LEPTON DIPOLE MOMENTS

ADVANCED SERIES ON DIRECTIONS IN HIGH ENERGY PHYSICS

Published

Vol. 1 – High Energy Electron–Positron Physics (*eds. A. Ali and P. Söding*)
Vol. 2 – Hadronic Multiparticle Production (*ed. P. Carruthers*)
Vol. 3 – CP Violation (*ed. C. Jarlskog*)
Vol. 4 – Proton–Antiproton Collider Physics (*eds. G. Altarelli and L. Di Lella*)
Vol. 5 – Perturbative QCD (*ed. A. Mueller*)
Vol. 6 – Quark–Gluon Plasma (*ed. R. C. Hwa*)
Vol. 7 – Quantum Electrodynamics (*ed. T. Kinoshita*)
Vol. 9 – Instrumentation in High Energy Physics (*ed. F. Sauli*)
Vol. 10 – Heavy Flavours (*eds. A. J. Buras and M. Lindner*)
Vol. 11 – Quantum Fields on the Computer (*ed. M. Creutz*)
Vol. 12 – Advances of Accelerator Physics and Technologies (*ed. H. Schopper*)
Vol. 13 – Perspectives on Higgs Physics (*ed. G. L. Kane*)
Vol. 14 – Precision Tests of the Standard Electroweak Model (*ed. P. Langacker*)
Vol. 15 – Heavy Flavours II (*eds. A. J. Buras and M. Lindner*)
Vol. 16 – Electroweak Symmetry Breaking and New Physics at the TeV Scale (*eds. T. L. Barklow, S. Dawson, H. E. Haber and J. L. Siegrist*)
Vol. 17 – Perspectives on Higgs Physics II (*ed. G. L. Kane*)
Vol. 18 – Perspectives on Supersymmetry (*ed. G. L. Kane*)
Vol. 19 – Linear Collider Physics in the New Millennium (*eds. K. Fujii, D. J. Miller and A. Soni*)

Forthcoming

Vol. 8 – Standard Model, Hadron Phenomenology and Weak Decays on the Lattice (*ed. G. Martinelli*)

Advanced Series on
Directions in High Energy Physics — Vol. 20

LEPTON DIPOLE MOMENTS

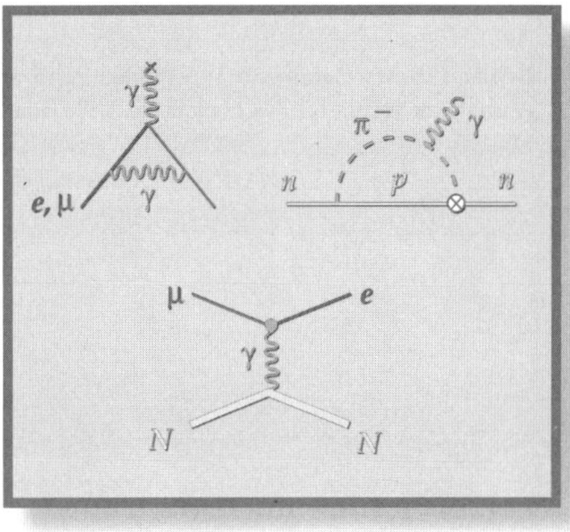

Editors

B Lee Roberts
Boston University, USA

William J Marciano
Brookhaven National Laboratory, USA

World Scientific

NEW JERSEY · LONDON · SINGAPORE · BEIJING · SHANGHAI · HONG KONG · TAIPEI · CHENNAI

Published by

World Scientific Publishing Co. Pte. Ltd.

5 Toh Tuck Link, Singapore 596224

USA office: 27 Warren Street, Suite 401-402, Hackensack, NJ 07601

UK office: 57 Shelton Street, Covent Garden, London WC2H 9HE

British Library Cataloguing-in-Publication Data
A catalogue record for this book is available from the British Library.

LEPTON DIPOLE MOMENTS
Advanced Series on Directions in High Energy Physics — Vol. 20

ISBN-13 978-981-4271-83-7
ISBN-10 981-4271-83-7

Printed in Singapore by World Scientific Printers

Preface

As the title suggests, lepton electromagnetic dipole moments, including anomalous magnetic, electric, and transition moments, are the main subject of this volume. Studies of these quantities test the Standard Model of elementary particle physics at the level of its quantum fluctuations, and search for New Physics effects.

Those searches fall into two categories. The first approach entails precision experimental measurements of the electron and muon anomalous magnetic moments, which can then be compared with theoretical Standard-Model predictions of comparable accuracy. A clear discrepancy would point to additional contributions of New Physics origin. The second approach involves searches for non-vanishing electric, and transition dipole moments (e.g. $\mu \to e\gamma$). The Standard Model predicts those quantities to be unobservably small. Hence, discovery of a non-zero value would be interpreted as direct evidence for New Physics.

The measurement and theory of the electron and muon magnetic moments has a long and distinguished history. The former was intimately intertwined with the development of quantum electrodynamics, and the calculation of the electron anomalous magnetic moment (anomaly) by Schwinger represented the very first quantum-loop computation. Its simple but elegant value is inscribed on the memorial marker located near his grave in the Mount Auburn Cemetery in Cambridge Massachusetts.

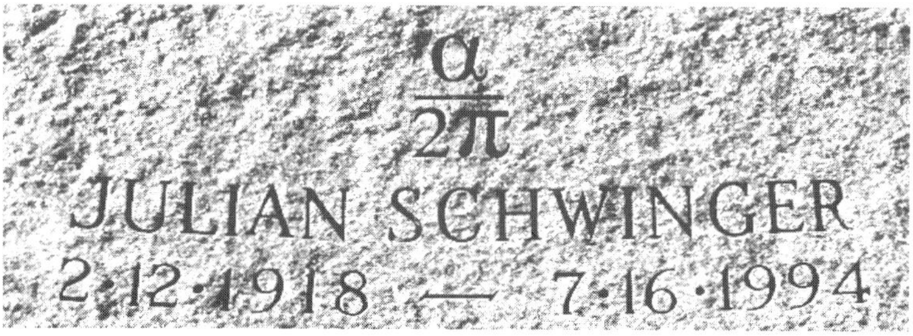

QED calculations of the electron anomaly have become an industry, with the sixth-order (3-loop) contribution having been calculated analytically by Laporta and Remiddi. The eighth- and tenth-order (4- and 5-loop) contributions have occupied a significant fraction of Kinoshita's career, and with his collaborators he continues these numerical calculations today. Meanwhile, the experiments by Gabrielse and his collaborators have reached the remarkable precision of 0.24 parts per billion on the electron anomaly, some 20 times more precise than independent measurements of the fine-structure constant α. Chapters by the above-mentioned experts, along with an historical introduction by BLR and a general overview of electromagnetic moments by A. Czarnecki and WJM, provide an up-to-date review of the status of the electron magnetic moment. We also include a brief discussion of the various measurements of α by G. Gabrielse and an article by K. Pachucki and J. Sapirstein on the theory necessary to extract α from helium fine structure. At present, the electron g-value along with the QED theory provides the best measure of α.

The relative sensitivity of the muon anomaly to higher mass scales compared to the electron goes as $(m_\mu/m_e)^2 \simeq 43,000$, which requires knowledge of the hadronic contribution arising from virtual hadrons in vacuum polarization loops (which dominate the uncertainty on the Standard-Model value of the muon anomaly), as well as the one- and two-loop contributions from the weak gauge bosons, fermions and Higgs scalar. Thus, at the present experimental precision for the muon anomaly of 0.54 ppm, there is significant sensitivity to the several-hundred GeV mass scale. The current Standard-Model prediction for the muon anomalous magnetic moment and potential effects due to New Physics are reviewed in chapters by Czarnecki and WJM; M. Davier; J. Prades, E. de Rafael and A. Vainshtein; K. Lynch; and D. Stöckinger, while its experimental status is described in a chapter by J. Miller, BLR and K. Jungmann.

Dedicated searches for electric dipole moments (EDMs) date back to the pioneering observation by Purcell and Ramsey in 1950, that a particle EDM would violate parity, but should nevertheless be searched for as a test of that symmetry. The experimental quest for an EDM of the electron, the neutron, and of atomic nuclei has become an important area in the search for physics beyond the Standard Model. The level of precision that has been reached, $< 1.6 \times 10^{-27}$ e-cm for the electron, $< 2.9 \times 10^{-26}$ e-cm for the neutron and $< 3.1 \times 10^{-29}$ e-cm for ^{199}Hg, is beginning to challenge models such as supersymmetry. There is substantial hope that the discovery of an EDM will come in the present generation of experiments. Reviews of all of these searches, along with the related theoretical issues, are covered in this volume by M. Pospelov and A. Ritz; E. Commins and D. DeMille; S. Lamoreaux and R. Golub; W.C. Griffith, M. Swallows and N. Fortson; all active experts in the field. The new idea of using storage rings to search for EDMs of charged particles is covered in a chapter by BLR, J. Miller and Y. Semertzidis.

The related process, the transition dipole moment that would permit lepton flavor (muon number) violation (LFV) in reactions such as $\mu^- \mathcal{N} \to e^- \mathcal{N}$ and $\mu^+ \to e^+ \gamma$ are complementary to the studies of electric and magnetic dipole moments. Since the Standard-Model predictions for such reactions are suppressed by $(m_\nu / M_W)^4 < 10^{-45}$ and thus experimentally unobservable, any observation of LFV in the charged sector would signal the presence of New Physics. Charged lepton transition moments due to New Physics and experimental searches are covered in the chapters by Y. Okada and Y. Kuno which complete the book.

The idea for this volume came about when after a seminar given at Imperial College, BLR was approached by an editor from Imperial College Press to write a monograph on muon physics. The counter proposal was a volume dedicated to the topics covered at the series of symposia on Lepton Moments started by Klaus Jungmann in Heidelberg in 1999 and continued by BLR on Cape Cod in 2003, 2006 and planned for 2010. We are indeed grateful that so many of our friends and colleagues have joined with us to create this volume. We gratefully acknowledge Kevin R. Lynch for his encyclopedic expertise in LaTeX, which he used to solve numerous issues in putting this document together.

We dedicate this volume to Norman Ramsey, and to the memory of Paul Dirac, Julian Schwinger, Polykarp Kusch and Edward Purcell, all pictured on the next page, who carried out the seminal work which began our modern journey through the field of magnetic and electric dipole moments.

B. Lee Roberts and William J. Marciano

Clockwise:

Julian Schwinger,
Polykarp Kusch,
Paul Dirac,
Norman Ramsey and
Edward Purcell

Courtesy AIP Emilio Segrè Visual Archives (full credits overleaf)

Photo credits: Schwinger memorial marker, photo by BLR; Schwinger photo from AIP Emilio Segrè Visual Archives; Kusch photo from National Archives and Records Administration (NARA), courtesy AIP Emilio Segrè Visual Archives, Physics Today Collection, W. F. Meggers Gallery of Nobel Laureates; Dirac photo from AIP Emilio Segrè Visual Archives; Ramsey photo from AIP Emilio Segrè Visual Archives, Ramsey Collection; Purcell photo from AIP Emilio Segrè Visual Archives, Physics Today Collection, W. F. Meggers Gallery of Nobel Laureates.

Contents

Chapter 1

Historical Introduction to Electric and Magnetic Moments

B. Lee Roberts
Department of Physics, Boston University
Boston, MA 01890 U.S.A.
roberts@bu.edu

The historical development of the discovery of spin and magnetic moments is reviewed, along with the development of searches for electric dipole moments.

Contents

1.1. The Discovery of Spin

As physics developed at the beginning of the 20th century, a number of intriguing puzzles existed that could only be explained by radically new ideas. In 1911 Rutherford proposed the nuclear atom [1]. This hypothesis, combined with Thompson's discovery of the electron [2] and Millikan's discovery that the electron charge is quantized [3], implied that electrons were somehow in orbit around the positive nucleus, leading to a neutral atom. Classically such a system is unstable, and in 1913 Bohr proposed his quantum theory [4]. Of course, many conceptual problems remained, which began to be understood once Schrödinger's wave equation [5] was published in 1926.

In 1921, two interesting proposals were published: Compton proposed [6] a spinning electron to explain ferromagnetism, which he realized was difficult to explain by any other means.[a] Stern proposed an experiment to study space quantization [7] to test the Sommerfeld quantum theory, where he presented the details of what we now call the Stern–Gerlach experiment. An atomic beam of silver atoms was to be projected through a gradient magnetic field where the net force on the magnetic dipole would separate the different magnetic quantum states. For a classical dipole the deflection would be continuous, since the direction of the dipole moment could have any value.[b]

Over the next two years the famous experiments were carried out [8], and the two-band structure observed. By 1924, Stern and Gerlach concluded that to within 10%, the magnetic moment of the silver atom in its ground state was one Bohr magneton [9]. Their papers made no reference to the developments in spectroscopy, and in their 1924 review article, no conclusions beyond the magnetic moment were drawn from the two-band structure.

Independently, in 1925 Uhlenbeck and Goudsmit [10] proposed the "spinning electron" to explain the fine-structure observed in the anomalous Zeeman effect in atomic spectra.[c] The fine-structure splitting can be understood as the interaction of the magnetic dipole moment of the electron with the magnetic field produced by the nuclear motion, which in the electron's rest frame appears to be orbiting about the electron. The electron's magnetic dipole moment is along its spin and is given by

$$\vec{\mu} = g\left(\frac{q}{2m}\right)\vec{s}, \tag{1.1}$$

where $q = \pm e$ is the charge of the particle in terms of the magnitude of the electron charge e, and the proportionality constant g is the g-factor for spin (which is sometimes written as g_s). In their second paper [11], Uhlenbeck and Goudsmit conclude that the g-factor for spin is twice that for orbital angular momentum, however the calculated fine-structure splitting was then twice as large as the observed splitting. Only later in 1926, when Thomas showed that the factor of 2 discrepancy between experiment and calculation was a kinematic effect [12], did spin start to become an accepted

[a] In his paper Compton acknowledges A.L. Parson (Smithsonian Misc. Collections, 1915) as first proposing that the electron was a spinning ring of charge. Compton modified this proposal to be a much smaller distribution "concentrated principally near its center." Compton's paper is almost unknown.

[b] See Allan Franklin, http://plato.stanford.edu/entries/physics-experiment/app5.html *Stanford Encyclopedia of Philosophy*, Appendix 5, for a nice discussion putting the Stern–Gerlach experiment into historical context.

[c] In their Nature paper [11] of 1926, they acknowledge Compton's independent suggestion of spin.

theory. Thomas later wrote to Goudsmit, indicating that Kronig had also suggested spin [13]:

> I think you and Uhlenbeck have been very lucky to get your spinning electron published and talked about before Pauli heard of it. It appears that more than a year ago, Kronig believed in the spinning electron and worked out something; the first person he showed it to was Pauli. Pauli ridiculed the whole thing so much that the first person became also the last and no one else heard anything of it. Which all goes to show that the infallibility of the Deity does not extend to his self-styled vicar on earth.

Incidentally, no mention is made of the Stern–Gerlach measurements in the Uhlenbeck and Goudsmit papers. However, the Stern–Gerlach result was noticed by Phipps and Taylor at the University of Illinois at Urbana, and they did draw the connection between the Stern–Gerlach experiment and the electron spin proposed by Uhlenbeck and Goudsmit. They repeated the Stern–Gerlach experiment with an atomic beam of hydrogen in 1926. While technically more challenging than the silver experiment, they reached a similar conclusion, viz. that the magnetic moment of the hydrogen atom was also one Bohr magneton [14].

Today, we understand that the magnetic moment measured in both of these atomic-beam experiments was that of the un-paired atomic electron. We can conclude that a magnetic moment of one Bohr magneton implies that the g-factor for spin is 2. Although in our undergraduate modern physics courses we emphasize that the Stern–Gerlach experiment showed clearly the existence of half-integer spin, historically it seems to have played a much less important role than spectroscopy did.[d] In his book, *The Story of Spin*, Tomonaga does not mention the Stern–Gerlach result [16].

1.2. Dirac's Theory and Beyond

It was not until Dirac's famous 1928 paper [17], where he introduced his relativistic wave equation for the electron, that the picture became clear. Dirac pointed out that an electron in external electric and magnetic fields has "the two extra terms[e]

$$\frac{e\hbar}{c}\left(\sigma, \mathbf{H}\right) + i\frac{e\hbar}{c}\rho_1\left(\sigma, \mathbf{E}\right), \tag{1.2}$$

[d]The recollections of Goudsmit agree with this assessment, see Ref. [15].
[e]Here we use Dirac's original notation.

...when divided by the factor $2m$ can be regarded as the additional potential energy of the electron due to its new degree of freedom." These terms represent the magnetic dipole (Dirac) moment and electric dipole moment interactions with the external magnetic and electric fields.[f] Dirac theory predicts that the electron magnetic moment is one Bohr-magneton (viz. $g = 2$), consistent with the value measured by the experiments.[g] Dirac later commented: "It gave just the properties that one needed for an electron. That was an unexpected bonus for me, completely unexpected [18]."

As an aside, Dirac had little use for the electric dipole moment (EDM), and stated "The electric moment, being a pure imaginary, we should not expect to appear in the model. It is doubtful whether the electric moment has any physical meaning, since the Hamiltonian ... that we started from is real, and the imaginary part only appeared when we multiplied it up in an artificial way in order to make it resemble the Hamiltonian of previous theories." We now understand that the presence of an electric dipole moment violates both parity (P) and time reversal (T) symmetries, and CP as well if CPT holds.

1.2.1. *The discovery of anomalous magnetic moments*

For some years, the experimental situation remained the same. The electron had $g = 2$, and the Dirac equation seemed to describe nature. Then a surprising and completely unexpected result was obtained. In 1933, against the advice of Pauli who believed that the proton was a pure Dirac particle [16], Stern and his collaborators [19] showed that the g-factor of the proton was ~ 5.5, a long way from the expected value of 2. Even more surprising was the discovery in 1940 by Alvarez and Bloch [20] that the neutron had a large magnetic moment (see Eq. (1.1)). These two results remained quite mysterious for many years, and are still not perfectly understood. With the advent of the quark model, one does get a 10 to 20% description of baryon magnetic moments, but given that experiments show that very little of the proton spin is carried by the quarks, the whole spin structure of baryons remains a topic of investigation.[h] It became convenient

[f]However, it appears that the Dirac complex phase is an artifact of his second-order formalism analysis rather than a real EDM.

[g]The Dirac equation also predicts that the g-factor associated with orbital angular momentum $g_\ell = 1$.

[h]A.W. Thomas claims that this crisis is resolved [21], but according to R.L. Jaffe [22] this is a minority view.

to break the magnetic moment into two pieces:

$$\mu = (1+a)\,\frac{q\hbar}{2m} \quad \text{where} \quad a = \frac{g-2}{2}. \tag{1.3}$$

The first piece, predicted by the Dirac equation and called the Dirac moment, is 1 in units of the appropriate magneton, $q\hbar/2m$. The second piece is the anomalous (Pauli) moment [23], where the dimensionless quantity a is sometimes referred to as the *anomaly.*

The development of radio frequency engineering and microwave technology during the Second World War was quickly put to use afterward in the laboratory. In 1947, motivated by measurements of the hyperfine structure in hydrogen that obtained splittings larger than expected from the Dirac theory [24–26], Schwinger [27] showed that from a theoretical viewpoint these "discrepancies can be accounted for by a small additional electron spin magnetic moment" that arises from the lowest-order radiative correction to the Dirac moment.[i] In his paper, Schwinger points out three important features of his new theory.

> The new Hamiltonian is superior to the original one in essentially three ways: it involves the experimental electron mass, rather than the unobservable mechanical mass; an electron now interacts with the radiation field only in the presence of an external field ... the interaction of an electron with an external field is now subject to a finite radiative correction.

In today's language, Schwinger pointed out that one replaces the bare mass and charge with the physical (dressed) mass and charge (see Chapter 3 for additional details).

The one-loop contribution to a is shown diagrammatically in Fig. 1.1(b) and has the value $a_e = \alpha/(2\pi) \simeq 0.00116\cdots$, which is independent of mass and is the same for a_μ and a_τ.

In the same year, Kusch and Foley [29] measured a_e with 4% precision, and found that the measured electron anomaly agreed well with Schwinger's prediction. They state that: "... the results can be described by $g_\ell = 1$ and $g_s = 2(1.00119 \pm 0.00005)$."[j]

[i] In response to Nafe, *et al.* [24], Breit [28] conjectured that this discrepancy could be explained by the presence of a small Pauli moment. It's not clear whether this paper influenced Schwinger's work, but in a footnote Schwinger states: "However, Breit has not correctly drawn the consequences of his empirical hypothesis."

[j] The choice that $g_\ell = 1$ and $g_s > 2$ was guided by theoretical prejudice. The modern experiments, which confine a single electron in a Penning trap, measure g_s directly and fully justify this assumption.

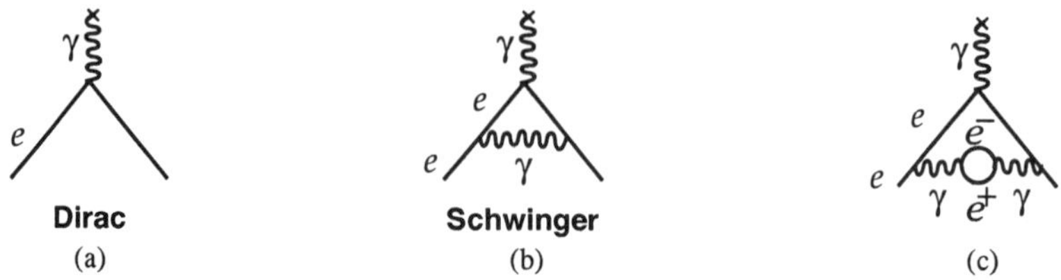

Fig. 1.1. The Feynman graphs for: (a) $g = 2$; (b) the lowest-order radiative correction first calculated by Schwinger; and (c) the vacuum polarization contribution, which is one of five fourth-order, $(\alpha/\pi)^2$, terms.

In the intervening time since the Kusch and Foley paper, many improvements have been made in the precision of the electron anomaly [30–32], as well as in the theory (see Chapters 3 and 4). Most recently, a_e has been measured to a relative precision of 0.24 ppb (parts per billion) [32], and the comparison with theory is limited by the knowledge of the fine-structure constant, α. See Chapters 3 and 6 for the most recent theory and experimental values of a_e.

The ability to calculate the higher-order QED contributions to the anomaly has gone well beyond what could have been imagined by the inventors. In response to a question about how the QED pioneers viewed the theory Freeman Dyson said [33]:

> The main point was that all of us who put QED together, including especially Feynman, considered it a jerry-built and provisional structure which would either collapse or be replaced by something more permanent within a few years. So I find it amazing that it has lasted for fifty years and still agrees with experiments to twelve significant figures. It seems that Nature is telling us something. Perhaps she is telling us that she loves sloppiness.

The muon anomaly has been measured to a precision of 0.54 ppm [34]. Naively, this level of precision would seem to limit the physics reach of the muon anomaly when compared to that of the electron. However, since the relative sensitivity of the anomaly to higher mass scales goes as $(m_\mu/m_e)^2 \simeq 43,000$, the muon anomaly has measurable sensitivity up to the several hundred GeV scale, as discussed in the Chapter 2.

1.3. The Search for Electric Dipole Moments

Dirac [17] discovered an electric dipole moment (EDM) term in his relativistic electron theory. Like the magnetic dipole moment, the electric dipole moment must be along the spin. We can write an expression similar to Eq. (1.1),

$$\vec{d} = \eta \left(\frac{q}{2mc} \right) \vec{s}, \tag{1.4}$$

where η is a dimensionless constant that is analogous to g in Eq. (1.1). While magnetic dipole moments (MDMs) are a natural property of charged particles with spin, electric dipole moments (EDMs) are forbidden both by parity and by time reversal symmetry.

The search for an EDM dates back to the suggestion of Purcell and Ramsey [35] in 1950, well in advance of the paper by Lee and Yang [36], that a measurement of the neutron EDM would be a good way to search for parity violation in the nuclear force. An experiment was mounted at Oak Ridge [37] soon thereafter which placed a limit on the neutron EDM of $d_n < 5 \times 10^{-20}$ e-cm, although the result was not published until after the discovery of parity violation.

Once parity violation was established, Landau [38] and Ramsey [39] pointed out that an EDM would violate both P and T symmetries. This can be seen by examining the Hamiltonian for a spin one-half particle in the presence of both an electric and magnetic field,

$$\mathcal{H} = -\vec{\mu} \cdot \vec{B} - \vec{d} \cdot \vec{E}. \tag{1.5}$$

The transformation properties of $\vec{E}$, $\vec{B}$, $\vec{\mu}$ and $\vec{d}$ are given in Table 1.1, and we see that while $\vec{\mu} \cdot \vec{B}$ is even under all three symmetries, $\vec{d} \cdot \vec{E}$ is odd under both P and T. Thus the existence of an EDM implies that both P and T are not good symmetries of the interaction Hamiltonian, Eq. (1.5). In the context of CPT symmetry, an EDM implies CP violation.

Table 1.1. Transformation properties of the magnetic and electric fields and dipole moments.

	$\vec{E}$	$\vec{B}$	$\vec{\mu}$ or $\vec{d}$
P	-	+	+
C	-	-	-
T	+	-	-

The Standard Model value for the electron (muon) EDM is $\leq 10^{-38}$ e-cm ($\leq 2 \times 10^{-36}$ e-cm), well beyond the reach of experiments (which are at the 1.6×10^{-27} (1.8×10^{-19}) e-cm level). Likewise, the Standard-Model value for the neutron is 10^{-32} e-cm, with the present experimental limit of $2.9 \times 10^{-26} e$-cm.

Concerning these symmetries, Ramsey states [39]:

> However, it should be emphasized that while such arguments are appealing from the point of view of symmetry, they are not necessarily valid. Ultimately the validity of all such symmetry arguments must rest on experiment.

Fortunately this advice has been followed by many experimental investigators during the intervening 50 years. Today the searches for a (*CP* violating) permanent electric dipole moment of the electron, neutron, and of an atomic nucleus have become an important part of the search for physics beyond the Standard Model. Since the Standard Model *CP* violation observed in the neutral kaon and B-meson systems is inadequate to explain the predominance of matter over antimatter in the universe, the search for new sources of *CP* violation beyond that embodied in the CKM formalism takes on a certain urgency. These searches, along with the relevant theoretical framework, form a major portion of this volume.

References

[1] E. Rutherford, Proc. of the Manch. Lit. and Phil. Soc., **IV, 55**, (1911) 18, and Phil. Mag., Series 6, **21** (1911) 669.
[2] J.J. Thompson, Phil. Mag. 44 (1897) 293.
[3] R.A. Millikan, Phys. Mag. XIX, **6** (1910) 209.
[4] N. Bohr, Phil. Mag. 26, 1 (1913).
[5] E. Schrödinger, Ann. Phys. **79** (1926) 361.
[6] A.K. Compton, Jour. Franklin. Inst., **192** Aug. (1921) 145.
[7] O. Stern, Z. Phys. **7**, 249 (1921).
[8] W. Gerlach and O. Stern, , Z. Phys. **8**, 110 (1922), Z. Phys. **9** and 349(1922), Z. Phys. **9**, 353 (1924).
[9] W. Gerlach and O. Stern, Ann. Phys. **74**, 673 (1924).
[10] G.E. Uhlenbeck and S. Goudsmit, Naturwissenschaften **47**, 953 (1925).
[11] G.E. Uhlenbeck and S. Goudsmit, Nature **117** (1926) 264.
[12] L.H. Thomas, Nature **117**, (1926) 514 and Phil. Mag. **3** (1927) 1.
[13] From a letter by L.H. Thomas to Goudsmit (25 March 1926). A reproduction from a transparency shown by Goudsmit during his 1971 lecture at Leiden [15]. The original is presumably in the Goudsmit archive kept by the American Institute of Physics Center for History of Physics.
[14] T.E. Phipps and J.B. Taylor, Phys. Rev. **29**, 309 (1927).
[15] http://www.lorentz.leidenuniv.nl/history/spin/goudsmit.htm
[16] Sin-itiro Tomonaga, *The Story of Spin*, translated by Takeshi Oka, U. Chicago Press, 1997.

[17] P.A.M. Dirac, Proc. R. Soc. (London) **A117**, 610 (1928), and **A118**, 351 (1928). See also, P.A.M. Dirac, *The Principles of Quantum Mechanics*, 4th edition, Oxford University Press, London, 1958.
[18] Abraham Pais in *Paul Dirac: The Man and His Work*, P. Goddard, ed., Cambridge U. Press, New York (1998).
[19] R. Frisch and O. Stern, Z. Phys. **85**, 4 (1933), and I. Estermann and O. Stern, Z. Phys. **85**, 17 (1933).
[20] Luis W. Alvarez and F. Bloch, Phys. Rev. **57**, 111 (1940).
[21] A.W. Thomas, Prog. Part. Nucl. Phys. **61**, 219 (2008), (arXiv:0805.4437v1).
[22] R.L. Jaffe, private communication, Nov. 2008 and http://www.bnl.gov/gbunce/talks.asp
[23] Hans A. Bethe and Edwin E. Salpeter, *Quantum Mechanics of One- and Two-Electron Atoms*, Springer-Verlag, (1957), p. 51.
[24] J.E. Nafe, E.B. Nelson and I.I. Rabi Phys. Rev. **71**, 914(1947).
[25] D.E. Nagel, R.S. Julian and J.R. Zacharias, Phys. Rev. **72**, 971 (1947).
[26] P. Kusch and H.M Foley, Phys. Rev **72**, 1256 (1947).
[27] J. Schwinger, Phys. Rev. **73**, 416L (1948), and Phys. Rev. **76** 790 (1949). The former paper contains a misprint in the expression for a_e that is corrected in the longer paper.
[28] G. Breit, Phys. Rev. **72** 984, (1947).
[29] P. Kusch and H.M Foley, Phys. Rev. **73**, 250 (1948).
[30] See Arthur Rich and John Wesley, Reviews of Modern Physics **44**, 250 (1972) for a nice historical overview of the lepton g - factors.
[31] R.S. Van Dyck *et al.*, Phys. Rev. Lett., **59**, 26(1987) and in *Quantum Electrodynamics*, (Directions in High Energy Physics Vol. 7) T. Kinoshita d., World Scientific, 1990, p. 322.
[32] D. Hanneke, S. Fogwell and G. Gabrielse, Phys. Rev. Lett. **100**, 120801, (2008).
[33] F. Dyson, private communication to BLR, December 2006.
[34] G. Bennett, *et al.*, (Muon $(g-2)$ Collaboration), Phys. Rev. **D73**, 072003 (2006).
[35] E.M. Purcell and N.F. Ramsey, Phys. Rev. **78**, 807 (1950).
[36] T.D. Lee and C.N. Yang, Phys. Rev. **104** (1956) 254.
[37] J.H. Smith, E.M. Purcell and N.F. Ramsey, Phys. Rev. **108**, 120 (1957).
[38] L. Landau, Nucl. Phys. **3**, 127 (1957).
[39] N.F. Ramsey Phys. Rev. **109**, 225 (1958).

Chapter 2

Electromagnetic Dipole Moments and New Physics

Andrzej Czarnecki

Department of Physics, University of Alberta,
Edmonton, AB, Canada T6G 2G7

William J. Marciano

Physics Department, Brookhaven National Laboratory,
Upton, NY 11973, USA

As an introduction to the more detailed chapters that follow, we present a general overview of spin 1/2 fermion electromagnetic dipole moments produced by quantum loop effects. Standard Model predictions are given and possible New Physics contributions are parameterized in terms of the mass scale responsible for anomalous magnetic, electric, and transition dipole moments. Experimental measurements and bounds are discussed. The muon anomalous magnetic moment is covered in some detail because it may already be exhibiting signs of New Physics. Electron and neutron electric dipole moments along with $\mu \to e\gamma$ transition moments are shown to have New Physics sensitivities extending up to $\mathcal{O}$ (1000 TeV) mass scales, modulo CP and flavor violation suppressions. Various other less constraining fermion dipole moments are discussed.

Contents

2.1. The Dirac Equation and Electron Dipole Moments

The Dirac equation [1, 2],

$$i\left(\partial_\mu - ieA_\mu\left(x\right)\right)\gamma^\mu\psi\left(x\right) = m_e\psi\left(x\right), \tag{2.1}$$

introduced in 1928 is a cornerstone of modern physics. Using the now famous four-by-four Dirac γ^μ matrices, it succinctly describes a four component (spinor) electron wave function, $\psi\left(x\right)$, in an electromagnetic potential $A_\mu\left(x\right)$. That elegant equation combined quantum mechanics, special relativity, spin and electromagnetic gauge invariance in one simple expression and laid the foundation for later developments in quantum electrodynamics (QED). Today, it provides a basis for our $SU\left(3\right)_c \times SU\left(2\right)_L \times U\left(1\right)_Y$ Standard Model of elementary particle physics.

The Dirac equation is primarily acclaimed for its (later realized [3]) prediction of antimatter, corresponding to negative energy solutions. Subsequent discovery of the positron, the electron's antimatter partner, was thus its crowning glory. However, it left us with a modern day puzzle as to why Nature chose to populate our Universe with matter and not antimatter, i.e. why is it so matter-antimatter asymmetric? Resolving that puzzle will likely require New Physics beyond Standard Model expectations. One of the necessary ingredients [4] is expected to be a new source of CP violation that differentiates the properties of particles and antiparticles. As we shall see, a signature of that New Physics could be the existence of particle electric dipole moments (EDMs) [5], one of the main topics of this chapter and book.

The immediate success of the Dirac equation was not, however, to predict antimatter. It was the explanation [6] as to why the gyromagnetic ratio, g_e, of the electron is equal to 2. That parameter, which expresses the relationship between the electron's magnetic moment, $\vec{\mu}_e$, and its spin $\vec{s}$,

$$\vec{\mu}_e = g_e \frac{Qe}{2m_e}\vec{s} \tag{2.2}$$

would be 1 if it were relating atomic orbital angular momentum and its associated magnetic moment.[a] Of course, the need for $g_e = 2$ was already well established by atomic fine structure spectroscopy before 1928. Nevertheless, the Dirac equation provided a natural explanation and strong underpinning for that fundamental value.

A deviation from $g_e = 2$ can be easily accommodated, if necessary, by adding a so-called Pauli interaction term [7, 8],

$$\frac{e}{4m_e} a_e F_{\mu\nu}(x)\sigma^{\mu\nu}\psi(x) \tag{2.3}$$

$$F_{\mu\nu} = \partial_\mu A_\nu - \partial_\nu A_\mu, \tag{2.4}$$

$$\sigma^{\mu\nu} = \frac{i}{2}\left[\gamma^\mu, \gamma^\nu\right], \tag{2.5}$$

to Eq. (2.1), where a_e is called the anomalous magnetic moment because it leads to

$$g_e = 2\left(1 + a_e\right) \tag{2.6}$$

i.e. an increase in the intrinsic magnetic dipole moment by $a_e \frac{e}{2m_e}$. Such an addition is very much required for the proton, where one finds [9] $g_p \simeq 5.59$ rather than 2 (see Section 2.2) due to its underlying quark substructure. However, Dirac had no need for a Pauli term, since it was known in 1928 that $g_e = 2$ with rather good certainty.

What forbids the addition of a Pauli term for an elementary spin 1/2 fermion such as the electron? That term respects Lorentz covariance and local gauge invariance; however, it runs counter to Dirac's principles of elegance and simplicity as well as his use of minimal coupling (the replacement of ∂_μ by the covariant derivative $\partial_\mu - ieA_\mu$ in the non-interacting Dirac equation).

Today, we would automatically exclude Pauli terms at the level of our fundamental classical interaction Lagrangian because they correspond to what are called dimension 5 operators which are known to spoil renormalizability. However, such dimension 5 terms can and do arise in quantum field theories as a result of virtual loop fluctuations. In that respect, their existence is to be expected and they can be viewed as a window to quantum loops including effects due to heavy new particles with masses well above direct experimental accessibility. That feature forms the main theme of

[a]We define $e > 0$ and $Q = -1$ for electrons, $Q = +1$ for positrons. Many field theory texts employ $e < 0$ as the electron charge and express all derived results in terms of that negative quantity. Such an approach is a little awkward for EDMs where the unit $e{\cdot}$cm is conventionally used, since negative units can lead to sign inconsistencies.

this chapter, how measurements of various quantum induced dimension 5 dipole operators can be used to provide indirect evidence for New Physics or at least constrain speculations regarding its properties.

2.1.1. *Electron anomalous magnetic moment*

In 1947, small anomalous effects at about the 0.1 percent level began to be observed in high precision atomic hyperfine spectroscopy [10, 11]. Breit suggested [12], on empirical grounds, that such observations could be explained if g_e deviated slightly from 2. Schwinger then demonstrated [13] the power of QED and his own computational prowess by calculating the leading quantum contribution to a_e,

$$a_e = \frac{g_e - 2}{2} = \frac{\alpha}{2\pi} \simeq 0.00116, \tag{2.7}$$

where $\alpha = e^2/4\pi \simeq 1/137$ is the fine structure constant.

His result agreed with experiment [14] and ushered in an era of precision measurements that tested the validity of QED to high order in α and searched for deviations that might indicate the presence of New Physics.

Today, as a result of many pioneering efforts, including the Nobel prize winning experiments of H. Dehmelt and his collaborators [15], the electron anomalous magnetic moment has been measured with phenomenal accuracy (see Chapter 5). The most precise value, due to Hanneke, Fogwell and Gabrielse [16] is currently

$$a_e^{\exp} = \frac{g_e - 2}{2} = 0.001\,159\,652\,180\,73\,(28)\,, \tag{2.8}$$

where the numbers in parenthesis represent the one sigma uncertainty in the last two decimal places. That result is truly impressive. It can be compared with the four-loop QED prediction and estimated five-loop uncertainty (due to the heroic work of many theorists, see Chapters 3, 4 and Ref. [17])

$$\begin{aligned} a_e^{\rm SM} &= \frac{\alpha}{2\pi} - 0.328\,478\,444\,003\left(\frac{\alpha}{\pi}\right)^2 + 1.181\,234\,016\,8\left(\frac{\alpha}{\pi}\right)^3 \\ &\quad -1.9144(35)\left(\frac{\alpha}{\pi}\right)^4 + 0.0(4.6)\left(\frac{\alpha}{\pi}\right)^5 + 1.71\times 10^{-12} \end{aligned} \tag{2.9}$$

where we have truncated the two and three loop numerical coefficients at the level of their uncertainty (due to uncertainties in the muon and tau lepton masses m_μ and m_τ) and have included a small Standard Model correction $\left(1.71 \times 10^{-12}\right)$ due to hadronic loops $\left(1.68 \times 10^{-12}\right)$ and electroweak effects $\left(0.03 \times 10^{-12}\right)$.

Equations (2.8) and (2.9) can be compared in two different ways. First, assuming no New Physics, they can be equated to give the world's most precise determination of the fine structure constant

$$\alpha^{-1}(a_e) = 137.035\,999\,084\,(51)\,, \tag{2.10}$$

where the uncertainty comes from Eq. (2.8) and the error in Eq. (2.9). Alternatively, one can take a more direct low energy atomic physics or condensed matter determination of α and obtain a numerical prediction from Eq. (2.9) which can be compared with Eq. (2.8). Using the recent Rydberg based value [18] (which is next best after Eq. (2.10))

$$\alpha^{-1}(\text{Rydberg}) = 137.035\,999\,450\,(620) \tag{2.11}$$

leads to

$$a_e^{\text{SM}}(\text{Rydberg}) = 0.001\,159\,652\,177\,60\,(520)\,. \tag{2.12}$$

That prediction agrees with Eq. (2.8) but its error is almost 20 times larger. If New Physics is contributing to a_e^{exp}, its contribution must satisfy

$$|a_e(\text{New Physics})| = \left|a_e^{\text{exp}} - a_e^{\text{SM}}\right| < 10^{-11}. \tag{2.13}$$

That bound could be improved by more than an order of magnitude if α were much better independently determined [19].

How large a value of a_e (New Physics) might be expected from new short distance interactions parametrized by the mass scale Λ? Because anomalous magnetic moments change chirality ($R \leftrightarrow L$), we expect New Physics contributions to $a_e \frac{e}{2m_e}$ to vanish in the chiral limit $m_e \to 0$. Therefore, one anticipates the quadratic dependence[b]

$$a_e(\text{New Physics}) = C\left(\frac{m_e}{\Lambda}\right)^2 \tag{2.14}$$

where C could be $\mathcal{O}(1)$ (see Section 2.3) or smaller, e.g. $\mathcal{O}(\alpha)$ in weak coupling loop scenarios. Taking $C \simeq 1$, we find from Eq. (2.13), that $\Lambda \lesssim 160$ GeV is currently being probed by a_e^{exp} measurements. For $C \simeq \alpha/\pi$, that sensitivity is reduced to about 8 GeV, a finding that is consistent with the fact that (due to the uncertainty in α) a_e^{SM} in Eq. (2.12) is still about two orders of magnitude away from electroweak contributions $\left(3 \times 10^{-14}\right)$ which correspond to a physics mass scale of about $m_W \simeq 80$ GeV.

From the above exercise, we conclude that although a_e provides a stringent test of QED at the four-loop quantum level, it is not a particularly good probe of high mass scale New Physics. Indeed, as we subsequently illustrate, other experiments are potentially sensitive to Λ in the multi-TeV region and in some cases future efforts could probe beyond 1000 TeV!

[b] One could tune m_e and a_e such that a_e exhibits a linear dependence on m_e; however, we do not consider such scenarios here. For a discussion of that case see Chapter 5.

2.1.2. *Electron electric dipole moment*

If instead of adding the Pauli term to the Dirac equation, we were to append a

$$\frac{i}{2} d_e F_{\mu\nu}(x)\,\sigma^{\mu\nu}\gamma_5\psi(x) \tag{2.15}$$

interaction, it would correspond to an electron electric dipole moment (EDM), d_e [20], interacting with the external electromagnetic fields $F_{\mu\nu}(x)$. Apparently, Dirac noted the possibility of EDM effects (see Chapter 1) but dismissed them as unphysical.

EDMs violate the discrete symmetries of P (parity) and T (time reversal) [21–24]. Of course, we now know that both symmetries are violated by weak interactions [25]; so, we should expect at some level $d_e \neq 0$ due to Standard Model loop effects. It has been estimated, that such an effect arising from quark mixing via the CKM matrix [25, 26] (from four-loop order) is roughly

$$\left|d_e^{\rm SM}\right| \simeq 10^{-38}\, e\cdot{\rm cm} \qquad \text{Standard Model.} \tag{2.16}$$

In other words, $d_e^{\rm SM}$ is unobservably small, since current experiments probe $d_e \sim \mathcal{O}\left(10^{-27}\right)\, e\cdot$cm and it is hard to imagine improvements in sensitivity by more than ten orders of magnitude. However, New Physics EDM effects that violate P and T could arise from one or two loop order and be much larger than the tiny Standard Model prediction even if they stem from high mass scales.

Parameterizing the effect of New Physics (NP) on a_e and d_e by the relationship (see Section 2.2 for a discussion)

$$d_e\,(\text{New Physics}) = a_e\,(\text{New Physics})\,\frac{e}{2m_e}\tan\phi_e^{\rm NP} \tag{2.17}$$

with $\phi_e^{\rm NP}$ a new physics model dependent phase, we can relate a_e and d_e sensitivities. Using the experimental constraint from atomic physics [27],

$$|d_e| < 2\times 10^{-27}\, e\cdot{\rm cm} \tag{2.18}$$

or in units of $e/2m_e$ (electron Bohr magneton)

$$|d_e| < 1\times 10^{-16}\,\frac{e}{2m_e} \tag{2.19}$$

we find by comparing Eqs. (2.13) and (2.19) and employing Eq. (2.17) that d_e provides a better constraint on New Physics than a_e by about $10^5\tan\phi_e^{\rm NP}$, i.e. it already explores scales of about $\Lambda \sim 50\,{\rm TeV}\times\sqrt{C\tan\phi_e^{\rm NP}}$. If $C\tan\phi_e^{\rm NP} \sim \mathcal{O}(1)$, that represents extremely good sensitivity. Even for $C\tan\phi_e^{\rm NP} \simeq 0.01$, $\Lambda \sim 5$ TeV is competitive with the scale of physics being directly explored at the LHC (Large Hadron Collider). See [28] for a recent example. That simple comparison suggests that the electron EDM is a particularly good place to look for a new source of P and T (CP) violation. One that may, in fact, be linked with the matter-antimatter asymmetry of our Universe and thus responsible for our existence. Indeed, in some supersymmetric models, a non-zero d_e is often predicted to be close at hand (see Chapter 13). Since the Standard Model prediction for d_e is currently negligible and does not present a background problem, searches for a $d_e \neq 0$ should be pushed as far as technologically possible. It is expected that planned experiments will improve d_e sensitivity by more than two orders of magnitude, reaching for $C \tan \phi_e^{\rm NP} \sim 1$ scales of New Physics approaching $\Lambda \sim \mathcal{O}\,(1000\,{\rm TeV})$. Alternatively, for low scale New Physics scenarios with $\Lambda \simeq 200$ GeV, such as supersymmetry, $C \tan \phi^{\rm NP}$ as small as 4×10^{-8} will be probed.

2.2. Spin 1/2 Electromagnetic Form Factors

Having described the sensitivity of electron anomalous magnetic and electric dipole moments for probing New Physics via dimension 5 induced operators, we now present a general field theory based analysis applicable to dipole moments of arbitrary spin 1/2 fermions, elementary or composite. We also discuss flavor-changing, dimension 5, electromagnetic transition dipole moments that allow for the decay $\mu \to e\gamma$ and related reactions.

Our discussion begins with the matrix element of the electromagnetic current $J_\mu^{em} = e\sum_f Q_f \bar{f}\gamma_\mu f$, between initial and final states of an arbitrary spin 1/2 fermion f, with momenta p and p' respectively (so that $q = p' - p$)

$$\langle f\,(p')\,|J_\mu^{em}|\,f\,(p)\rangle = \bar{u}_f\,(p')\,\Gamma_\mu u_f\,(p) \tag{2.20}$$

where $\bar{u}_f$ and u_f are Dirac spinor fields and Γ_μ has the general Lorentz structure

$$\begin{aligned}\Gamma_\mu = {} & F_1\left(q^2\right)\gamma_\mu + iF_2\left(q^2\right)\sigma_{\mu\nu}q^\nu - F_3\left(q^2\right)\sigma_{\mu\nu}q^\nu\gamma_5 \\ & + F_A\left(q^2\right)\left(\gamma_\mu q^2 - 2m_f q_\mu\right)\gamma_5 .\end{aligned} \tag{2.21}$$

Hermiticity of J_μ^{em} requires that the form factors in Eq. (2.21) be real (modulo unstable particle effects).

The three $F_i\left(q^2\right)$, $i = 1, 2, 3$ in Eq. (2.21) are the charge, anomalous magnetic dipole and electric dipole form factors. $F_A\left(q^2\right)$ is called the anapole form factor. Anapole effects violate parity and are generally a component of electroweak loop physics. Although interesting, we will not discuss anapole induced interactions in this article.

The static charge and dipole moments are defined at $q^2 = 0$,

$$F_1(0) = Q_f e = \text{electric charge}, \tag{2.22}$$

$$F_2(0) = a_f Q_f \frac{e}{2m_f} = \text{anomalous magnetic moment}, \tag{2.23}$$

$$F_3(0) = d_f Q_f = \text{electric dipole moment}. \tag{2.24}$$

The effective (quantum loop induced) Hamiltonian that gives rise to F_2 and F_3 interactions is

$$H_{\text{dipole}} = -\frac{1}{2}\left(F_2 \bar{f}(x)\sigma_{\mu\nu} f(x) + iF_3 \bar{f}(x)\sigma_{\mu\nu}\gamma_5 f(x)\right) F^{\mu\nu}(x), \tag{2.25}$$

$$F_{\mu\nu}(x) = \partial_\mu A_\nu(x) - \partial_\nu A_\mu(x). \tag{2.26}$$

In the case of neutral spin 1/2 particles with $Q_f = 0$, such as neutrons or (Dirac) neutrinos, $F_{2,3}(0) \neq 0$ and Q_f parameterization in Eqs. (2.23, 2.24) is not appropriate; so, we take $Q_f \to \pm 1$ depending on the charge of their isospin partner, e.g. $Q_f \to 1$ for the neutron and -1 for the neutrino. In the non-relativistic limit, the electric dipole interaction reduces to $-d_f \vec{s}\cdot\vec{E}$. That term is odd (changes sign) under P and T transformations, hence, it violates both symmetries [21].

In modern terminology, the interactions in Eq. (2.25) are called dimension 5 operators. That nomenclature stems from the fact that spinor fields have dimension 3/2 while the dimension of $F_{\mu\nu}$ is 2. Hence, the field products in Eq. (2.25) have dimension 5. Since the Hamiltonian has overall dimension 4, the form factors F_2 and F_3 are necessarily of dimension -1 (they behave like $1/M$).

Dimension 5 operators are generally not allowed in fundamental classical Lagrangians because they spoil renormalizability at the quantum field theory level. They will, nevertheless, arise at the quantum loop level, if no symmetry forbids them. As such, both a_f and d_f must be finite and calculable in terms of other parameters of the theory. Unfortunately, they can often be difficult to reliably compute because they can be due to high orders in loop perturbation theory, may be clouded by strong interaction uncertainties or, in the case of EDMs, depend on unknown model dependent phases.

2.2.1. *Lepton anomalous magnetic and electric dipole moments*

In Table 2.1, we list the current measured values of the electron and muon anomalous magnetic moments. Note that a_e is more precisely determined

Table 2.1. Measured values and bounds for charged lepton anomalous magnetic and electric dipole moments. EDM constraints are given in e·cm as well as $e/2m_l$ magneton units. The muon EDM bound has recently been submitted for publication [29].

Lepton (l)	a_l	$\lvert d_l \rvert$
electron	$115\ 965\ 218\ 073(28) \times 10^{-14}$	$< 2 \times 10^{-27}\, e \cdot \mathrm{cm} \simeq 1 \times 10^{-16} \frac{e}{2m_e}$
muon	$116\ 592\ 080(63) \times 10^{-11}$	$< 1.8 \times 10^{-19}\, e \cdot \mathrm{cm} \simeq 2 \times 10^{-6} \frac{e}{2m_\mu}$
tau	$< 2 \times 10^{-2}$	$< 10^{-16}\, e \cdot \mathrm{cm} \simeq 2 \times 10^{-2} \frac{e}{2m_\tau}$

than a_μ roughly by a factor of 2300. However, New Physics contributions are expected to scale as m_l^2 for both $l = e$ and μ. So, in general, a_μ should be about $\left(\frac{m_\mu}{m_e}\right)^2 \simeq 43\,000$ times[c] more sensitive to New Physics than a_e. In addition, the Standard Model prediction for a_e, as noted in Section 2.1, falls short of a_e^{exp} in precision by about a factor of 20 primarily due to the uncertainty in α. So, overall a_μ is currently about 400 times more sensitive to New Physics than a_e and probes mass scales about 20 times higher. Indeed, already a sizable difference between a_μ^{exp} and the Standard Model prediction, a_μ^{SM}, exists and can easily be identified with various reasonable examples of New Physics with Λ extending into the TeV region. That possible hint of New Physics will be discussed in Section 2.3 in some detail. There, we also present a rather up to date discussion of the Standard Model prediction for a_μ and its underlying uncertainties.

In the case of the tau anomalous magnetic moment, as well as its EDM, the current bounds on a_τ and d_τ in Table 2.1 come from the good agreement between theory and experiment for $e^+e^- \to \tau^+\tau^-$. For $s = q^2 \gg 4m_\tau^2$, the cross section for an unexpectedly large F_2 or F_3 is increased to

$$\sigma\left(e^+e^- \to \tau^+\tau^-\right) = \frac{4\pi\alpha^2}{3s} + \frac{\alpha}{6}\left(\left|F_2\left(s\right)\right|^2 + \left|F_3\left(s\right)\right|^2\right) + \ldots \tag{2.27}$$

The agreement between $\sigma\left(e^+e^- \to \tau^+\tau^-\right)^{\mathrm{exp}}$ at $q^2 \simeq (35\,\mathrm{GeV})^2$ with expectations then leads to the bounds on a_τ and d_τ in Table 2.1. Those constraints are not competitive for probing New Physics. In fact the a_τ constraint is more than an order of magnitude larger than the Standard Model prediction, $a_\tau \simeq \frac{\alpha}{2\pi} \simeq 0.001$. To reach even that level of sensitivity will require a more direct experimental approach to a_τ. For a discussion of tau dipole moments see [30].

[c]Of course, there is always the possibility that New Physics violates electron-muon universality and that simple quadratic mass relationship fails. We do not consider such cases here.

In the case of lepton EDMs, the Standard Model predictions are far below experimental capabilities (e.g. $d_e^{\rm SM} \simeq 10^{-38}\, e{\cdot}{\rm cm}$). Hence, the observation of a non-vanishing lepton EDM would be heralded as a CP violating New Physics effect. Since the New Physics is naively expected to scale as m_l (or in some models as m_l^3), we normally estimate[d]

$$|d_e| : |d_\mu| : |d_\tau| :: 1 : 200 : 3500. \tag{2.28}$$

Assuming that relationship, the muon $d_\mu^{\rm exp}$ sensitivity would have to reach about $4 \times 10^{-25}\, e{\cdot}{\rm cm}$ to become competitive with the existing $d_e^{\rm exp}$. As illustrated in Table 2.1, the current bound is far short of that level. Nevertheless, an experiment with $|d_\mu| \simeq 10^{-24}\, e{\cdot}{\rm cm}$ as its goal has been proposed at JPARC in Japan. Such an effort is extremely well motivated. It would bring the muon to a New Physics sensitivity similar to that of the present electron and neutron bounds, our current best probes.

If neutrinos are four-component Dirac particles, i.e. they have left- and right-handed components, they can have magnetic and electric dipole moments. Those moments are expected to be very small since they are likely to be proportional to the neutrino masses which are tiny < 1 eV. For example, if one merely adds singlet sterile right-handed neutrinos to the Standard Model, they can give rise to Dirac neutrino masses, m_{ν_i}, $i = 1, 2, 3$, via the Higgs mechanism as well as loop-induced magnetic dipole moments and EDMs. The leading contribution to $a_{\nu_i} \frac{e}{2m_e}$ (i.e. units of electron Bohr magneton) is given by [32–34]

$$a_{\nu_i} = -\frac{3G_\mu m_e m_{\nu_i}}{4\sqrt{2}\pi^2} = -3 \times 10^{-19} m_{\nu_i}\,({\rm eV}) \tag{2.29}$$

where $G_\mu = 1.166\,37 \times 10^{-5}$ GeV^{-2} and the neutrino magnetic moment is $\vec{\mu}_{\nu_i} = -\frac{e}{m_e} a_{\nu_i} \vec{s}_{\nu_i}$. Such small values are well below current experimental sensitivities. However, in some New Physics scenarios, such as left-right symmetric models [35] or extended Higgs models, it is possible to have larger dipole moments [36]. Therefore, it is interesting to examine what direct laboratory, astrophysical and cosmological bounds can be placed on neutrino dipole moments (magnetic or electric), independent of theory (see [37, 38] for a recent detailed review). In Table 2.2, we list various bounds.

Because the neutrinos considered in Table 2.2 are relativistic, we have cited bounds for the combined moment $\sqrt{\left|\frac{a_{\nu_i} e}{2m_e}\right|^2 + |d_{\nu_i}|^2}$ rather than just $\left|\frac{a_{\nu_i} e}{2m_e}\right|$ as is customarily done. Of course, if one assumes $|d_{\nu_i}| \ll \left|\frac{a_{\nu_i} e}{2m_e}\right|$,

[d]For examples where d_μ is enhanced beyond Eq. (2.28) see [31].

Table 2.2. Some bounds on magnetic and electric dipole moments $a_{\nu_i}e/2m_e$ and d_{ν_i}, $i = 1, 2, 3$, for neutrino mass eigenstates.

Neutrino	$\sqrt{\left\|\frac{a_{\nu_i}e}{2m_e}\right\|^2 + \|d_{\nu_i}\|^2}$	Source
ν_1	$< 2\times 10^{-10}\frac{e}{2m_e} \simeq 4\times 10^{-21}\, e\cdot\mathrm{cm}$	$\bar{\nu}_e e \to \bar{\nu}_e e$
ν_2	$< 3\times 10^{-10}\frac{e}{2m_e} \simeq 6\times 10^{-21}\, e\cdot\mathrm{cm}$	$\bar{\nu}_e e \to \bar{\nu}_e e$
ν_3	$< 5\times 10^{-10}\frac{e}{2m_e} \simeq 3\times 1\times 10^{-20}\, e\cdot\mathrm{cm}$	$\nu_\mu e \to \nu_\mu e$
ν_i	$< 3\times 10^{-12}\frac{e}{2m_e} \simeq 6\times 10^{-23}\, e\cdot\mathrm{cm}$	Astrophysics
ν_1	$< 6\times 10^{-13}\frac{e}{2m_e} \simeq 1.2\times 10^{-23}\, e\cdot\mathrm{cm}$	Supernova 1987A
ν_2	$< 8\times 10^{-13}\frac{e}{2m_e} \simeq 1.6\times 10^{-23}\, e\cdot\mathrm{cm}$	Supernova 1987A
ν_i	$< 10^{-16}\frac{e}{2m_e} \simeq 2\times 10^{-27}\, e\cdot\mathrm{cm}$	Nucleosynthesis

they can be taken as simply magnetic moment bounds. Also, we apply the bounds to neutrino mass eigenstates rather than flavor states (as customarily done). So, for example, the best direct laboratory bounds on dipole moments currently come from $\bar{\nu}_e e \to \bar{\nu}_e e$ and $\nu_\mu e \to \nu_\mu e$ scattering cross section measurements. Those cross sections are increased by [39]

$$\frac{\Delta d\sigma}{dy} \simeq |\kappa_\nu|^2 \frac{\pi\alpha^2}{m_e^2}\left(\frac{1}{y} - 1\right) \qquad (2.30)$$

$$y = \frac{E_e' - m_e}{E_\nu}$$

if neutrinos have dipole moments $\kappa_\nu e/2m_e$,

$$|\kappa_\nu|^2 = |a_\nu|^2 + \left|d_\nu \frac{2m_e}{e}\right|^2 . \qquad (2.31)$$

Converting from a flavor to a mass eigenstate basis then dilutes somewhat the bounds. For example, using the three-by-three mixing matrix angles θ_{12}, θ_{13}, and θ_{23} (see Section 2.4), one finds (ignoring oscillations)

$$|\kappa_{\nu_e}|^2 = |\kappa_{\nu_1}|^2 \cos^2\theta_{12}\cos^2\theta_{13} + |\kappa_{\nu_2}|^2 \sin^2\theta_{12}\cos^2\theta_{13} + |\kappa_{\nu_3}|^2 \sin^2\theta_{13} \qquad (2.32)$$

so, using $\sin^2\theta_{12} \simeq 0.3$ and $\sin^2\theta_{13} \simeq 0$

$$|\kappa_{\nu_e}|^2 = 0.7\,|\kappa_{\nu_1}|^2 + 0.3\,|\kappa_{\nu_2}|^2 . \qquad (2.33)$$

Similarly, using $\sin^2\theta_{23} \simeq 0.5$ implies

$$|\kappa_{\nu_\mu}|^2 = 0.14\,|\kappa_{\nu_1}|^2 + 0.36\,|\kappa_{\nu_2}|^2 + 0.5\,|\kappa_{\nu_3}|^2 . \qquad (2.34)$$

Interestingly, ν_3 gives the largest contribution to κ_{ν_μ}.

A larger than expected $\nu e \to \nu e$ scattering cross section, particularly one exhibiting the distinctive $1/y$ dependence in Eq. (2.30), would be evidence for a non-vanishing neutrino dipole moment. Consistency of the current $\bar{\nu}_e e$ and $\nu_\mu e$ cross sections with Standard Model expectations, along with solar neutrino measurements, gives the constraints [37, 38]

$$|\kappa_{\nu_e}| < 6 \times 10^{-11}, \qquad |\kappa_{\nu_\mu}| < 1.5 \times 10^{-10} \tag{2.35}$$

used to derive the first three bounds in Table 2.2. Those rather direct bounds could be improved by new dedicated low energy scattering experiments, perhaps by one or two orders of magnitude. It would be very difficult to do much better because of the (squared) $|\kappa_\nu|^2$ dependence in Eq. (2.30). So, it seems unlikely that neutrino dipole moments are directly observable in scattering experiments.

Tighter constraints on (Dirac) neutrino magnetic and electric dipole moments can be obtained from astrophysics. If neutrinos couple to photons via such moments, then the decay, plasmon $\to \nu\bar{\nu}$, can occur in stellar interiors [35, 40–42]. (A plasmon is an effectively massive photon with mass ω_p (plasma frequency) due to electron-hole excitations in a plasma.) Because neutrinos immediately escape and carry away energy, that decay process would speed up stellar evolution. Since evidence for such a speed up is not seen, the bound of $6 \times 10^{-23}\, e{\cdot}\mathrm{cm}$ [43], independent of neutrino species, in Table 2.2 is obtained.

An independent bound on κ_{ν_e} (or κ_{ν_1} and κ_{ν_2} via Eq. (2.33)) can be obtained from the detection of $\bar{\nu}_e$ neutrinos from supernova 1987A at about the expected flux level [44–46]. A dipole moment would lead to the production of sterile right-handed neutrinos (left-handed antineutrinos) in the very dense pre-supernova core. A significant flux loss would alter the supernova collapse dynamics and subsequent explosion. Those considerations lead to the bound $|\kappa_{\nu_e}| < 5 \times 10^{-13}$ and the slightly diluted constraints in Table 2.2.

The most constraining bound on neutrino dipole moments currently comes from primordial nucleosynthesis under the assumption of a background magnetic field at those early times [47]. Spin precession in such a field would have resulted in sterile neutrinos and a change in the expansion rate of the Universe. The good agreement between the observed helium abundance and the usual "Big Bang" prediction then leads to the bound $|\kappa_{\nu_i}| < 10^{-16}$ in Table 2.2. Note that even that most stringent bound when phrased in terms of a New Physics scale Λ seems to suggest $C\frac{m_\nu m_e}{\Lambda^2} \simeq 10^{-16}$ is at best being probed. Even for $C \sim \mathcal{O}(1)$, that corresponds (assuming $m_\nu < 1$ eV) to $\Lambda \lesssim \mathcal{O}(70\text{ GeV})$ or smaller; not a competitive approach.

In the case of Majorana neutrinos, they are self-conjugate and, therefore, must have vanishing magnetic and electric dipole moments. They can have,

Table 2.3. Comparison of the neutron and proton anomalous magnetic moments (in units of $e/2m_N$, $m_N = (m_n + m_p)/2$). Also given are the current bounds on neutron [51] and proton [52] EDMs in units of $e{\cdot}$cm and $e/2m_N$.

N	F_{2N} $(e/2m_N)$	$\|F_{3N}\|$
n	$-1.913\,042\,7\,(5)$	$< 0.29 \times 10^{-25}\, e{\cdot}\text{cm} \simeq 1 \times 10^{-13} e/2m_N$
p	$+1.792\,158\,142\,(28)$	$< 7.9 \times 10^{-25}\, e{\cdot}\text{cm} \simeq 3 \times 10^{-12} e/2m_N$

however, transition moments between different mass eigenstates which lead to $\nu_i \to \nu_j\gamma$ [32] or spin-flavor precession in magnetic fields [48–50]. The direct bounds in Table 2.2 from νe scattering, solar neutrino oscillations and the astrophysical plasmon decay constraint can be applied to such transition moments, but the astrophysical and cosmological constraints are significantly diluted (or rendered inapplicable) because the final state neutrinos are active rather than sterile.

2.2.2. *Nucleon dipole moments*

The neutron and proton are known to have large anomalous magnetic dipole moments due to their composite structure. Current values are listed in Table 2.3 in units of the nucleon Bohr magneton, $e/2m_N$.

Since P and T symmetries are violated in Nature, it is quite likely (almost certain) that nucleons also have electric dipole moments $= d_N$ in their spin direction. In fact, all spin 1/2 Dirac particles should have, albeit tiny, EDMs. Indeed, the Standard Model predicts, on the basis of CKM quark mixing [53, 54],

$$|d_N| \simeq 10^{-32}\, e \cdot \text{cm} \qquad \text{Standard Model} \tag{2.36}$$

which (unfortunately) is more than six orders of magnitude below the existing neutron EDM bound in Table 2.3 and experimentally unobservable in the foreseeable future. Of course, our inability to confirm the Standard Model prediction can be viewed as fortuitous. It means that any discovery of a nucleon EDM with $|d_N| \gg 10^{-32}\, e{\cdot}$cm is providing direct evidence for New Physics. Furthermore, that New Physics would be an additional source of CP violation and might help explain the matter-antimatter asymmetry of our Universe, an exciting possibility.

We see from Table 2.3 that the nucleon magnetic dipole moments (particularly the proton's) are very precisely measured. In fact, the uncertainty in the proton's anomalous magnetic moment, $\pm 28 \times 10^{-9}$, is only a factor of 44 worse than the muon's a_μ^{exp} $\left(\pm 63 \times 10^{-11}\right)$. If the proton were an elementary particle and if its hadronic contributions were under theoretical control, a_p because of the proton's larger mass, $(m_p/m_\mu)^2 \simeq 77$, would be a slightly better probe than the muon a_μ for New Physics. However, that is not the case. The proton is not elementary and there are very large hadronic uncertainties in the Standard Model prediction for a_p, at least at the several percent level, making its high precision useless as a window to New Physics.

We can, however, learn some valuable lessons from $F_{2p}(0)$ and $F_{2n}(0)$ that may be useful in trying to anticipate and motivate possible measurements of $F_{3p}(0)$ and $F_{3n}(0)$, the nucleon EDMs.

We first observe that when written as isovector and isoscalar combinations

$$F_{2N}^{(I=1)} = \frac{F_{2p} - F_{2n}}{2} \simeq 1.85,$$
$$F_{2N}^{(I=0)} = \frac{F_{2p} + F_{2n}}{2} \simeq -0.06, \tag{2.37}$$

it is clear that the isovector component dominates, i.e. $F_{2n} \simeq -F_{2p}$.

Will the nucleon EDMs also be isovector? A priori, nothing tells us that must be the case. Therefore, we should strive to measure both d_n and d_p in order to determine the isospin dependence and relate it to the predictions of New Physics scenarios. Some models may predict nearly pure isovector nucleon EDMs. Others may have a relatively large isoscalar part. In other words, we really have to consider d_n and d_p as independent quantities and should try to determine both with similar precision, if possible. Here, we might remark that current lattice techniques are better for determining the isovector EDM combination

$$F_{3N}^{(I=1)} = \frac{F_{3p} - F_{3n}}{2}, \tag{2.38}$$

rather than the isoscalar

$$F_{3N}^{(I=0)} = \frac{F_{3p} + F_{3n}}{2}. \tag{2.39}$$

Hence, an experimental determination of $d_N^{(I=1)} = (d_p - d_n)/2$ will be very important if we are to use lattice gauge theory to unfold the underlying New Physics responsible for nucleon EDMs. That requires a measurement of both d_n and d_p.

The values of the anomalous nucleon magnetic moments in Eq. (2.37) also illustrate a success of the $SU(6)$ constituent quark model. ($SU(6)$ is a largely unsuccessful attempt to combine $SU(3)$ flavor and $SU(2)$ spin into a spin-flavor symmetry.) It predicts for the full nucleon magnetic dipole moments $\vec{\mu}_N$

$$\begin{aligned}\vec{\mu}_n &= \frac{4}{3}\vec{\mu}_d - \frac{1}{3}\vec{\mu}_u \\ \vec{\mu}_p &= \frac{4}{3}\vec{\mu}_u - \frac{1}{3}\vec{\mu}_d\end{aligned} \tag{2.40}$$

where $\vec{\mu}_d$ and $\vec{\mu}_u$ are the constituent down and up quark dipole moments. Then employing

$$\vec{\mu}_u = -2\vec{\mu}_d, \qquad |\vec{\mu}_u| = \frac{2}{3}\frac{e}{2m_q} = \frac{e}{m_N} \tag{2.41}$$

where $m_q = \frac{m_N}{3}$ has been employed, one finds

$$\begin{aligned}F_{2n}(0) &= a_n = -2\frac{e}{2m_N} \\ F_{2p}(0) &= a_p = 2\frac{e}{2m_N}\end{aligned} \tag{2.42}$$

which implies pure isovector anomalous nucleon magnetic moments, with values in very good accord with Eq. (2.37). For some things, that naive quark model approach works remarkably well and Eq. (2.42) was one of its early successes. As a result, the $SU(6)$ constituent quark model is often used to predict nucleon EDMs from quark EDMs $\vec{d}_d$ and $\vec{d}_u$. Following Eq. (2.40) one expects

$$\begin{aligned}\vec{d}_n &= \frac{4}{3}\vec{d}_d - \frac{1}{3}\vec{d}_u \\ \vec{d}_p &= \frac{4}{3}\vec{d}_u - \frac{1}{3}\vec{d}_d.\end{aligned} \tag{2.43}$$

In such a scenario, the isovector EDM will dominate if $\vec{d}_d \simeq -\vec{d}_u$ while the isoscalar dominates if $\vec{d}_d \simeq \vec{d}_u$. Without a specific model it is hard to favor one over the other. Indeed, underlying calculations of quark EDMs are unlikely to satisfy simple isospin relations. On the basis of Eq. (2.43), one might, very roughly, expect

$$\frac{1}{4} < \left|\frac{d_p}{d_n}\right| < 4 \tag{2.44}$$

for the relative magnitudes, with either relative sign possible. Of course, cancellations could take the ratio outside these limits.

To study further the relationship between d_n and d_p requires one to examine specific New Physics models of CP violation. Such a detailed discussion is beyond the scope of this chapter. Instead, we defer to Chapter 13 by Pospelov and Ritz and Ref. [55] where a much more thorough discussion of various models is given. Some of their examples suggest isovector EDM dominance, while others (particularly if heavy quark or gluonic CP violation is dominant) have relatively large isoscalar components. We will end this subsection with a review of QCD $\bar{\theta}$ parameter contributions to nucleon EDMs. Then in Subsection 2.2.3, a complex phase formalism that combines magnetic and electric dipole moments is introduced. As we shall see, that approach provides a convenient means of comparing the New Physics sensitivity of different EDM searches. In the case of leptons, it also allows us to compare anomalous magnetic and electric dipole moments in a more unified manner.

The $\bar{\theta}$ parameter The vacuum state of QCD (Quantum Chromodynamics) can be parameterized by an angle θ, $0 \leq \theta \leq 2\pi$. The effect of that angle is to modify the QCD Lagrangian such that

$$\mathcal{L}_{\rm QCD}^{\rm eff} = \mathcal{L}_{\rm QCD} + \theta \frac{g_{\rm QCD}^2}{32\pi^2} F^{a\mu\nu} \widetilde{F}_{\mu\nu}^a, \qquad a = 1, 2, \ldots, 8 \tag{2.45}$$

where $\widetilde{F}_{\mu\nu}^a$ is the dual tensor, $\widetilde{F}_{\mu\nu}^a = \frac{1}{2}\epsilon_{\mu\nu\alpha\beta}F^{a\alpha\beta}$. The appended term violates P and T symmetries and appears capable of giving rise to strong interaction CP violation. However, that term can be written as a total divergence and as such will not have perturbative consequences. It will, nevertheless, have non-perturbative implications via surface effects due to long distance gluon field configurations called instantons. Instantons are known to play an important role in QCD. For example, they break what appears to be an extra $U_A(1)$ symmetry in the theory and provide the η' pseudoscalar meson with a relatively large mass [56]. So, instanton effects combined with a non-vanishing θ will lead to nucleon EDMs.

One can set $\theta = 0$ by demanding that the QCD Lagrangian conserve CP. However, the problem returns when we add CP violation back into the theory via electroweak physics, particularly a relatively large New Physics CP violating source. The New Physics can contribute to the quark mass matrix, M, via loop effects such that its rediagonalization and elimination of complex phases replaces θ in Eq. (2.45) by $\bar{\theta}$ where

$$\bar{\theta} = \theta + \arg(\det M) \tag{2.46}$$

via the chiral $U_A(1)$ anomaly. In that way, new sources of CP violation can slip into QCD. To naturally remove $\bar{\theta}$ requires an axion (very light pseudoscalar particle) or massless quark, neither of which appears to exist in Nature. Alternatively, one can require that $\bar{\theta}$ be very small and examine its non-vanishing consequences.

What are the most dramatic implications of $\bar{\theta} \neq 0$? It will give rise to non-vanishing d_n and d_p. To reliably compute those induced EDMs as a function of $\bar{\theta}$ is somewhat complicated and must address the $U_A(1)$ symmetry breaking due to instantons, if it is to be meaningful.

A realistic leading log (in $\ln(m_p/m_\pi)$) current algebra calculation by Crewther *et al.* [57] gives

$$|d_n| \simeq 3.6 \times 10^{-16}\bar{\theta}\, e \cdot \mathrm{cm} \tag{2.47}$$

which implies from the bound on $|d_n|$ in Table 2.3

$$\bar{\theta} \lesssim 10^{-10}. \tag{2.48}$$

Although that constraint seems quite severe, one should remember that it likely represents the effect of new electroweak physics and is, therefore, expected to be small relative to strong interactions.

We note that for the proton, leading log contributions give

$$d_p \simeq -d_n \qquad \text{(leading logs)} \tag{2.49}$$

i.e. an isovector nucleon EDM. That should not be too surprising since the calculation of those EDMs involves effective pion loops. So, if a non-vanishing measurement of d_n is made, one of the next steps should be to determine d_p and see if Eq. (2.49) is respected. Of course, non leading-log $\bar{\theta}$ contributions will alter the relationship in Eq. (2.49) somewhat and give an isoscalar contribution. Reliably computing the non-leading $\bar{\theta}$ contribution to nucleon EDMs, particularly the isoscalar part is challenging but should be undertaken.

What are the prospects for measuring d_n and d_p or improving their bounds? Searches for a neutron EDM have a long distinguished history. New efforts currently getting underway expect to reach about $|d_n| \sim 10^{-28}\, e \cdot \mathrm{cm}$ sensitivity or better, i.e. an improvement of more than two orders of magnitude.

In the case of the proton, bounds on d_p have traditionally been a byproduct of atomic EDM searches. However, recently the idea of directly measuring a proton (or deuteron) EDM using a storage ring (discussed in Chapter 17) has been explored. It seems that reaching $10^{-28}\, e \cdot \mathrm{cm}$ or better for $|d_p|$ is technically feasible. Hopefully, some of the discussion given here reinforces the need to push both d_n and d_p as far as possible. The two measurements are complementary and both will be needed to properly interpret a non-vanishing result, including its underlying New Physics explanation. Nucleon EDMs are considered in detail in Chapters 15 and 16.

2.2.3. *Complex formalism*

The anomalous magnetic and electric dipole form factor operators in Eq. (2.21) differ by an $i\gamma_5$ factor. That difference is suggestive of a chiral phase effect. It can, therefore, be useful to combine the two real form factors, F_2 and F_3 into a single complex dipole form factor $F_D\left(q^2\right)$ such that

$$F_D\left(q^2\right) = F_2\left(q^2\right) + iF_3\left(q^2\right). \tag{2.50}$$

In that way, the interactions in Eq. (2.25) can be recast as

$$H_{\text{dipole}} = -\frac{1}{2}\left(F_D \bar{f}_L \sigma_{\mu\nu} f_R + F_D^* \bar{f}_R \sigma_{\mu\nu} f_L\right) F^{\mu\nu} \tag{2.51}$$

where $f_R = \frac{1+\gamma_5}{2} f$ and $f_L = \frac{1-\gamma_5}{2} f$ are the right and left-handed chiral spinor projections. At $q^2 = 0$, we have

$$F_D(0) = \left(a_f \frac{e}{2m_f} + id_f\right) Q_f \tag{2.52}$$

which can be written as

$$F_D(0) = |F_D(0)|\, e^{i\phi_f} \tag{2.53}$$

where

$$|F_D(0)| = \sqrt{\left(a_f \frac{e}{2m_f}\right)^2 + d_f^2}$$

$$\tan\phi_f = \frac{d_f}{a_f \frac{e}{2m_f}}. \tag{2.54}$$

The phase ϕ_f merely parameterizes the relative size of a fermion's EDM and its anomalous magnetic moment. It can be regarded as a measure of CP violation.

Using the prescription in Eq. (2.54), we can turn EDM bounds into ϕ_f constraints. The resulting bounds on $\tan\phi_f$ are illustrated in Table 2.4 for $f = e, \mu, p, n$. Those bounds tell us that the phases are indeed very small. Of course, that is to be expected since for charged leptons the anomalous magnetic moment $a_l \simeq \alpha/2\pi$ is of QED origin and for nucleons $|a_N| \simeq 2$ is due to strong interactions, while EDMs are generally induced by some high scale, possibly multi-loop weak effect. For example, in the case of the nucleons, the Standard Model (SM) expectation would be $\tan\phi_N^{\text{SM}} \simeq 10^{-19}$.

At this point, it might appear that the phase, ϕ_f, in Eq. (2.53) can be removed from Eq. (2.51) by a chiral phase rotation that differs for f_R and f_L. That is possible; however, such a rotation would induce a complex

Table 2.4. Bounds on the CP violating dipole phase, $\tan\phi_f = \frac{d_f}{a_f \frac{e}{2m_f}}$ based on values in Tables 2.1 and 2.3.

Fermion f	$\lvert\tan\phi_f\rvert$ bound
electron	$< 9 \times 10^{-14}$
muon	$< 2 \times 10^{-3}$
neutron	$< 5 \times 10^{-14}$
proton	$< 1 \times 10^{-12}$

chiral phase in the fermion mass which has a similar chiral (changing) structure as Eq. (2.51). The phase ϕ_f should be viewed as the relative phase between a fermion's dipole moment and its mass. (For an unstable particle, there should be an imaginary phase associated with its width which will be common to both the mass and dipole moment.)

The phases, ϕ_f, defined above are all expected to be very small (except possibly in the case of a Dirac neutrino where the anomalous magnetic moment and electric dipole moment are both tiny). As such, they are not very useful except as a reminder that CP violation is generally a very small effect.

A more practical application of the complex formalism is to use it for New Physics (NP) contributions separately [58–60]. Calling $a_f^{\rm NP}$ and $d_f^{\rm NP}$ the contributions coming from a new physics source, one defines the relation

$$\tan\phi_f^{\rm NP} = \frac{d_f^{\rm NP}}{a_f^{\rm NP}\frac{e}{2m_f}}. \tag{2.55}$$

In that way, $\phi_f^{\rm NP}$ parameterizes the relative size of CP violating and non-violating NP effects in dipole moments. Here, again, we assume a relative phase in which the fermion mass is real.

The New Physics phase, $\phi_f^{\rm NP}$, can be useful for discussing the reach or sensitivity of EDM experiments relative to anomalous magnetic moment precision measurements. So, for example, if the NP at mass scale Λ_f contributes to an anomalous magnetic moment

$$a_f^{\rm NP} = C_f \frac{m_f^2}{\Lambda_f^2}, \tag{2.56}$$

it will also lead to

$$d_f^{\rm NP} = C_f \frac{m_f^2}{\Lambda_f^2}\frac{e}{2m_f}\tan\phi_f^{\rm NP}. \tag{2.57}$$

So, $\tan\phi_f^{\rm NP}$ simply parameterizes the degree of suppression for CP violation in the New Physics. It can be $\tan\phi_f^{\rm NP} \sim \mathcal{O}(1)$ if both $a_f^{\rm NP}$ and $d_f^{\rm NP}$ are of the same order in perturbation theory and large phases in the NP couplings are present. However, more likely $\tan\phi_f^{\rm NP} \ll 1$. For example, if a fermion mass and its dipole moments are due to the same NP (see Section 2.3.3.2) it is quite possible that $C_f \sim \mathcal{O}(1)$ but $\tan\phi_f^{\rm NP} \simeq 0$ because the same chiral rotation that renders the fermion mass real also removes the phase between $a_f^{\rm NP}$ and $d_f^{\rm NP}$. So, one should take any assumption regarding the value of $\tan\phi_f^{\rm NP}$ with a grain of salt. It is very model dependent.

Consider the constraints on New Physics from the leptons e and μ. We find from Table 2.1 and Standard Model predictions for the a_l, $l = e, \mu$ that d_e is about $10^5 \tan\phi_e^{\rm NP}$ more sensitive to NP than a_e. On the other hand, d_μ is currently about $4 \times 10^{-4} \tan\phi_\mu^{\rm NP}$ less constraining than a_μ and not competitive. (A proposed five orders of magnitude improvement in $d_\mu^{\rm exp}$ could alter the situation.)

Taking $C_e \simeq C_\mu \simeq 1$, one finds the NP scales probed by lepton dipole moment experiments are

$$\begin{aligned}
\Lambda_e &\lesssim 160 \text{ GeV from } a_e \\
\Lambda_e &\lesssim 50 \text{ TeV} \times \sqrt{\tan\phi_{\rm e}^{\rm NP}} \text{ from } d_e \\
\Lambda_\mu &\lesssim 4 \text{ TeV from } a_\mu .
\end{aligned} \tag{2.58}$$

Those values should be considered upper bounds on the scale of physics probed, since we assumed $C_e \simeq C_\mu \simeq 1$. From that perspective, it seems that only d_e and a_μ are currently sensitive to the types of New Physics we hope to explore at the LHC. If $\tan\phi_{\rm e}^{\rm NP} \gtrsim 0.006$, it appears that d_e is the most constraining. That is very encouraging, since $d_{\rm e}^{\rm exp}$ probes are expected to further improve by several orders of magnitude in the near future, pushing Λ_e sensitivity to $\mathcal{O}\left(1000\,\text{TeV} \cdot \sqrt{\tan\phi_{\rm e}^{\rm NP}}\right)$.

In the case of the muon, $a_\mu^{\rm exp}$ already seems to disagree with the Standard Model prediction $a_\mu^{\rm SM}$. New Physics scenarios with $a_\mu^{\rm NP} \sim 3 \times 10^{-9}$ that can explain the disagreement are discussed in Section 2.3 and in Chapter 12. They would suggest $d_\mu \simeq 3 \times 10^{-22} \tan\phi_\mu^{\rm NP}$ e·cm.

Nucleon EDMs are also sensitive probes of New Physics. It is, however, harder to parameterize their dependence on the underlying NP scale. One expects

$$d_N \sim C_N \frac{m}{\Lambda_N^2} \tan\phi_N^{\rm NP} \tag{2.59}$$

where m represents a quark or hadron mass scale. Taking $m \simeq 15$ MeV and $C_n \simeq 1$, we find from the d_n bound in Table 2.3

$$\Lambda_n > 70 \text{ TeV} \times \sqrt{\tan \phi_n^{\text{NP}}} \qquad \text{from } d_n. \tag{2.60}$$

Anticipated improvements in d_n^{exp} sensitivity by more than two orders of magnitude will probe $\Lambda \sim 1000 \text{ TeV} \sqrt{\tan \phi_n^{\text{NP}}}$. So, d_n and d_e bounds currently give roughly similar sensitivity to New Physics. However, there is no clear relationship between $\tan \phi_n^{\text{NP}}$ and $\tan \phi_e^{\text{NP}}$ and even their sources of new CP violation could be very different. It is, therefore, extremely important that d_n improvements by more than two orders of magnitude, currently planned, be carried out concurrent with new d_e^{exp} efforts.

2.2.4. *Transition dipole moments*

Flavor-changing transition amplitudes between distinct fermions can result from flavor off-diagonal matrix elements of the electromagnetic current and lead to $f_i \to f_j + \gamma$ decays ($i \neq j$). We can parameterize those amplitudes in analogy with Eqs. (2.20) and (2.21), but in terms of transition electric and magnetic form factors

$$\langle f_j(p') | J_\mu^{\text{em}} | f_i(p) \rangle = \bar{u}_j(p') \Gamma_\mu^{ij} u_i(p), \tag{2.61}$$

$$\begin{aligned} \Gamma_\mu^{ij} = {} & \left(q^2 g_{\mu\nu} - q_\mu q_\nu\right) \gamma^\nu \left[F_{E0}^{ij}(q^2) + \gamma_5 F_{M0}^{ij}(q^2)\right] \\ & + i\sigma_{\mu\nu} q^\nu \left[F_{M1}^{ij}(q^2) + \gamma_5 F_{E1}^{ij}(q^2)\right]. \end{aligned} \tag{2.62}$$

The first two form factors, F_{E0}^{ij} and F_{M0}^{ij} (transition analogs of the charge radius and anapole moment), contribute to chiral conserving flavor-changing amplitudes at $q^2 \neq 0$ and are part of more general dimension six operators. They can be important, for example in the strangeness changing decay $K^+ \to \pi^+ e^+ e^-$, or in $\mu^+ \to e^+ e^+ e^-$, but are outside the scope of this chapter and will not be discussed further here. We briefly mention them again in Section 2.4.

The transition dipole moments, F_{M1}^{ij} and F_{E1}^{ij}, change chirality and are flavor-changing analogs of magnetic and electric dipole moments. They give rise to gauge invariant dimension five operators and decays $f_i \to f_j + \gamma$ such as $b \to s\gamma$, $\mu \to e\gamma$, $\tau \to \mu\gamma$, etc.

In terms of those form factors, one finds the decay rate

$$\Gamma(f_i \to f_j + \gamma) = \frac{1}{8\pi} \left(\frac{m_{f_i}^2 - m_{f_j}^2}{m_{f_i}}\right)^3 \left(\left|F_{M1}^{ij}\right|^2 + \left|F_{E1}^{ij}\right|^2\right). \tag{2.63}$$

We mention in passing that if the fermions $f_{i,j}$ are charged, there can be an unusually large negative QED correction to this decay rate [61]. Because of the anomalous dimension of the dimension five operators in Eq. (2.62), a virtual photon loop connecting the parent and daughter charged fermions is enhanced by the logarithm of the New Physics mass responsible for the flavor off-diagonal coupling. For sub-TeV mass scales this correction is universal; it does not depend on the details of the New Physics model. It results in a multiplicative factor

$$\Gamma\left(f_i \to f_j + \gamma\right) \to \Gamma\left(f_i \to f_j + \gamma\right)\left(1 - \frac{8 Q_{f_i} Q_{f_j} \alpha}{\pi} \ln \frac{\Lambda}{m_{f_i}}\right), \tag{2.64}$$

where $Q_{f_{i,j}}$ denote charges of the fermions involved and Λ is the characteristic mass scale responsible for their coupling. For example, for the decay $\mu \to e\gamma$ and for $\Lambda = 250$ GeV, this correction is -15%. For comparison, the QED correction to the Standard Model muon decay is only about -0.4%.

We note that the same correction with $Q_i = Q_j$ (and half the coefficient since they are amplitudes) also applies to the anomalous magnetic and electric dipole moment contributions of high scale New Physics effects. An illustration is given for supersymmetry in Subsection 2.3.3.1.

Sometimes, it is convenient to separate the transition dipoles into right and left components,

$$\begin{aligned} D_R^{ij} &= F_{M1}^{ij} + F_{E1}^{ij} \\ D_L^{ij} &= F_{M1}^{ij} - F_{E1}^{ij}. \end{aligned} \tag{2.65}$$

Then, Eq. (2.63) becomes

$$\Gamma\left(f_i \to f_j + \gamma\right) = \frac{1}{16\pi}\left(\frac{m_{f_i}^2 - m_{f_j}^2}{m_{f_i}}\right)^3 \left(\left|D_R^{ij}\right|^2 + \left|D_L^{ij}\right|^2\right). \tag{2.66}$$

Of course, both sets of form factors must be very small for several reasons. First, they are dimension five and must stem from quantum loops. Second, they change quark or lepton flavor and are likely to be suppressed by mixing effects and unitarity cancellations. Third, they are expected to be proportional to $\frac{m_{f_i}}{\Lambda^2}$, where Λ is the scale of New Physics responsible for those quantum loop generated amplitudes. With those suppressions in mind, we write the chiral transition dipoles in Eq. (2.65) as

$$\begin{aligned} D_R^{ij} &= \frac{eQ_{f_i}}{2} \frac{m_{f_i}}{\left(\Lambda_R^{ij}\right)^2} \mathcal{E}_{ij}^R, \\ D_L^{ij} &= \frac{eQ_{f_i}}{2} \frac{m_{f_i}}{\left(\Lambda_L^{ij}\right)^2} \mathcal{E}_{ij}^L, \end{aligned} \tag{2.67}$$

where $\mathcal{E}_{ij}^{R,L}$ parameterize possible coupling and mixing suppression factors. The factor of 1/2 in Eq. (2.67) has been included to make the New Physics scale normalization similar to that used in anomalous magnetic and electric dipole moments. We later take $\mathcal{E}_{ij}^{R}$ and $\mathcal{E}_{ij}^{L}$ to be $\mathcal{O}(1)$ for the purpose of crudely estimating the largest scale of physics probed by various flavor-changing reactions. For that purpose, we will use (for $m_{f_i} \gg m_{f_j}$)

$$\Gamma(f_i \to f_j + \gamma) \simeq \frac{\alpha}{16} Q_{f_i}^2 m_{f_i}^5 \left(\left| \frac{\mathcal{E}_{ij}^R}{\left(\Lambda_R^{ij}\right)^2} \right|^2 + \left| \frac{\mathcal{E}_{ij}^L}{\left(\Lambda_L^{ij}\right)^2} \right|^2 \right). \tag{2.68}$$

Of course, New Physics scale sensitivity described in that way is highly subjective. However, it can be useful for comparison of different searches for rare processes and other probes of New Physics.

2.3. Muon Anomalous Magnetic Moment

2.3.1. *Introduction*

The muon's anomalous magnetic moment a_μ provides a particularly sensitive probe for New Physics, as mentioned in Section 2.2.3. Three factors are particularly relevant [62]:

- Muons can be copiously produced in a fully polarized state and live sufficiently long for precise measurements of their precession frequency in a magnetic field [63];
- The Standard Model value of a_μ has been precisely evaluated through efforts of several groups of theorists [64]; and finally
- The muon is sufficiently heavy to be relatively sensitive to New Physics phenomena.

In this Section we briefly review these three aspects of a_μ, emphasizing its sensitivity to New Physics.

Precise measurements of the muon's anomalous magnetic moment began with two decades of dedicated experiments at CERN, completed in 1977, that found [65]

$$a_\mu^{\text{exp}} = 116\,592\,300(840) \times 10^{-11} \qquad \text{(CERN 1977)}. \tag{2.69}$$

Between 1994 and 2001, the experiment E821 at Brookhaven National Laboratory (BNL) ran with much higher statistics and a very stable, well measured magnetic field in its storage ring. It resulted in a 14-fold improvement over the CERN result and, based on data taken with μ^+ and μ^-, lead to [66]

$$a_\mu^{\rm exp} = 116\,592\,080(63) \times 10^{-11} \qquad \text{(BNL final)}, \tag{2.70}$$

a 0.5 ppm determination. Further improvement of this result has been proposed [67]. With an upgrade of E821, a new experiment would aim for a 2–5-fold reduction of the experimental error.

As discussed in Section 2.2.1, although $a_\mu^{\rm exp}$ is currently about 2300 times less precise than $a_e^{\rm exp}$, it is still much more sensitive to hadronic and electroweak quantum loops as well as New Physics effects, since such contributions [68] are generally proportional to m_l^2. The $m_\mu^2/m_e^2 \simeq 43\,000$ enhancement more than compensates for the reduced experimental precision and makes $a_\mu^{\rm exp}$ a better probe of short-distance phenomena. Indeed, as we later illustrate, a deviation in $a_\mu^{\rm exp}$ from the Standard Model prediction, $a_\mu^{\rm SM}$, even at its current level of sensitivity can quite naturally be interpreted as the appearance of New Physics such as supersymmetry at 200-500 GeV, or other even higher scale phenomena. Such an interpretation hinges on a reliable theoretical prediction for $a_\mu^{\rm SM}$ with which to compare, an issue that we address in the next subsection.

2.3.2. a_μ *in the Standard Model*

2.3.2.1. *QED contribution*

99.993 percent of the value of a_μ is due to QED (see Chapters 3 and 4). Similar to the case of the electron, Eq. (2.9), the QED contribution can be described as a series in the fine structure constant α. The difference for the muon is that effects due to virtual electron loops are enhanced by powers of large logarithms $\ln(m_\mu/m_e) \simeq 5.3$ and/or factors of π [69, 70]. Thus, even though a_μ is measured less accurately than a_e, it is necessary to compute the enhanced effects through five loops [71, 72] (see also [73]).

Coefficients of the perturbative expansion depend on two ratios of lepton masses which we take from CODATA 2006 recommended values [74],

$$\frac{m_\mu}{m_e} = 206.768\,2823(52), \qquad \frac{m_\mu}{m_\tau} = 5.945\,92(97) \cdot 10^{-2}. \tag{2.71}$$

With these values one obtains [75, 76]

$$a_\mu^{\rm QED} = \frac{\alpha}{2\pi} + 0.765\,857\,410(27)\left(\frac{\alpha}{\pi}\right)^2 + 24.050\,509\,64(43)\left(\frac{\alpha}{\pi}\right)^3 \\ +130.8055(80)\left(\frac{\alpha}{\pi}\right)^4 + 663(20)\left(\frac{\alpha}{\pi}\right)^5. \tag{2.72}$$

The first three terms are known analytically as discussed in Chapter 4. The four-loop coefficient is a sum of the universal mass-independent $-1.9144(35)$ the same as in the electron a_e in Eq. (2.9); the large electron-loop contribution 132.6823(72) [72]; and the small tau-mass dependent part 0.0376(1) [77]. The last, five-loop coefficient, results from 32 gauge-invariant subsets of diagrams. 20 subsets have been evaluated so far [76], and they already include those diagrams that are enhanced by large logarithms of the electron to muon mass ratio (see also [78]). Employing the value of α from a_e in Eq. (2.10) leads to

$$a_\mu^{\mathrm{QED}} = 116\,584\,718.1(2) \times 10^{-11}. \tag{2.73}$$

The current QED uncertainty is far below the $\pm 63 \times 10^{-11}$ experimental error from E821 and plays no essential role in the confrontation between theory and experiment.

2.3.2.2. *Hadronic loop corrections*

Beginning with $\mathcal{O}\left(\alpha^2\right)$, hadronic loop effects contribute to a_μ via vacuum polarization (see Fig. 2.1(a)). A first principles QCD calculation of that effect does not exist. Fortunately, it is possible to evaluate the leading effect via the dispersion integral [79]

$$a_\mu^{\mathrm{Had}}(\text{vac. pol.}) = \frac{1}{4\pi^3}\int_{4m_\pi^2}^{\infty} \mathrm{d}s\, K(s)\, \sigma^0(s)_{e^+e^-\to\mathrm{hadrons}}, \tag{2.74}$$

where $\sigma^0(s)_{e^+e^-\to\mathrm{hadrons}}$ means QED vacuum polarization and some other extraneous radiative corrections (e.g. initial state radiation) have been

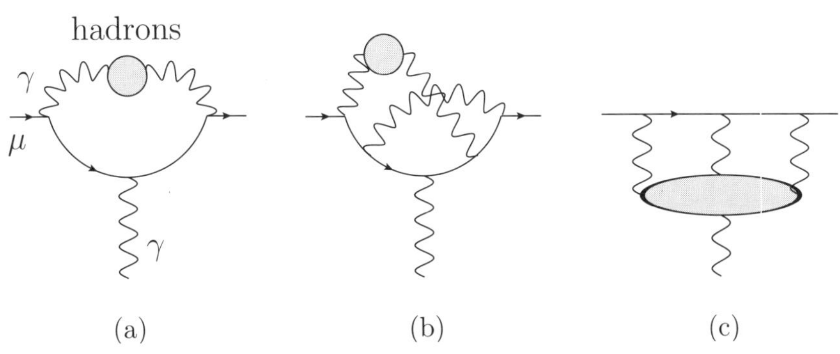

Fig. 2.1. Hadronic contributions to a_μ: (a,b) leading and an example of next-to-leading vacuum polarization diagrams; and (c) light-by-light scattering.

subtracted from measured $e^+e^- \to$ hadrons cross sections, and

$$K(s) = x^2\left(1-\frac{x^2}{2}\right) + (1+x)^2\left(1+\frac{1}{x^2}\right)\left[\ln(1+x) - x + \frac{x^2}{2}\right]$$
$$+\frac{1+x}{1-x}x^2\ln x$$
$$x = \frac{1-\sqrt{1-4m_\mu^2/s}}{1+\sqrt{1-4m_\mu^2/s}}. \tag{2.75}$$

Detailed studies of Eq. (2.74) have been carried out by a number of authors [80–92] and discussed by Davier in Chapter 8.

For the present analysis we adopt a recent value based on published experimental e^+e^- annihilation data [90]

$$a_\mu^{\text{Had}}(\text{vac. pol.}) = 6894(46) \times 10^{-11}. \tag{2.76}$$

This result and its uncertainty are dominated by the low energy region. In fact, the $\rho(770$ MeV) resonance provides about 72% of the total hadronic contribution to a_μ^{Had}(vac. pol.).

To reduce the uncertainty in the ρ resonance region one sometimes employs $\Gamma(\tau \to \nu_\tau\pi^-\pi^0)/\Gamma(\tau \to \nu_\tau\bar{\nu}_e e^-)$ data to supplement or replace $e^+e^- \to \pi^+\pi^-$ cross sections. In the $I = 1$ channel they are related by isospin. Unfortunately, that relation is not exact because of isospin-breaking effects [93–95], quark mass and charge differences [96–98]. Those corrections introduce a theoretical uncertainty that is at present difficult to fully assess [99, 100]. Using tau data [101] in place of e^+e^- increases the Standard Model prediction for a_μ in Eq. (2.76) by about 150×10^{-11} which, as we will see, would bring it much closer to the measured value.

Another way to augment the e^+e^- annihilation data is the radiative return method [102]. Collision energy can be reduced in some events at medium-energy facilities designed to study e^+e^- collisions at the ϕ (~ 1 GeV) or Υ (~ 10 GeV) resonance, due to initial state radiation. Such events can be used to map out the $e^+e^- \to$ hadrons cross section throughout the energy spectrum. Preliminary BaBar results using this approach seem to agree with the tau data [103]; however, radiative return data from KLOE at the ϕ resonance confirm the results in Eq. (2.76) [104, 105].

Smaller but important hadronic effects occur also at three loops: photonic corrections to the diagram with a hadronic vacuum polarization (see Fig. 2.1(b)), and light-by-light scattering, Fig. 2.1(c). The former contributes [85, 106, 107]

$$\Delta a_\mu^{\text{Had}}(\text{vac. pol.}) = -98(1) \times 10^{-11}. \tag{2.77}$$

Light-by-light hadronic diagrams pose a whole new level of theoretical difficulties. Like the hadronic vacuum polarization, they cannot be evaluated from first principles in QCD, although lattice efforts are being undertaken [108]. Chiral perturbation theory and models of light hadron interactions have been employed to estimate their effect [109–114]. A compromise value accepted [116] by the authors of Chapter 9 is

$$\Delta a_\mu^{\rm Had}(\text{light-by-light}) = 105(26) \times 10^{-11}. \tag{2.78}$$

Adding those contributions to Eqs. (2.76) leads to the total hadronic contribution

$$a_\mu^{\rm Had} = 6901(53) \times 10^{-11} \tag{2.79}$$

which we will subsequently use in comparison of theory and experiment.

However, one should be mindful of the differences among the various e^+e^- and τ decay results. The disagreement between those studies represents the main theoretical issue in $a_\mu^{\rm SM}$ which we have not attempted to quantify. It would be very valuable to supplement the above evaluation of $a_\mu^{\rm Had}$ with lattice calculations (for the light-by-light contribution), as well as further improved e^+e^- data and tau decay studies.

2.3.2.3. *Electroweak corrections*

The original goal of E821 at Brookhaven was to measure the smallest among the Standard Model contributions to a_μ, the electroweak radiative corrections (see Fig. 2.2). The leading, one-loop electroweak effects are [117–123]

$$\begin{aligned} a_\mu^{\rm EW}(1\,\text{loop}) &= \frac{5}{3}\frac{\mathrm{G}_\mu \mathrm{m}_\mu^2}{8\sqrt{2}\pi^2} \\ &\times \left[1 + \frac{1}{5}(1-4\sin^2\theta_W)^2 + \mathcal{O}\left(\frac{m_\mu^2}{M^2}\right)\right] \\ &\approx 195 \times 10^{-11} \end{aligned} \tag{2.80}$$

where $G_\mu = 1.16637(1) \times 10^{-5}$ GeV^{-2}, $\sin^2\theta_W \equiv 1 - M_W^2/M_Z^2 \simeq 0.223$. and $M = M_W$ or $M_{\rm Higgs}$.

The muon $g-2$ was the first observable to which full two-loop electroweak corrections were calculated [124, 125]. Those corrections, described by about 200 Feynman diagrams, include a number of interesting effects. For example, the Higgs boson contribution at the two-loop order is much larger than at one loop. The one-loop Higgs boson diagram, shown in Fig. 2.2, is suppressed by the two scalar couplings to a relatively light lepton. At two loops, if the Higgs boson couples only once to the muon, for

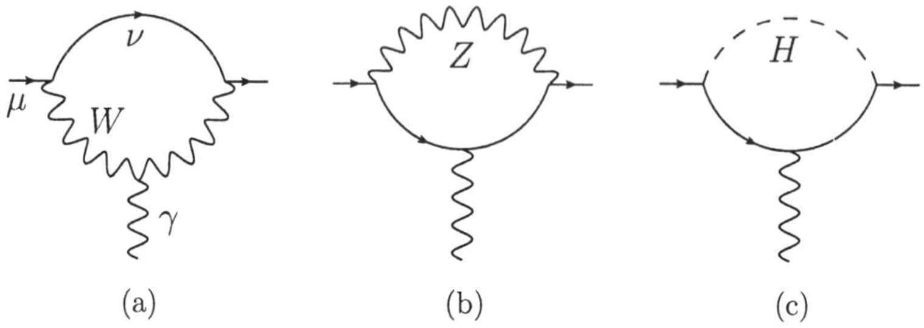

Fig. 2.2. One-loop electroweak radiative corrections to the muon anomalous magnetic moment.

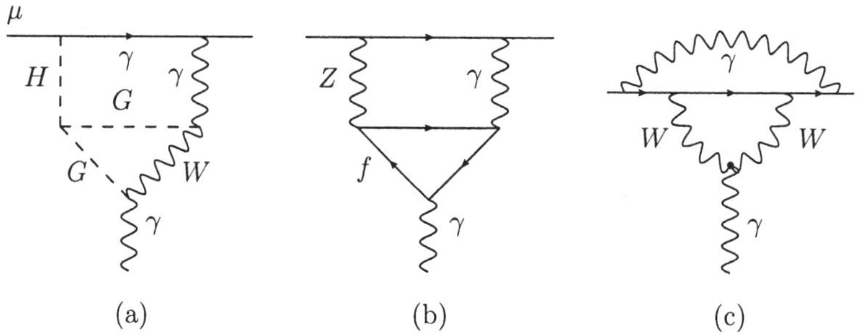

Fig. 2.3. Examples of two-loop electroweak corrections to the muon $g-2$.

example as in Fig. 2.3(a), this single scalar coupling does not introduce any relative suppression, since one factor of the muon mass is needed for the chirality-flipping effect of the dipole coupling. Thus the two-loop Higgs contribution is larger than one loop by $\mathcal{O}\left(\frac{\alpha M_W^2}{\pi m_\mu^2}\right)$, about 1000 times. This is a realization of the Bjorken-Weinberg two-loop/one-loop enhancement mechanism originally pointed out for the $\mu \to e$ transition dipole moment [126].

An important enhancement of the two-loop electroweak contributions is due to the presence of $\ln m_Z^2/m_\mu^2 \simeq 13.5$ terms, as first pointed out by [127]. They arise from the same electromagnetic anomalous dimension of the dipole operator that suppresses the decay $\mu \to e\gamma$, discussed in Section 2.2.4. Those logarithms appear in two-loop diagrams with a virtual photon exchange, examples of which are shown in Fig. 2.3(a-c). Among those contributions, diagram (b) is particularly interesting. For a single fermion f, it is gauge dependent and even ultraviolet divergent in the unitary gauge, due to the axial-vector triangle anomaly in the Zff coupling.

However, when all charged fermions in a single generation are summed over, the anomaly cancellation results in a finite gauge independent contribution for which the large logs also cancel but leave a residual long distance effect [68, 125, 128–130].

The evaluation of the non-logarithmic contributions of the light hadrons in these diagrams is somewhat subtle and relies in part on models of the hadronic interactions [128, 129]. Fortunately, those effects are small and do not contribute significantly to the theoretical uncertainty.

In total, the two-loop corrections decrease the electroweak effect by

$$a_\mu^{\rm EW}(2\text{ loop}) = -41(1)(2)\times 10^{-11}, \tag{2.81}$$

where the first error is an estimate of hadronic uncertainty and the second corresponds to the allowed Higgs mass range 114 GeV $\lesssim M_H \lesssim$ 250 GeV, the current top mass uncertainty, and higher-order corrections.

The central value in Eq. (2.81) is obtained with $m_H \simeq 150$ GeV. It is quite insensitive to the exact value of m_H for $m_H > 114$ GeV, since the Higgs contribution is either logarithmic (reflecting non-renormalizability of the Standard Model without the Higgs boson) and thus slowly varying, or suppressed by its mass squared and thus small if the Higgs boson is heavy. The residual sensitivity arises primarily from the bosonic two-loop diagrams and is illustrated in Fig. 2.4.

Combining Eqs. (2.80) and (2.81) gives the electroweak contribution

$$a_\mu^{\rm EW} = 154(1)(2)\times 10^{-11}. \tag{2.82}$$

Higher-order (three loop and beyond) leading logs of the form $(\alpha \ln m_Z^2/m_\mu^2)^n$, $n = 2, 3, \ldots$ can be computed via renormalization group techniques [128, 131]. Due to cancellations between the running of α and anomalous dimension effects, they give a relatively negligible $\sim 0.1\times 10^{-11}$ contribution to $a_\mu^{\rm EW}$. It is safely included in the uncertainty of Eq. (2.82).

In the case of electroweak contributions to the electron anomalous magnetic moment, a large -35% reduction found in [124] from the 2-loop electroweak radiative corrections leads to the contribution $a_e^{EW} = 3\times 10^{-14}$ employed in Eq. (2.9).

2.3.2.4. *Comparison with experiment*

The complete Standard Model prediction for a_μ is

$$a_\mu^{\rm SM} = a_\mu^{\rm QED} + a_\mu^{\rm Had} + a_\mu^{\rm EW}, \tag{2.83}$$

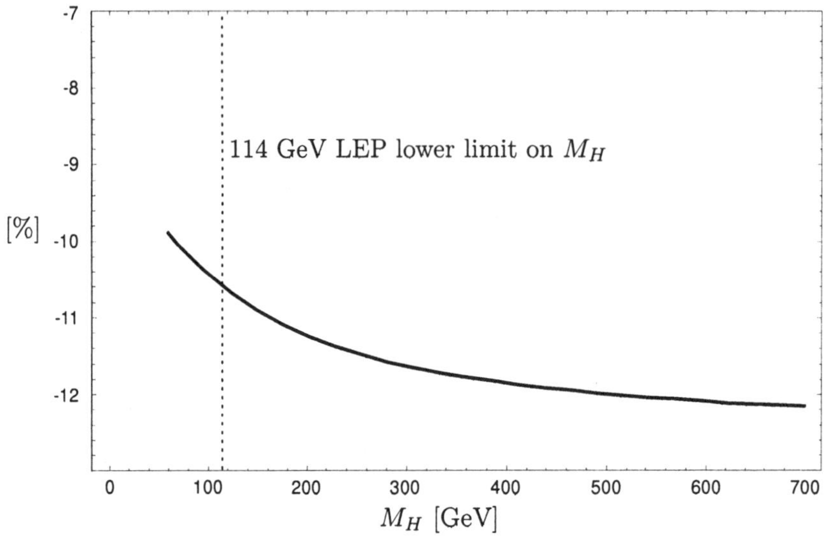

Fig. 2.4. Higgs mass dependence of the two-loop bosonic correction to a_μ expressed in percents of the one-loop effect, $\frac{a_\mu^{\rm EW,bos}(\text{two-loop})}{a_\mu^{\rm EW}(\text{one-loop})}$ [132]. The vertical dotted line shows the lower limit for the Higgs boson mass from direct searches, 114 GeV.

with the errors added in quadrature. Combining Eqs. (2.73), (2.79) and (2.82), one finds

$$a_\mu^{\rm SM} = 116\,591\,773(53) \times 10^{-11}. \tag{2.84}$$

Comparing Eq. (2.84) with the experimental result in Eq. (2.70) gives

$$a_\mu^{\rm exp} - a_\mu^{\rm SM} = 307 \pm 82 \times 10^{-11}. \tag{2.85}$$

The roughly 3.7σ difference is potentially very exciting. It may be an indicator or harbinger of contributions from New Physics beyond the Standard Model. At 90% CL, one finds

$$172 \times 10^{-11} \leq a_\mu(\text{New Physics}) \leq 440 \times 10^{-11}, \tag{2.86}$$

which suggests a relatively large New Physics effect, even larger than the predicted 154×10^{-11} Standard Model electroweak contribution, is starting to be seen. As we show in the next Section, several realistic examples of New Physics could quite easily lead to $a_\mu(\text{New Physics}) \sim \mathcal{O}\left(300 \times 10^{-11}\right)$ and might be responsible for the apparent deviation.

We caution, however, that tau decays and preliminary radiative return BaBar results suggest a reduction in Eq. (2.85) by about 150×10^{-11} which would leave only about a two sigma deviation. In fact, depending on exactly how one chooses to treat experimental input into the hadronic vacuum polarization correction, the discrepancy can reasonably range between 2 and 4 sigma. Clearly, further studies are needed to resolve that ambiguity.

Nevertheless, we should point out that if larger hadronic vacuum polarization corrections to a_μ are, in fact, responsible for the current disagreement between theory and experiment, they will have other serious implications for precision electroweak physics that also depends on e^+e^- annihilation data via dispersion relations. For example, it has been shown [133] that an increase in the hadronic cross section would likely reduce the Standard Model Higgs mass prediction below the current 150GeV (95%CL) upper bound, and could potentially lead to a conflict with the direct experimental constraint $m_H > 114.4$ GeV.

2.3.3. *New Physics effects*

Since the anomalous magnetic moment comes from a dimension 5 operator, New Physics (i.e. beyond the Standard Model expectations) will contribute to a_μ via induced quantum loop effects (rather than tree level). Whenever a new model or Standard Model extension is proposed, such effects are examined and $a_\mu^{\rm exp} - a_\mu^{\rm SM}$ is often employed to constrain or rule it out.

Here we describe several examples mainly taken from our work in ref. [134] of interesting New Physics probed by $a_\mu^{\rm exp} - a_\mu^{\rm SM}$. Rather than attempting to be inclusive, we concentrate on two general scenarios: 1) Supersymmetric loop effects which can be substantial and would be heralded as the most likely explanation if the deviation in $a_\mu^{\rm exp}$ is confirmed and 2) Models of radiative muon mass generation which predict $a_\mu(\text{New Physics}) \sim m_\mu^2/M^2$ where M is the scale of New Physics. Either case is capable of explaining the apparent deviation in $a_\mu^{\rm exp} - a_\mu^{\rm SM}$ exhibited in Eq. (2.85). Both examples can be cast in the form $a_\mu^{\rm NP} \simeq C_\mu m_\mu^2/\Lambda^2$, the first with $C_\mu \sim \mathcal{O}\left(\frac{\alpha}{\pi}\right)$ and the second with $C_\mu \sim \mathcal{O}\left(1\right)$. Other types of potential New Physics contributions to a_μ are only briefly discussed.

2.3.3.1. *Supersymmetry*

The supersymmetric contributions to a_μ stem from sneutrino-chargino and smuon–neutralino loops (see Fig. 2.5). They include 2 chargino and 4 neutralino states and could in principle entail slepton mixing and phases. Depending on SUSY masses, mixing and other parameters, the contribution of $a_\mu^{\rm SUSY}$ can span a broad range of possibilities. Studies have been carried out for a variety of models where the parameters are specified. Here we give a discussion primarily intended to illustrate the strong likelihood that evidence for supersymmetry can be inferred from $a_\mu^{\rm exp}$ and may in fact be the natural explanation for the apparent deviation from SM theory reported by E821.

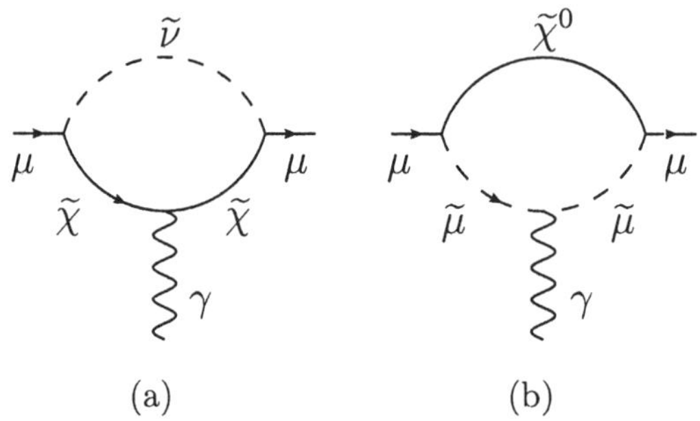

Fig. 2.5. Supersymmetric loops contributing to the muon anomalous magnetic moment.

Early studies of the supersymmetric contributions $a_\mu^{\rm SUSY}$ were carried out in the context of the minimal SUSY Standard Model (MSSM) [135–142], in an E_6 string-inspired model [143, 144], and in an extension of the MSSM with an additional singlet [145, 146]. An important observation made in [147], namely that some of the contributions are enhanced due to mixing by the ratio of Higgs' vacuum expectation values, $\tan\beta \equiv \langle\Phi_2\rangle/\langle\Phi_1\rangle$, which in some models is large (in some cases of order $m_t/m_b \approx 40$). In addition, larger values of $\tan\beta \gtrsim 2$ are generally in better accord with the recent LEP II Higgs mass bound $m_H \gtrsim 114$ GeV and, therefore, currently favored. The main contribution is generally due to the chargino-sneutrino diagram (Fig. 2.5(a)), which is enhanced by a Yukawa coupling in the muon-sneutrino-Higgsino vertex (charginos are admixtures of Winos and Higgsinos).

The leading effect from Fig. 2.5(a) is approximately given in the large $\tan\beta$ limit by

$$|a_\mu^{\rm SUSY}| \simeq \frac{\alpha(m_Z)}{8\pi\sin^2\theta_W}\frac{m_\mu^2}{\widetilde{m}^2}\tan\beta\left(1-\frac{4\alpha}{\pi}\ln\frac{\widetilde{m}}{m_\mu}\right), \tag{2.87}$$

where $\alpha(m_Z) \simeq 1/128$, and $\widetilde{m} = m_{\rm SUSY}$ represents a typical SUSY loop mass. SUSY mass scales are actually assumed degenerate in Eq. (2.87) [148]. (For a detailed discussion of degeneracy conditions see [149, 150].)

Also, we have included a 7–8% suppression factor due to leading two-loop EW effects. Like most New Physics effects, SUSY loops contribute directly to the dimension 5 magnetic dipole operator. From the calculation in Ref. [124, 128, 131], one finds a leading log suppression factor

$$1 - \frac{4\alpha}{\pi} \ln \frac{M}{m_\mu} \tag{2.88}$$

where M is the characteristic New Physics scale. For $M \sim 200$ GeV, that factor corresponds to about a 7% reduction. That reduction factor has the same source as the correction given for electromagnetic transition rates in Eq. (2.64). Note, Eq. (2.88) also applies to New Physics induced EDMs.

Numerically, one expects in the large $\tan\beta$ regime (after a small negative contribution from Fig. 2.5(b) is included, again assuming degenerate SUSY mass scales)

$$\left|a_\mu^{\mathrm{SUSY}}\right| \simeq 130 \times 10^{-11} \left(\frac{100 \text{ GeV}}{\widetilde{m}}\right)^2 \tan\beta, \tag{2.89}$$

where a_μ^{SUSY} generally has the same sign as the μ-parameter in SUSY models. Eq. (2.89) represents the leading effect up to corrections of $\mathcal{O}\left(m_W/\widetilde{m}\right)$ and $\mathcal{O}\left(1/\tan\beta\right)$.

Supersymmetric effects in a_μ have been computed in a variety of models [148, 151–170]. Also two-loop effects have been determined in various scenarios [171–175]. For a detailed review of supersymmetry contributions to a_μ, see Chapter 12 and Ref. [149].

Rather than focusing on a specific model, we simply employ for illustration the large $\tan\beta$ approximate formula in Eq. (2.89) with degenerate SUSY mass scales and the current constraint in Eq. (2.85). Then we find (for positive $\mathrm{sgn}(\mu)$)

$$\tan\beta \left(\frac{100 \text{ GeV}}{\widetilde{m}}\right)^2 \simeq 2.4 \pm 0.6, \tag{2.90}$$

or

$$\widetilde{m} \simeq (65 \pm 10 \text{ GeV})\sqrt{\tan\beta}. \tag{2.91}$$

(Of course, in specific models with non-degenerate SUSY mass scales, a more detailed analysis is required, but here we only want to illustrate roughly the scale of supersymmetry being probed.) Negative μ models give the opposite sign contribution to a_μ and are strongly disfavored by current $a_\mu^{\mathrm{exp}} - a_\mu^{\mathrm{SM}}$ results.

For large $\tan\beta$ in the range $4 \sim 40$, where the approximate results given above should be valid, one finds (assuming $\widetilde{m} > 200$ GeV from other experimental constraints and the region of Eq. (2.89) validity)

$$\widetilde{m} \simeq 200 - 500 \text{ GeV} \tag{2.92}$$

precisely the range where SUSY particles are often expected. If supersymmetry in the mass range of Eq. (2.92) with relatively large $\tan\beta$ is responsible for the apparent $a_\mu^{\text{exp}} - a_\mu^{\text{SM}}$ difference, it will have many dramatic consequences. Besides expanding the known symmetries of Nature and our fundamental notion of space-time, it will impact other new exploratory experiments. Indeed, for $\widetilde{m} \simeq 200 - 500$ GeV, one can expect a plethora of new SUSY particles to be discovered soon, either at the Fermilab 2 TeV $p\bar{p}$ collider or certainly at the LHC 14 TeV pp collider which is expected to start dedicated running in 2009.

Large $\tan\beta$ supersymmetry can also have other interesting loop-induced low energy consequences beyond a_μ. For example, it can affect the $b \to s\gamma$ branching ratio. Even for the muon, New Physics in a_μ is likely to suggest potentially observable $\mu \to e\gamma$, $\mu^- N \to e^- N$ and a muon electric dipole moment, depending on the degree of flavor mixing and CP violating phases. Searches for these phenomena are now entering an exciting phase, with a new generation of experiments being proposed or constructed. The decay $\mu \to e\gamma$ is currently being searched for with 2×10^{-13} (later to improve to 2×10^{-14}) single event sensitivity (SES) at the Paul Scherrer Institute (PSI) [176]. The mu2e experiment at Fermilab [177] will search for the muon-electron conversion, $\mu^-\text{Al} \to e^-\text{Al}$, with 2×10^{-17} SES. A proposal has been made [178] to search for the muon's EDM with sensitivity of about 10^{-24} e·cm. Certainly, the hint of supersymmetry suggested by a_μ^{exp} will provide strong additional motivation to extend such studies both theoretically and experimentally.

2.3.3.2. *Radiative muon mass models*

The relatively light masses of the muon and most other known fundamental fermions could suggest that they are radiatively loop induced by New Physics beyond the Standard Model. Although no compelling model exists, the concept is very attractive as a natural way to explain the flavor mass hierarchy, i.e. why most fermion masses are so much smaller than the electroweak scale ~ 250 GeV.

The basic idea, described in [179], is to start off with a naturally zero bare fermion mass due to an underlying chiral symmetry. The symmetry is broken in the fermion 2-point function by quantum loop effects. They lead to a finite calculable mass which depends on the mass scales, coupling strengths and dynamics of the underlying symmetry breaking mechanism. In such a scenario, one generically expects for the muon

$$m_\mu \propto \frac{g^2}{16\pi^2} M_F, \tag{2.93}$$

where g is some new interaction coupling strength and $M_F \sim 100 - 1000$ GeV is a heavy scale associated with chiral symmetry breaking and perhaps electroweak symmetry breaking. Of course, there may be other suppression factors at work in Eq. (2.93) that keep the muon mass small.

Whatever source of chiral symmetry breaking is responsible for generating the muon's mass will also give rise to non-Standard Model contributions in a_μ. Indeed, fermion masses and anomalous magnetic moments are intimately connected chiral symmetry breaking operators. Remarkably, in such radiative scenarios, the additional contribution to a_μ is quite generally given by [179, 180]

$$a_\mu(\mathrm{N}P) \simeq C \frac{m_\mu^2}{M^2}, \qquad C \simeq \mathcal{O}(1), \tag{2.94}$$

where M is some physical high mass scale associated with the New Physics and C is a model-dependent number roughly of order 1. M need not be the same scale as M_F in Eq. (2.93). In fact, M is usually a somewhat larger gauge or scalar boson mass responsible for mediating the chiral symmetry breaking interaction. The result in Eq. (2.94) is remarkably simple in that it is largely independent of coupling strengths, dynamics, etc. Furthermore, rather than exhibiting the usual $g^2/16\pi^2$ loop suppression factor, $a_\mu(\mathrm{N}P)$ is related to m_μ^2/M^2 by a (model dependent) constant, C, roughly of $\mathcal{O}(1)$, thus exhibiting the m_f^2/Λ^2 possibility we discussed earlier.

Toy model example To demonstrate how the relationship in Eq. (2.94) arises, we first review a toy model example [179] for muon mass generation which is graphically depicted in Fig. 2.6.

If the muon is massless in lowest order (i.e. no bare m_μ^0 is possible due to a symmetry), but couples to a heavy fermion F via scalar, S, and pseudoscalar, P, bosons with couplings g and $g\gamma_5$ respectively, then the diagrams in Fig. (2.6) give rise to

$$m_\mu \simeq \frac{g^2}{16\pi^2} M_F \left(\frac{M_S^2}{M_S^2 - M_F^2} \ln \frac{M_S^2}{M_F^2} - \frac{M_P^2}{M_P^2 - M_F^2} \ln \frac{M_P^2}{M_F^2} \right) \tag{2.95}$$

$$\rightarrow \frac{g^2}{16\pi^2} M_F \ln \left(\frac{M_S^2}{M_P^2} \right) \quad (M_{S,P} \gg M_F). \tag{2.96}$$

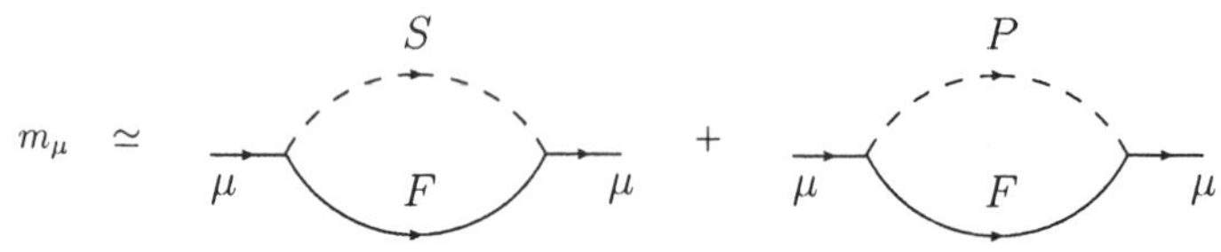

Fig. 2.6. One-loop diagrams which can induce a finite radiative muon mass.

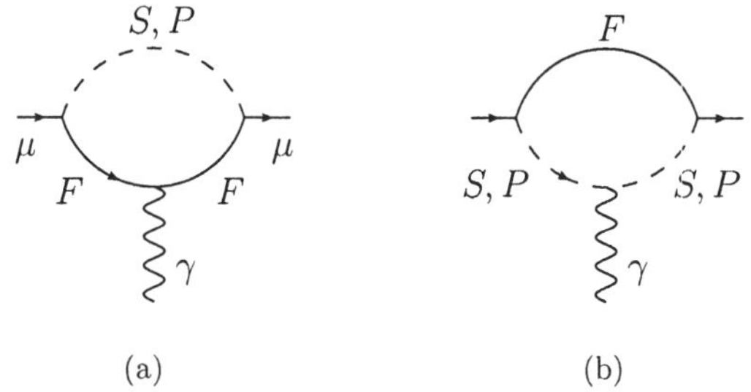

Fig. 2.7. Diagrams that could potentially contribute to the anomalous magnetic moment in radiative muon mass models.

Note that short-distance ultraviolet divergences have canceled and the induced mass vanishes in the chirally symmetric limit $M_S = M_P$.

If we attach a photon to the heavy internal fermion, F, or boson S or P (assumed to carry fractions Q_F and $1 - Q_F$ of the muon charge, respectively), then a new contribution to a_μ is also induced (see Fig. 2.7). In the limit $M_{S,P} \gg M_F$ and $Q_F = 1$, one finds [179]

$$a_\mu(\mathrm{N}P) \simeq \frac{g^2}{8\pi^2}\frac{m_\mu M_F}{M_P^2}\left(\frac{M_P^2}{M_S^2}\ln\frac{M_S^2}{M_F^2} - \ln\frac{M_P^2}{M_F^2}\right), \tag{2.97}$$

while for $Q_F = 0$

$$a_\mu(\mathrm{N}P) \simeq \frac{g^2}{8\pi^2}\frac{m_\mu M_F}{M_P^2}\left(1 - \frac{M_P^2}{M_S^2}\right). \tag{2.98}$$

The induced $a_\mu(\mathrm{N}P)$ also vanishes in the $M_S = M_P$ chiral symmetry limit. Interestingly, $a_\mu(\mathrm{N}P)$ exhibits a linear rather than quadratic dependence on m_μ at this point.

Although Eqs. (2.96) and (2.97) both depend on unknown parameters such as g and M_F, those quantities largely cancel when we combine both expressions. One finds

$$a_\mu(\mathrm{N}P) \simeq C\frac{m_\mu^2}{M_P^2},$$
$$C = 2\left[1 - \left(1 - \frac{M_P^2}{M_S^2}\right)\ln\frac{M_S^2}{M_F^2}/\ln\frac{M_S^2}{M_P^2}\right] \quad \text{for } Q_F = 1,$$
$$C = \left(1 - \frac{M_P^2}{M_S^2}\right)/\ln\frac{M_S^2}{M_P^2} \quad \text{for } Q_F = 0, \tag{2.99}$$

where C is very roughly $\mathcal{O}(1)$. It actually spans a broad range and take on either sign, depending on the M_S/M_P ratio and Q_F. A loop produced $a_\mu(\mathrm{N}P)$ effect that started out at $\mathcal{O}\left(g^2/16\pi^2\right)$ has effectively been promoted to $\mathcal{O}(1)$ by absorbing the couplings and M_F factor into m_μ. Along the way, the linear dependence on m_μ has been replaced by a more natural quadratic dependence.

Technicolor An alternative procedure for radiatively generating fermion masses involves new strong dynamics, e.g. extended technicolor. In such scenarios, technifermions acquire, via new strong dynamics, dynamical self-energies

$$\Sigma_F(p) \simeq m_F\left(\frac{\Lambda^2}{\Lambda^2 - p^2}\right)^{1-\frac{\gamma}{2}}, \tag{2.100}$$

where $0 < \gamma < 2$ is an anomalous dimension, $m_F \simeq \mathcal{O}(300\ \mathrm{GeV})$, and Λ is the new strong interaction scale $\sim \mathcal{O}(1\ \mathrm{TeV})$.

Ordinary fermions such as the muon receive loop induced masses via the diagram in Fig. 2.8.

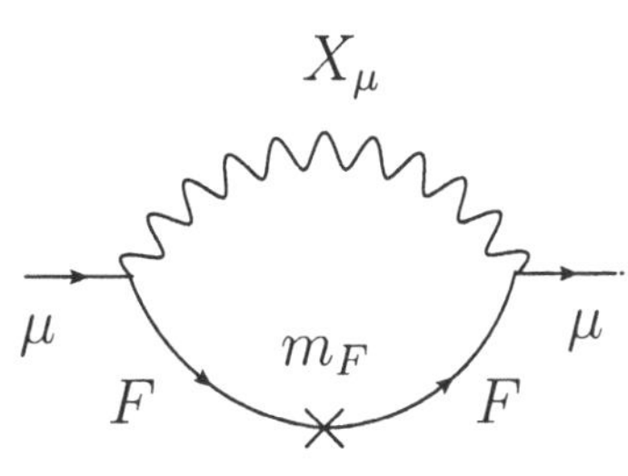

Fig. 2.8. Extended technicolor-like diagram responsible for generating the muon mass.

The extended gauge boson X_μ links μ and F via the non-chiral coupling

$$g\gamma_\mu\left(a\frac{1-\gamma_5}{2} + b\frac{1+\gamma_5}{2}\right) \tag{2.101}$$

and gives rise to a mass [179, 180]

$$m_\mu \simeq \frac{g^2 ab}{4\pi^2} m_F \left(\frac{\Lambda}{m_{X_\mu}}\right)^{2-\gamma} \Gamma\left(\frac{\gamma}{2}\right)\Gamma\left(1-\frac{\gamma}{2}\right), \tag{2.102}$$

where $\Gamma(x)$ is the Gamma function. The possible ultraviolet divergence at $\gamma = 2$ corresponds to a non-dynamical m_F.

If we attach a photon to one of the internal propagators of Fig. 2.8 one obtains an anomalous magnetic moment of the form

$$a_\mu(\text{New Dynamics}) \simeq \frac{g^2 ab}{2\pi^2} \frac{m_\mu m_F}{m_{X_\mu}^2} \left(\frac{\Lambda}{m_{X_\mu}}\right)^{2-\gamma} F(\gamma), \tag{2.103}$$

where $F(\gamma)$ is a model dependent dynamics factor. Again, we see a linear dependence on m_μ. However, when Eq. (2.102) and (2.103) are combined, one finds for $\gamma \lesssim 1$

$$a_\mu(\text{New Dynamics}) \simeq \mathcal{O}(1) \frac{m_\mu^2}{m_{X_\mu}^2}, \tag{2.104}$$

i.e. the generic result $\mathcal{O}(1)\, m_\mu^2/M^2$ where M is the New Physics scale (here the extended-techniboson mass) emerges [181].

A similar relationship, $a_\mu(NP) \simeq Cm_\mu^2/M^2$, has been found in more realistic multi-Higgs models [182], SUSY with soft masses [183], etc. It is also a natural expectation in composite models [184–186] or some models with large extra dimensions [187, 188], although studies of such cases have not necessarily made that same connection. Basically, the requirement that m_μ remain relatively small in the presence of new chiral symmetry breaking interactions forces a_μ(New Physics) to effectively exhibit a quadratic m_μ^2 dependence.

For models of the above variety, where $|a_\mu(\text{New Physics})| \simeq m_\mu^2/M^2$, the current constraint in Eq. (2.86) suggests (very roughly)

$$M \simeq 2 \text{ TeV}. \tag{2.105}$$

Of course, for a specific model, one must check that the sign of the induced $a_\mu^{\rm NP}$ is in accord with experiment (i.e. it should be positive).

Such a scale of New Physics could be quite natural in multi-Higgs radiative mass models, including large extra dimensions, and soft SUSY mass scenarios [133]. It would be somewhat low for dynamical symmetry breaking and compositeness, however, confirmation of an $a_\mu^{\rm exp}$ deviation from $a_\mu^{\rm SM}$ will certainly lead to all possibilities being revisited.

2.3.3.3. *Other New Physics examples*

Anomalous W boson properties Anomalous W boson magnetic dipole and electric quadrupole moments can also lead to a deviation in a_μ from SM expectations. Generalizing the γWW coupling, the W boson magnetic dipole moment is given by

$$\mu_W = \frac{e}{2m_W}(1+\kappa+\lambda) \tag{2.106}$$

and electric quadrupole moment by

$$Q_W = -\frac{e}{2m_W}(\kappa-\lambda) \tag{2.107}$$

where $\kappa = 1$ and $\lambda = 0$ in the Standard Model, i.e. the gyromagnetic ratio $g_W = \kappa + 1 = 2$. For non-standard couplings, one obtains the additional one loop contribution to a_μ given by [189–193]

$$a_\mu(\kappa,\lambda) \simeq \frac{G_\mu m_\mu^2}{4\sqrt{2}\pi^2}\left[(\kappa-1)\ln\frac{\Lambda^2}{m_W^2} - \frac{1}{3}\lambda\right], \tag{2.108}$$

where Λ is the high momentum cutoff required to give a finite result. It presumably corresponds to the onset of New Physics such as the W compositeness scale, or new strong dynamics. Higher order electroweak loop effects reduce that contribution by roughly the suppression in Eq. (2.88), i.e. $\sim 9\%$.

For $\Lambda \simeq 1$ TeV, the deviation in Eq. (2.85) corresponds to

$$\kappa - 1 = 0.28 \pm 0.07. \tag{2.109}$$

Such a large deviation from Standard Model expectations, $\kappa = 1$, is already ruled out by $e^+e^- \to W^+W^-$ data at LEP II which gives [194, 195]

$$\kappa - 1 = 0.04 \pm 0.08 \qquad \text{(LEP II)}. \tag{2.110}$$

One could reduce the requirement in Eq. (2.109) somewhat by assuming a much larger Λ cutoff in Eq. (2.108). However, it is generally felt that $\kappa - 1$ and Λ should be inversely correlated. For example $\kappa - 1 \sim m_W/\Lambda$ or $(m_W/\Lambda)^2$. So, the rather substantial $\kappa - 1$ needed to accommodate $a_\mu^{\rm exp}$ would argue against a much larger Λ. Similarly, the large value of the anomalous W electric quadrupole moment $\lambda \simeq -4$ needed to reconcile $a_\mu^{\rm exp} - a_\mu^{\rm SM}$ is also ruled out by collider data (which implies $|\lambda| \lesssim 0.1$). Hence, it appears that anomalous W boson properties cannot be the source of the discrepancy in $a_\mu^{\rm exp}$.

We note that the existence of a W boson EDM would induce fermion EDMs in a manner very similar to the anomalous magnetic moment discussion given above. Indeed, for a W EDM, $d_W = e\lambda_W/2m_W$, one finds analogous to Eq. (2.108) the fermion induced EDM [196]

$$d_f = \frac{eT_{3L}G_F m_f \lambda_W}{4\sqrt{2}\pi^2}\left(\ln\frac{\Lambda^2}{m_W^2} + \mathcal{O}(1)\right) \tag{2.111}$$

where T_{3L} is the third component of the weak $SU(2)_L$ isospin of the fermion f.

New gauge bosons The local $SU(3)_C \times SU(2)_L \times U(1)_Y$ symmetry of the Standard Model can be easily expanded to a larger gauge group with additional charged and neutral gauge bosons. Here, we consider effects due to a charged $W_R^\pm$ which couples to right-handed charged currents in generic left-right symmetric models and a neutral gauge boson, Z', which can naturally arise in higher rank GUT models such as $SO(10)$ or E_6. A general analysis of one-loop contributions to a_μ from extra gauge bosons has been carried out by Leveille [197] and summarized in Chapter 10.2. The specific examples considered here were illustrated in [68] (see also [198]). Therefore, we will only discuss the likelihood of such bosons being the source of the apparent $a_\mu^{\rm exp} - a_\mu^{\rm SM}$ discrepancy.

For the case of a W_R coupled to μ_R and a (very light) ν_R with gauge coupling g_R, one finds

$$a_\mu(W_R) \simeq (390 \times 10^{-11})\frac{g_R^2}{g_2^2}\frac{m_W^2}{m_{W_R}^2}. \tag{2.112}$$

To accommodate the discrepancy in Eq. (2.85) requires $m_{W_R} \simeq m_W = 80.4$ GeV for $g_R \simeq g_2$, which is clearly ruled out by direct searches and precision measurements which give $m_{W_R} \gtrsim 715$ GeV. Hence, $W_R^\pm$ is not a viable candidate for explaining the $a_\mu^{\rm exp}$ discrepancy.

Extra neutral gauge bosons (with diagonal $\bar{\mu}\mu$ couplings) do much worse in trying to explain $a_\mu^{\rm exp} - a_\mu^{\rm SM}$, partly because they often tend to give a contribution with opposite sign. For example, the Z_χ of $SO(10)$ leads to

$$a_\mu(Z_\chi) \simeq -6 \times 10^{-11}\frac{m_Z^2}{m_{Z_\chi}^2}. \tag{2.113}$$

Given the collider constraint $m_{Z_\chi} \gtrsim 720$ GeV, that effect would be much too small to observe in $a_\mu^{\rm exp}$. Most other Z' scenarios give similar results.

An exception to the small effects from gauge bosons illustrated above is provided by non-chiral coupled bosons which connect μ and a heavy fermion F. In those cases, $\Delta a_\mu \simeq \frac{g^2}{16\pi^2}\frac{m_\mu m_F}{M^2}$, where M is the gauge boson mass. However, loop effects then give $\delta m_\mu \sim g^2 m_F$ (see the discussion in Section 2.3.3.2) and we have argued that in such scenarios Δa_μ should actually turn out to be $\sim m_\mu^2/M^2$. As previously pointed out in Eq. (2.105), $a_\mu^{\rm exp} - a_\mu^{\rm SM}$ then corresponds to $M \sim 2$ TeV.

Many other examples of New Physics contributions to a_μ have been considered in the literature. A general analysis in terms of effective interactions was presented in [199]. Specific other examples include effects due to muon compositeness [186, 200, 201], extra Higgs bosons [202–206], leptoquarks [207–209], bileptons [210], two-loop pseudoscalar effects [211], compact and large extra dimensions [212–215], extended family models [216], brane models [217–219], unparticles [220], etc.

2.4. Flavor Violating Transition Dipole Moments

Searches for flavor-changing weak neutral current effects in the quark sector of the Standard Model have had a rich and glorious history. The need to theoretically suppress $s \to d$ transitions in decays such as $K_L \to \mu^+\mu^-$ led to the introduction of charm and the GIM (Glashow-Iliopoulos-Maiani) mechanism [221] of loop cancellations. That mechanism was also instrumental in suggesting that a third generation of quarks may explain CP violation via CKM mixing. More recently, the accurate measurement of the $b \to s\gamma$ branching ratio which occurs via transition dipole moments confirmed the Standard Model top quark loop prediction and has been used [222, 223] to constrain possible New Physics effects such as in supersymmetry. Indeed, that branching ratio currently provides sensitivity to supersymmetry [224–227] competitive with and complementary to the muon anomalous magnetic dipole moment discussion in Section 2.3.

In the case of charged lepton decays, searches for flavor-changing neutral current effects such as $\mu \to e\gamma$ or $\tau \to \mu\gamma$ have all, so far, proven to be negative (see Table 2.5). Only experimental bounds on transition dipole moments exist (see Table 2.6) in spite of the fact that we now know from neutrino oscillation studies that lepton flavor is not conserved. Their suppression in charged lepton processes is accidentally due to the smallness of neutrino masses rather than from intrinsically small flavor mixing. Indeed, we have found from oscillations that the flavor basis states ν_e, ν_μ, and ν_τ produced in weak interaction reactions are related to the neutrino mass

Table 2.5. Current bounds on various flavor-changing charged lepton processes along with future expected or possible improvements [228].

Reaction	Current bound	Expected	Possible
$B(\mu^+ \to e^+\gamma)$	$< 1.2 \times 10^{-11}$	2×10^{-13}	2×10^{-14}
$B(\mu^+ \to e^+e^-e^+)$	$< 1.0 \times 10^{-12}$	—	10^{-15}
$R(\mu^- \text{Au} \to e^- \text{Au})$	$< 7 \times 10^{-13}$		
$R(\mu^- \text{Al} \to e^- \text{Al})$	—	10^{-16}	10^{-18}
$B(\tau \to \mu\gamma)$	$< 5.9 \times 10^{-8}$		$\mathcal{O}(10^{-9})$
$B(\tau \to e\gamma)$	$< 8.5 \times 10^{-8}$		$\mathcal{O}(10^{-9})$
$B(\tau \to \mu\mu^+\mu^-)$	$< 2.0 \times 10^{-8}$		$\mathcal{O}(10^{-10})$

eigenstates ν_1, ν_2, and ν_3 via the mixing matrix

$$\begin{pmatrix} |\nu_e\rangle \\ |\nu_\mu\rangle \\ |\nu_\tau\rangle \end{pmatrix} = U \begin{pmatrix} |\nu_1\rangle \\ |\nu_2\rangle \\ |\nu_3\rangle \end{pmatrix}, \text{ where} \tag{2.114}$$

$$U = \begin{pmatrix} c_{12}c_{13} & s_{12}c_{13} & s_{13}e^{-i\delta} \\ -s_{12}c_{23} - c_{12}s_{23}s_{13}e^{i\delta} & c_{12}c_{23} - s_{12}s_{23}s_{13}e^{i\delta} & s_{23}c_{13} \\ s_{12}s_{23} - c_{12}c_{23}s_{13}e^{i\delta} & -c_{12}s_{23} - s_{12}c_{23}s_{13}e^{i\delta} & c_{23}c_{13} \end{pmatrix},$$

$$c_{ij} = \cos\theta_{ij}, \quad s_{ij} = \sin\theta_{ij}, \quad i, j = 1, 2, 3$$

with

$$\sin^2 2\theta_{23} \simeq 1, \qquad \sin^2 2\theta_{12} \simeq 0.8, \qquad \sin^2 2\theta_{13} < 0.15 \tag{2.115}$$

and a completely undetermined phase $0 \le \delta < 2\pi$.

The measured mixing angles θ_{23} and θ_{12} are quite large and give rise to some near maximal neutrino oscillation effects (such as solar ν_e flux reaching the earth as roughly $\frac{1}{3}\nu_e + \frac{1}{3}\nu_\mu + \frac{1}{3}\nu_\tau$). However, the measured neutrino mass squared differences found from oscillations are very small,

$$\begin{aligned} \Delta m_{32}^2 &= m_3^2 - m_2^2 \simeq \pm 2.5 \times 10^{-3}\ \text{eV}^2 \\ \Delta m_{21}^2 &= m_2^2 - m_1^2 \simeq 8 \times 10^{-5}\ \text{eV}^2. \end{aligned} \tag{2.116}$$

Since charged lepton flavor violation in the Standard Model must vanish in the limit where both $\Delta m_{ij}^2 = 0$, decay amplitudes for $l_1 \to l_2 + \gamma$ will be proportional to the Δm_{ij}^2 and, therefore, highly suppressed. For example, in the case of $\tau \to \mu\gamma$, one finds for the transition dipole moments given in Section 2.2.4, the leading Δm_{32}^2 contribution

$$\begin{aligned} D_R^{\tau\mu} &= -\frac{eG_F m_\tau}{16\sqrt{2}\pi^2} \sin\theta_{23}\cos\theta_{23}\cos^2\theta_{13}\frac{\Delta m_{32}^2}{m_W^2}, \\ D_L^{\tau\mu} &\simeq 0. \end{aligned} \tag{2.117}$$

Table 2.6. Current experimental bounds on the transition dipole moments defined in Eq. (2.62) along with possible future sensitivities. The bounds are based on the constraints in Table 2.5 with the best future $\mu - e$ sensitivity expected to come from $\mu\text{Al} \to e\text{Al}$ conversion.

Transition moment	Current bound	Possible future sensitivity
$\sqrt{\lvert F_{M1}^{\mu e}\rvert^2 + \lvert F_{E1}^{\mu e}\rvert^2}$	$< 2 \times 10^{-26}\, e{\cdot}\text{cm}$	$< 1 \times 10^{-28}\, e{\cdot}\text{cm}$
$\sqrt{\lvert F_{M1}^{\tau\mu}\rvert^2 + \lvert F_{E1}^{\tau\mu}\rvert^2}$	$< 5 \times 10^{-23}\, e{\cdot}\text{cm}$	$< 6 \times 10^{-24}\, e{\cdot}\text{cm}$
$\sqrt{\lvert F_{M1}^{\tau e}\rvert^2 + \lvert F_{E1}^{\tau e}\rvert^2}$	$< 6 \times 10^{-23}\, e{\cdot}\text{cm}$	$< 6 \times 10^{-24}\, e{\cdot}\text{cm}$

The left-hand transition moment is non-vanishing, but rendered extremely small because of the $\frac{\Delta m_{32}^2}{m_W^2} \simeq 4 \times 10^{-25}$ suppression factor. Using the branching ratio relation,

$$B(\tau \to \mu\gamma) \simeq \frac{12\pi^2}{G_F^2 m_\tau^2} \left(|D_R^{\tau\mu}|^2 + |D_L^{\tau\mu}|^2 \right) B(\tau \to \mu\nu\bar{\nu}) , \tag{2.118}$$

leads to the prediction $B(\tau \to \mu\gamma) \simeq 2 \times 10^{-54}$, i.e. more than 10^{-47} below the current bound in Table 2.5. (A similar size $B(\mu \to e\gamma)$ is expected.) Such tiny levels of lepton flavor violation are clearly unobservable.

So, the bad news is that we cannot measure neutrino mass and mixing in flavor-changing charged lepton decays such as $\mu \to e\gamma$ or $\tau \to \mu\gamma$. The good news is that the observation of such decays would, therefore, provide direct evidence for New Physics beyond neutrino mass effects. Also, there are many models that predict potentially observable charged lepton flavor violation reaction and planned experiments will probe (as we shall see) $\mathcal{O}(1000\ \text{TeV})$ mass scales, a significant window to New Physics.

2.4.1. *Muon flavor violation*

Experiments to find the rare processes $\mu \to e\gamma$, $\mu \to ee^+e^-$, and $\mu^- N \to e^- N$ have been at the forefront of searches for lepton flavor violation (see Table 2.5). With regard to flavor transition electromagnetic dipole moments, the decay $\mu \to e\gamma$ provides the most direct probe and currently the best constraint on its related New Physics [229, 230] as discussed in Chapters 18 and 19. Phrased in terms of the right- and left-handed transition dipole moments in Eq. (2.65) and the total muon decay rate

$$\Gamma(\mu \to \text{all}) \simeq \Gamma(\mu \to e\nu\bar{\nu}) \simeq \frac{G_F^2 m_\mu^5}{192\pi^3} \tag{2.119}$$

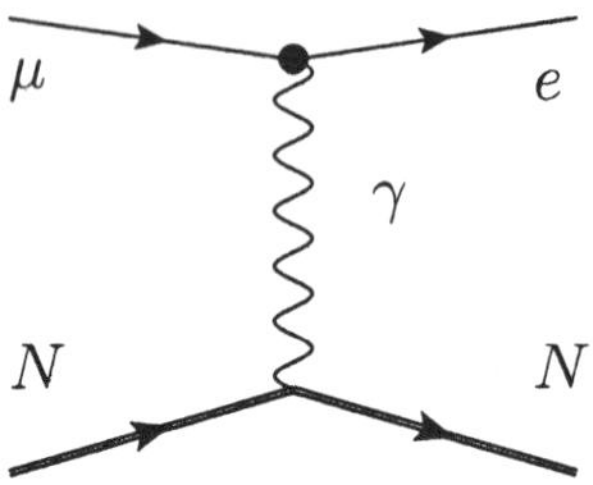

Fig. 2.9. Coherent muon-electron conversion in the field of the nucleus N, induced by the dipole operator coupling μ, e, and γ.

one expects a branching ratio

$$B(\mu \to e\gamma) \simeq \frac{12\pi^2}{G_F^2 m_\mu^2}\left(|D_R^{\mu e}|^2 + |D_L^{\mu e}|^2\right). \tag{2.120}$$

These same dipole moments also give rise to $\mu \to ee^+e^-$ and $\mu^- N \to e^- N$ (coherent conversion in the field of a nucleus) via virtual photon effects at $q^2 \neq 0$, but at a somewhat reduced rate [231]. One finds

$$B\left(\mu \to ee^+e^-\right) \simeq 0.006 B(\mu \to e\gamma). \tag{2.121}$$

In the case of $R(\mu^- N \to e^- N) \equiv \Gamma(\mu^- N \to e^- N)/\Gamma(\mu^- N \to \nu_\mu N')$, the rate for conversion is traditionally compared with ordinary weak charged current capture and both depend on the specific nucleus, N, employed. So, for example, one finds from a detailed analysis [232, 233]

$$R\left(\mu^- \mathrm{Al} \to e^- \mathrm{Al}\right) \simeq 2.6 \times 10^{-3} B(\mu \to e\gamma). \tag{2.122}$$

However, both $\mu \to ee^+e^-$ and $\mu^- N \to e^- N$ could turn out to be much larger if they are dominated by chiral conserving dimension six four fermion operators rather than dimension five transition dipole moments. Those dimension six operators provide a primary motivation for $\mu \to ee^+e^-$ and $\mu^- N \to e^- N$ searches; but, we will not discuss them further here because they are outside the scope of our dimension five New Physics theme.

If the dipole transition moments dominate, one expects

$$B(\mu \to e\gamma) : B\left(\mu \to ee^+e^-\right) : R\left(\mu^- \mathrm{Al} \to e^- \mathrm{Al}\right) :: 389 : 2.3 : 1. \tag{2.123}$$

However, from the last column in Table 2.5, it seems that $\mu \to ee^+e^-$ and $\mu^- \mathrm{Al} \to e^- \mathrm{Al}$ can be experimentally pushed much farther than $\mu \to e\gamma$ because backgrounds are easier to control. One could expect long term future experimental sensitivities to go roughly as

$$\begin{aligned} &B(\mu \to e\gamma)^{\exp} : B\left(\mu \to ee^+e^-\right)^{\exp} : R\left(\mu^- \mathrm{Al} \to e^- \mathrm{Al}\right)^{\exp} \\ &:: 5 \times 10^{-5} : 10^{-3} : 1. \end{aligned} \tag{2.124}$$

So, in the long term $\mu^- N \to e^- N$ may become the best probe of transition dipole moments as well as dimension six muon flavor-changing amplitudes. However, currently $B(\mu \to e\gamma)$ provides the best constraint on $D_R^{\mu e}$ and $D_L^{\mu e}$ and it should continue to do so for some time; so, we concentrate our discussion on it.

To estimate the New Physics sensitivity of $B(\mu \to e\gamma)$, we apply the parameterization of Eq. (2.67) and find

$$B(\mu \to e\gamma) \simeq \frac{12\alpha\pi^3}{G_F^2}\left(\left|\frac{\mathcal{E}_{\mu e}^R}{(\Lambda_R^{\mu e})^2}\right|^2 + \left|\frac{\mathcal{E}_{\mu e}^L}{(\Lambda_L^{\mu e})^2}\right|^2\right), \tag{2.125}$$

or

$$B(\mu \to e\gamma) \simeq 2 \times 10^{10}\ \text{GeV}^4 \left(\left|\frac{\mathcal{E}_{\mu e}^R}{(\Lambda_R^{\mu e})^2}\right|^2 + \left|\frac{\mathcal{E}_{\mu e}^L}{(\Lambda_L^{\mu e})^2}\right|^2\right). \tag{2.126}$$

Then using $B(\mu \to e\gamma) < 1.2 \times 10^{-11}$ and assuming $\frac{\mathcal{E}_{\mu e}^R}{(\Lambda_R^{\mu e})^2}$ or $\frac{\mathcal{E}_{\mu e}^L}{(\Lambda_L^{\mu e})^2} \equiv \frac{\mathcal{E}_{\mu e}}{(\Lambda^{\mu e})^2}$ (but not both), we find the New Physics constraint

$$\Lambda^{\mu e} \gtrsim 200\ \text{TeV} \times \sqrt{\mathcal{E}_{\mu e}}. \tag{2.127}$$

That bound is already quite stringent. It will increase to about

$$\Lambda^{\mu e} \gtrsim 1000\ \text{TeV} \times \sqrt{\mathcal{E}_{\mu e}}, \tag{2.128}$$

if the long term goal, 2×10^{-14}, of an experiment at Paul Scherrer Institute is realized. (Better yet, a non-null value may result.)

Even for $\mathcal{E}_{\mu e} \sim \mathcal{O}\left(\frac{\alpha}{\pi}\right)$, that bound becomes ~ 50 TeV which is well beyond collider direct discovery capabilities.

Of course, one can turn the constraint around and ask for $\Lambda^{\mu e} \sim 1$ TeV, the scale of LHC New Physics, what size $\mathcal{E}_{\mu e}$ is probed? Then, one finds that the current bound of 1.2×10^{-11} requires

$$\mathcal{E}_{\mu e} \lesssim 2.5 \times 10^{-5} \text{ for } \Lambda^{\mu e} \sim 1\,\text{TeV} \tag{2.129}$$

and that sensitivity will improve to $\mathcal{E}_{\mu e} \lesssim 10^{-6}$ if the new PSI experimental goals are met. The constraint in Eq. (2.129) is very interesting from a model independent perspective; but, it becomes even more impressive if viewed in terms of a New Physics explanation for the discrepancy between the anomalous muon magnetic moment experiment and theory discussed in Section 2.3.

2.4.2. *The New Physics connection between a_μ and $\mu \to e\gamma$*

If there is New Physics, such as supersymmetry [60], responsible for the $\Delta a_\mu\,(\mathrm{NP}) = a_\mu^{\exp} - a_\mu^{\mathrm{SM}} = 307\,(82) \times 10^{-11}$ discrepancy, it could be expected to also contribute to $\Gamma\,(\mu \to e\gamma)$ (examples of Feynman diagrams are shown in Fig. 2.10). Indeed, the latter can be viewed as a flavor off-diagonal or transition version of a_μ but suppressed by $\mathcal{E}_{\mu e}$. Assuming $\Delta a_\mu\,(\mathrm{NP}) = m_\mu^2/\Lambda^2$ (which corresponds to an effective $\Lambda \simeq 1.9$ TeV) and applying $B\,(\mu \to e\gamma) < 1.2 \times 10^{-11}$ in Eq. (2.127), one finds

$$\mathcal{E}_{\mu e} < 10^{-4}, \tag{2.130}$$

a constraint similar to Eq. (2.129) for which $\Lambda \simeq 1$ TeV was assumed.

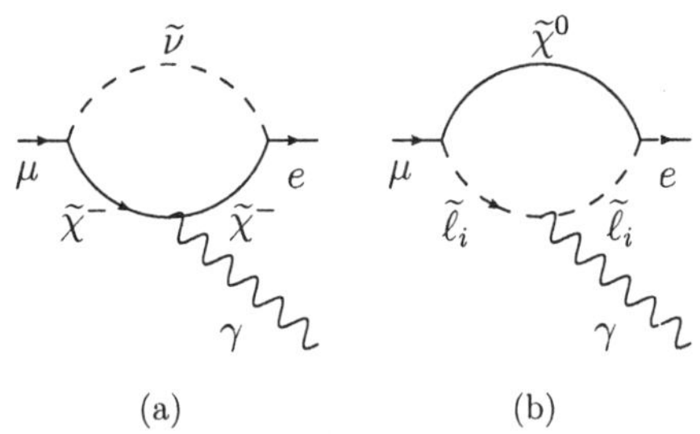

Fig. 2.10. Diagrams which might give rise to the decay $\mu \to e\gamma$ in supersymmetric theories. Note the similarity to the anomalous magnetic moment, Fig. 2.5.

The small suppression factor in Eq. (2.130) has an interesting origin in supersymmetry scenarios. It is either due to very small sparticle flavor mixing or nearly degenerate sparticle masses. For the latter case (a super GIM mechanism), that suppression corresponds to (including a factor of 1/2 for mixing)

$$\frac{m_{\tilde{1}}^2 - m_{\tilde{2}}^2}{m_{\tilde{1}}^2} < 2 \times 10^{-4}, \tag{2.131}$$

or for $m_1 \simeq m_2 \simeq 200$ GeV,

$$\Delta m = m_1 - m_2 \simeq 20 \text{ MeV}. \tag{2.132}$$

Such a near degeneracy, to 100 ppm, may be difficult to arrange or appear to be fine tuned. Its further reduction by improvements in the $B\,(\mu \to e\gamma)$ bound (by more than an order of magnitude) would be even more contrived. Turning that reasoning around, suggests a $\mu \to e\gamma$ discovery may be close at hand. So, the discrepancy in Δa_μ may not only be a harbinger of New Physics but may also be heralding a possible discovery of $\mu \to e\gamma$.

2.4.3. *Tau flavor violation*

Tau flavor-changing transition dipole moments of the form in Eq. (2.65) give rise to the radiative decays $\tau \to \mu\gamma$ and $e\gamma$ as well as other rare decays such as $\tau \to \mu\mu^+\mu^-$, μe^+e^-, etc.

Recently, searches for those processes have been extended at the SLAC and KEK B factories which are also tau factories, producing similar amounts of $\tau^+\tau^-$ and $b\bar{b}$ pairs. Indeed, they have started to reach branching ratio sensitivities approaching 10^{-8} (see Table 2.5). Future SuperB factories, currently under design, are expected to attain $10^{-9} - 10^{-10}$ levels, making them competitive, in some models, with $\mu \to e\gamma$ for unveiling New Physics.

Parameterizing New Physics at a scale $\Lambda^{\tau\ell}_{R,L}$ by the $D^{\tau\ell}_{R,L}$, $\ell = \mu, e$, in Eq. (2.67) and setting for simplicity $D^{\tau\ell}_R$ or $D^{\tau\ell}_L = \mathcal{E}_{\tau\ell}/\left(\Lambda^{\tau\ell}\right)^2$ leads to the decay branching ratios

$$B\left(\tau \to \ell\gamma\right) \simeq 3.4 \times 10^9\ \mathrm{GeV}^4 \frac{\mathcal{E}^2_{\tau\ell}}{\left(\Lambda^{\tau\ell}\right)^4}. \tag{2.133}$$

Then, using the current bounds in Table 2.5 implies

$$\begin{aligned} \Lambda^{\tau\mu} &\gtrsim 15\ \mathrm{TeV} \times \sqrt{\mathcal{E}_{\tau\mu}}, \\ \Lambda^{\tau e} &\gtrsim 14\ \mathrm{TeV} \times \sqrt{\mathcal{E}_{\tau e}}. \end{aligned} \tag{2.134}$$

Those constraints are more than an order of magnitude below the $\Lambda^{\mu e}$ bound in Eq. (2.127) for $\mathcal{E}_{\mu e} \sim \mathcal{E}_{\tau\mu} \sim \mathcal{E}_{\tau e}$. However, it is possible that $\mathcal{E}_{\tau\mu}$ and $\mathcal{E}_{\tau e}$ are much larger than $\mathcal{E}_{\mu e}$, rendering rare tau decays competitive as probes of New Physics. For example, in supersymmetry scenarios with a super-GIM mechanism, the first two generations of slepton partners may be very nearly degenerate in mass while the third generation sleptons need not be. Of course, their mixing might then be smaller. All things considered, one guesstimates $\mathcal{E}_{\tau\mu}$ and $\mathcal{E}_{\tau e}$ could be $10 \sim 100$ times larger than $\mathcal{E}_{\mu e}$. So, $\tau \to \mu\gamma$ and $e\gamma$ are starting to become interesting and will even be competitive with improved future $\mu \to e\gamma$ searches if rare tau decays can reach $10^{-9} - 10^{-10}$ sensitivities.

2.4.4. *Neutrino transition dipole moments*

As we discussed in Section 2.2, massive Dirac neutrinos can have magnetic and electric dipole moments induced by quantum loops. Bounds on those quantities and predictions were also given there. For all practical purposes, those dipole moments are expected not to have important consequences unless they are significantly enhanced by New Physics effects.

Both Dirac and Majorana neutrinos can have transition dipole moments that connect different mass eigenstates $\nu_i \to \nu_j + \gamma$, $i \neq j$. Those moments are also dimension five quantum loop induced effects. Since they require flavor mixing, they are expected (in the Dirac case) to be somewhat suppressed relative to the neutrino magnetic moments mentioned above. Also, as in the $\mu \to e\gamma, \tau \to \mu\gamma$, and $\tau \to e\gamma$ amplitudes, they are expected to be GIM suppressed.

Employing the notation of Eq. (2.65), one finds for Dirac neutrinos with masses m_{ν_i} such that $m_{\nu_3} > m_{\nu_2} > m_{\nu_1}$, the transition moments

$$\begin{aligned} D_R^{ij} &\simeq \frac{3eG_F}{16\sqrt{2}\pi^2} \frac{m_\tau^2}{m_W^2} m_{\nu_i} U_{\nu_i \tau}^* U_{\nu_j \tau}, \\ D_L^{ij} &\simeq \frac{3eG_F}{16\sqrt{2}\pi^2} \frac{m_\tau^2}{m_W^2} m_{\nu_j} U_{\nu_i \tau}^* U_{\nu_j \tau}, \end{aligned} \tag{2.135}$$

where we have neglected terms with relative suppression m_μ^2/m_τ^2 and m_e^2/m_τ^2. Assuming $m_{\nu_i} \gg m_{\nu_j}$, the hierarchy $m_{\nu_3} > m_{\nu_2} > m_{\nu_1}$ and $\theta_{13} \simeq 0$ in the U mixing matrix then leads to

$$\begin{aligned} D_R^{32} &\simeq \frac{3eG_F}{16\sqrt{2}\pi^2} \frac{m_\tau^2}{m_W^2} m_{\nu_3} \left(-c_{12}c_{13}c_{23}s_{23}\right), \\ D_R^{31} &\simeq \frac{3eG_F}{16\sqrt{2}\pi^2} \frac{m_\tau^2}{m_W^2} m_{\nu_3} \left(c_{13}c_{23}s_{12}s_{23}\right), \\ D_R^{21} &\simeq \frac{3eG_F}{16\sqrt{2}\pi^2} \frac{m_\tau^2}{m_W^2} m_{\nu_2} \left(-s_{12}c_{12}s_{23}^2\right). \end{aligned} \tag{2.136}$$

Those transition moments exhibit some interesting features. The mixing angle effects are significant for $\theta_{23} \simeq 45°$, $\theta_{12} \simeq 32°$. The moments are linear in the neutrino mass rather than quadratic, as found in the case of charged lepton transition moments. So, neutrino transition moments can be $\mathcal{O}\left(m_\tau/m_{\nu_3}\right) \simeq 3\times 10^{10}$ times larger than their charged lepton counterparts. Nonetheless, they are still very small unless enhanced by additional New Physics beyond neutrino mass effects.

The transition moments in Eq. (2.136) give rise to radiative decay rates

$$\Gamma\left(\nu_i \to \nu_j \gamma\right) \simeq \frac{1}{16\pi} m_{\nu_i}^3 \left(|D_R|^2 + |D_L|^2\right), \tag{2.137}$$

which because of the phase space suppression correspond to neutrino lifetimes $> 10^{37}$ yrs, very long indeed.

In the case of Majorana neutrinos, the situation is less definite, depending in part on how the Majorana mass is generated. However [234], in general $D_L = \pm D_R = D_R\,(\text{Dirac})$. Overall, that feature doubles the decay rate in Eq. (2.137); but, it is still tiny.

In magnitude, the above transition moments (module mixing angle effects) are roughly (in units of electron Bohr magnetons)

$$\frac{3eG_F}{16\sqrt{2}\pi^2}\frac{m_\tau^2}{m_W^2}m_{\nu_3} \simeq 4\times 10^{-24}\frac{e}{2m_e}. \tag{2.138}$$

The only place where such tiny moments could come into play might be supernova phenomena where the combination of high matter densities and very large magnetic fields, $\mathcal{O}\left(10^8 - 10^{11}\right)$ Gauss, may give rise to resonant neutrino spin-flavor transitions. However, even there the moment in Eq. (2.138) is too small by several orders of magnitude to have much of an effect. For neutrino transition dipole moments to play any significant role in astrophysics or cosmology, there must be relatively large additional New Physics contributions that give enhancements beyond the neutrino mass effects in Eq. (2.138).

2.5. Conclusion

Starting with the one loop calculation of Schwinger and the experimental discovery $a_e \neq 0$ by Kusch and Foley [10, 14], anomalous magnetic dipole moments have played a leading role in testing QED and constraining New Physics. Recent experimental advances in measurements of a_e^{exp} and a_μ^{exp} by factors of 15 and 14 respectively have interesting consequences. The electron value of a_e provides the world's best determination of α, the fine structure constant, and is pushing QED calculations to five loops. The muon a_μ is less precise, but 43,000 times more sensitive to hadronic, electroweak and New Physics loop effects. A current discrepancy between experiment and a_μ Standard Model theory could be a strong hint of New Physics at scales $\lesssim 2$ TeV, with supersymmetry the leading candidate explanation.

Electric dipole moments of fermions provide a different type of dimension five New Physics probe. Any non-vanishing value would be direct evidence for New Physics at high scales and could be related to the matter-antimatter asymmetry of our Universe, since EDMs require CP violation, an important ingredient needed to generate such an asymmetry. Significant experimental advances in the electron and neutron EDM searches are expected in the next few years. A major discovery could soon be at hand.

Table 2.7. New Physics predictions for electromagnetic dipole moments and charged lepton flavor violating radiative decays assuming $a_\mu^{\rm NP} = a_\mu^{\rm exp} - a_\mu^{\rm SM} \simeq 3 \times 10^{-9}$ and the scaling relations discussed in the text. Also given are expected (in progress or proposed) experimental sensitivities.

Quantity	NP prediction	Future sensitivity
$a_\mu^{\rm NP}$	3×10^{-9}	$\pm 3 \times 10^{-10}$
$a_e^{\rm NP}$	7×10^{-14}	$\pm 13 \times 10^{-14}$
$d_\mu^{\rm NP}$	$3 \times 10^{-22} \tan\phi_\mu^{\rm NP}$ $e{\cdot}$cm	$\left\|\tan\phi_\mu^{\rm NP}\right\| \sim 3 \times 10^{-3}$
$d_e^{\rm NP}$	$1.5 \times 10^{-24} \tan\phi_e^{\rm NP}$ $e{\cdot}$cm	$\left\|\tan\phi_e^{\rm NP}\right\| \sim 10^{-6}$
$\left\|d_n^{\rm NP}\right\|$	$4 \times 10^{-23} \tan\phi_n^{\rm NP}$ $e{\cdot}$cm	$\left\|\tan\phi_n^{\rm NP}\right\| \sim 10^{-6}$
$\left\|d_p^{\rm NP}\right\|$	$4 \times 10^{-23} \tan\phi_p^{\rm NP}$ $e{\cdot}$cm	$\left\|\tan\phi_p^{\rm NP}\right\| \sim 10^{-6}$
$B(\mu \to e\gamma)$	$1.5 \times 10^{-3} \left\|\epsilon_{\mu e}\right\|^2$	$\left\|\epsilon_{\mu e}\right\| \sim 4 \times 10^{-6}$
$B(\tau \to \mu\gamma)$	$3 \times 10^{-4} \left\|\epsilon_{\tau\mu}\right\|^2$	$\left\|\epsilon_{\tau\mu}\right\| \sim 2 \times 10^{-3}$
$B(\tau \to e\gamma)$	$3 \times 10^{-4} \left\|\epsilon_{\tau e}\right\|^2$	$\left\|\epsilon_{\tau e}\right\| \sim 2 \times 10^{-3}$

New ideas for pushing proton and muon EDM sensitivities by many orders of magnitude using storage rings have been proposed that could make them competitive with n and e EDM searches.

Perhaps the most promising place to look for New Physics induced dimension five operators is in the flavor-changing transition dipole moments. Searches for $\mu \to e\gamma$ and related reactions $\mu^- N \to e^- N$ and $\mu \to ee^+e^-$ will probe New Physics scales beyond 1000 TeV if flavor-changing loop suppressions are not too severe. A new generation of rare tau decay studies, $\tau \to \mu\gamma$, $e\gamma$, $\mu\mu^+\mu^-$ etc. at SuperB factories promises to make them potentially competitive with rare muon decays.

To illustrate possible relationships among different dipole moments, we give in Table 2.7 New Physics predictions under the assumption that the same underlying Cm_μ^2/Λ^2 physics responsible for $a_\mu^{\rm NP} \simeq 3 \times 10^{-9}$ is also contributing to other induced dipole moments. If that is the case, we see that small CP violating phases, $\left|\tan\phi^{\rm NP}\right| \sim 10^{-6}$, and flavor-changing mixing effects, $|\epsilon_{\mu e}| \sim 4 \times 10^{-6}$, will be explored by the coming generation of EDM and rare muon decay experiments. At that level of sensitivity, a discovery is certainly possible.

Efforts to push experimental sensitivities for anomalous magnetic, electric and transition dipole moments are well motivated and complementary to direct searches for New Physics at high energy colliders such as the LHC. They should be extended as far as possible.

Acknowledgments

This research was supported by the United States DOE grant DE-AC02-76CH00016, and by Natural Sciences and Engineering Research Canada.

References

[1] P. A. M. Dirac, Proc. Roy. Soc. Lond. **A117**, 610 (1928).
[2] P. A. M. Dirac, Proc. Roy. Soc. Lond. **A118**, 351 (1928).
[3] P. A. M. Dirac, Proc. Roy. Soc. Lond. **A126**, 360 (1930).
[4] A. D. Sakharov, Pisma Zh. Eksp. Teor. Fiz. **5**, 32 (1967), translation in JETP Lett. **5**, 24 (1967).
[5] E. M. Purcell and N. F. Ramsey, Phys. Rev. **78**, 807 (1950).
[6] *Proceedings of the Dirac Centennial Symposium, Tallahassee, USA, December 6-7, 2002*, edited by H. Baer and A. Belyaev (World Scientific, Singapore, 2003).
[7] W. Pauli, in *Handbuch der Physik*, 2 ed., edited by H. Geiger and K. Scheel (Springer, Berlin, 1933), Vol. 24/1, p. 83, revised version in S. Flügge (ed.), Handbuch der Physik 5/1, p. 1 (1958).
[8] W. Pauli, Rev. Mod. Phys. **13**, 203 (1941).
[9] C. Amsler *et al.*, Phys. Lett. **B667**, 1 (2008).
[10] J. E. Nafe, E. B. Nelson, and I. I. Rabi, Phys. Rev. **71**, 914 (1947).
[11] D. E. Nagle, R. S. Julian, and J. R. Zacharias, Phys. Rev. **72**, 971 (1947).
[12] G. Breit, Phys. Rev. **72**, 984 (1947).
[13] J. Schwinger, Phys. Rev. **73**, 416 (1948).
[14] P. Kusch and H. M. Foley, Phys. Rev. **74**, 250 (1948).
[15] R. S. van Dyck Jr., P. B. Schwinberg, and H. G. Dehmelt, Phys. Rev. Lett. **59**, 26 (1987).
[16] D. Hanneke, S. Fogwell, and G. Gabrielse, Phys. Rev. Lett. **100**, 120801 (2008).
[17] T. Aoyama, M. Hayakawa, T. Kinoshita, and M. Nio, Phys. Rev. **D77**, 053012 (2008).
[18] M. Cadoret *et al.*, Phys. Rev. Lett. **101**, 230801 (2008).
[19] A. Czarnecki, Nature **442**, 516 (2006).
[20] W. Bernreuther and M. Suzuki, Rev. Mod. Phys. **63**, 313 (1991), erratum: *ibid.* **64**, 633 (1992).
[21] W. J. Marciano, in *CP Violation 1990*, edited by S. Dawson and A. Soni (World Scientific, Singapore, 1991), p. 35, proceedings of the BNL Summer Study, Upton, USA, May 21 – June 22, 1990.
[22] S. M. Barr and W. J. Marciano, Adv. Ser. Direct. High Energy Phys. **3**, 455 (1989).
[23] I. B. Khriplovich and S. K. Lamoreaux, *CP Violation Without Strangeness: Electric Dipole Moments of Particles, Atoms, and Molecules* (Springer, Berlin, 1997).
[24] E. D. Commins, J. Phys. Soc. Jap. **76**, 111010 (2007).

[25] M. Kobayashi and T. Maskawa, Prog. Theor. Phys. **49**, 652 (1973).
[26] N. Cabibbo, Phys. Rev. Lett. **10**, 531 (1963).
[27] B. C. Regan, E. D. Commins, C. J. Schmidt, and D. DeMille, Phys. Rev. Lett. **88**, 071805 (2002).
[28] S. Jung and J. D. Wells, arXiv:0811.4140 (unpublished).
[29] G.W. Bennett, *et al.*, Muon $(g-2)$ Collaboration), arXiv:0811.1207v1, Nov. 2008 and submitted to Phys. Rev. **D**.
[30] A. G. Grozin, I. B. Khriplovich, and A. S. Rudenko, arXiv:0811.1641 (unpublished).
[31] T. Ibrahim and P. Nath, Phys. Rev. **D64**, 093002 (2001).
[32] W. J. Marciano and A. I. Sanda, Phys. Lett. **B67**, 303 (1977).
[33] B. W. Lee and R. E. Shrock, Phys. Rev. **D16**, 1444 (1977).
[34] K. Fujikawa and R. Shrock, Phys. Rev. Lett. **45**, 963 (1980).
[35] M. A. B. Beg, W. J. Marciano, and M. Ruderman, Phys. Rev. **D17**, 1395 (1978).
[36] M. B. Voloshin, Sov. J. Nucl. Phys. **48**, 512 (1988).
[37] A. Studenikin, arXiv:0812.4716 (unpublished).
[38] C. Giunti and A. Studenikin, arXiv:0812.3646 (unpublished).
[39] A. V. Kyuldjiev, Nucl. Phys. **B243**, 387 (1984).
[40] P. Sutherland *et al.*, Phys. Rev. **D13**, 2700 (1976).
[41] M. Fukugita and S. Yazaki, Phys. Rev. **D36**, 3817 (1987).
[42] G. G. Raffelt and D. S. P. Dearborn, Phys. Rev. **D37**, 549 (1988).
[43] G. G. Raffelt, Phys. Rev. Lett. **64**, 2856 (1990).
[44] J. M. Lattimer and J. Cooperstein, Phys. Rev. Lett. **61**, 23 (1988).
[45] R. Barbieri and R. N. Mohapatra, Phys. Rev. Lett. **61**, 27 (1988).
[46] M. B. Voloshin, Phys. Lett. **B209**, 360 (1988).
[47] M. Fukugita, D. Notzold, G. Raffelt, and J. Silk, Phys. Rev. Lett. **60**, 879 (1988).
[48] C.-S. Lim and W. J. Marciano, Phys. Rev. **D37**, 1368 (1988).
[49] E. K. Akhmedov, Phys. Lett. **B213**, 64 (1988).
[50] S. Davidson, M. Gorbahn, and A. Santamaria, Phys. Lett. **B626**, 151 (2005).
[51] C. A. Baker *et al.*, Phys. Rev. Lett. **97**, 131801 (2006).
[52] W. C. Griffith, M. D. Swallows, T. H. Loftus, M. V. Romalis, B. R. Heckel and E. N. Fortson, Phys. Rev. Lett. **102**, 101601 (2009).
[53] M. Pospelov and A. Ritz, Phys. Rev. Lett. **83**, 2526 (1999).
[54] M. Pospelov and A. Ritz, Phys. Rev. **D63**, 073015 (2001).
[55] M. Pospelov and A. Ritz, Annals Phys. **318**, 119 (2005).
[56] G. 't Hooft, Phys. Rev. Lett. **37**, 8 (1976).
[57] R. J. Crewther, P. Di Vecchia, G. Veneziano, and E. Witten, Phys. Lett. **B88**, 123 (1979), erratum: Phys. Lett. **B91**, 341 (1980).
[58] J. L. Feng, K. T. Matchev, and Y. Shadmi, Nucl. Phys. **B613**, 366 (2001).
[59] M. Graesser and S. D. Thomas, Phys. Rev. **D65**, 075012 (2002).
[60] Z. Chacko and G. D. Kribs, Phys. Rev. **D64**, 075015 (2001).
[61] A. Czarnecki and E. Jankowski, Phys. Rev. **D65**, 113004 (2002).
[62] J. P. Miller, E. de Rafael, and B. L. Roberts, Rept. Prog. Phys. **70**, 795

(2007).
[63] G. W. Bennett *et al.*, Phys. Rev. **D73**, 072003 (2006).
[64] K. Melnikov and A. Vainshtein, *Theory of the Muon Anomalous Magnetic Moment*, No. 216 in *Springer tracts in modern physics* (Springer, Berlin, 2006).
[65] F. Combley, F. J. M. Farley, and E. Picasso, Phys. Rept. **68**, 93 (1981).
[66] G. W. Bennett *et al.*, Phys. Rev. Lett. **92**, 161802 (2004).
[67] D. W. Hertzog *et al.*, arXiv:0705.4617 (unpublished).
[68] T. Kinoshita and W. J. Marciano, in *Quantum Electrodynamics*, edited by T. Kinoshita (World Scientific, Singapore, 1990), pp. 419–478.
[69] J. Calmet, S. Narison, M. Perrottet, and E. de Rafael, Rev. Mod. Phys. **49**, 21 (1977).
[70] A. Czarnecki and W. J. Marciano, Nucl. Phys. B (Proc. Suppl.) **76**, 245 (1999).
[71] T. Kinoshita and M. Nio, Phys. Rev. **D73**, 053007 (2006).
[72] M. Nio, T. Aoyama, M. Hayakawa, and T. Kinoshita, Nucl. Phys. Proc. Suppl. **169**, 238 (2007).
[73] J.-P. Aguilar, D. Greynat, and E. de Rafael, Phys. Rev. **D77**, 093010 (2008).
[74] P. J. Mohr, B. N. Taylor, and D. B. Newell, Rev. Mod. Phys. **80**, 633 (2008).
[75] M. Passera, Phys. Rev. **D75**, 013002 (2007).
[76] T. Aoyama, M. Hayakawa, T. Kinoshita, and M. Nio, Phys. Rev. **D78**, 113006 (2008).
[77] T. Kinoshita and M. Nio, Phys. Rev. **D70**, 113001 (2004).
[78] A. L. Kataev, Phys. Rev. **D74**, 073011 (2006).
[79] M. Gourdin and E. De Rafael, Nucl. Phys. **B10**, 667 (1969).
[80] R. Alemany, M. Davier, and A. Höcker, Eur. Phys. J. **C2**, 123 (1998).
[81] M. Davier and A. Höcker, Phys. Lett. **B435**, 427 (1998).
[82] M. Davier, hep-ex/9912044 (unpublished).
[83] M. Davier, Nucl. Phys. B (Proc. Suppl.) **76**, 327 (1999).
[84] S. Eidelman and F. Jegerlehner, Z. Phys. **C67**, 585 (1995).
[85] T. Kinoshita, B. Nizic, and Y. Okamoto, Phys. Rev. **D31**, 2108 (1985).
[86] E. de Rafael, Phys. Lett. **B322**, 239 (1994).
[87] J. Erler and M. Luo, hep-ph/0101010 (unpublished).
[88] M. Davier, Nucl. Phys. Proc. Suppl. **169**, 288 (2007).
[89] S. Eidelman, Acta Phys. Polon. **B38**, 3499 (2007).
[90] K. Hagiwara, A. D. Martin, D. Nomura, and T. Teubner, Phys. Lett. **B649**, 173 (2007).
[91] F. Jegerlehner, Acta Phys. Polon. **B38**, 3021 (2007).
[92] J. F. de Troconiz and F. J. Yndurain, Phys. Rev. **D71**, 073008 (2005).
[93] M. Davier and W. J. Marciano, Ann. Rev. Nucl. Part. Sci. **54**, 115 (2004).
[94] W. J. Marciano, Precision electroweak measurements and the Higgs mass, 2004, hep-ph/0411179, talk at 32nd SLAC Summer Institute on Particle Physics4.
[95] W. J. Marciano and B. L. Roberts (unpublished).

[96] W. J. Marciano and A. Sirlin, Phys. Rev. Lett. **61**, 1815 (1988).
[97] V. Cirigliano, G. Ecker, and H. Neufeld, Phys. Lett. **B513**, 361 (2001).
[98] V. Cirigliano, G. Ecker, and H. Neufeld, JHEP **08**, 002 (2002).
[99] S. Eidelman, private communication.
[100] F. Jegerlehner, in *Radiative Corrections*, edited by J. Solà (World Scientific, Singapore, 1999), pp. 75–89.
[101] M. Fujikawa *et al.*, Phys. Rev. **D78**, 072006 (2008).
[102] S. Binner, J. H. Kuhn, and K. Melnikov, Phys. Lett. **B459**, 279 (1999).
[103] M. Davier, talk given at the 10th International Worshop on Tau Lepton Physics, Novisibirsk, September 2008; http://tau08.inp.nsk.su/talks/24/Davier.ppt.
[104] F. Ambrosino *et al.*, Phys. Lett. **B670**, 285 (2009).
[105] W. Kluge, Nucl. Phys. Proc. Suppl. **181-182**, 280 (2008).
[106] B. Krause, Phys. Lett. **B390**, 392 (1997).
[107] K. Hagiwara, A. D. Martin, D. Nomura, and T. Teubner, Phys. Rev. **D69**, 093003 (2004).
[108] C. Aubin and T. Blum, Nucl. Phys. Proc. Suppl. **162**, 251 (2006).
[109] J. Bijnens, E. Pallante, and J. Prades, Nucl. Phys. **B474**, 379 (1996).
[110] M. Hayakawa and T. Kinoshita, Phys. Rev. **D57**, 465 (1998), erratum Phys. Rev. D66, 019902 (2002).
[111] K. Melnikov and A. Vainshtein, Phys. Rev. **D70**, 113006 (2004).
[112] M. Knecht and A. Nyffeler, Phys. Rev. **D65**, 073034 (2002).
[113] M. Knecht, A. Nyffeler, M. Perrottet, and E. de Rafael, Phys. Rev. Lett. **88**, 071802 (2002).
[114] J. Erler and G. T. Sanchez, Phys. Rev. Lett. **97**, 161801 (2006).
[115] J. Prades, E. de Rafael, and A. Vainshtein, Hadronic Light-by-Light Scattering Contribution to the Muon Anomalous Magnetic Moment, 2009, in this book.
[116] J. Prades, E. de Rafael, and A. Vainshtein, arXiv:0901.0306 (unpublished).
[117] S. J. Brodsky and J. D. Sullivan, Phys. Rev. **156**, 1644 (1967).
[118] T. Burnett and M. J. Levine, Phys. Lett. **24B**, 467 (1967).
[119] R. Jackiw and S. Weinberg, Phys. Rev. **D5**, 2396 (1972).
[120] K. Fujikawa, B. W. Lee, and A. I. Sanda, Phys. Rev. **D6**, 2923 (1972).
[121] I. Bars and M. Yoshimura, Phys. Rev. **D6**, 374 (1972).
[122] G. Altarelli, N. Cabibbo, and L. Maiani, Phys. Lett. **B40**, 415 (1972).
[123] W. A. Bardeen, R. Gastmans, and B. E. Lautrup, Nucl. Phys. **B46**, 315 (1972).
[124] A. Czarnecki, B. Krause, and W. Marciano, Phys. Rev. Lett. **76**, 3267 (1996).
[125] A. Czarnecki, B. Krause, and W. Marciano, Phys. Rev. **D52**, 2619 (1995).
[126] J. D. Bjorken and S. Weinberg, Phys. Rev. Lett. **38**, 622 (1977).
[127] T. V. Kukhto, E. A. Kuraev, A. Schiller, and Z. K. Silagadze, Nucl. Phys. **B371**, 567 (1992).
[128] A. Czarnecki, W. J. Marciano, and A. Vainshtein, Phys. Rev. **D67**, 073006 (2003).
[129] A. Czarnecki, W. J. Marciano, and A. Vainshtein, Acta Phys. Polon. **B34**,

5669 (2003).
[130] S. Peris, M. Perrottet, and E. de Rafael, Phys. Lett. **B355**, 523 (1995).
[131] G. Degrassi and G. F. Giudice, Phys. Rev. **D58**, 053007 (1998).
[132] T. Gribouk and A. Czarnecki, Phys. Rev. **D72**, 053016 (2005).
[133] M. Passera, W. J. Marciano, and A. Sirlin, Phys. Rev. **D78**, 013009 (2008); AIP Conf. Proc. **1078**, 378-381 (2209).
[134] A. Czarnecki and W. J. Marciano, Phys. Rev. **D64**, 013014 (2001).
[135] P. Fayet, in *Unification of the Fundamental Particle Interactions*, edited by S. Ferrara, J. Ellis, and P. van Nieuwenhuizen (Plenum, New York, 1980), p. 587.
[136] J. A. Grifols and A. Mendez, Phys. Rev. **D26**, 1809 (1982).
[137] J. Ellis, J. Hagelin, and D. V. Nanopoulos, Phys. Lett. **B116**, 283 (1982).
[138] R. Barbieri and L. Maiani, Phys. Lett. **B117**, 203 (1982).
[139] J. C. Romao, A. Barroso, M. C. Bento, and G. C. Branco, Nucl. Phys. **B250**, 295 (1985).
[140] D. A. Kosower, L. M. Krauss, and N. Sakai, Phys. Lett. **B133**, 305 (1983).
[141] T. C. Yuan, R. Arnowitt, A. H. Chamseddine, and P. Nath, Z. Phys. **C26**, 407 (1984).
[142] I. Vendramin, Nuovo Cim. **A101**, 731 (1989).
[143] J. A. Grifols, J. Sola, and A. Mendez, Phys. Rev. Lett. **57**, 2348 (1986).
[144] D. A. Morris, Phys. Rev. **D37**, 2012 (1988).
[145] M. Frank and C. S. Kalman, Phys. Rev. **D38**, 1469 (1988).
[146] R. M. Francis, M. Frank, and C. S. Kalman, Phys. Rev. **D43**, 2369 (1991).
[147] J. L. Lopez, D. V. Nanopoulos, and X. Wang, Phys. Rev. **D49**, 366 (1994).
[148] T. Moroi, Phys. Rev. **D53**, 6565 (1996), erratum *ibid.* **D56**, 4424 (1997).
[149] D. Stockinger, J. Phys. **G34**, R45 (2007).
[150] W.-S. Hou, F.-F. Lee, and C.-Y. Ma, arXiv:0812.0064 (unpublished).
[151] T. Ibrahim and P. Nath, Phys. Rev. **D62**, 015004 (2000).
[152] A. Brignole, E. Perazzi, and F. Zwirner, JHEP **09**, 002 (1999).
[153] M. Carena, G. F. Giudice, and C. E. M. Wagner, Phys. Lett. **B390**, 234 (1997).
[154] K. T. Mahanthappa and S. Oh, Phys. Rev. **D62**, 015012 (2000).
[155] U. Chattopadhyay and P. Nath, Phys. Rev. **D53**, 1648 (1996).
[156] T. Goto, Y. Okada, and Y. Shimizu, hep-ph/9908499 (unpublished).
[157] T. Blazek, hep-ph/9912460 (unpublished).
[158] U. Chattopadhyay, D. K. Ghosh, and S. Roy, hep-ph/0006049 (unpublished).
[159] E. Ma, hep-ph/0109249 (unpublished).
[160] G. Belanger *et al.*, Phys. Lett. **B519**, 93 (2001).
[161] D. G. Cerdeno *et al.*, Phys. Rev. **D64**, 093012 (2001).
[162] H. Baer, C. Balazs, J. Ferrandis, and X. Tata, Phys. Rev. **D64**, 035004 (2001).
[163] K.-m. Cheung, C.-H. Chou, and O. C. W. Kong, Phys. Rev. **D64**, 111301 (2001).
[164] L. L. Everett, G. L. Kane, S. Rigolin, and L.-T. Wang, Phys. Rev. Lett. **86**, 3484 (2001).

[165] E. A. Baltz and P. Gondolo, Phys. Rev. Lett. **86**, 5004 (2001).
[166] M. Byrne, C. Kolda, and J. E. Lennon, Phys. Rev. **D67**, 075004 (2003).
[167] S. P. Martin and J. D. Wells, Phys. Rev. **D67**, 015002 (2003).
[168] K. S. Babu and J. C. Pati, Phys. Rev. **D68**, 035004 (2003).
[169] V. Barger, C. Kao, P. Langacker, and H.-S. Lee, Phys. Lett. **B614**, 67 (2005).
[170] M. J. Ramsey-Musolf and S. Su, Phys. Rept. **456**, 1 (2008).
[171] S. Heinemeyer, D. Stockinger, and G. Weiglein, Nucl. Phys. **B690**, 62 (2004).
[172] S. Heinemeyer, D. Stockinger, and G. Weiglein, Nucl. Phys. **B699**, 103 (2004).
[173] D. Stockinger, Nucl. Phys. Proc. Suppl. **135**, 311 (2004).
[174] T.-F. Feng *et al.*, Phys. Rev. **D73**, 116001 (2006).
[175] T.-F. Feng, L. Sun, and X.-Y. Yang, Phys. Rev. **D77**, 116008 (2008).
[176] T. Mori *et al.*, Search for $\mu^+ \to e^+\gamma$ down to 10^{-14} branching ratio, 1999, proposal to PSI. http://meg.psi.ch/doc.
[177] R. M. Carey *et al.*, letter of intent: A muon to electron conversion experiment at Fermilab; FERMILAB-TM-2396-AD-E-TD (unpublished).
[178] Y. K. Semertzidis *et al.*, hep-ph/0012087 (unpublished).
[179] W. Marciano, in *Particle Theory and Phenomenology*, edited by K. Lassila *et al.* (World Scientific, Singapore, 1996), p. 22.
[180] W. J. Marciano, in *Radiative Corrections: Status and Outlook*, edited by B. F. L. Ward (World Scientific, Singapore, 1995), pp. 403–414.
[181] Z.-H. Xiong and J. M. Yang, Phys. Lett. **B508**, 295 (2001).
[182] K. S. Babu and E. Ma, Mod. Phys. Lett. **A4**, 1975 (1989).
[183] F. Borzumati, G. R. Farrar, N. Polonsky, and S. Thomas, Nucl. Phys. **B555**, 53 (1999).
[184] S. J. Brodsky and S. D. Drell, Phys. Rev. **D22**, 2236 (1980).
[185] G. L. Shaw, D. Silverman, and R. Slansky, Phys. Lett. **B94**, 57 (1980).
[186] M. C. Gonzalez-Garcia and S. F. Novaes, Phys. Lett. **B389**, 707 (1996).
[187] H. Davoudiasl, J. L. Hewett, and T. G. Rizzo, hep-ph/0006097 (unpublished).
[188] R. Casadio, A. Gruppuso, and G. Venturi, hep-th/0010065 (unpublished).
[189] P. Mery, S. E. Moubarik, M. Perrottet, and F. M. Renard, Z. Phys. **C46**, 229 (1990).
[190] F. Herzog, Phys. Lett. **148B**, 355 (1984).
[191] M. Suzuki, Phys. Lett. **153B**, 289 (1985).
[192] A. Grau and J. A. Grifols, Phys. Lett. **154B**, 283 (1985).
[193] M. Beccaria, F. M. Renard, S. Spagnolo, and C. Verzegnassi, Phys. Lett. **B448**, 129 (1999).
[194] D. E. Groom *et al. (Particle Data Group)*, Eur. Phys. J. **C15**, 1 (2000).
[195] H. Przysiezniak, in *Intersections of Particle and Nuclear Physics*, edited by Z. Parsa and W. J. Marciano (AIP, Melville, NY, 2000), p. 1.
[196] W. J. Marciano and A. Queijeiro, Phys. Rev. **D33**, 3449 (1986).
[197] J. P. Leveille, Nucl. Phys. **B137**, 63 (1978).
[198] H. Chavez and J. A. Martins Simoes, Nucl. Phys. **B783**, 76 (2007).

[199] R. Escribano and E. Masso, Eur. Phys. J. **C4**, 139 (1998).
[200] Y.-B. Dai, C.-S. Huang, and A. Zhang, J. Phys. **G28**, 139 (2002).
[201] C. S. Kim, J. D. Kim, and J.-H. Song, Phys. Lett. **B511**, 251 (2001).
[202] M. Krawczyk and J. Zochowski, Phys. Rev. **D55**, 6968 (1997).
[203] A. Dedes and H. E. Haber, JHEP **05**, 006 (2001).
[204] M. Krawczyk, Acta Phys. Polon. **B33**, 2621 (2002).
[205] R. A. Diaz, R. Martinez, and J. A. Rodriguez, Phys. Rev. **D67**, 075011 (2003).
[206] K. Cheung and O. C. W. Kong, Phys. Rev. **D68**, 053003 (2003).
[207] G. Couture and H. König, Phys. Rev. **D53**, 555 (1996).
[208] S. Davidson, D. Bailey, and B. A. Campbell, Z. Phys. **C61**, 613 (1994).
[209] K.-m. Cheung, Phys. Rev. **D64**, 033001 (2001).
[210] F. Cuypers and S. Davidson, Eur. Phys. J. **C2**, 503 (1998).
[211] D. Chang, W.-F. Chang, C.-H. Chou, and W.-Y. Keung, hep-ph/0009292 (unpublished).
[212] M. L. Graesser, Phys. Rev. **D61**, 074019 (2000).
[213] P. Nath and M. Yamaguchi, Phys. Rev. **D60**, 116006 (1999).
[214] G. Cacciapaglia, M. Cirelli, and G. Cristadoro, Nucl. Phys. **B634**, 230 (2002).
[215] U. Chattopadhyay and P. Nath, Phys. Rev. **D66**, 093001 (2002).
[216] T. W. Kephart and H. Pas, Phys. Rev. **D65**, 093014 (2002).
[217] E. Kiritsis and P. Anastasopoulos, JHEP **05**, 054 (2002).
[218] S. C. Park and H. S. Song, Phys. Lett. **B523**, 161 (2001).
[219] K. Sawa, Phys. Rev. **D73**, 025010 (2006).
[220] A. Hektor, Y. Kajiyama, and K. Kannike, Phys. Rev. **D78**, 053008 (2008).
[221] S. L. Glashow, J. Iliopoulos, and L. Maiani, Phys. Rev. **D2**, 1285 (1970).
[222] M. Misiak, Acta Phys. Polon. **B38**, 2879 (2007).
[223] B. Grzadkowski and M. Misiak, Phys. Rev. **D78**, 077501 (2008).
[224] S. Heinemeyer, X. Miao, S. Su, and G. Weiglein, JHEP **08**, 087 (2008).
[225] J. R. Ellis *et al.*, JHEP **08**, 083 (2007).
[226] S. Heinemeyer, W. Hollik, and G. Weiglein, Phys. Rept. **425**, 265 (2006).
[227] G. Degrassi, P. Gambino, and P. Slavich, Comput. Phys. Commun. **179**, 759 (2008).
[228] W. J. Marciano, T. Mori, and J. M. Roney, Ann. Rev. Nucl. Part. Sc. **58**, 35 (2008).
[229] M. L. Brooks *et al.*, Phys. Rev. Lett. **83**, 1521 (1999).
[230] Y. Kuno and Y. Okada, Rev. Mod. Phys. **73**, 151 (2001).
[231] W. J. Marciano and A. I. Sanda, Phys. Rev. Lett. **38**, 1512 (1977).
[232] A. Czarnecki, W. J. Marciano, and K. Melnikov, in *Physics at the First Muon Collider*, edited by S. Geer and R. Raja (AIP, Woodbury, 1998), pp. 409–418.
[233] A. Czarnecki, W. J. Marciano, and K. Melnikov, in *High Intensity Muon Sources*, edited by Y. Kuno and T. Yokoi (World Scientific, Singapore, 2001), p. 61.
[234] P. B. Pal and L. Wolfenstein, Phys. Rev. **D25**, 766 (1982).

Chapter 3

In Search of the Breakdown of QED: Study of Lepton $g-2$ from 1947 to Present

Toichiro Kinoshita

Laboratory of Elementary-Particle Physics, Cornell University
Ithaca, NY, U.S.A. 14853
tk42@cornell.edu

This article is a revised and expanded version of a talk presented at the *Third International Symposium on Lepton Moments*, Cape Cod, June 19 - 22, 2006. It is expanded considerably to provide a historical perspective on the study of lepton $g-2$, focusing on the numerical integration method which consists of two steps: (1) Analytic construction of FORTRAN codes of renormalized amplitudes for the lepton $g-2$. (2) Numerical evaluation of the codes obtained in step (1). A systematic formulation was developed by 1974 to deal with the sixth-order case, later extended to the eighth-order case. To handle the far more complicated tenth-order case, we developed an algorithm that enables us to automatically construct fully renormalized FORTRAN codes representing a large fraction of 12672 Feynman diagrams. While applying this automation code to the eighth-order a_e for the purpose of debugging, we found an inconsistency in the previous handling of linear infrared divergence. With this error corrected we now have two independent evaluations of the eighth-order term of a_e, which agree with each other within the estimated uncertainty of numerical integration. The possible existence of further algebraic error, which might have eluded detection being smaller than the uncertainty of numerical integration, is eliminated by an extensive test evaluation of the *integrands* (not integrals themselves) in double precision, which shows that the old and new integrands agree to the first 15 digits at arbitrarily chosen points in the domain of integration. The current status of the calculation of the muon $g-2$ is reviewed briefly. Numerical evaluation of the tenth-order a_e is in an advanced stage.

Contents

3.1. Introduction

According to Dirac's theory [1], the electron has an intrinsic magnetic moment accompanying its spin, whose value, when expressed in the form $\frac{geh}{4\pi mc}$, is given by $g = 2$, in good agreement with the available experiments.[a] However, the Dirac equation itself does not rule out the possibility that the electron has an additional interaction with the magnetic field which might cause the g value to deviate from 2. It was an intriguing question why the observed g was so close to 2. An answer was obtained in 1947 when a tiny but non-vanishing deviation from the prediction of Dirac equation was discovered experimentally [2]. Furthermore, Schwinger showed that the deviation can be explained as an effect caused by the interaction of the electron with photons [3]. Together with the discovery of the Lamb shift of the hydrogen atom in the same year [4], this provided strong experimental support for the renormalization theory of quantum electrodynamics (QED) which was just being developed [5]. Schwinger's calculation of a_e to the order α was one of the most important milestones in the development of QED. This also meant that the non-QED contribution to $g-2$ is extremely small, if it exists at all.

In spite of the spectacular success of QED, because of mathematically dubious treatment of ultraviolet (UV) divergences, there has been widespread suspicion from the beginning that the renormalization is just a

[a] e is the charge carried by the electron, m is the rest mass of the electron, h is the Planck constant, and c is the velocity of light in vacuum.

temporal patch to hide or circumvent the real problem and is far from the satisfactory solution of the divergence problem [6, 7].

To find out where QED actually breaks down became a challenge to both theorists and experimentalists. Since then, the measurement and theory of a_e have been improved by seven orders of magnitude. But no sign of failure of QED is yet in sight. Instead it keeps providing the most precise and rigorous verification for the validity of QED.

It looks plausible that the UV divergence is somehow related to the assumption that the electron is a point particle with no internal structure. One plausible way to eliminate divergence is thus to give a finite size to the electron, an idea explored since ancient times. String theories and brane theories may be regarded as its modern incarnation.

The discovery of a large number of new particles since the good old days of 1947 made it clear that QED by itself cannot be the comprehensive theory of Nature. To accommodate new particles QED was extended to the Standard Model, a *renormalizable* gauge theory unifying the electromagnetic, weak, and strong interactions.

At present the test of the Standard Model is in a fair shape, consistent with all measurements within the precision of hadronic and electroweak data available. Furthermore, in processes where the electromagnetic interaction is dominant and the effect of other forces is small and known moderately well, it is possible to call it a test of the validity of QED. Thus, more precisely speaking, the question is "How good is QED within the context of the Standard Model ?"

Other questions often asked are:

(a) Will the perturbative expansion in the elementary charge e (or α), on which the success of QED is based, converge?
(b) Even if it diverges, could it be an asymptotic expansion?

It is impossible to answer these questions without calculating high-order terms. Unfortunately, this is easy to say but not easy to perform because of the enormous complexity of such calculations. The simplest system in which such a calculation might be feasible to sufficiently high orders is a free electron in a constant magnetic field. Of course there is no guarantee that a breakdown occurs at some finite order. However, we will never know unless we try. This is why the electron's magnetic moment has become the target of intense scrutiny both experimentally and theoretically.

In section 3.2 we review the history of the test of $g-2$ of the electron and the muon, including the new Harvard measurement of a_e and the revised QED calculation up to the order α^4.

In section 3.3 we discuss the status of the work in progress on the tenth-order radiative correction to the lepton $g-2$, followed in section 3.4 by the discussion of the prospect of future tests of QED.

3.2. QED Test by Lepton $g-2$: Interplay of Theory and Experiment

3.2.1. *Pre-1947 era*

The history of the study of the magnetic moment of the electron goes all the way back to early measurements of atomic spectra, in particular the Stern-Gerlach experiment [8], in which a collimated beam of neutral silver atoms was found to split into just two separated beams (instead of the expected three) when they were passed through an inhomogeneous magnetic field. A critical examination of this anomalous Zeeman effect led Pauli [9] to propose (December 1924) that the doublet structure of the atomic spectra is caused by a two-valuedness of some quantum property of the electron. This was followed by the formulation of the exclusion principle in January 1925 [10]. In October 1925, Uehlenbeck and Goudsmit suggested [11] that it is associated with a "proper" rotation of the electron with the intrinsic angular momentum $h/4\pi$, or the spin.[b]

The quantitative understanding of the atomic fine structure required further work. A relativistic effect (Thomas precession) had to be understood [13]. Also, the value e/mc, twice the value expected classically, had to be assigned to the gyromagnetic ratio of the electron. In May 1927 Pauli succeeded in describing the spinning electron by a two-component wave function introducing 2×2 matrices in an *ad hoc* fashion [14]. Darwin proposed a similar equation in July 1927 [15].

The origin of these features became clear only when Dirac proposed, in January 1928, a relativistically covariant 4-component wave equation of the electron based on a few general ansaetze [1]. This equation incorporated in a natural way the spin angular momentum $\frac{h}{4\pi}$ and the gyromagnetic ratio $g = 2$. Furthermore, the spectrum of the hydrogen atom given by the Dirac

[b]The spin as an internal angular momentum responsible for the electron's fourth quantum number was first mentioned by R. Kronig [12] in January 1925.

equation was in excellent agreement with atomic experiments, including the fine structure.

At the same time, however, Dirac's equation raised a new and profound paradox, namely, the apparent existence of negative energy states. This was resolved only when it was recognized that Dirac's wave function ψ was not the probability amplitude in the sense of the Schrödinger equation, but must be reinterpreted as an operator which destroys an electron and creates a positron. In other words, the electron had to be treated as a quantized field just as the photon was, although the exclusion principle had to be invoked to quantize the electron field [16].

Relativistic quantum theory incorporating all these features, namely, quantum electrodynamics (QED), was formulated around 1929 [17, 18]. This theory was in good agreement with experiments in the lowest order of perturbation theory. However, QED was not yet satisfactory: It suffered from a very severe problem that higher order corrections to the prediction of QED are divergent. The resolution of this difficulty had to wait for nearly 20 years until it was solved in 1947 by the renormalization of mass and charge of the electron [5].

3.2.2. *Early tests of QED*

In spite of serious theoretical difficulties, the "naive" prediction $g = 2$ of the Dirac equation was in excellent agreement with the available experiments for 20 years within the experimental precision. Only in 1947 was a tiny but unambiguous deviation of the electron g value from 2 discovered by an accurate measurement of the Zeeman splitting of the gallium atom in a magnetic field [2]

$$a_e = (g_e - 2)/2 = 0.001\ 15\ (4). \tag{3.1}$$

Schwinger showed that it can be explained as a QED effect [3]:

$$a_e = \frac{\alpha}{2\pi} = 0.001\ 161... \tag{3.2}$$

It is important to note that, at the present stage of development of QED (or its generalization including the Standard Model of electroweak and strong interactions), neither the mass nor the charge of the electron is calculable from the theory itself. They must be treated as input parameters whose values have to be determined experimentally. Because of this the simplest quantity that can be actually calculated from first principles is the anomalous magnetic moment of the electron. This is why the study of a_e occupies a particularly important niche in the high precision test of QED.

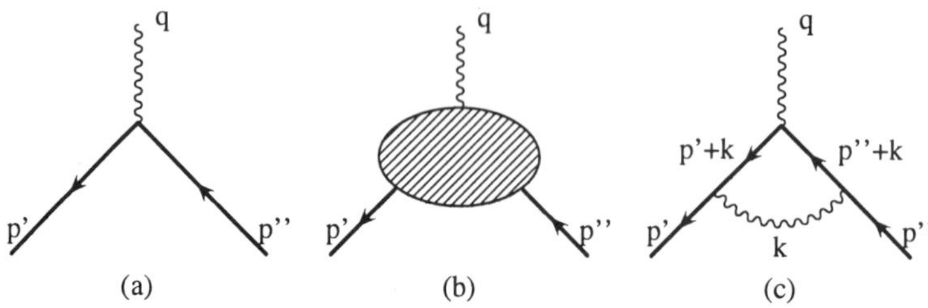

Fig. 3.1. (a) Lowest-order Feynman diagram describing scattering of an electron by an external magnetic field. (b) Schematic diagram representing an infinite set of Feynman diagrams contributing to a_e. (c) Second-order vertex diagram.

The magnetic property of the electron can be studied most conveniently by examining the theoretical idealization, namely, scattering of the electron by a static magnetic field. If the interaction with virtual photons is ignored, this can be expressed, in the limit of weak magnetic field (which is the case under normal experimental conditions) by the Feynman diagram shown in Fig. 3.1(a). Application of Feynman–Dyson rules to this diagram leads to the scattering amplitude (apart from a factor $-2\pi i\delta({p'}_0 - {p''}_0)$ for energy conservation)[c]

$$e\bar{u}(p')\gamma^\mu u(p'')A^e_\mu(\vec{q}), \tag{3.3}$$

with

$$A^e_\mu(\vec{q}) = \frac{1}{(2\pi)^3}\int d^3x e^{-i\vec{q}\cdot\vec{x}}A^e_\mu(\vec{x}), \tag{3.4}$$

where $A^e_\mu(\vec{x})$ is the vector potential of the external static magnetic field. For $u(p'')$ and $\bar{u}(p')$ satisfying the Dirac equation the electric current $\bar{u}(p')\gamma^\mu u(p'')$ can be decomposed into convection and spin currents

$$\bar{u}(p')\gamma^\mu u(p'') = \frac{1}{2m}\bar{u}(p')(p'+p'')^\mu u(p'') + \frac{i}{2m}\bar{u}(p')q_\nu\sigma^{\mu\nu}u(p''), \tag{3.5}$$

where $q = p' - p''$ and $\sigma^{\mu\nu} = \frac{i}{2}(\gamma^\mu\gamma^\nu - \gamma^\nu\gamma^\mu)$. The second term exhibits that the Landé g-factor of a free electron is equal to 2 in Dirac's theory of the electron.

Because of the interaction with the virtual photon field surrounding the charge, however, the diagram in Fig. 3.1(a) must be replaced by an infinite set of Feynman diagrams, all having the structure schematically represented

[c]We put $c = 1$ and $h = 2\pi$ for the rest of this article.

by Fig. 3.1(b). Taking account of Lorentz, C, P, and T invariances, the corresponding amplitude can be written as a sum of two terms

$$e\bar{u}(p')\left[\gamma^\mu F_1(q^2)+\frac{i}{2m}\sigma^{\mu\nu}q_\nu F_2(q^2)\right]u(p'')A^e_\mu(\vec{q}). \tag{3.6}$$

F_1 and F_2 are the charge and magnetic form factors, respectively. The charge form factor is normalized so that $F_1(0)=1$. Thus the first term reduces to the amplitude Eq. (3.3) in the static limit and contributes a factor 2 to the g factor. The magnetic moment anomaly a_e is the static limit of $F_2(q^2)$, and, rewriting p' and p'' as $p+\frac{1}{2}q$ and $p-\frac{1}{2}q$, can be expressed as

$$a_e = F_2(0) = Z_2 M, \tag{3.7}$$

with

$$M=\lim_{q\to 0}\frac{m}{4p^2q^2}Tr\left[(m\gamma^\nu p^2-(m^2+\frac{1}{2}q^2)p^\nu)\right.$$
$$\left.(\gamma^\alpha(p_\alpha+\frac{1}{2}q_\alpha)+m)\Gamma_\nu(\gamma^\beta(p_\beta-\frac{1}{2}q_\beta)+m)\right], \tag{3.8}$$

where $p^2=m^2-\frac{1}{4}q^2, p\cdot q=0$. Γ^ν is the proper vertex part represented by Fig. 3.1(b), and Z_2 is the wave function renormalization constant.

3.2.2.1. *Early electron tests*

Taking the presence of the muon and tau particle into account the QED contribution to the electron $g-2$ can be written in the general form

$$a_e(QED)=A_1+A_2(m_e/m_\mu)+A_2(m_e/m_\tau)+A_3(m_e/m_\mu,m_e/m_\tau), \tag{3.9}$$

where A_i can be expanded into power series in $\frac{\alpha}{\pi}$

$$A_i=A_i^{(2)}\left(\frac{\alpha}{\pi}\right)+A_i^{(4)}\left(\frac{\alpha}{\pi}\right)^2+A_i^{(6)}\left(\frac{\alpha}{\pi}\right)^3+\ldots,\ i=1,2,3, \tag{3.10}$$

whose coefficients are finite calculable quantities, which is guaranteed by the renormalizability of QED.

The second-order coefficient $A_1^{(2)}$ can be calculated from the Feynman diagram of Fig. 3.1(c). The scattering amplitude corresponding to this diagram is readily given by the Feynman–Dyson rules:

$$\Gamma^\mu=\frac{-1}{(2\pi)^4}\int d^4k\frac{1}{k^2}\bar{u}(p')$$
$$\gamma^\lambda\frac{1}{\gamma^\alpha(p'_\alpha+k_\alpha)-m}\gamma^\mu\frac{1}{\gamma^\beta(p''_\beta+k_\beta)-m}\gamma_\lambda u(p'')A^e_\mu(\vec{q}). \tag{3.11}$$

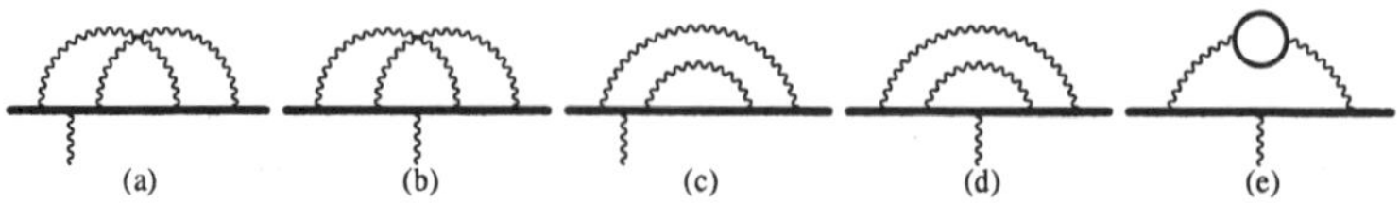

Fig. 3.2. Feynman diagrams contributing to a_e of fourth-order. Two more diagrams related by time-reversal are not shown.

Substituting this into Eq. (3.8) and carrying out the integration over the 4-momentum k, one finds

$$A_1^{(2)} = \frac{1}{2} \quad \text{or} \quad a_e^{(2)} = \frac{\alpha}{2\pi}. \tag{3.12}$$

This is Schwinger's result [3].

The early attempt by Karplus and Kroll to calculate the α^2 term $A_1^{(4)}$, contributed by 7 Feynman diagrams of Fig. 3.2 [19], had an unfortunate error which was corrected[d] by Petermann [20] and, independently, by Sommerfield [21] in 1957.[e] The corrected result is

$$\begin{aligned} A_1^{(4)} &= \frac{197}{144} + \left(\frac{1}{2} - 3\ln 2\right)\zeta(2) + \frac{3}{4}\zeta(3) \\ &= -0.328\ 478\ 965\ 579\ldots, \end{aligned} \tag{3.13}$$

where $\zeta(n)$ is the Riemann zeta function of argument n.

Meanwhile experimental effort has been going on to improve the initial result of Kusch and Foley [2] by measurement of μ_p/μ_0, where μ_0 is the Bohr magneton and μ_p is the proton magnetic moment. Combined with the measurement of μ_e/μ_p, this led to $a_e = 0.001\ 165\ (11)$ in 1956 [22], which disagreed with the calculation of Karplus and Kroll by 1.6 standard deviations. This problem was resolved soon afterwards by the correct calculation given in Eq. (3.13).

A far more substantial improvement in precision was achieved by Michigan group in measurement of the electron $g-2$, not g itself, by means of the precession of the electron spin in a uniform magnetic field [23]. The final value obtained by this method is [24]:

$$a_{e^-}[\text{UM71}] = 1\ 159\ 657\ 7\ (35)\times\ 10^{-10}. \tag{3.14}$$

Since this uncertainty is only 3.6 times smaller than

$$\left(\frac{\alpha}{\pi}\right)^3 \simeq 125 \times 10^{-10}, \tag{3.15}$$

[d]Examining the Karplus-Kroll article [19] Petermann discovered a sign error in one of the integrals by a numerical method which led him to re-evaluate the entire $A_1^{(4)}$.
[e]Schwinger asked his graduate student, Sommerfield, to solve the electron $g-2$ problem exactly to all orders in α. Sommerfield solved it to the order α^2.

it is necessary to evaluate the coefficient $A_1^{(6)}$ of the α^3 term to match the precision of theory with the experiment.

3.2.2.2. *Early muon tests*

The first observation that the muon has a spin rotation consistent with $g = 2$ was reported from the Columbia–Nevis cyclotron [25]. In a subsequent paper [26] they reported the measurement

$$g_\mu = 2(1.00113^{+0.00016}_{-0.00012}), \tag{3.16}$$

which shows that the Schwinger's radiative correction given by Eq. (3.2) applies to the muon, too. This is the first convincing experimental evidence that the muon behaves like the electron, unlike the proton whose magnetic moment is 2.792 847 351 (28) times the nuclear magneton [27]. In other words, experimentally, the muon seems to be identical with the electron in all respects except for the rest mass.[f]

However, the mass difference between the muon and the electron affects the muon anomaly a_μ in the fourth-order, as was first pointed out by Suura, Wichman [28] and by Petermann [29]. Bouchiat, Michel [30] and Durand [31] pointed out that a_μ has also an important contribution from hadronic vacuum-polarization, because of the strong enhancement effect caused by the ρ-resonance.

Farley proposed [32] a high precision muon $g-2$ experiment at CERN in 1962, which was followed by second and third measurements with steadily improving methods and precision [33, 34]. See section 3.2.9 for more details.

These experiments measure the spin precession of the muon in a magnetic field, which is similar to the Michigan experiment for the electron [23]. However, the muon has a great advantage that it has a built-in spin polarizer and analyzer because of the parity non-conservation while the Michigan electron spin measurement had to rely on a tiny spin dependence of the elastic electron-nucleus scattering cross section (Mott scattering). On the other hand, the ultimate precision achievable by the muon measurement of this type is constrained by the short lifetime of the μ-e decay.

3.2.3. *Back to theory*

Inspired by these developments I started in 1966 a serious effort to evaluate the sixth-order QED contribution. The QED contribution to a_μ can be

[f] I. I. Rabi once quipped "Who ordered the muon?"

written in the general form:

$$a_\mu(QED) = A_1 + A_2(m_\mu/m_e) + A_2(m_\mu/m_\tau) + A_3(m_\mu/m_e, m_\mu/m_\tau), \quad (3.17)$$

taking account of the presence of other leptons. A_i can be expanded into power series in $\frac{\alpha}{\pi}$ as in Eq. (3.10). A_1 is mass-independent so that it is common to a_e and a_μ. Suura, Wichmann [28], and Petermann [29] found that the fourth-order term contributing to $a_\mu - a_e$, namely $A_2^{(4)}(m_\mu/m_e)$, has a logarithmic dependence on m_μ/m_e:

$$A_2^{(4)}(m_\mu/m_e) = \frac{1}{3}\ln(m_\mu/m_e) - \frac{25}{36} + \cdots . \quad (3.18)$$

The analytic result was obtained later as a series expansion in r, where $r = m_e/m_\mu$. The first few terms are [35, 36]:

$$\begin{aligned} A_2^{(4)}(m_\mu/m_e) = & -\frac{1}{3}\ln r - \frac{25}{36} + \frac{\pi^2}{4}r + (3 + 4\ln r)r^2 - \frac{5}{4}\pi^2 r^3 \\ & + \left(\frac{\pi^2}{3} + \frac{44}{9} - \frac{14}{3}\ln r + 2(\ln r)^2\right) r^4 \\ & + \left(-\frac{109}{225} + \frac{8}{15}\ln r\right) r^6 + \cdots . \end{aligned} \quad (3.19)$$

Note that the original (unrenormalized) amplitude does not have a logarithmic mass-singularity. Namely, the appearance of the logarithmic term is nothing but a consequence of charge renormalization [37]. Once this was realized, it was straightforward to reproduce Eq. (3.18) using only the renormalization group idea and a theorem on mass singularity [38] without carrying out any integration at all. What was more interesting, the same argument immediately led to the derivation of leading logarithmic terms of sixth-order diagrams (diagrams (a), (b), (c) of Fig. 3.3) by an algebraic manipulation of lower order terms, which in fact was the first application of the renormalization group method [37].

The sixth-order term $A_2^{(6)}(m_\mu/m_e)$ has also contributions from six diagrams (represented by Fig. 3.3(d)) that contain a light-by-light scattering subdiagram. If one can show that these diagrams have no logarithmic contribution, then the leading $\ln(m_\mu/m_e)$ contribution to the muon $g-2$ would come only from the diagrams (a), (b), and (c) of Fig. 3.3, which I had solved already. Having tried unsuccessfully to prove this *no-log* conjecture [39], I decided to examine it by numerical means, and persuaded my graduate student J. Aldins to work it out. S. Brodsky and his graduate student A. J. Dufner were also working on the same problem. We decided to check each other's calculation and write a joint paper [40]. At that time

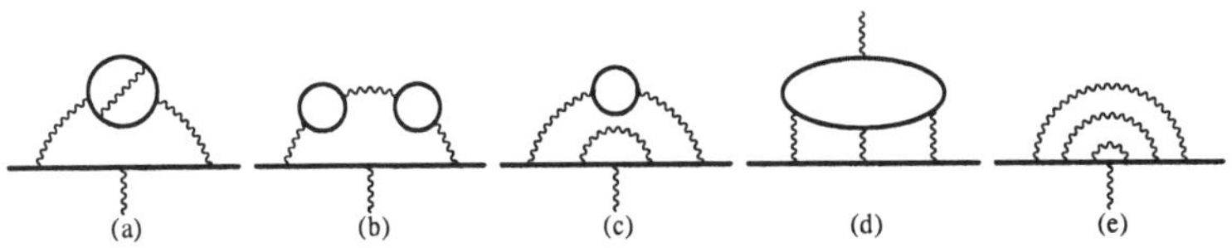

Fig. 3.3. Representatives of 72 Feynman diagrams contributing to a_e of sixth-order. Diagrams (a), (b), and (c) represent 16 diagrams containing various vacuum-polarization loops. The diagram (d) represents 6 diagrams with a light-by-light-scattering subdiagram. The diagram (e) represents 50 diagrams without closed lepton loop, which are called *q-type* diagrams. Diagrams (a) – (d), in which open lines are muon lines and closed lines are electron lines, give mass-dependent contributions to a_μ.

no algebraic manipulation program was available so that the integrands had to be worked out by hand, which required a very nontrivial effort. Fortunately, for carrying out the 7-dimensional integration, an adaptive-iterative Monte-Carlo integration routine SHEPPY [41] has become available.

The initial naive expectation was that the contribution of Fig. 3.3(d) to a_μ would be small because the light-by-light scattering cross-section calculated from the Euler–Heisenberg Lagrangian is very small [42]. Thus we were quite surprised when we found by numerical integration [40] that it is actually very large (~ 18). This suggested the presence of a $\ln(m_\mu/m_e)$ term, which was readily confirmed by calculating it numerically for several values of m_μ/m_e [40]. Lautrup and Samuel [43] obtained later the leading term of $A_2^{(6)}(l\!-\!l)$ analytically

$$A_2^{(6)}(l\!-\!l) = \frac{2\pi^2}{3}\ln(m_\mu/m_e) + \cdots , \tag{3.20}$$

which established the presence of a large numerical factor $2\pi^2/3$.

The reason why my initial expectation was wrong is that I failed to pay attention to the fact that the Euler–Heisenberg interaction Lagrangian is valid only in the low energy region where it is suppressed by the fourth-power of photon momentum (a consequence of gauge invariance). For the diagrams of Fig. 3.3(d), in which the light-by-light-scattering subdiagram receives contributions from photons of all energies, an exact formula must be used instead of the low energy approximation.

It would be instructive here to examine the origin of $\ln(m_\mu/m_e)$ and its large coefficient. Since $A_2^{(6)}(l\!-\!l)$ is UV-finite, the term $\ln m_\mu$ comes from the scale set by the largest physical mass of the system, m_μ. The $\ln(m_\mu/m_e)$ term arises from the integration over the domain $D_1(m_e < |k| < m_\mu,\ |p_i| \leq m_e)$ in which the loop momentum k of the light-by-light subdiagram covers the large range $m_e < |k| < m_\mu$ while the momenta $p_i, i = 1, 2, 3$, of photons exchanged between the electron and the muon are restricted to the small domains as shown. This is the type of mass singularity which appears for the first time in Feynman diagrams containing closed lepton loops with 4 or more photons attached. Other domains such as $D_2(\text{any } k,\ |p_i| > m_e)$ do not contribute to $\ln m_e$, since the lower end of k integration does not reach m_e.

What makes $A_2^{(6)}(l\text{–}l)$ really large, however, is the presence of the coefficient π^2 in Eq. (3.20). A nice physical explanation for the appearance of this coefficient was given by Elikhovskii [44]. He pointed out that, in the large m_μ/m_e limit, in the subdomain $D_3(m_e < |k| < m_\mu,\ |p_i| < \alpha m_e)$ of D_1, where α is the fine structure constant, the muon is nearly at rest and may be regarded as a static source of Coulomb photon as well as the hyperfine spin-spin interaction. Of the three photons exchanged between the muon line and the electron loop of Fig. 3.3(d), one photon is responsible for the hyperfine spin-spin interaction while the other two act essentially like the static Coulomb potential. In this limit it is easy to carry out integration over the Coulomb photon momenta. Each integration gives a factor $i\pi$ so that two such integrations give a factor $\pi^2(\sim 10)$ to the leading term of Eq. (3.20).

The full analytic result of $A_2^{(6)}(l\text{–}l)$ was obtained in 1993 by Laporta and Remiddi [45]. The first few terms in the expansion in r, where $r = m_e/m_\mu$, are

$$\begin{aligned}
A_2^{(6)}(l\text{–}l) &= \frac{2\pi^2}{3}\ln(1/r) + \frac{59\pi^4}{270} - 3\zeta(3) - \frac{10\pi^2}{3} + \frac{2}{3} \\
&+ r\left(\frac{4\pi^2}{3}\ln(1/r) - \frac{196\pi^2}{3}\ln 2 + \frac{424\pi^2}{9}\right) \\
&+ r^2\left(-\frac{2}{3}\ln^3(1/r) + (\frac{\pi^2}{9} - \frac{20}{3})\ln^2(1/r)\right. \\
&- (\frac{16\pi^4}{135} + 4\zeta(3) - \frac{32\pi^2}{9} + \frac{61}{3})\ln(1/r) \\
&+ \left.\frac{4}{3}\zeta(3)\pi^2 - \frac{61\pi^4}{270} + 3\zeta(3) + \frac{25\pi^2}{18} - \frac{283}{12}\right) \\
&+ r^3\left(\frac{10\pi^2}{9}\ln(1/r) - \frac{11\pi^2}{9}\right) + \cdots . \qquad (3.21)
\end{aligned}$$

For $1/r = 206.768\ 283\ 8\ (54)$ [27] we obtain [46]

$$A_2^{(6)}(l\text{–}l) = 20.947\ 924\ 89\ (16). \qquad (3.22)$$

In this way all diagrams containing vacuum-polarization loop and/or light-by-light-scattering loop were evaluated numerically [37, 40, 47–49] or analytically [37, 50] by 1975 for both a_μ and a_e. In order to match the improving precision of the electron $g-2$ measurement, however, it was necessary to evaluate the mass-independent term $A_1^{(6)}$ of a_e, to which all 72 Feynman diagrams contribute including those of Fig. 3.3(e).

3.2.4. *Feynman-parametric integral for numerical integration*

Although some simpler sixth-order diagrams of Fig. 3.3(e) had been evaluated analytically by 1974 [51], others looked so formidable that my graduate student Cvitanovic and I decided to tackle them by numerical integration. Of course this was not unrelated to our successful calculation by the numerical integration of diagrams of Fig. 3.3(d) described in section 3.2.3. Many of the techniques used later were developed already in Ref. [40].

Our approach starts from an exact analytic construction of renormalized amplitudes for lepton $g-2$ in the Feynman parametric form. This step involves no approximation except for the expansion in powers of α.

Suppose G is a $2n$-th order contribution to the proper electron vertex part of the form given by Fig. 3.1(b). Feynman–Dyson rules assign the propagators

$$i\frac{\gamma_\mu p_i^\mu + m_i}{p_i^2 - m_i^2}, \qquad \frac{-ig^{\mu\nu}}{p_i^2 - m_i^2}, \tag{3.23}$$

to the electron lines and photon lines, besides various factors to the vertices. The momentum p_i may be decomposed as $k_i + q_i$, where k_i is a linear combination of integration variables and q_i is a linear combination of (fixed) external momenta p' and p''. m_i is the mass associated with the line i, which is temporarily distinguished from each other.

Before carrying out the momentum integration we replace $p_j = k_j + q_j$ of the numerator of lepton propagator by an operator D_j^μ defined by [19]

$$D_j^\mu \equiv \frac{1}{2}\int_{m_j^2}^{\infty} dm_j^2 \frac{\partial}{\partial q_{j\mu}}. \tag{3.24}$$

Since D_j^μ does not depend explicitly on the integration variables k_j, the numerators can be pulled out in front of the momentum integration as far as the integrand is adequately regularized.

The product of denominators are then combined into one using the Feynman formula

$$\prod_{i=1}^{N} \frac{1}{a_i} = (N-1)! \int (dz)_G \frac{1}{(\sum_{i=1}^{N} z_i a_i)^N}, \tag{3.25}$$

where $N = 3n$ and

$$(dz)_G \equiv \prod_{i=1}^{N} dz_i \delta \left(1 - \sum_{i=1}^{N} z_i \right). \tag{3.26}$$

The sum $\sum_i z_i a_i$ is a quadratic form of loop momenta so that it can be integrated analytically. As a consequence the amplitude is converted into an integral over the Feynman parameters z_i

$$\left(\frac{\alpha}{\pi}\right)^n \Gamma_\nu^{(2n)} = \left(-\frac{\alpha}{4\pi}\right)^n (n-1)!\, \mathbb{F} \int \frac{(dz)_G}{U^2 V^n}, \tag{3.27}$$

where U is the Jacobian of mapping of the momentum space onto the Feynman-parametric space, and $\mathbb{F}$ is an operator consisting of γ^μ from vertices and $\gamma_\mu D_j^\mu + m_j$ from electron lines. An explicit form of V is given later. The action of $\mathbb{F}$ on $1/V^n$ produces terms of the form

$$\mathbb{F}\frac{1}{V^n} = \frac{F_0}{V^n} + \frac{F_1}{V^{n-1}} + \frac{F_2}{V^{n-2}} + \dots, \tag{3.28}$$

where the subscript k of F_k stands for the number of contractions. By contraction we mean picking a pair of $\gamma_\mu D_i^\mu + m_i$ and $\gamma_\nu D_j^\nu + m_j$ from $\mathbb{F}$, making the substitution

$$\gamma_\mu D_i^\mu + m_i, \quad \gamma_\nu D_j^\nu + m_j \Rightarrow \gamma_\mu, \quad \gamma_\nu, \tag{3.29}$$

multiplying it by a factor $-\frac{1}{2U} g^{\mu\nu} B_{ij}$, and summing the results over all distinct pairs. (Actually this corresponds to carrying out the momentum integration involving $\gamma_\mu k^\mu$ and $\gamma_\nu k^\nu$.) The uncontracted parts of D_i^μ (which involves the constant momenta q_i^μ introduced in Eq. (3.24)) are then replaced by

$$Q_i^{'\mu} = -\frac{1}{U} \sum_{j=1}^{2n} q_j^\mu z_j \left(B_{ij} - \delta_{ij} \frac{U}{z_i} \right). \tag{3.30}$$

For $k \geq 1$, F_k includes an overall factor $(m-1)^{-1}(m-2)^{-1} \cdots (m-k)^{-1}$. The magnetic moment projection of Eq. (3.27) is an integral M_G of Feynman parameters z_i, and "symbolic" building blocks U, V, and A_i, B_{ij}, where i, j are restricted to the indices of electron lines [52]

$$M_G = \int (dz)_G \left[\frac{F_0(A_i)}{U^2 V^{n-1}} + \frac{F_1(B_{ij}, A_i)}{U^3 V^{n-2}} + \cdots \right]. \tag{3.31}$$

Here the factor $(\alpha/\pi)^n$ is suppressed for simplicity.

The conversion of the momentum integral into the Feynman-parametric integral was achieved using an algebraic manipulation program SCHOONSCHIP [53]. $Q_i^{'\mu}$ is a linear combination of $p' \equiv p + q/2$ and

$p'' \equiv p - q/2$, or equivalently a linear combination of p and q. If the external momentum p is chosen to flow through the set E of continuous electron lines, the coefficient of p for $i \in E$ is given by $A_i \gamma^\mu$, where[g]

$$A_i = -\frac{1}{U} \sum_{j \in E} z_j \left(B_{ij} - \delta_{ij} \frac{U}{z_i} \right), \tag{3.32}$$

and V has a form common for all diagrams in the limit $q = 0$:

$$V = \sum_{i \in E} z_i (m_i^2 - A_i p_i^2) + \sum_{j}^{photons\ only} z_j \lambda^2. \tag{3.33}$$

Here m_i is the mass of the electron line i, $p_i = p$ if the external momentum p flows through the line i, and $p_i = 0$ otherwise. λ is the infrared cutoff. The next step is to express B_{ij} and U as homogeneous polynomials of $z_1, z_2, ..., z_N$. This was not difficult to carry out by hand in the sixth-order case.

The integral Eq. (3.31) has in general UV divergences coming from subdiagrams of vertex type and/or self-energy type. In order to deal with these divergences by renormalization we first regularize each relevant photon propagator by introducing the Feynman cutoff

$$\frac{1}{k^2 - \lambda^2} \rightarrow \frac{1}{k^2 - \lambda^2} - \frac{1}{k^2 - \Lambda^2} = - \int_{\lambda^2}^{\Lambda^2} \frac{dL}{(k^2 - L)^2}, \tag{3.34}$$

where Λ is the UV-cutoff. Throughout this paper let us assume that all UV-divergent integrals are regularized by the Feynman cutoff. This requires just a minor modification of Eq. (3.31). Of course the Feynman cutoff is not needed for convergent integrals, and the limit $\Lambda \rightarrow \infty$ must be taken after the renormalization is carried out.

3.2.5. *K-operation and I-operation*

Finally, we have to renormalize the integrals. Our approach is a subtractive renormalization. This is carried out by construction of subtraction integrals by K-operation defined as follows [52].

Suppose we want to find out whether M_G diverges when all loop momenta of a subdiagram S consisting of N_S lines and n_S closed loops go to infinity. In the parametric formulation this limit corresponds to the vanishing of U when all $z_i \in S$ vanish simultaneously. To find the criterion for

[g]The coefficient of q has a slightly different form. It is not included explicitly in Eq. (3.31) to avoid cluttering.

a UV divergence from S, consider the part of the integration domain where $z_i \in S$ satisfy $\sum_{i\in S} z_i \leq \epsilon$. In the limit $\epsilon \to 0$, one finds

$$\begin{aligned} V &= \mathcal{O}(1), \quad U = \mathcal{O}(\epsilon^{n_S}), \\ B_{ij} &= \mathcal{O}(\epsilon^{n_S - 1}) \text{ if } i, j \in S, \\ B_{ij} &= \mathcal{O}(\epsilon^{n_S}) \qquad \text{otherwise.} \end{aligned} \tag{3.35}$$

For a vertex subdiagram S the K_S-operation is defined as follows.

(a) In the limit Eq. (3.35) keep only terms with lowest power of ϵ in U, B_{ij}, A_i. (Then U factorized as $U_S U_R$. Similarly for B_{ij}. V is reduced to V_R, where R is obtained from G by shrinking S to a point in G.)
(b) Replace V_R by $V_R + V_S$, where V_S is the V function defined on S.
(c) Rewrite the integrand of M_G in terms of parametric function redefined in (a) and (b), drop all terms except those with the largest number of contractions (see Eq. (3.29)) within S, and call the result $\mathbb{K}_S M_G$ (which means $\mathbb{K}_S$ operating on M_G).

Since $\mathbb{K}_S M_G$ have the same (logarithmic) UV divergence as M_G in the common domain of Feynman-parametric space it can be used for *point-wise* subtraction of the UV singularity of M_G. Furthermore, by construction, $\mathbb{K}_S M_G$ factorizes exactly into a product of a part of renormalization constant and magnetic moment of lower orders:

$$\mathbb{K}_S M_G = \widehat{L}_S M_{G/S}, \tag{3.36}$$

where $\widehat{L}_S$ is the overall UV-divergent part of the vertex renormalization constant L_S. It is important to note that the factorization in Eq. (3.36) does not work unless both sides are well-defined integrals (made finite by the Feynman cutoff or some other regularization). Note also that $\widehat{L}_S$ and $M_{G/S}$ can be separately constructed as integrals representing lower-order diagrams. The naive product of these integrals, however, has a singularity different from that of M_G so that it cannot be used for point-wise subtraction.

This procedure is also applicable to self-energy-type subdiagrams S although it leads to a somewhat more complicated factorization [52]:

$$\mathbb{K}_S M_G = \delta\widehat{m}_S M_{G/S(i^*)} + \widehat{B}_S M_{G/[S,i]}, \tag{3.37}$$

where S is an electron self-energy part inserted between consecutive lines i and j of G, and $\delta\widehat{m}_S$ and $\widehat{B}_S$ are the overall UV-divergent parts of renormalization constants δm_S and B_S. (See [52] for definitions of $M_{G/S(i^*)}$ and $M_{G/[S,i]}$.)

An infrared divergence, which has its origin in the vanishing of virtual photon momenta, arises from the part of the integration domain where the z_i's assigned to these photons take the largest possible values under the constraint $\sum_i z_i = 1$. This means that all other z_i's are pushed to zero in the IR limit. This is, however, not the sufficient condition. In order that the IR singularity actually becomes divergent, it must be enhanced by vanishing of two or more denominators of electron propagators which share a 3-point-vertex with the infrared photons and external (on-shell) electron lines. This corresponds to the vanishing of the denominator V in the integration domain characterized by

$$\begin{aligned} z_i &= \mathcal{O}(\delta) && \text{if } i \text{ is an electron line in } R, \\ z_i &= \mathcal{O}(1) && \text{if } i \text{ is a photon line in } R, \\ z_i &= \mathcal{O}(\epsilon),\ \epsilon \sim \delta^2 && \text{if } i \in S, \end{aligned} \tag{3.38}$$

where $R = G/S$. (The last condition of Eq. (3.38) is actually an artifact of the constraint $\sum_i z_i = 1$ which can be readily lifted.) In this limit V behaves as $\mathcal{O}(\delta^2)$.

If two electron propagators participate in the enhancement, we obtain a *logarithmic* IR-divergence. In this case we can construct an IR subtraction term by a simple power counting rule and an I-operation similar to the K-operation of the UV case [52] . For the subdiagram $R = G/S$ the I_R-operation is defined as follows:[h]

(a) In the limit Eq. (3.38) keep only terms with lowest power of ϵ and δ in U, B_{ij}, A_i.

(b) Make the following replacements:

$$U \to U_S U_R, \quad V \to V_S + V_R, \quad F \to F_0[L_R] F_S, \tag{3.39}$$

where $F_0[L_R]$ is the no-contraction part of the vertex renormalization constant defined on R, and F_S is the product of γ matrices and D_i^μ operators for the diagram S.

(c) Rewrite the integrand of M_G in terms of redefined parametric functions, keep only the IR-divergent terms [52].

3.2.6. *Sixth-order calculation*

By the time we started evaluating the sixth-order term $A_1^{(6)}$ of Fig. 3.3, algebraic manipulation programs optimized for the QED calculation, such

[h]The following rule works for the sixth-order case but must be replaced by an extended rule for diagrams of eighth- and higher-orders in which some IR singularities are enhanced strongly by more than two electron propagators. See section 3.3.1 for details.

as SCHOONSCHIP [53] and REDUCE [54], became available. Thus we no longer had to carry out the trace calculation and momentum integration by hand, although we still prepared numerous small components of the integrand manually. The resulting FORTRAN programs, which were made finite by subtraction of all UV and IR divergences, were evaluated by the second-generation Monte-Carlo numerical integration routine RIWIAD [55] and, a few years later, by VEGAS [56].

Construction of Feynman-parametric integrals of sixth-order diagrams, in particular those diagrams that have no closed lepton loops, which will be called *q-type*, was carried out in two independent ways. One is a straightforward evaluation of individual vertex diagram Γ_ν using the magnetic moment projection Eq. (3.8). Another is to first combine five vertex diagrams, all generated from a self-energy diagram Σ by insertion of a magnetic vertex in all electron lines, into one using the equation

$$\Lambda^\nu(p,q) \simeq -q_\mu [\frac{\partial \Lambda_\mu(p,q)}{\partial q_\nu}]_{q=0} - \frac{\partial \Sigma(p)}{\partial p_\nu} \tag{3.40}$$

which is derived from the Ward–Takahashi identity. Using this identity and time reversal invariance of QED, we can cut down the number of independent integrals of *q-type* from 50 to 8. This reduces the amount of computing time substantially since the size of each of these eight integrals is not much larger than that of individual vertex diagrams.

Construction of Feynman-parametric integral and its magnetic moment projection is slightly more complicated for the Ward–Takahashi-summed amplitude than for a single vertex diagram. The major difference is that we can take the limit $q = 0$ from the outset and that new building block C_{ij} appears. For details see [52].

The availability of two independent methods was crucial for eliminating algebraic and programming errors. To make sure that these programs were free of further error, they were derived by two people working independently of each other [57].

A crude numerical evaluation of the complete sixth-order term by RIWIAD [55] was obtained by 1974 [57]:

$$A_1^{(6)} = 1.195\ (26), \tag{3.41}$$

which is the weighted average of the results obtained by the two independent methods described above. This led to the value of a_e which is in fair agreement with the Michigan experiment [24]. Somewhat different numerical approaches of Refs. [58, 59] gave results consistent with ours.

Meanwhile, the rapidly increasing computing power enabled us to improve Eq. (3.41) substantially. The best numerical value of the most difficult sixth-order diagram M_{6H}, combined with analytical results of other diagrams obtained by 1995, led to [60]:[i]

$$A_1^{(6)} = 1.181\ 259\ (40). \tag{3.42}$$

Analytic evaluation of $A_1^{(6)}$ was completed in 1996 by Laporta and Remiddi after many years of hard work [61]:

$$\begin{aligned} A_1^{(6)} &= \frac{83}{72}\pi^2\zeta(3) - \frac{215}{24}\zeta(5) - \frac{239}{2160}\pi^4 + \frac{139}{18}\zeta(3) \\ &\quad - \frac{298}{9}\pi^2 \ln 2 + \frac{17101}{810}\pi^2 + \frac{28259}{5184} \\ &\quad + \frac{100}{3}\left[\left(a_4 + \frac{1}{24}\ln^4 2\right) - \frac{1}{24}\pi^2 \ln^2 2\right] \\ &= 1.181\ 241\ 456\cdots, \end{aligned} \tag{3.43}$$

where $a_4 = \sum_{n=1}^{\infty} \frac{1}{2^n n^4} = 0.517\ 479\ 061\cdots$. It is reassuring that the numerical result Eq. (3.42) and the analytic result Eq. (3.43) agree to 5 digits, which is within the uncertainty of the numerical work.

Obviously, our numerical approach must pay constant attention to two distinct sources of error. One is possible algebraic and analytic error in the derivation of the FORTRAN codes, and the other is associated with the method of numerical integration itself. Let me first discuss the latter.

3.2.7. *How reliable is VEGAS?*

Numerical integration of our integral is carried out mostly by an adaptive-iterative Monte-Carlo integration routine VEGAS [56]. Let me describe briefly how VEGAS works for our problem since the reliability of the results of integration is critically dependent on that of VEGAS.

[i]Our algebraic work on the sixth-order term $A_1^{(6)}$ benefited greatly from SCHOONSCHIP, which was an excellent program for algebraic manipulation. One minor problem was that it was written in the Pauli metric in which the time coordinate was purely imaginary. Thus we had to pay close attention to the difference between i of time and i of quantum mechanics. Unfortunately, SCHOONSCHIP was written in the machine-specific language (which was Veltman's idiosyncrasy) so that it could not be easily ported to other computers. By the time we started working on the eighth-order term $A_1^{(8)}$, the machines on which SCHOONSCHIP ran began to disappear. FORM was created by Vermaseren as a successor of SCHOONSCHIP. Since it is written in the C language, it works on a wide variety of computers. Also, FORM is more flexible than SCHOONSCHIP and accepts the Bjorken-Drell metric which is widely used by high energy physicists.

Usually, the first iteration of VEGAS begins with evaluation of the integrand at randomly chosen points uniformly distributed throughout the integration domain. This gives an approximate value of the integral and its variance. Furthermore, it gives information on where important contributions to the integral come from. In the next iteration, the distribution of random points is adjusted to reflect the information obtained in the last iteration. This process (*adaptation*) is repeated until the combined result of all iterations reaches the desired precision.

Normally VEGAS is formulated on a unit hypercube. Thus it is necessary to map the n-dimensional hyperplane $\sum_{i=1}^{n+1} z_i = 1$ in the $(n+1)$-dimensional space onto a n-dimensional unit cube. There are infinitely many possible choices of such mapping. This gives us an opportunity to choose several different mappings. They are equivalent analytically but different from the viewpoint of numerical integration, responding differently to random sampling conducted by VEGAS.

For technical reasons, however, it is desirable to choose a mapping such that the singular behavior of the (original) unrenormalized part of the integrand is confined to a surface of smallest-dimension. (By construction singularities of renormalized integrands contributing to a_e are confined to the boundary surface of the unit cube and are integrable.) This is because randomly selected points, after just a few iterations, tend to accumulate towards the end point (0 or 1) of one of the axes on such a surface.

Frequently, the integrand blows up after several iterations. This is because our integrand is renormalized point-wise (namely, UV- or IR-counterterm is subtracted point by point throughout the domain of integration) so that the result (difference of two large and nearly equal numbers) is sensitive to the rounding of the number of digits available. Eventually, the integral may be overwhelmed by the noise caused by the round-off error in the vicinity of a singularity.

This problem can be alleviated by "stretching", namely, by expanding the neighborhood of the singular surface by further (nonlinear) mapping. This takes advantage of the fact that our integrand is analytically well-defined and integrable so that it can be made better-behaved after such a mapping, even if the exact nature of the singularity is unknown. In some difficult cases, however, it is necessary to go from the (usual) double precision arithmetic to higher precision such as quadruple precision, to minimize the round-off error problem.

Iterations performed in a well-chosen mapping may converge more rapidly than others even though they approach the same limit eventually.

This flexibility enables us to evaluate each diagram in several different ways, thereby greatly enhancing the reliability of the final numerical result. All integrals have been thoroughly tested in this manner. For instance this method gave us full confidence in our sixth-order numerical result Eq. (3.42), up to five decimal points, even before the analytic result Eq. (3.43) was published.

3.2.8. *Current status of a_e test*

In 1987 the measurement of the electron $g-2$ was improved over the previous value Eq. (3.14) by three orders of magnitude in a Penning trap experiment by Dehmelt *et al.* at University of Washington [62]. They reported

$$\begin{aligned} a_{e^-}[\text{UW87}] &= 1\ 159\ 652\ 188.4\ (4.3) \times 10^{-12}, \\ a_{e^+}[\text{UW87}] &= 1\ 159\ 652\ 187.9\ (4.3) \times 10^{-12}, \end{aligned} \tag{3.44}$$

for the electron and positron, respectively.

The uncertainty of the measurement Eq. (3.44) was dominated by the cavity shift due to the interaction of the electron with the hyperboloid cavity which has a complicated resonance structure. Several ways to reduce this error have been examined: (a) Use a cavity with smaller Q [63].

(b) Study the cavity shift by many ($\sim$ 1000)-electron cluster which magnifies the shift [64].
(c) Use a cylindrical cavity, whose property is calculated analytically [65].

Recent Harvard measurements [66, 67] determine a_e up to 18 times more accurately than the 1987 measurement Eq. (3.44), and shift the measured value by 1.7 standard deviations:

$$a_{e^-}[\text{HV06}] = 1\ 159\ 652\ 180.85\ (0.76) \times 10^{-12} \qquad [0.66\ \text{ppb}], \tag{3.45}$$

$$a_{e^-}[\text{HV08}] = 1\ 159\ 652\ 180.73\ (0.28) \times 10^{-12} \qquad [0.24\ \text{ppb}]. \tag{3.46}$$

These experiments deal with cavity shifts using a calculable cylindrical cavity geometry [68, 69] (method (c) above), along with two different methods to measure needed cavity properties that cannot be calculated, using parametrically pumped stored electrons [70] and using one suspended electron [67]. Also crucial to the new measurements is realizing and resolving the quantization of the electron cyclotron motion [71], using cavity-inhibited spontaneous emission, a very low cavity temperature of 100 mK and feedback detection [72]. Thanks to these innovations the 2008 measurement was not limited by cavity shift uncertainties but mostly by the need for further study of the observed resonance lineshapes.

Since the experimental uncertainty in Eq. (3.46) is less than 1% of

$$\left(\frac{\alpha}{\pi}\right)^4 \simeq 29 \times 10^{-12}, \tag{3.47}$$

it is necessary to know the actual value of the coefficient $A_1^{(8)}$ of the α^4 term to match the precision of theory with experiment. This requires evaluation of 891 Feynman diagrams, which can be classified into 13 gauge-invariant sets, representatives of which are shown in Fig. 3.4. In anticipation of high precision measurements which may become available some day, we began the effort to evaluate $A_1^{(8)}$ in early 1980s [73]. The algebraic work to express the integrands as functions of Feynman parameters was carried out initially by SCHOONSCHIP and later by FORM [74] , following the procedures outlined in section 3.2.4 and section 3.2.5. This gives us an algebraically exact fully renormalized integral. No approximation is involved at all in the preparation of the Feynman-parametric integral. For the *'spot check'* test which confirms this statement by numerical means, see the discussion preceding Eq. (3.49).

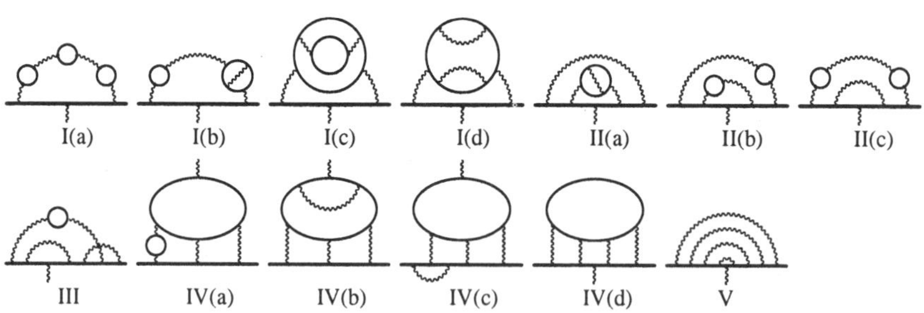

Fig. 3.4. Representative of 891 Feynman diagrams contributing to a_e of eighth-order.

Evaluation of Groups I, II, and III shown in Fig. 3.4 is relatively easy. The Group IV was much more difficult but was evaluated in two independent ways. Their latest numerical values are given in [75]. The value of I(a) is also known analytically [76, 77]. An alternative evaluation of I(c) is carried out using the photon spectral function of order α^3 derived from the QCD spectral function given in [78]. The value of I(d) is also evaluated using the Padé approximant of the sixth-order photon spectral function given in [79–82].

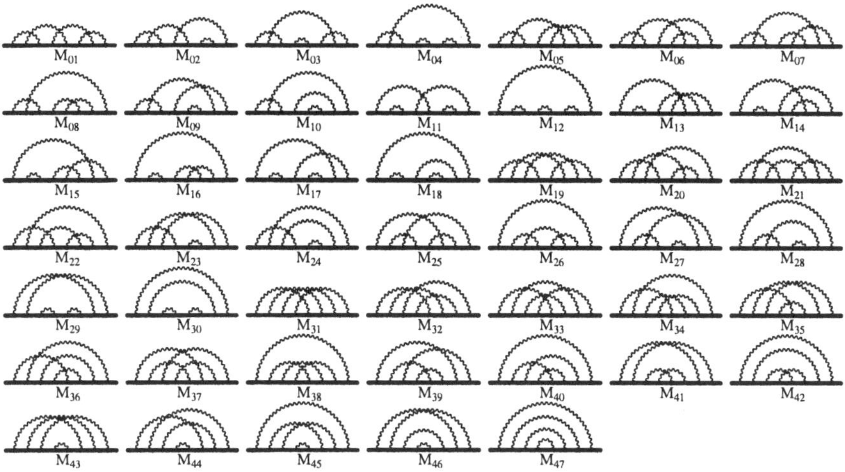

Fig. 3.5. Eighth-order Group V diagrams. 47 self-energy-like diagrams of $M_{01} - M_{47}$ represent 518 vertex diagrams.

The diagrams of Group V are far more complicated than the rest of the eighth-order diagrams. However, the general approach developed for the sixth-order case was found to be applicable to them with some modification. In view of the enormous demand on the computing power, we decided to proceed only with the approach which utilizes Eq. (3.40), forgoing the double-checking opportunity provided by the alternative method. Unfortunately, this left our calculation vulnerable to a possible programming error. Only very recently were we able to carry out a second independent calculation using FORTRAN codes generated by an automatic code-generating algorithm "gencodeN" [83, 84] described in section 3.3.

Although "gencodeN" was developed primarily to handle the tenth-order term, we have applied it to fourth-, sixth-, and eighth-order *q-type* diagrams as part of the debugging effort. With the help of "gencodeN" eighth-order FORTRAN codes are generated very easily. However, their numerical evaluation by VEGAS [56] is quite nontrivial and requires a huge computational resource. Numerical work has thus far reached a relative uncertainty of about 3% [85]. Although this is more than an order of magnitude less accurate than the uncertainty of the old calculation [75], it is good enough for checking the algebra of the old calculation.

UV divergences of vertex and self-energy subdiagrams are removed by K-operation (see section 3.2.5), which is identical with the old approach.

For diagrams containing self-energy subdiagrams, however, "gencodeN" treats UV-finite parts of self-energy subdiagrams and IR-divergences differently from the old approach.

To our dismay the comparison of the new and old calculation has revealed a subtle inconsistency in the treatment of infrared divergence in the latter. After correcting this programming error in the diagrams M_{16} and M_{18} of Group V (see Fig. 3.5), we now have two independent and consistent calculations of Group V diagrams.

Fortunately, the analytic form of the correction terms themselves can be obtained easily and evaluated precisely, yielding [85]]

$$A_1^{(8)}(\text{correction}) = -0.186\ 104\ (21). \tag{3.48}$$

The problem with the old calculation arose from the treatment of two eighth-order diagrams M_{16} and M_{18} which have linear IR divergence. Since this case was not covered by the method developed for the sixth-order case, we handled it by improvising an *ad hoc* subtraction method. What we found is that some inconsistencies remained undetected in the old treatment of M_{16} and M_{18} causing some finite shift in the value of $A_1^{(8)}$.

Note that all integrals of the new calculation have been generated from the same master code "gencodeN". If there were an error in any one of them, all others would suffer from the same error. On the other hand, integrals of the old version were constructed semi-manually one by one so that they might not be completely free from some undetected errors, even though the good numerical agreement between all 47 Group-V integrals of the new and old version is reassuring.

Of course, much more numerical work is required for the new version to reach the precision comparable to that of the old calculation. There is, however, an alternative and powerful way to prove or disprove the algebraic equivalence of old and new versions. It is to evaluate corresponding integrands, *not integrals*, of old and new versions numerically (at equivalent but not necessarily identical points) with high precision (real*8 or higher) by a *'spot check'* method.[j] It evaluates the integrand at as many arbitrarily chosen sets of points as is needed. It enables us to test not only the whole integrand but also individual components, such as the unrenormalized part and UV-divergence subtraction parts, separately. IR-divergence subtraction parts are defined differently in the old and new versions, but they can also be compared by the *'spot check'* method after some adjustment. We

[j]The *'spot check'* method was originally introduced around 1992 in order to debug UV- and IR- subtraction terms. However, it was not quoted explicitly until 2006 [75, 88].

can thus establish an algebraic equivalence of every corresponding part of the old and new integrals as precisely as we wish. Typically it is easy to get equivalence of the first 15 digits in real*8, with a small difference coming from possible differences in the treatment of round-off error. Any algebraic error may not escape detection by this very powerful method.

As a matter of fact, the algebraic error of the old M_{16} was discovered by comparison with the new M_{16} by means of the *'spot check'* method [85].

Lately we applied the *'spot check'* method to all diagrams of Group V, including residual renormalization terms. The result shows the complete algebraic equivalence of the old and new integrals.[k] We are thus fully convinced that all FORTRAN programs of Group V diagrams are indeed free from any algebraic error. The uncertainty in the value of $A_1^{(8)}$ given in Eq. (3.49) is therefore purely statistical and can be improved steadily by accumulating more and more sampling statistics.

As for the purely analytic integration of the eighth-order diagrams, it seems that, although analytic techniques developed for integrating $A_1^{(6)}$, in particular, that of integration by part, have been useful for the study of some of the eighth-order terms, further development seems to be necessary for a complete analytic integration of $A_1^{(8)}$ [86]. Thus far only a small number of eighth-order diagrams of *q-type* have been evaluated by analytic means [87].

Let us summarize here the values of mass-independent terms $A_1^{(n)}$, up to $n = 8$:

$$
\begin{aligned}
A_1^{(2)} &= 0.5 && \text{1 diagram (analytic)} \\
A_1^{(4)} &= -0.328\,478\,965\,\ldots && \text{7 diagrams (analytic)} \\
A_1^{(6)} &= 1.181\,241\,456\,\ldots && \text{72 diagrams (numerical, analytic)} \\
A_1^{(8)} &= -1.914\,4\,(35) && \text{891 diagrams (numerical).} \qquad (3.49)
\end{aligned}
$$

Note that the value of $A_1^{(8)}$ is different from the value $(-1.7283(35))$ published in [75] because of the correction mentioned above [85]. Let us emphasize that all terms of Eq. (3.49) have now been checked by two or more different methods. One may still ask whether the numerical value of $A_1^{(8)}$ given in Eq. (3.49) can be trusted or not. Of course it will never be exact, being a product of numerical integration. A more relevant question is how reliable is the estimation of error-bar quoted. Besides the absence of algebraic error described above I may emphasize that

[k]The author wishes to thank M. Nio and T. Aoyama for carrying out the *'spot check'* on very short notice.

(i) the uncertainty estimated by VEGAS internally is very reliable for all functions tested whose integrals are known exactly.

(ii) the freedom of choice of hyperplane-to-hypercube mapping enables us to experiment several choices of mapping and gives us strong confidence that we are not misled about the magnitude of uncertainty. (See discussions in section 3.2.7.)

As a consequence of all these tests we are sure that the value and error-bars of $A_1^{(8)}$ given in Eq. (3.49) are very solid and need not be replaced by a better one until an extensive new calculation is carried out in the future.

Mass-dependent terms $A_2^{(4)}(m_e/m_\mu)$ and $A_2^{(4)}(m_e/m_\tau)$ are known exactly [35, 36]. Recent evaluation of these contributions to a_e are [46]

$$A_2^{(4)}(m_e/m_\mu) = 5.197\ 386\ 70\ (28) \times 10^{-7},$$
$$A_2^{(4)}(m_e/m_\tau) = 1.837\ 62\ (60) \times 10^{-9}, \tag{3.50}$$

where the errors are only due to the uncertainty in the measured mass ratios [27].

Mass-dependent terms $A_2^{(6)}(m_e/m_\mu)$ and $A_2^{(6)}(m_e/m_\tau)$ are also known exactly [89]. Recent evaluation of their contributions to a_e, including both vacuum-polarization and light-by-light-scattering contributions, are [46]

$$A_2^{(6)}(m_e/m_\mu) = -7.373\ 941\ 64\ (29) \times 10^{-6},$$
$$A_2^{(6)}(m_e/m_\tau) = -6.581\ 9\ (19) \times 10^{-8}, \tag{3.51}$$

where the errors are only due to the uncertainty in the measured mass ratios [27].

Thus the total contribution of A_2 to a_e is small ($\sim 2.72 \times 10^{-12}$) but not negligible compared with the measurement uncertainty of Eq. (3.46). That of A_3 is even smaller ($\sim 2.4 \times 10^{-21}$) and completely negligible at present.

The direct evaluations of the leading and next-to-leading contributions of the hadronic vacuum-polarization to a_e yield [90, 91][1]

$$a_e(\text{had.vp.}) = 1.875\ (18) \times 10^{-12},$$
$$a_e(\text{had.NLO}) = -0.225\ (5) \times 10^{-12}. \tag{3.52}$$

The contribution of the hadronic light-by-light-scattering term, obtained by scaling down from $a_\mu(\text{had.ll})$ [93–95] by a factor $(m_e/m_\mu)^2$, is

$$a_e(\text{had.ll}) = 0.0257\ (94) \times 10^{-12}. \tag{3.53}$$

[1]Note, however, that $a_e(\text{had.vp.})$ must be replaced by $1.906\ (16) \times 10^{-12}$ if the preliminary measurement of $a_\mu(\text{had.vp.})$ reported in [92] is confirmed by further work. This will cause some changes in the rest of section 3.2.8 as is noted at the end of the section.

A direct evaluation of a_e(had.ll), without scaling down approximation, leads to[m]

$$a_e(\text{had.ll}) = 0.039\ (5) \times 10^{-12}. \tag{3.54}$$

The contribution of the electroweak effect to 2-loop order, scaled down from the electroweak effect on a_μ, is [97, 98]

$$a_e(\text{weak}) = 0.0297\ (5) \times 10^{-12}. \tag{3.55}$$

To summarize, the total non-QED contribution of the Standard Model to a_e is $1.72(2) \times 10^{-12}$. It is small but not negligible compared with the measurement uncertainty of Eq. (3.46). It will play an important role when better non-QED values of α become available in the future. Currently, the information gained for a_μ plays an indirect but important role in controlling the uncertainty in a_e arising from the hadronic and electroweak interactions (within the context of the Standard Model).

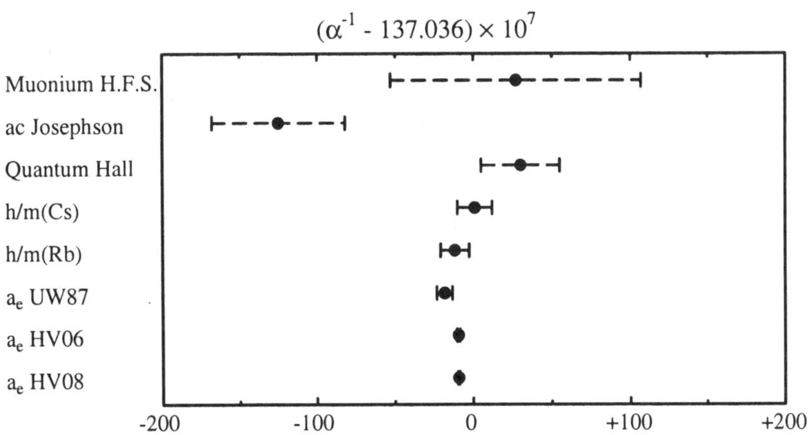

Fig. 3.6. Comparison of various α^{-1} of high precision.

To compare the theory with the measured a_e one needs an α obtained by an independent measurement. Some of the most precise α^{-1} are shown in Fig. 3.6 and Fig. 3.7. The best α's independent of the electron $g-2$ are

$$\alpha^{-1}(h/M_{Rb}) = 137.035\ 998\ 84\ (91), \qquad [6.7\ \text{ppb}] \tag{3.56}$$

$$\alpha^{-1}(h/M_{Cs}) = 137.036\ 000\ 00\ (110), \qquad [8.0\ \text{ppb}] \tag{3.57}$$

[m]See footnote 7 of [96].

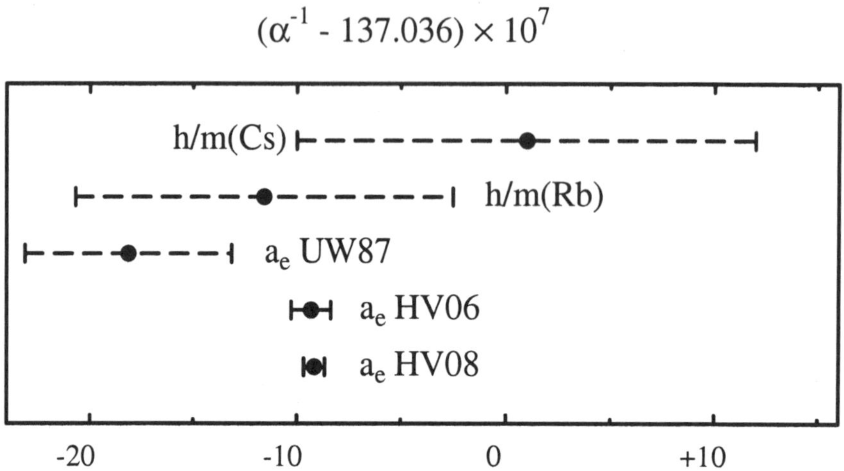

Fig. 3.7. Magnification of the lower half of Fig. 3.6 by a factor 10.

obtained by an optical lattice method [99] and atom interferometry [100], respectively. Assuming $|A_1^{(10)}| < x$, we find

$$a_e(h/M_{Rb}) = 1\ 159\ 652\ 182.79\ (0.11)(0.08x)(7.72) \times\ 10^{-12},$$
$$a_e(h/M_{Cs}) = 1\ 159\ 652\ 172.99\ (0.11)(0.08x)(9.33) \times\ 10^{-12}, \quad (3.58)$$

where 0.11 and $0.08x$ are the uncertainties arising from the eighth-order and (unknown) tenth-order terms, respectively, and 7.72 and 9.33 are from the measurements of α in Eqs. (3.56) and (3.57), respectively. The uncertainty arising from the hadronic and electroweak terms is not listed in Eq. (3.58) to avoid overcrowding. It is about 0.02×10^{-12}.

Clearly, not knowing $A_1^{(10)}$ is not a serious source of concern as far as $0.08x \ll 8$. For $x = 4.6$[n], which satisfies this criterion, theory and experiment are in good agreement:

$$a_e[\text{HV08}] - a_e(h/M_{Rb}) = -2.06\ (7.72) \times\ 10^{-12},$$
$$a_e[\text{HV08}] - a_e(h/M_{Cs}) = +7.74\ (9.33) \times\ 10^{-12}, \quad (3.59)$$

where a_e[HV08] is from Eq. (3.46).

Note that errors in $a_e(h/M_{Rb})$ and $a_e(h/M_{Cs})$ listed in Eq. (3.58) are mostly from the measurement of α. In other words, the non-QED α, even

[n]Ref. [27] gave $x = 3.8$ determined by a formula which depends on the value of $A_1^{(8)}$. The same formula applied to the revised value of $A_1^{(8)}$ leads to $x = 4.6$.

the best ones, is too crude to test QED to the precision achieved by theory and measurement of a_e. For testing QED it makes more sense to get α from a_e assuming that QED is valid, and compare it with other α's. This gives

$$\alpha^{-1}(a_e, x) = 137.035\ 999\ 085\ (12)(8x)(33), \qquad (3.60)$$

where 33 is the uncertainty of a_e measurement in Eq. (3.46). For $x = 4.6$ we obtain

$$\alpha^{-1}(a_e[\text{HV08}]) = 137.035\ 999\ 085\ (51), \qquad [0.37\ \text{ppb}] \qquad (3.61)$$

This is the most precise value of α available at present [67, 101, 102].

Note that the new measurement of $a_\mu(\text{had.vp.})$ [92], if confirmed by further work, would shift the results Eq. (3.58) to

$$\begin{aligned} a_e(h/M_{Rb}) &= 1\ 159\ 652\ 182.83\ (0.11)(0.08x)(7.72) \times\ 10^{-12}, \\ a_e(h/M_{Cs}) &= 1\ 159\ 652\ 173.03\ (0.11)(0.08x)(9.33) \times\ 10^{-12}, \end{aligned} \qquad (3.62)$$

and Eq. (3.61) to

$$\alpha^{-1}(a_e[\text{HV08}]) = 137.035\ 999\ 088\ (51). \qquad [0.37\ \text{ppb}] \qquad (3.63)$$

3.2.9. *Current status of a_μ test*

Due to the different sensitivity to the mass dependence the muon $g-2$ is about 4×10^4 times more sensitive to the hadronic and electroweak effects than the electron $g-2$. In particular,the hadronic contribution to a_μ is about 60 ppm. Thus, a pure QED is already in disagreement with experiment at the level of the α^3 correction. On the other hand, this means that the muon $g-2$ provides one of the most sensitive probes of the validity of the Standard Model, or possible physics beyond the Standard Model.

The final result of the CERN experiment is [34]:

$$\begin{aligned} a_\mu(\text{exp}) &= 11\ 659\ 23\ (8.5) \times 10^{-9} \quad [7\ \text{ppm}], \\ a_\mu(\text{exp}) - a_e(\text{exp}) &= 5271\ (8.5) \times 10^{-9}. \end{aligned} \qquad (3.64)$$

After years of preparation the muon $g-2$ experiment at the Brookhaven National Laboratory [103] has come close to the goal, which is to improve the CERN value by a factor 20. Including these results the current world average is [103]

$$a_\mu(\text{exp}) = 116\ 592\ 080\ (63) \times 10^{-11} \quad [0.5\ \text{ppm}]. \qquad (3.65)$$

Besides the QED corrections of up to the order α^4 an accurate hadronic correction and electroweak correction within the context of the Standard

Model are needed to test theory at the experimental precision. Of course the QED contribution is by far the largest. Let us first summarize the QED contribution to a_μ.

Mass-dependent terms $A_2^{(4)}$, $A_2^{(6)}$, and $A_3^{(6)}$ have been evaluated by numerical integration, analytic integration, asymptotic expansion in m_μ/m_e, or power series expansion in m_μ/m_τ [36, 37, 104]. Recent re-evaluation [46] using the values $m_\mu/m_e = 206.768\,283\,8\,(54)$ and $m_\mu/m_\tau = 5.945\,92\,(97) \times 10^{-2}$ [27] gives

$$\begin{aligned} A_2^{(4)}(m_\mu/m_e) &= 1.094\,258\,311\,1\,(84), \\ A_2^{(4)}(m_\mu/m_\tau) &= 7.8064\,(25) \times 10^{-5}, \\ A_2^{(6)}(m_\mu/m_e) &= 22.868\,380\,02\,(20), \\ A_2^{(6)}(m_\mu/m_\tau) &= 36.051\,(21) \times 10^{-5}, \\ A_3^{(6)}(m_\mu/m_e, m_\mu/m_\tau) &= 0.527\,66\,(17) \times 10^{-3}. \end{aligned} \tag{3.66}$$

Note that $A_2^{(6)}(m_\mu/m_e)$ is very large. As was discussed earlier [40] this is dominated by the contribution of diagrams containing light-by-light-scattering subdiagrams, which has a $\ln(m_\mu/m_e)$ term with a large numerical coefficient (see the discussion in section 3.2.3).

Thus far $A_2^{(8)}(m_\mu/m_e)$ and $A_2^{(8)}(m_\mu/m_e, m_\mu/m_\tau)$ have been evaluated by numerical method only, using the Monte-Carlo integration code VEGAS [56]. The latest results are [105]

$$\begin{aligned} A_2^{(8)}(m_\mu/m_e) &= 132.682\,3\,(72), \\ A_3^{(8)}(m_\mu/m_e, m_\mu/m_\tau) &= 0.037\,6\,(1). \end{aligned} \tag{3.67}$$

Diagrams I(a–d), II(a–c), III, IV(a) of Fig. 3.4 have vacuum-polarization loops so that they have leading $\ln(m_\mu/m_e)$ terms arising from charge renormalization. Their next-to-leading terms can be studied by the renormalization group method [106]. For latest developments see [107]. The diagrams I(c) of Fig. 3.4 have a second-order vacuum-polarization loop within another vacuum-polarization loop. They must be treated with care because naive application of the renormalization group method can lead to a wrong next-to-leading term [108], which was first discovered by comparison with the numerical integration result [109].

We also have a partial (but dominant) value of the tenth-order term [110, 111]

$$A_2^{(10)}(m_\mu/m_e) = 663\,(20), \tag{3.68}$$

obtained from 2604 vertex diagrams which include most of important diagrams, and replaces the previous crude estimate [106, 112]. This will be discussed in more detail in section 3.3.

The total QED contribution to a_μ is [110]

$$a_\mu(QED) = \ 116\ 584\ 718.30\ (0.02)(0.14)(0.78) \times\ 10^{-11}, \qquad (3.69)$$

where 0.02 is the calculated uncertainty of the eighth-order term, 0.14 is the estimate based on the partly calculated uncertainty of the tenth-order contribution, and 0.78 is the uncertainty of the fine structure constant given in Eq. (3.56). Thus the uncertainty in the QED contribution to a_μ is much smaller than the current experimental uncertainty.

The largest uncertainty in a_μ comes from the hadronic vacuum-polarization term. Unfortunately, this has not yet been evaluated from first principles, namely QCD. At present this contribution is evaluated using the experimental information. Three types of measurements are available for this purpose: (1) $e^+e^- \rightarrow hadrons$, (2) $\tau^\pm \rightarrow \nu + \pi^\pm + \pi^0$, (3) $e^+e^- \rightarrow \gamma + hadrons$, called *radiative return* process.

Of these three the process (1) yields the most detailed information at present. Since this is discussed in chapter 8, (see also Ref. [113] and references therein.), let us just give the summary here.

$$a_\mu(\text{had.vp.}) = 6901\ (42)_{exp}(19)_{rad}(7)_{QCD} \times 10^{-11}\ . \qquad (3.70)$$

The NLO hadronic contribution summarized in [113] is

$$a_\mu(\text{had.NLO}) = -97.9\ (0.9)_{exp}(0.3)_{rad} \times 10^{-11}\ . \qquad (3.71)$$

The hadronic light-by-light scattering contribution is of similar size as a_μ(had.NLO) [93–95], but has a much larger theoretical uncertainty, as discussed in chapter 9.

$$a_\mu(\text{had.ll}) = 110\ (40) \times 10^{-11}\ . \qquad (3.72)$$

Recent evaluation [96] of a_μ(had.ll), taking into account a new short distance constraint on pion exchange and including other hadronic light-by-light-scattering contributions leads to

$$a_\mu(\text{had.ll}) = 116\ (40) \times 10^{-11}\ . \qquad (3.73)$$

Finally, the electroweak contribution to 2-loop order is [98]

$$a_\mu(\text{weak}) = 154\ (2) \times 10^{-11}\ . \qquad (3.74)$$

Including the QED contribution, the hadronic vacuum-polarization, the hadronic light-by-light scattering term Eq. (3.73), and the electroweak contribution, the theoretical value of a_μ in the Standard Model is

$$\begin{aligned} a_\mu(SM) &= 116\ 591\ 791\ (62) \times\ 10^{-11}, \\ a_\mu(\text{exp}) - a_\mu(SM) &= 289\ (86) \times\ 10^{-11}, \end{aligned} \tag{3.75}$$

where the uncertainty in "theory" is mostly due to the hadronic terms.

It is important to note that the hadronic contribution is still far from being settled. For instance, the recent preliminary value based on the radiative recoil process (3) increases the value given in Eq. (3.70) by 135×10^{-11}, far outside of the uncertainty given in Eq. (3.70) [92]. If this is confirmed by further work, the values listed in Eq. (3.75) must be replaced by

$$\begin{aligned} a_\mu(SM) &= 116\ 591\ 926\ (62) \times\ 10^{-11}, \\ a_\mu(\text{exp}) - a_\mu(SM) &= 154\ (86) \times\ 10^{-11}. \end{aligned} \tag{3.76}$$

Both theory and experiment must be improved before we can decide whether the 3.4 s.d. for Eq. (3.75) (or 1.8 s.d. for Eq. (3.76)) is really an indicator of physics beyond the Standard Model.

3.3. Tenth-Order Term

In the derivation of α from the electron $g-2$ described in Eq. (3.60) [101, 102]

$$\alpha^{-1}(a_e, x) = 137.035\ 999\ 057\ (12)(8x)(33), \tag{3.77}$$

it is to be noted that the uncertainty 33 from the measurement of a_e is smaller than that of theory, assuming $x = 4.6$. Thus, an actual value of the tenth-order coefficient is needed to match the experimental precision, which requires evaluation of 12672 Feynman diagrams of tenth-order.

Besides their gigantic size, none of the contributing Feynman diagrams is dominant so that every one of them must be evaluated accurately. Of course, since

$$\left(\frac{\alpha}{\pi}\right)^5 \simeq 0.068 \times 10^{-12}, \tag{3.78}$$

the precision of the numerical evaluation itself does not have to be very high at present. Thus the primary question is whether it is feasible to obtain FORTRAN codes which are analytically correct. One may naturally ask:

Is such an attempt realistic? Our answer turns out to be yes! But only if it is highly automated.

The first step is to classify all Feynman diagrams of tenth-order into gauge-invariant sets. They consists of 32 gauge-invariant sets within 6 supersets as shown in Figs. 3.8–3.13 [110].

Actually, many (but not all) of diagrams of Sets I, II, III, IV, and VI (see Figs. 3.8, 3.9, 3.10, 3.11, and 3.13, respectively) can be evaluated with relative ease by simple modification of integrals obtained in lower-order calculations. By far the hardest is the evaluation of the Set V of Fig. 3.12, which consists of 6354 Feynman diagrams of *q-type* that have no closed lepton loops. This means that their integrals cannot be derived from lower-order integrals.

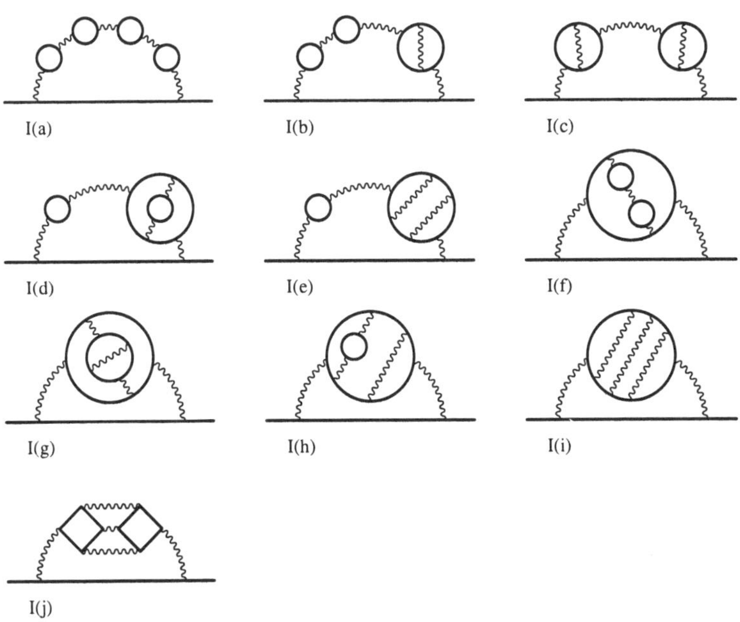

Fig. 3.8. Diagrams of Set I are built from a second-order vertex. This set contributes 208 diagrams to $A_1^{(10)}$ and 498 diagrams to $A_2^{(10)}$.

3.3.1. *Automated evaluation of the set V contribution to a_e*

Obviously, a complete evaluation of the tenth-order contribution $A_1^{(10)}$ to a_e is not achievable until a way is found to deal with Set V. Fortunately, this

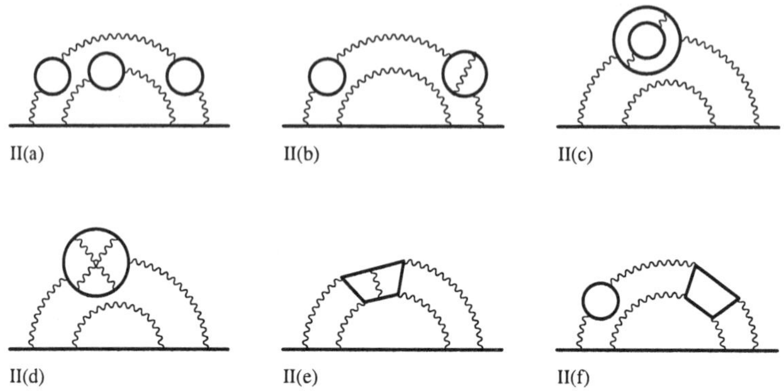

Fig. 3.9. Diagrams of Set II are built from fourth-order proper vertices. This set contributes 600 diagrams to $A_1^{(10)}$ and 1176 diagrams to $A_2^{(10)}$.

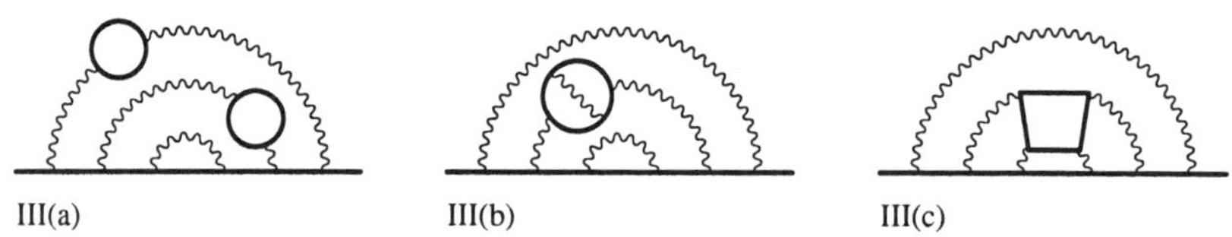

Fig. 3.10. Diagrams of Set III are built from sixth-order proper vertices. This set contributes 1140 diagrams to $A_1^{(10)}$ and 1740 diagrams to $A_2^{(10)}$.

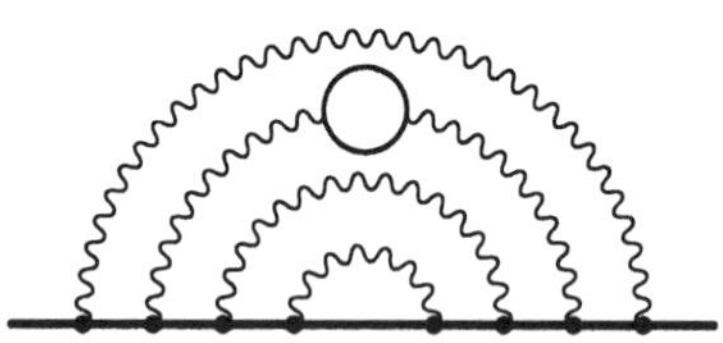

Fig. 3.11. Diagrams of Set IV are built from eighth-order proper vertices. This set contributes 2072 diagrams to both $A_1^{(10)}$ and $A_2^{(10)}$.

set has the simplifying feature that a set of nine vertex diagrams is related to a self-energy-like diagram by Eq. (3.40) derived from the Ward–Takahashi identity. Using this identity we can cut the number of independent integrals to 706. The time-reversal invariance reduces it further to 389.

Analytic evaluation of these integrals is likely to be far in the future. At present numerical integration is the only viable option. In view of the

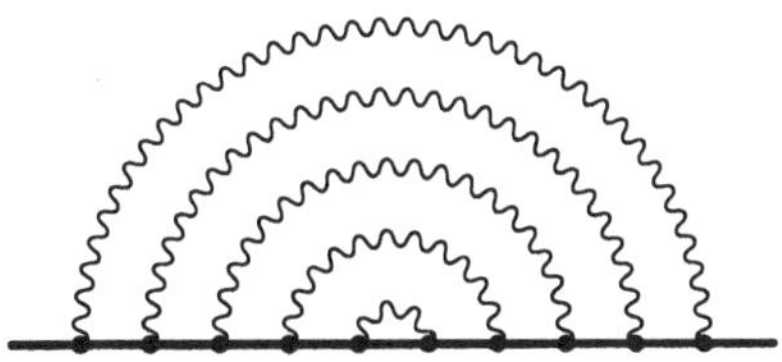

Fig. 3.12. Diagrams of Set V consists of 10th-order proper vertices of *q-type*, namely diagrams which have no closed lepton loops. This set contributes 6354 Feynman diagrams only to $A_1^{(10)}$.

gigantic size of integrals and an enormous number of renormalization terms (more than 10,000 terms for Set V), however, it is practically impossible to carry out such a calculation without committing errors unless some way is found to make it fully automated. In order to solve this problem we developed an algorithm "gencodeN" which carries out the entire calculation automatically [83, 84]. It consists of several steps:

(A) *Diagram generation.*
A *q-type* diagram G of Set V is specified uniquely by the pattern of parings of vertices connected by virtual photons. The complete set of distinct diagrams are thus generated in a combinatorial manner, which are named as X_{abc}, $abc = 001, 002, \ldots, 389$. Note that the pairing pattern specifies the form of a diagram completely. In particular, it specifies all UV- and IR-divergent subdiagrams.

(B) *Construction of unrenormalized integrands.*
The diagram "X_{abc}" is expressed as a momentum integral by the Feynman–Dyson rule. The momentum integration is carried out analytically, which leads to an integral of the form Eq. (3.31) which is a function of Feynman parameters z_i, "symbolic" building blocks U, V, A_i, B_{ij}, and C_{ij}, $i, j = 1, 2, \ldots, N$. (See section 3.2.4 for notation.) Recall that these integrals have UV-divergent subdiagrams, which must be regularized by the Feynman cutoff Eq. (3.34).

(C) *Construction of building blocks.*
The building blocks A_i, B_{ij}, C_{ij}, U, and V are expressed as homogeneous functions of $z_1, z_2, \ldots, z_N$. V has a form given by Eq. (3.33). (See section 3.2.4 for notation.)

(D) *Construction of UV subtraction terms.*
We deal with the UV renormalization by subtractive approach. The subtracting integrand is derived from the original integrand by

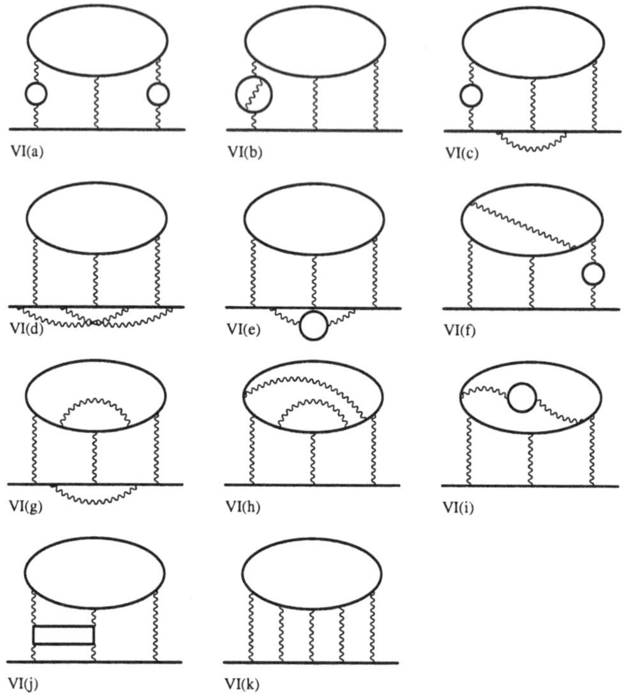

Fig. 3.13. Set VI consists of diagrams containing various light-by-light scattering subdiagrams. This set contributes 2298 diagrams to $A_1^{(10)}$ and 3594 diagrams to $A_2^{(10)}$.

K-operations [52] for Zimmermann's forest of subdiagrams. Each K-operation is defined for a divergent subdiagram in terms of a simple power-counting rule. (See Eq. (3.35)).

The K-operation has the following properties:

(i) It generates an integral which subtracts the UV divergence point by point in the Feynman parametric space.

(ii) The subtraction term factorizes analytically into a product of known lower-order quantities.

(iii) It gives only the leading UV-divergent parts (denoted by $\widehat{\delta m}_S$, $\widehat{L}_S$, $\widehat{B}_S$) of renormalization constants δm_S, L_S, B_S. Thus, an additional (finite) renormalization is required to attain the standard on-shell renormalization. We shall call this a residual renormalization.

Fig. 3.14. 389 self-energy-like diagrams that represent 6354 vertex diagrams of Set V.

(E) *Construction of IR subtraction terms.*

As was noted in section 3.2.5 the IR divergence has its origin in vanishing of virtual photon momenta. This is, however, just a necessary condition but not a sufficient condition. In order that the IR singularity actually becomes logarithmically divergent, it must be enhanced by vanishing of denominators of two electron propagators each of which shares a 3-point-vertex with the infrared photons and external (on-shell) electron lines, which we shall call "enhancers". This corresponds to the vanishing of the denominator V as $\mathcal{O}(\epsilon^2)$ in the corner of integration domain characterized by Eq. (3.38). This is the case where the (W-T-summed) diagram has just one self-energy-like subdiagram (of any order). When a diagram has two self-energy-like subdiagrams, however, the number of "enhancers" becomes three, and the unrenormalized integral M_G develops a linear IR divergence.

Suppose S is one of these subdiagrams. As is readily seen from the analysis of Feynman diagrams, this divergence is not the source of a real problem since it is canceled exactly by the mass-renormalization term $\delta m_S M_{G/S(i^*)}$, where δm_S is the (UV-divergent) self-mass associated with the subdiagram S. The reduced diagram $M_{G/S(i^*)}$ is the one that has a linear IR divergence. As a consequence

$$M_G - \delta m_S M_{G/S(i^*)} \tag{3.79}$$

is free from linear IR divergence. Although this cancellation is analytically correct, it is not a point-wise cancellation in the domain of M_G. Our problem is thus to translate the second term into a form which is defined in the same domain as that of M_G and cancels the IR divergence of M_G point-by-point. (In order to avoid excessive notations let us ignore divergences coming from subdiagrams of S.) Now, we know that the K_S-operation acting on M_G creates

$$\mathbb{K}_S M_G = \widehat{\delta m}_S M_{G/S(i^*)} + \widehat{B}_S M_{G/[S,i]}, \tag{3.80}$$

which may be rewritten as

$$\mathbb{K}_S M_G + \widetilde{\delta m}_S M_{G/S(i^*)} = \delta m_S M_{G/S(i^*)} + \widehat{B}_S M_{G/[S,i]}, \tag{3.81}$$

where $\widetilde{\delta m_S} = \delta m_S - \widehat{\delta m_S}$. Thus, if an operator $\mathbb{R}_S$ is found that causes point-wise cancellation of linear IR divergence in the domain of M_G and also produces the factorization on the right-hand side

$$\mathbb{R}_S M_G = \widetilde{\delta m}_S M_{G/S(i^*)}, \tag{3.82}$$

then we will have

$$(\mathbb{K}_S + \mathbb{R}_S) M_G = \delta m_S M_{G/S(i^*)} + \widehat{B}_S M_{G/[S,i]}. \tag{3.83}$$

It turned out that it is not difficult to find such an operator. Furthermore, it can be readily incorporated in our automation algorithm.

After the K- and R-operations are carried out the amplitude defined by $(1 - \mathbb{K}_S - \mathbb{R}_S)M_G$ is free from the UV divergence due to S and has only a logarithmic IR divergence from $M_{G/S(i^*)}$ so that it can be handled by the I-operation. (Note, however, that we found it useful to define the I-operation differently from the old I-operation in the way the IR-finite terms is handled [84]. This simplifies the treatment of IR subtraction considerably.)[o]

[o]In the previous work [52] the problem arising from the linear IR divergence was handled, not by an R-operation, but by an *ad hoc* subtraction of the IR divergent term. Although this is not incorrect by itself, it caused some complication which contributed to the unfortunate failure to detect an inconsistency in the treatment of the IR divergence [75].

(F) *Residual renormalization.*
The output of the above steps produces finite integrals. However, as is mentioned in Step (D), it is not the standard renormalized integral. Thus the additional finite renormalization is required to obtain the on-shell-renormalized result.

The Step (A) is performed by a separate program. The information describing diagrams in single-line representation is stored in a plain text file.

The steps (B), (C), (D), and (E) are implemented as separate Perl programs that use FORM and Maple internally. These symbolic manipulation programs take traces, project out the magnetic moment, perform analytic integration over momentum variables by means of home-made integration tables written in FORM, carry out inversion of $(3n-1) \times (3n-1)$ matrices which creates B_{ij} and U, where $2n$ is the order of diagrams, and execute K-operations. The programs for the IR subtraction part are integrated with the programs that generate the codes for UV-finite amplitudes developed previously [83], to form the automated code generation system which creates FORTRAN codes free from both UV and IR divergences.

The Perl program takes the name of the diagram and finds the corresponding single-line expression of the diagram from the table prepared in Step (A). Then it generates the numerical integration code in the FORTRAN format that is readily integrated by Monte-Carlo integration routine VEGAS [56]. All the steps are controlled by the **make** utility and a shell script. Note that some steps are independent of each other so that they can be carried out simultaneously. The residual renormalization, Step (F), is executed separately at the last stage.

In order to debug the automation code, we applied it first to the α^3 case evaluating the FORTRAN output by numerical integration. The result was in good agreement with the previous numerical result [60] as well as the analytic result [61]. While testing by numerical integration the FORTRAN codes generated by "gencodeN" for the α^4 case, however, we ran into a disagreement with the previous result [75], which is significantly larger than the precision of the numerical evaluation. A detailed comparison of the old and new calculations has uncovered a subtle inconsistencies in the handling of the non-divergent parts of IR subtraction terms of two diagrams (M_{16} and M_{18} of Group V (see Fig. 3.4)) in the old calculation [75]. After correcting this error, the old and new results agree within the numerical uncertainty of the new (although still tentative) calculation [85].

It is to be noted that, until now, the old calculation [75] was the only complete evaluation of the eighth-order term of a_e. In view of its enormous size and complexity no one has attempted to perform an independent check until now. Luckily the development of the automation code (that can handle diagrams of any order) has provided us with the opportunity to treat *q-type* eighth-order diagrams independently and expeditiously. As a consequence we now have two independent evaluations of the complete eighth-order term, which agree with each other after the error of the old calculation is corrected.

The work on the eighth-order term has given us the opportunity to examine the automation code itself thoroughly, enhancing our confidence in its mechanism, particularly in its handling of infrared divergence. The evaluation of the Set V contribution to the α^5 term is now in an advanced stage [116].

3.3.2. *Evaluation of other tenth-order diagrams*

At present only a small fraction of tenth-order integrals, from the subsets I(a), I(b), I(c), II(a), and II(b), are known analytically for an arbitrary mass ratio [114]. The numerical values of their contributions to the muon anomaly a_μ, expanded in the ratio m_e/m_μ, are quoted from Table 2 of [114]:

$$\begin{aligned} a_\mu[I(a)] &= 22.566\,973\ (3), \\ a_\mu[I(b)] &= 30.667\,091\ (3), \\ a_\mu[I(c)] &= 5.141\,395\ (1), \\ a_\mu[II(a_s)] &= -36.174\,859\ (2), \\ a_\mu[II(b_s)] &= -23.426\,173\ (1), \end{aligned} \tag{3.84}$$

where the uncertainties come from the measurement uncertainty of m_e/m_μ only [27]. The subscript s in $II(a_s)$ and $II(b_s)$ indicates that these diagrams are subsets of $II(a)$ and $II(b)$ in which vacuum polarization loops are inserted only in the same photon line. The contributions of these diagrams to the electron anomaly a_e are also given in [114].

Before we began working on the tenth-order diagrams of Set V, which contributes only to a_e, we evaluated numerically many other tenth-order diagrams that contribute to both a_μ and a_e. Thus far the contributions of 17 gauge-invariant subsets I(a, b, c, d, e, f), II(a, b, f), VI(a, b, c, e, f, i, j, k) have been evaluated. Altogether these sets contain 2958 vertex diagrams, which include all numerically dominant terms contributing to a_μ. Identification of such diagrams is not difficult in view of the discussion given in section 3.2.3. Namely, they are primarily diagrams that contain light-by-light-scattering subdiagrams, which has a logarithmic dependence on m_μ/m_e with a large numerical factor [40]. They are further enhanced by vacuum-polarization subdiagrams, which contribute additional $\ln(m_\mu/m_e)$ factors through charge renormalization.

Based on these considerations, it is obvious that the most important contribution will come from the Set VI(a), followed by the Set VI(b). We confirmed this expectation by the numerical integration which yielded [110]:

$$\begin{aligned} A_2[VI(a)] &= 629.1407\ (118), \\ A_2[VI(b)] &= 181.1285\ (51). \end{aligned} \tag{3.85}$$

Another set of interest is Set VI(k) whose leading term in the large m_μ/m_e limit was obtained by Elikhovskii [44]

$$A_2[VI(k))] = \pi^4(0.438...\ln(m_\mu/m_e) + \ldots), \tag{3.86}$$

where the large factor $\pi^4 \sim 97$ comes from integrations over the momenta of four Coulomb photons exchanged between the closed electron loop and the muon. (See discussion in section 3.2.3.) The numerical coefficient 0.438... was evaluated in Ref. [115]. Based on this result Karshenboim estimated that $A_2[VI(k)]$ will be about 180. In order to check this estimate we evaluated the Set VI(k) numerically and obtained [110]

$$A_2[VI(k)] = 97.123\ (62), \tag{3.87}$$

which shows that Karshenboim overestimated it by about 100. Actually, this is not surprising since the leading log term is usually followed by fairly large negative term.

We have also evaluated numerically the contributions of sets of secondary importance. The results are [110]

$$\begin{aligned} A_2[I(a)] &= 22.567\ 05\ (25), \\ A_2[I(b)] &= 30.667\ 54\ (33), \\ A_2[I(c)] &= 5.141\ 38\ (15), \\ A_2[I(d)] &= 8.892\ 07\ (102), \\ A_2[I(e)] &= -1.219\ 20\ (71), \\ A_2[I(f)] &= 3.685\ 10\ (13), \end{aligned}$$

$$\begin{aligned}
A_2[II(a)] &= -70.471\,7\ (38),\\
A_2[II(b)] &= -34.771\,5\ (26),\\
A_2[II(f)] &= -77.464\,8\ (120),\\
A_2[VI(c)] &= -36.576\,3\ (1141),\\
A_2[VI(e)] &= -4.321\,5\ (1341),\\
A_2[VI(f)] &= -38.159\,8\ (1488),\\
A_2[VI(i)] &= -27.337\,3\ (1147),\\
A_2[VI(j)] &= -25.505\ (20).
\end{aligned} \tag{3.88}$$

$A_2[I(a)]$, $A_2[I(b)]$, $A_2[I(c)]$, and $II(a_s)$ and $II(b_s)$ parts of $A_2[II(a)]$ and $A_2[II(b)]$ are in good agreement with the analytic results given in Eq. (3.84).

Recently we also evaluated the contribution of the set I(j). This set is an interesting one whose eighth-order vacuum-polarization diagrams consist of two light-by-light-scattering subdiagrams connected by three photon lines [111]. Numerical integration gives

$$A_2[I(j)] = -1.263\,44\ (14). \tag{3.89}$$

The sum of contributions of Eq. (3.85), Eq. (3.87), Eq. (3.88), and Eq. (3.89) is

$$A_2^{(10)}(m_\mu/m_e)[\mathbf{part}] = 661.24\ (27). \tag{3.90}$$

Since the contribution of remaining diagrams is not likely to be large, we may choose as the best provisional estimate the value

$$A_2^{(10)}(m_\mu/m_e)[\mathbf{estimate}] = 661\ (20). \tag{3.91}$$

This is 8.5 times more precise than the old estimate [112]

$$A_2^{(10)}(m_\mu/m_e)[\mathbf{old\ estimate}] = 930\ (170), \tag{3.92}$$

and downgrades $A_2^{(10)}$ as a serious source of theoretical uncertainty.

In parallel with the calculation of a_μ, we have obtained the contribution to the electron $g-2$ from 964 vertex diagrams [110, 111]:

$$A_1^{(10)}[\mathbf{part}] = -1.823\,5\ (63). \tag{3.93}$$

Since this comes from less than 8% of the entire diagrams contributing to a_e, it is not significant numerically except that it is not excessively large.

3.3.3. *Remaining task*

The automation method developed in [83, 84] can be readily applied for a speedy evaluation of the Sets III(a), III(b), and IV. The work on these sets are in advanced stages. All diagrams of Set V have been evaluated except for the residual renormalization terms [116]. A somewhat different automation algorithm is required to evaluate the Set I(i), which contains vacuum-polarization subdiagrams of eighth order. This work is almost finished, waiting for the evaluation of residual renormalization terms. The report on the sets I(g) and I(h) has now been published [117]. Works on the sets II(c), II(d), III(a), III(b), and IV are also in advanced stages. The Set II(e), which contain a light-by-light-scattering subdiagram of sixth order, is next on the agenda. The remaining sets III(c) and VI(d, g, h) do not seem to present particular complication. It is thus probable that we can complete this project within a year or two.

3.4. How Far Can We Go?

The successful calculations of a_e and the Lamb shift established QED as the theory of electromagnetic interaction by 1948. In spite of initial doubt expressed by many people, it has survived rigorous tests for nearly 60 years. Of course, physics has become much more complex since 1947. The pure QED is not the theory of all physics: It does not describe other interactions, weak and strong. Fortunately, these interactions turned out to be renormalizable within the framework of the Standard Model. As is seen from Eqs. (3.70), (3.71), (3.72), and [92], the hadronic effect is very large for a_μ:

$$7048(62) \times 10^{-11} \simeq 5.62(5) \left(\frac{\alpha}{\pi}\right)^3 , \tag{3.94}$$

so that the naive QED fails already at the level of sixth order. The contribution of weak interaction to a_μ is about [98]

$$154(2) \times 10^{-11} \simeq 0.123(2) \left(\frac{\alpha}{\pi}\right)^4 , \tag{3.95}$$

which cannot be ignored in comparison with the eighth-order term. For a_μ the comparison of theory and experiment actually test the validity of the Standard Model, and it tests the QED only to the extent that the non-QED effects are under reasonable control.

In contrast, the hadronic and weak effects on the electron are about 1.65×10^{-12} and 0.03×10^{-12}, respectively, so that they become significant only at the levels of $(\alpha/\pi)^4$ and $(\alpha/\pi)^5$, respectively. For a_e QED still plays the dominant role, and the test of QED makes sense, as far as short-distance effects due to other forces are controlled within small uncertainties.

Thus far, we have not discussed the possibility that comparing experiment and theory of a_e probes for possible electron substructure. An electron whose constituents would have mass $m^* \gg m$ has a natural size scale $R = h/(2\pi m^* c)$. This would lead to an addition to a_e of $\delta \sim (m/m^*)^2$ in a chirally invariant model [118]. This would lead to $m^* > 130$ Gev/c^2 and $R < 1 \times 10^{-18} m$. If this test was limited only by the experimental uncertainty of a_e, and not by the precision of α, then one could set a stricter limit $m^* > 600$ Gev/c^2. We have to wait for the next generation experiments at LHC to see whether such a substructure, if it exists, can be identified with the "physics beyond the Standard Model" or something else.

Finally, let me ask "How far can the test of QED go by means of a_e?"

On the experimental side:
Uncertainty in the measurement of a_e is likely to be reduced from 0.28×10^{-12} to at least 0.1×10^{-12} before long.

On the theoretical side:
Uncertainty in a_e caused by $A_1^{(8)}$ may be reduced from 0.08×10^{-12} to 0.01×10^{-12}, although it requires a large scale computation.

Our work on $A_1^{(10)}$ will soon reach the precision of 1%, or even 0.1%, corresponding to uncertainty $\Delta A_1^{(10)} (\alpha/\pi)^5 \sim 0.008 \times 10^{-12}$. $A_1^{(12)}$ will not be needed for a while since

$$(\alpha/\pi)^6 \simeq 0.00016 \times 10^{-12}. \tag{3.96}$$

The real obstacle is likely to be the hadronic effect. At present the uncertainties in a_e and $\alpha(a_e)$ due to the hadronic effect are $\sim 0.02 \times 10^{-12}$ and $\sim 2.7 \times 10^{-12}$, respectively. Unless this is improved, it will become a serious barrier which is already encountered by the muon $g-2$. Testing QED (or the Standard Model) beyond 0.003 ppb in $\alpha(a_e)$ will then run into a brick wall very difficult to penetrate.

At this point let me clarify the meaning of testing QED. It used to mean checking QED calculation against experiment using high precision α obtained from the quantum Hall effect, etc. Now that $\alpha(a_e)$ is more accurate than any other α, however, this does not make sense any longer.

As a matter of fact non-QED α is actually a QED α. This is because it is based on quantum mechanics, which is the non-relativistic limit of QED in the sense that it requires physical mass and charge, whose justification depends on the renormalizability of QED. From this viewpoint comparison of other α with $\alpha(a_e)$ is really checking the *internal consistency* of QED, in the guise of atomic physics, condensed matter physics, laser physics, etc.

Thus, the discrepancy between these α's by itself does not mean the breakdown of QED. Instead it may indicate shortcoming of some theories on which measurements of non-QED α are based.

An intriguing possibility is that theories and measurements of these non-QED α are error-free and measured very precisely, but still they disagree with $\alpha(a_e)$. Could it possibly be an indication of internal inconsistency of quantum mechanics?

On the other hand, if $\alpha(a_e)$ and other α's are in agreement up to 0.003 ppb, or its future improvement, we may never be able to observe the actual breakdown of QED (or quantum mechanics).

Finally, let me emphasize that a_e provides the most stringent test for any theory beyond the Standard Model in the sense that such a theory must be able to calculate the measured value of the electron mass (which is just an external parameter in the Standard Model) and a_e at least to the precision achieved by the Harvard experiment [67].

Acknowledgments

The author thanks M. Nio, M. Hayakawa, and T. Aoyama for their careful reading of the manuscript and valuable comments. Thanks are due to G. Gabrielse who elucidated to me the new Harvard measurement of a_e. This work is supported in part by the U. S. National Science Foundation under Grant No. PHYS-0355005.

References

[1] P. A. M. Dirac, *Proc. Roy. Soc.* **A117**, 610 (1928).
[2] P. Kusch and H. M. Foley, *Phys. Rev.* **72**, 1256 (1947).
[3] J. Schwinger, *Phys. Rev.* **73**, 416L (1948). An error in this paper is corrected in *Phys. Rev.* **75**, 898 (1949).
[4] W. E. Lamb and R. C. Retherford, *Phys. Rev.* **72**, 241 (1947).
[5] S. Tomonaga, *Prog. Theor. Phys.* **1**, 27 (1946); Z. Koba and S. Tomonaga, *Prog. Theor. Phys.* **2**, 218 (1947); S. Tomonaga, *Phys. Rev.* **74**, 224 (1948); J. Schwinger, *Phys. Rev.* **74**, 1439 (1948); R. P. Feynman, *Phys. Rev.* **76**, 749 (1949), **76**, 769 (1949); F. J. Dyson, *Phys. Rev.* **75**, 486 (1949), **75**, 1736 (1949).
[6] S. Tomonaga, private communication.
[7] F. J. Dyson, in *Physics Today*, 15, August 2006.

[8] O. Stern, *Z. f. Phys.* **7**, 249 (1921); W. Gerlach and O. Stern, *Z. f. Phys.* **8**, 110 (1922); **9**, 349 (1922).
[9] W. Pauli, *Z. f. Phys.* **31**, 373 (1927).
[10] W. Pauli, *Z. f. Phys.* **31**, 765 (1927).
[11] G. Uehlenbeck and S. Goudsmit, *Naturwiss.* **13**, 593 (1925); *Nature* **117**, 264 (1926).
[12] B. L. van der Waerden, *Sources of Quantum Mechanics* (Dover Publications, Inc., New York, 1967).
[13] L. H. Thomas, *Nature* **117**, 514 (1926).
[14] W. Pauli, *Z. f. Phys.* **43**, 601 (1927).
[15] C. G. Darwin, *Proc. Roy. Soc.* **A116**, 227 (1927).
[16] P. Jordan and E. P. Wigner, *Zeits. fur Phys.* **47**, 631 (1928).
[17] W. Heisenberg, and W. Pauli, *Zeits. fur Phys.* **56**, 1 (1929).
[18] P. A. M. Dirac, *Proc. Roy. Soc.* **A136**, 453 (1932); P. A. M. Dirac, V. Fock, and B. Podolsky, *Phys. U.S.S.R.* **2**, 468 (1932).
[19] R. Karplus and N. M. Kroll, *Phys. Rev.* **77**, 536 (1950).
[20] A.Petermann, *Helv. Phys. Acta* **30**, 407 (1957).
[21] C. Sommerfield, *Phys. Rev.* **107**, 328 (1957); *Ann. Phys. (NY)*, **5**, 26 (1958).
[22] J. H. Gardner and E. M. Purcell, *Phys. Rev.* **76**, 1262 (1949); P. A. Franken and S. Liebes, *Phys. Rev.* **104**, 1197 (1956); S. Liebes and P. A. Franken, *Phys. Rev.* **116**, 633 (1959).
[23] W. H. Louisell, R. W. Pidd, and H. R. Crane, *Phys. Rev.* **91**, 475 (1953).
[24] J. C. Wesley and A. Rich, *Phys. Rev. A* **4**, 1341 (1971).
[25] R. L. Garwin, L. M. Lederman, and M. Weinrich, *Phys. Rev.* **105**, 1415 (1957).
[26] See "Note added in proof" in R. L. Garwin, D. P. Hutchinson, S. Penman, and G. Shapiro, *Phys. Rev.* **118**, 271 (1960).
[27] P. J. Mohr and B. N. Taylor, *Rev. Mod. Phys.* **77**, 1 (2005).
[28] H. Suura and E. Wichman, *Phys. Rev. Lett.* **105**, 1930 (1957).
[29] A. Petermann, *Phys. Rev. Lett.* **105**, 1931 (1957).
[30] C. Bouchiat and L. Michel, *J. Phys. Radium* **22**, 121 (1961).
[31] L. Durand, III, *Phys. Rev.* **128**, 441 (1962).
[32] F. J. M Farley, *CERN Internal report* NP/4733 (1962).
[33] J. Bailey *et al.*, *Phys. Lett. B* **55**, 420 (1975).
[34] F. J. M. Farley and E. Picasso, in *Quantum Electrodynamics*, edited by T. Kinoshita (World Scientific, Singapore, 1990), p. 479.
[35] H. H. Elend, *Phys. Lett.* **20**, 682 (1966); *Phys. Lett.* **21**, 720(E) (1966). G. W. Erickson and H. H. T. Liu, UCD-CNL-81 report (1968).
[36] M. A. Samuel and G. Li, *Phys. Rev. D* **44**, 3935 (1991); *Phys. Rev. D* **48**, 1879(E) (1991); G. Li, R. Mendell, and M. A. Samuel, *Phys. Rev. D* **47**, 1723 (1993).
[37] T. Kinoshita, *Nuovo Cimento* **51B**, 140 (1967).
[38] T. Kinoshita, *J. Math. Phys.* **3**, 650 (1962).
[39] T. Kinoshita, in "Cargèse Lectures in Physics, Vol. 2" (Gordon and Breach, New York, 1968), p. 209.
[40] J. Aldins, T. Kinoshita, S. J. Brodsky, and A.J. Dufner, *Phys. Rev. Lett.*

23, 441 (1969); *Phys. Rev. D* **1**, 2378 (1970).
[41] W. Czyz, G. C. Sheppy, and J. D. Walecka, *Nuovo Cimento* **34**m 404 (1970).
[42] H. Euler, *Ann. der Phys.* **26**, 398 (1936).
[43] B. E. Lautrup, and M. A. Samuel, *Phys. Lett.* B **72**, 114 (1977).
[44] A. S. Elikhovskii, *Yad. Fiz.* **49**, 1059 (1989) [*Sov. J. Nucl. Phys.* **49**, 656 (1989)].
[45] S. Laporta and E. Remiddi, *Phys. Lett. B* **301**, 440 (1993).
[46] M. Passera, *Phys. Rev. D* **75**, 013002 (2007).
[47] B. E. Lautrup and E. de Rafael, *Phys. Rev.* **174**, 1835 (1968); B. E. Lautrup and E. de Rafael, *Nuovo Cimento A* **64**, 322 (1969); B. E. Lautrup, *Phys. Lett. B* **32**, 627 (1970); S. J. Brodsky and T. Kinoshita, *Phys. Rev. D* **3**, 356 (1971); J. Calmet and M. Perrottet, *Phys. Rev. D* **3**, 356 (1971).
[48] J. Calmet and A. Petermann, CERN preprint TH.1724 (1973).
[49] C. T. Chang and M. J. Levine, unpublished (1973).
[50] J. A. Mignaco and E. Remiddi, *Nuovo Cimento* **60A**, 519 (1969); R. Barbieri and E. Remiddi, *Nucl. Phys.B* **90**, 233 (1975); R. Barbieri, M. Caffo and E. Remiddi, *Phys. Lett. B* **57**, 460 (1975).
[51] K. A. Milton, W. Y. Tsai, and L. L. De Raad, Jr., *Phys. Rev. D* **9**, 1809 (1974); L. L. De Raad, Jr., K. A. Milton, and W. Y. Tsai, *Phys. Rev. D* **9**, 1814 (1974).
[52] T. Kinoshita, in *Quantum Electrodynamics*, edited by T. Kinoshita (World Scientific, Singapore 1990), p. 218.
[53] Originally written by M. Veltman. Reported by H. Strubbe, *Compt. Phys. Commun.* **8**, 1 (1974); **18**, 1 (1979).
[54] A. C. Hearn,"REDUCE 2 User's Manual" Stanford Artificial Intelligence Project Memo AIM-133.
[55] RIWIAD is a version of SHEPPY [41] modified by Lautrup, Sheppy, and Dufner. For a description of RIWIAD see, for instance, B. E. Lautrup, in *Proceedings of the Second Colloquium on Advanced Computer Methods in Theoretical Physics, Marseilles*, Marseilles, 1971, edited by A. Visconti (Univ. of Marseilles, Marseilles, 1971).
[56] G. P. Lepage, *J. Comput. Phys.* **27**, 192 (1978).
[57] P. Cvitanovic and T. Kinoshita, *Phys. Rev. D* **10**, 4007 (1974).
[58] M. J. Levine and J. Wright, *Phys. Rev. D* **8**, 3171 (1973).
[59] R. Carroll, *Phys. Rev. D* **12**, 2344 (1975).
[60] T. Kinoshita, *Phys. Rev. Lett.* **75**, 4728 (1995).
[61] S. Laporta and E. Remiddi, *Phys. Lett. B* **379**, 283 (1996), and references quoted in this paper.
[62] R. S. Van Dyck, Jr., P. B. Schwinberg, and H. G. Dehmelt, *Phys. Rev. Lett.* **59**, 26 (1987).
[63] R. S. Van Dyck, Jr. *et al.*, 1991, unpublished.
[64] R. Mittleman, H. Dehmelt, and S. Kim, *Phys. Rev. Lett.* **75**, 2839 (1995).
[65] L. S. Brown and G. Gabrielse, *Rev. Mod. Phys.* **58**, 233 (1986).
[66] B. Odom, D. Hanneke, B. D'Urso, G. Gabrielse, *Phys. Rev. Lett.* **97**, 030801 (2006).
[67] D. Hanneke, S. Fogwell, and G. Gabrielse, *Phys. Rev. Lett.* **100**, 120801

(2008).
[68] G. Gabrielse and F. Colin MacKintosh, *J. of Mass Spec.* **57**, 1 (1984).
[69] G. Gabrielse, J. N. Tan, and L. S. Brown, in *Quantum Electrodynamics*, edited by T. Kinoshita (World Scientific, Singapore 1990), p. 389.
[70] J. Tan and G. Gabrielse, *Phys. Rev. Lett.* **67**, 3090 (1991).
[71] S. Peil and G. Gabrielse, *Phys. Rev. Lett.* **83**, 1287 (1999).
[72] B. D'Urso, R. Van Handel, B. Odom, D. Hanneke, and G. Gabrielse, *Phys. Rev. Lett.* **94**, 113002 (2005).
[73] T. Kinoshita and W. B. Lindquist, *Phys. Rev. Lett.* **47**, 1573 (1981).
[74] J. A. M. Vermaseren, FORM ver. 2.3 (1998). The first version was written in 1984.
[75] T. Kinoshita and M. Nio, *Phys. Rev. D* **73**, 013003 (2006).
[76] M. Caffo, E. Remiddi, and S. Turrini, *Nucl. Phys. B* **141**, 302 (1978); J. A. Mignaco and E. Remiddi, 1969 (unpublished).
[77] J.-P. Aguilar, E. de Rafael, and D.Greynat, Phys. Rev. **D 77**, 093010 (2009).
[78] A. H. Hoang *et al.*, *Nucl. Phys. B* **452**, 175 (1995).
[79] D. J. Broadhurst, A. L. Kataev, and O. V. Tarasov, *Phys. Lett. B* **298**, 445 (1993); P. A. Baikov and D. J. Broadhurst, in Proceedings of the 4th International Workshop on Software Engineering and Artificial Intelligence for High Energy and Nuclear Physics (AIHENP95), Pisa, Italy, 1995 (unpublished), p. 167.
[80] K. G. Chetyrkin, R. Harlander, J. H. Kühn, and M. Steinhauser, Nucl. Phys. **B503**, 339 (1997).
[81] K. G. Chetyrkin, J. H. Kühn, and M. Steinhauser, Nucl. Phys. **B505**, 40 (1997).
[82] T. Kinoshita and M. Nio, *Phys. Rev. D* **60**, 053008 (1999).
[83] T. Aoyama, M. Hayakawa, T. Kinoshita, and M. Nio, *Nucl. Phys. B* **740**, 138 (2006).
[84] T. Aoyama, M. Hayakawa, T. Kinoshita, and M. Nio, *Nucl. Phys. B* **796**, 184 (2008).
[85] T. Aoyama, M. Hayakawa, T. Kinoshita, and M. Nio, *Phys. Rev. Lett.* **99**, 110406 (2007); *Phys. Rev. D* **77**, 053012 (2008).
[86] S. Laporta, *Int. J. Mod. Phys. A* **15**, 5087 (2000); C. Anastasiou and A. Lazopoulos, *J. High Energy Phys.* **07** 046 (2004); A. Smirnov and M. Steinhauser, *Nucl. Phys. B* **672**, 199 (2003).
[87] M. Caffo, S. Turrini, and E. Remiddi, *Phys. Rev. D* **30**, 483 (1984); E. Remiddi and S. P. Sorrella, *Lett. Nuovo Cimento* **44**, 231 (1985).
[88] T. Kinoshita, in *An Isolated Atomic Particle at Rest in Free Space. A Tribute to Hans Dehmelt, Nobel Laureate*, edited by E. N. Fortson, E. M. Henley, and W. G. Nagourney (Alpha Science International Ltd., Oxford, U. K.), pp. 77 - 90.
[89] S. Laporta, *Nuovo Cimento Soc. Ital. Fis. A* **106**, 675 (1993); S. Laporta and E. Remiddi, *Phys. Lett. B* **301**, 440 (1993).
[90] M. Davier and A. Höcker, *Phys. Lett.* **B435**, 427 (1998).
[91] B. Krause, *Phys. Lett.* **B390**, 392 (1997).
[92] M. Davier, talk presented at Tau08, 22–25/9/2008.

[92] M. Davier, talk presented at Tau08, 22–25/9/2008.
[93] K. Melnikov and A. Vainshtein, *Phys. Rev. D* **70**, 113006 (2004).
[94] M. Davier and W. J. Marciano, *Annu. Rev. Nucl. Part. Sci.***54**, 115 (2004).
[95] J. Bijnens and J. Prades, *Mod. Phys. Lett.* **A22**, 767 (2007).
[96] A. Nyffeler, *arXiv:0901.1172* [hep-ph] 9 Jan 2009.
[97] A. Czarnecki, B. Krause, and W. J. Marciano, *Phys. Rev. Lett.* **76**, 3267 (1996).
[98] M. Knecht, S. Peris, M. Perrottet,and E. de Rafael, *J. High Energy Phys.* **11**, 003 (2002); A. Czarnecki, W. M. Marciano, and A. Vainshtein, *Phys. Rev. D* **67**, 073006 (2003).
[99] P. Cladé *et al.*, *Phys. Rev. A* **74**, 052109 (2006).
[100] A. Wicht *et al.*, *Physica Scripta* **T102**, 82-88, 2002; V. Gerginov *et al.*, *Phys. Rev. A* **73**, 032504 (2006).
[101] G. Gabrielse, D. Hanneke, T. Kinoshita, M. Nio, B. Odom, *Phys. Rev. Lett.* **97**, 030802 (2006).
[102] G. Gabrielse, D. Hanneke, T. Kinoshita, M. Nio, B. Odom, *Phys. Rev. Lett.* **99**, 039902 (2007).
[103] Bennett *et al.*, *Phys. Rev. Lett.* **92**, 161802 (2004).
[104] S. Laporta, *Nuovo Cimento* **B106**, 675 (1993); A. Czarnecki and M. Skrzypek, *Phys. Lett. B* **449**, 354 (1999).
[105] T. Kinoshita and M. Nio, *Phys. Rev. D* **70**, 113001 (2004).
[106] T. Kinoshita and W. J. Marciano, in *Quantum Electrodynamics*, edited by T. Kinoshita (World Scientific, Singapore 1990), p. 419.
[107] A. L. Kataev, *Phys. Rev. D* **74**, 073011 92006).
[108] H. Kawai, T. Kinoshita, and Y. Okamoto, *Phys. Lett. B* **260**, 193 (1991); T. Kinoshita, H. Kawai, and Y. Okamoto, *Phys. Lett. B* **254**, 235 (1991); R. N. Faustov, A. L. Kataev, S. A. Larin, and V. V. Starshenko, *Phys. Lett. B* **254**, 241 (1991).
[109] T. Kinoshita, B. Nizic, and Y. Okamoto, *Phys. Rev. D* **41**, 593 (1990).
[110] T. Kinoshita and M. Nio, *Phys. Rev. D* **73**, 053007 (2006).
[111] T. Aoyama, M. Hayakawa, T. Kinoshita, M. Nio, and N. Watanabe, *Phys. Rev. D* **78**, 053005 (2008).
[112] S. G. Karshenboim, *Yad. Fiz.* **56**, 252 (1993) [*Phys. At. Nucl.* **56**, 857 (1993).
[113] J. P. Miller, E. de Rafael, and B. L. Roberts, *Rept. Prog. Phys.* **70**, 795 (2007).
[114] S. Laporta, *Phys. Lett. B* **328**, 522 (1994).
[115] A. I. Milstein and A. S. Yelkhovsky, *Phys. Lett. B* **233**, 11 (1989).
[116] T. Aoyama, M. Hayakawa, T. Kinoshita, and M. Nio, in preparation.
[117] T. Aoyama, M. Hayakawa, T. Kinoshita, and M. Nio, *Phys. Rev. D* **78**, 113006 (2008).
[118] S. J. Brodsky and S. D. Drell, *Phys. Rev. D* **22**, 2236 (1980).

Chapter 4

Analytic QED Calculations of the Anomalous Magnetic Moment of the Electron

Stefano Laporta

Museo Storico della Fisica e Centro Studi e Ricerche Enrico Fermi, Roma
Dipartimento di Fisica, Università di Bologna
INFN, Sezione di Bologna

Ettore Remiddi

Dipartimento di Fisica, Università di Bologna and INFN,
Sezione di Bologna, via Irnerio 46, I-40126 Bologna, Italy

Contents

4.1. Introduction

The analytic evaluation of the anomalous magnetic moment of the electron at three loops in perturbative QED was carried out in almost 28 years, from the initial result of Ref. [1] to the completion of Ref. [2], with the successive use of a great number of increasingly more powerful computational techniques. An account of the status of the calculation in 1990 can be found in Ref. [3]. Rather than discussing the chronological evolution of the techniques, we will describe the resulting algorithm, which could be used now for restarting the calculation from scratch, and which was in fact essentially used in Ref. [4] for obtaining a strictly related static quantity, namely the three loop slope of the Dirac form factor of the electron.

The main ingredients of the resulting algorithm are:

- the d-continuous dimension regularization, used consistently and systematically through all the algebra and in the evaluation of all the loop integrals, for dealing in particular with all the *ultraviolet* (*UV*) and *infrared* (*IR*) divergences;
- the extraction of the considered scalar quantities (the electromagnetic form factors or their static limiting values) from the Feynman graphs, as given by perturbative quantum field theory, by means of suitable projectors by evaluating traces of Dirac gamma matrices in d-continuous dimension. When that is done, the contribution of each Feynman graph to the electron anomaly becomes the sum of several (up to one thousand or more) scalar integrals;
- the exploitation of the integration by parts (and related) identities [5] for expressing all the occurring scalar integrals in terms of the Master Integrals (MI's) of the problem (17 MI's are sufficient for the complete three-loop anomaly);
- the analytic evaluation of the MI's; that is obtained by writing the MI's as multiple integrals involving 4-dimensional hyperspherical variables and suitable dispersive representations; the actual integration is then carried out within the formalism of Euler's dilogarithm [8] and its generalizations [9, 10].

The above points will be discussed in the various sections which follow.

Radiative correction calculations are notoriously very demanding from the algebraic point of view; even if the final results are usually relatively simple, all the algorithms found so far generate large algebraic expressions in the intermediate steps, which cannot be processed without a powerful computer algebra program. For the authors of this review it was essential to use, in various occasions, the programs `SCHOONSCHIP` [11] by M. Veltman, `ASHMEDAI` [12] by M. Levine and `FORM` [13] by J. Vermaseren, who have been all extremely kind and helpful also with their personal advice for the installation and use of their programs on the various hardware platforms used along the years.

4.2. Regularization and Renormalization

The electromagnetic vertex with the electron on mass shell and momentum transfer t contains two scalar form factors, referred to as the Dirac or electric form factor $F_1(t)$ and the Pauli or magnetic form factor $F_2(t)$. The electron magnetic anomaly, $a_e = F_2(0)$, is both *UV* and *IR* finite at any order in perturbation theory; the electron charge slope, $F_1'(0)$, is also *UV* finite at any order and *IR* finite from 2-loop included on (at 1 loop $F_1'(0)$ is *IR* divergent). Therefore both a_e and $F_1'(0)$ at 3 loops are *UV* and *IR* finite, but the contributions to them from individual unrenormalized Feynman graphs can develop *UV* or *IR* divergences. An *UV* regularization procedure is therefore needed (in particular to carry out the renormalization of the inserted subgraphs), as well as an *IR* regularization for properly parametrizing the contributions from separate graphs.

Let us recall that the *UV* divergences of the subgraphs to be regularized and renormalized in perturbative QED are notoriously the divergence associated to vacuum polarization, the two divergences of the electron self-mass (namely those corresponding to electron mass and electron wave function), and the divergence of the electric charge or $F_1(0)$ (while the slope $F_1'(0)$, as remarked above, is *UV* finite); let us recall also that the counterterm (*c.t.*) for the wave function (*w.f.*), usually dubbed Z_2 and the *c.t.* Z_1 for the electric charge must be taken equal, $Z_1 = Z_2$ (the celebrated Ward–Takahashi identity) to enforce the gauge invariance of QED. Due to $Z_1 = Z_2$, the contributions to a_e and $F_1'(0)$ from Z_1 and Z_2 compensate exactly, so that electric charge and wave function renormalization are in fact not really needed, provided that all the vertex graphs (including also the self-mass insertions on the external legs) are properly accounted for.

However, wave function and charge renormalization can be carried out with various prescriptions (provided that the relation $Z_1 = Z_2$ is maintained). When the so-called on-shell renormalization scheme is used (as in the approach which we are describing), the graphs with self-mass insertions in the external electron legs, once renormalized (with the mass *c.t.* and a *w.f.* subtraction proportional to Z_2) give identically vanishing contributions. For that reason those unrenormalized graphs and their *c.t.*'s are usually neglected altogether; when that is done, in the remaining graphs the contributions of Z_1 and Z_2 do not compensate any more, so that wave function and charge renormalization must be carried out explicitly.

In the on-shell renormalization scheme Z_1 and Z_2 are of course *UV* divergent (as they are meant to compensate the *UV* divergences of charge and wave function), but they turn out to be also *IR* divergent; it can therefore happen that an unrenormalized graph, which is on itself *UV* divergent but *IR* finite, gives an *IR* divergent contribution when renormalized. (All the *IR* divergences cancel out, as already repeatedly remarked, in the final result for a_e).

The initial (and most of the following) calculations of a_e, where carried out by modifying the photon propagator of momentum k as

$$\frac{1}{k^2} \rightarrow \frac{1}{k^2+\lambda^2} - \frac{1}{k^2+A^2} . \tag{4.1}$$

The mass A, taken to be much larger than the electron mass m, $A \gg m$, regularises the *UV* divergences. That prescription is known as Pauli–Villars (PV) regularization [14]; loop integrations are carried out neglecting systematically terms of order m/A in the result, and the *UV* divergences of the unregularized graphs show up as powers of $\ln(A/m)$ (typically up to one power for each loop). Similarly, the mass λ is taken to be much smaller than m, $\lambda \ll m$, terms of order λ/m are systematically neglected in the results and the *IR* divergences of the unregularized graphs show up as powers of $\ln(\lambda/m)$.

The PV-prescription Eq. (4.1) is not sufficient to deal with closed electron loops, as those occurring in graphs with vacuum polarization insertions. Let us indicate by $L(p,q;m)$ the integrand of some vacuum-polarization electron loop, where p stands for the external vector, q for the electron loop momentum and m is again the electron mass; the PV regularization then consists in the replacement

$$L(p,q;m) \rightarrow L(p,q;m) - \sum_i c_i L(p,q;M_i) , \tag{4.2}$$

where the M_i are (large) regulator masses, $M_i \gg m$, the c_i suitable coefficients, and the actual number of the M_i and the values of the corresponding c_i are chosen so that the loop integrals converge. After carrying out the renormalization the vacuum polarization amplitude can be written in the form of a subtracted dispersion relation, in which any reference to the regularizators has disappeared, see Sec. 4.6.

The case of the light-light graphs is different; they must be *UV* regularized even if renormalization is not needed, see Sec. 4.7 for more details.

In 1972 't Hooft and Veltman proposed the d-continuous dimension scheme for regularizing Quantum Field Theory, and showed the power of the method using it for renormalizing non-Abelian gauge theories [15]. The main ingredient of the method consists in replacing the 4-dimensional momenta by d-dimensional momenta, where d is continuous; all the calculations are carried out for arbitrary (continuous) d, in the final result the $d \to 4$ limit is taken (see however Section 4.11 for more proper definitions and examples of typical results).

Both the *UV* and the *IR* divergences present in the original 4-dimensional expressions show up, in the $d \to 4$ limit, as polar singularities in $(d-4)$, *i.e.* powers of $1/(d-4)$.

The d-continuous regularization proved extremely powerful also in the actual analytic evaluation of loop integrals; indeed, in d-continuous dimensions the loop integrals are always well defined, without convergence problems, so that a very wide set of formal manipulations on loop integrals can be carried out without ambiguities, allowing in particular to establish the integration by parts identities (ibp-id's) of Ref. [5]. Those give a very interesting set of relations between loop integrals, which can be used in particular to express all the scalar integrals occurring in a calculation in terms of a smaller set of "reference" integrals, usually dubbed "Master Integrals" (MI's).

Somewhat ironically, the new regularization scheme was ignored for many years in the analytic evaluation of a_e. That was due, besides the usual obvious inertia in switching from older to newer techniques, to the (wrong) perception that continuous d-dimension was better tailored to massless theories (the first applications of ibp-id's were indeed in massless QCD) than to massive QED. As matter of fact, only the latest (but most demanding) a_e calculations were actually carried out in the d-continuous regularization scheme.

Assume that some simple quantity with an *UV* divergence, when evaluated in the PV regularization takes the form

$$c \ln \frac{A}{m} + f ,$$

where the numbers c, f are independent of A; in the limit $A \to \infty$ terms proportional to m/A can indeed be dropped. In the d-continuous scheme that same quantity is a function of d, which in the $d \to 4$ limit can be expanded as

$$C \frac{1}{d-4} + F + (d-4)G + \dots .$$

The correspondence between the coefficients of the leading singularities, *i.e.* c, the coefficient of $\ln(A/m)$, in the first scheme and C, the coefficient of

$1/(d-4)$ is immediate, $C = -c$, but no simple relation exists, in general, between the finite terms f and F of the two schemes (if the considered quantity is finite, i.e. $c = C = 0$, then its value is of course the same in the two schemes, i.e. $f = F$).

Further, when carrying out renormalization, one has often to consider terms equal to the product of some subtraction constant, say Z, containing an UV divergence, times some finite quantity, say Q. In the PV scheme, in the limit $A \gg m$, Z could be (in a simple case) something like

$$Z = Z(A) = z\ \ln\frac{A}{m} + z_1\ ;$$

as everything depends, in principle, on A, one might write something like

$$Q = Q(A) = q + \left(\frac{m}{A}\right)^2 q_1 + \ldots\ ;$$

but in the $A \to \infty$ limit

$$\left(\frac{m}{A}\right)^2 \ln\frac{A}{m} \to 0\ ,$$

giving

$$QZ = Q(A)Z(A) \to \left(z\ \ln\frac{A}{m} + z_1\right) q\ ,$$

so that q_1 does not appear in the result and does not need to be evaluated.

In the d-continuous scheme, on the contrary, one expects something like

$$Z = Z(d) = -z\ \frac{1}{d-4} + z'_1\ ,$$
$$Q = Q(d) = q + (d-4)q'_1 + \ldots\ ,$$

so that when evaluating the product QZ in the $d \to 4$ limit one has

$$QZ = Q(d)Z(d) \to \left(-z\ \frac{1}{d-4} + z'_1\right) q - zq'_1\ ,$$

where the term q'_1 cannot be dropped – and must be evaluated explicitly.

4.3. The Projectors

Let M_μ be a QED vertex amplitude for on mass shell electrons and momentum transfer t; its decomposition in form factors is

$$\bar{u}(p_2)M_\mu u(p_1) = \bar{u}(p_2)\left[F_1(t)\gamma_\mu - \frac{i}{4m}F_2(t)\left(\gamma_\mu \not{\Delta} - \not{\Delta}\gamma_\mu\right)\right] u(p_1)\ , \quad (4.3)$$

where p_1, p_2 are the momenta of the initial and final electrons, the mass shell condition reads $p_1^2 = p_2^2 = -m^2$, the spinors satisfy the equation

$$(-i\not{p}_i + m)u(p_i) = 0 \ ,$$

$\Delta = p_1 - p_2$ and the momentum transfer is $t = -\Delta^2$; define further $p = (p_1 + p_2)/2$, so that $(p \cdot \Delta) = 0$, $p^2 = -m^2 + t/4$.

No γ_5 appears in the formulae, so that the extension to d-continuous dimensions is straightforward, and one easily finds that in d-continuous dimensions the form factors can be extracted by means of the formulae

$$F_1(t) = \frac{1}{2(d-2)(t-4m^2)} \mathrm{Tr}\left[\left(\gamma_\mu - 4i\frac{(d-1)m}{t-4m^2}p_\mu\right)\right.$$
$$\left.(-i\not{p}_2 + m)M_\mu(-i\not{p}_1 + m)\right]$$
$$F_2(t) = \frac{2m^2}{(d-2)t(t-4m^2)} \mathrm{Tr}\left[\left(-\gamma_\mu + i\frac{(d-2)t + 4m^2}{m(t-4m^2)}p_\mu\right)\right.$$
$$\left.(-i\not{p}_2 + m)M_\mu(-i\not{p}_1 + m)\right] \ , \tag{4.4}$$

which hold for arbitrary t.

We are actually interested only in the static $t \to 0$ limit. That limit is trivial in all the terms not containing the factor $1/t$; in the terms multiplied by $1/t = -1/\Delta^2$, which turn out to have already a factor proportional to Δ when the traces are explicitly evaluated, it is sufficient to expand M_μ up to first order in Δ_μ

$$M_\mu = M_\mu(p, \Delta) \simeq M_\mu(p, 0) + \Delta_\nu \frac{\partial}{\partial \Delta_\nu} M_\mu(p, \Delta)|_{\Delta=0}$$
$$\equiv V_\mu(p) + \Delta_\nu T_{\nu\mu}(p) \tag{4.5}$$

and then to average over the solid angle $\Omega(d-1)$ of the $(d-1)$ space dimensions of Δ, orthogonal to p. The terms linear in Δ vanish for symmetry, while for the quadratic terms the average is given by

$$\frac{1}{\Omega(d-1)} \int d\Omega(d-1)\ \Delta_\mu \Delta_\nu = \frac{\Delta^2}{d-1}\left(\delta_{\mu\nu} - \frac{p_\mu p_\nu}{p^2}\right) \ ,$$

after which the $1/t = -1/\Delta^2$ factors disappear. The final results for the static quantities $F_1(0)$, relevant for obtaining the charge renormalization *c.t.* and the magnetic anomaly $a_e = F_2(0)$ are

$$F_1(0) = -\frac{i}{4m^2}\mathrm{Tr}\left[(-i\not{p}+m)\,p_\mu V_\mu\right]$$

$$F_2(0) = \frac{1}{4m^2(d-1)}\mathrm{Tr}\left[\left(m^2\gamma_\mu + i(d-1)mp_\mu + d\not{p}p_\mu\right)V_\mu \right.$$

$$\left. -\frac{im}{2(d-2)}(-i\not{p}+m)(\gamma_\mu\gamma_\nu - \gamma_\nu\gamma_\mu)(-i\not{p}+m)T_{\mu\nu})\right] . \quad (4.6)$$

A similar formula can be established for the slope $F_1'(0)$, which however requires one more derivative in Δ_μ.

4.4. The ibp (and Other) Identities

Consider an arbitrary Feynman graph depending on n external momenta $p_i, i = 1,..,n$, on l integration loop momenta $k_i, i = 1,..,l$, and containing P different propagators. The number of the scalar products depending on the loop momenta, namely of type $(k_i \cdot k_j)$ and $(p_i \cdot k_j)$, is $N = l(l+1)/2 + nl$; let us also observe that, in any case, $N \geq P$. For a_e at 3 loops, $l = 3$; in the static limit of Eq. (4.6) $n = 1$, so that $N = 9$, while for the 3-loop graphs P, the number of different propagators in a Feynman graph, may vary (in the $\Delta \to 0$ limit) from 6 to 8 depending on the considered graph.

Call $D_i, i = 1,..,P$, the scalar denominators of the propagators; P of the N scalar products can then be expressed as linear combinations of the denominators D_i, while the remaining $S = N - P$ scalar products $S_i, i = 1,..,S$, can be dubbed *independent* of the D_i (the actual choice of the S_i has some arbitrariness, but their number S is anyhow fixed).

By using the projectors described in the previous section, one can extract the contribution of the graph to any of the desired static quantity as a sum of (products of) scalar products divided the D_i; note that some of the denominators D_i may occur in the original expression raised to powers higher than 1, due to the nature of the graph or to the algebra of the projection, Eq. (4.5), which can imply a differentiation.

Some of the scalar products in the numerator can be expressed in terms of the D_i; after obvious simplifications of the numerators against the denominators, only independent scalar products remain in the numerator, and one is left with a sum of scalar integrals of the form

$$\int \left(\prod_{i=1}^{l} d^d k_i\right) \frac{\prod_{j=1}^{S} S_j^{y_j}}{\prod_{r=1}^{P} D_r^{1+x_r}} , \quad (4.7)$$

where the (y_j, x_r) are non-negative integer numbers, $y_j, x_r \geq 0$, plus other integrals where some of the D_r are missing as a consequence of the simplifications (recall that in the d-continuous regularization scheme all integrals remain separately well defined or "convergent"). In the terms with fewer denominators the bookkeeping of scalar products in the numerator against denominators is different (as the set of denominators has shrunk), so that the procedure must be repeated with the smaller set of D_r; the process will anyhow come to an end, as in the d-continuous regularization scheme loop integrals with less denominators than loops vanish regardless of the numerator.

As a consequence of the "convergence" of all the integrals, in d-continuous dimension for any scalar loop integral, say of the kind of Eq. (4.7), one can write the (celebrated) integration by parts identity (ibp-id) [5]

$$\int \left(\prod_{i=1}^{l} d^d k_i \right) \frac{\partial}{\partial k_{a,\mu}} \left(v_\mu \frac{\prod_{j=1}^{S} S_j^{y_j}}{\prod_{r=1}^{P} D_r^{1+x_r}} \right) = 0 \ , \tag{4.8}$$

where a is any of the l loops, while v_μ stands for any of the external or loop momenta. For any given integral like Eq. (4.7) one can therefore write $l(l+n)$ such identities, all formally different from each other (see below for their actual independence); in the case of a_e at 3 loop, $l(l+n) = 3(3+1) = 12$, so that for any integral like Eq. (4.7) there are 12 such different identities.

Let us now discuss the explicit structure of the identity Eq. (4.8). To start with, let us give to each integral of the form of Eq. (4.7) a "weight" (or more precisely a set of weights) (P, X, Y), where P is the number of the different propagators, $X = \sum_r x_r$ is the sum of the extra powers of the denominators and $Y = \sum_j y_j$, is the sum of the powers of the independent scalar products in the numerator. Acting on the numerator multiplying by a vector v_μ and then taking the derivative with respect to a loop momentum will modify the structure of the scalar products, but not their total power, which will therefore remain Y. On the contrary, when acting with the derivative on the denominator one obtains a sum of terms in which the power of one of the denominators has increased by one, times a term linear in the momenta (the denominators are quadratic in the momenta!), which combined with the vector v_μ generates a sum of scalar products (or a single scalar product in the simplest cases).

Some of the extra scalar products in the numerator may simplify against some of the denominators, but some can remain; summarizing, the identities originated by an integrand with weight (P, X, Y) will involve, in general, integrals with the same P propagators, but additional extra powers, i.e. integrals of weight $(P, X+1, Y+1)$, plus a combination of terms with equal or lower values of X, Y, i.e. weight $(P, X+1, Y)$, $(P, X, Y+1)$ or (P, X, Y), and finally also terms with $P-1$ propagators etc. The number of integrals with weight $(P, X+1, Y+1)$ is surely larger than the number of those with weight (P, X, Y), so at first sight each identity seems to involve an increasingly larger number of new integrals with higher and higher extra powers, in a kind of runaway situation.

Fortunately, it is not so. The number of ways in which X objects can be distributed in P boxes can be obtained by considering a set of $(P+X-1)$ lined up points and choosing among them $(P-1)$ "separators"; the points from the beginning to the first separator excluded will give the number of the objects in the first box, the points between the first and the second separator those in the second box, etc. Clearly, the way in which $(P-1)$ separators can be chosen among $(P+X-1)$ points is

$$\binom{P+X-1}{P-1} = \binom{P+X-1}{X}, \tag{4.9}$$

which grows at most as a polynomial of order $(P-1)$ in X for large X, (but for $X < P$ it is just a polynomial of degree X; P is a constant in these consideration). Eq. (4.9) is therefore the number of integrals with a given numerator and X extra powers of the P denominators. Similarly, the number of integrals with a same denominator and Y extra powers of the independent scalar products in the numerator is

$$\binom{S+Y-1}{S-1} = \binom{S+Y-1}{Y}, \tag{4.10}$$

which for large Y grows polynomially in Y (S is a constant here). For large enough X, Y, the ratio of the integrals of weight $(P, X+1, Y+1)$ to the integrals of weight (P, X, Y) will approach 1, while the number of the identities which can be written is always equal to the product of $l(l+n)$ times the number of the integrals of weight (P, X, Y). As a conclusion: if $l(l+n) > 4$ (in the case we are interested the actual value is 12), for large enough X, Y,

$$
\begin{aligned}
l(l+n) \times & \binom{P+X-1}{P-1}\binom{S+Y-1}{S-1} \\
& > \left[\binom{P+X}{P-1}+\binom{P+X-1}{P-1}\right] \\
& \times \left[\binom{S+Y}{S-1}+\binom{S+Y-1}{S-1}\right] , \qquad (4.11)
\end{aligned}
$$

so that (for large enough X, Y) the number of the generated identities is larger than the number of the involved scalar integrals with P different propagators – *i.e.* there are more identities than integrals, or, in other words, the system of all the equations generated by evaluating explicitly the identities of Eqs. (4.8) for all the sets (x_r, y_j) corresponding to a given pair (X, Y) is apparently overconstrained! That is by no means contradictory, but simply a clear evidence that as (X, Y) grow the obtained identities are not all independent.

For completeness, let us recall that there are also other possible sources of identities between scalar integrals, such as symmetry identities, and let us consider for the following the enlarged system of all the available identities.

As varying the numbers x_r, y_j in Eq. (4.8) one obtains a redundant system of an infinite number of equations for an infinite number of unknown scalar integrals of the type of Eq. (4.7), it is convenient to work out an appropriate algorithm [2, 6, 7] for the actual solution of the system.

To that aim, let us complete the set of weights (P, X, Y) already introduced by giving additional weights to the occurring integrals, so that each integral has a different set of weights, assigning, to be definite, higher weights to integrals which are more laborious to evaluate analytically.

The first weight remains P, the number of propagators; terms with a smaller number of propagators (as those terms sometimes arising when simplifying in an identity numerator and denominator) have a smaller P, and will be considered simpler than terms with higher P. The second weight is X, the number of extra powers of the denominators; again, terms with the same number of propagators P but more extra powers are considered more complicated than those with smaller X; similarly, one can keep Y as a third weight. The set of the weights can then be completed with the numbers (x_r, y_j) (or something equivalent) etc., until each integral is eventually given a different weight.

One can now consider the set of all the identities (such as the ibp-identities like Eq. (4.8) and the symmetry identities, if any) which can be written explicitly by using all the integrals with a given P up to some convenient value of (X, Y), large enough to guarantee that there are more identities than involved integrals. In any case, the identities must contain all the scalar integrals occurring in the original expression for the considered physical quantity obtained by the projection procedure of Section 4.3. Typically, the contribution of a 3-loop Feynman graph to a_e consists of several hundred scalar integrals, and the corresponding system to be solved may contain a few thousand identities.

To solve the system, take one of the identities as the first identity. Identify among the scalar integrals which appear in that identity the scalar integral with the highest weight, and solve the identity by expressing that integral in terms of the other integrals of lower weight occurring in the identity. Then look at all the other identities, and replace in them the expression just obtained for the integral with the highest weight in the first identity (Gauss substitution rule).

Consider the next equation, solve it for the integral with the highest weight which occurs in the identity, substitute the result into the other identities and so on until all the identities are worked out. Due to the redundance of the system of identities discussed above, some of the identities are automatically satisfied once the solutions of the previous identities are substituted, and therefore they will not give any new information. The order in which the equations are considered and solved is in principle irrelevant, even if in practice the intermediate results can depend heavily on it.

The final result of the procedure will be a (long) list of relations, expressing almost all the scalar integrals appearing in the identities in terms of a few independent integrals, dubbed the *Master Integrals* (MI's) of the problem. In the simplest cases, all the integrals with a given value of P (the number of different propagators) can be expressed in terms of integrals with strictly lower P, corresponding to subgraphs (sometimes called also subtopologies in this context) of the original graph (or topology), which is then said to possess no MI on its own.

It is to be noted that each graph generates a large number of subtopologies, but a same subtopology can be generated by several graphs, and as a rule of thumb the number of all the subtopologies is of the order of the original number of graphs. The number of the MI's, finally, is smaller than the total number of topologies and subtopologies (in the case of a_e at 3-loop some topology has 2 MI's, but many others have no MI at all). As already anticipated, the whole calculation of the 3-loop a_e and $F_1'(0)$ involves 17 independent MI's only; see Section 4.11 for the complete list.

It is to be observed that, strictly speaking, there is no proof that no relations exist between the obtained MI's, but that does not spoil the correctness of the final results; the discovery of a new relation between MI's might indeed further simplify the final analytic expression of the concerned physical quantities, but would not change its actual numerical value. To make an example, it is known that the Riemann ζ-function of even argument is related to the even powers of π, (one has for instance $\zeta(2) = \pi^2/6$), while nothing similar seems to exist for odd arguments; but the hypothetical discovery that, say, $\zeta(3)$ can be expressed as a combination of other mathematical constants times simple rational coefficients would not change the numerical value of any of the current results which are now written in terms of $\zeta(3)$.

Finally, the actual choice of the MIs is related to the specific choice of the weights, which is to some extent arbitrary, but once the result is established in terms of a given set of MIs, moving to another set of MIs is an almost trivial task. In an analytic calculation in d-continuous dimensions, integrals with less propagators are easier to deal with and are therefore preferred, even if their analytic expression contains several $1/(d-4)$ factors, divergent in the $d \to 4$ limit. Furthermore, additional $1/(d-4)$ factors may appear in the expression of the anomaly in terms of to MIs, a few terms of the expansions of the MIs in $(d-4)$ (besides the singular and the finite term) are also needed (needless to say, all the singularities cancel out in the final result).

In a numerical approach to the calculation of the same quantity, the preference may go to well convergent integrals with smooth integrands, so that a different set of weights leading to a different set of MI's might be more convenient.

4.5. The Feynman Graphs

At 3 loops, there are in total 72 different Feynman graphs, of which 40 are actually different when accounting for mirror symmetry; 12 of them involve electron loops and are shown in Fig. 4.1, the other 28 without electron loops are shown in Fig. 4.2, following the numbering of Ref. [3]. The label $\times 2$ appearing in some of them, as for instance in graph J2 of Fig. 4.1, accounts for the multiplicity.

For the actual evaluation of the contributions of the graphs, and in particular in view of the static limit $\Delta_\mu \to 0$, it is convenient to group together the vertex graphs corresponding to the insertions of the external

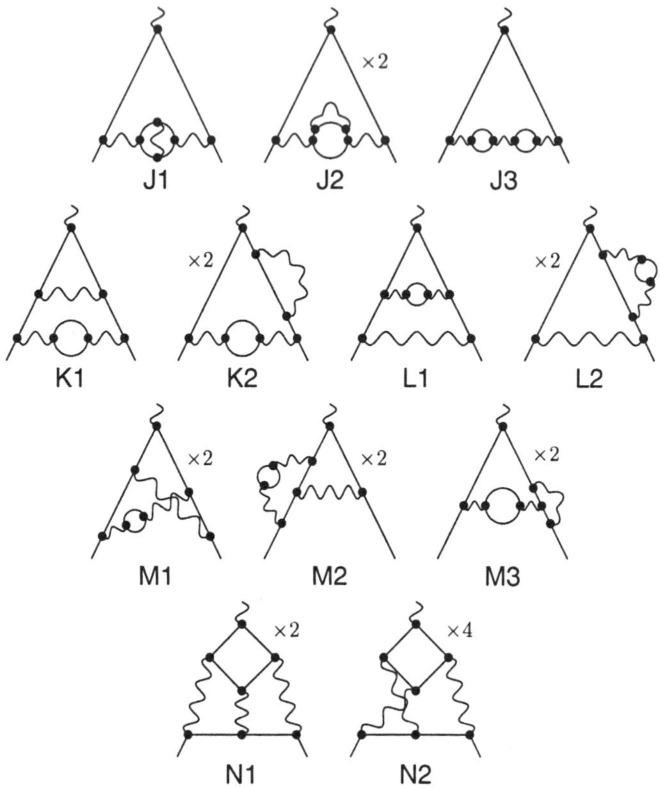

Fig. 4.1. Graphs with electron loops.

field line with the momentum Δ_μ in a same self-mass graph, as shown for instance for the graphs H1, H2 and H3 in Fig. 4.3. As a further comment, note that *IR* divergences *do not* cancel, in general, among vertex graphs corresponding to a same self-mass.

4.6. Graphs with a Closed Electron Loop: Vacuum Polarization Insertions

The vacuum polarization tensor of a photon of momentum k can be written as

$$\Pi_{\mu\nu}(k) = i\ (k_\mu k_\nu - k^2\delta_{\mu\nu})\Pi(-k^2)\ , \tag{4.12}$$

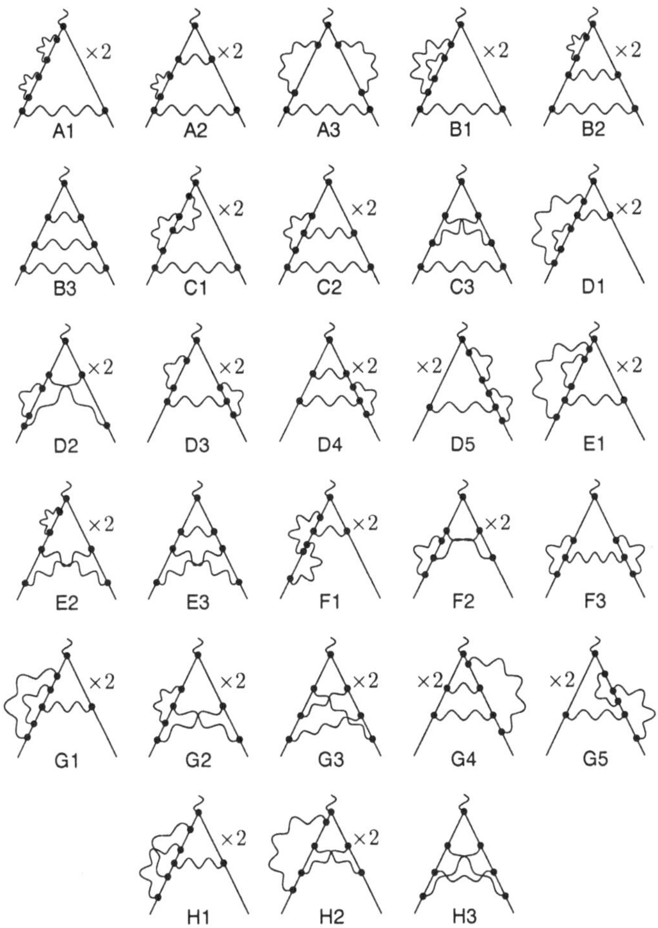

Fig. 4.2. Graphs without electron loops.

where the renormalized amplitude $\Pi(-k^2)$ satisfies the subtracted dispersion relation

$$\Pi(-k^2) = -k^2 \int_{4m^2}^{\infty} \frac{dt}{t(k^2 + t - i\epsilon)} \frac{1}{\pi} Im\Pi(t) \,. \tag{4.13}$$

Inserting a vacuum polarization tensor into a photon line of momentum k amounts to the replacement

$$\frac{-i}{k^2 - i\epsilon}\, \delta_{\mu\nu} \rightarrow \int_{4m^2}^{\infty} \frac{dt}{t} \frac{1}{\pi} Im\Pi(t) \, \frac{-i}{k^2 + t - i\epsilon}\, \delta_{\mu\nu} \,, \tag{4.14}$$

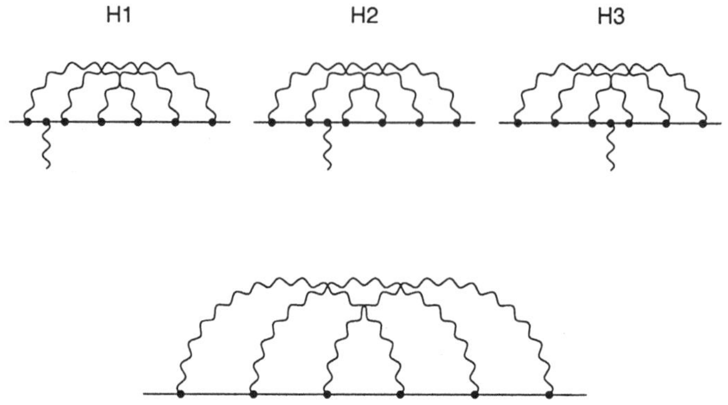

Fig. 4.3. The vertex graphs H1, H2, H3 and the corresponding self-mass.

where the term in $k_\mu k_\nu$ of Eq. (4.12), which cancels out when summed on all the Feynman graphs (gauge invariance), has been dropped. According to the previous equation, for evaluating the contribution of graphs with vacuum polarization insertions one can evaluate the electron anomaly due to a massive photon of arbitrary mass $\sqrt{t}$, i.e. propagator $-i\delta_{\mu\nu}/(k^2+t-i\epsilon)$, and then integrate it on t with the weighting function $(1/t\pi)Im\Pi(t)$.

Let us call $K(t)$ such an anomaly; the actual vacuum polarization contribution to the electron $(g-2)$ is then

$$a_e(\mathrm{vp}) = \int_{4m^2}^{\infty} \frac{dt}{t}\, \frac{1}{\pi} Im\Pi(t)\; K(t). \tag{4.15}$$

In perturbative QED, both $Im\Pi(t)$ and $K(t)$ can be expanded in powers of (α/π)

$$Im\Pi(t) = \sum_{n=1}^{\infty} \left(\frac{\alpha}{\pi}\right)^n Im\Pi^{(n)}(t) \ ,$$

$$K(t) = \sum_{n=1}^{\infty} \left(\frac{\alpha}{\pi}\right)^n K^{(n)}(t) \ ,$$

and one can define the corresponding contributions to a_e as

$$a_e^{(i,j)}(\mathrm{vp}) = \int_{4m^2}^{\infty} \frac{dt}{t}\, \frac{1}{\pi} Im\Pi^{(i)}(t)\; K^{(j)}(t) \ . \tag{4.16}$$

Note that $a_e^{(i,j)}(\mathrm{vp})$ comes therefore from a Feynman graph with $(i+j)$ loops altogether.

One has for instance, at 1-loop QED,

$$\frac{1}{\pi} Im\Pi^{(1)}(t) = \frac{t+2m^2}{3t}\sqrt{\frac{t-4m^2}{t}} \,,$$

$$K^{(1)}(t) = \frac{1}{2} - \frac{t}{m^2} + \frac{t(t-2m^2)}{2m^4}\ln\frac{t}{m^2} + \frac{t(t^2-4m^2t+2m^4)}{2m^4\sqrt{t(t-4m^2)}}\ln\frac{\sqrt{t}-\sqrt{t-4m^2}}{\sqrt{t}+\sqrt{t-4m^2}} \,,$$

from which one easily obtains the well-known 2-loop contribution [16]

$$a_e^{(1,1)}(\mathrm{vp}) = \frac{119}{36} - \frac{\pi^2}{3} = 0.015\,687\,421\ldots. \quad (4.17)$$

$Im\Pi^{(2)}(t)$, first evaluated by Källen and Sabry [17], and $K^{(2)}(t)$, which can be found in Ref. [18], are too long to be listed here. By using $Im\Pi^{(2)}(t)$ and $K^{(1)}(t)$ one obtains the 3-loop contributions

$$a_e^{(2,1)} = a_e(J1) + 2a_e(J2) + a_e(J3)$$

from the graphs J1, J2, J3 of Fig. 4.1, (see Ref. [1]), and by using $Im\Pi^{(1)}(t)$ and $K^{(2)}(t)$ the contributions

$$a_e^{(1,2)} = a_e(K1) + 2a_e(K2) + a_e(L1) + 2a_e(L2) + 2a_e(M1) + 2a_e(M2) + 2a_e(M3) \quad (4.18)$$

from the remaining graphs [18] K1, K2, L1, L2, M1, M2, M3.

The explicit results follow: $a_e(J1), a_e(J2)$ etc. stand for the contributions of the corresponding graphs of Fig. 4.1, note that we write explicitly the multiplicity of the various graphs:

$$a_e(J1) = +\frac{32}{3}\left(a_4 + \frac{1}{24}\ln^4 2\right) - \frac{4}{9}\pi^2\ln^2 2 - \frac{7}{270}\pi^4 + \frac{49}{18}\zeta_3 - \frac{22}{9}\pi^2\ln 2 + \frac{161}{162}\pi^2 + \frac{1145}{432}$$
$$= -\,0.001\,804\,385\,803\ldots \quad (4.19)$$

$$2a_e(J2) = -\,2\zeta_3 + 2\pi^2\ln 2 - \frac{3}{2}\pi^2 + \frac{1547}{432}$$
$$= 0.054\,675\,038\,279\ldots \quad (4.20)$$

$$a_e(J3) = +\frac{8}{3}\zeta_3 - \frac{4}{135}\pi^2 - \frac{943}{324} = 0.002\,558\,524\,936\ldots \quad (4.21)$$

$$a_e^{(1,2)} = +\frac{1}{18}\pi^4 - \frac{35}{8}\zeta_3 - \frac{4}{3}\pi^2\ln 2 + \frac{31}{54}\pi^2 + \frac{227}{72}$$
$$= -\,0.150\,172\,282\,099\ldots \quad (4.22)$$

In the above equations

$$a_4 = \sum_{n=1}^{\infty} 1/(2^n n^4) = 0.517479061\ldots,$$

while $\zeta_k = \sum_{n=1}^{\infty} 1/n^k$ denotes as usual the Riemann zeta function of argument k.

The contributions of the graphs where the closed electron loop is replaced by a closed μ-meson loop, say $a_e(\mathrm{vp};\mu)$ is obtained by simply rescaling the vacuum polarization amplitude, (in the rest of this section we write the electron mass as m_e, to avoid confusion with other masses)

$$a_e(\mathrm{vp};\mu) = \int_{4m_\mu^2}^{\infty} \frac{dt}{t}\,\frac{1}{\pi} Im\Pi\left(\frac{m_e^2}{m_\mu^2}t\right)\; K(t)\;. \tag{4.23}$$

It is convenient to expand the integral in powers of (m_e^2/m_μ^2), which is a small parameter; at 2-loop one finds [19]

$$\begin{aligned} a_e^{(1,1)}(\mathrm{vp};\mu) &\simeq \frac{1}{45}\left(\frac{m_e}{m_\mu}\right)^2 + \left[-\frac{1}{70}\ln\frac{m_\mu}{m_e} + \frac{9}{19600}\right]\left(\frac{m_e}{m_\mu}\right)^4 + \ldots \\ &\simeq 5.1973 \times 10^{-7}\;. \end{aligned} \tag{4.24}$$

The contribution is very small, but relevant at the 1 ppb level; indeed, 10^{-7} in the numerical value of a 2-loop contribution to a_e, due to the accompanying $(\alpha/\pi)^2$ factor, corresponds to a relative contribution of 0.46 ppb to a_e, as the order of magnitude of a_e is $\frac{1}{2}\left(\frac{\alpha}{\pi}\right)$.

For completeness, we list also the leading terms of the 3-loop contributions due to muon vacuum polarization insertions [20], expanded again in powers of (m_e/m_μ). In an almost obvious notation, with $a_e(J1;\mu)$ referring to the graph $J1$ of Fig. 4.1 with the electron loop replaced by a a muon loop etc., one finds

$$a_e(J1;\mu) + 2a_e(J2;\mu) \simeq \left(\frac{m_e}{m_\mu}\right)^2\left[\frac{41}{486}\right]\;, \tag{4.25}$$

$$a_e^{(1,2)}(\mathrm{vp};\mu) \simeq \left(\frac{m_e}{m_\mu}\right)^2\left[-\frac{23}{135}\ln\frac{m_\mu}{m_e} - \frac{2}{135}\pi^2 + \frac{229}{8100}\right]. \tag{4.26}$$

The graph $J3$ of Fig. 4.1 contains two electron vacuum polarizations loops; one of them or both can be substituted with muon loops, the corresponding contributions are

$$a_e(J3;e,\mu) \simeq \left(\frac{m_e}{m_\mu}\right)^2\left[-\frac{4}{135}\pi^2 + \frac{41}{135}\right]\;, \tag{4.27}$$

$$a_e(J3;\mu,\mu) \simeq \left(\frac{m_e}{m_\mu}\right)^4\left[\frac{2}{225}\ln\frac{m_\mu}{m_e} - \frac{161}{54\,000}\right]\;. \tag{4.28}$$

The contribution of all the 3-loop muon vacuum polarization graphs is -2.17×10^{-5}. Note that 10^{-5} in the numerical value of a three loop contribution to a_e, due to the $(\alpha/\pi)^3$ factor, corresponds to a relative contribution of 0.1 ppb to a_e.

As a last remark, let us recall that the contribution to the anomaly of the μ-meson due to electron vacuum polarization loops is

$$a_\mu(\text{vp}; e) = \int_{4m_\mu^2}^{\infty} \frac{dt}{t}\,\frac{1}{\pi} Im\Pi(t)\; K\left(\frac{m_\mu^2}{m_e^2}t\right) \;; \tag{4.29}$$

in this case, the expansion in (m_μ^2/m_e^2) gives rise to leading logarithmic terms in $\ln(m_\mu/m_e)$. One has for instance [21]

$$a_\mu^{(1,1)}(\text{vp}; e) \simeq \frac{1}{3}\ln\frac{m_\mu}{m_e} - \frac{25}{36} + \ldots .$$

4.7. Graphs with a Closed Electron Loop: Light-Light Scattering

The 1-loop light-light scattering amplitude $T_{\mu\nu\rho\sigma}$ present in the graphs N1, N2 of Fig. 4.1 is, strictly speaking, *UV* divergent: the naive power counting gives, for the electron loop, 8 powers of the loop momentum q in the numerator (4 due to the integration on the 4 components d^4q and 4 due to the numerators of the 4 electron propagators) and 8 powers in the denominator (due to the scalar part of the 4 propagators). When evaluated naively, just taking the trace on the closed electron loop, the term with 4 powers of the loop momentum q in the numerator drops out and the remaining terms give convergent integrals.

On the other hand, when the momenta of the 4 external photons vanish, the naive amplitude tends to a finite, non-vanishing value (proportional to the symmetric product of the Kronecker δ-functions in the photon polarization indices); the result is therefore wrong, because for vanishing photon momenta the light-light amplitude should also vanish (due to gauge invariance, the amplitude should couple the electromagnetic fields, which are linear in the the associated momenta, while the Kronecker δ's of the naive calculation couple directly the polarizations). Only when properly regularized, $T_{\mu\nu\rho\sigma}$ does vanish for vanishing momenta of the external photons, so that it does not couple anymore the polarizations. It is to be recalled here that the light-light amplitude $T_{\mu\nu\rho\sigma}$ is in this respect unique, because even if it requires regularization, it does not need to be renormalized (at variance with vacuum polarization, electron self-mass etc., which all require both regularization and renormalization).

As already observed in Ref. [22] the regularized $T_{\mu\nu\rho\sigma}$ satisfies

$$\Delta_\mu T_{\mu\nu\rho\sigma} = 0 \; ,$$

as required by gauge invariance; differentiating with respect to Δ_μ gives

$$T_{\mu\nu\rho\sigma} = -\Delta_\alpha \frac{\partial}{\partial \Delta_\mu} T_{\alpha\nu\rho\sigma} \; .$$

Carrying out the derivative $\partial/\partial\Delta_\mu$ on the integrand of the Feynman graph gives a safely convergent integral, so that the above equation provides for $T_{\mu\nu\rho\sigma}$ an expression which does not need anymore to be regularized, and was indeed used also in Ref. [23].

The results for the two independent light-light graphs of Fig. 4.1 (note the multiplicity of N2) are

$$\begin{aligned} 2a_e(N1) + 4a_e(N2) &= -\frac{5}{18}\pi^2\zeta_3 + \frac{5}{6}\zeta_5 + 16\left(a_4 + \frac{1}{24}\ln^4 2\right) - \frac{2}{3}\pi^2\ln^2 2 \\ &\quad -\frac{41}{540}\pi^4 - \frac{4}{3}\zeta_3 - 24\pi^2\ln 2 + \frac{931}{54}\pi^2 + \frac{5}{9} \\ &= 0.371\,005\,292\ldots \end{aligned} \tag{4.30}$$

The contribution of the light-light diagrams with an internal muon loop is [24]

$$2a_e(N1;\mu) + 4a_e(N2;\mu) \simeq \left(\frac{m_e}{m_\mu}\right)^2 \left[\frac{3}{2}\zeta_3 - \frac{19}{16}\right] \; ; \tag{4.31}$$

the corresponding numerical value is about 1.44×10^{-5}.

4.8. Graphs without Closed Electron Loops

Many of the 28 graphs without electron loops of Fig. 4.2 are *IR* divergent; we present their contributions by grouping them in 14 *IR* finite combinations. The results for the various sets of graphs are taken from the references which follow (we repeat, for completeness, also the references to the sets J, N already seen in the previous sections):

- for set A from Ref. [25] and Ref. [26],
- for set B from Ref. [25],
- for set C from Ref. [27], Ref. [28] and Ref. [29],
- for set D from Ref. [29],
- for set E from Ref. [30],
- for set F from Ref. [31] and Ref. [32],
- for set G from Ref. [33],

- for set H from Ref. [2],
- for set J from Ref. [1],
- for set K from Ref. [34], Ref. [35] and Ref. [36],
- for set L from Ref. [35] and Ref. [37],
- for set M from Ref. [38], and
- for set N from Ref. [23].

As in previous sections, $2a_e(D3)$ stands for the contribution of the graph $D3$ of Fig. 4.2, which has multiplicity 2, etc.; the results are:

$$a_e(A3) + 2a_e(D3) + a_e(F3) = -\frac{17}{36}\pi^2\zeta_3 + \frac{5}{3}\zeta_5 - \frac{23}{180}\pi^4 + \frac{37}{24}\zeta_3 + 3\pi^2\ln 2 - \frac{143}{144}\pi^2 + \frac{25}{6} = 0.421\,171\,047\ldots \tag{4.32}$$

$$2a_e(D1) + 2a_e(F1) = +\frac{85}{36}\pi^2\zeta_3 - \frac{140}{3}\zeta_5 - \frac{1}{9}\pi^4 + \frac{101}{2}\zeta_3 + \frac{7}{6}\pi^2\ln 2 - \frac{67}{18}\pi^2 - \frac{9}{8} = -0.378\,099\,956\ldots \tag{4.33}$$

$$2a_e(B1) + 2a_e(D5) + 2a_e(G1) = +\frac{5}{9}\pi^2\zeta_3 - \frac{5}{2}\zeta_5 - 28\left(a_4 + \frac{1}{24}\ln^4 2\right) + 2\pi^2\ln^2 2 + \frac{13}{240}\pi^4 - \frac{305}{36}\zeta_3 + \frac{53}{18}\pi^2\ln 2 - \frac{517}{324}\pi^2 + \frac{1123}{864} = -0.489\,778\,473\ldots \tag{4.34}$$

$$2a_e(E1) + 2a_e(G5) = -\frac{43}{72}\pi^2\zeta_3 + \frac{95}{24}\zeta_5 - 16\left(a_4 + \frac{1}{24}\ln^4 2\right) - \frac{1}{3}\pi^2\ln^2 2 + \frac{277}{1080}\pi^4 - \frac{31}{2}\zeta_3 + \frac{20}{9}\pi^2\ln 2 - \frac{103}{108}\pi^2 + \frac{109}{48} = 1.417\,302\,845\ldots \tag{4.35}$$

$$2a_e(A1) + 2a_e(C1) + 2a_e(H1) = -\frac{2}{3}\pi^2\zeta_3 + \frac{215}{12}\zeta_5 + \frac{44}{3}\left(a_4 + \frac{1}{24}\ln^4 2\right) - \frac{11}{18}\pi^2\ln^2 2 - \frac{181}{2160}\pi^4 - \frac{1025}{72}\zeta_3 - \frac{17}{6}\pi^2\ln 2 + \frac{2051}{648}\pi^2 - \frac{1813}{864} = -0.016\,069\,834\ldots \tag{4.36}$$

$$
\begin{aligned}
2a_e(D2) + 2a_e(F2) &= -\frac{43}{36}\pi^2\zeta_3 + \frac{235}{6}\zeta_5 - 4\left(a_4 + \frac{1}{24}\ln^4 2\right) \\
&\quad - \frac{4}{3}\pi^2\ln^2 2 + \frac{79}{240}\pi^4 - \frac{345}{8}\zeta_3 - \frac{119}{18}\pi^2\ln 2 \\
&\quad + \frac{2117}{432}\pi^2 + \frac{1}{6} \\
&= 1.541\,648\,949\ldots
\end{aligned} \tag{4.37}
$$

$$
\begin{aligned}
2a_e(H2) &= +\frac{29}{18}\pi^2\zeta_3 - \frac{275}{12}\zeta_5 - \frac{5}{3}\pi^2\ln^2 2 + \frac{43}{540}\pi^4 + \frac{623}{72}\zeta_3 \\
&\quad + \frac{70}{9}\pi^2\ln 2 - \frac{1951}{324}\pi^2 - \frac{493}{432} \\
&= -1.757\,936\,343\ldots
\end{aligned} \tag{4.38}
$$

$$
\begin{aligned}
2a_e(E2) + 2a_e(G2) &= -\frac{2}{3}\pi^2\zeta_3 + \frac{25}{6}\zeta_5 + \frac{224}{9}\left(a_4 + \frac{1}{24}\ln^4 2\right) \\
&\quad - \frac{65}{54}\pi^2\ln^2 2 - \frac{253}{1080}\pi^4 + \frac{407}{24}\zeta_3 - \frac{32}{3}\pi^2\ln 2 + \frac{9607}{1296}\pi^2 - \frac{485}{432} \\
&= 0.455\,451\,856\ldots
\end{aligned} \tag{4.39}
$$

$$
\begin{aligned}
2a_e(G4) &= -\frac{3}{8}\pi^2\zeta_3 + \frac{95}{24}\zeta_5 - \frac{28}{3}\left(a_4 + \frac{1}{24}\ln^4 2\right) + \frac{37}{18}\pi^2\ln^2 2 \\
&\quad - \frac{43}{432}\pi^4 - \frac{635}{72}\zeta_3 + \frac{83}{18}\pi^2\ln 2 - \frac{4777}{2592}\pi^2 + \frac{1835}{864} \\
&= -0.334\,695\,103\ldots
\end{aligned} \tag{4.40}
$$

$$
\begin{aligned}
&2a_e(A2) + 2a_e(B2) + 2a_e(C2) + 2a_e(D4) = \\
&\quad + \frac{52}{3}\left(a_4 + \frac{1}{24}\ln^4 2\right) + \frac{5}{18}\pi^2\ln^2 2 - \frac{347}{2160}\pi^4 \\
&\quad + \frac{491}{72}\zeta_3 - \frac{29}{18}\pi^2\ln 2 + \frac{3025}{2592}\pi^2 - \frac{3371}{864} \\
&\quad = -0.402\,717\,114\ldots
\end{aligned} \tag{4.41}
$$

$$
\begin{aligned}
a_e(B3) &= +\frac{7}{18}\zeta_3 + \frac{59}{648}\pi^2 + \frac{733}{1728} \\
&= 1.790\,277\,776\ldots
\end{aligned} \tag{4.42}
$$

$$
\begin{aligned}
a_e(C3) + a_e(E3) &= -\frac{40}{3}\left(a_4 + \frac{1}{24}\ln^4 2\right) + \frac{37}{18}\pi^2\ln^2 2 - \frac{49}{1080}\pi^4 \\
&\quad - 10\zeta_3 - \frac{71}{18}\pi^2\ln 2 + \frac{3209}{864}\pi^2 + \frac{251}{288} \\
&= -3.176\,684\,762\ldots
\end{aligned} \tag{4.43}
$$

$$2a_e(G3) = +\frac{95}{72}\pi^2\zeta_3 - \frac{215}{24}\zeta_5 + \frac{160}{9}\left(a_4 + \frac{1}{24}\ln^4 2\right) - \frac{137}{27}\pi^2\ln^2 2 + \frac{41}{180}\pi^4 + \frac{69}{4}\zeta_3 - \frac{101}{18}\pi^2\ln 2 + \frac{2401}{2592}\pi^2 - \frac{3017}{864} = 1.861\,907\,872\ldots \quad (4.44)$$

$$a_e(H3) = -\frac{4}{9}\pi^2\zeta_3 + \frac{5}{12}\zeta_5 + \frac{8}{3}\left(a_4 + \frac{1}{24}\ln^4 2\right) + \frac{32}{9}\pi^2\ln^2 2 - \frac{161}{1080}\pi^4 + \frac{97}{12}\zeta_3 + \frac{20}{9}\pi^2\ln 2 - \frac{1043}{432}\pi^2 - \frac{1}{48} = -0.026\,799\,490\ldots \quad (4.45)$$

4.9. 3-Loop Results

Summing up the contributions from the various graphs described in the previous sections we obtain the total electron anomaly at 3 loops in perturbative QED [2]

$$F_2(0) = \frac{83}{72}\pi^2\zeta_3 - \frac{215}{24}\zeta_5 + \frac{100}{3}\left[\left(a_4 + \frac{1}{24}\ln^4 2\right) - \frac{1}{24}\pi^2\ln^2 2\right] - \frac{239}{2160}\pi^4 + \frac{139}{18}\zeta_3 - \frac{298}{9}\pi^2\ln 2 + \frac{17101}{810}\pi^2 + \frac{28259}{5184} = 1.181\,241\,456\ldots. \quad (4.46)$$

We recall for the convenience of the reader that $\zeta_k = \sum_{n=1}^{\infty} 1/n^k$ denotes as usual the Riemann zeta function of argument k and $a_4 = \sum_{n=1}^{\infty} 1/(2^n n^4)$. For completeness, we give here also $F_1'(0)$ at 3 loops from Ref. [4]

$$F_1'(0) = -\frac{17}{24}\pi^2\zeta_3 + \frac{25}{8}\zeta_5 - \frac{217}{9}\left(a_4 + \frac{1}{24}\ln^4 2\right) - \frac{103}{1080}\pi^2\ln^2 2 + \frac{3899}{25920}\pi^4 - \frac{2929}{288}\zeta_3 + \frac{41671}{2160}\pi^2\ln 2 - \frac{454979}{38880}\pi^2 - \frac{77513}{186624} = 0.171\,720\,018\ldots. \quad (4.47)$$

4.10. Analytic Integration Techniques

It is well known that logarithms, the Euler dilogarithm (or the equivalent Spence function), the Nielsen PolyLogarithms [8, 9] and their further generalization, the Harmonic PolyLogarithms (introduced, however, a few years later in Ref. [10], therefore not yet directly available for the calculation of the 3-loop anomaly), play a special role in the evaluation of the (simplest) Feynman loop integrals. But at any fixed order in perturbation theory the electron anomaly is just a number, the value of a multi-dimensional definite integral, not a function, so that strictly speaking one cannot refer to a specific set of mathematical functions for expressing the result, and the explicit class of functions actually encountered in carrying out the integration depends heavily on the path chosen for attempting the integration.

Historically, almost all calculations were carried out for scalar integrals which are finite in $d = 4$ dimensions, and the results for most of the MI's listed in section 4.11 were in fact obtained by inverting the relations between the already evaluated scalar integrals and their expression in terms of MI's.

Various techniques of increasing power have of course been implemented for the various sets of graphs; we will illustrate in some detail only the technique used for one of the most demanding scalar integrals, evaluated in Ref. [39],

$$
\begin{aligned}
M &= \frac{m^4}{(2\pi)^6} \int \mathcal{D}k \frac{1}{D_1..D_8} \\
&= \frac{m^4}{(2\pi)^6} \int \frac{d^4q}{(q^2+m^2)\,(p-q)^2} \int \frac{d^4r}{(r^2+m^2)\,(p-r)^2} \\
&\times \frac{d^4k}{k^2[(q-k)^2+m^2][(p-q-r+k)^2+m^2][(r-k)^2+m^2]}
\end{aligned}
\tag{4.48}
$$

($\mathcal{D}k$ refers to the 3-loop Euclidean integration variables, the mass shell is at $p^2 = -m^2$, and the D_i stand for the 8 denominators appearing in the formula), which occurs in the triple cross graphs H3 of Fig. 4.2. To be precise, the main Master Integral, occurring in the triple cross graphs only, is

$$
\int \mathcal{D}k \frac{(p\cdot k)}{D_1..D_8} \, , \tag{4.49}
$$

while the integral Eq. (4.48) is not even a Master Integral; but they can be both evaluated with the same technique, which will be discussed here, the integral Eq. (4.48) being marginally simpler and therefore easier to describe than the integral Eq. (4.49).

As a first step, rewrite Eq. (4.48) as

$$M = \frac{1}{(2\pi)^2}\int \frac{d^4q}{(q^2+m^2)\,(p-q)^2} V((p-q)^2, q^2) \,, \tag{4.50}$$

with

$$V((p-q)^2,q^2) = \frac{m^4}{(2\pi)^4}\int \frac{d^4r}{(r^2+m^2)\,(p-r)^2} \times \frac{d^4k}{k^2[(q-k)^2+m^2][(p-q-r+k)^2+m^2][(r-k)^2+m^2]} \; ; \tag{4.51}$$

$V((p-q)^2, q^2)$ is the scalar part of the 2-loop off-mass-shell vertex with external momenta q, p and $(p-q)$, where q^2 corresponds to an electron leg out of mass shell, $p^2 = -m^2$ to the leg on mass shell, so that the dependence of $V((p-q)^2, q^2)$ on p^2 does not need to be recalled explicitly.

At this point one writes a dispersion relation in the variable $(p-q)^2$ at fixed q^2 and p^2,

$$V((p-q)^2,q^2) = \frac{1}{\pi}\int_{4m^2}^{\infty} \frac{dt}{t+(p-q)^2} ImV(-t,q^2) \,, \tag{4.52}$$

so that Eq. (4.50) reads

$$M = \frac{1}{(2\pi)^2}\int_{4m^2}^{\infty} dt \int \frac{d^4q}{(q^2+m^2)(p-q)^2\,[t+(p-q)^2]} \frac{1}{\pi} ImV(-t,q^2) \,.$$

Note that $ImV(-t, q^2)$ depends only on the two variables t and q^2, while all the dependence on $(p \cdot q)$ is explicitly shown in the denominators; it is then convenient to use 4-dimensional hyperspherical coordinates for q,

$$d^4q = \frac{1}{2} q^2 dq^2 \; d\Omega_4(\hat{q})$$

and perform the angular integration. For more complicated hyperspherical integrals one can use the properties of Gegenbauer polynomials, but in our case the angular integration is elementary. For spacelike p one has at once

$$\int \frac{d\Omega_4(\hat{q})}{t+(p-q)^2} = 2\pi^2 \, \frac{p^2+q^2+t-R(t,-q^2,-p^2)}{2p^2q^2} \,,$$

where

$$R(a,b,c) = \sqrt{a^2+b^2+c^2-2ab-2ac-2bc} \tag{4.53}$$

is the familiar square root of the two body relativistic kinematics. When $t > 0$ the continuation $p^2 \to -m^2$ is trivial, while in the limiting case $t \to 0$ the continuation requires some more care, involving also a deformation of the contour of the q^2 integration (see for instance Sec. 3.2 of Ref. [3]). The result can be written as

$$\int_0^\infty q^2\, dq^2 \int \frac{d\Omega_4(\hat{q})}{(p-q)^2} f(q^2) = 2\pi^2 \int_{-m^2}^\infty q^2\, dq^2 \frac{q^2+m^2}{m^2 q^2} f(q^2) \\ -2\pi^2 \frac{1}{m^2} \int_0^\infty q^2\, dq^2 f(q^2) \; , \qquad (4.54)$$

so that M becomes

$$\begin{aligned} M &= \frac{1}{4} M_1 - \frac{1}{8} M_2 - \frac{1}{8} M_3 + \frac{1}{8} M_4 \; , \\ M_1 &= \frac{1}{m^2} \int_{-m^2}^\infty dq^2 \int_{4m^2}^\infty \frac{dt}{t} \frac{1}{\pi} ImV(-t,q^2) \; , \\ M_2 &= \frac{1}{m^2} \int_0^\infty dq^2 \int_{4m^2}^\infty \frac{dt}{t} \frac{1}{\pi} ImV(-t,q^2) \; , \\ M_3 &= \frac{1}{m^2} \int_0^\infty \frac{dq^2}{q^2+m^2} \int_{4m^2}^\infty \frac{dt}{t} R(t,-q^2,m^2) \frac{1}{\pi} ImV(-t,q^2) \; , \\ M_4 &= \frac{1}{m^2} \int_0^\infty \frac{dq^2}{q^2+m^2} \int_{4m^2}^\infty dt \frac{1}{\pi} ImV(-t,q^2) \; . \end{aligned} \qquad (4.55)$$

$ImV(-t,q^2)$, the discontinuity in $(p-q)^2$ of $V((p-q)^2,q^2)$ of Eq. (4.51), consists of 3 contributions, namely a 2-body cut, obtained by cutting the two propagators $[(p-q-r+k)^2+m^2]$ and $[(r-k)^2+m^2]$, and two 3-body cuts, obtained by cutting respectively $[(q-k)^2+m^2], (p-r)^2, [(r-k)^2+m^2]$ and $[(p-q-r+k)^2+m^2], k^2, (r^2+m^2)$. The 3-body contribution and the 2-body contribution in which $[(r-k)^2+m^2]$ is cut are both infrared divergent, but their sum is IR finite (in Ref. [39] the IR divergence was regulated by a small photon mass).

For the analytic integration algorithm to be described here, the essential point is that $ImV(-t,q^2)$ can be written as

$$\frac{1}{\pi} ImV(-t,q^2) = \frac{m^4\; H(t,q^2)}{R(t,-q^2,m^2)\sqrt{(t+q^2+m^2)(t+q^2-3m^2)}} \; , \qquad (4.56)$$

where the (dimensionless) function $H(t,q^2)$ is a *polylogarithmic function of weight 3* (or *a 3-logarithm function* in the terminology of Ref. [39]; the factor m^4 has been introduced for convenience).

By *polylogarithmic function of weight n*, or n-polylogarithm for short, we mean here a function of a set of variables x_i, whose derivatives with respect to any of the x_i is an $(n-1)$-polylogarithm times an *algebraic fraction*, i.e. a fraction whose numerator and denominator are in general two algebraic functions (polynomials and square roots) of the x_i.

The usual logarithm is then an 1-polylogarithm, as its derivative is in general an algebraic fraction; the Euler's dilogarithm $\mathrm{Li}_2(x)$ satisfies

$$\frac{d}{dx}\mathrm{Li}_2(x) = -\frac{1}{x}\ln(1-x)$$

and is therefore a 2-polylogarithm in the present terminology; the product of 2 logarithms is also a 2-polylogarithm etc.

Going back to the structure of the discontinuity of $V((p-q)^2, q^2)$, the relativistic 3-body phase space has 5 independent variables; the integrations on 3 of those variables are easily carried out (thanks to double radicals which simplify as perfect squares of a simple radicals!) the result being a combination of logarithms times algebraic factors (involving square roots), and one is left with the 3-body cut contributions expressed as a combination of double definite integrals. The contribution of the 2-body cut is also expressed in the form of similar double definite integrals.

It turns out that a typical term contributing to $H(t, q^2)$, say $A(x)$ (x can be any of the two variables t, q^2, and we drop the dependence on the other variable for ease of typing), can be written in the form

$$A(x) = \int_{y_1}^{y_2} dy \frac{\partial L(x,y)}{\partial y} B(x,y)\ , \tag{4.57}$$

$$B(x,y) = \int_{z_1}^{z_2} dz \frac{\partial M(x,y,z)}{\partial z} C(x,y,z)\ , \tag{4.58}$$

where $C(x, y, z)$ is a 1-polylogarithm (*i.e.* a logarithm of suitable argument), while $\partial L(x, y)/\partial y$ and $\partial M(x, y, z)/\partial z$ are algebraic fractions (in the sense specified above), equal to the y or z-derivatives of two functions, $L(x, y)$ and $M(x, y, z)$, which are logarithms whose arguments are in turn suitable algebraic fractions. As an example, consider the definite integral

$$D(x) = \int_{y_1}^{y_2} dy \frac{1}{R(x,y,z)(y+a)} f(x,y)\ , \tag{4.59}$$

involving the algebraic fraction $1/[R(x, y, z)(y + a)]$, where $R(x, y, z)$ given as usual by Eq. (4.53), and the (unspecified) function $f(x, y)$. One finds

$$\frac{R(x,-a,z)}{R(x,y,z)(y+a)} = \frac{\partial N(x,y,z)}{\partial y}\ , \tag{4.60}$$

where

$$N(x,y,z) = \frac{1}{2}\ln\frac{ay + (x+z)(y-a) - (x-z)^2 + R(x,-a,z)R(x,y,z)}{ay + (x+z)(y-a) - (x-z)^2 - R(x,-a,z)R(x,y,z)}\ ,$$

so that one can write the algebraic fraction $1/[R(x,y,z)(y+a)]$ in terms of $\partial N(x,y,z)/\partial y$, obtaining

$$R(x,-a,z)\ D(x) = \int_{y_1}^{y_2} dy \frac{\partial N(x,y,z)}{\partial y} f(x,y) \ .$$

The integral Eq. (4.59) has been rewritten in the form of Eq. (4.57) or (4.58) thanks to the introduction of the factor $R(x,-a,z)$, independent of the integration variable y, in front of the integral. (Needless to say, it is not always possible to write an algebraic fraction as the derivative of a logarithm of suitable argument!)

The direct integration of Eq.s(4.57), and (4.58) would be extremely hard, and in fact useless for the rest of the calculation; but one can obtain a convenient expression for the derivative of $A(x)$ directly from the integral representation Eq. (4.57), without evaluating it explicitly. Assuming for simplicity that y_1, y_2 do not depend on x (the generalization would be immediate) one differentiates in x, and after an integration by parts in y one obtains

$$\begin{aligned}\frac{\partial}{\partial x}A(x) &= \frac{\partial L(x,y_2)}{\partial x}\ B(x,y_2) - \frac{\partial L(x,y_1)}{\partial x}\ B(x,y_1) \\ &+ \int_{y_1}^{y_2} dy \left(\frac{\partial L(x,y)}{\partial y}\frac{\partial B(x,y)}{\partial x} - \frac{\partial L(x,y)}{\partial x}\frac{\partial B(x,y)}{\partial y} \right) , \qquad (4.61)\end{aligned}$$

which involves, among other known things, the derivatives of the still unknown function $B(x,y)$. But from the integral representation Eq. (4.58) one obtains directly, as for the derivative of $A(x)$,

$$\begin{aligned}\frac{\partial}{\partial x}B(x,y) &= \frac{\partial M(x,y,z_2)}{\partial x}\ C(x,y,z_2) - \frac{\partial M(x,y,z_1)}{\partial x}\ C(x,y,z_1) \\ &+ \int_{z_1}^{z_2} dz \left(\frac{\partial M(x,y,z)}{\partial z}\frac{\partial C(x,y,z)}{\partial x} \right. \\ &\left. - \frac{\partial M(x,y,z)}{\partial x}\frac{\partial C(x,y,z)}{\partial z} \right) , \qquad (4.62)\end{aligned}$$

and a similar expression for the other derivative $\partial B(x,y)/\partial y$. The explicit evaluation of Eq. (4.62) is much easier than the evaluation of Eq. (4.58), as the integrand contains algebraic functions only (the function $C(x,y,z)$ is, by assumption, a 1-polylogarithm). In the cases which were worked out for the electron anomaly the explicit results of the z-integration are sums of various terms all equal to an algebraic fraction times a 1-polylogarithm. The end-point values are also of the same kind, so that the term $B(x,y)$ is found to be a a 2-polylogarithm; note that the z integration does not need to be carried out in the definition Eq. (4.58) but only in the much simpler Eq. (4.62).

One can then go back to Eq. (4.61). The end-point values $B(x,y_1)$ and $B(x,y_2)$ are usually easy to evaluate; the explicit results obtained for the derivatives of $B(x,y)$ can then be inserted in Eq. (4.61); after some algebra one finds that $\partial A(x)/\partial x$ is the product of an algebraic fraction (in the variable x) times an integral in y, say $A_1(x)$, again of the form of Eq. (4.57), but containing, instead of the 2-polylogarithm $B(x,y)$, just the 1-logarithmic functions coming from its derivatives, while the integration variable z has obviously disappeared at this stage of the calculation.

The whole procedure can be iterated for the x-derivative of $A_1(x)$, which can then be evaluated explicitly and turns out to be a sum of products of algebraic fractions in x times 1-polylogarithms depending also only on x (the integration variables y and z have both disappeared at this point). From the x-derivatives one might try to reconstruct the original functions by quadrature (the simpler the argument of an n-polylogarithm, the easier to work out the quadrature); but it was in fact more convenient to evaluate the original integrals of Eqs. 4.55) and (4.56), integrating repeatedly by parts in t or q^2, so that only the corresponding derivatives of $H(t,q^2)$ are actually required. Integrating by parts the various terms of Eqs. (4.55) and (4.56) is however not immediate. Consider for instance the first term, M_1; substituting Eq. (4.56) indeed gives

$$M_1 = \int_{-m^2}^{\infty} dq^2 \int_{4m^2}^{\infty} \frac{dt}{t} \frac{m^2 H(t,q^2)}{R(t,-q^2,m^2)\sqrt{(t+q^2+m^2)(t+q^2-3m^2)}} ,$$

which contains the product of the two square roots, i.e. the square root of a polynomial of 4th order in both the variables q^2 and t, which cannot be expressed as the derivative of a suitable logarithmic function as in Eq. (4.60).

To proceed, one can change the integration variables (q^2, t) into the pair (u,t), with $u = t + q^2 + m^2$, so that $\sqrt{(t+q^2+m^2)(t+q^2-3m^2)} \to \sqrt{u(u-4m^2)}$ and $R(t,-q^2,m^2) \to \sqrt{u^2-4m^2t}$, with a simpler dependence of the square roots on t. One finds, writing for short $H(t,q^2)$ instead of

$H(t, u - t - m^2)$,

$$M_1 = \int_{4m^2}^{\infty} \frac{du\ m^2}{\sqrt{u(u-4m^2)}} \int_{4m^2}^{u} dt\ \frac{1}{t\sqrt{u^2-4m^2t}} H(t,q^2)$$
$$= \int_{4m^2}^{\infty} du \frac{m^2}{u\sqrt{u(u-4m^2)}} \int_{4m^2}^{u} dt \left(\frac{u}{t\sqrt{u^2-4m^2t}} \right) H(t,q^2)$$
$$= \int_{4m^2}^{\infty} du \frac{1}{2} \left(\frac{\partial}{\partial u} \sqrt{\frac{u-4m^2}{u}} \right) \int_{4m^2}^{u} dt \left(\frac{\partial}{\partial t} K_1(t,u) \right) H(t,q^2) \ ,$$

where

$$K_1(t,u) = -\ln \frac{u+\sqrt{u^2-4m^2t}}{u+\sqrt{u^2-4m^2t}} \ .$$

The t-integral has the form of Eqs.(4.57) and (4.58); integrating by parts in u is trivial, and according to the above discussion one is left with the derivatives of $H(t,q^2)$. Similarly one obtains

$$M_2 = \int_{5m^2}^{\infty} du \frac{1}{2} \left(\frac{\partial}{\partial u} \sqrt{\frac{u-4m^2}{u}} \right) \int_{4m^2}^{u-m^2} dt \left(\frac{\partial}{\partial t} K_1(t,u) \right) H(t,q^2) \ ,$$
$$M_3 = \int_{5m^2}^{\infty} du \frac{1}{2} \left(\frac{\partial}{\partial u} \sqrt{\frac{u-4m^2}{u}} \right) \int_{4m^2}^{u-m^2} dt \left(\frac{\partial}{\partial t} \ln \frac{t}{u-t} \right) H(t,q^2) \ .$$

A similar approach could be followed also for M_4; it is however simpler to observe that

$$\int_{4m^2}^{\infty} dt \frac{1}{\pi} Im V(-t,q^2)$$

is nothing but the coefficient of the $1/(p-q)^2$ term in the expansion of $V((p-q)^2,q^2)$ for large $(p-q)^2$; but that coefficient vanishes, because $V((p-q)^2,q^2)$ itself vanishes in that limit faster than $1/(p-q)^2$, as can be seen from the definition Eq. (4.51).

For completeness we recall here that the analytic value of M, Eq. (4.55), is [39]

$$M = \frac{3}{8}\zeta_2 \ln^2 2 - \frac{3}{32}\zeta_2^2 \ ,$$

while the value of the Master Integral Eq. (4.49) corresponds to the integral I_1 of the next section.

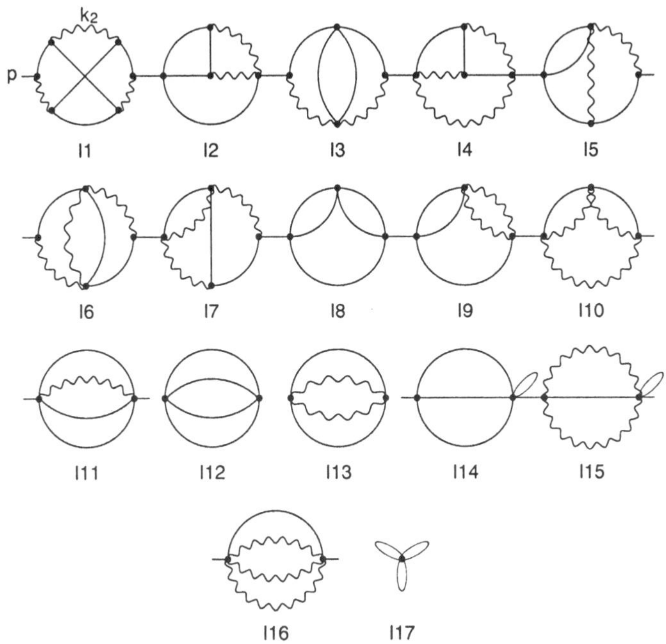

Fig. 4.4. The 17 Master Integrals.

4.11. The Master Integrals

The 17 MI's requested for the evaluation of the magnetic anomaly a_e and the slope $F_1'(0)$ at 3-loop QED, represented in Fig. 4.4 in the form of scalar Feynman graphs, are listed in this section. The conventions and notations used here (which may differ from the rest of the paper) are the following: d is the continuous dimension, the physical limit is $d = 4$ and ϵ is defined as $\epsilon = (4 - d)/2$; as explained at the end of section 4.4, several $1/\epsilon$ factors are present in the expression of the relevant physical quantities in terms of the MI's, so that the terms of the corresponding order in ϵ of the MI's must also be evaluated; $C(\epsilon)$, defined as

$$C(\epsilon) = \left(\pi^{\epsilon}\Gamma(1+\epsilon)\right)^3 \ ,$$

is an overall normalization factor, whose limiting value at $\epsilon = 0$ is 1; as the final physical results have no $1/\epsilon$ singularities, the expansion of $C(\epsilon)$ in ϵ is not needed; C_1, C_2 are two constants, entering in the analytic expression of MI's, but disappearing in the final result, corresponding to

$$C_1 = -\frac{49}{4}\zeta_5 + \frac{25}{12}\pi^2\zeta_3 - \frac{49}{180}\pi^4 + \zeta_3 + 2\pi^2\ln 2 - \frac{\pi^2}{3}\ , \tag{4.63}$$

$$C_2 = -\frac{173}{4}\zeta_5 + \frac{53}{12}\pi^2\zeta_3 - \frac{2}{15}\pi^4 + 18\zeta_3 + 2\pi^2\ln 2 - 3\pi^2\ ; \tag{4.64}$$

the d-dimensional integration loop momenta k_i are Minkoskian; the denominators D_i are, in $m_e = 1$ units,

$$\begin{aligned} D_1 &= (p-k_1)^2 + 1 - i\epsilon\ , & D_2 &= (p-k_1-k_2)^2 + 1 - i\epsilon\ , \\ D_3 &= (p-k_1-k_2-k_3)^2 + 1 - i\epsilon, & D_4 &= (p-k_2-k_3)^2 + 1 - i\epsilon\ , \\ D_5 &= (p-k_3)^2 + 1 - i\epsilon\ , & D_6 &= k_1^2 - i\epsilon\ , \\ D_7 &= k_2^2 - i\epsilon\ , & D_8 &= k_3^2 - i\epsilon\ . \end{aligned}$$

The MI's then are

$$\begin{aligned} I_1 &= \left(\frac{-i}{\pi^{d-2}}\right)^3 \int d^dk_1 d^dk_2 d^dk_3\ \frac{p\cdot k_2}{D_1D_2D_3D_4D_5D_6D_7D_8} \\ &= C(\epsilon)\left[5\zeta_5 - \frac{1}{2}\pi^2\zeta_3 + O(\epsilon)\right]\ , \end{aligned}$$

$$\begin{aligned} I_2 &= \left(\frac{-i}{\pi^{d-2}}\right)^3 \int d^dk_1 d^dk_2 d^dk_3\ \frac{1}{D_1D_2D_3D_4D_7D_8} \\ &= C(\epsilon)\Bigg[\ 2\frac{\zeta_3}{\epsilon} - \frac{13}{90}\pi^4 - \frac{1}{3}\pi^2 + 10\zeta_3 + \epsilon\Bigg(\frac{385}{2}\zeta_5 \\ &\quad -\frac{85}{6}\pi^2\zeta_3 - \frac{7}{15}\pi^4 - 82\zeta_3 - 4\pi^2\ln 2 + 16\pi^2 - 2C_1 \\ &\quad +6C_2\Bigg) + O(\epsilon^2)\Bigg]\ , \end{aligned}$$

$$\begin{aligned} I_3 &= \left(\frac{-i}{\pi^{d-2}}\right)^3 \int d^dk_1 d^dk_2 d^dk_3\ \frac{1}{D_1D_2D_4D_5D_6D_8} \\ &= C(\epsilon)\Bigg[\frac{1}{3\epsilon^3} + \frac{7}{3\epsilon^2} + \frac{31}{3\epsilon} - \frac{2}{15}\pi^4 - \frac{4}{3}\zeta_3 + \frac{103}{3} \\ &\quad +\epsilon\Bigg(95\zeta_5 - \frac{25}{3}\pi^2\zeta_3 - \frac{1}{15}\pi^4 - \frac{184}{3}\zeta_3 - 8\pi^2\ln 2 \\ &\quad +\frac{44}{3}\pi^2 + \frac{235}{3} + 4C_2\Bigg) + O(\epsilon^2)\Bigg]\ , \end{aligned}$$

$$
\begin{aligned}
I_4 &= \left(\frac{-i}{\pi^{d-2}}\right)^3 \int d^dk_1 d^dk_2 d^dk_3 \, \frac{1}{D_2D_3D_4D_6D_7D_8} \\
&= C(\epsilon)\Bigg[2\frac{\zeta_3}{\epsilon} - \frac{7}{90}\pi^4 + 2\zeta_3 + \frac{1}{3}\pi^2 + \epsilon\bigg(\frac{385}{2}\zeta_5 \\
&\quad -\frac{85}{6}\pi^2\zeta_3 - \frac{7}{15}\pi^4 - 82\zeta_3 - 4\pi^2\ln 2 + 16\pi^2 - 2C_1 \\
&\quad +4C_2\bigg) + O(\epsilon^2)\Bigg] ,
\end{aligned}
$$

$$
\begin{aligned}
I_5 &= \left(\frac{-i}{\pi^{d-2}}\right)^3 \int d^dk_1 d^dk_2 d^dk_3 \, \frac{1}{D_1D_3D_4D_5D_7D_8} \\
&= C(\epsilon)\Bigg[\frac{1}{6\epsilon^3} + \frac{3}{2\epsilon^2} + \frac{1}{\epsilon}\left(-\frac{1}{3}\pi^2 + \frac{55}{6}\right) - \frac{4}{45}\pi^4 - \frac{14}{3}\zeta_3 \\
&\quad -\frac{7}{3}\pi^2 + \frac{95}{2} + \epsilon\bigg(-\frac{2}{9}\pi^4 - 44\zeta_3 - \frac{29}{3}\pi^2 + \frac{1351}{6} \\
&\quad +2C_1\bigg) + O(\epsilon^2)\Bigg] ,
\end{aligned}
$$

$$
\begin{aligned}
I_6 &= \left(\frac{-i}{\pi^{d-2}}\right)^3 \int d^dk_1 d^dk_2 d^dk_3 \, \frac{1}{D_1D_3D_5D_6D_7D_8} \\
&= C(\epsilon)\Bigg[\frac{1}{3\epsilon^3} + \frac{7}{3\epsilon^2} + \frac{31}{3\epsilon} - \frac{4}{45}\pi^4 + \frac{2}{3}\zeta_3 + \frac{1}{3}\pi^2 + \frac{103}{3} \\
&\quad +\epsilon\bigg(\frac{45}{2}\zeta_5 - \frac{7}{2}\pi^2\zeta_3 + \frac{11}{45}\pi^4 + \frac{14}{3}\zeta_3 - 4\pi^2\ln 2 \\
&\quad +\frac{14}{3}\pi^2 + \frac{235}{3} + 2C_1\bigg) + O(\epsilon^2)\Bigg] ,
\end{aligned}
$$

$$
\begin{aligned}
I_7 &= \left(\frac{-i}{\pi^{d-2}}\right)^3 \int d^dk_1 d^dk_2 d^dk_3 \, \frac{1}{D_2D_4D_5D_6D_7D_8} \\
&= C(\epsilon)\Bigg[\frac{1}{6\epsilon^3} + \frac{3}{2\epsilon^2} + \frac{1}{\epsilon}\left(-\frac{1}{3}\pi^2 + \frac{55}{6}\right) - \frac{1}{15}\pi^4 - \frac{8}{3}\zeta_3 \\
&\quad -2\pi^2 + \frac{95}{2} + \epsilon\bigg(\frac{45}{2}\zeta_5 - \frac{17}{6}\pi^2\zeta_3 - \frac{7}{9}\pi^4 - 50\zeta_3
\end{aligned}
$$

$$-4\pi^2\ln 2 + \frac{1}{3}\pi^2 + \frac{1351}{6} + 2C_2\Big) + O(\epsilon^2)\Big] \, ,$$

$$I_8 = \left(\frac{-i}{\pi^{d-2}}\right)^3 \int d^dk_1 d^dk_2 d^dk_3 \; \frac{1}{D_1D_2D_3D_4D_5}$$

$$= C(\epsilon)\Big[-\frac{1}{\epsilon^3} - \frac{16}{3\epsilon^2} - \frac{16}{\epsilon} + 2\zeta_3 - \frac{8}{3}\pi^2 - 20$$

$$+\epsilon\Big(-\frac{3}{10}\pi^4 - \frac{200}{3}\zeta_3 + 16\pi^2\ln 2 - 28\pi^2 + \frac{364}{3}\Big)$$

$$+\epsilon^2\Big(-126\zeta_5 + 21\pi^2\zeta_3 + \frac{46}{15}\pi^4 - 512\left(a_4 + \frac{1}{24}\ln^4 2\right)$$

$$-\frac{80}{3}\pi^2\ln^2 2 - 776\zeta_3 + 168\pi^2\ln 2 - 188\pi^2 + 1244\Big)$$

$$+O(\epsilon^3)\Big] \, ,$$

$$I_9 = \left(\frac{-i}{\pi^{d-2}}\right)^3 \int d^dk_1 d^dk_2 d^dk_3 \; \frac{1}{D_2D_3D_5D_6D_7}$$

$$= C(\epsilon)\Big[-\frac{2}{3\epsilon^3} - \frac{10}{3\epsilon^2} + \frac{1}{\epsilon}\Big(-\frac{1}{3}\pi^2 - \frac{26}{3}\Big) - \frac{16}{3}\zeta_3$$

$$-\frac{11}{3}\pi^2 - 2 + \epsilon\Big(-\frac{13}{45}\pi^4 - \frac{248}{3}\zeta_3 + 16\pi^2\ln 2 - \frac{73}{3}\pi^2$$

$$+\frac{398}{3}\Big) + \epsilon^2\Big(-96\zeta_5 - \frac{8}{3}\pi^2\zeta_3 + \frac{3}{5}\pi^4$$

$$-512\left(a_4 + \frac{1}{24}\ln^4 2\right) - \frac{128}{3}\pi^2\ln^2 2 - \frac{1888}{3}\zeta_3$$

$$+160\pi^2\ln 2 - 129\pi^2 + 1038\Big) + O(\epsilon^3)\Big] \, ,$$

$$I_{10} = \left(\frac{-i}{\pi^{d-2}}\right)^3 \int d^dk_1 d^dk_2 d^dk_3 \; \frac{1}{D_2D_4D_6D_7D_8} \tag{4.65}$$

$$= C(\epsilon)\Big[-\frac{1}{3\epsilon^3} - \frac{5}{3\epsilon^2} + \frac{1}{\epsilon}\Big(-\frac{2}{3}\pi^2 - 4\Big) - \frac{26}{3}\zeta_3 - \frac{7}{3}\pi^2$$

$$+\frac{10}{3}+\epsilon\left(-\frac{35}{18}\pi^4-\frac{94}{3}\zeta_3-\pi^2+\frac{302}{3}\right)$$

$$-\epsilon^2\left(-734+\frac{76}{3}\pi^2\zeta_3-\frac{101}{3}\pi^2+20\zeta_3+\frac{551}{90}\pi^4+462\zeta_5\right)+O(\epsilon^3)\Bigg] ,$$

$$I_{11}=\left(\frac{-i}{\pi^{d-2}}\right)^3\int d^dk_1d^dk_2d^dk_3\,\frac{1}{D_1D_3D_5D_7}$$

$$=C(\epsilon)\left[\frac{1}{\epsilon^3}+\frac{7}{2\epsilon^2}+\frac{253}{36\epsilon}+\frac{2501}{216}+\epsilon\left(-\frac{64}{9}\pi^2+\frac{59437}{1296}\right)\right.$$

$$+\epsilon^2\left(-\frac{1792}{9}\zeta_3+\frac{256}{3}\pi^2\ln 2-\frac{2272}{27}\pi^2+\frac{2831381}{7776}\right)$$

$$+\epsilon^3\left(\frac{2752}{135}\pi^4-\frac{8192}{3}\left(a_4+\frac{1}{24}\ln^4 2\right)-\frac{3584}{9}\pi^2\ln^2 2\right.$$

$$\left.-\frac{63616}{27}\zeta_3+\frac{9088}{9}\pi^2\ln 2-\frac{49840}{81}\pi^2+\frac{117529021}{46656}\right)$$

$$\left.+O(\epsilon^4)\right] , \tag{4.66}$$

$$I_{12}=\left(\frac{-i}{\pi^{d-2}}\right)^3\int d^dk_1d^dk_2d^dk_3\,\frac{1}{D_1D_2D_4D_5}$$

$$=C(\epsilon)\left[\frac{2}{\epsilon^3}+\frac{23}{3\epsilon^2}+\frac{35}{2\epsilon}+\frac{275}{12}+\epsilon\left(\frac{112}{3}\zeta_3-\frac{189}{8}\right)\right.$$

$$+\epsilon^2\left(-\frac{136}{45}\pi^4+256\left(a_4+\frac{1}{24}\ln^4 2\right)-\frac{32}{3}\pi^2\ln^2 2\right.$$

$$\left.\left.+280\zeta_3-\frac{14917}{48}\right)+O(\epsilon^3)\right] ,$$

$$I_{13}=\left(\frac{-i}{\pi^{d-2}}\right)^3\int d^dk_1d^dk_2d^dk_3\,\frac{1}{D_3D_5D_6D_7}$$

$$=C(\epsilon)\left[\frac{1}{3\epsilon^3}+\frac{7}{6\epsilon^2}+\frac{25}{12\epsilon}+\frac{8}{3}\zeta_3-\frac{5}{24}+\epsilon\left(-\frac{2}{15}\pi^4\right.\right.$$

$$\left.+\frac{28}{3}\zeta_3-\frac{959}{48}\right)+\epsilon^2\left(48\zeta_5-\frac{7}{15}\pi^4+\frac{50}{3}\zeta_3\right.$$

$$-\frac{10493}{96}\Big) + O(\epsilon^3)\Bigg] ,$$

$$I_{14} = \left(\frac{-i}{\pi^{d-2}}\right)^3 \int d^dk_1 d^dk_2 d^dk_3 \; \frac{1}{D_2 D_3 D_4 D_5}$$

$$= C(\epsilon)\Bigg[\frac{3}{2\epsilon^3} + \frac{23}{4\epsilon^2} + \frac{105}{8\epsilon} + \frac{4}{3}\pi^2 + \frac{275}{16} + \epsilon\Big(28\zeta_3$$

$$-8\pi^2\ln 2 + 10\pi^2 - \frac{567}{32}\Big) + \epsilon^2\Big(-\frac{62}{45}\pi^4$$

$$+192\left(a_4 + \frac{1}{24}\ln^4 2\right) + 16\pi^2\ln^2 2 + 210\zeta_3 - 60\pi^2\ln 2$$

$$+\frac{145}{3}\pi^2 - \frac{14917}{64}\Big) + O(\epsilon^3)\Bigg] ,$$

$$I_{15} = \left(\frac{-i}{\pi^{d-2}}\right)^3 \int d^dk_1 d^dk_2 d^dk_3 \; \frac{1}{D_3 D_4 D_7 D_8}$$

$$= C(\epsilon)\Bigg[\frac{1}{2\epsilon^3} + \frac{7}{4\epsilon^2} + \frac{1}{\epsilon}\left(\frac{1}{3}\pi^2 + \frac{25}{8}\right) + 4\zeta_3 + \frac{7}{6}\pi^2$$

$$-\frac{5}{16} + \epsilon\left(\frac{16}{45}\pi^4 + 14\zeta_3 + \frac{25}{12}\pi^2 - \frac{959}{32}\right)$$

$$+\epsilon^2\Big(72\zeta_5 + \frac{8}{3}\pi^2\zeta_3 + \frac{56}{45}\pi^4 + 25\zeta_3 - \frac{5}{24}\pi^2$$

$$-\frac{10493}{64}\Big) + O(\epsilon^3)\Bigg] ,$$

$$I_{16} = \left(\frac{-i}{\pi^{d-2}}\right)^3 \int d^dk_1 d^dk_2 d^dk_3 \; \frac{1}{D_3 D_6 D_7 D_8}$$

$$= C(\epsilon)\Bigg[-\frac{1}{6\epsilon^2} - \frac{35}{36\epsilon} - \frac{1}{3}\pi^2 - \frac{559}{216} + \epsilon\Big(-\frac{16}{3}\zeta_3$$

$$-\frac{35}{18}\pi^2 + \frac{2737}{1296}\Big) + \epsilon^2\Big(-\frac{37}{45}\pi^4 - \frac{280}{9}\zeta_3$$

$$-\frac{559}{108}\pi^2 + \frac{552041}{7776}\Big) + O(\epsilon^3)\Bigg] .$$

$$
\begin{aligned}
I_{17} &= \left(\frac{-i}{\pi^{d-2}}\right)^3 \int d^dk_1 d^dk_2 d^dk_3 \, \frac{1}{D_1 D_4 D_5} \, . \\
&= C(\epsilon)\left(-\frac{1}{\epsilon^3} - \frac{3}{\epsilon^2} - \frac{6}{\epsilon} - 10 - 15\epsilon - 21\epsilon^2 - 28\epsilon^3 + O(\epsilon^4)\right) \, .
\end{aligned}
$$

References

[1] J. A. Mignaco and E. Remiddi, Nuovo Cimento A **60**, 519 (1969).

[2] S. Laporta and E. Remiddi, Phys. Lett. B **379** (1996) 283 [arXiv:hep-ph/9602417].

[3] R. Z. Roskies, E. Remiddi and M. Levine, in *Quantum Electrodynamics*, edited by T. Kinoshita, Advanced Series on Directions in High Energy Physics, Vol. 7, (World Scientific, Singapore, 1990) 162.

[4] K. Melnikov and T. van Ritbergen, Phys. Rev. Lett.84:1673-1676 (2000).

[5] F. V. Tkachov, Phys. Lett. B **100**, 65 (1981); K. G. Chetyrkin and F. V. Tkachov, Nucl. Phys. B **192**, 159 (1981).

[6] T. Gehrmann and E. Remiddi, Nucl. Phys. B **580**, 485 (2000) [arXiv:hep-ph/9912329].

[7] S. Laporta, Int. J. Mod. Phys. A **15** (2000) 5087 [arXiv:hep-ph/0102033].

[8] N. Nielsen, *Der Eulersche Dilogarithmus und seine Verallgemeinerungen*, Nova Acta Leopoldina (Halle) **90**, 123 (1909).

[9] L. Lewin, it Polylogarithms and Associated Functions, North Holland 1981.

[10] E. Remiddi and J. A. M. Vermaseren, Int. J. Mod. Phys. A **15** (2000) 725 [arXiv:hep-ph/9905237].

[11] M. Veltman, SCHOONSCHIP a CDC 6600 Program for Symbolic Evaluation of Algebraic Expressions, CERN report (1967) unpublished;
M. J. G. Veltman and D. N. Williams, Univ. Michigan preprint UM–TH–91–18 (1991).

[12] M. J. Levine, U.S. AEC Report No. CAR-882-25 (1971), unpublished.

[13] J. A. M. Vermaseren, "New features of FORM," arXiv:math-ph/0010025.

[14] W. Pauli and F. Villars, Rev. Mod. Phys. **21**, 434 (1949).

[15] G. 't Hooft and M. J. G. Veltman, Nucl. Phys. B **44** (1972) 189.

[16] R. Karplus and N. M. Kroll, Phys. Rev. **77**, 536 (1950), where $a_e^{(1,1)}(\text{vp})$ of our Eq. (4.17) is referred to as μ^{IIe}, Eq.(53); note that in the last section of the paper, the summary, the same result is copied with by a typing error. Let us recall here that contributions $\mu^I + \mu^{IIc}$ were on the contrary incorrect, as pointed out by C. M. Sommerfield, Phys. Rev. **107**, 328 (1957) and A. Petermann, Helv. Phys. Acta **30**, 407 (1957).

[17] G. Källen and A. Sabry, *Mat. Fys. Medd. Dans. Vid. Selsk.* **29**, no.17 (1955).

[18] R. Barbieri and E. Remiddi, Nucl. Phys. B **90** (1975) 233.

[19] The leading term of the expansion was first given in B. E. Lautrup and E. de Rafael, Phys. Rev. **174**, 1835 (1968); an formula exact in (m_e/m_μ) is contained in Glen W. Erickson and Henry H. T. Liu, UCD-CNL-81 report (1968).

[20] S. Laporta, Nuovo Cim. A **106** (1993) 675.
[21] H. Suura and E. Wichmann, Phys. Rev. **105** 1930 (1957), A. Petermann, Phys. Rev. **105** 1931 (1957).
[22] J. Aldins, T. Kinoshita, S. J. Brodsky and A. J. Dufner, Phys. Rev. D **1** (1970) 2378.
[23] S. Laporta and E. Remiddi, Phys. Lett. B **265**, 182 (1991).
[24] S. Laporta and E. Remiddi, Phys. Lett. B **301**, 440 (1993).
[25] M. J. Levine and R. Roskies, Phys. Rev. D **9**, 421 (1974).
[26] K. A. Milton, W. Tsai and L. L. De Raad, Jr., Phys. Rev. D **9**, 1809 (1974).
[27] R. Barbieri, M. Caffo and E. Remiddi, Phys. Lett. B**57**, 460 (1975).
[28] R. Barbieri, M. Caffo, E. Remiddi, S. Turrini and D. Oury, Nuclear Physics B **144**, 329 (1978).
[29] M. J. Levine, R. C. Perisho and R. Roskies, Phys. Rev. D **13**, 997 (1976).
[30] M. J. Levine and R. Roskies, Phys. Rev. D **14**, 2191 (1976).
[31] M. J. Levine, E. Remiddi and R. Roskies, Phys. Rev. D**20**, 2068 (1979).
[32] S. Laporta, Phys. Rev. D **47**, 4793 (1993).
[33] S. Laporta, Phys. Lett. B **343**, 421 (1995).
[34] D. Billi, M. Caffo and E. Remiddi, Nuovo Cimento Lett. **4**, 657 (1972).
[35] R. Barbieri, M. Caffo and E. Remiddi, Nuovo Cimento Lett. **9**, 690 (1974).
[36] K. A. Milton, W. Tsai and L. L. De Raad, Jr., Phys. Rev. D **9**, 1814 (1974).
[37] R. Barbieri, M. Caffo and E. Remiddi, Nuovo Cimento Lett. **5**, 769 (1972).
[38] R. Barbieri and E. Remiddi, Phys. Lett. B **49**, 468 (1974).
[39] S. Laporta and E. Remiddi, Phys. Lett. B **356** (1995) 390.

Chapter 5

Measurements of the Electron Magnetic Moment

G. Gabrielse

Department of Physics, Harvard University
17 Oxford Street, Cambridge, MA 02138

New measurements determine the electron magnetic moment in Bohr magnetons, $g/2 = 1.001\,159\,652\,180\,73\,(28)$ [0.28 ppt]. The uncertainty is 15 times smaller than for the measurement that had stood for 20 years, and the value is shifted by 1.7 standard deviations. The cyclotron and spin states of a single trapped electron are fully resolved thanks to a cylindrical Penning trap cavity at 100 mK, cavity-modified radiation fields, inhibited spontaneous emission, and a one-particle self-excited oscillator. The new $g/2$ and QED theory determine the fine structure constant, $\alpha^{-1} = 137.035\ 999\ 084\ (51)$ [0.37 ppb], more than an order of magnitude more accurately than any independent determination.

Contents

5.1. Introduction and Overview

Measurements of the electron magnetic moment ($\boldsymbol{\mu}$) probe the electron's interaction with the fluctuating vacuum described by quantum electrodynamics (QED). They also probe for possible electron substructure that is not part of the Standard Model of particle physics. As an eigenstate of spin $\mathbf{S}$, the electron (charge $-e$ and mass m) has $\boldsymbol{\mu} \propto \mathbf{S}$,

$$\boldsymbol{\mu} = -\frac{g}{2}\frac{e\hbar}{2m}\frac{\mathbf{S}}{\hbar/2}. \tag{5.1}$$

The g-value is a dimensionless measure of the moment, with the dimensions and approximate size given by the Bohr magneton, $e\hbar/(2m)$. Thus $g/2$ is the magnetic moment in units of Bohr magnetons for a spin 1/2 particle like an electron or muon.

If the electron was a mechanical system, and spin was an orbital angular momentum, then g would characterize the relative distributions of the rotating charge and mass, with $g = 1$ for identical distributions. (Cyclotron motion of a charge in a magnetic field B, at frequency $\nu_c = eB/(2\pi m)$, is one example.) A Dirac point particle has $g = 2$, the leading term in

$$g/2 = 1 + a_{QED}(\alpha) + a_{hadronic} + a_{weak} + a_{new}. \tag{5.2}$$

QED predicts that vacuum fluctuations and polarization slightly increase this value by the small "anomaly" $a_{QED}(\alpha) \approx 10^{-3}$ that is a function of the fine structure constant α. Hadronic and weak interactions are calculated within the Standard Model to be very small and negligible, respectively. Electron substructure (or other deviations from the Standard Model) would make $g/2$ deviate by a_{new} from the Dirac/QED prediction, as quark-gluon substructure does for a proton.

Why measure the electron $g/2$? The motivations include:

(1) The electron $g/2$ is the property that can be most accurately measured for an important ingredient of our universe, an unusual particle that is predicted to have no internal structure.

(2) The most stringent test of QED comes from measuring $g/2$ and comparing to the value $g(\alpha)$ calculated using an independently determined α in Eq. (5.2).
(3) The most accurate determination of the fine structure constant, by more than an order of magnitude, comes from solving Eq. (5.2) for α in terms of the measured $g/2$. (No physics beyond the Standard Model, i.e. $a_{new} = 0$, is assumed.)
(4) A search for physics beyond the Standard Model (e.g. electron substructure) comes from using the best measurement of $g/2$ and the best independent α (with calculated values of $a_{hadronic}$ and a_{weak}) in Eq. (5.2) to set a limit on a_{new}.
(5) Comparing $g/2$ for an electron and a positron is the most stringent test of CPT invariance with leptons.

Owing to the great importance of the dimensionless magnetic moment, there have been many measurements of the electron $g/2$. A long list of measurements of this fundamental quantity has been compiled [1]. Worthy of special mention is a long series of measurements at the Univ. of Michigan [2]. The spin precession relative to the cyclotron rotation of keV electrons was measured. Also worthy of special mention is the series of measurements at the Univ. of Washington [3]. In the end these measurements [4] used a single electron trapped in a hyperbolic Penning trap.

New Harvard measurements determine the electron magnetic moment [5, 7] to a much higher accuracy than do previous measurements. The most recent in the long history of applying new methods to measuring $g/2$, they supersede the UW measurement that stood for about 20 years [4]. The uncertainty is 15 times lower and the measured value is shifted by 1.7 standard deviations (Fig. 5.1).

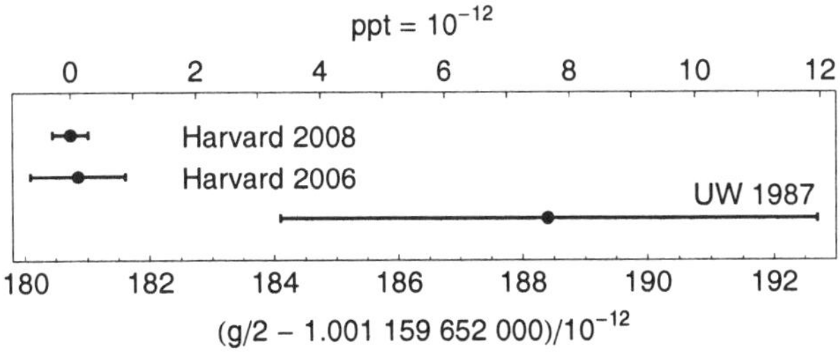

Fig. 5.1. Most accurate measurements of the electron $g/2$.

The substantially higher accuracy of the new measurements was the result of new experimental methods, developed and demonstrated one thesis at a time over 20 years by a string of excellent Ph.D. students – C. H. Tseng, D. Enzer, J. Tan, S. Peil, B. D'Urso, B. Odom and D. Hanneke. Progress continues in the ongoing work of Ph.D. students S. Fogwell and J. C. Dorr. The unifying idea for the new methods was that of a one-electron quantum cyclotron – with fully resolved cyclotron and spin energy levels, and a detection sensitivity sufficient to detect one quantum transitions. The new methods included:

(1) A cylindrical Penning trap was used to suspend the electron. The cylindrical trap was invented to form a microwave cavity that could inhibit spontaneous emission. The calculable cavity shape made it possible to understand and correct for cavity shifts of the measured cyclotron frequency.
(2) Cavity-inhibited spontaneous emission (by a factor of up to 250) narrowed measured linewidths and gave us the crucial averaging time that we needed to resolve one-quantum changes in the electron's cyclotron state.
(3) The cavity was cooled to 100 mK rather than to 4.2 K so that in thermal equilibrium the electron's cyclotron motion would be in its ground state.
(4) Detection with good signal-to-noise ratio came from feeding back a signal derived from the electron's motion along the magnetic field to the electron to cancel the damping due to the detection impedance. The "classical measurement system" for the quantum cyclotron motion was this large self-excited motion of the electron, with a quantum nondemolition coupling between the classical and quantum systems.
(5) A silver trap cavity avoided the magnetic field variations due to temperature fluctuations of the paramagnetism of conventional copper trap electrodes.
(6) The measurement was entirely automated so that the best data could be taken at night, when the electrical, magnetic and mechanical disturbances were lowest, with no person present.
(7) A parametric excitation of electrons suspended in the trap was used to measure the radiation modes of the radiation field in the trap cavity.

(8) The damping rate of a single trapped electron was used as a second probe of the radiation fields within the trap cavity.

5.2. One-Electron Quantum Cyclotron

5.2.1. *A homemade atom*

A one-electron quantum cyclotron is a single electron suspended within a magnetic field, with the quantum structure in its cyclotron motion fully resolved. Accurate measurements of the resonant frequencies of driven transitions between the energy levels of this homemade atom – an electron bound to our trap – reveals the electron magnetic moment in units of Bohr magnetons, $g/2$. The energy levels and what must be measured to determine $g/2$ are presented in this section. The experimental devices and methods needed to realize the one-electron quantum cyclotron are discussed in following sections.

A nonrelativistic electron in a magnetic field has energy levels

$$E(n, m_s) = \tfrac{g}{2} h\nu_c m_s + (n + \tfrac{1}{2}) h\nu_c. \tag{5.3}$$

These depend in a familiar way upon the electron's cyclotron frequency ν_c and its spin frequency $\nu_s \equiv (g/2)\nu_c$. The electron $g/2$ is thus specified by the two frequencies,

$$\frac{g}{2} = \frac{\nu_s}{\nu_c} = 1 + \frac{\nu_s - \nu_c}{\nu_c} = 1 + \frac{\nu_a}{\nu_c}, \tag{5.4}$$

or equivalently by their difference (the anomaly frequency $\nu_a \equiv \nu_s - \nu_c$) and ν_c. Because ν_s and ν_c differ by only a part-per-thousand, measuring ν_a and ν_c to a precision of 1 part in 10^{10} gives $g/2$ to 1 part in 10^{13}.

Although one electron suspended in a magnetic field will not remain in one place long enough for a measurement, two features of determining $g/2$ by measuring ν_a and ν_c are apparent in Eq. (5.4).

(1) Nothing in physics can be measured more accurately than a frequency (the art of timekeeping being so highly developed) except for a ratio of frequencies.
(2) Although both of these frequencies depend upon the magnetic field, the field dependence drops out of the ratio. The magnetic field thus needs to be stable only on the time scale on which both frequencies can be measured, and no absolute calibration of the magnetic field is required.

To confine the electron for precise measurements, an ideal Penning trap includes an electrostatic quadrupole potential $V \sim z^2 - \frac{1}{2}\rho^2$ with a magnetic field $B\hat{\mathbf{z}}$ [7]. This potential shifts the cyclotron frequency from the free-space value ν_c to $\bar{\nu}_c$. The latter frequency is also slightly shifted by the unavoidable leading imperfections of a real laboratory trap – a misalignment of the symmetry axis of the electrostatic quadrupole and the magnetic field, and quadratic distortions of the electrostatic potential.

The lowest cyclotron energy levels (with quantum numbers $n = 0, 1, \ldots$) and the spin energy levels (with quantum numbers $m_s = \pm 1/2$) are given by

$$E(n, m_s) = \frac{g}{2} h\nu_c m_s + (n + \tfrac{1}{2})h\bar{\nu}_c - \tfrac{1}{2}h\delta(n + \tfrac{1}{2} + m_s)^2. \qquad (5.5)$$

The lowest cyclotron and spin energy levels are represented in Fig. 5.2.

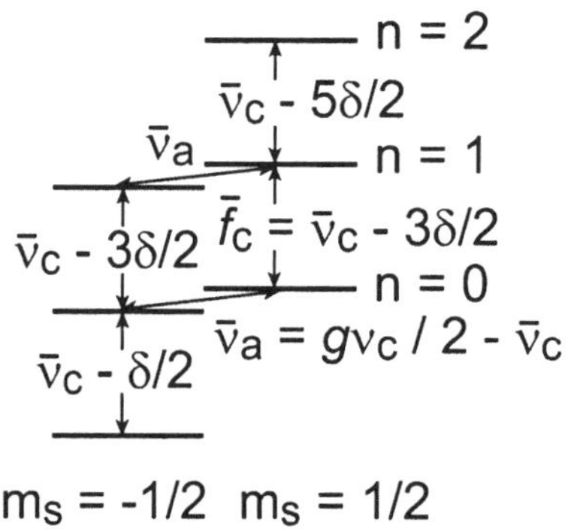

Fig. 5.2. Lowest cyclotron and spin levels of an electron in a Penning trap.

Special relativity is important for even the lowest quantum levels. The third term in Eq. (5.5) is the leading relativistic correction [7] to the energy levels. Special relativity makes the transition frequency between two cyclotron levels $|n, m_s\rangle \leftrightarrow |n+1, m_s\rangle$ decrease from $\bar{\nu}_c$ to $\bar{\nu}_c + \Delta\bar{\nu}_c$, with the shift

$$\Delta\bar{\nu}_c = -\delta(n + 1 + m_s) \qquad (5.6)$$

depending upon the spin state and cyclotron state. This very small shift, with

$$\delta/\nu_c \equiv h\nu_c/(mc^2) \approx 10^{-9}, \qquad (5.7)$$

is nonetheless significant at our precision. An important new feature of our measurement is that special relativity adds no uncertainty to our measurements. Quantum transitions between identified energy levels with a precisely known relativistic contribution to the energy levels are resolved. When only the average cyclotron frequency of an unknown distribution of cyclotron states was all that can be measured [4], figuring out the size of the relativistic frequency shift was difficult.

We have seen how $g/2$ is determined by the anomaly frequency ν_a and the free-space cyclotron frequency $\nu_c = eB/(2\pi m)$. However, neither of these frequencies is an eigenfrequency of the trapped electron. We actually measure the transition frequencies

$$\bar{f}_c \equiv \bar{\nu}_c - \frac{3}{2}\delta \tag{5.8}$$

$$\bar{\nu}_a \equiv \frac{g}{2}\nu_c - \bar{\nu}_c \tag{5.9}$$

represented by the arrows in Fig. 5.5 for an electron initially prepared in the state $|n = 0, m_s = 1/2\rangle$.

The needed $\nu_c = eB/(2\pi m)$ is deduced from the three observable eigenfrequencies of an electron bound in the trap by the Brown–Gabrielse invariance theorem [8],

$$(\nu_c)^2 = (\bar{\nu}_c)^2 + (\bar{\nu}_z)^2 + (\bar{\nu}_m)^2. \tag{5.10}$$

The three measurable eigenfrequencies on the right include the cyclotron frequency $\bar{\nu}_c$ for the quantum cyclotron motion we have been discussing. The second measurable eigenfrequency is the axial oscillation frequency $\bar{\nu}_z$ for the nearly-harmonic, classical electron motion along the direction of the magnetic field. The third measurable eigenfrequency is the magnetron oscillation frequency for the classical magnetron motion along the circular orbit for which the electric field of the trap and the motional electric field exactly cancel.

The invariance theorem applies for a perfect Penning trap, but also in the presence of the mentioned imperfection shifts of the eigenfrequencies for an electron in a trap. This theorem, together with the well-defined hierarchy of trap eigenfrequencies, $\bar{\nu}_c \gg \bar{\nu}_z \gg \bar{\nu}_m \gg \delta$, yields an approximate expression that is sufficient at our accuracy. We thus determine the electron $g/2$ using

$$\frac{g}{2} = \frac{\bar{\nu}_c + \bar{\nu}_a}{\nu_c} \simeq 1 + \frac{\bar{\nu}_a - \bar{\nu}_z^2/(2\bar{f}_c)}{\bar{f}_c + 3\delta/2 + \bar{\nu}_z^2/(2\bar{f}_c)} + \frac{\Delta g_{cav}}{2}. \tag{5.11}$$

The cavity shift $\Delta g_{cav}/2$ that arises from the interaction of the cyclotron motion and the trap cavity is presently discussed in detail.

5.2.2. *Cylindrical penning trap cavity*

A cylindrical Penning trap (Fig. 5.3) is the key device that makes these measurements possible. It was invented [9] and demonstrated [10] to provide boundary conditions that produce a controllable and understandable radiation field within the trap cavity, along with the needed electrostatic quadrupole potential. Spontaneous emission can be significantly inhibited at the same time as corresponding shifts of the electron's oscillation frequencies are avoided. We shall see that this is critical to the new Harvard measurements in several ways.

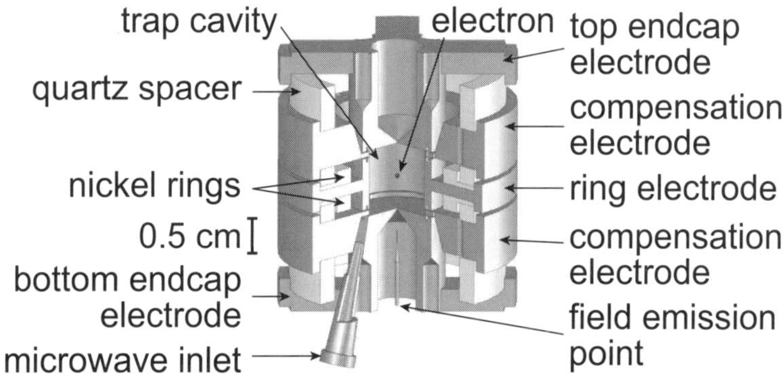

Fig. 5.3. Cylindrical Penning trap cavity used to confine a single electron and inhibit spontaneous emission.

A necessary function of the trap electrodes is to produce a very good approximation to an electrostatic quadrupole potential. This is possible with cylindrical electrodes but only if the relative geometry of the electrodes is carefully chosen [9]. The electrodes of the cylindrical trap are symmetric under rotations about the center axis ($\hat{\mathbf{z}}$), which is parallel to the spatially uniform magnetic field ($B\hat{\mathbf{z}}$). The potential (about 100 V) applied between the endcap electrodes and the ring electrode provides the basic trapping potential and sets the axial frequency $\bar{\nu}_z$ of the nearly harmonic oscillation of the electron parallel to the magnetic field. The potential applied to the compensation electrodes is adjusted to tune the shape of the potential, to make the oscillation as harmonic as possible. The tuning does not change $\bar{\nu}_z$ very much owing to an orthogonalization [11, 30] that arises from the geometry choice. What we found was that one electron could be observed within a cylindrical Penning trap with as good or better signal-to-noise ratio than was realized in hyperbolic Penning traps.

Table 5.1. Properties of the trapped electron.

Cyclotron frequency	$\omega_c/(2\pi)$	150 GHz
Trap-modified cyc. freq.	$\omega_+/(2\pi)$	150 GHz
Axial frequency	$\omega_z/(2\pi)$	200 MHz
Magnetron frequency	$\omega_-/(2\pi)$	133 kHz
Cyclotron damping (free space)	τ_+	0.09 s
Axial damping	τ_z	30 ms
Magnetron damping	τ_-	10^9 yr

The principle motivation for the cylindrical Penning trap is to form a microwave cavity whose radiation properties are well understood and controlled – the best possible approximation to a perfect cylindrical trap cavity. (Our calculation attempts with a hyperbolic trap cavity were much less successful [12].) The modes of the electromagnetic radiation field that are consistent with this boundary condition are the well-known transverse electric TE_{mnp} and transverse magnetic TM_{mnp} modes (see e.g. [13, Sec. 8.7]).

For a right circular cylinder of diameter $2\rho_0$ and height $2z_0$ the TE and TM modes have characteristic resonance frequencies,

$$TE: \quad \omega_{mnp} = c\sqrt{\left(\frac{x'_{mn}}{\rho_0}\right)^2 + \left(\frac{p\pi}{2z_0}\right)^2} \tag{5.12a}$$

$$TM: \quad \omega_{mnp} = c\sqrt{\left(\frac{x_{mn}}{\rho_0}\right)^2 + \left(\frac{p\pi}{2z_0}\right)^2}. \tag{5.12b}$$

They are indexed with integers

$$m = 0, 1, 2, \cdots \tag{5.13}$$

$$n = 1, 2, 3, \cdots \tag{5.14}$$

$$p = 1, 2, 3, \cdots, \tag{5.15}$$

and are functions of the nth zeros of Bessel functions and their derivatives

$$J_m(x_{mn}) = 0 \tag{5.16}$$

$$J'_m(x'_{mn}) = 0. \tag{5.17}$$

The zeros force the boundary conditions at the cylindrical wall. All but the $m = 0$ modes are doubly degenerate.

Of primary interest is the magnitude of the cavity electric fields that couple to the cyclotron motion of an electron suspended in the center of the trap. For both TE and TM modes, the transverse components of **E** are proportional to

$$\sin(\tfrac{p\pi}{2}(\tfrac{z}{z_0}+1)) = \begin{cases} (-1)^{p/2}\sin(\frac{p\pi z}{2z_0}) & \text{for even } p, \\ (-1)^{(p-1)/2}\cos(\frac{p\pi z}{2z_0}) & \text{for odd } p. \end{cases} \tag{5.18}$$

For an electron close to the cavity center, ($z \approx 0$), only modes with odd p thus have any appreciable coupling.

The transverse components of the electric fields are also proportional to either the order-m Bessel function times m/ρ for the TE modes, or to the derivative of the order-m Bessel function for the TM modes. Close to the cavity center ($\rho \approx 0$),

$$\frac{m}{\rho}\ J_m(x_{mn}^{(\prime)}\tfrac{\rho}{\rho_0}) \sim \begin{cases} \dfrac{\rho^{m-1}}{(m-1)!}\left(\dfrac{x_{mn}^{(\prime)}}{2\rho_0}\right)^m & \text{for } m>0 \\ 0 & \text{for } m=0 \end{cases} \tag{5.19a}$$

$$\frac{x_{mn}^{(\prime)}}{\rho_0} J_m'(x_{mn}^{(\prime)}\tfrac{\rho}{\rho_0}) \sim \begin{cases} \dfrac{\rho^{m-1}}{(m-1)!}\left(\dfrac{x_{mn}^{(\prime)}}{2\rho_0}\right)^m & \text{for } m>0 \\ -\dfrac{x_{0n}^{(\prime)2}}{2\rho_0^2}\rho & \text{for } m=0. \end{cases} \tag{5.19b}$$

In the limit $\rho \to 0$, all but the $m = 1$ modes vanish.

For a perfect cylindrical cavity the only radiation modes that couple to an electron perfectly centered in the cavity are $\mathrm{TE}_{1n(\mathrm{odd})}$ and $\mathrm{TM}_{1n(\mathrm{odd})}$. If the electron is moved slightly off-center axially it will begin to couple to radiation modes with $mnp = 1n(\mathrm{even})$. If the electron is moved slightly off-center radially it similarly begins to couple to modes with $m \neq 1$.

In the real trap cavity, the perturbation caused by the small space between the electrodes is minimized by the use of "choke flanges" – small channels that tend to reflect the radiation leaking out of the trap back to cancel itself, and thus to minimize the losses from the trap. The measured radiation modes, discussed later, are close enough to the calculated frequencies for a perfect cylindrical cavity that we have been able to identify more than 100 different radiation modes for such trap cavities [14–16]. The spatial properties of the electric and magnetic field for the radiation that builds up within the cavity are thus quite well understood. Some of the modes couple to cyclotron motion of an electron centered in the cavity, others couple to the spin of a centered electron, and still others have the symmetry that we hope will one day allow us to sideband-cool the axial motion.

5.2.3. *100 mK and 5 T*

Detecting transitions between energy levels of the quantum cyclotron requires that the electron-bound-to-the-trap system be prepared in a definite quantum state. Two key elements are a high magnetic field, and a low temperature for the trap cavity. A high field makes the spacing of the cyclotron energy levels to be large. A high field and low temperature make a very large Boltzmann probability to be in the lowest cyclotron state, $P \propto \exp[-h\bar{\nu}_c/(kT)]$, which is negligibly different from unity.

The trap cavity is cooled to 0.1 K or below via a thermal contact with the mixing chamber of an Oxford Instruments Kelvinox 300 dilution refrigerator (Fig. 5.4). The electrodes of this trap cavity are housed within a separate vacuum enclosure that is entirely at the base temperature. Measurements on an apparatus with a similar design but at 4.2 K found the vacuum in the enclosure to be better than 5×10^{-17} torr [17]. Our much lower temperature make our background gas pressure much lower. We are able

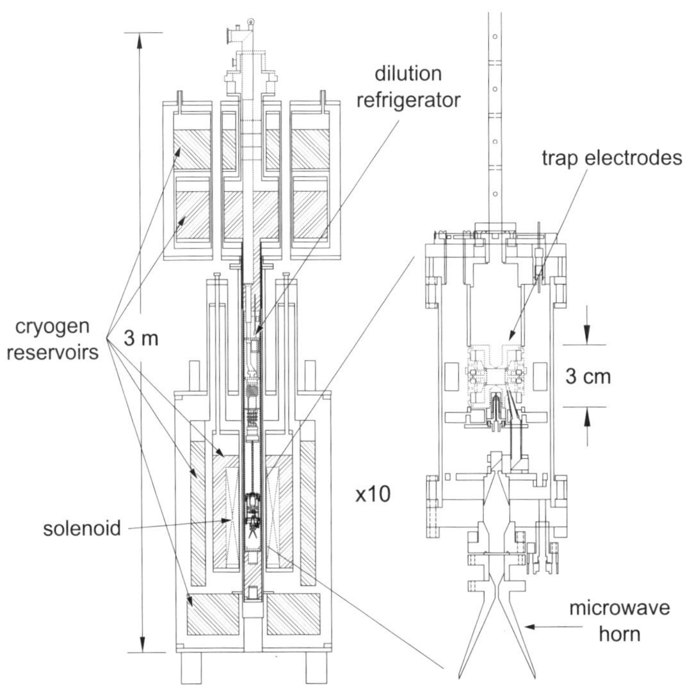

Fig. 5.4. The apparatus includes a trap electrodes near the central axis, surrounded by a superconducting solenoid. The trap is suspended from a dilution refrigerator.

to keep one electron suspended in our apparatus for as long as desired – regularly months at a time. Substantial reservoirs for liquid helium and liquid nitrogen make it possible to keep the trap cold for five to seven days before the disruption of adding more liquid helium or nitrogen is required.

The trap and its vacuum container is located within a superconducting solenoid (Fig. 5.4) that makes a very homogeneous magnetic field over the interior volume of the trap cavity. A large dewar sitting on top of the solenoid dewar provides the helium needed around the dilution refrigerator below. The superconducting solenoid is entirely self-contained, with a bore that can operate from room temperature down to 77 K. It possesses shim coils capable of creating a field homogeneity better than a part in 10^8 over a 1 cm diameter sphere and has a passive "shield" coil that reduces fluctuations in the ambient magnetic field [18, 19]. When properly energized (and after the steps described in the next section have been taken) it achieves field stability better than a part in 10^9 per hour. We regularly observe drifts below 10^{-9} per night.

5.2.4. *Stabilizing the energy levels*

Measuring the electron $g/2$ with a precision of parts in 10^{13} requires that the energy levels of our homemade atom, an electron bound to a Penning trap, be exceptionally stable. The energy levels depend upon the magnetic field and upon the potential that we apply to the trap electrodes. The magnetic field must be stable at least on the timescale that is required to measure the two frequencies, $\bar{f}_c$ and $\bar{\nu}_a$, that are both proportional to the magnetic field.

One defense against external field fluctuations is a high magnetic field. This makes field fluctuations due to outside sources to be relatively smaller. The largest source of ambient magnetic noise is a subway that produces 50 nT (0.5 mG, 10 ppb) fluctuations in our lab and that would limit us to four hours of data taking per day (when the subway stops running) if we did not shield the electron from them. Eddy currents in the high-conductivity aluminum and copper cylinders of the dewars and the magnet bore shield high-frequency fluctuations [20]. For slower fluctuations, the aforementioned shelf-shielding solenoid [18] has the correct geometry to make the central field always equal to the average field over the solenoid cross-section. This translates flux conservation into central-field conservation, shielding external fluctuations by more than a factor of 150 [19].

Stabilizing the field produced by the solenoid requires that care is taken when the field value is changed, since changing the current in the solenoid alters the forces between windings. Resulting stresses can take months to stabilize if the coil is not pre-stressed by "over-currenting" the magnet. Our recipe is to overshoot the target value by a few percent of the change, undershoot by a similar amount, and then move to the desired field.

The apparatus in Fig. 5.4 evolved historically rather than being designed for maximum magnetic field stability in the final configuration. Because the solenoid and the trap electrodes are suspended from widely separated support points, temperature and pressure changes can cause the electrodes to move relative to the solenoid. Apparatus vibrations can do the same insofar as the magnetic field is not perfectly homogeneous, despite careful adjusting of the persistent currents in ten superconducting shim coils. Any relative motion of the electron and solenoid changes the field seen by the electron.

To counteract this, we regulate the five He and N_2 pressures in the cryostats to maintain the temperature of both the bath and the solenoid itself [21, 22]. Recently we also relocated the dilution refrigerator vacuum pumps to an isolated room at the end of a 12 m pipe run. This reduced vibration by more than an order of magnitude at frequencies related to the pump motion and reduced the noise level for the experimenters but did not obviously improve the $g/2$ data.

Because some of the structure establishing the relative location of the trap electrodes and the solenoid is at room temperature, changes in room temperature can move the electron in the magnetic field. The lab temperature routinely cycles 1–2 K daily, so we house the apparatus in a large, insulated enclosure within which we actively regulate the air temperature to 0.1 K. A refrigerated circulating bath (ThermoNeslab RTE-17) pumps water into the regulated zone and through an automobile transmission fluid radiator, heating and cooling the water to maintain constant air temperature. Fans couple the water and air temperatures and keep a uniform air temperature throughout.

The choice of materials for the trap electrodes and its vacuum container is also crucial to attaining high field stability [5, 23]. Copper trap electrodes, for example, have a nuclear paramagnetism at 0.1 mK that makes the electron see a magnetic field that changes at an unacceptable level with very small changes in trap temperature. We thus use only low-Curie-constant materials such as silver, quartz, titanium, and molybdenum at the refrigerator base temperature and we regulate the mixing chamber temperature to 1 mK or better.

A stable axial frequency is also extremely important since small changes in the measured axial frequency reveal one-quantum transitions of the cyclotron and spin energy (as will be discussed in Sec. 5.3.1). A trapping potential without thermal fluctuations is provided by a charged capacitor ($10\mu F$) that has a very low leakage resistance at low temperature. We add to or subtract from the charge on the capacitor using 50 ms current pulses sent to the capacitor through a 100 $M\Omega$ resistor as needed to keep the measured axial frequency constant. Because of the orthogonalized trap design [9] already discussed, the potential applied to the compensation electrodes (to make the electron see as close to a pure electrostatic quadrupole potential as possible) has little effect upon the axial frequency.

5.2.5. *Motions and damping of the suspended electron*

We load a single electron using an electron beam from a sharp tungsten field emission tip. A hole in the bottom endcap electrode admits the beam, which hits the top endcap electrode and releases gas atoms cryopumped on the surface. Collisions between the beam and gas atoms eventually cause an electron to fall into the trap. Adjusting the beam energy and the time it is on determines the number of electrons loaded.

The electron has three motions in the Penning trap formed by the $B = 5.4$ T magnetic field, and the electrostatic quadrupole potential. The cyclotron motion in the trap has a cyclotron frequency $\bar{\nu}_c \approx 150$ GHz. The axial frequency, for the harmonic oscillator parallel to the magnetic field direction, is $\bar{\nu}_z \approx 200$ MHz. A circular magnetron motion, perpendicular to $\mathbf{B}$, has an oscillation frequency, $\bar{\nu}_m \approx 133$ kHz. The spin precession frequency, which we do not measure directly, is slightly higher than the cyclotron frequency. The frequency difference is the anomaly frequency, $\bar{\nu}_a \approx 174$ MHz, which we do measure directly.

The undamped spin motion is essentially uncoupled from its environment [7]. The cyclotron motion is only weakly damped. By controlling the cyclotron frequency relative to that of the cavity radiation modes, we alter the density of radiation states and inhibit the spontaneous emission of synchrotron radiation [7, 24] by 10 to 50 times the $(90\text{ ms})^{-1}$ free-space rate. Blackbody photons that could excite from the cyclotron ground state are eliminated because the trap cavity is cooled by the dilution refrigerator to 100 mK [25]. The axial motion is cooled by a resonant circuit at a rate $\gamma_z \approx (0.2\text{ s})^{-1}$ to as low as 230 mK (from 5 K) when the detection amplifier is off. The magnetron radius is minimized with axial sideband cooling [7].

5.3. Non-destructive Detection of One-Quantum Transitions

5.3.1. *QND detection*

Quantum nondemolition (QND) detection has the property that repeated measurements of the energy eigenstate of the quantum system do not change the state of the quantum system [26, 27]. This is crucial for detecting one-quantum transitions in the cyclotron motion insofar as it avoids transitions produced by the detection system. In this section we discuss the QND coupling, and in the next section the self-excited oscillator readout system.

Detecting a single 150 GHz photon from the decay of one cyclotron energy level to the level below would be very difficult – because the frequency is so high and because it is difficult to cover the solid angle into which the photon could be emitted. Instead we get the one-quantum sensitivity by coupling the cyclotron motion to the orthogonal axial motion at 200 MHz, a frequency at which we are able to make sensitive detection electronics [28]. The QND detection keeps the thermally driven axial motion of the electron from changing the state of the cyclotron motion.

We use a magnetic bottle gradient that is familiar from plasma physics and from earlier electron measurements [9, 29],

$$\mathbf{\Delta B} = B_2 \left[\left(z^2 - \rho^2/2 \right) \hat{\mathbf{z}} - z\rho\hat{\boldsymbol{\rho}} \right], \tag{5.20}$$

with $B_2 = 1540$ T/m^2. This is the lowest order gradient that is symmetric under reflections $z \to -z$ and is cylindrically symmetric about $\hat{\mathbf{z}}$. The gradient arises from a pair of thin nickel rings (Fig. 5.3) that are completely saturated in the strong field from the superconducting solenoid. To lowest order the rings modify B by $\approx -0.7\%$ – merely changing the magnetic field that the electron experiences without affecting our measurement.

The formal requirement for a QND measurement is that the Hamiltonian of the quantum system (i.e. the cyclotron Hamiltonian) and the Hamiltonian describing the interaction of the quantum system and the classical measurement system must commute. The Hamiltonian that couples the quantum cyclotron and spin motions to the axial motion does so. It has the form $-\mu B$, where μ is the magnetic moment associated with the cyclotron motion or the spin. The coupling Hamiltonian thus has a term that goes as μz^2. This term has the same spatial symmetry as does the axial Hamiltonian, $H = \frac{1}{2}m(2\pi\bar{\nu}_z)^2 z^2$. A change in the magnetic moment that takes place from a one-quantum change in the cyclotron or spin magnetic

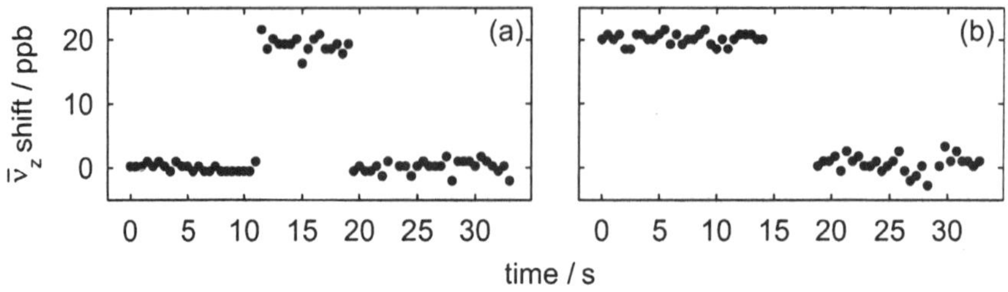

Fig. 5.5. Two quantum jumps: A cyclotron jump (a) and spin flip (b) measured via a QND coupling to shifts in the axial frequency.

moment thus changes the observed axial frequency of the suspended electron.

The result is that the frequency of the axial motion $\bar{\nu}_z$ shifts by

$$\Delta\bar{\nu}_z = \delta_B(n + m_s), \tag{5.21}$$

in proportion to the cyclotron quantum number n and the spin quantum number m_s. Fig. 5.5 shows the $\Delta\bar{\nu}_z = 4$ Hz shift in the 200 MHz axial frequency that takes place for one-quantum changes in cyclotron (Fig. 5.5(a)) and spin energy (Fig. 5.5(b)). The 20 ppb shift is easy to observe with an averaging time of only 0.5 s. We typically measure with an averaging time that is half this value.

5.3.2. *One-electron self-excited oscillator*

The QND coupling makes small changes in the electron's axial oscillation frequency, the signature of one-quantum cyclotron transitions and spin flips. Measuring these small frequency changes is facilitated by a large axial oscillation amplitude. To this end we use electrical feedback which we demonstrated could be used effectively to either cool the axial motion [30] or to make a large self-excited axial oscillation [31]. Cyclotron excitations and spin flips are generally induced while the detection system is off, as will be discussed. After an attempt to excite the cyclotron motion or to flip the spin has been made, the detection system is then turned on. The self-excited oscillator rapidly reaches steady state, and its oscillation frequency is then measured by fourier transforming the signal.

The 200 MHz axial frequency lies in the radio-frequency (RF) range which is more experimentally accessible than the microwave range of the 150 GHz cyclotron and spin frequencies, as mentioned. Nevertheless, standard RF techniques must be carefully tailored for our low-noise, cryogenic experiment. The electron axial oscillation induces image currents in the trap electrodes that are proportional to the axial velocity of the electron [7, 32]. An inductor (actually the inductance of a cryogenic feedthrough) is placed in parallel with the capacitance between two trap electrodes to cancel the reactance of the capacitor which would otherwise short out the induced signal. The RF loss in the tuned circuit that is formed is an effective resistance that damps the axial motion.

The voltage that the electron motion induces across this effective resistance is amplified with two cryogenic detection amplifiers. The heart of each amplifier is a single-gate high electron mobility transistor (Fujitsu FHX13LG).

The first amplifier is at the 100 mK dilution refrigerator base temperature. Operating this amplifier without crashing the dilution refrigerator requires operating with a power dissipation in the FET that is three orders of magnitude below the transistor's 10 mW design dissipation. The effective axial temperature for the electron while current is flowing through the FET is about 5 K, well above the ambient temperature. Very careful heat sinking makes it possible for the effective axial temperature of the electron to cool to below 350 mK in several seconds after the amplifier is turned off, taking the electron axial motion to this temperature. Cyclotron excitations and spin flips are induced only when the axial motion is so cooled, with the detection amplifiers off, since the electron is then making the smallest possible excursion in the magnetic bottle gradient.

The second cryogenic amplifier is mounted on the nominally 600 mK still of the dilution refrigerator. This amplifier counteracts the attenuation of a thermally-isolating but lossy stainless steel transmission line that carries the amplified signal out of the refrigerator. The second amplifier boosts the signal above the noise floor of the first room-temperature amplifier.

Because the induced image-current signal is proportional to the electron's axial velocity, feeding this signal back to drive the electron alters the axial damping force, a force that is also proportional to the electron velocity. Changing the feedback gain thus changes the damping rate. As the gain increases, the damping rate decreases as does the effective axial temperature of the electron, in accord with the fluctuation dissipation theorem [33]. Feedback cooling of the one-electron oscillator from 5.2 K to 850 mK was demonstrated [30]. The invariant ratio of the separately measured damping rate and the effective temperature was also demonstrated, showing that the amplifier adds very little noise to the feedback.

Setting the feedback gain to make the feedback drive exactly cancel the damping in the attached circuit could sustain a large axial oscillation amplitude, in principle. However, since the gain cannot be perfectly adjusted, noise fluctuations will always drive the axial oscillation exponentially away from equilibrium. We thus stabilize the oscillation amplitude using a digital signal processor (DSP) that Fourier transforms the signal in real time, and adjusts the feedback gain to keep the signal size at a fixed value.

The one-particle self-excited oscillator is turned on after an attempt has been made to excite the cyclotron energy up one level, or to flip the spin. The frequency of the axial oscillation that rapidly stabilizes at a large and easily detected amplitude is then measured. Small shifts in this frequency reveal whether the cyclotron motion has been excited by one quantum or whether the spin has flipped, as illustrated in Fig. 5.5.

5.3.3. *Inhibited spontaneous emission*

The spontaneous emission of synchrotron radiation in free space would make the damping time for an electron's cyclotron motion to be less than 0.1 s. This is not enough time to average down the noise in our detection system to the level that would allow the resolution of one-quantum transitions between cyclotron states. Also, to drive cyclotron transitions "in the dark", with the detection system off, requires that the cyclotron excitations persist long enough for the detection electronics to be turned on. Cavity-inhibition of the spontaneous emission gives us the averaging time that we need.

One of the early papers in what has come to be known as cavity QED was an observation of inhibited spontaneous emission within a Penning trap [24] – the first time that inhibited spontaneous emission was observed within a cavity and with only one particle – as anticipated earlier [34, 35]. As already mentioned, the cylindrical Penning trap [9] was invented to provide understandable boundary conditions to control the spontaneous emission rate with only predictable cavity shifts of the electron's cyclotron frequency.

The spontaneous emission rate is measured directly, by making a histogram of the time the electron spends in the first excited state after being excited by a microwave drive injected into the trap cavity with the detector left on. Fig. 5.6 shows a sample histogram which fits well to an exponential (solid curve) with a lifetime of 0.41 s in this example.

Stimulated emission is avoided by making these observations only when the cavity is at low temperature so that effectively no blackbody photons are present. The detector makes thermal fluctuations of the axial oscillation

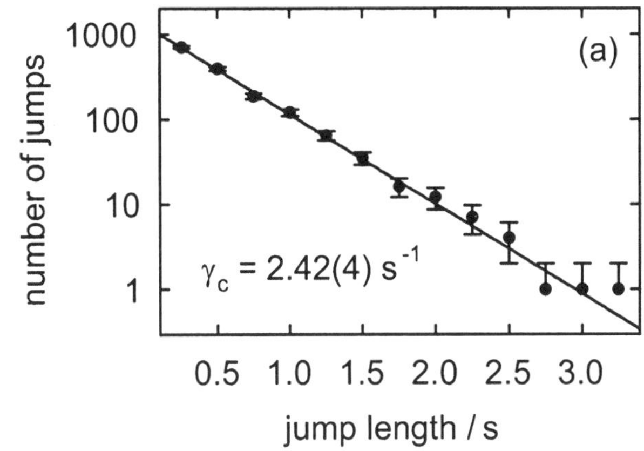

Fig. 5.6. A histogram of the time that the electron spends in the first excited state that is fit to an exponential reveals the substantial inhibition of the spontaneous emission of synchrotron radiation. The decay time, 0.41 s in this example, depends on how close the cyclotron frequency is to neighboring radiation modes of the trap cavity. Lifetimes as long as 16 s have been observed.

amplitude, and these in turn make the cyclotron frequency fluctuate. For measuring the cyclotron decay time, however, this does not matter as long as the fluctuations in axial amplitude are small compared to the 2 mm wavelength of the radiation that excites the cyclotron motion.

The spontaneous emission rate into free space is [7]

$$\gamma_+ = \frac{4}{3}\frac{r_e}{c}(\omega_c)^2 \approx \frac{1}{0.89 \text{ ms}} . \tag{5.22}$$

The measured rate in this example is thus suppressed by a factor of 4.5. The density of states within the cylindrical trap cavity is not that of free space. Instead the density of states for the radiation is peaked at the resonance frequencies of the radiation modes of the cavity, and falls to very low values between the radiation modes. We attain the inhibited spontaneous emission by tuning the magnetic field so that the cyclotron frequency is as far as possible from resonance with the cavity radiation modes. With the right choice of magnetic field we have increased the lifetime to 16 s, which is a cavity suppression of spontaneous emission by a factor of 180.

In a following section we report on using the direct measurements of the radiation rate for electron cyclotron motion to probe the radiation modes of the cavity, with the radiation rate increasing sharply at frequencies that approach a resonant mode of the cavity.

5.4. Elements of an Electron $g/2$ Measurement

5.4.1. *Quantum jump spectroscopy*

We determine the cyclotron and anomaly frequencies using quantum jump spectroscopy, in which a near resonance drive attempts to either excite the cyclotron motion or flip the spin. After each attempt we check whether a one-quantum transition has taken place, and build up a histogram of transitions per attempt. Fig. 5.7 shows the observed quantum jump lineshapes upon which our 2008 measurement is based.

A typical data run consists of alternating scans of the cyclotron and anomaly lines. The runs occur at night, with daytime runs only possible on Sundays and holidays when the ambient magnetic field noise is lower. Interleaved every three hours among these scans are periods of magnetic field monitoring to track long-term drifts using the electron itself as the magnetometer. In addition, we continuously monitor over fifty environmental parameters such as refrigerator temperatures, cryogen pressures and flows, and the ambient magnetic field in the lab so that we may screen data for abnormal conditions and troubleshoot problems.

Cyclotron transitions are driven by injecting microwaves into the trap cavity. The microwaves originate as a 15 GHz drive from a signal generator (Agilent E8251A) whose low-phase-noise, 10 MHz oven-controlled crystal oscillator serves as the timebase for all frequencies in the experiment. After passing through a waveguide that removes all subharmonics, the signal enters a microwave circuit that includes an impact ionization avalanche transit-time (IMPATT) diode, which multiplies the frequency by ten and outputs the $\bar{f}_c$ drive at a power of 2 mW. Voltage-controlled attenuators reduce the strength of the drive, which is broadcast from a room temperature horn through teflon lenses to a horn at 100 mK (Fig. 5.4) and enters the trap cavity through an inlet waveguide (Fig. 5.3).

Anomaly transitions are driven by potentials, oscillating near $\bar{\nu}_a$, applied to electrodes to drive off-resonant axial motion through the magnetic bottle gradient (Eq. (5.20)). The gradient's $z\rho\hat{\rho}$ term mixes the driven oscillation of z at $\bar{\nu}_a$ with that of ρ at $\bar{f}_c$ to produce an oscillating magnetic field perpendicular to $\mathbf{B}$ as needed to flip the spin. The axial amplitude required to produce the desired transition probability is too small to affect the lineshape (Sec. 5.4.4); nevertheless, we apply a detuned drive of the same strength during cyclotron attempts so the electron samples the same magnetic gradient.

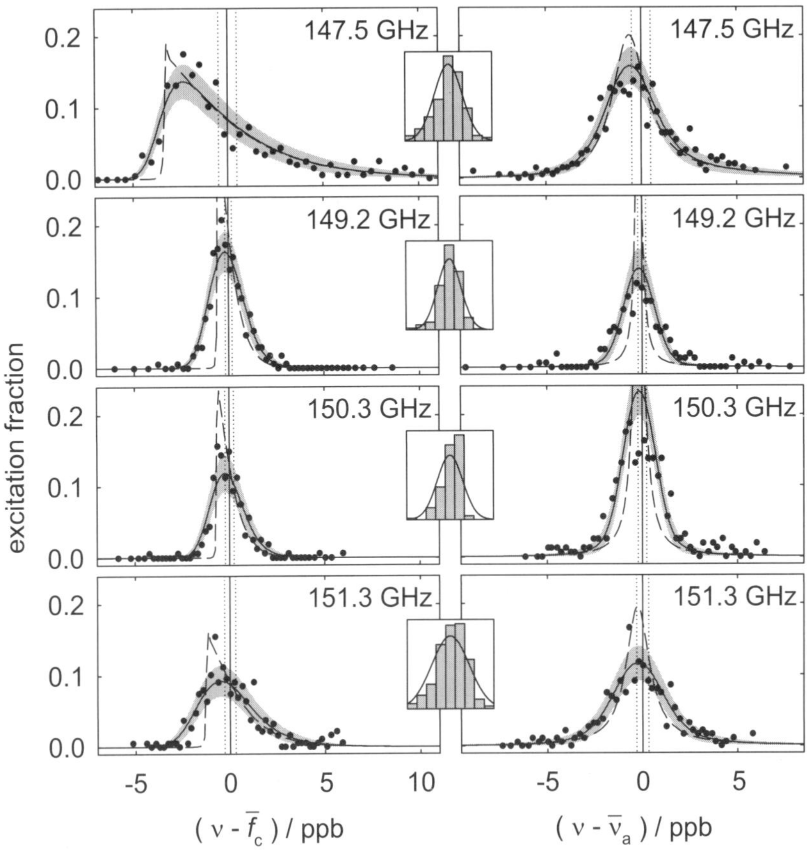

Fig. 5.7. Quantum-jump spectroscopy lineshapes for cyclotron (left) and anomaly (right) transitions with maximum-likelihood fits to broadened lineshape models (solid) and inset resolution functions (solid) and edge-tracking data (histogram). Vertical lines show the 1-σ uncertainties for extracted resonance frequencies. Corresponding unbroadened lineshapes are dashed. Gray bands indicate 1-σ confidence limits for distributions about broadened fits. All plots share the same relative frequency scale.

Quantum jump spectroscopy of each resonance follows the same procedure. With the electron prepared in the spin-up ground state $|0, \frac{1}{2}\rangle$, the magnetron radius is reduced with 1.5 s of strong sideband cooling at $\bar{\nu}_z + \bar{\nu}_m$ with the SEO turned off immediately and the detection amplifiers turned off after 0.5 s. After an additional 1 s to allow the axial motion to thermalize with the tuned circuit, we apply a 2 s pulse of either a cyclotron drive near $\bar{f}_c$ or an anomaly drive near $\bar{\nu}_a$ with the other drive applied simultaneously but detuned far from resonance. The detection electronics and SEO are turned back on; after waiting 1 s to build a steady-state axial amplitude, we measure $\bar{\nu}_z$ and look for a 20 ppb shift up (from a cyclotron transition) or down (from an anomaly transition followed by a spontaneous decay to $|0, -\frac{1}{2}\rangle$) in frequency. Cavity-inhibited spontaneous emission provides the time needed to observe cyclotron transitions before decay. The several-cyclotron-lifetimes wait for a spontaneous decay after an anomaly attempt is the rate-limiting step in the spectroscopy. After a successful anomaly transition and decay, simultaneous cyclotron and anomaly drives pump the electron back to $|0, \frac{1}{2}\rangle$. All timing is done in hardware. We probe each resonance line with discrete excitation attempts spaced in frequency by approximately 10% of the linewidth. We step through each drive frequency on the $\bar{f}_c$ line, then each on the $\bar{\nu}_a$ line, and repeat.

5.4.2. *The electron as magnetometer*

Slow drifts of the magnetic field are corrected using the electron itself as a magnetometer. Accounting for these drifts allows the combination of data taken over many days, giving a lineshape signal-to-noise that allows the systematic investigation of lineshape uncertainties at each field. For a half-hour at the beginning and end of a run and again every three hours throughout, we alter our cyclotron spectroscopy routine by applying a stronger drive at a frequency below $\bar{f}_c$. Using the same timing as above but a ten-times-finer frequency step, we increase the drive frequency until observing a successful transition. We then jump back 60 steps and begin again.

We model the magnetic field drift by fitting a polynomial to these "edge" points (so-called because the ideal cyclotron lineshape has a sharp low-frequency edge). Since we time-stamp every cyclotron and anomaly attempt, we use the smooth curve to remove any field drift. This edge-tracking adds a 20% overhead, but allows the use of data from nights with a larger than usual field drift, and the combination of data from different nights.

5.4.3. *Measuring the axial frequency*

In addition to $\bar{f}_c$ and $\bar{\nu}_a$, measuring $g/2$ requires a determination of the axial frequency $\bar{\nu}_z$ (Eq. (5.11)). To keep the relative uncertainty in $g/2$ from $\bar{\nu}_z$ below 0.1 ppt, we must know $\bar{\nu}_z$ to better than 50 ppb (10 Hz). This is easily accomplished. We routinely measure $\bar{\nu}_z$ when determining the cyclotron and spin states. However, the large self-excited oscillation amplitude in the slightly anharmonic axial potential typically results in a

10 ppb shift, compared to the $\bar{\nu}_z$ for the thermally-excited amplitude during the cyclotron and anomaly pulses. We cannot directly measure the axial frequency under the pulse conditions because the amplifiers are off. We come close when measuring $\bar{\nu}_z$ with the amplifiers on and all axial drives off. This thermal axial resonance appears as a dip on the amplifier noise resonance [32], and we use it as our measurement. The difference in $\bar{\nu}_z$ with the amplifiers on and off is negligible. A second shift comes from the interaction between the axial motion and the amplifier, which both damps the motion and shifts $\bar{\nu}_z$. The maximum shift of $\bar{\nu}_z$ is 1/4 of the damping rate, which at ≈ 1 ppb is negligible at our current precision. A third shift of $\bar{\nu}_z$ comes from the anomaly drive, which induces both a frequency-pulling from the off-resonant axial force and a Paul-trap shift from the change in effective trapping potential [36]; based on extrapolation from measured shifts at higher powers, we estimate these shifts combine to -1 ppb at the highest anomaly power used for the measurement—too small to affect $g/2$.

5.4.4. *Frequencies from lineshapes*

The cyclotron frequency $\bar{f}_c$ and anomaly frequency $\bar{\nu}_a$ (Fig. 5.2) must be determined from their respective quantum jump spectroscopy lineshapes (Fig. 5.7). The observed lineshapes are much broader than the natural linewidth that arises because the excited cyclotron state decays by the cavity-inhibited spontaneous emission of synchrotron radiation. The shape arises because the electron experiences a magnetic field that varies during the course of a measurement. Variations arise because of the electron's thermal axial motion within the magnetic bottle gradient, for example. Other possible variations could arise because the magnetic field for the Penning trap fluctuates in time, or because of a distribution of magnetron orbit sizes for the quantum jump trials.

Once the slow drift of the magnetic field (p. 178) has been removed, there is no reason for the electron to sample a different distribution of magnetic field values while the anomaly frequency is being measured compared to when the trap-modified cyclotron frequency is being measured. Each resonance shape converts the distribution of sampled magnetic field values into the corresponding distribution of frequency values. Dividing the quantum jump lineshapes into frequency bins, we obtain average cyclotron and anomaly frequencies by weighting the frequency of each bin by the number of quantum jumps in the bin, and use these average frequencies in Eq. (5.11).

Using the weighted average frequencies will remove shifts to $g/2$ caused by the thermal axial motion of the electron within the magnetic bottle gradient, the largest source of the observed linewidth. The use of weighted average frequencies should also account for temporal fluctuations in the magnetic field of the Penning trap on the measurement time scale for the frequencies. If there is a distribution of magnetron radii for the quantum jump trials, the weighted average method should account for the resulting distribution of magnetic field values as well.

To verify the weighted averages method, and to assign safe uncertainties to the average frequencies that we deduce using it, we also analyze our measured lineshapes in a very different way. We start with an analytic calculation of the lineshape for thermal Brownian motion of the axial motion for a given axial temperature T_z [37]. We then fit the measured cyclotron and anomaly lineshapes (Fig. 5.7) to the ideal lineshape convolved with a Gaussian broadening function to take into account other sources of the magnetic field distribution. The analytically calculated lineshapes are the dashed curves in Fig. 5.7, the maximum-likelihood fits to the broadened lineshapes are solid curves, and the gray bands indicate where we would expect 68% of the measured points to lie. The insets to Fig. 5.7 show the best-fit resolution functions. We assign a lineshape uncertainty that is the size of the differences between the $g/2$ value determine from the fitting and our preferred weighted averages method.

The linewidths are wider for two of the four measurements in Fig. 5.7, and they remained reproducible over the weeks required to take each data point. A wider cyclotron linewidth indicates a higher axial temperature. We know of no reason why the axial temperatures should be different for different values of the Penning trap field; this is one reason that we assigned the larger uncertainties that reflect the difference between the two methods. The narrower lineshapes have better agreement between the weighted average method and the fit method, and hence the assigned lineshape uncertainties are smaller. Not surprisingly, the narrower lines better determine the corresponding frequencies.

For the 2008 measurement the lineshape uncertainty is larger than any other. Future efforts will focus upon understanding and reducing the lineshape broadening and uncertainty.

5.4.5. *Cavity shifts*

Despite the precision reached in this measurement, one correction to the directly measured $g/2$ value is required, the $\Delta g_{cav}/2$ included in Eq. (5.11).

The correction is a cavity shift correction that depends upon interaction of the electron with nearby cavity radiation modes. The trap cavity modifies the density of states of the radiation modes of free space, though not enough to significantly affect QED calculations of g [38]. Since the cavity shift correction depends upon the electron cyclotron frequency, we measure $g/2$ at four different cyclotron frequencies to make sure that the same $g/2$ is deduced when cavity shifts of different sizes are applied.

The cavity-inhibited spontaneous emission narrows the cyclotron resonance line, giving the time in the excited state that is needed to turn on the self-excited oscillator, and to average its signal long enough to determine the cyclotron state. Cavity shifts are the unfortunate downside of the cavity, arising because the cyclotron oscillator has its frequency pulled by its coupling to nearby radiation modes of the cavity.

The cylindrical Penning trap was invented to make a microwave cavity with a calculable geometry. Section 5.2.2 describes a perfect cylindrical trap cavity and the radiation fields that it can support. However, the trap is not perfectly machined, it changes its size as it cools from room temperature down to 0.1 K, and it has small slits that make it possible to bias sections to form a Penning trap.

The shape of the radiation fields near the center of the trap cavity are not greatly altered for the real cavity, but the resonant frequencies of the modes are slightly shifted. The frequency shifts are not enough to keep us from identifying most modes by comparing to calculated frequencies, but are large enough that we must measure the mode frequencies if we are to characterize the interaction of the cavity and an electron. The mode quality factors (resonant frequencies divided by energy damping rates) must also be determined. The decay of the radiation field within the cavity depends upon power dissipated by currents (induced in the electrodes and modified by the slits), and upon the loss of microwave power that escapes the trap despite the choke flanges in the slits.

We developed two methods to learn the resonant frequencies of the radiation modes of the real trap cavity:

(1) A cloud of electrons near the center of the trap is heated using a parametric driving force. The electrons cool via synchrotron radiation with a rate that is highest when their cyclotron frequency is resonant with a cavity radiation mode, and that is very small far from resonance [14–16, 39]. Fig. 5.8(a) shows the peaks in the signal from the electrons that correspond to resonance with cavity radiation modes that are labeled as described earlier.

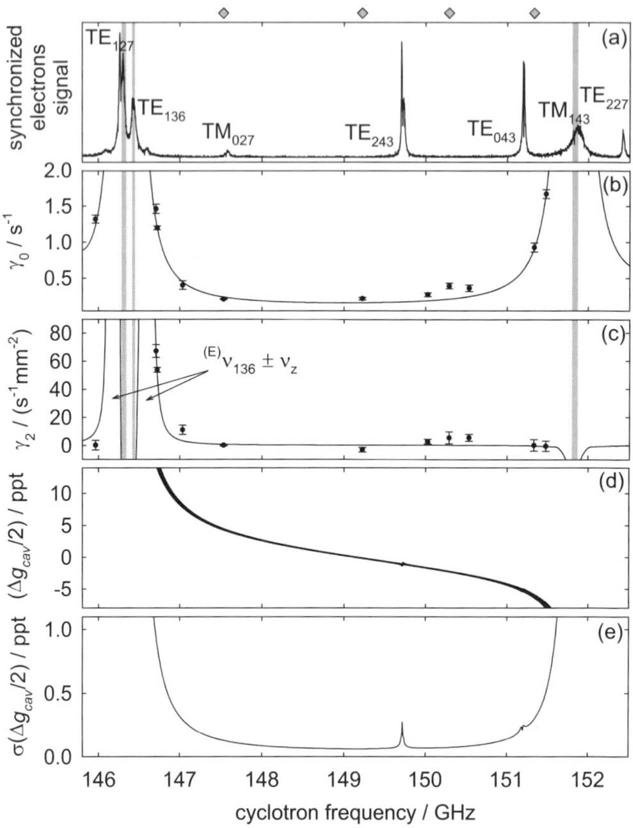

Fig. 5.8. Cavity shift results come from synchronized electrons (a) and from direct measurements with one electron of γ_c (b) and its dependence on axial amplitude (c). Together, they provide uncertainties in the frequencies of coupled cavity radiation modes (gray) that translate into an uncertainty band of cavity shifts $\Delta g_{cav}/2$ (d) whose half-width, i.e., the cavity shift uncertainty, is plotted in (e). The diamonds at the top indicate the cyclotron frequencies of the four $g/2$ measurements.

(2) The measured spontaneous emission rate for a single electron near the center of the trap cavity (Fig. 5.8b), and the dependence of this rate upon the amplitude of the axial oscillation of the electron (Fig. 5.8c), both depend upon the proximity of the electron cyclotron frequency to cavity radiation modes that couple to a nearly centered electron. Fig. 5.9 illustrates how the one-electron damping rate and dependence upon axial oscillation amplitude are measured.

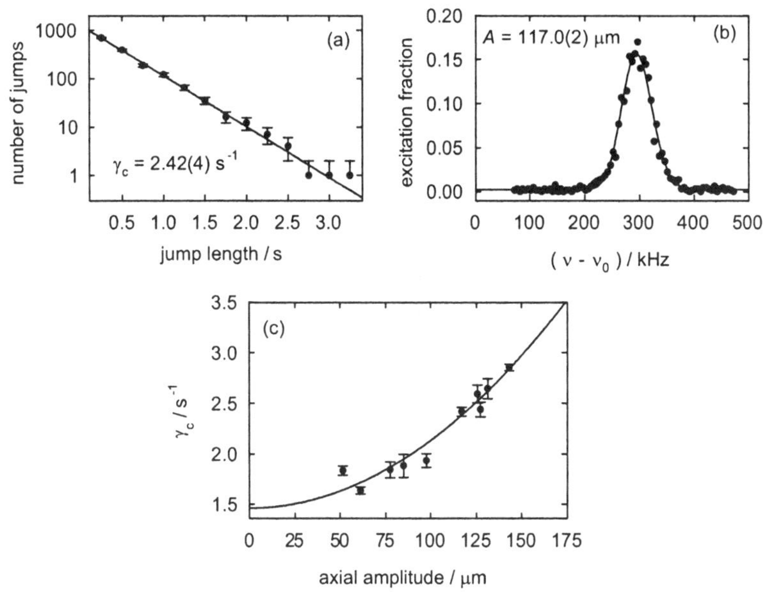

Fig. 5.9. Measurement of the cyclotron damping rate at 146.70 GHz, near the upper sideband of TE_{136}. The cyclotron damping rate as a function of axial amplitude (c) extrapolates to the desired lifetime. Each point in (c) consists of a damping rate measured from a fit to a histogram of cyclotron jump lengths (a) as well as an axial amplitude measured from a driven cyclotron line (b).

From the cavity spectra in Figs. 5.8(a-c) we deduce the mode frequencies and uncertainties represented by the gray bands in these figures. Our identification of the modes is aided by several features of the spectra. Modes that are strongly coupled to the electrons (the coupling increases with electron number) can split into two normal modes. A large axial oscillation during measurements of the cavity spectrum produces sidebands at the axial frequency for modes with a node at the trap center, and at twice the sideband frequency for radiation modes with an antinode at the center. Modes which would not couple to a perfectly centered electron will couple more strongly to the electrons as their number is increased so that they occupy a larger volume. From 2006 to 2008 our understanding of the cavity improved when we became aware of, and were able to measure, a small displacement of the electrostatic center of the trap (where the electron resides), and the center for the cavity radiation modes.

So far we have used the calculable cylindrical trap geometry to know which radiation modes can couple to an electron near the center of the trap, and we have recognized these modes in measured cavity spectra by comparing their measured frequencies to what is calculated for a perfect cavity. Next we use the measured radiation mode frequencies and quality factors as input to a calculation of the cavity shift of the electron cyclotron frequency as a function of the electron cyclotron frequency (Fig. 5.8d).

A calculation of the shifts [37, 40] must carefully distinguish and remove the electron self-energy from the electron-cavity interaction. The uncertainty in the measured inputs gives a cavity shift uncertainty (Fig. 5.8e) that is small between the resonance frequencies of modes that couple strongly to a centered electron, and then increases strongly closer to the resonant frequencies of these modes. The diamonds at the top of the figure show how, in our four measurements of $g/2$, we avoid the electron cyclotron frequencies for which the uncertainty is the largest. Fig. 5.10 shows the good agreement attained between the four measurements when the cavity shifts are applied.

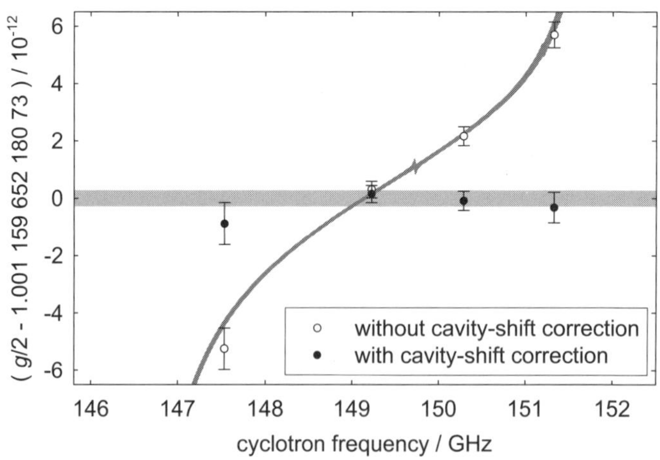

Fig. 5.10. Four measurements of $g/2$ without (open) and with (filled) cavity-shift corrections. The light gray uncertainty band shows the average of the corrected data. The dark gray band indicates the expected location of the uncorrected data given our result in Eq. (5.23) and including only the cavity shift uncertainty.

Table 5.2. Measurements and shifts with uncertainties multiplied by 10^{12}. The cavity-shifted "$g/2$ raw" and corrected "$g/2$" are offset from our result in Eq. (5.23).

f_c	147.5 GHz	149.2 GHz	150.3 GHz	151.3 GHz
$g/2$ raw	-5.24 (0.39)	0.31 (0.17)	2.17 (0.17)	5.70 (0.24)
Cav. shift	4.36 (0.13)	-0.16 (0.06)	-2.25 (0.07)	-6.02 (0.28)
Lineshape				
correlated	(0.24)	(0.24)	(0.24)	(0.24)
uncorrelated	(0.56)	(0.00)	(0.15)	(0.30)
$g/2$	-0.88 (0.73)	0.15 (0.30)	-0.08 (0.34)	-0.32 (0.53)

5.5. Results and Applications

5.5.1. *Most accurate electron $g/2$*

The measured values, shifts, and uncertainties for the four separate measurements of $g/2$ are in Table 5.2. The uncertainties are lower for measurements with smaller cavity shifts and smaller linewidths, as might be expected. Uncertainties for variations of the power of the $\bar{\nu}_a$ and $\bar{f}_c$ drives are estimated to be too small to show up in the table. A weighted average of the four measurements, with uncorrelated and correlated errors combined appropriately, gives the electron magnetic moment in Bohr magnetons,

$$g/2 = 1.001\,159\,652\,180\,73\,(28) \qquad [0.28 \text{ ppt}]. \tag{5.23}$$

The uncertainty is 2.7 and 15 times smaller than the 2006 and 1987 measurements, and 2300 times smaller than has been achieved for the heavier muon lepton [41].

5.5.2. *Most accurate determination of α*

The new measurement determines the fine structure constant, $\alpha = e^2/(4\pi\epsilon_0\hbar c)$, much more accurately than does any other method. The fine structure constant is the fundamental measure of the strength of the electromagnetic interaction in the low energy limit, and it is also a crucial ingredient of our system of fundamental constants [42]. A full discussion of α, its importance, the quantum electrodynamics theory used to determine it from the measured $g/2$, and alternative methods to determine α is in Chapter 6. Only the bare essentials of what is needed to determine α from $g/2$ are summarized here.

The Standard Model relates g and α by

$$\frac{g}{2} = 1 + C_2\left(\frac{\alpha}{\pi}\right) + C_4\left(\frac{\alpha}{\pi}\right)^2 + C_6\left(\frac{\alpha}{\pi}\right)^3 + C_8\left(\frac{\alpha}{\pi}\right)^4 + C_{10}\left(\frac{\alpha}{\pi}\right)^5 + \ldots + a_{\text{hadronic}} + a_{\text{weak}}, \tag{5.24}$$

with the asymptotic series and the values of the C_k coming from QED. Very small hadronic and weak contributions are included, along with the assumption that there is no significant modification from electron substructure or other physics beyond the Standard Model.

QED calculations (summarized more extensively in Chapter 6) give the constants C_k,

$$C_2 = \quad 0.500\,000\,000\,000\,00 \text{ (exact)} \tag{5.25}$$
$$C_4 = -\,0.328\,478\,444\,002\,90\,(60) \tag{5.26}$$
$$C_6 = \quad 1.181\,234\,016\,827\,(19) \tag{5.27}$$
$$C_8 = -\,1.914\,4\,(35) \tag{5.28}$$
$$C_{10} = \quad 0.0\,(4.6). \tag{5.29}$$

The QED theory for C_2 [43], C_4 [11, 13, 44], and C_6 [47] is exact, with no uncertainty, except for an essentially negligible uncertainty in C_4 and C_6 that comes from a weak functional dependence upon the lepton mass ratios, m_μ/m_e and m_τ/m_e. Numerical QED calculations [48] give the value and uncertainty for C_8. The hadronic anomaly a_{hadronic}, calculated within the context of the Standard Model,

$$a_e^{\text{hadronic}} = 1.682(20) \times 10^{-12}\,, \tag{5.30}$$

contributes at the level of several times the current experimental uncertainty, but the calculation uncertainty in the hadronic anomaly is not important [38, 42]. See Chapters 8 and 9 for further details. The weak anomaly is completely negligible.

The most accurately determined fine structure constant is given by

$$\begin{aligned} \alpha^{-1} &= 137.035\,999\,084\,(33)\,(39) \quad [0.24 \text{ ppb}]\,[0.28 \text{ ppb}], \\ &= 137.035\,999\,084\,(51) \qquad [0.37 \text{ ppb}]. \end{aligned} \tag{5.31}$$

The first line shows experimental (first) and theoretical (second) uncertainties that are nearly the same. The theory uncertainty contribution to α is divided as (12) and (37) for C_8 and C_{10}. It should decrease when a calculation underway [48] replaces the crude estimate $C_{10} = 0.0\,(4.6)$ [42, 50]. The α^{-1} of Eq. (5.31) will then shift by $2\alpha^3\pi^{-4}C_{10}$, which is $8.0\,C_{10} \times 10^{-9}$. A change Δ_8 in the calculated C_8 would add $2\alpha^2\pi^{-3}\Delta_8$.

The total 0.37 ppb uncertainty in α is 12 and 21 times smaller than for the next most precise independent methods (Fig. 5.11). These so-called atom recoil methods (see Chapter 6) utilize measurements of the Rydberg constant [17, 51], transition frequencies [21, 53], mass ratios [19, 55], and either a Rb [54] or Cs [57] recoil velocity measured in an atom interferometer.

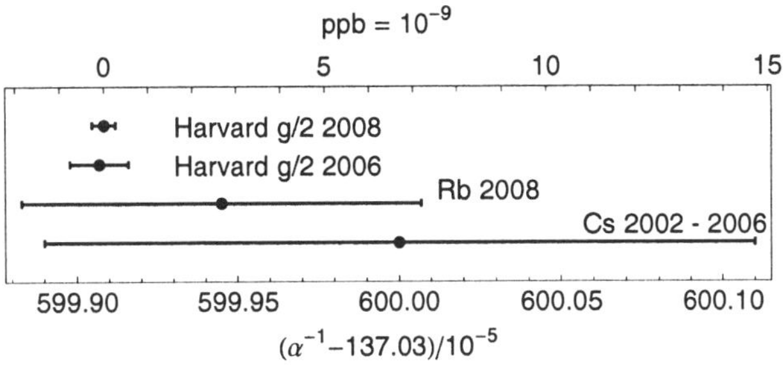

Fig. 5.11. The most accurate determinations of α are determined from the measured electron $g/2$. These are compared to the best independently measured values.

5.5.3. *Testing the standard model and QED*

The dimensionless electron magnetic moment g that is measured can be compared to the $g(\alpha)$ that is predicted by the Standard Model of particle physics. The input needed to calculate $g(\alpha)$ is the measured fine structure constant α (that is determined without the use of the electron magnetic moment). The most accurately measured and calculated values of $g/2$ are currently given by

$$g/2 = 1.001\ 159\ 652\ 180\ 73\ \ (28)\ [0.28\ \text{ppt}], \qquad (5.32)$$

$$g(\alpha)/2 = 1.001\ 159\ 652\ 177\ 60\ (520)\ [5.2\ \ \text{ppt}]. \qquad (5.33)$$

The measurement is our one-electron quantum cyclotron measurement [6]. The calculated value $g(\alpha)/2$ comes from using the Rb value of α(Rb08) in Eq. (5.24). The large uncertainty in this "calculated" value actually comes from the large uncertainty in the Rb α; the theoretical uncertainty is believed to be much smaller, comparable to the measurement uncertainty for $g/2$. The Standard Model prediction is thus tested and verified to about 5 ppt. The much smaller 0.3 ppt uncertainty in the measured $g/2$, along with the comparable uncertainty in the QED calculation, would allow a much better test of QED.

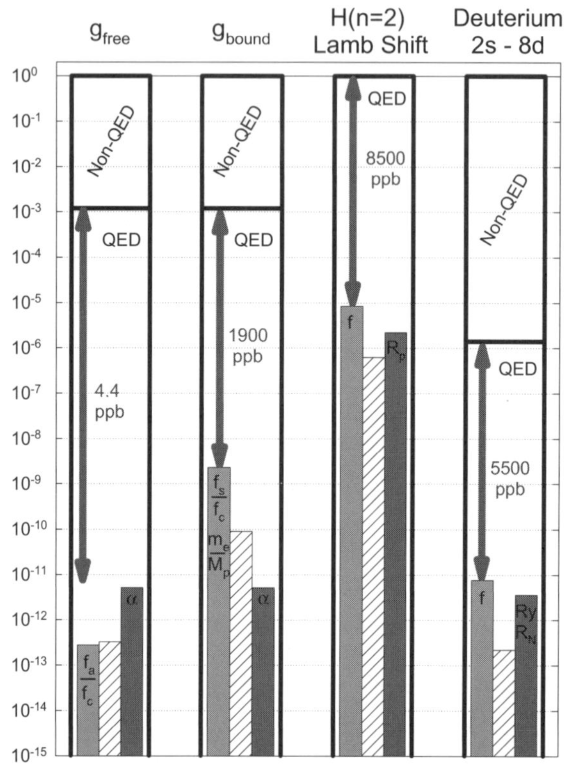

Fig. 5.12. Comparisons of precise tests of QED. The arrows represent the fractional accuracy to which the QED contribution to the measured g values and frequencies that are measured.

About 1 part per thousand of the electron $g/2$ comes from the unavoidable interaction of the electron with the virtual particles of "empty space", as described by quantum electrodynamics (QED) and represented in Fig. 5.12. Where testing QED is the primary focus, measured and calculated values of the so-called anomalous magnetic moment of the electron (defined by $a = g/2 - 1$ so that the Dirac contribution is subtracted out) are often compared. The measured and calculated values of a that correspond to the $g/2$ values above are

$$a = 0.001\ 159\ 652\ 180\ 73\ (\ 28)\ [0.24\ \text{ppb}], \tag{5.34}$$

$$a(\alpha) = 0.001\ 159\ 652\ 177\ 60\ (520)\ [4.4\ \ \text{ppb}]. \tag{5.35}$$

At the one standard deviation level, the difference of the measured and calculated values is

$$\delta a = a - a(\alpha) \tag{5.36}$$
$$= g/2 - g(\alpha)/2 \tag{5.37}$$
$$= 3.1(5.2) \times 10^{-12}. \tag{5.38}$$

The possible difference between the measurement and calculation is thus bounded by

$$|\delta a| < 8.3 \times 10^{-12}, \tag{5.39}$$

at the one standard deviation level, with this bound arising almost entirely from the uncertainty in the measurement of α from Rb.

Some of the most precise tests of bound-state QED are compared in Fig. 5.12 with the electron $g/2$. The QED test based upon the measurements [56] and calculation [58] of $g/2$ for an electron bound in an ion provides a test of the QED contribution to the electron magnetic moment at the 4.4 ppb level. Bound electron $g/2$ measurements test QED less precisely. In fact, the calculated value of the bound g values depends upon the mass of the electron strongly enough that this measurement is now being used to determine the electron mass in amu, much as we determine α from our measurements of the magnetic moment of the free electron. The n=2 Lamb shift in hydrogen is essentially entirely due to QED. However, the measurements are much less precise so that QED is again tested less precisely.

The last example in the figure is a QED test based upon a number of measurements of hydrogen and deuterium transition frequencies – the QED contribution to which are typically at the ppm level. Theoretical calculations that depend upon the Rydberg constant, the fine structure constant, the ratio of the electron and proton masses and the size of the nucleus are fit to a number of accurately measured transition frequencies for hydrogen and deuterium. The fit determines values for the mentioned constants. The QED test comes from removing one of the measured lines from the fit, and using the best fit to predict the value of the transition frequency that was omitted. This process tests the Standard Model prediction at a comparable precision to that provided using the magnetic moment of the electron. However, it tests the size of the QED contribution to a much lower fractional precision.

The QED tests described so far test QED predictions to the highest precision and the highest order in α. There are many other tests of QED with a much lower precision. Although these tests are outside of the scope of this work it is worth mentioning that it is interesting to probe QED in other ways. For example, it seems interesting to test QED for systems whose binding energy is very large, even comparable to the electron rest mass energy as can be done in high Z systems. Another example is probing the QED of positronium, the bound state of an electron and a positron, insofar as annihilation and exchange effects are quite different than what must be calculated for normal atoms.

5.5.4. *Probe for electron substructure*

Comparing experiment and theory probes for possible electron substructure at an energy scale one might only expect from a large accelerator. An electron whose constituents would have mass $m^* \gg m$ has a natural size scale, $R = \hbar/(m^* c)$. The simplest analysis of the resulting magnetic moment [59] gives $\delta a \sim m/m^*$, suggesting that $m^* > 61,000$ TeV/c^2 and $R < 3 \times 10^{-24}$ m. This would be an incredible limit, since the largest e^+e^- collider (LEP) probes for a contact interaction at an $E = 10.3$ TeV [60], with $R < (\hbar c)/E = 2 \times 10^{-20}$ m.

However, the simplest argument also implies that the first-order contribution to the electron self-energy goes as m^* [59]. Without heroic fine tuning (e.g. the bare mass canceling this contribution to produce the small electron mass) some internal symmetry of the electron model must suppress both mass and moment. For example, a chirally invariant model [59], leads to $\delta a \sim (m/m^*)^2$. In this case, $m^* > 177$ GeV/c^2 and $R < 1 \times 10^{-18}$ m. These limits seem remarkable for an experiment carried out at 100 mK, although they do not compete with LEP. If this test was limited only by the experimental uncertainty in a then we could set a limit $m^* > 1$ GeV.

5.5.5. *Comparison to the muon g/2*

The electron $g/2$ is measured about 2300 times more accurately than is $g/2$ for its heavier muon sibling [7, 41]. Because the electron is stable there is time to isolate one electron, cool it so that it occupies a very small volume within a magnetic field, and to resolve the quantum structure in its cyclotron and spin motions. The short-lived muon must be studied before it decays in a very small fraction of a second, during which time it orbits in a very large orbit over which the same magnetic field homogeneity realized with a nearly motionless electron cannot be maintained.

Why then measure the muon $g/2$? The compelling reason is that the muon $g/2$ is expected to be more sensitive to physics beyond the Standard

Model by about a factor of 4×10^4, which is the square of the ratio of the muon to the electron mass. In terms of Eq. (5.2) this means that a_{new} is expected to be bigger for the muon than for the electron by this large factor, making the muon a more attractive probe for New Physics.

Unfortunately, the other Standard Model contribution, $a_{\text{hadronic}} + a_{\text{weak}}$, is also bigger by approximately the same large factor, rather than being essentially negligible as in the electron case. Correctly calculating these terms is a significant challenge to detecting New Physics. These large terms, and the much lower measurement precision, also make the muon an unattractive candidate (compared to the electron) for determining the fine structure constant and for testing QED.

The measured electron $g/2$ makes two contributions to using the muon system for probing for physics beyond the Standard Model. Both relate to determining the muon QED anomaly $a_{\text{QED}}(\alpha)$

(1) The electron measurement of $g/2$ makes possible the most accurate determination of the fine structure constant (discussed in the previous section and in Chapter 6) as is needed to calculate $a_{\text{QED}}(\alpha)$.
(2) The electron measurement of $g/2$ and an independently measured value of α test QED calculations of the very similar $a_{\text{QED}}(\alpha)$ terms in the electron system.

The QED contribution must be accurately subtracted from the measured muon $g - 2$ if the much smaller possible contribution from New Physics is to be observed.

5.6. Prospects and Conclusion

In conclusion, our 2008 measurement of the electron $g/2$ is 15 times more accurate than the 1987 measurement that provided $g/2$ and α for nearly 20 years, and 2.7 times more accurate than our 2006 measurement that superseded it. Achieving the reported electron $g/2$ uncertainty with a positron seems feasible, to make the most stringent lepton CPT test.

With QED and the assumption of no New Physics beyond the Standard Model of particle physics, the new measurement determines α 12 times more accurately than any independent method. The measured $g/2$ makes it possible to test QED and probe for electron size. In fact, the sensitivity of all of these applications would immediately be improved by the factor of 12 if a more accurate independent measurement of α, at our level of precision, is realized.

Several experimental items warrant further study. First is the broadening of the expected lineshapes which limits the splitting of the resonance lines. Second, the variation in axial temperatures in the observed resonance lineshapes is not understood, and a larger uncertainty comes from the wider lineshapes. Third, cavity sideband cooling could cool the axial motion to near its quantum ground state for a more controlled measurement. Fourth, a new apparatus should be much less sensitive to vibration and other variations in the laboratory environment. A more accurate measurement of the electron $g/2$ is the expected result.

Acknowledgments

It has been a pleasure and privilege to collaborate with a string of excellent graduate students – C. H. Tseng, D. Enzer, J. Tan, S. Peil, B. D'Urso, B. Odom and D. Hanneke – to develop the new method and apparatus that made it possible to measure the electron $g/2$ and α so much more accurately than had been possible. Progress continues in the ongoing work of Ph.D students S. Fogwell and J. C. Dorr who also provided useful comments.

References

[1] B. Lautrup and H. Zinkernagel, *Stud. Hist. Phil. Mod. Phys.* **30**, 85 – 110, (1999).
[2] A. Rich and J. C. Wesley, *Rev. Mod. Phys.* **44**, 250, (1972).
[3] R. S. Van Dyck Jr. , P. B. Schwinberg, and H. G. Dehmelt, *The Electron*, pp. 239–293. Kluwer Academic Publishers, Netherlands, (1991).
[4] R. S. Van Dyck, Jr., P. B. Schwinberg, and H. G. Dehmelt, *Phys. Rev. Lett.* **59**, 26–29, (1987).
[5] B. Odom, D. Hanneke, B. D'Urso, and G. Gabrielse, *Phys. Rev. Lett.* **97**, 030801, (2006).
[6] D. Hanneke, S. Fogwell, and G. Gabrielse, *Phys. Rev. Lett.* **100**, 120801, (2008).
[7] L. S. Brown and G. Gabrielse, *Rev. Mod. Phys.* **58**, 233–311, (1986).
[8] L. S. Brown and G. Gabrielse, *Phys. Rev. A.* **25**, 2423–2425, (1982).
[9] G. Gabrielse and F. C. MacKintosh, *Intl. J. of Mass Spec. and Ion Proc.* **57**, 1–17, (1984).
[10] J. N. Tan and G. Gabrielse, *Appl. Phys. Lett.* **55**, 2144–2146, (1989).
[11] G. Gabrielse, *Phys. Rev. A.* **27**, 2277–2290, (1983).
[12] L. S. Brown, G. Gabrielse, J. N. Tan, and K. C. D. Chan, *Phys. Rev. A.* **37**, 4163–4171, (1988).
[13] J. D. Jackson, *Classical Electrodynamics, 2nd Edition.* (John Wiley and Sons, Inc., New York, 1975).

[14] J. Tan and G. Gabrielse, *Phys. Rev. Lett.* **67**, 3090–3093, (1991).
[15] J. N. Tan and G. Gabrielse, *Phys. Rev. A.* **48**, 3105–3122, (1993).
[16] G. Gabrielse, J. N. Tan, and L. S. Brown, *Cavity Shifts of Measured Electron Magnetic Moments*, In (ed.) T. Kinoshita, *Quantum Electrodynamics*, pp. 389–418. World Scientific, Singapore, (1990).
[17] G. Gabrielse, X. Fei, L. A. Orozco, R. L. Tjoelker, J. Haas, H. Kalinowsky, T. A. Trainor, and W. Kells, *Phys. Rev. Lett.* **65**, 1317–1320, (1990).
[18] G. Gabrielse and J. Tan, *J. Appl. Phys.* **63**, 5143–5148, (1988).
[19] G. Gabrielse, J. N. Tan, L. A. Orozco, S. L. Rolston, C. H. Tseng, and R. L. Tjoelker, *J. Mag. Res.* **91**, 564–572, (1991).
[20] E. S. Meyer, I. F. Silvera, and B. L. Brandt, *Rev. Sci. Instrum.* **60**, 2964–2968, (1989).
[21] D. F. Phillips. *A Precision Comparison of the $\bar{p}-p$ Charge-to-Mass Ratios.* Ph.D. thesis, Harvard University, (1996).
[22] R. S. Van Dyck, Jr., D. L. Farnham, S. L. Zafonte, and P. B. Schwinberg, *Rev. Sci. Instrum.* **70**, 1665–1671, (1999).
[23] B. Odom. *Fully Quantum Measurement of the Electron Magnetic Moment.* Ph.D. thesis, Harvard University, (2004).
[24] G. Gabrielse and H. Dehmelt, *Phys. Rev. Lett.* **55**, 67–70, (1985).
[25] S. Peil and G. Gabrielse, *Phys. Rev. Lett.* **83**, 1287–1290, (1999).
[26] K. S. Thorne, R. W. P. Drever, and C. M. Caves, *Phys. Rev. Lett.* **40**(11), 667–670, (1978).
[27] V. B. Braginsky and F. Y. Khalili, *Rev. Mod. Phys.* **68**(1-11), 1, (1996).
[28] B. D'Urso. *Cooling and Self-Excitation of a One-Electron Oscillator.* Ph.D. thesis, Harvard Univ., (2003).
[29] R. Van Dyck, Jr., P. Ekstrom, and H. Dehmelt, *Nature.* **262**, 776, (1976).
[30] B. D'Urso, B. Odom, and G. Gabrielse, *Phys. Rev. Lett.* **90**(4), 043001, (2003).
[31] B. D'Urso, R. Van Handel, B. Odom, D. Hanneke, and G. Gabrielse, *Phys. Rev. Lett.* **94**, 113002, (2005).
[32] D. J. Wineland and H. G. Dehmelt, *J. Appl. Phys.* **46**, 919–930, (1975).
[33] R. Kubo, *Rep. Prog. Phys.* **29**(1), 255–284, (1966).
[34] E. M. Purcell, *Phys. Rev.* **69**, 681, (1946).
[35] D. Kleppner, *Phys. Rev. Lett.* **47**, 233, (1981).
[36] F. L. Palmer, *Phys. Rev. A.* **47**, 2610, (1993).
[37] L. S. Brown, *Ann. Phys. (N.Y.).* **159**, 62–98, (1985).
[38] D. G. Boulware, L. S. Brown, and T. Lee, *Phys. Rev. D.* **32**, 729–735, (1985).
[39] G. Gabrielse and J. N. Tan, *One Electron in a Cavity*, In ed. P. Berman, *Cavity Quantum Electrodynamics*, pp. 267–299. Academic Press, New York, (1994).
[40] L. S. Brown, *Phys. Rev. Lett.* **52**, 2013–2015, (1984).
[41] G. W. Bennett and *et al.*, *Phys. Rev. D.* **73**, 072003, (2006).
[42] P. J. Mohr and B. N. Taylor, *Rev. Mod. Phys.* **77**, 1 – 107, (2005).
[43] J. Schwinger, *Phys. Rev.* **73**, 416L, (1948).
[44] A. Petermann, *Helv. Phys. Acta.* **30**, 407, (1957).
[45] C. M. Sommerfield, *Phys. Rev.* **107**, 328, (1957).

[46] C. M. Sommerfield, *Ann. Phys. (N.Y.)*. **5**, 26, (1958).
[47] S. Laporta and E. Remiddi, *Phys. Lett. B*. **379**, 283, (1996).
[48] T. Aoyama, M. Hayakawa, T. Kinoshita, and M. Nio, *Phys. Rev. Lett.* **99**, 110406, (2007).
[49] A. Czarnecki, B. Krause, and W. J. Marciano, *Phys. Rev. Lett.* **76**, 3267–3270, (1996).
[50] G. Gabrielse, D. Hanneke, T. Kinoshita, M. Nio, and B. Odom, *Phys. Rev. Lett.* **97**, 030802, (2006). *ibid.* **99**, 039902 (2007).
[51] T. Udem, A. Huber, B. Gross, J. Reichert, M. Prevedelli, M. Weitz, and T. W. Hänsch, *Phys. Rev. Lett.* **79**, 2646–2649, (1997).
[52] C. Schwob, L. Jozefowski, B. de Beauvoir, L. Hilico, F. Nez, L. Julien, F. Biraben, O. Acef, J. J. Zondy, and A. Clairon, *Phys. Rev. Lett.* **82**, 4960–4963, (1999).
[53] V. Gerginov, K. Calkins, C. E. Tanner, J. J. McFerran, S. Diddams, A. Bartels, and L. Hollberg, *Phys. Rev. A*. **73**, 032504, (2006).
[54] M. Cadoret, E. de Mirandes, P. Cladé, S. Guellati-Hkélifa, C. Schwob, F. Nez, L. Julien, and F. Biraben, *Phys. Rev. Lett.* **101**, 230801, (2008).
[55] M. P. Bradley, J. V. Porto, S. Rainville, J. K. Thompson, and D. E. Pritchard, *Phys. Rev. Lett.* **83**, 4510–4513, (1999).
[56] G. Werth, J. Alonso, T. Beier, K. Blaum, S. Djekic, H. Häffner, N. Hermanspahn, W. Quint, S. Stahl, J. Verdú, T. Valenzuela, and M. Vogel, *Int. J. Mass Spectrom.* **251**, 152, (2006).
[57] A. Wicht, J. M. Hensley, E. Sarajlic, and S. Chu, *Phys. Scr.* **T102**, 82–88, (2002).
[58] K. Pachucki, A. Czarnecki, U. Jentschura, and V. A. Yerokhin, *Phys. Rev. A*. **72**, 022108, (2005).
[59] S. J. Brodsky and S. D. Drell, *Phys. Rev. D*. **22**, 2236–2243, (1980).
[60] D. Bourilkov, *Phys. Rev. D*. **64**, 071701R, (2001).

Chapter 6

Determining the Fine Structure Constant

G. Gabrielse

Department of Physics, Harvard University
17 Oxford Street, Cambridge, MA 02138
gabrielse@physics.harvard.edu

The most accurate determination of the fine structure constant α is $\alpha^{-1} = 137.035\,999\,084\,(51)\,[0.37\,\text{ppb}]$. This value is deduced from the measured electron $g/2$ (the electron magnetic moment in Bohr magnetons) using the relationship of α and $g/2$ that comes primarily from Dirac and QED theory. Less accurate by factors of 12 and 21 are determinations of α from combined measurements of the Rydberg constant, two mass ratios, an optical frequency, and a recoil shift for Rb and Cs atoms. Helium fine structure intervals have been measured well enough to determine α with nearly the same precision – if two-electron QED calculations can be sorted out. Less accurate measurements are also compared.

Contents

6.1. Introduction

The fundamental and dimensionless fine structure constant α is defined (in SI units) by

$$\alpha = \frac{1}{4\pi\epsilon_0}\frac{e^2}{\hbar c}. \tag{6.1}$$

The well known value $\alpha^{-1} \approx 137$ is not predicted within the Standard Model of particle physics.

The most accurate determination of α comes from a new Harvard measurement [7, 8] of the dimensionless electron magnetic moment, $g/2$, that is 15 times more accurate than the measurement that stood for twenty years [9]. The fine structure constant is obtained from $g/2$ using the theory of a Dirac point particle with QED corrections [10–15]. The most accurate α, and the two most accurate independent values, are given by

$$\alpha^{-1}(\mathrm{H08}) = 137.035\,999\,084\,(51) \qquad [0.37\ \mathrm{ppb}] \tag{6.2}$$

$$\alpha^{-1}(\mathrm{Rb08}) = 137.035\,999\,45\,(62) \qquad [4.5\ \mathrm{ppb}] \tag{6.3}$$

$$\alpha^{-1}(\mathrm{Cs06}) = 137.036\,000\,0\,(11) \qquad [8.0\ \mathrm{ppb}]. \tag{6.4}$$

Fig. 6.1 compares the most accurate values.

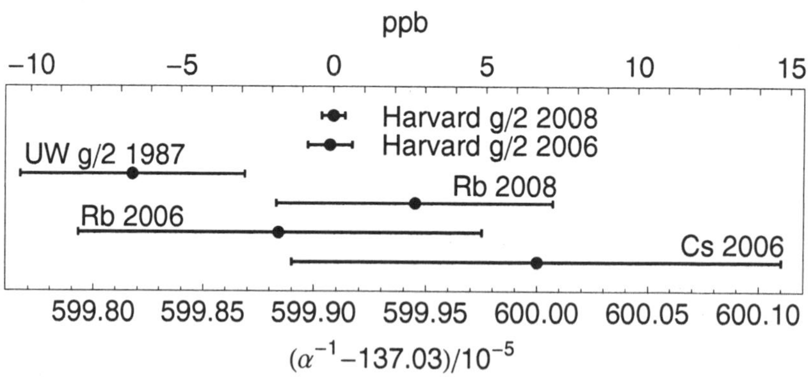

Fig. 6.1. The most precise determinations of α.

The uncertainties in the two independent determinations of α are within a factor of 12 and 21 of the α from $g/2$. They rely upon separate measurements of the Rydberg constant [16, 17], mass ratios [18, 19], optical frequencies [20, 21], and atom recoil [21, 22]. Theory also plays an important role for this method, to determine the Rydberg constant (reviewed in Ref. [23]) and one of the mass ratios [24].

In what follows, the importance of the fine structure constant is discussed first. Determining α from the measured electron $g/2$ comes next, starting with an operational summary of how this is done, and finishing with an overview of the status and reliability of the theory. Determining α from the combined measurements mentioned above is the next topic. The possibility to determine α with nearly the same precision from atomic fine structure is then considered. Helium fine structure intervals have been measured with enough accuracy to do so, [1–4, 25] if inconsistencies in the needed two-electron QED theory [5, 6] can be cleared up. Other methods that are important for historical reasons are mentioned, and followed by a conclusion.

6.2. Importance of the Fine Structure Constant

The fine structure constant appears in many contexts and is important for many reasons.

(1) The fine structure constant is the low energy electromagnetic coupling constant, the measure of the strength of the electromagnetic interaction in the low energy limit.

(2) The fine structure constant is the basic dimensionless constant of atomic physics, distinguishing the energy scales that are important for atoms. In terms of the electron rest energy, $m_e c^2$:

   (a) The binding energy of an atom is approximately $\alpha^2 m_e c^2$.
   (b) The fine structure energy splitting in atoms goes as $\alpha^4 m_e c^2$.
   (c) The hyperfine structure energy splitting goes as $(m_e/M)\,\alpha^4 m_e c^2$, like the fine structure splitting except reduced by an additional ratio of an electron mass to the nucleon mass (M).
   (d) The lamb shift in an atom goes as $\alpha^5 m_e c^2$.

(3) The fine structure constant is also important for condensed matter physics, the condensed matter and atomic energy scales being similar. Important examples include the quantum hall resistance and the oscillation frequency of a Josephson junction.

(4) The fine structure constant is important and central to our interlinked system of fundamental constants [23]. Its role will be enhanced if a contemplated redefinition of the SI system of units

(to remove the dependence upon an artifact mass standard) is adopted [27].

(5) Measurements of the muon magnetic moment [28], made to test for possible breakdowns of the Standard Model of particle physics, require a value for α. Small departures from the Standard Model would only be visible once the large α-dependent QED contribution to the muon g value is subtracted out.

(6) Comparing α values from methods that depend differently upon QED theory is a test of the QED theory.

6.3. Most Accurate α Comes from Electron $g/2$

6.3.1. *New Harvard measurement and QED theory*

The most accurate determination of the fine structure constant utilizes a new measurement of the electron magnetic moment, measured in Bohr magnetons [7],

$$g/2 = 1.001\,159\,652\,180\,73\,(28)\,[0.28\,\text{ppt}]. \tag{6.5}$$

This 2008 measurement of $g/2$ (Chapter 5) is 15 times more precise than the 1987 measurement [9] that had stood for about twenty years. The high precision and accuracy came from new methods that made it possible to resolve the quantum cyclotron levels [29], as well as the spin levels, of one electron suspended for months at a time in a cylindrical Penning trap [30].

The electron $g/2$ is essentially the ratio of the spin and cyclotron frequencies. This ratio is deduced from measurable oscillation frequencies in the trap using an invariance theorem [31]. These frequencies are measured using quantum jump spectroscopy of one-quantum transitions between the lowest energy levels [8]. The cylindrical Penning trap electrodes form a microwave cavity that shapes the radiation field in which the electron is located, narrowing resonance linewidths by inhibiting spontaneous emission [29, 32], and providing boundary conditions which make it possible to identify the symmetries of cavity radiation modes [7, 33]. A QND (quantum nondemolition) coupling, of the cyclotron and spin energies to the frequency of an orthogonal and nearly harmonic electron axial oscillation, reveals the quantum state [29]. This harmonic oscillation of the electron is self-excited [34], by a feedback signal [35] derived from its own motion, to produce the large signal-to-noise ratio needed to quickly read out the quantum state without ambiguity.

Within the Standard Model of particle physics the measured electron $g/2$ is related to the fine structure constant by

$$g/2 = 1 + C_2\left(\frac{\alpha}{\pi}\right) + C_4\left(\frac{\alpha}{\pi}\right)^2 + C_6\left(\frac{\alpha}{\pi}\right)^3 + C_8\left(\frac{\alpha}{\pi}\right)^4 + C_{10}\left(\frac{\alpha}{\pi}\right)^5 + \ldots + a_{\text{hadronic}} + a_{\text{weak}}. \tag{6.6}$$

Dirac theory of the electron provides the leading term on the right. Fig. 6.2 compares the size of the measured $g/2$ (gray) with its measurement uncertainty (black) to size of this leading Dirac term and other theoretical contributions (gray). The uncertainties (black) of the theoretical contributions arise from the uncertainty for the coefficients.

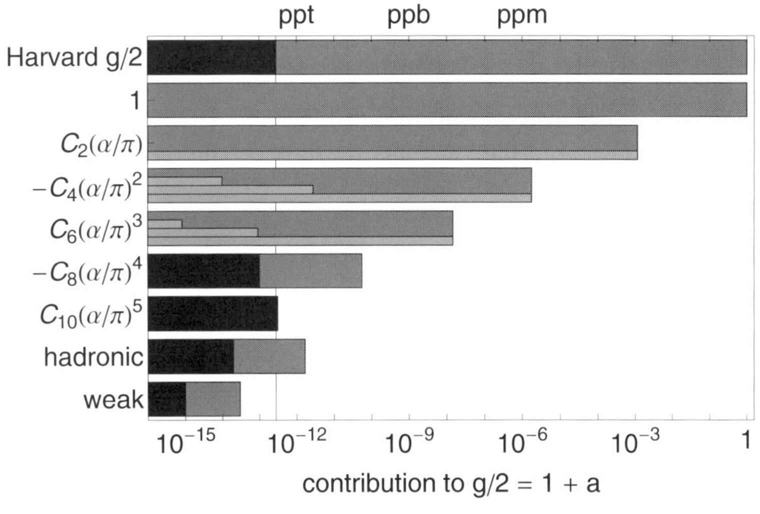

Fig. 6.2. Contributions to $g/2$ for the experiment (top bar), terms in the QED series (below), and from small distance physics (below). Uncertainties are black. The inset light gray bars represent the magnitude of the larger mass-independent terms (A_1) and the smaller A_2 terms that depend upon either m_e/m_μ or m_e/m_τ. The even smaller A_3 terms, functions of both mass ratios, are not visible on this scale.

Quantum electrodynamics (QED) provides the expansion in the small ratio $\alpha/\pi \approx 2\times10^{-3}$, and the values of the coefficients C_k. The first three of these, C_2 [10], C_4 [11–13], C_6 [14] are exactly known functions which have no theoretical uncertainty. The small uncertainties in C_4 and C_6, completely negligible at the current level of experimental precision (Fig. 6.2), arise because C_4 and C_6 depend slightly upon lepton mass ratios.

$$C_2 = 0.500\,000\,000\,000\,00 \text{ (exact)} \tag{6.7}$$

$$C_4 = -\,0.328\,478\,444\,002\,90\,(60) \tag{6.8}$$

$$C_6 = 1.181\,234\,016\,827\,(19) \tag{6.9}$$

$$C_8 = -\,1.914\,4\,(35) \tag{6.10}$$

$$C_{10} = 0.0\,(4.6). \tag{6.11}$$

There is no analytic solution for C_8 yet but this coefficient has been calculated numerically [15]. Unfortunately, C_{10} has not yet been calculated; the quoted bound is a simple extrapolation from the lower-order C_k [36].

Very small additional contributions due to short distance physics have also been evaluated [37, 38],

$$a_{\text{hadronic}} = 0.000\,000\,000\,001\,682\,(20) \tag{6.12}$$

$$a_{\text{weak}} = 0.000\,000\,000\,000\,030\,(01). \tag{6.13}$$

The hadronic contribution is important at the current level of experimental precision, but the reported uncertainty for this contribution is much smaller than is currently needed to determine α from $g/2$. See Chapters 8 and 9 for further details.

The most precise value of the fine structure constant comes from using the very accurately measured electron $g/2$ (Eq. (6.5)) in the Standard Model relationship between $g/2$ and α (Eq. (6.6)). The result is

$$\begin{aligned}
\alpha^{-1}(\text{H08}) &= 137.035\,999\,084\,(33)\,(39) \qquad [0.24 \text{ ppb}]\,[0.28 \text{ ppb}],\\
&= 137.035\,999\,084\,(33)\,(12)\,(37) \quad [0.24 \text{ ppb}]\,[0.09 \text{ ppb}]\,[0.27 \text{ ppb}],\\
&= 137.035\,999\,084\,(51) \qquad [0.37 \text{ ppb}].
\end{aligned} \tag{6.14}$$

The first line shows experimental (first) and theoretical (second) uncertainties that are nearly the same. The second line separates the theoretical uncertainty into two parts, the numerical uncertainty in C_8 (second) and the estimated uncertainty for C_{10} (third). The third line gives the total 0.37 ppb uncertainty. A graphical comparison of the experimental and theoretical uncertainties in determining α from $g/2$ is in Fig. 6.3.

The crudely estimated theoretical uncertainty in the uncalculated C_{10} currently adds more to the uncertainty in α more than does the measurement uncertainty for $g/2$. As a result, the factor of 15 reduction in the measurement uncertainty for $g/2$ results in only a factor of 10 reduction in the uncertainty in α.

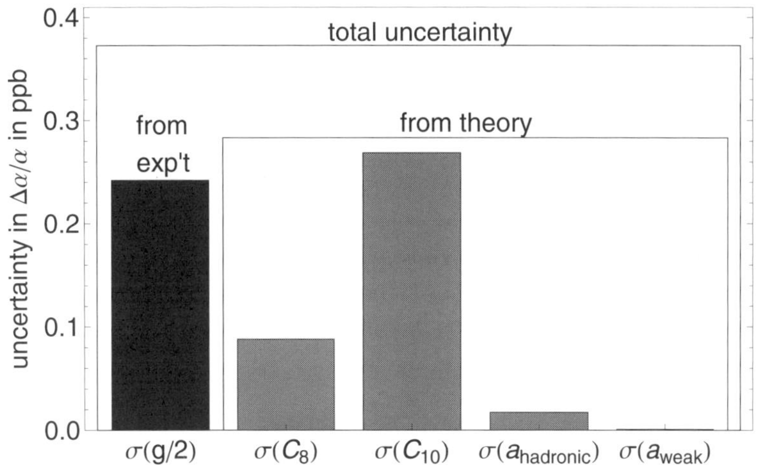

Fig. 6.3. Experimental uncertainty (black) and theoretical uncertainties (gray) that determine the uncertainty in the α that is determined from the measured electron $g/2$.

Figure 6.1 compares our α^{-1}(H08) to other accurate determinations of α. The fine structure constant is currently determined about 12 and 21 times more precisely from $g/2$ than from the best Cs and Rb measurements (to be discussed). No other α determination has error bars small enough to fit in this figure. Comparing our α with the most accurate independent determinations is a test of the Standard Model prediction in Eq. (6.6), along with the theoretical assumptions used for the other determinations. More accurate independent α values would improve upon what is already the most stringent test of QED theory.

6.3.2. *Status and reliability of the QED theory*

The electron $g/2$ differs from 1 by about one part in 10^3 as a result of the QED corrections to the Dirac theory. How uncertain and how reliable is the QED theory that is needed to accurately determine α from $g/2$? Given the complexity of the theory, and mistakes that have been discovered in the past, how likely is it that additional mistakes will either appreciably change α in the future, or go undetected?

In this section we summarize the status of calculations of the C_k coefficients, the current values of which are already listed in Eqs. 6.7–6.11. The history and method of the calculations are discussed in Chapters 3 and 4. We illustrate how impressive analytic calculations have made it easy to now evaluate the lowest order coefficients (C_2, C_4 and C_6) to an arbitrary precision with no theoretical uncertainty, provided that no mistakes have been made. Numerical calculations and verifications of C_8, and the prospects for numerical calculations of C_{10}, are also summarized.

There is no theoretical uncertainty in the Dirac unit contribution to $g/2$ in Eq. (6.6). There is also no theoretical uncertainty in the leading QED correction, $C_2(\alpha/\pi)$, insofar as long ago a single Feynman diagram was evaluated analytically to determine C_2 exactly [10].

The C_4 coefficient is the sum of a mass-independent term and two much smaller terms that are functions of lepton mass ratios,

$$C_4 = A_1^{(4)} + A_2^{(4)}(\frac{m_e}{m_\mu}) + A_2^{(4)}(\frac{m_e}{m_\tau}). \tag{6.15}$$

The mass-independent term is larger by many orders of magnitude. This pure number, involving 7 Feynman diagrams, is given by [11–13, 39]

$$A_1^{(4)} = \frac{197}{144} + \frac{\pi^2}{12} + \frac{3}{4}\zeta(3) - \frac{\pi^2}{2}\ln(2) \tag{6.16}$$
$$= -0.328\,478\,965\,579\,193\ldots \tag{6.17}$$

where $\zeta(s)$ is the Riemann zeta function (Zeta[s] in Mathematica).

There is no theoretical uncertainty in this contribution, which can easily be evaluated to any desired precision. Of course, this is only true if there are no mistakes in the analytic derivation. The original result [40] had an error in the evaluation of an integral. This was corrected some years later [12] (and then confirmed independently [11]) after the initial result did not agree with a numerical calculation. This was the first of several instances where independent evaluations allowed the elimination of mistakes, as we shall see.

The mass-dependent function $A_2^{(4)}(x)$ is an analytical evaluation of one Feynman diagram [41]. In a convenient form [42] it is given by

$$\begin{aligned} A_2^{(4)}(1/x) = &-\frac{25}{36} - \frac{\ln(x)}{3} + x^2[4 + 3\ln(x)] + \frac{x}{2}(1 - 5x^2) \\ &\times \left[\frac{\pi^2}{2} - \ln(x)\ln(\frac{1-x}{1+x}) - \mathrm{Li}_2(\mathrm{x}) + \mathrm{Li}_2(-\mathrm{x})\right] \\ &+ x^4\left[\frac{\pi^2}{3} - 2\ln(x)\ln(\frac{1}{x} - x) - \mathrm{Li}_2(\mathrm{x}^2)\right]. \end{aligned} \tag{6.18}$$

The dilogarithm function is a special case of the polylogarithm (PolyLog[n,x] in Mathematica); it has a series expansion $\mathrm{Li_n(x)} = \sum_{\mathrm{k=1}}^{\infty} \mathrm{x^k/k^n}$

that converges for the cases we need. The exactly calculated mass-dependent function is evaluated as a function of two lepton mass ratios [23, 43],

$$m_\mu/m_e = 206.768\,276\,(24) \tag{6.19}$$

$$m_\tau/m_e = 3\,477.48\,(57). \tag{6.20}$$

There is no theoretical uncertainty in the mass-dependent terms

$$A_2^{(4)}(m_e/m_\mu) = 5.197\,387\,71\,(12) \times 10^{-7}, \tag{6.21}$$

$$A_2^{(4)}(m_e/m_\tau) = 1.837\,63\,(60) \times 10^{-9}. \tag{6.22}$$

The uncertainties are from the uncertainties in the measured mass ratios. When multiplied by $(\alpha/\pi)^2$ these are very small contributions to $g/2$. The first of these two contributions is larger than the current experimental precision (Fig. 6.2) while the second is not. The uncertainties in both terms are so small as to not even be visible in Fig. 6.2.

The higher order coefficients, C_k with $k = 6, 8, 10, \ldots$, are each the sum of a constant and functions of mass ratios,

$$C_k = A_1^{(k)} + A_2^{(k)}(\frac{m_e}{m_\mu}) + A_2^{(k)}(\frac{m_e}{m_\tau}) + A_3^{(k)}(\frac{m_e}{m_\mu}, \frac{m_e}{m_\tau}). \tag{6.23}$$

The leading mass-independent term, $A_1^{(k)}$, is much larger than the small mass-dependent corrections. In fact, for $k \geq 8$, the mass-dependent corrections should not be needed to determine α from $g/2$ at the current or foreseeable measurement precision in $g/2$ owing to their very small values.

For sixth order the mass-independent term requires the evaluation of 72 Feynman diagrams. An analytic evaluation of this term, mostly by Remiddi and Laporta [14], is

$$\begin{aligned} A_1^{(6)} &= \frac{83}{72}\pi^2\zeta(3) - \frac{215}{24}\zeta(5) - \frac{239}{2160}\pi^4 + \frac{28259}{5184} \\ &+ \frac{139}{18}\zeta(3) - \frac{298}{9}\pi^2\ln(2) + \frac{17101}{810}\pi^2 \\ &+ \frac{100}{3}\left[\mathrm{Li}_4(\frac{1}{2}) + \frac{\ln^4(2)}{24} - \frac{\pi^2\ln^2(2)}{24}\right] \qquad (6.24) \\ &= 1.181\,241\,456\,587\,\ldots. \qquad (6.25) \end{aligned}$$

This remarkable analytic expression, easily evaluated to any desired numerical precision with no theoretical error, is very significant for determining α from $g/2$ insofar as it completely removes what otherwise would be a significant numerical uncertainty.

Is the remarkable analytic expression free of mistakes? The best confirmation is the good agreement between the extremely complicated analytic derivation and a simpler but computation-intensive numerical calculation, $A_1^{(6)} = 1.181\ 259\ (40)$ [44]. This result used the best computers available many years ago; it could (and should) now be greatly improved. An earlier numerical evaluation led to the discovery and correction of a mistake made in an earlier analytic derivation of a renormalization term [44]. This further illustrates the importance of checking analytic derivations numerically.

An exact analytic calculation of the 48 Feynman diagrams that determine the mass-dependent function $A_2^{(6)}$ has also been completed [45, 46]. However, the resulting expressions are apparently too lengthy to publish in a printed form. Instead, expansions for small mass ratios are made available

$$A_2^{(6)}(r) = \sum_{k=1}^{4} r^{2k} f_{2k}(r). \tag{6.26}$$

The expansions make it easy to calculate the two most important mass dependent contributions to the precision at which the measurement uncertainty in the mass ratios is important for any foreseeable improvements in the mass ratio uncertainties. Functions f_2 and f_4 are from Ref. [46], f_6 is from Refs. [45] and [47], and f_8 is from Ref. [42].

$$f_2(r) = \frac{23\ln(r)}{135} + \frac{3\zeta(3)}{2} - \frac{2\pi^2}{45} - \frac{74957}{97200}, \tag{6.27}$$

$$f_4(r) = -\frac{4337\ln^2(r)}{22680} + \frac{209891\ln(r)}{476280} + \frac{1811\zeta(3)}{2304} - \frac{1919\pi^2}{68040} - \frac{451205689}{533433600}, \tag{6.28}$$

$$f_6(r) = -\frac{2807\ln^2(r)}{21600} + \frac{665641\ln(r)}{2976750} + \frac{3077\zeta(3)}{5760} - \frac{16967\pi^2}{907200} - \frac{246800849221}{480090240000}, \tag{6.29}$$

$$f_8(r) = -\frac{55163\ln^2(r)}{594000} + \frac{24063509989\ln(r)}{172889640000} + \frac{9289\zeta(3)}{23040} - \frac{340019\pi^2}{24948000} - \frac{896194260575549}{2396250410400000}. \tag{6.30}$$

These expansions have been compared to the exact calculations to verify the claim that their accuracy is much higher than any experimental uncertainty that will likely be reached [42].

With the current values of the mass ratios,

$$A_2^{(6)}(m_e/m_\mu) = -7.373\ 941\ 58\ (28) \times 10^{-6}, \tag{6.31}$$

$$A_2^{(6)}(m_e/m_\tau) = -6.581\ 9\ (19) \times 10^{-8}. \tag{6.32}$$

The uncertainties arise from the measurement imprecision in the mass ratios, not from any theoretical uncertainty. The term that depends upon both mass ratios [42],

$$A_3^{(6)}(m_e/m_\mu, m_e/m_\tau) = 1.909\ 45\ (62) \times 10^{-13}, \qquad (6.33)$$

is too small to be important for the electron $g/2$ in the foreseeable future, or to even have its uncertainty visible in Fig. 6.2.

For the current and foreseeable experimental precisions, only the mass-independent term is required in eighth order. Kinoshita and his collaborators have reduced the 891 Feynman diagrams to a much smaller number of master integrals, which were then evaluated by Monte Carlo integrations over the course of ten years. The latest result is [15]

$$C_8 = A_1^{(8)} = -1.9144\ (35). \qquad (6.34)$$

The uncertainty is that of the numerical integration as evaluated by an integration routine [48], limited by the computer time available for the integrations. A calculation of this coefficient to 0.2% is a remarkable result that is critical for determining α from $g/2$.

Checking the eighth-order coefficient to make sure that it is correctly evaluated is a formidable challenge. There is no analytic result to compare (yet). Only the collaborating groups of Kinoshita and Nio have had the courage and tenacity needed to complete such a challenging calculation. The complexity of the calculation makes it very difficult to avoid mistakes. The strategy has been to check each part of the calculation by using more than one independent formulation [49].

Our 2006 measurement of $g/2$ came while the theoretical checking was underway. At this point we published a value of α along with a warning that the theoretical checking for eighth order was not yet complete [50]. In 2007, a calculation using an independent formulation reached a precision sufficient to reveal a mistake [15] in how infrared divergences were handled in two master integrals. When the mistake in C_8 was corrected, the α determined from $g/2$ shifted a bit [50].

One could take the moral of the 2007 adjustment to be that the sheer complexity of the high order QED calculation makes it impossible to be certain that they are done correctly. I take the opposite conclusion, choosing to be reassured that the theory is checked so carefully that even a very small mishandling of divergences can be identified and corrected. Now that the eighth order calculation is completely checked by an independent formulation, to a level of precision that the theorists deem is sufficient to detect mistakes, it seems much less likely that another substantial change in α will be necessary. The check will be even better when the new calculation reaches the numerical precision of the calculation being checked.

An evaluation of, or at least a reasonable bound on, the tenth-order coefficient, $C_{10} \approx A_1^{(10)}$, is needed as a result of the level of accuracy of our 2008 measurement of $g/2$. A calculation is not easy given that 12 672 Feynman diagrams contribute. The estimated bound suggested in the meantime [37],

$$C_{10} = 0.0\,(4.6), \tag{6.35}$$

takes the uncalculated coefficient to be zero with an uncertainty that is an extrapolation of the size of the lower order coefficients. This crude estimate is not so convincing. It is especially unsatisfying given that it now limits the accuracy with which α can be determined from the measured $g/2$, as illustrated in Fig. 6.3.

6.3.3. *How much better can α be determined?*

Fig. 6.3 shows the experimental and theoretical contributions to the uncertainty in the α determined from $g/2$. This uncertainty is currently divided nearly equally between measurement uncertainty in $g/2$ and theoretical uncertainty in the Standard Model relation between $g/2$ and α. The largest theoretical uncertainty is from the uncalculated C_{10}, followed by numerical uncertainty in C_8.

The first calculation of $C_{10} \approx A_1^{(10)}$ is now underway [15, 51, 52]. It has already produced an automated code that was checked by recomputing the eighth-order coefficient. (This is the independent calculation that in 2007 reached the precision needed to expose a mistake in the calculation of C_8 [15].) No limit or bound will apparently be available until the impressive calculation is completed at some level of precision because many contributions with similar magnitudes sum to make a smaller result. A completed calculation of C_{10} will likely reduce the theoretical uncertainty enough so that the uncertainty in α would approach the 0.26 ppb uncertainty that comes from the measurement uncertainty in $g/2$.

The uncertainty in C_8 can be reduced once the uncertainty in C_{10} has been reduced enough to warrant this. More computation time would reduce the numerical integration uncertainty in C_8. A better hope is that parts or all of this coefficient will eventually be calculated analytically. Efforts in this direction are underway [53].

It thus seems likely that the theoretical uncertainty that limits the accuracy to which α can be determined from $g/2$ can and will be reduced below

0.1 ppb. The corresponding good news is that it also seems likely that the uncertainty in α from the measurement of $g/2$ can also be reduced below 0.1 ppb. With enough experimental and theoretical effort it may well be possible to do even better.

6.4. Determining α from the Rydberg, Two Mass Ratios and $\hbar/M$ for an Atom

All the determinations of α whose uncertainty is not much larger than 20 times the uncertainty of the α from $g/2$ are compared in Fig. 6.1. The values not from $g/2$ in this figure do not come from a single measurement. Instead, each requires the determination of four quantities from a minimum of six precise measurements, each measurement contributing to the uncertainty in the α that is determined. Theory, including QED theory, is essential to determining two of the measured quantities.

The definitions for α and the Rydberg constant R_∞ taken together yield

$$\alpha^2 = \frac{4\pi}{c} R_\infty \frac{\hbar}{m_e}. \tag{6.36}$$

No accurate measurement of $\hbar/m_e$ for the electron is available. However, a precisely measured $\hbar/M_x$ for a Cs or Rb atom (of mass M_x) can be used along with two measurable mass ratios, $A_r(e)$ and $A_r(x)$,

$$\alpha^2 = \frac{4\pi}{c} R_\infty \frac{A_r(x)}{A_r(e)} \frac{\hbar}{M_x}. \tag{6.37}$$

The speed of light, c, is defined in the SI system of units.

The first of the needed mass ratios, $A_r(e) = 12 m_e / M(^{12}C)$, is the electron mass in atomic mass units (amu). The second is the mass of Cs or Rb in amu, $A_r(x) = 12M(x)/M(^{12}C)$. Determining the Rydberg constant accurately requires the precise measurements of two hydrogen transition frequencies (and less accurate measurements of other quantities). Determining $\hbar/M_x$ for Cs and Rb requires the measurement of an optical frequency ω and an atom recoil velocity v_r, or equivalent recoil frequency shift, ω_r.

The fractional uncertainties that contribute to the uncertainty in α are listed in Table 6.1 for Cs, and in Table 6.2 for Rb, in order of increasing precision. Owing to the square in Eq. (6.37) the fractional uncertainty in α is half the fractional uncertainty of the contributing measurements.

The Rydberg constant describes the structure of a non-relativistic hydrogen atom in the limit of an infinite proton mass. Real hydrogen atoms, of course, have fine structure, Lamb shifts, and hyperfine structure. The proton has a finite mass. The Dirac energy eigenvalues must be corrected for

Table 6.1. Measurements determining $\alpha(Cs)$.

Measurement quantity	ppb	$\Delta\alpha/\alpha$ ppb	References
ω_r	15.	7.7	[[22]]
$A_r(e)$	0.4	0.2	[[23, 54]]
$A_r(Cs)$	0.2	0.1	[[18]]
ω	0.007	0.007	[[20]]
R_∞	0.007	0.004	[[16, 17, 23]]
Best α(Cs)		8.0	[[22]]

Table 6.2. Measurements determining $\alpha(Rb)$.

Measurement quantity	ppb	$\Delta\alpha/\alpha$ ppb	References
ω_r	9.1	4.6	[[21]]
$A_r(e)$	0.4	0.2	[[23, 54]]
$A_r(Rb)$	0.2	0.1	[[18]]
ω	0.4	0.4	[[21]]
R_∞	0.007	0.004	[[16, 17, 23]]
Best α(Rb)		4.6	[[21]]

relativistic recoil, QED self-energy effects, and QED vacuum polarization. Corrections for nuclear polarization, nuclear size and nuclear self-energy are important at the precision with which transition energies can be measured.

The theory needed to determine the Rydberg constant from measurements is described in a seven-page section of Ref. [23] entitled "Theory relevant to the Rydberg constant." The accepted value of the Rydberg comes from a best fit of the measurements of a number of accurately measured hydrogen transitions [16, 17], the proton-to-electron mass ratio [19], the size of the proton, etc. to the intricate hydrogen theory for each of the hydrogen transitions, using more precisely measured values for every quantity that is not determined best by fitting. A full discussion of this process and a complete bibliography for all the measurements and calculations that make important contributions is beyond the scope of this work. Tables 6.1 and 6.2 thus show the currently accepted uncertainty for the Rydberg constant [23] rather than the uncertainties from all the contributing measurements.

The measured electron mass in amu, $A_r(e)$, relies equally upon precise measurements [19, 54] and upon bound state QED theory [24], using

$$\frac{m_e}{M} = \frac{g_{bound}}{2} \frac{1}{q/e} \frac{\omega_c}{\omega_s}, \tag{6.38}$$

where q/e is the integer charge of the ion in terms of one quantum of charge. Measurements are made using a $^{12}C^{5+}$ (or $^{16}O^{7+}$) ion trapped in a pair of open access Penning traps [55], a type of trap we developed for accurate measurements of q/m for an antiproton. Spin flips and cyclotron excitations are made in one trap and then transferred to the other for detection in a strong magnetic gradient. The spin frequency ω_s of the electron bound in an ion is measured. The cyclotron frequencies ω_c of the ion is deduced from the measurable oscillation frequencies of the trapped ion using the Brown–Gabrielse invariance theorem [31, 56]. This determination of the electron mass in amu could not take place without an extensive QED calculation of the g value of an electron bound into an ion [24]. A less accurate measurement of the electron mass in amu does not rely on QED theory [57]; it agrees with the more accurate method.

The needed mass ratios, $A_r(x)$, are from measurements [18] using isolated ions in a orthogonalized hyperbolic Penning trap [58], a trap design we developed to facilitate precise measurements. Ion cyclotron frequencies are deduced from oscillation frequencies of the ions in the trap using the same invariance theorem [31, 56]. Ion cyclotron energy is transferred to the axial motion using a sideband method that allows cyclotron information to be read out by a SQUID detector that is coupled to the axial motion of an ion in a trap. Ratios of ion frequencies give the ratios of masses in a simple and direct way that is insensitive to theory. Ratios to of M_x to the carbon mass, as needed to get amu, came from using ions like CO_2^+ and several hydrocarbon ions as reference ions.

The basic idea of the $\hbar/M_x$ measurements for Cs and Rb is that when an atom absorbs a quantum of light from a laser field, or is stimulated to emit a quantum of light into a laser field, then the atom recoils with a momentum $M_x v_r = \hbar k$, where for a laser field with angular frequency ω we have $k = \omega/c$. Thus $\hbar/M_x$ is determined by the measured optical frequency of the laser radiation, ω, and by the atom recoil velocity v_r. The latter can be accurately measured from the recoil shift ω_r in the resonance frequency caused by the recoil of the atom.

The laser frequency is measured a bit differently for Cs and Rb. For Cs the needed frequencies are measured with a precision of 0.007 ppb, much more accurately than will likely be needed for some time, using an optical comb to directly measure the frequency with respect to hydrogen maser and a Cs fountain clock [20]. For Rb, a diode laser is locked to a stable cavity, and its frequency is compared using an intermediate reference laser to that of a two-photon Rb standard [59].

The largest uncertainty in determining α using Eq. (6.37) is the uncertainty in measuring the atom recoil velocity v_r, or equivalently the recoil shift $\omega_r = \frac{1}{2}M_x v_r^2/\hbar$. This measurement uncertainty is much larger than the measurement uncertainty in R_∞, $A_r(e)$, $A_r(x)$, and ω, and is thus the limit to the accuracy with which α can currently be determined by this method. The availability of extremely cold laser-cooled atoms has led to significant progress by two different research groups. First came a Cs measurement at Stanford [22] in 2002. More recently came 2006 and 2008 measurements of slightly higher precision with Rb atoms at the LKB in Paris [21, 59].

The Cs recoil measurement [22] and the most accurate of the Rb measurement [21] both measure the atom recoil using atom interferometry. The so-called Ramsey–Bordé spectrometer [60] configuration that is used in both cases was developed to apply Ramsey separated oscillator field methods at optical frequencies. Pairs of stimulated Raman $\pi/2$ pulses produced by counter-propagating laser beams [61] split the wave packet of a cold atom into two phase-coherent wave packets with different atom velocities. A series of N Raman π pulses then add recoil kicks to both parts of the atom wave packet. When a final pair of Raman $\pi/2$ pulses make it possible for the previously separated parts of the wave packet to interfere, the interference pattern reveals the energy difference, and hence the recoil frequency difference, for the wave packets in the two arms of the interferometer.

The measured phase difference that reveals v_r and ω_r goes basically as N, where N is the number of additional recoil kicks given to the wave packets in both arms of the spectrometer. The experiments differ in the way that they seek to make N as large as possible. The initial Cs measurement used a sequence of π pulses to achieve $N = 30$. The most accurate of the Rb measurements achieved $N = 1600$ using a series of Raman transitions with the frequency difference between the counter-propagating laser beams being swept linearly in time. This can equivalently be regarded as a type of Bloch oscillation within an accelerating optical lattice [62].

An improved apparatus is under construction in the hope of improving the 2008 measurement of the Rb recoil shift on the time scale of a year or two. Although no Cs recoil measurement has been reported since 2002, an improved apparatus has been built. A goal of soon measuring the Cs recoil shift accurately enough to determine α to sub-ppb accuracy was mentioned in a recent report on improved beam splitters for a Cs atom interferometer [63].

6.5. Other Measurements to Determine α

6.5.1. *Determining α from He fine structure*

Surprisingly none of the accurate measurements determine α by measuring atomic fine structure intervals. Helium fine structure intervals have been measured precisely enough so that two-electron QED theory could determine α from the interval at about the same precision as do the combined Rydberg, mass ratios and atom recoil measurements. Helium is a better candidate for such measurements than is hydrogen because the fine structure splittings are larger, and the radiation lifetimes of the levels are longer so that narrow resonance lines can be measured. Unfortunately, theoretical inconsistencies need to be resolved.

The most accurate measurements of three 2^3P ^{4}He fine structure intervals [1–4, 25] are in good agreement as illustrated in Fig. 6.4. Our Harvard laser spectroscopy measurements [25] have the smallest uncertainties,

$$f_{12} = 2\,291\,175.59 \pm 0.51 \text{ kHz} \qquad [220 \text{ ppb}] \qquad (6.39)$$

$$f_{01} = 29\,616\,951.66 \pm 0.70 \text{ kHz} \qquad [\,24 \text{ ppb}] \qquad (6.40)$$

$$f_{02} = 31\,908\,126.78 \pm 0.94 \text{ kHz} \qquad [\,29 \text{ ppb}]. \qquad (6.41)$$

The figure shows good agreement between measurements of the largest intervals; these are best for determining α. The measurements of the small interval also agree well. This interval is less useful for determining α but is a useful check on the theory.

Because a fine structure interval frequency f goes as $R_\infty \alpha^2$ to lowest order, and the Rydberg is known much more accurately than α, a fractional uncertainty in f translates into a fractional uncertainty for α that is smaller by half – if the theory would contribute no additional uncertainty. The 24 ppb fractional uncertainty in the f_{01} that we reported back in 2005 would then suffice to determine α to 12 ppb, a small enough uncertainty to allow this value to be plotted with the most precise measurements in Fig. 6.1.

A big disappointment is that Fig. 6.4 reveals two serious problems with calculations done independently by two different groups [5, 6]. (See Chapter 7.) The calculated interval frequencies (using α from $g/2$) are plotted below the measurements in the figure. The first problem is that the two

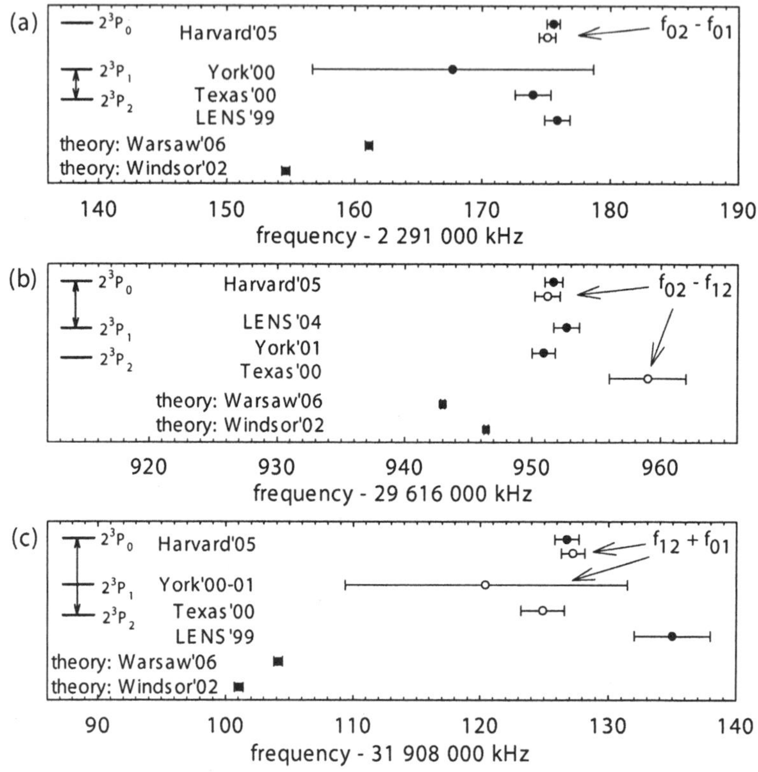

Fig. 6.4. Most accurate measured [1–4] and calculated [5, 6] ^{4}He fine structure intervals with standard deviations. Directly measured intervals (black filled circles) are compared to indirect values (open circles) deduced from measurements of the other two intervals. Uncorrelated errors are assumed for the indirect values for other groups.

calculations do not agree, raising questions as to whether mistakes have been made. It is not hard to imagine mistakes given that the two-electron QED theory gives interval frequencies that are the sum of a series in powers of both α and $\ln \alpha$. The convergence is not rapid, and the many terms to be summed present a significant bookkeeping challenge. The second problem is that both theories disagree with the measurements, for both the large and small intervals. The measurements from 2005 and earlier, though they have an accuracy that would suffice to be one of the most precise determinations of α, cannot be used until the theory issues are resolved.

A serious difficulty with two-electron QED theory seems surprising given how successful one-electron QED theory has been in its predictions. Is there a fundamental problem or is this a case of mistakes? Until the two calculations agree the latter explanation is hard to discount, and neither calculation agrees with experiments.

A problem with the measurements is the other possibility, though the good agreement between measurements with very different systematic effects would suggest otherwise. One caution is that the most accurate measurements determine to 700 Hz the center of resonance lines that are slightly bigger than 1.6 MHz natural linewidths. "Splitting the line" to a few parts in 10^4 of the linewidth is challenging, requiring as it does that systematic shifts and distortions of the measured resonance lines be either insignificant or well understood. It is hard to believe that a helium fine structure measurement could ever approach the accuracy of the current α from $g/2$.

After we published our measurement of the helium fine structure intervals we narrowed our laser linewidth to below 5 kHz and stabilized it to an iodine clock using an optical comb that we built to bridge between the very different frequencies of our clock and the 1.08 μm optical transitions that we measured. We also greatly improved the signal-to-noise ratio in our measured resonance lines. Within a couple of hours we could get close to 100 Hz resolution for all three intervals, and we could do this in an automated way during the mechanically and electrically quiet nighttimes with none of us present.

However, at the new level of precision that we were exploring we encountered systematic frequency shifts that suggested to us that we had pushed saturated absorption measurements in a discharge cell as far as they should reasonably be pushed. Given the large amount of line splitting already being done, and the theoretical inconsistencies, we decided not to replace the cell with a helium beam. Instead, several years ago we shut the experiment down – perhaps the first discontinued optical comb experiment – and decided to pursue measurements of the electric dipole moment of the electron instead.

6.5.2. *Historically important methods*

In Fig. 6.1 there is a factor of more than 20 between the sizes of the uncertainties for the most accurate determinations of α that have already been discussed above. All other measurements of α have larger error bars that will not fit on this scale. Several additional measurements fit on the 8 times expanded scale of Fig. 6.5, though the error bars for the most accurate determinations of α from $g/2$ are then too small to be visible.

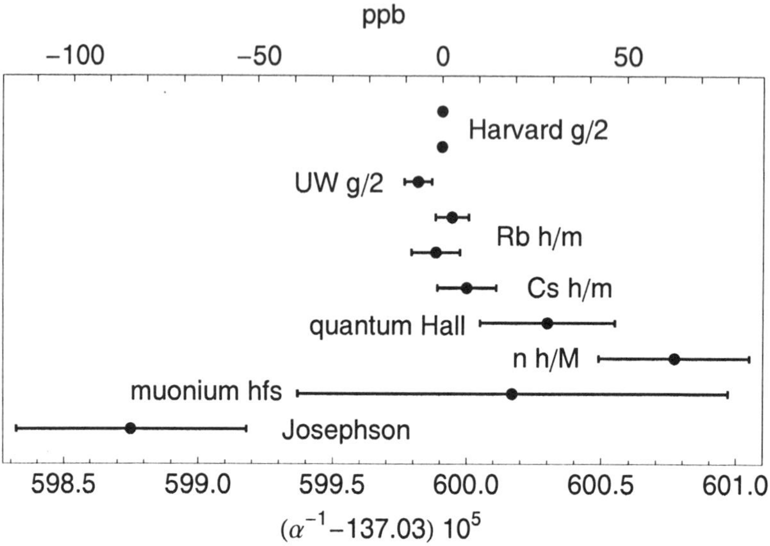

Fig. 6.5. Less accurate measurements of α compared upon an expanded scale. The uncertainties in the two most accurate determinations of α are too small to be visible on this large scale.

A summary and discussion of traditional measurements of α is in Ref. [23]. The work includes the value deduced from the quantum Hall resistance [64], a value that essentially agrees with the more accurate determinations of α insofar as these lie almost within its one standard deviation error bars.

A measurement using neutrons [65] that is similar in spirit to the described Cs and Rb measurements is also plotted. Different mass ratios are required, of course, but an even more important difference is that $\hbar/M_n$ is deduced from the diffraction of cold neutrons from a Si crystal. The lattice spacing in Si is thus crucial, and there is an impressive range of differing values for this lattice constant [23]. A recommended value [23] is used for the figure but given the range of measured lattice constants it is not so surprising that this value of α does not agree so well with more accurate measurements.

Values from muonium hyperfine structure measurements [23, 43] and from measurements of the AC Josephson effect (with related measurements [23]) are also plotted because of their importance in the past. It is not clear why the latter solid state measurement disagree so much with the more accurate values.

6.6. Conclusion

Combined measurements of the Rydberg constant, two mass ratios, a laser frequency, and an atom recoil frequency together determine α using Cs atoms to 8.0 ppb, and using Rb atoms to 4.6 ppb. Efforts are underway to improve both sets of measurements enough to determine α to 1 ppb.

Helium fine structure measurements are now accurate enough to determine α at nearly the same precision, but with completely different systematic uncertainties. Unfortunately, the two-electron QED theory needed to relate fine structure intervals to α heeds to be clarified before this can happen.

New measurements of the electron magnetic moment $g/2$, along with QED calculations, determine the fine structure constant much more accurately than ever before, to 0.4 ppb. The uncertainty in α will be reduced, without the need for a more accurate measurement of $g/2$, when a first calculation of the tenth-order QED coefficient is completed. It seems reasonable to reduce the experimental and theoretical contribution to determinations of α from $g/2$ to 0.1 ppb or better in efforts now underway, though this will take some time.

Acknowledgments

Useful comments on this manuscript from F. Biraben, D. Hanneke, T. Kinoshita, S. Laporta, W. Marciano, P. Mohr, H. Mueller, M. Nio, M. Passera, E. de Rafael, E. Remiddi, and B. L. Roberts are gratefully acknowledged. Support for this work came from the NSF, the AFOSR, and from the Humboldt Foundation.

References

[1] G. Giusfredi, P. de Natale, D. Mazzotti, P. C. Pastor, C. de Mauro, L. Fallani, G. Hagel, V. Krachmalnicoff, and M. Inguscio, *Can. J. Phys.* **83**, 301–309, (2005).

[2] J. Castillega, D. Livingston, A. Sanders, and D. Shiner, *Phys. Rev. Lett.* **84**, 4321–4324, (2000).

[3] C. H. Storry, M. C. George, and E. A. Hessels, *Phys. Rev. Lett.* **84**, 3274–3277, (2000).

[4] M. C. George, L. D. Lombardi, and E. A. Hessels, *Phys. Rev. Lett.* **87**, 173002, (2001).

[5] G. W. F. Drake, *Can. J. Phys.* **80**, 1195–1212, (2002).

[6] K. Pachucki, *Phys. Rev. Lett.* **97**, 013002, (2006).

[7] D. Hanneke, S. Fogwell, and G. Gabrielse, *Phys. Rev. Lett.* **100**, 120801, (2008).
[8] B. Odom, D. Hanneke, B. D'Urso, and G. Gabrielse, *Phys. Rev. Lett.* **97**, 030801, (2006).
[9] R. S. Van Dyck, Jr., P. B. Schwinberg, and H. G. Dehmelt, *Phys. Rev. Lett.* **59**, 26–29, (1987).
[10] J. Schwinger, *Phys. Rev.* **73**, 416L, (1948).
[11] C. M. Sommerfield, *Phys. Rev.* **107**, 328, (1957).
[12] A. Petermann, *Helv. Phys. Acta.* **30**, 407, (1957).
[13] C. M. Sommerfield, *Ann. Phys. (N.Y.).* **5**, 26, (1958).
[14] S. Laporta and E. Remiddi, *Phys. Lett. B.* **379**, 283, (1996).
[15] T. Aoyama, M. Hayakawa, T. Kinoshita, and M. Nio, *Phys. Rev. Lett.* **99**, 110406, (2007).
[16] T. Udem, A. Huber, B. Gross, J. Reichert, M. Prevedelli, M. Weitz, and T. W. Hänsch, *Phys. Rev. Lett.* **79**, 2646–2649, (1997).
[17] C. Schwob, L. Jozefowski, B. de Beauvoir, L. Hilico, F. Nez, L. Julien, F. Biraben, O. Acef, J. J. Zondy, and A. Clairon, *Phys. Rev. Lett.* **82**, 4960–4963, (1999).
[18] M. P. Bradley, J. V. Porto, S. Rainville, J. K. Thompson, and D. E. Pritchard, *Phys. Rev. Lett.* **83**, 4510–4513, (1999).
[19] G. Werth, J. Alonso, T. Beier, K. Blaum, S. Djekic, H. Häffner, N. Hermanspahn, W. Quint, S. Stahl, J. Verdú, T. Valenzuela, and M. Vogel, *Int. J. Mass Spectrom.* **251**, 152, (2006).
[20] V. Gerginov, K. Calkins, C. E. Tanner, J. J. McFerran, S. Diddams, A. Bartels, and L. Hollberg, *Phys. Rev. A.* **73**, 032504, (2006).
[21] M. Cadoret, E. de Mirandes, P. Cladé, S. Guellati-Hkélifa, C. Schwob, F. Nez, L. Julien, and F. Biraben, *Phys. Rev. Lett.* **101**, 230801, (2008).
[22] A. Wicht, J. M. Hensley, E. Sarajlic, and S. Chu, *Phys. Scr.* **T102**, 82–88, (2002).
[23] P. J. Mohr, B. N. Taylor, and D. B. Newall, *Rev. Mod. Phys.* **80**, 633, (2008).
[24] K. Pachucki, A. Czarnecki, U. Jentschura, and V. A. Yerokhin, *Phys. Rev. A.* **72**, 022108, (2005).
[25] T. Zelevinsky, D. Farkas, and G. Gabrielse, *Phys. Rev. Lett.* **95**, 203001, (2005).
[26] G. Giusfredi, P. de Natale, D. Mazzotti, P. C. Pastor, C. de Mauro, L. Fallani, G. Hagel, V. Krachmalnicoff, and M. Inguscio, *Can. J. Phys.* **83**, 301–309, (2005).
[27] I. M. Mills, P. J. Mohr, T. J. Quinn, B. N. Taylor, and E. R. Williams, *Metrologia.* **43**, 227, (2006).
[28] G. W. Bennett and *et al.*, *Phys. Rev. D.* **73**, 072003, (2006).
[29] S. Peil and G. Gabrielse, *Phys. Rev. Lett.* **83**, 1287–1290, (1999).
[30] G. Gabrielse and F. C. MacKintosh, *Intl. J. of Mass Spec. and Ion Proc.* **57**, 1–17, (1984).
[31] L. S. Brown and G. Gabrielse, *Phys. Rev. A.* **25**, 2423–2425, (1982).
[32] G. Gabrielse and H. Dehmelt, *Phys. Rev. Lett.* **55**, 67–70, (1985).
[33] J. Tan and G. Gabrielse, *Phys. Rev. Lett.* **67**, 3090–3093, (1991).

[34] B. D'Urso, R. Van Handel, B. Odom, D. Hanneke, and G. Gabrielse, *Phys. Rev. Lett.* **94**, 113002, (2005).
[35] B. D'Urso, B. Odom, and G. Gabrielse, *Phys. Rev. Lett.* **90** (4), 043001, (2003).
[36] P. J. Mohr and B. N. Taylor, *Rev. Mod. Phys.* **72**, 351–495, (2000).
[37] P. J. Mohr and B. N. Taylor, *Rev. Mod. Phys.* **77**, 1–107, (2005).
[38] A. Czarnecki, B. Krause, and W. J. Marciano, *Phys. Rev. Lett.* **76**, 3267–3270, (1996).
[39] A. Petermann, *Nucl. Phys.* **5**, 677, (1958).
[40] R. Karplus and N. M. Kroll, *Phys. Rev.* **77**, 536, (1950).
[41] H. H. Elend, *Phys. Rev. Lett.* **20**, 682, (1966). **21**, 720(E) (1966).
[42] M. Passera, *Phys. Rev. D.* **75**, 013002, (2007).
[43] W. Liu, M. G. Boshier, O. v. D. S. Dhawan, P. Egan, X. Fei, M. G. Perdekamp, V. W. Hughes, M. Janousch, K. Jungmann, D. Kawall, F. G. Mariam, C. Pillai, R. Prigl, G. z. Putlitz, I. Reinhard, W. Schwarz, P. A. Thompson, and K. A. Woodle, *Phys. Rev. Lett.* **82**, 711–714, (1999).
[44] T. Kinoshita, *Phys. Rev. Lett.* **75**, 4728, (1995).
[45] S. Laporta, *Nuovo Cim. A.* **106A**, 675–683, (1993).
[46] S. Laporta and E. Remiddi, *Phys. Lett. B.* **301**, 440–446, (1993).
[47] J. H. Kühn, *et al.*, *Phys. Rev. D.* **68**, 033018, (2003).
[48] G. P. Lepage, *J. Comput. Phys.* **27**, 192–203, (1978).
[49] T. Kinoshita and M. Nio, *Phys. Rev. Lett.* **90**, 021803, (2003).
[50] G. Gabrielse, D. Hanneke, T. Kinoshita, M. Nio, and B. Odom, *Phys. Rev. Lett.* **97**, 030802, (2006). *ibid.* **99**, 039902 (2007).
[51] T. Aoyama, M. Hayakawa, T. Kinoshita, and M. Nio, *Nucl. Phys.* **B740**, 138, (2006).
[52] T. Aoyama, M. Hayakawa, T. Kinoshita, and M. Nio, *Phys. Rev. D.* **78**, 113006, (2008).
[53] S. Laporta and E. Remiddi, (private communication).
[54] T. Beier, H. Häffner, N. Hermanspahn, S. G. Karshenboim, H.-J. Kluge, W. Quint, S. Stahl, J. Verdú, and G. Werth, *Phys. Rev. Lett.* **88**, 011603, (2002).
[55] G. Gabrielse, L. Haarsma, and S. L. Rolston, *Intl. J. of Mass Spec. and Ion Proc.* **88**, 319–332, (1989). ibid. 93, 121 1989.
[56] G. Gabrielse, *Int. J. Mass Spectrom.* **279**, 107, (2009).
[57] D. L. Farnham, R. S. Van Dyck, Jr., and P. B. Schwinberg, *Phys. Rev. Lett.* **75**, 3598–3601, (1995).
[58] G. Gabrielse, *Phys. Rev. A.* **27**, 2277–2290, (1983).
[59] P. Cladé, E. de Mirandes, M. Cadoret, S. Guellati-Khélifa, C. Schwob, F. Nez, L. Julien, and F. Biraben, *Phys. Rev. Lett.* **96**, 033001, (2006). Phys. Rev. A **74**, 052109 (2006).
[60] C. Bordé, *Phys. Lett. A.* **140**, 10, (1989).
[61] D. S. Weiss, B. C. Young, and S. Chu, *Phys. Rev. Lett.* **70**, 2706–2709, (1993).
[62] E. Peik, M. B. Dahan, I. Bouchoule, Y. Castin, and C. Salomon, *Phys. Rev. D.* **55**, 2289, (1997).

[63] H. Müller, S. Chiow, Q. Long, S. Herrmann, and S. Chu, *Phys. Rev. Lett.* **100**, 180405, (2008).
[64] A. M. Jeffery, R. E. Elmquist, L. H. Lee, J. Q. Shields, and R. F. Dziuba, *IEEE Trans. Instrum. Meas.* **46**, 264, (1997).
[65] E. Krüger, W. Nistler, and W. Weirauch, *Metrologia.* **36** (2), 147–148, (1999).

Chapter 7

Helium Fine Structure Theory for the Determination of α

Krzysztof Pachucki
Institute of Theoretical Physics, University of Warsaw
Hoża 69, 00-681 Warsaw, Poland
krp@fuw.edu.pl

Jonathan Sapirstein
Department of Physics, University of Notre Dame
Notre Dame, IN 46556
jsapirst@nd.edu

Recent advances in the application of effective field theory to the helium atom have allowed the calculation of all contributions up to order $m\alpha^7$ to the fine structure of 2^3P_J states along with recoil corrections up to order $m\alpha^6 \frac{m}{M}$. Combined with very precise experiments these calculations allow a determination of the fine structure constant α. The derivation of α from helium fine structure, while not at present competitive in accuracy with the value of α available from electron g-2, has a very different dependence on theory. A discrepancy with another calculation and directions for future progress are discussed.

Contents

7.1. Introduction

The fine structure constant, $\alpha = \frac{e^2}{4\pi\epsilon_0 \hbar c} \approx 1/137$, was introduced in 1916 by Sommerfeld [1], who made Bohr's derivation of hydrogen energy levels consistent with relativistic theory. He showed that the nonrelativistic degeneracy of energy levels for a given principal quantum number n is partially removed by relativistic effects. Specifically, the first two terms in the Taylor expansion in $Z\alpha$ of his formula for energy levels, later rigorously derived from the Dirac equation, are

$$E(n,j) = -\frac{m(Z\alpha)^2}{2n^2} - \frac{m(Z\alpha)^4}{2n^3}\left[\frac{1}{j+1/2} - \frac{3}{4n}\right]. \tag{7.1}$$

We note they depend only on the total angular momentum j, but not on spin or orbital angular momentum. Here we introduce a convention of keeping the nuclear charge Z general, which is useful for distinguishing loop corrections, which go as α^n, from "binding corrections", which scale as powers of $Z\alpha$. If this were the exact equation for the energy, and one had an accurate measurement of a splitting, one could use that measurement to deduce the value of α from atomic spectroscopy as the Rydberg constant

$$R_\infty = \frac{\alpha^2 m_e c}{2h} \tag{7.2}$$

is known with very high precision. An example we will concentrate on in the following is the frequency associated with the splitting ΔE between the $2p_{1/2}$ and $2p_{3/2}$ states in hydrogen, $\nu_0 \equiv \alpha^2 R_\infty c/16$. The most accurate direct experimental determination of this frequency is [2]

$$\nu(\text{exp}) = \frac{\Delta E}{h} = 10\ 969.13(10)\,\text{MHz}. \tag{7.3}$$

One can then solve for α_0, the value of the fine structure constant in the infinite nuclear mass limit and with the neglect of relativistic and quantum electrodynamic (QED) corrections, through

$$\alpha_0^2 = \frac{16\nu(\text{exp})}{R_\infty c}. \tag{7.4}$$

Using the CODATA value [3]

$$R_\infty c = 3.289\ 841\ 960\ 361(22)\ 10^9\ \text{MHz} \tag{7.5}$$

then gives

$$\alpha_0^{-1}(\mathrm{H}) = 136.911\ 97(62). \tag{7.6}$$

In the first part of this chapter we describe an effective field theory approach, later generalized to the helium problem, that allows the systematic calculation of corrections associated with the finite mass of the nucleus, relativistic, and QED corrections up to order $\alpha^3\nu_0$, and show that their inclusion shifts this value of the fine structure constant to

$$\alpha^{-1}(\mathrm{H}) = 137.035\ 45(62). \tag{7.7}$$

In general there are both theory and experimental sources of errors, but in this case the uncertainty associated with uncalculated theoretical terms is negligible compared to the error coming from the experiment. This value of the fine structure constant is consistent with the most accurately known value of α at present,

$$\alpha^{-1}(g{-}2) = 137.035\ 999\ 084(51). \tag{7.8}$$

This value is determined from the measurement of the electron anomalous magnetic moment a_e by Hanneke *et al.* [4],

$$a_e = 0.001\ 159\ 652\ 180\ 73(28), \tag{7.9}$$

together with the recently revised four-loop calculation by Kinoshita and collaborators [5]. The 4.5 ppm determination of α from hydrogen is much less precise than the 0.37 ppb value from g-2 because of the short lifetime of hydrogenic $2p$ states, of order 10^{-9} seconds. We note in passing that determinations of α that require much less QED calculation, and are approaching the accuracy of the g-2 measurement, are available from recoil measurements on rubidium [6] and cesium [7].

We now describe the corrections for hydrogen fine structure in some detail, as a similar approach will be used for the helium calculation, and in addition some contributions carry over to the latter case. We begin by describing how the corrections are categorized by size, and will then describe the actual calculation using an effective field theory. There are three expansion parameters in the bound state problem, the ratio of the electron mass to the nuclear mass, $\frac{m}{M}$, the fine structure constant α, and $Z\alpha$. Any correction of order $\frac{m}{M}$ or higher will be called a recoil correction. As mentioned above, there are two sources of corrections of order α^n, one without a factor of Z that is associated with the number of photon-electron vertices, and one with that factor, associated with photon-nucleus vertices. A characteristic difficulty of applying QED to bound states is the fact that one sometimes encounters infinite sets of Feynman diagrams that all contribute to the same order in $Z\alpha$. Technically this comes from diagrams in which adding a photon exchange between the electron and nucleus, which nominally is down by a factor $(Z\alpha)^2$, stays the same order because the extra electron propagator is almost on mass shell and gives an inverse factor of $(Z\alpha)^2$. This is the case with the leading QED correction, which begins in order $m\alpha(Z\alpha)^4$. While the term "Lamb shift" is sometimes used only for the $2s_{1/2} - 2p_{1/2}$ splitting in hydrogen, in the following we will refer to any QED contribution that starts in this order as a Lamb shift. While part of the Lamb shift is associated with one loop diagrams with free electron propagators, the so-called "Bethe log" term involves an infinite set of diagrams which fortunately can be rewritten as a closed form expression with a bound electron propagator. Binding corrections to the one-loop Lamb shift of order $m\alpha(Z\alpha)^n$ have been carried out to $n = 6$ [8] and of the two-loop Lamb shift of order $m\alpha^2(Z\alpha)^n$ to $n = 5$ [9]. In addition the three-loop Lamb shift without binding corrections, which is of order $m\alpha^3(Z\alpha)^4$, has been evaluated [10]. At $Z = 1$ these orders are of course all $m\alpha^7$. Significant progress has been made on even higher orders from numerical evaluations of the one-loop Lamb shift [11] , effective field theory treatment of the two-loop Lamb shift [12], and numerical evaluation of the two-loop Lamb shift [13], but at this point we simply wish to stress that it has been a major effort to reach the $m\alpha^7$ level in hydrogen. It is only very recently that the same level has been reached for fine structure in helium using effective field theory [14], and the main purpose of this chapter is to describe this calculation. We note that a Bethe–Salpeter approach that claims the same accuracy has been used by Zhang [15], but our results differ from his significantly, as will be discussed in the conclusion.

As we will be using the tool of effective field theory to carry out the more challenging helium calculation, it is useful to introduce it in the simpler hydrogenic case. The basic idea of effective field theory is to represent various corrections to energies in terms of effective operators, the expectation values of which are evaluated with Schrödinger wave functions. There are various ways to obtain these operators. A particularly powerful method used for calculating the $m\alpha^7$ terms involves the Foldy–Wouthuysen transformation [16], but for the derivation of lower order terms we use a method based on considerations involving the scattering of free particles.

In this approach we carry out a matching procedure, where we first calculate scattering amplitudes of low velocity on-shell particles using QED. We then introduce an effective theory that has perturbations added to a nonrelativistic starting point that are determined by requiring that the two scattering amplitudes agree up to a given order in the velocities. We work in the center of mass system, so that if the incoming electron has momentum $\vec{p}_{\rm i}$ the incoming nucleus has momentum $-\vec{p}_{\rm i}$, and similarly if the outgoing electron is taken to have momentum $\vec{p}_{\rm f}$ the outgoing nucleus has momentum $-\vec{p}_{\rm f}$. The hydrogen fine structure problem is complicated by the presence of the proton spin, but at the level of accuracy of the experiment it is a valid approximation to ignore it, so in the following we will drop any terms depending on $\vec{\sigma}_p$ and write $\vec{\sigma}$ for the electron spin $\vec{\sigma}_e$.

The most appropriate gauge for this problem is Coulomb gauge, though Feynman gauge can be used for the QED part of the calculation providing a gauge invariant set of graphs is considered. If we start with an electron scattering on a nucleus, use of free Dirac spinors, normalized so that $u^\dagger u = 1$,

$$u(\vec{p}) = \sqrt{\frac{E+m}{2E}}\begin{pmatrix} 1 \\ \frac{\vec{\sigma}\cdot\vec{p}}{E+m}\end{pmatrix} \tag{7.10}$$

with $E = \sqrt{\vec{p}^{\,2}+m^2}$ gives for Coulomb photon exchange the following scattering amplitude,

$$\begin{aligned}\mathcal{M}_C(\vec{p}_{\rm f},\vec{p}_{\rm i}) &= -\frac{4\pi Z\alpha}{\vec{q}^{\,2}}u^\dagger(p_{\rm f})u(p_{\rm i}) \\ &= -\frac{4\pi Z\alpha}{\vec{q}^{\,2}}\left[1-\frac{\vec{q}^{\,2}}{8m^2}+i\frac{\vec{\sigma}\cdot\vec{p}_{\rm f}\times\vec{p}_{\rm i}}{4m^2}+\frac{5}{128m^4}({p_{\rm f}}^2-{p_{\rm i}}^2)^2\right. \\ &\quad \left.+\frac{3}{64m^4}({p_{\rm f}}^2+{p_{\rm i}}^2)(\vec{q}^{\,2}-2i\vec{\sigma}\cdot\vec{p}_{\rm f}\times\vec{p}_{\rm i})\right]+\ldots\,. \end{aligned} \tag{7.11}$$

Here $\vec{q}=\vec{p}_{\rm f}-\vec{p}_{\rm i}$ and we have made a Taylor expansion assuming $|\vec{p}_{\rm i}|/m << 1$ and $|\vec{p}_{\rm f}|/m << 1$. The nucleus has similar Dirac spinors, but their effect for Coulomb photon exchange enters in order $1/M^2$ and can be neglected. We can also account for the leading QED correction by making the replacement

$$\gamma_\mu \to \gamma_\mu F_1(q^2)+i\frac{\sigma_{\mu\nu}q^\nu}{2m}F_2(q^2), \tag{7.12}$$

with $q_0 \equiv E(\vec{p}_{\rm f})-E(\vec{p}_{\rm i})$ and $\sigma_{\mu\nu}\equiv\frac{i}{2}[\gamma_\mu,\gamma_\nu]$. Note that q_0 is of order $\vec{p}^{\,2}/m$ so that $q^2\approx-\vec{q}^{\,2}$. The one-loop form factor F_2 has the low q^2 expansion

$$F_2(q^2)=\frac{\alpha}{2\pi}+\frac{\alpha}{\pi}\frac{q^2}{12m^2}+\ldots\,. \tag{7.13}$$

The expansion of F_1 involves an infrared divergence that will be treated later, but its lowest order value is of course one. If we make the approximation $F_2(q^2)=\frac{\alpha}{2\pi}$ and again consider only Coulomb photon exchange, a simple calculation shows that the F_1 term reproduces Eq. (7.11), and that the F_2 term gives

$$\mathcal{M}^{(0)}_{C(a_e)}(\vec{p}_{\rm f},\vec{p}_{\rm i}) = -\frac{4\pi Z\alpha}{\vec{q}^{\,2}}\bar{u}(p_{\rm f})\frac{-i\sigma_{0j}q_j}{2m}u(p_{\rm i})\frac{\alpha}{2\pi} = \frac{4\pi Z\alpha}{\vec{q}^{\,2}}\left[\frac{\vec{q}^{\,2}}{4m^2} - i\frac{\vec{\sigma}\cdot\vec{p}_{\rm f}\times\vec{p}_{\rm i}}{2m^2}\right]\frac{\alpha}{2\pi}. \quad (7.14)$$

Comparing this contribution with the lowest order one, we see that we can combine them by inserting a factor $1+2\frac{\alpha}{2\pi}$ in both the spin-independent part, which is the Darwin term, and in the spin-dependent part, the spin-orbit term. In fact, we can automatically account for an infinite set of terms by making the replacement $\frac{\alpha}{2\pi} \to a_e$, since we know $F_2(0) = a_e$, and in the following we will always follow this convention. We observe that this practice eliminates a small α dependence of the theoretical fine structure that would arise if one used the theoretical expansion of a_e in powers of α instead of the accurately known [4] experimental value a_e. We also note two higher-order corrections from the F_2 vertex,

$$\mathcal{M}^{(2)}_{C(a_e)}(\vec{p}_{\rm f},\vec{p}_{\rm i}) = \frac{4\pi Z\alpha}{\vec{q}^{\,2}}\frac{a_e}{32m^4}\Big[-3p_{\rm i}^4 - 3p_{\rm f}^4 + 4(p_{\rm i}^2+p_{\rm f}^2)\vec{p}_{\rm f}\cdot\vec{p}_{\rm i} - 2p_{\rm i}^2p_{\rm f}^2 + 4i(p_{\rm i}^2+p_{\rm f}^2)\vec{\sigma}\cdot\vec{p}_{\rm f}\times\vec{p}_{\rm i}\Big] - \frac{\alpha Z\alpha}{3m^2}\left[\frac{\vec{q}^{\,2}}{4m^2} - \frac{i\vec{\sigma}\cdot\vec{p}_{\rm f}\times\vec{p}_{\rm i}}{2m^2}\right], \quad (7.15)$$

where the first correction comes from expanding the Dirac spinors further in velocities and the second from the second term in $F_2(q^2)$. In the bound state problem $p/m \propto Z\alpha$, so both these corrections will lead to $\alpha(Z\alpha)^2$ corrections to fine structure, that is, they will contribute in order $m\alpha(Z\alpha)^6$.

Turning now to transverse photon exchange, we note that the Dirac spinors of the nucleus must now be treated, and the net result will be a recoil effect, as a single power of the nuclear mass M is present in the denominator. The diagram leads to a scattering amplitude

$$\mathcal{M}_T(\vec{p}_{\rm f},\vec{p}_{\rm i}) = -\frac{2\pi Z\alpha}{Mq^2}(p_{\rm i}+p_{\rm f})_i\left(\delta_{ij} - \frac{q_iq_j}{\vec{q}^{\,2}}\right)\bar{u}(p_{\rm f})\gamma_j u(p_{\rm i}) \equiv -\frac{2\pi Z\alpha}{Mq^2}(p_{\rm i}+p_{\rm f})_i P_{ij}(q)\bar{u}(p_{\rm f})\gamma_j u(p_{\rm i}), \quad (7.16)$$

where, as discussed above, terms involving the spin of the proton have been suppressed. Again expanding in powers of momentum divided by mass and keeping only the leading term gives

$$\mathcal{M}^{(0)}_T(\vec{p}_{\rm f},\vec{p}_{\rm i}) = -\frac{4\pi Z\alpha}{Mm\vec{q}^{\,2}}\left(\frac{i}{2}\vec{\sigma}\cdot\vec{p}_{\rm f}\times\vec{p}_{\rm i} + \frac{\vec{q}^{\,2}\vec{p}_{\rm f}\cdot\vec{p}_{\rm i} - \vec{q}\cdot\vec{p}_{\rm f}\,\vec{q}\cdot\vec{p}_{\rm i}}{\vec{q}^{\,2}}\right). \quad (7.17)$$

After carrying out another calculation with the F_2 form factor, the spin-dependent part of the above gets multiplied by $1 + a_e$ rather than $1 + 2a_e$, while the spin independent term is unaffected. Our results up to second order can be summarized as

$$\mathcal{M}^{(2)}(\vec{p}_{\rm f}, \vec{p}_{\rm i}) = -\frac{4\pi Z\alpha}{\vec{q}^{\,2}}\left[1 - \frac{\vec{q}^{\,2}}{8m^2}(1+2a_e) + i\frac{\vec{\sigma}\cdot\vec{p}_{\rm f}\times\vec{p}_{\rm i}}{4m^2}(1+2a_e) + \right.$$
$$\left. i\frac{\vec{\sigma}\cdot\vec{p}_{\rm f}\times\vec{p}_{\rm i}}{2mM}(1+a_e) + \frac{\vec{q}^{\,2}\vec{p}_{\rm f}\cdot\vec{p}_{\rm i} - \vec{q}\cdot\vec{p}_{\rm f}\vec{q}\cdot\vec{p}_{\rm i}}{Mm\vec{q}^{\,2}}\right]. \quad (7.18)$$

So far we have been doing a QED calculation of the scattering of low velocity particles in momentum space. In coordinate space we have corresponding operators defined by

$$O(\vec{r})\delta^3(\vec{r}-\vec{r}\,') = \frac{1}{(2\pi)^6}\int d^3p_{\rm f}d^3p_{\rm i}e^{i\vec{p}_{\rm f}\cdot\vec{r}}\mathcal{M}(\vec{p}_{\rm f},\vec{p}_{\rm i})e^{-i\vec{p}_{\rm i}\cdot\vec{r}\,'}. \quad (7.19)$$

If we consider only the first term in Eq. (7.18), the coordinate space form of the operator is $-Z\alpha/r$, as can be seen by changing variables from $\vec{p}_{\rm f}$ to $\vec{q}$. In general, whenever a scattering amplitude depends only on $\vec{q}$ an operator that is a function only of $\vec{r}$ results, but if additional factors of momentum are present in the numerator terms involving derivatives, which of course can be written as momentum operators, will be present.

The only spin-dependent operators needed to make the nonrelativistic theory agree with QED are the spin-orbit operator, the coordinate space version of the third term in Eq. (7.18),

$$H_{SO} = \frac{Z\alpha}{4m^2}\vec{\sigma}\cdot\frac{\vec{r}\times\vec{p}}{r^3}(1+2a_e), \quad (7.20)$$

along with the recoil term

$$H_R = \frac{Z\alpha}{2mM}\vec{\sigma}\cdot\frac{\vec{r}\times\vec{p}}{r^3}(1+a_e). \quad (7.21)$$

While not contributing to fine structure in lowest order, we note the presence of three spin-independent terms, the Darwin term,

$$H_D = \frac{\pi Z\alpha}{2m^2}\delta(\vec{r})(1+2a_e), \quad (7.22)$$

a second term frequently called the Breit term,

$$H_B = -\frac{Z\,\alpha}{2\,M m\,r}\left(\delta_{ij} + \frac{r_i r_j}{r^2}\right)p_i p_j, \quad (7.23)$$

and finally the standard relativistic mass increase operator

$$H_{\rm RM} = -\frac{p^4}{8m^3}. \quad (7.24)$$

If we add these operators as perturbations to a nonrelativistic starting point, the theory will agree with QED to order $(v/c)^2$ for the scattering of slowly moving particles.

The Hamiltonian of the effective theory without relativistic corrections is the familiar one from elementary quantum mechanics with the electron mass replaced with the reduced mass, since the Hamiltonian in the system we have chosen is

$$\begin{aligned} H_{NR} &= \frac{\vec{p}^{\,2}}{2m} + \frac{\vec{p}^{\,2}}{2M} - \frac{Z\alpha}{r} \\ &\equiv \frac{\vec{p}^{\,2}}{2m_r} - \frac{Z\alpha}{r}. \end{aligned} \tag{7.25}$$

When used to solve the bound state problem this Hamiltonian gives the usual Schrödinger energies and wave functions in terms of the reduced mass. A crucial point in effective field theories is that the extra operators derived above from scattering considerations can also be treated as perturbations to this Hamiltonian for the bound state problem, so one can use standard Rayleigh–Schrödinger perturbation theory to evaluate their contributions, with the operator defined in Eq. (7.19) giving the energy shift

$$\Delta E = \int d^3r \phi^*(\vec{r}) O(\vec{r}) \phi(\vec{r}). \tag{7.26}$$

Scaling arguments ($p \propto m_r\alpha$, $r \propto 1/m_r\alpha$) show that the reduced mass then enters as $(m_r/m)^3$ for both operators contributing to fine structure. Evaluating their expectation value gives a much more accurate replacement for our original formula $\nu_0 = \alpha^2 R_\infty c/16$, specifically

$$\begin{aligned} \bar{\nu}_0 &= \frac{\alpha^2 R_\infty c}{16}\left(\frac{m_r}{m}\right)^3\left(1 + 2a_e + 2\frac{m}{M}(1 + a_e)\right) \\ &\equiv A\,\alpha^2 \\ A &= 205\ 979\ 506.129(1)\ \text{MHz}. \end{aligned} \tag{7.27}$$

When we turn to helium, we will start with a form analogous to this. By including these dominant corrections we find

$$\bar{\alpha}_0^{-1} = 137.033\ 24(62), \tag{7.28}$$

now only 20 ppm smaller than the g-2 value of α^{-1}.

We include higher order theoretical contributions by defining

$$\begin{aligned} \nu_{th} &\equiv \bar{\nu}_0 + \delta\nu_{th} \\ &= \bar{\nu}_0 + \nu^{(4)} + \nu^{(4r)} + \nu^{(5)} + \ldots, \end{aligned} \tag{7.29}$$

where $\nu^{(4)}$ accounts for corrections of order $\alpha^4 R_\infty$, $\nu^{(4r)}$ for corrections of order $\frac{m}{M}\alpha^4 R_\infty$, and $\nu^{(5)}$ of order $\alpha^5 R_\infty$. These corrections are of course of order $\alpha^2\nu_0$, $\frac{m}{M}\alpha^2\nu_0$, and $\alpha^3\nu_0$. Once they are computed, the formula for determining α is

$$\alpha^2 = \frac{\nu_{exp} - \delta\nu_{th}}{A}, \tag{7.30}$$

where we are assuming the validity of QED to identify ν_{th} with the measured value ν_{exp}. In the following we will show that

$$\begin{aligned}
\nu^{(4)} &= \frac{5m(Z\alpha)^6}{256} \\
\nu^{(4r)} &= -\frac{m^2(Z\alpha)^6}{64M} \\
\nu^{(5)} &= \frac{\alpha m(Z\alpha)^6}{8\pi}\left[\frac{1}{9} + \frac{1}{4}\ln(Z\alpha)^2 + 0.38442 + \frac{1}{20}\right],
\end{aligned} \tag{7.31}$$

which contribute 0.364 415 MHz, -0.000 159 MHz, and -0.008 495 MHz respectively (the factor 1/20 in $\nu^{(5)}$ is from vacuum polarization). Since the experimental error is 0.10 MHz, only $\nu^{(4)}$ plays a significant role for hydrogen. Using it in Eq. (7.30) then allows one to derive the value of α given in Eq. (7.7). While much more complex, the helium calculation presented in the next section follows the same pattern, but in this case the experiment is far more precise, so that the accurate determination of all of the terms in $\delta\nu_{th}$ is essential.

We now present the theory for $\delta\nu_{th}$ in hydrogen, starting with $\nu^{(4)}$. Because we treat $\nu^{(4r)}$ separately, $\nu^{(4)}$ is to be evaluated in the nonrecoil limit, so the Dirac equation can be used. The next term in the expansion of Sommerfeld's formula,

$$\begin{aligned}
E(n,j) &= mf(n,j) \\
&= m\left[1 + \left(\frac{Z\alpha}{n - (j+\frac{1}{2}) + \sqrt{(j+\frac{1}{2})^2 - (Z\alpha)^2}}\right)^2\right]^{-1/2}
\end{aligned} \tag{7.32}$$

contributes a relative factor $5/8\alpha^2$ to $n = 2$ hydrogen fine structure. However, since we are going to use an effective field theory approach for helium, there being no counterpart of the Dirac equation in that system [17], we show below how the $5/8\alpha^2$ correction is calculated using this approach. In Ref. [18] this procedure was carried out for the more difficult case of s-state energies, and we briefly describe it here to introduce techniques that will be used extensively in the helium calculation.

The use of effective field theory described above is, with the exception of the inclusion of the anomalous magnetic moment of the electron and carrying out some Taylor expansions to fourth rather than second order, very well known, with the derivation of the fine structure formula a standard exercise in quantum mechanics textbooks. However, in textbooks the discussion stops at first-order perturbation theory. To obtain terms of order $m(Z\alpha)^6$ one must go to second-order perturbation theory with the operators already derived, and in addition new operators two orders of α smaller must be found by carrying out the matching procedure to higher orders in the velocities of the particles. At this point the fact that the operators are nearly singular becomes important. To show the problem, we first consider the effect of second order perturbation theory in the Darwin interaction. The general expression for second-order perturbation theory is

$$\begin{aligned}\Delta E_2 &= \sum_{m\neq 0} \frac{\langle 0|V|m\rangle\langle m|V|0\rangle}{E_0 - E_m} \\ &= \sum_{m\neq 0} \int d^3r_1 d^3r_2 \frac{\phi_0^*(\vec{r}_1)V(\vec{r}_1)\phi_m(\vec{r}_1)\phi_m^*(\vec{r}_2)V(\vec{r}_2)\phi_0(\vec{r}_2)}{E_0 - E_m} \\ &\equiv \int d^3r_1 d^3r_2 \phi_0^*(\vec{r}_1)V(\vec{r}_1)G_R(\vec{r}_1,\vec{r}_2;E_0)V(\vec{r}_2)\phi_0(\vec{r}_2),\end{aligned} \tag{7.33}$$

where G_R is a reduced Green's function that builds in the restriction $m \neq 0$. If this is transformed into momentum space we get

$$\Delta E_2 = \int \frac{d^3q_1 d^3q_2 d^3q_3 d^3q_4}{(2\pi)^{12}} \phi_0^*(\vec{q}_4)V(\vec{q}_4 - \vec{q}_3)G_R(\vec{q}_3,\vec{q}_2;E_0)V(\vec{q}_2 - \vec{q}_1)\phi(\vec{q}_1). \tag{7.34}$$

The part of the reduced propagator that gives rise to the strongest singularity is the free propagator,

$$G_0(\vec{q}_3,\vec{q}_2;E_0) = (2\pi)^3 \frac{\delta^3(\vec{q}_3 - \vec{q}_2)}{E_0 - \frac{\vec{q}_3^{\,2}}{2m}}. \tag{7.35}$$

Replacing G_R with G_0 when both potentials are Darwin terms, which are simply constants in momentum space, leads to a linearly divergent integral of the form

$$\begin{aligned}\Delta E_2(G_0) &\propto \int d^3q_1 d^3q_2 d^3q_4 \phi_0^*(\vec{q}_4) \frac{1}{\vec{q}_2^{\,2} - 2mE_0} \phi_0(\vec{q}_1) \\ &\propto |\phi_0(0)|^2 \int d^3p \frac{1}{\vec{p}^{\,2} - 2mE_0}.\end{aligned} \tag{7.36}$$

We note that this term vanishes for p states, but for s states it must be dealt with. To do so one must first regularize the linear divergence, which can be done using either dimensional regularization [19] or by cutting off momentum integrals, either by multiplying by factors like $\Lambda^2/(\vec{p}^{\,2}+\Lambda^2)$ or simply restricting $|\vec{p}| < \Lambda$.

Another source of divergences is the Hamiltonian $H^{(6)}$ that comes from the part of Eq. (7.11) that is quartic in the momenta. This is to be treated in the first order of perturbation theory, and when s-states are treated linearly divergent terms arise. These can be shown to cancel the linear divergences discussed above that arise from second-order perturbation theory, summing to the known Dirac equation result [18]. However, since we are treating fine structure, the only contributing part of $H^{(6)}$ comes from the last term in Eq. (7.11),

$$H^{(6)} = -\frac{3Z\alpha}{16m^4}\frac{1}{r^3}\vec{\sigma}\cdot(\vec{r}\times\vec{p})\vec{p}^{\,2}, \tag{7.37}$$

which has a finite expectation value for the $2p_{3/2} - 2p_{1/2}$ splitting, $-21/512\ m(Z\alpha)^6$. This disagrees with the value expected from the Dirac equation because the calculation is incomplete. To finish it, the second-order perturbation theory expression must be evaluated, specifically

$$\sum_{m\neq 2p}\frac{\langle 2p|(H_{\rm RM}+H_{SO})|m\rangle\langle m|(H_{\rm RM}+H_{SO})|2p\rangle}{E_{2p}-E_m}, \tag{7.38}$$

where we have dropped the Darwin term, which vanishes for p states. The contribution of this term to fine structure, which is also finite, can be evaluated with the method of Dalgarno and Lewis, and one finds $31/512\ m(Z\alpha)^6$, which gives the Dirac equation result, $5/256\ m(Z\alpha)^6$, when summed with the expectation value of $H^{(6)}$. We note that the factor of 31/512 is the sum of 49/4608 from two spin-orbit interactions and 230/4608 from one spin-orbit and one relativistic mass increase interaction, with two mass increase operators not contributing to fine structure.

We next turn to the evaluation of $\nu^{(4r)}$. We have already accounted for recoil corrections in the previous order, $\frac{m^2}{M}(Z\alpha)^4$, but note the interesting fact that in this order all corrections of order $m(\frac{m}{M})^n(Z\alpha)^4$ can be accounted for through the formula [20, 21]

$$\begin{aligned} E &= (M+m) + m_r(f(n,j)-1) - \frac{m_r^2}{2M}(f(n,j)-1)^2 \\ &\quad + (1-\delta_{l0})\frac{(Z\alpha)^4 m_r^3}{2n^3M^2}\left[\frac{1}{j+\frac{1}{2}} - \frac{1}{l+\frac{1}{2}}\right], \end{aligned} \tag{7.39}$$

where $f(n,j)$ is the Dirac energy function given above. We refer to this as the Barker–Glover contribution in the following. For the fine structure we are treating it has the $Z\alpha$ expansion

$$E_{fs} = \frac{m_r(Z\alpha)^4}{32}\left(1 - \frac{m_r^2}{M^2}\right) + \frac{5m_r(Z\alpha)^6}{256}\left(1 + \frac{m_r}{5M}\right), \tag{7.40}$$

the first term of which reproduces the result shown in Eq. (7.27) when a_e is dropped.

One method of studying the recoil corrections in order $(Z\alpha)^6$ is to carry out calculations to all orders in $Z\alpha$ using numerical methods, an approach originated by Shabaev [22] and applied to $2p_{1/2}$ states in Ref. [23] and $2p_{3/2}$ states in Ref. [24]. These results were recently confirmed by Adkins et al. [25]. Carrying out a numerical fit to the exact calculation at low Z reproduces the Barker–Glover result above, even though that formula is supposed to be valid only to order $m^2/M(Z\alpha)^4$. There are in fact terms of order $m^2/M(Z\alpha)^6$, but we now show that they cancel for fine structure, leading to

$$\nu^{(4r)} = -\frac{m^2(Z\alpha)^6}{64M}. \tag{7.41}$$

The Barker–Glover result follows from working with the Hamiltonian

$$H_{\text{eff}} = \vec{\alpha}\cdot\vec{p} + \beta m + \frac{\vec{p}^{\,2}}{2M} + V_{\text{eff}}, \tag{7.42}$$

where the Breit interaction is included in V_{eff},

$$V_{\text{eff}} = -\frac{Z\alpha}{r}\left[1 + \frac{1}{2M}\left(\delta_{ij} + \frac{r_i r_j}{r^2}\right)\alpha^i p^j\right]. \tag{7.43}$$

This is corrected by three terms coming from loop diagrams with either 0, 1, or 2 transverse photons present. All three terms involve the Dirac-Coulomb propagator, though when considering fine structure to the order we are interested in it suffices to use a free electron propagator.

The first, all-Coulomb term does not contribute to fine structure. The one transverse photon term is

$$E_T = -\frac{Z\alpha}{M}\int\frac{d\omega}{2\pi i}\frac{1}{\omega - i\epsilon}\int\frac{d^3k}{(2\pi)^3}\frac{P_{ij}(k)}{\omega^2 - k^2} \langle\phi|\gamma^i e^{ikr} S_F(E-\omega)\gamma^0[p^j, V]|\phi\rangle + \text{c.c.} \tag{7.44}$$

A short analysis gives a fine structure contribution of

$$E_T(fs) = \frac{m^2}{2M}(Z\alpha)^6\langle\phi|\frac{\vec{L}\cdot\vec{S}}{r^4}|\phi\rangle. \tag{7.45}$$

Finally, the two-transverse photon term is

$$E_{TT} = -\frac{Z^2\alpha^2}{M}\int\frac{d\omega}{2\pi i}\int\frac{d^3k_1}{(2\pi)^3}\int\frac{d^3k_2}{(2\pi)^3}\frac{P_{ik}(k_1)}{\omega^2-k_1^2}\frac{P_{jk}(k_2)}{\omega^2-k_2^2}$$
$$\langle\phi|\gamma^i e^{-ik_1r}S_F(E-\omega)\gamma^j e^{ik_2r}|\phi\rangle. \qquad (7.46)$$

Another short analysis that relies on the integral

$$\int\frac{d^3k}{(2\pi)^3}\frac{1}{k^2}P_{ij}(k)e^{-i\vec{k}\cdot\vec{r}} = \frac{\delta_{ij}r^2+r_ir_j}{8\pi r^3} \qquad (7.47)$$

gives

$$E_{TT}(fs) = -\frac{m^2}{2M}(Z\alpha)^6\langle\phi|\frac{\vec{L}\cdot\vec{S}}{r^4}|\phi\rangle, \qquad (7.48)$$

which cancels $E_T(fs)$, so that there is in fact no contribution to fine structure in this order beyond the Barker–Glover formula.

We now turn to radiative corrections, which are the most difficult part of the calculation [28]. The leading order radiative correction has already been accounted for through our use of the $(1+a_e)$ and $(1+2a_e)$ factors in Eq. (7.27). There are other contributions of order $m\alpha(Z\alpha)^4$, but they cancel in the difference. An additional simplification is the fact that the first binding corrections to the Lamb shift, of order $m\alpha(Z\alpha)^5$, vanish for p states. In terms of scattering diagrams, this is connected with the behavior of one-loop corrections to two-Coulomb exchange diagrams. These diagrams were studied by Baranger, Bethe, and Feynman [26] in the early days of QED, and found to behave as a delta function of position, which contributes only for s states. We note that this order is the same when $Z=1$ as the $m(Z\alpha)^6$ correction discussed above. However, the next term in the binding expansion, of order $m\alpha(Z\alpha)^6$, not only does not vanish for p states, but also has a nonvanishing contribution to the fine structure that contains a logarithm of $Z\alpha$. It in fact changes α by about 1 ppm, and its counterpart in the helium calculation will play an important role. We now turn to a discussion of the evaluation of this logarithmic term along with the associated constant term using effective field theory.

Before evaluating the logarithmic term, we calculate the two terms associated with using the F_2 form factor beyond first order given in Eq. (7.15). The part of the first equation that contributes to fine structure can be seen to be proportional to $H^{(6)}$, with the constant of proportionality being $\frac{4}{3}a_e$. Thus no new calculation is needed, and in fact the calculation could be subsumed into the $H^{(6)}$ calculation as was done for the basic formula Eq. (7.27). When we turn to the helium calculation, where there are a set of operators $H_i^{(6)}$ first derived by Douglas and Kroll (D–K) [27], we will see that the effect of F_2 is to multiply each operator by $1 + A_i a_e$, where A_i is always a simple integer or rational fraction. The other contribution comes from the known q^2 behavior of the form factor, and gives a spin-dependent effective operator

$$\frac{\alpha(Z\alpha)}{6m^4}\vec{\sigma} \cdot (\vec{\nabla}\delta^3(\vec{r})) \times \vec{p}, \tag{7.49}$$

that contributes $-\frac{m\alpha(Z\alpha)^6}{64\pi}$ to fine structure. We now discuss the other form factor.

We introduced the form factor F_1 above, but put off defining it. This is because it has an infrared divergence. This can be regularized either by introducing a photon mass, using dimensional regularization, or making a cutoff. Making the last choice, we introduce a parameter ϵ that divides the calculation into a low energy part, with photon energies $\omega < \epsilon$, and a high energy part, with $\omega > \epsilon$. In the high energy part we can account for QED effects by evaluating F_1 with this cutoff, and find

$$F_1(q^2) = 1 + \frac{\alpha}{3\pi}\frac{q^2}{m^2}\left[\ln\left(\frac{m}{2\epsilon}\right) + \frac{11}{24}\right]. \tag{7.50}$$

With this regularized form of F_1 we can continue the matching procedure discussed above to derive the effective operator arising from the q^2 term in F_1,

$$\frac{\alpha(Z\alpha)}{3m^4}\left[\ln\frac{m}{2\epsilon} + \frac{11}{24}\right]\vec{\sigma} \cdot (\vec{\nabla}\delta^3(\vec{r})) \times \vec{p}, \tag{7.51}$$

which leads to the energy shift

$$E_H{}^1 = -\frac{m\alpha(Z\alpha)^6}{32\pi}\left[\ln\frac{m}{2\epsilon} + \frac{11}{24}\right]. \tag{7.52}$$

To show how the regularized divergence cancels, we have to now consider the effective theory in second-order perturbation theory. Now, the matching procedure we have carried out so far has not yet included diagrams in which a radiated photon is present, but nonrelativistically we of course know that such photons couple to the current $\vec{p}/m$ in the dipole approximation. Using this interaction in second order gives a nonrelativistic form of the Lamb shift,

$$\Delta E = \alpha \int_0^\epsilon \frac{d^3k}{(2\pi)^3 2k} P_{ij}(k) \langle\phi|\frac{p^i}{m}\frac{1}{E - H_{NR} - k}\frac{p^j}{m}|\phi\rangle, \tag{7.53}$$

which is linearly divergent because diagrams with an electron-positron pair, the so-called "Z" graphs, which cancel that divergence are not present in the nonrelativistic theory. The cutoff is required because we have chosen to regularize the effective theory by not allowing photons to have energy greater than ϵ. The linear divergence of this integral is reduced to a logarithmic divergence after mass renormalization, which process replaces

$$\frac{1}{E-H-k} \to \frac{E-H}{k(E-H-k)}. \tag{7.54}$$

In full QED one starts with a logarithmic divergence that is made finite by renormalization, but here a logarithmic divergence remains that is canceled by the F_1 term discussed above.

As mentioned above, most of the Lamb shift, which the above integral is representing the nonrelativistic part of, does not contribute to the fine structure splitting. To get a splitting we must include the effect of the spin-orbit operator. This can occur in three ways; through the shift of the energy, the shift of the propagator, and the shift of the wave function. This requires the evaluation of

$$\begin{aligned} E_{L1} &= \frac{2\alpha}{3\pi}\int_0^\epsilon kdk[2\langle\delta\phi|\frac{p^i}{m}\frac{1}{E-H-k}\frac{p^i}{m}|\phi\rangle \\ &+ \langle\phi|\frac{p^i}{m}\frac{1}{E-H-k}(H_{SO}-\langle\phi|H_{SO}|\phi\rangle)\frac{1}{E-H-k}\frac{p^i}{m}|\phi\rangle. \end{aligned} \tag{7.55}$$

In addition to this calculation, we must also consider corrections to the current the photon couples to. In QED this is of course $\bar{u}(\vec{p}_{\rm f})\gamma_i u(\vec{p}_{\rm i})$ instead of $(\vec{p}_{\rm f}+\vec{p}_{\rm i})/m$. The difference has already been discussed in connection with transverse photon exchange. Including corrections to the dipole approximation, the current changes to

$$\begin{aligned} \frac{\vec{p}}{m} &\to \frac{\vec{p}}{m} - \frac{Z\alpha}{2m^2r^3}\vec{r}\times\vec{\sigma} + \frac{1}{2m}(\vec{k}\cdot\vec{r})\vec{k}\times\vec{\sigma} \\ &\equiv \frac{\vec{p}}{m} + \delta\vec{j}. \end{aligned} \tag{7.56}$$

The first part of $\delta\vec{j}$ can be thought of as arising from a Z graph, and the second comes from the first term in the expansion of $e^{i\vec{k}\cdot\vec{r}}$ together with a magnetic interaction. $\delta\vec{j}$ gives rise to the additional shift

$$E_{L2} = 2\alpha\int_0^\infty \frac{d^3k}{(2\pi)^3 2k}P_{ij}(k)\langle\phi|\delta j^i\frac{1}{E-H-k}\frac{p^j}{m}|\phi\rangle. \tag{7.57}$$

The calculation of E_{L1} and E_{L2} must be done numerically, and the result is

$$E_L = m\frac{\alpha}{\pi}\frac{(Z\alpha)^6}{8}\left[-\frac{1}{4}\ln\frac{2\epsilon}{m(Z\alpha)^2} + 0.384\,42(1)\right]. \tag{7.58}$$

Finally putting everything together the complete correction is cutoff independent and equal to

$$E_{FS} = \frac{m\alpha(Z\alpha)^6}{8\pi}\left[\frac{1}{9} + \frac{1}{4}\ln(Z\alpha)^2 + 0.38442\right]. \tag{7.59}$$

The derivation of this $m(Z\alpha)^6$ correction is seen to be quite complicated even in hydrogen. While it is a sort of "proof of principle" that effective field theory methods can be applied to the bound state problem, and means that the helium calculation is possible, it also indicates that the calculation will be very complex, and in fact the approach here was completed only recently [14]. We now turn to this more challenging problem.

7.2. Helium Fine Structure

Our interest here is in determining α from the simplest many-electron atom, helium. This program was initiated in 1964 by Schwartz [29]. The problems faced in carrying out calculations of the same accuracy as has been only recently achieved for hydrogen are considerable, but the much higher experimental precision achieved in helium means these calculations can lead to an accurate determination of α that depends on the low energy scales characteristic of atomic physics. There are certain simplifying features of the calculation that stem from the form of the helium wave functions, which we now describe.

Helium fine structure is associated with states that can be thought of as having one electron in the ground state and the other in a $2p$ state. However, while in hydrogen the fine structure would be associated with the energy difference between the $2p_{1/2}$ and $2p_{3/2}$ states, in this two-electron system LS coupling instead of jj coupling is obeyed, so one first sums the electron spins to form either a singlet (S=0) or triplet (S=1), and then adds this spin to the orbital angular momentum, which in this case is L=1. The fine structure studied here is associated with triplets, so the allowed values of total angular momentum are J=0,1, and 2. The spectroscopic notation for the three states is 2^3P_J, with a onefold, threefold, and fivefold degeneracy for J=0,1, and 2 respectively. Nonrelativistically they are degenerate, but relativistic effects lead to a frequency splitting of order $\alpha^2 R_\infty c$. We will be concerned in particular with two intervals, the "large interval"

$$\nu_{01} = \frac{E(2^3P_0) - E(2^3P_1)}{h} \tag{7.60}$$

and the "small interval"

$$\nu_{12} = \frac{E(2^3P_1) - E(2^3P_2)}{h}. \tag{7.61}$$

The former will be used for determining α and the latter as a test of theory. The energy diagram is shown in Fig. 7.1. The linewidths of these states are approximately 100 times narrower than hydrogen, which is why Schwartz [29] proposed their study for an accurate fine structure constant determination. The experiments are still extremely challenging, but have reached accuracies such that, were there no theoretical uncertainties, a value of α accurate to 10 ppb could be determined. We further note that while we will use M to refer to the mass of the nucleus of an arbitrary isotope of helium, in practice we are interested in the nucleus being an alpha particle, with [3]

$$Mc^2 = 3727.379\ 109(93)\text{MeV}. \tag{7.62}$$

Since the nucleus has spin 0 the degeneracy mentioned above remains in the absence of external fields.

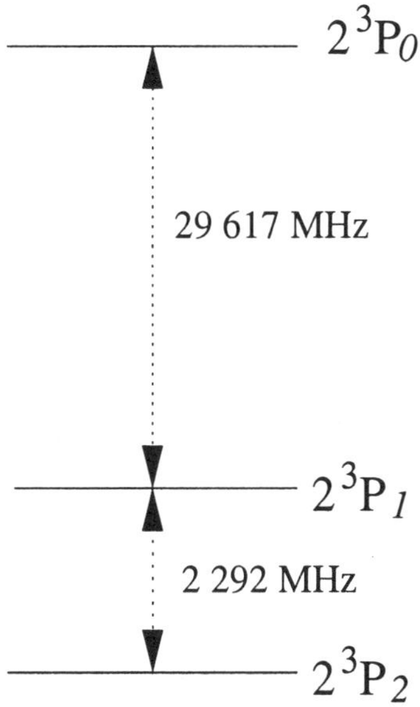

Fig. 7.1. Energy levels of the 2P triplet states in helium.

The theoretical simplicity of these states comes from the fact that the helium wave function can be written as a product of spatial and spin wave functions. Since the spin part of the wave function for triplets is symmetric, the spatial part is antisymmetric, so any operator that behaves as $\delta^3(\vec{r}_1 - \vec{r}_2)$ can be dropped. It was for this reason that Douglas and Kroll [27] were able to derive operators contributing at the level $m\alpha^6$ to helium fine structure more than two decades before it became possible to carry out calculations of equivalent accuracy for energies of singlet states [30].

To illustrate how one determines α from helium fine structure, we use a recent measurement of ν_{01} [31],

$$\nu_{01}(\text{exp}) = 29\ 616.951\ 66(70)\text{MHz}, \tag{7.63}$$

which we note has an accuracy of 24 ppb. As with hydrogen, the leading recoil corrections together with the leading radiative corrections can be built into the lowest order formula, where we replace the constant A defined in Eq. (7.27) for hydrogen with B, defined in Eq. (7.71). In terms of this constant,

$$\begin{aligned} \bar{\nu}_{01} &= B\alpha^2, \\ B &= 556\ 200\ 289.46\ \text{MHz}. \end{aligned} \tag{7.64}$$

The coefficient above has to be obtained numerically instead of analytically, as was the case for hydrogen. We note that highly accurate wave functions are required, which was a major issue at the beginning of Schwartz program [29], but since then this problem has been solved [32, 33], and all digits shown are correct. Were this all there was to the theoretical calculation, one would solve for α_0^2 by setting the approximated theory equal to the experiment, giving

$$\alpha_0^2 = \frac{\nu_{01}(\text{exp})}{B}, \tag{7.65}$$

which leads to

$$\alpha_0^{-1} = 137.039\ 9791(2). \tag{7.66}$$

As with hydrogen, inclusion of the leading recoil and radiative corrections gives a starting value of α within 30 ppm of the g-2 value, and we will see that the bulk of the difference comes from the order α^2 corrections that have not yet been included. It is the accurate and complete determination of these and higher-order corrections using effective field theory together with accurate helium wave functions that will be described in the rest of this chapter.

7.3. Organization of Helium Fine Structure Calculation

The analog of the factor 1/16 in ν_0 for hydrogen involves evaluating the expectation value of three operators with accurate nonrelativistic helium wave functions. One of the operators is the same spin-orbit coupling between the electron and nucleus that gives rise to the 1/16 in hydrogen, the second is a spin-orbit coupling where the electron spin interacts with the magnetic field associated with the motion of the other electron, and the third involves the magnetic interaction of the electrons' spins with each other. The explicit form of these operators will be given in the following section. There are three corresponding expectation values denoted $\tilde{E}_1$, $\tilde{E}_2$, and $\tilde{E}_3$, where the tilde indicates that the mass of the nucleus is taken to infinity, and they shift the energy of the three states to

$$\begin{aligned}
\nu(2^3P_0) &= \alpha^2 R_\infty c\left[\frac{\tilde{E}_1}{2}+\frac{\tilde{E}_2}{2}+\frac{\tilde{E}_3}{2}\right],\\
\nu(2^3P_1) &= \alpha^2 R_\infty c\left[-\frac{\tilde{E}_1}{4}+\frac{\tilde{E}_2}{4}+\frac{\tilde{E}_3}{4}\right],\\
\nu(2^3P_2) &= \alpha^2 R_\infty c\left[\frac{\tilde{E}_1}{20}-\frac{\tilde{E}_2}{4}-\frac{\tilde{E}_3}{4}\right].
\end{aligned} \tag{7.67}$$

We note that $\tilde{E}_1$, the expectation value of the spin-spin operator, contributes to $J = 0, 1, 2$ as $(1/2, -1/4, 1/20)$, while $\tilde{E}_2$ and $\tilde{E}_3$, the spin-orbit operators, contribute as $(1/2, 1/4, -1/4)$, which follows from an angular momentum analysis of the three states. The leading QED correction to these formulas can be obtained by including the electron anomalous magnetic moment operator in effective field theory in a manner analogous to the hydrogen calculation. Again, as with hydrogen, there is another operator that is pure recoil which we will also describe in the next section, with expectation value $\tilde{E}_4$. However, to fully account for recoil, in addition to introducing scaling factors of m_r/m account must be taken of an additional term known as the mass polarization operator, given in Eq. (7.81), in the nonrelativistic Hamiltonian. This can be accounted for by simply generating wave functions with Hamiltonians including mass polarization, and we denote such constants as E_i. The net result is the somewhat more complicated looking, but much more accurate set of formulas,

$$\begin{aligned}
\nu(2^3P_0) &= (\frac{m_r}{m})^3\alpha^2 R_\infty c\left[\frac{E_1}{2}(1+a_e)^2+\frac{E_2}{2}(1+2a_e)\right.\\
&\quad\left.+\frac{E_3}{2}(1+\frac{4}{3}a_e)+\frac{m}{M}\frac{E_4}{2}(1+a_e)\right],\\
\nu(2^3P_1) &= (\frac{m_r}{m})^3\alpha^2 R_\infty c\left[-\frac{E_1}{4}(1+a_e)^2+\frac{E_2}{4}(1+2a_e)\right.\\
&\quad\left.+\frac{E_3}{4}(1+\frac{4}{3}a_e)+\frac{m}{M}\frac{E_4}{4}(1+a_e)\right],
\end{aligned}$$

$$\nu(2^3P_2) = (\frac{m_r}{m})^3\alpha^2 R_\infty c\bigg[\frac{E_1}{20}(1+a_e)^2 - \frac{E_2}{4}(1+2a_e) - \frac{E_3}{4}(1+\frac{4}{3}a_e) - \frac{m}{M}\frac{E_4}{4}(1+a_e)\bigg]. \tag{7.68}$$

While the dominant QED and recoil corrections are included in the above, QED corrections of relative order α^2, α^3, and α^4 are not completely included, and in addition recoil corrections of relative order $\alpha^2\frac{m}{M}$ need to be calculated. It is important to emphasize the fact that many higher order QED and recoil corrections *are* included in these formulas through the inclusion of a_e, which can be written as a power series in α with the first four terms known, along with recoil effects that could be treated separately, for example by using $\tilde{E}_i$ and including mass polarization perturbatively. Therefore any future calculations designed to check the two discrepant results that presently exist need to avoid double counting if they adopt the approach of this chapter.

We then write the four corrections to the large interval, following the treatment of hydrogen in the previous section, as

$$\delta\nu_{01} \equiv \nu_{01}^{(4)} + \nu_{01}^{(4r)} + \nu_{01}^{(5)}. \tag{7.69}$$

The complete equation that determines α from a measurement of ν_{01} is

$$\alpha^2 = \frac{\nu_{01}(\exp) - \delta\nu_{01}}{B}, \tag{7.70}$$

where B, the numerical value of which has been given above in Eq. (7.64), is the helium analog of Eq. (7.27),

$$B = (\frac{m_r}{m})^3 R_\infty c\bigg[\frac{3E_1}{4}(1+a_e)^2+\frac{E_2}{4}(1+2a_e)+\frac{E_3}{4}(1+\frac{4}{3}a_e)+\frac{m}{M}\frac{E_4}{4}(1+a_e)\bigg]. \tag{7.71}$$

The numerical values of the E_i are given in Eq. (7.94). The values of the first three contributions to $\delta\nu_{01}$ are

$$\begin{aligned} {\nu_{01}}^{(4)} &= -1.547\ 06(10)\,\text{MHz} \\ {\nu_{01}}^{(4r)} &= -0.010\ 38\quad\ \ \text{MHz} \\ {\nu_{01}}^{(5)} &= \ \ 0.081\ 72(16)\,\text{MHz}. \end{aligned} \tag{7.72}$$

Putting these in the above formula gives our final answer for α as determined by helium fine structure,

$$\alpha^{-1}(\text{He fs}) = 137.035\ 979(2). \tag{7.73}$$

This is in poor agreement with other fine structure constant determinations, and also in significant disagreement with another theoretical calculation: this unsatisfactory situation will be described in the concluding section. We now derive the operators whose expectation values give the constants E_i.

7.4. Lowest-Order Contributions

As with hydrogen, we begin by studying the scattering of free particles, in this case two electrons and an alpha particle nucleus. The α particle has spin 0, so our previous neglect of spin as an approximation for the hydrogen problem is not an issue. The two electrons have incoming momenta $\vec{p}_1, \vec{p}_2$, and outgoing momenta $\vec{p}_3, \vec{p}_4$. The nucleus then has in the center of mass system incoming momentum $-\vec{p}_1 - \vec{p}_2$ and outgoing momentum $-\vec{p}_3 - \vec{p}_4$. The general formula that connects an operator to an energy shift is

$$E_i = \frac{1}{(2\pi)^{12}} \int d^3p_1 d^3p_2 d^3p_3 d^3p_4 \phi^*(\vec{p}_3, \vec{p}_4) O_i(\vec{p}_3, \vec{p}_4; \vec{p}_1, \vec{p}_2) \phi(\vec{p}_1, \vec{p}_2). \tag{7.74}$$

When the electrons interact only with each other momentum conservation allow us to write

$$O_i^{ee}(\vec{p}_3, \vec{p}_4; \vec{p}_1, \vec{p}_2) = (2\pi)^3 \delta^3(\vec{p}_3 + \vec{p}_4 - \vec{p}_1 - \vec{p}_2) \mathcal{M}_i^{e-e}(\vec{p}_3; \vec{p}_1, \vec{p}_2). \tag{7.75}$$

If the interaction is through the exchange of a Coulomb photon one has

$$\mathcal{M}_i^{ee-C} = \frac{4\pi\alpha}{|\vec{p}_3 - \vec{p}_1|^2}, \tag{7.76}$$

which corresponds in coordinate space to the interaction

$$\begin{aligned} H_{eeC} &= \frac{\alpha}{|\vec{r}_1 - \vec{r}_2|} \\ &\equiv \frac{\alpha}{r}. \end{aligned} \tag{7.77}$$

When one of the electrons interacts only with the nucleus and the other electron is a spectator we have

$$O_i^{eN}(\vec{p}_3, \vec{p}_4; \vec{p}_1, \vec{p}_2) = (2\pi)^3 \delta^3(\vec{p}_4 - \vec{p}_2) \mathcal{M}_i^{e-N}(\vec{p}_3; \vec{p}_1). \tag{7.78}$$

In this case a Coulomb photon exchange is

$$\mathcal{M}_i^{eN_1} = -\frac{4\pi Z\alpha}{|\vec{p}_3 - \vec{p}_1|^2} \tag{7.79}$$

for one electron, with a similar expression involving $\vec{p}_2$ and $\vec{p}_4$ for the other, which corresponds in coordinate space to

$$\begin{aligned} H_{eN} &= -\frac{Z\alpha}{|\vec{r}_1|} - \frac{Z\alpha}{|\vec{r}_2|} \\ &\equiv -\frac{Z\alpha}{r_1} - \frac{Z\alpha}{r_2}. \end{aligned} \tag{7.80}$$

In addition the kinetic energy is

$$\begin{aligned} &\frac{\vec{p}_1^{\,2}}{2m} + \frac{\vec{p}_2^{\,2}}{2m} + \frac{(-\vec{p}_1 - \vec{p}_2)^2}{2M} \\ &= \frac{\vec{p}_1^{\,2}}{2m_r} + \frac{\vec{p}_2^{\,2}}{2m_r} + \frac{\vec{p}_1 \cdot \vec{p}_2}{M}. \end{aligned} \tag{7.81}$$

While the electron mass is replaced by the reduced mass, as in the hydrogen case, helium is complicated by the presence of the last term, which is the mass polarization correction mentioned above. However, in practice this term is simple to deal with when we solve the Schrödinger equation for helium, and in the following it is always assumed to be built into the nonrelativistic wave functions. It is still the case that momenta scale proportionally to m_r and distances to ${m_r}^{-1}$, so that lowest order fine structure comes with a multiplicative factor $(m_r/m)^3$.

To get the lowest order operators contributing to fine structure for helium, we generalize the above discussion to relativistic corrections to $\mathcal{M}^{eN}$ and $\mathcal{M}^{ee}$. The non-recoil part of the former is essentially identical to the hydrogen case, and consists of the sum of the spin-orbit interactions of the two electrons,

$$H_{SL}(eN) = \frac{Z\alpha}{4m^2}\left[\frac{1}{r_1^3}\vec{\sigma}_1 \cdot (\vec{r}_1 \times \vec{p}_1) + \frac{1}{r_2^3}\vec{\sigma}_2 \cdot (\vec{r}_2 \times \vec{p}_2)\right]. \tag{7.82}$$

The expectation value of this operator is E_2 times the appropriate J-dependent factor.

Continuing with $\mathcal{M}^{eN}$, we next consider exchange of a transverse photon. While in hydrogen the spin-dependent part of this effect is simply proportional to the spin-orbit interaction, here the presence of the second electron, even though it does not have a direct interaction, leads to an additional term because of its momentum. Specifically, since the initial nuclear momentum is $-\vec{p}_1 - \vec{p}_2$ and the final $-\vec{p}_3 - \vec{p}_2$, the convection current has an additional term $-2\vec{p}_2$. Thus, while E_4 has one contribution from $2\frac{m}{M}H_{SL}(eN)$, it has another that can be written

$$H_{SL}(eN)(\text{recoil}) = \frac{Z\alpha}{2mM}\left[\frac{1}{r_1^3}\vec{\sigma}_1 \cdot (\vec{r}_1 \times \vec{p}_2) + \frac{1}{r_2^3}\vec{\sigma}_2 \cdot (\vec{r}_2 \times \vec{p}_1)\right]. \tag{7.83}$$

The expectation value of the sum of these two gives E_4.

The remaining terms contributing to lowest order fine structure are specific to helium, as they involve exchange of a photon between the electrons. The exchange of a Coulomb photon has the scattering amplitude

$$\begin{aligned}\mathcal{M}^{ee}(C) &= \frac{4\pi\alpha}{|\vec{q}|^2} u^\dagger(\vec{p}_3)u(\vec{p}_1)u^\dagger(\vec{p}_4)u(\vec{p}_2) \\ &\rightarrow \frac{\pi\alpha}{m^2|\vec{q}|^2}[i\vec{\sigma}_1\cdot(\vec{p}_3\times\vec{p}_1) + i\vec{\sigma}_2\cdot(\vec{p}_4\times\vec{p}_2)], \end{aligned} \tag{7.84}$$

where we have kept only the lowest order spin-dependent terms. In the same approximation transverse photon exchange leads to the scattering amplitude ($\vec{q} = \vec{p}_3 - \vec{p}_1 = \vec{p}_2 - \vec{p}_4$)

$$\begin{aligned}\mathcal{M}^{ee}(T) &= -\frac{4\pi\alpha}{|\vec{q}|^2} u^\dagger(\vec{p}_3)\alpha_i u(\vec{p}_1)u^\dagger(\vec{p}_4)\alpha_j u(\vec{p}_2)P_{ij}(q) \\ &= -\frac{\pi\alpha}{m^2|\vec{q}|^2}[2i\vec{\sigma}_1\cdot(\vec{p}_2\times\vec{p}_4) + 2i\vec{\sigma}_2\cdot(\vec{p}_1\times\vec{p}_3) \\ &\quad +\vec{\sigma}_1\cdot(\hat{i}\times\vec{q})\vec{\sigma}_2\cdot(\hat{i}\times\vec{q})]. \end{aligned} \tag{7.85}$$

The constant E_3 is associated with the terms in the above two expressions with a single Pauli matrix, being the expectation value of the coordinate space form

$$H_{SL} = \frac{\alpha}{4\,m^2\,r^3}\left[(\vec{\sigma}_2 + 2\,\vec{\sigma}_1)\cdot\vec{r}\times\vec{p}_2 - (\vec{\sigma}_1 + 2\,\vec{\sigma}_2)\cdot\vec{r}\times\vec{p}_1\right]. \tag{7.86}$$

Finally, the constant E_1 comes from the spin-spin interaction in the transverse photon term,

$$H_{SS} = \frac{\alpha}{4\,m^2}\left(\frac{\vec{\sigma}_1\cdot\vec{\sigma}_2}{r^3} - 3\,\frac{\vec{\sigma}_1\cdot\vec{r}\,\vec{\sigma}_2\cdot\vec{r}}{r^5}\right). \tag{7.87}$$

In addition to these spin-dependent effective Hamiltonians we will later need a spin-independent effective Hamiltonian, denoted H_{SI}, for higher-order calculations. Its explicit form is

$$\begin{aligned}H_{SI} &= -\frac{1}{8\,m^3}\left(p_1^4 + p_2^4\right) + \frac{Z\,\alpha\,\pi}{2\,m^2}\left(\delta^3(r_1) + \delta^3(r_2)\right) \\ &- \frac{\alpha}{2\,m^2}\,p_1^i\left(\frac{\delta^{ij}}{r} + \frac{r^i\,r^j}{r^3}\right)p_2^j - \frac{\pi\,\alpha}{m^2}\,\delta^3(r) - \frac{2\pi\alpha}{3m^2}\vec{\sigma}_1\cdot\vec{\sigma}_2\delta^3(r). \end{aligned} \tag{7.88}$$

As with the hydrogenic case, relativistic mass increase operators are included in the above along with two kinds of Darwin operator, one familiar from the hydrogenic case and the other a new one involving the electron-electron interaction.

7.4.1. *Leading QED corrections*

As with hydrogen, making the replacement

$$\gamma_\mu \to i\frac{\sigma_{\mu\nu}q^\nu}{2m}a_e, \tag{7.89}$$

allows the leading QED corrections to helium fine structure to be included with the lowest order corrections. The manipulations are identical for E_2 and E_4, which get multiplied by factors of $1+2a_e$ and $1+a_e$ respectively, as in Eq. (7.18). Further discussion is needed for E_1 and E_3. The factor $(1+a_e)^2$ in the former comes from four diagrams, the lowest order, two with one uncorrected vertex and one corrected vertex, and finally one with both vertices corrected. The factor of $(1+4/3\ a_e)$ in E_3 arises because this term has contributions both from Coulomb and transverse photon exchange, and the first has a factor $1+2a_e$ and the second $1+a_e$. Because for triplet states one can replace a single occurrence of $\vec{\sigma}_2$ with $\vec{\sigma}_1$, expressing the operator giving rise to E_3 in terms only of $\vec{\sigma}_1$ and putting in the appropriate anomalous magnetic moment corrections leads to the overall correction factor of $1+4/3\ a_e$. There are also a_e corrections to H_{SI}, with the electron-nucleus Darwin term multiplied by $1+2a_e$, the electron-electron Darwin term by $1+a_e$, and the spin-spin term by $(1+a_e)^2$. The net result is that the spin-dependent Hamiltonian including leading radiative corrections, which we denote H_{SD}, is the following combination of Eqs. (7.87), (7.82), and (7.86):

$$\begin{aligned} H_{SD} &= \frac{\alpha}{4\,m^2}\left(\frac{\vec{\sigma}_1\cdot\vec{\sigma}_2}{r^3} - 3\,\frac{\vec{\sigma}_1\cdot\vec{r}\,\vec{\sigma}_2\cdot\vec{r}}{r^5}\right)(1+a_e)^2 \\ &+ \frac{Z\alpha}{4m^2}\Big[\frac{1}{r_1^3}\vec{\sigma}_1\cdot(\vec{r}_1\times\vec{p}_1) + \frac{1}{r_2^3}\vec{\sigma}_2\cdot(\vec{r}_2\times\vec{p}_2)\Big](1+2a_e) \\ &+ \frac{\alpha}{4\,m^2\,r^3}\left[(\vec{\sigma}_2+2\,\vec{\sigma}_1)\cdot\vec{r}\times\vec{p}_2 - (\vec{\sigma}_1+2\,\vec{\sigma}_2)\cdot\vec{r}\times\vec{p}_1\right](1+\frac{4}{3}a_e). \end{aligned} \tag{7.90}$$

To accurately evaluate the expectation value of the spin-dependent operators, and thus determine E_1, E_2, E_3, and E_4, a high quality numerical representation of the wave function is needed, and we now turn to the description of a particularly simple and accurate method of determining it.

7.5. Helium Wave Functions

The helium wave function used here has been described in Ref. [33]. We represent the spatial part of the triplet P states we are concerned with here as

$$\vec{\phi}(\vec{r}_1,\vec{r}_2) = \sum_i C_i(\vec{r}_1 e^{-\alpha_i r_1-\beta_i r_2-\gamma_i r} - \vec{r}_2 e^{-\alpha_i r_2-\beta_i r_1-\gamma_i r}). \tag{7.91}$$

By choosing the constants α_i, β_i, and γ_i to lie randomly in certain ranges, and then minimizing the energy with respect to these ranges and the linear coefficients C_i, energies accurate to 18 digits or more are obtained with as few as 1200 basis functions. While this may seem "overkill", energies accurate to order δ^2 are associated with wave functions accurate only to order δ, so this accuracy is required to calculate matrix elements accurate to 9 digits. The nonrelativistic triplet P state energy for infinite nuclear mass, obtained in this way is

$$E(2^3P)_{NR} = 2.133\ 164\ 190\ 779\ 283\ 205\ 147(1), \tag{7.92}$$

which is in agreement with the less accurate result of Drake [32], who uses a modified Hylleraas basis set. Manipulations with this basis set are particularly straightforward, since all integrals encountered in the calculation can be related to the basic integral

$$\int d^3r_1 \int d^3r_2 \frac{e^{-\alpha r_1-\beta r_2-\gamma r}}{r_1 r_2 r} = \frac{16\pi^2}{(\alpha+\beta)(\alpha+\gamma)(\beta+\gamma)}. \tag{7.93}$$

Clearly any integral with a factor $r_1^m r_2^n r^t$ with m, n, t greater than or equal to -1 can be obtained by differentiating this formula. When a power is less than -1 one can integrate the formula, getting terms involving logarithms and dilogarithms. With the use of this wave function we determine for the constants E_i, $i = 1 \ldots 4$, the values

$$\begin{aligned} E_1 &= 0.180\ 220\ 618\ 633(1) \\ E_2 &= -0.277\ 401\ 358\ 713(1) \\ E_3 &= 0.411\ 999\ 963\ 626(1) \\ E_4 &= 0.241\ 945\ 125\ 694(9). \end{aligned} \tag{7.94}$$

These determinations are so accurate that they are essentially exact, in the sense that the experimental uncertainty is 100 times greater. If experimental precision were to increase, there is no difficulty in determining the constants to even greater accuracy. We have already used these values along with reduced mass and the anomalous magnetic moment of the electron in the above to determine α to be within 30 ppm of the CODATA value. As with the hydrogenic case, the next step is to determine the α^2 corrections, to which we now turn.

7.6. Determination of $\nu^{(4)}$

For hydrogen we noted that this contribution could be directly obtained from the Dirac eigenvalue, but instead derived it with effective field theory, noting that a Dirac equation for helium does not exist [17]. The calculation involved two parts, given in Eqns. (7.37) and (7.38), each of which generalizes to helium. The first part of the calculation involves deriving the spin-dependent parts of $H^{(6)}$ for helium. This was first carried out in a Bethe–Salpeter framework by Douglas and Kroll [27]. A much more compact derivation using effective field theory was given in Ref. [34], and will be summarized below. The second part involves evaluating second-order perturbation theory with the full Breit Hamiltonian, and was first carried out by Lewis and Serafino [35]. The total result can be written as

$$\nu^{(4)} = \nu_A^{(4)} + \nu_B^{(4)}, \tag{7.95}$$

with

$$\nu_A^{(4)} = \langle H^{(4)} \frac{1}{(E-H)'} H^{(4)} \rangle \tag{7.96}$$

and

$$\nu_B^{(4)} = \sum_{i=1}^{15} \langle H_i^{(6)} \rangle, \tag{7.97}$$

where the single $H^{(6)}$ for hydrogen given in Eq. (7.37) becomes 15 terms. We now give a brief description of how these terms can be derived using old-fashioned perturbation theory [34], followed by a briefer discussion of $\nu_A^{(4)}$.

For the calculation of the effective Hamiltonian we use time-ordered perturbation theory to derive the Douglas–Kroll operators, with the relevant diagrams being given in Fig. 7.2. We will classify the operators in terms of six terms, $V_1 - V_6$, each associated with one of the diagrams, but note that all 15 operators found by Douglas and Kroll are contained in them. The rules for deriving an expression from the corresponding diagram are as follows. We continue to use Coulomb gauge, with a dashed line representing a Coulomb photon and a wavy line a transverse photon; the dotted line is a Coulomb photon together with a kinetic energy term as discussed below. Three-momentum is conserved at each vertex, and there are projection operators between the vertices on electron lines onto positive or negative energy states, depending on the time arrow. Any diagram with more than one vertex on an electron line has energy denominators associated with it,

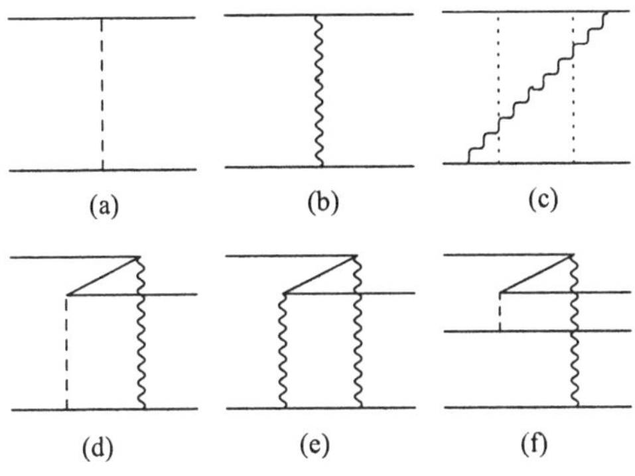

Fig. 7.2. Time-ordered diagrams contributing to fine structure splitting in order $m\,\alpha^6$. The middle horizontal line corresponds to helium, two outer horizontal lines correspond to two electrons. The wavy line is a transverse photon, the dashed line is a Coulomb photon, and the dotted line denotes retardation.

where each denominator is $E - \Sigma_i E_i$, where the sum goes over all intermediate states. There is also a factor -1 for any electron positron pair creation. When a transverse photon is involved, the leading term in the denominator is the photon energy, and for most cases one can neglect other terms in the denominator. The transverse photon in this approximation is denoted by a vertical wavy line. In some cases the denominator is expanded to account for the electron kinetic energy, and this effect is grouped together with Coulomb photon exchange and described graphically with the dotted line mentioned above. Since either electron could first emit a photon with the other electron absorbing it, an overall factor of $-2/q$ is present. The whole expression is enclosed by $u^\dagger(p')$ and $u(p)$, the free fermion bispinors, which correspond to the external lines. The simplifying features coming from working with 2^3P_J states mentioned above, specifically the fact that all expectation values with $(\vec{\sigma}_1 - \vec{\sigma}_2)$ vanish and that the wave function vanishes when $\vec{r}_1 = \vec{r}_2$, play an important role in the following.

The electron–nucleus part V_0 can be obtained from the diagram shown in in Fig. 7.2(a), with the lower line indicating in this case the nucleus, in a manner basically identical to that given for Coulomb photon exchange in hydrogen, the only difference being that a second electron is present. We have

$$V_0 = -\frac{4\pi Z\,\alpha}{q^2}\,\bar{u}(p_1')\gamma^0 u(p_1)\,, \tag{7.98}$$

where $\vec{q}$ is given by $\vec{q} = \vec{p_1}\,' - \vec{p}_1$ and $u(p)$ is the positive energy solution of the free Dirac equation previously given in Eq. (7.10). Using the expansion given in Eq. (7.11) leads to

$$V_0 = \frac{4\pi Z\,\alpha}{q^2}\,\frac{3\,i}{32\,m^4}(p_1^2 + p_1'^2)\,\vec{\sigma}_1 \cdot \vec{p}_1\,' \times \vec{p}_1. \tag{7.99}$$

Fourier transforming to coordinate space gives

$$V_0 = -\frac{3}{32\,m^4}\left(p_1^2\,\frac{Z\,\alpha}{r_1^3}\,\vec{\sigma}_1 \cdot \vec{r}_1 \times \vec{p}_1 + \frac{Z\,\alpha}{r_1^3}\,p_1^2\,\vec{\sigma}_1 \cdot \vec{r}_1 \times \vec{p}_1\right). \tag{7.100}$$

We can account for the second electron by simply doubling the result, and the final result is

$$V_0 = -\frac{3\,Z\,\alpha}{8\,m^4}\,p_1^2\,\frac{1}{r_1^3}\,\vec{\sigma}_1 \cdot \vec{r}_1 \times \vec{p}_1. \tag{7.101}$$

Now interpreting the lower line of Fig. 7.2(a) as an electron, we have the simplest electron–electron term. The initial expression is

$$V_1 = \frac{4\pi\alpha}{q^2}\,u^\dagger(p_1')u(p_1)\;u^\dagger(p_2')u(p_2)\,, \tag{7.102}$$

where

$$\vec{q} = \vec{p}_1\,' - \vec{p}_1 = \vec{p}_2 - \vec{p}\,'_2. \tag{7.103}$$

As usual, V_1 is expanded in momenta up to p^4, and we keep only the spin-dependent terms that give an $m\alpha^6$ correction to fine structure. After this procedure we find

$$\begin{aligned} V_1 = &-\frac{\pi}{4\,m^4}\,\frac{\alpha}{q^2} \\ &\left\{\frac{i}{2}\,q^2\,(\vec{\sigma}_1 \cdot \vec{p}\,'_1 \times \vec{p}_1 + \vec{\sigma}_2 \cdot \vec{p}\,'_2 \times \vec{p}_2) + \vec{\sigma}_1 \cdot \vec{p}\,'_1 \times \vec{p}_1\,\vec{\sigma}_2 \cdot \vec{p}\,'_2 \times \vec{p}_2 \right. \\ &\left. + \frac{3}{2}\,(p_1^2 + p_1'^2)\;i\,\vec{\sigma}_1 \cdot \vec{p}\,'_1 \times \vec{p}_1 + \frac{3}{2}\,(p_2^2 + p_2'^2)\;i\,\vec{\sigma}_2 \cdot \vec{p}\,'_2 \times \vec{p}_2\right\}. \end{aligned} \tag{7.104}$$

It is convenient at this point to introduce what we call a D–K simplification, in which we compress an expression by changing the index 2 to 1, so long as it gives the same matrix element. In coordinate space we then have

$$\begin{aligned} V_1 = &\frac{3\,\alpha}{8\,m^4}\,\vec{\sigma}_1 \cdot \frac{\vec{r}}{r^5}\,(\vec{r} \cdot \vec{p}_2) \times \vec{p}_1 + \frac{3\,\alpha}{8\,m^4}\,\vec{p}_1\,^2\,\vec{\sigma}_1 \cdot \frac{\vec{r}}{r^3} \times \vec{p}_1 \\ &- \frac{3\,\alpha}{16\,m^4\,r^5}\,\vec{r}(\vec{r} \times \vec{p}_1 \cdot \vec{\sigma}_1) \times \vec{p}_2 \cdot \vec{\sigma}_2 \\ &- \frac{\alpha}{16\,m^4\,r^3}\,(\vec{\sigma}_1 \cdot \vec{p}_2)\,(\vec{\sigma}_2 \cdot \vec{p}_1). \end{aligned} \tag{7.105}$$

If a transverse photon is exchanged, we first set the energy transfer in its propagator to zero, which is called the no retardation limit, with the correction to this procedure accounted for in Fig. 7.2(c). In this limit Fig. 7.2(b) gives

$$V_2 = -\frac{4\pi\alpha}{q^2}\,[u^\dagger(p_1')\,\alpha^i\,u(p_1)]\,[u^\dagger(p_2')\,\alpha^j\,u(p_2)]\,P_{ij}(q). \qquad (7.106)$$

The matrix element of α^i is expanded as

$$\begin{aligned} u^\dagger(p')\,\alpha^i\,u(p) = {} & \frac{1}{2\,m}\,(p^i + p'^i) + \frac{i}{2\,m}(\vec{\sigma}\times\vec{q})^i \\ & -\frac{1}{16\,m^3}\left[p^i\,(p'^2 + 3\,p^2) + p'^i\,(p^2 + 3\,p'^2)\right] \\ & +\frac{i}{16\,m^3}\left[(\vec{p}\,'\times\vec{\sigma})^i\,(p^2 + 3\,p'^2) - (\vec{p}\times\vec{\sigma})^i\,(p'^2 + 3\,p^2)\right], \end{aligned} \qquad (7.107)$$

which leads to (using the D–K simplification)

$$\begin{aligned} V_2 = {} & -\frac{\pi\alpha}{2\,m^4\,q^2}\Big\{4\,i\,p_1'^2\,\vec{\sigma}_1\times\vec{q}\cdot\vec{p}_1 + p_1'^2\,\vec{\sigma}_1\cdot\vec{q}\,\vec{\sigma}_2\cdot(\vec{p}\,'_1 + \vec{p}_1) \\ & +2\,p_1'^2\,(\vec{\sigma}_1\cdot\vec{q})\,(\vec{\sigma}_2\cdot\vec{q}) \\ & +2\,i\,p_1'^2\,P_{ij}(q)p_2^j\,[2\,\vec{p}_1\times\vec{\sigma}_1 + 3\,\vec{q}\times\vec{\sigma}_1]^i\,. \end{aligned} \qquad (7.108)$$

In configuration space V_2 takes the form

$$\begin{aligned} V_2 = {} & \frac{\alpha}{2\,m^4}\,p_1^2\,\vec{\sigma}_1\cdot\frac{\vec{r}}{r^3}\times\vec{p}_1 + \frac{9\,\alpha}{8\,m^4}\,p_1^2\,\frac{1}{r^5}\,(\vec{\sigma}_1\cdot\vec{r})\,(\vec{\sigma}_2\cdot\vec{r}) \\ & -\frac{i\,\alpha}{4\,m^4}\,p_1^2\,\frac{1}{r^3}\,(\vec{\sigma}_1\cdot\vec{r})\,(\vec{\sigma}_2\cdot\vec{p}_1) + \frac{i\,\alpha}{4\,m^2}\,p_1^2\,\frac{1}{r}\,\vec{\sigma}_1\cdot\vec{p}_1\times\vec{p}_2 \\ & -\frac{i\,\alpha}{4\,m^2}\,p_1^2\,\frac{\vec{r}}{r^3}\,(\vec{\sigma}_1\cdot\vec{r}\times\vec{p}_1)\vec{p}_2 - \frac{3\,\alpha}{4\,m^2}p_1^2\vec{\sigma}_1\cdot\frac{\vec{r}}{r^3}\times\vec{p}_2. \end{aligned} \qquad (7.109)$$

Figure 7.2(c), representing the retardation part of single transverse photon exchange, requires the most care. It is given by

$$\begin{aligned} V_3 = 4\pi\alpha & \int\frac{d^3k}{(2\,\pi)^3\,2\,k}\,P_{ij}(k)\left(\frac{p_1^i}{m} + \frac{i\,(\vec{\sigma}_1\times\vec{k})^i}{2\,m}\right) \\ & \left\{e^{i\,\vec{k}\cdot\vec{r}_1}\left(\frac{1}{E_0 - k - H_0} + \frac{1}{k}\right)e^{-i\,\vec{k}\cdot\vec{r}_2}\right\} \\ & \left(\frac{p_2^j}{m} - \frac{i\,(\vec{\sigma}_2\times\vec{k})^j}{2\,m}\right) + (1\leftrightarrow 2). \end{aligned} \qquad (7.110)$$

Only the second term in the retardation expansion contributes in order $m\,\alpha^6$ (notice two dotted lines in the corresponding diagram):

$$\frac{1}{E_0 - k - H_0} + \frac{1}{k} = \frac{(H_0 - E_0)}{k^2} - \frac{(H_0 - E_0)^2}{k^3} + \ldots, \qquad (7.111)$$

and the retardation part V_3 becomes

$$V_3 = -4\pi\alpha \int \frac{d^3k}{(2\,\pi)^3\, 2\,k}\, P_{ij}(k) \left(\frac{p_1^i}{m} + \frac{i\,(\vec{\sigma}_1 \times \vec{k})^i}{2\,m}\right)$$
$$\left\{e^{i\,\vec{k}\cdot\vec{r}_1}\, \frac{(H_0 - E_0)^2}{k^3} e^{-i\,\vec{k}\cdot\vec{r}_2}\right\} \left(\frac{p_2^j}{m} - \frac{i\,(\vec{\sigma}_2 \times \vec{k})^j}{2\,m}\right) + (1 \leftrightarrow 2)$$
$$= V_3^A + V_3^B\,, \tag{7.112}$$

where we have split the spin-dependent part of the calculation into two parts. Considering first the spin-spin term V_3^A we have

$$V_3^A = \frac{\pi\alpha}{2\,m^2} \int \frac{d^3k}{(2\,\pi)^3}\, \frac{\vec{\sigma}_1\cdot\vec{k}\,\vec{\sigma}_2\cdot\vec{k}}{k^4} e^{i\,\vec{k}\cdot\vec{r}_1}\,(H_0 - E_0)^2 e^{-i\,\vec{k}\cdot\vec{r}_2}$$
$$+ (1 \leftrightarrow 2). \tag{7.113}$$

Because $(H_0 - E_0)$ gives zero if next to a wave function, we can carry out the manipulation

$$e^{i\,\vec{k}\,\vec{r}_1}\,(H_0 - E_0)^2 e^{-i\,\vec{k}\cdot\vec{r}_2} + (1 \leftrightarrow 2)$$
$$\to \left[e^{i\,\vec{k}\cdot\vec{r}_1}\,,\, \left[(H_0 - E_0)^2\,,\, e^{-i\,\vec{k}\cdot\vec{r}_2}\right]\right]$$
$$= -\frac{1}{2\,m^2} \left[p_1^2\,,\, \left[p_2^2\,,\, e^{i\,\vec{k}\cdot\vec{r}}\right]\right] \tag{7.114}$$

to show

$$V_3^A = -\frac{\pi\alpha}{4\,m^4}\, \sigma_1^i\, \sigma_2^j \left[p_1^2\,,\, \left[p_2^2\,,\, \int \frac{d^3k}{(2\,\pi)^3}\, \frac{k^i\, k^j}{k^4}\, e^{i\,\vec{k}\cdot\vec{r}}\right]\right]. \tag{7.115}$$

Performing the k–integral with

$$\int \frac{d^3k}{(2\,\pi)^3}\, \frac{1}{k^2} \left(\frac{\delta^{ij}}{3} - \frac{k^i\, k^j}{k^2}\right) e^{i\,\vec{k}\cdot\vec{r}} = \frac{1}{8\,\pi\, r} \left(\frac{r^i\, r^j}{r^2} - \frac{\delta^{ij}}{3}\right) \tag{7.116}$$

then leads to, after some further manipulations, to

$$V_3^A = \frac{\alpha}{16\,m^4}\, p_1^2 \left[p_2^2, \frac{r^i\, r^j}{r^3}\right] {\sigma_1}^i\, {\sigma_2}^j \tag{7.117}$$
$$= \frac{3\,\alpha}{8\,m^4}\, p_1^2\, \frac{\vec{\sigma}_1\cdot\vec{r}\,\vec{\sigma}_2\cdot\vec{r}}{r^5}$$
$$+ \frac{\alpha}{8m^4}\, p_1^2\, \frac{i}{r^2}\, [(\hat{r}\cdot\vec{\sigma}_2)\vec{\sigma}_1 + (\hat{r}\cdot\vec{\sigma}_1)\vec{\sigma}_2 - 3(\hat{r}\cdot\vec{\sigma}_1)(\hat{r}\cdot\vec{\sigma}_2)\hat{r}]\cdot\vec{p}_2. \tag{7.118}$$

The second part, the spin-orbit term V_3^B, is obtained from Eq. (7.112)

$$V_3^B = -\frac{i\pi\,\alpha}{m^2} \int \frac{d^3k}{(2\,\pi)^3}\, \frac{1}{k^4} \Big[e^{i\,\vec{k}\cdot\vec{r}_1}\,(H_0 - E_0)^2 e^{-i\,\vec{k}\cdot\vec{r}_2}\, \vec{\sigma}_1 \times \vec{k}\cdot\vec{p}_2$$
$$-\vec{\sigma}_2 \times \vec{k}\cdot\vec{p}_1\, e^{i\,\vec{k}\cdot\vec{r}_1}\,(H_0 - E_0)^2 e^{-i\,\vec{k}\cdot\vec{r}_2}\Big] + (1 \leftrightarrow 2). \tag{7.119}$$

Using

$$\begin{aligned} & e^{i\vec{k}\vec{r}_1}(H_0-E_0)^2 e^{-i\vec{k}\cdot\vec{r}_2} \\ &= (H_0-E_0)\, e^{i\vec{k}\cdot\vec{r}}(H_0-E_0) + (H_0-E_0)\left[\frac{p_2^2}{2m},\, e^{i\vec{k}\cdot\vec{r}}\right] \\ &+ \left[e^{i\vec{k}\cdot\vec{r}},\, \frac{p_1^2}{2m}\right](H_0-E_0) + \left[\frac{p_2^2}{2m},\, \left[e^{i\vec{k}\cdot\vec{r}},\, \frac{p_1^2}{2m}\right]\right] \end{aligned} \tag{7.120}$$

allows it to be rewritten as

$$\begin{aligned} V_3^B &= -\frac{4\pi i\,\alpha}{m^2}\int \frac{d^3k}{(2\pi)^3} X S \frac{1}{k^4}\left\{\left[e^{i\vec{k}\cdot\vec{r}},\, \frac{p_1^2}{2m}\right]\left[H_0\,,\, \vec{\sigma}_1\times\vec{k}\cdot\vec{p}_2\right]\right. \\ &\left. + \left[\frac{p_2^2}{2m},\, \left[e^{i\vec{k}\cdot\vec{r}},\, \frac{p_1^2}{2m}\right]\right]\vec{\sigma}_1\times\vec{k}\cdot\vec{p}_2\right\}. \end{aligned} \tag{7.121}$$

Carrying out the k–integral using

$$\int \frac{d^3k}{(2\pi)^3}\,\frac{4\pi}{k^4}\left(e^{i\vec{k}\cdot\vec{r}}-1\right) = -\frac{r}{2} \tag{7.122}$$

and evaluating the commutators gives

$$\begin{aligned} V_3^B &= \frac{\alpha^2}{2m^3}\,\vec{\sigma}_1\cdot\frac{\vec{r}}{r^4}\times\vec{p}_1 + \frac{Z\alpha^2}{2m^3 r}\,\vec{\sigma}_1\cdot\frac{\vec{r}_1}{r_1^3}\times\vec{p}_2 + \frac{Z\alpha^2}{2m^3}\,\vec{\sigma}_1\cdot\frac{\vec{r}\times\vec{r}_1}{r^3 r_1^3}\,\vec{r}\cdot\vec{p}_2 \\ &-\frac{\alpha}{2m^4}\,p_1^2\,\vec{\sigma}_1\cdot\frac{\vec{r}}{r^3}\times\vec{p}_2 + \frac{\alpha}{2m^2}\,p_1^2\,\vec{\sigma}_1\cdot\frac{i\,\vec{r}}{r^3}\,(\vec{r}\cdot\vec{p}_2)\times\vec{p}_2. \end{aligned} \tag{7.123}$$

Figures 7.2(d) and 7.2(e) are the time-ordered perturbation theory forms of the "box" and "crossed box" diagrams with one transverse and one Coulomb, or two transverse photons respectively. Beginning with Fig. 7.2(d), we first analyze the upper electron line. We use q_1 and q_2 to denote the momenta transfered by the transverse and Coulomb photons respectively,

$$\begin{aligned} & u^\dagger(p_1')\,\alpha^j\,\Lambda_-(p_1+q_2)\,u(p_1) + u^\dagger(p_1')\,\Lambda_-(p_1'-q_2)\,\alpha^j\,u(p_1) \\ &\qquad = -\frac{i}{m}\,(\vec{q}_2\times\vec{\sigma}_1)^j. \end{aligned} \tag{7.124}$$

The lower electron line is

$$\begin{aligned} & u^\dagger(p_2')\,\alpha^i\,\Lambda_+(p_2'+q_1)\,u(p_2) + u^\dagger(p_2')\,\Lambda_+(p_2-q_1)\,\alpha^i\,u(p_2) \\ &\qquad = \frac{1}{m}(p_2^i + p_2'^i) + \frac{i}{m}\,(\vec{q}_1\times\vec{\sigma}_2)^i. \end{aligned} \tag{7.125}$$

Combining these expressions we find the effective interaction V_4 to be

$$V_4 = 32\pi^2 \int \frac{d^3 q_1}{(2\,\pi)^3} \int \frac{d^3 q_2}{(2\,\pi)^3} (2\,\pi)^3\, \delta^3(\vec{q}_1 + \vec{q}_2 - \vec{q})\, \frac{\alpha}{q_2^2}\, \frac{\alpha}{2\, q_1^2}\, P_{ij}(q_1)$$
$$\frac{1}{2\,m}\, \frac{i}{m}\, (\vec{q}_2 \times \vec{\sigma}_1)^j \left[\frac{1}{m}(p_2^i + p_2'^i) + \frac{i}{m}\, (\vec{q}_1 \times \vec{\sigma}_2)^i\right] \tag{7.126}$$
$$= \frac{8\pi^2}{m^3} \int \frac{d^3 q_1}{(2\,\pi)^3} \int \frac{d^3 q_2}{(2\,\pi)^3} (2\,\pi)^3\, \delta^3(\vec{q}_1 + \vec{q}_2 - \vec{q})\, \frac{\alpha}{q_1^2}\, \frac{\alpha}{q_2^2}$$
$$\left[(\vec{\sigma}_1 \cdot \vec{q}_1)\, (\vec{\sigma}_2 \cdot \vec{q}_2) - i\, P_{ij}(q)\, (\vec{\sigma}_1 \times \vec{q}_2)^i\, (\vec{p}_2 + \vec{p}_2')^j\right]. \tag{7.127}$$

The factor 2 in front of the first integral comes from the fact that Z is on upper or on the lower electron line. In configuration space V_4 is

$$V_4 = \frac{\alpha^2}{2\, m^3\, r^4}\, (\vec{\sigma}_1 \cdot \vec{r} \times \vec{p}_2 - \vec{\sigma}_1 \cdot \hat{r}\, \vec{\sigma}_2 \cdot \hat{r}). \tag{7.128}$$

The next contribution is the exchange of two transverse photons with a single Z, Fig. 7.2(e). This contribution consists of twelve diagrams. They reduce to two diagrams in the non-retardation limit,

$$V_5 = 16\pi^2 \int \frac{d^3 q_1}{(2\,\pi)^3} \int \frac{d^3 q_2}{(2\,\pi)^3} (2\,\pi)^3\, \delta^3(\vec{q}_1 + \vec{q}_2 - \vec{q})$$
$$\frac{\alpha}{q_1^2}\, P_{ik}(q_1)\, \frac{\alpha}{q_2^2}\, P_{jk}(q_2)\, \frac{1}{2\,m}$$
$$\left[\frac{(2\,\vec{p}_2 + i\,\vec{q}_1 \times \vec{\sigma}_2)^i}{2\,m}\, \frac{(2\,\vec{p}\,'_2 + i\,\vec{q}_2 \times \vec{\sigma}_2)^j}{2\,m}\right.$$
$$\left. + \frac{(2\,\vec{p}_1 - i\,\vec{q}_1 \times \vec{\sigma}_1)^i}{2\,m}\, \frac{(2\,\vec{p}\,'_1 - i\,\vec{q}_2 \times \vec{\sigma}_1)^j}{2\,m}\right]. \tag{7.129}$$

In coordinate space

$$V_5 = -\frac{\alpha^2}{2\, m^2}\, \vec{\sigma}_1 \cdot \frac{\vec{r}}{r^4} \times \vec{p}_1. \tag{7.130}$$

The diagram in Fig. 7.2(f) is a three-body effect, with the horizontal line in the middle being the nucleus. We note its similarity to Fig. 7.2(d) diagram, and a similar analysis leads to

$$V_6 = \frac{4\pi^2}{m^3} \int \frac{d^3 q_1}{(2\,\pi)^3} \int \frac{d^3 q_2}{(2\,\pi)^3} (2\,\pi)^3\, \delta^3(\vec{q}_1 + \vec{q}_2 - \vec{q})\, \frac{\alpha}{q_1^2}\, \frac{Z\,\alpha}{q_2^2}$$
$$\left[-\vec{\sigma}_1 \cdot \vec{q}_1\, \vec{\sigma}_2 \cdot \vec{q}_2 + i\, P_{ij}(q_1)\, (\vec{\sigma}_1 \times \vec{q}_2)^i\, (\vec{p}_2 + \vec{p}_2\,')^j\right]$$
$$+(1 \leftrightarrow 2). \tag{7.131}$$

Table 7.1. Douglas–Kroll terms in the infinite nuclear mass limit $\nu_B^{(4)}$, reduced mass and mass polarization corrections $\nu_B^{(4r)}$, in units of α^4 Ry = 9.329027 MHz.

Contribution	$\nu_{01B}^{(4)}$	$\nu_{12B}^{(4)}$	$\nu_{01B}^{(4r)}$	$\nu_{12B}^{(4r)}$
$\frac{3Z}{8}\nabla_1^2 \frac{1}{r_1^3}\vec{\sigma}_1\cdot(\vec{r}_1\times\vec{p}_1)$	0.224 210	0.448 420	0.000 081	0.000 162
$-\frac{Z}{r^3 r_1^3}\vec{\sigma}_1\cdot(\vec{r}_1\times\vec{r})(\vec{r}\cdot\vec{p}_2)$	0.016 418	0.032 836	-0.000 013	-0.000 026
$\frac{Z}{2}\frac{1}{r^3 r_1^3}(\vec{\sigma}_1\cdot\vec{r})(\vec{\sigma}_2\cdot\vec{r}_1)$	0.010 203	-0.004 081	-0.000 045	0.000 018
$\frac{1}{2}\frac{1}{r^4}\vec{\sigma}_1\cdot(\vec{r}\times\vec{p}_2)$	0.034 557	0.069 115	-0.000 002	-0.000 003
$-\frac{1}{2}\frac{1}{r^6}(\vec{\sigma}_1\cdot\vec{r})(\vec{\sigma}_2\cdot\vec{r})$	0.081 157	-0.032 463	-0.000 009	0.000 003
$-\frac{5}{8}\nabla_1^2\frac{1}{r^3}\vec{\sigma}_1\cdot(\vec{r}\times\vec{p}_1)$	-0.093 084	-0.186 167	-0.000 009	-0.000 018
$\frac{3}{4}\nabla_1^2\frac{1}{r^3}\vec{\sigma}_1\cdot(\vec{r}\times\vec{p}_2)$	-0.171 228	-0.342 456	0.000 054	0.000 108
$\frac{i}{4}\nabla_1^2\frac{1}{r}\vec{\sigma}_1\cdot(\vec{p}_1\times\vec{p}_2)$	0.000 221	0.000 442	0.000 007	0.000 015
$\frac{3i}{4}\nabla_1^2\frac{1}{r^3}(\vec{r}\cdot\vec{p}_2)\vec{\sigma}_1\cdot(\vec{r}\times\vec{p}_1)$	0.006 456	0.012 912	-0.000 022	-0.000 044
$\frac{3i}{8}\frac{1}{r^5}\vec{\sigma}_1\cdot(\vec{r}\times(\vec{r}\cdot\vec{p}_2)\vec{p}_1)$	-0.008 671	-0.017 343	0.000 001	0.000 002
$-\frac{3}{16}\frac{1}{r^5}\vec{\sigma}_2\cdot(\vec{r}\times(\vec{\sigma}_1\cdot(\vec{r}\times\vec{p}_1))\vec{p}_2)$	-0.004 457	0.001 783	0.000 004	-0.000 001
$-\frac{1}{16}\frac{1}{r^3}(\vec{\sigma}_1\cdot\vec{p}_2)(\vec{\sigma}_2\cdot\vec{p}_1)$	-0.010 325	0.004 130	0.000 000	0.000 000
$-\frac{3}{2}\nabla_1^2\frac{1}{r^5}(\vec{\sigma}_1\cdot\vec{r})(\vec{\sigma}_2\cdot\vec{r})$	-0.786 253	0.314 501	0.000 235	-0.000 094
$\frac{i}{4}\nabla_1^2\frac{1}{r^3}(\vec{\sigma}_1\cdot\vec{r})(\vec{\sigma}_2\cdot\vec{p}_1)$	0.135 711	-0.054 284	-0.000 032	0.000 013
$-\frac{i}{8}\nabla_1^2\frac{1}{r^3}(\vec{\sigma}_1\cdot\vec{r})(\vec{\sigma}_2\cdot\vec{p}_2)$	0.213 103	-0.085 241	-0.000 061	0.000 024
Total	-0.351 982	0.162 103	0.000 190	0.000 158

Its coordinate space representation is

$$V_6 = \frac{Z\,\alpha^2}{2\,m^3}\left[\frac{\vec{\sigma}_1\cdot\vec{r}}{r^3}\,\frac{\vec{\sigma}_2\cdot\vec{r}_1}{r_1^3} - \frac{1}{r\,r_1^3}\,\vec{\sigma}_1\cdot\vec{r}_1\times\vec{p}_2 - \frac{1}{r^3\,r_1^3}\,\vec{\sigma}_1\cdot\vec{r}_1\times\vec{r}\,(\vec{r}\cdot\vec{p}_2)\right]. \tag{7.132}$$

We note that the second term of V_6 cancels the second term of V_{3B}: we will see later when including the effect of $F_2(q^2)$ that this cancellation is no longer present.

The effective Hamiltonian $H^{(6)}$ is the sum of the V_i's derived above,

$$H^{(6)} = \Sigma_{i=0}^{6} V_i. \tag{7.133}$$

This complete result agrees with the Bethe–Salpeter equation based derivation of Douglas and Kroll [27], though those authors included other operators that lead to the $a_e^2 H_{SS}$ part of $(1+a_e)^2 H_{SS}$ that we have already included in our organization of the calculation.

We now discuss the contribution $\nu_A^{(4)}$, which comes from second-order perturbation theory. As mentioned above, in this case the entire Breit Hamiltonian must be included, as one can get fine structure from terms

Table 7.2. Contributions to $\nu_A^{(4)}$ broken down contributing intermediate states: For each J, first row with and second row without a_e factors included in Breit Hamiltonian. Units α^4 Ry.

State	^3P	^1P	^3D	^1D	^3F
$J=0$	-0.653470(9)	0.0	0.0	0.0	0.0
	-0.651602(9)	0.0	0.0	0.0	0.0
$J=1$	-0.127293(5)	-0.709495(9)	-0.002831	0.0	0.0
	-0.126969(5)	-0.707000(9)	-0.002823	0.0	0.0
$J=2$	0.041297(5)	0.0	-0.008262	-0.002385	-0.005626(1)
	0.041278(5)	0.0	-0.008236	-0.002379	-0.005600(1)

with one Hamiltonian being H_{SI} and the other H_{SD} as well as when both are H_{SD}. While straightforward in principle, these calculations, first carried out by Lewis and Serafino [35], are algebraically complicated and numerically challenging. The operators one is dealing with are nearly singular, so without very high quality numerical techniques significant wave-function related errors can arise. While it is evaluated in the non-recoil limit, it is trivial to incorporate radiative corrections proportional to a_e, as they enter as simple algebraic factors. Results with and without the radiative corrections are given in Table 7.2, but we note that Eq. (7.72) includes the a_e terms.

7.7. Determination of $\nu^{(4r)}$

Recoil corrections of order $\frac{m^2}{M}(Z\alpha)^6$ have been treated by Drake in Ref. [32] and by us in Refs. [36, 37]. A first set of contributions of this order is obtained by evaluating the Douglas–Kroll operators with a nonrelativistic Hamiltonian that includes mass polarization, which is analogous to using E_i rather than $\tilde{E}_i$ in the lowest order calculation, in addition including scaling factors of m_r/m. Since the original evaluation of the expectation value of the Douglas–Kroll operators [38] was carried out in the infinite nuclear mass limit, we tabulate the difference in Table 7.1 (see Ref. [36] for more details), finding excellent agreement with Ref. [32]. Secondly, a set of operators analogous to those giving rise to E_4 must be derived for this order. These operators were derived in Ref. [37] using effective field theory, and also by Zhang in his Bethe–Salpeter approach [39], [40]. We note that we found four terms of the same form but with opposite sign in comparing with his results. As with the Douglas–Kroll operators there are a number of contributions, and we only illustrate them by showing how a transverse photon exchange with the nucleus leads to a recoil operator of order $\frac{m^2}{M}(Z\alpha)^6$. We recall that this exchange in lowest order led to a recoil operator characterized by the constant E_4: The scattering amplitude associated with this term was

$$\mathcal{M}^{eN}(1T) = \frac{Z\alpha}{2M\vec{q}^{\,2}}\bar{u}(\vec{p}_3)\gamma_i u(\vec{p}_1)P_{ij}(q)(-p_1 - 2p_2 - p_3)_j. \tag{7.134}$$

The order α^2 corrections to this come from

$$\mathcal{M}_6^{eN}(1T) = \frac{Z\alpha}{32Mm^3\vec{q}^{\,2}}P_{ij}(q)(p_1 + p_3 + 2p_2)_j$$
$$[\sigma_{1i}\vec{\sigma}_1\cdot\vec{p}_1(3\vec{p}_1^{\,2} + \vec{p}_3^{\,2}) + \vec{\sigma}_1\cdot\vec{p}_3\sigma_{1i}(3\vec{p}_3^{\,2} + \vec{p}_1^{\,2})], \tag{7.135}$$

which has a spin-dependent part

$$\begin{aligned}\mathcal{M}_6^{eN}(1T)(fs) = \frac{i(Z\alpha)}{32Mm^3\vec{q}^{\,2}}\Bigg[&4\vec{\sigma}_1\cdot(\vec{p}_3\times\vec{p}_1)(\vec{p}_1^{\,2} + \vec{p}_3^{\,2})\\ &+ 2\vec{\sigma}_1\cdot(\vec{p}_3\times\vec{p}_2)(\vec{p}_1^{\,2} + 3\vec{p}_3^{\,2}) - 2\vec{\sigma}_1\cdot(\vec{p}_1\times\vec{p}_2)(3\vec{p}_1^{\,2} + \vec{p}_3^{\,2})\\ &- 2\frac{\vec{q}\cdot(\vec{p}_1 + \vec{p}_3 + 2\vec{p}_2)}{\vec{q}^{\,2}}\vec{\sigma}_1\cdot(\vec{p}_3\times\vec{p}_1)(\vec{p}_3^{\,2} - \vec{p}_1^{\,2})\Bigg].\end{aligned} \tag{7.136}$$

Fourier transforming to coordinate space leads to the effective operator

$$\begin{aligned}V_{1\text{rec}} = &\frac{i(Z\alpha)}{4Mm^3}\vec{p}_1^{\,2}\frac{1}{r_1}\vec{\sigma}_1\cdot(\vec{p}_1\times\vec{p}_2) - \frac{i(Z\alpha)}{4Mm^3}\vec{p}_1^{\,2}\frac{\vec{r}_1}{r_1^3}(\vec{\sigma}_1\cdot(\vec{r}_1\times\vec{p}_1))\cdot(\vec{p}_1 + \vec{p}_2)\\ &- \frac{3Z\alpha}{4Mm^3}\vec{p}_1^{\,2}\vec{\sigma}_1\cdot\frac{\vec{r}_1}{r_1^3}\times(\vec{p}_1 + \vec{p}_2).\end{aligned} \tag{7.137}$$

While six other contributions have to be considered, it is remarkable that their sum can be given in the compact form, where now the operators are understood to be in dimensionless units,

$$\begin{aligned}V_{\text{rec}}^{DK} = {}&2\alpha^4\frac{m}{M}\text{Ry}_\infty c\\ &\Bigg[\frac{Z}{4}p_1^2\Bigg(\frac{i}{r_1}\vec{\sigma}_1\cdot\vec{p}_1\times\vec{p}_2 - i\frac{\vec{r}_1}{r_1^3}(\vec{\sigma}_1\cdot\vec{r}_1\times\vec{p}_1)\cdot(\vec{p}_1 + \vec{p}_2)\\ &\qquad - \frac{3}{r_1^3}\vec{\sigma}_1\cdot\vec{r}_1\times(\vec{p}_1 + \vec{p}_2)\Bigg)\\ &+ Z^2\Bigg(\frac{\vec{\sigma}_1\cdot\vec{r}_1\times\vec{r}_2}{r_1^3r_2^3}\vec{r}_1\cdot\vec{p}_1 - \frac{1}{2}\vec{\sigma}_1\cdot\frac{\vec{r}_1}{r_1^4}\times(\vec{p}_1 + \vec{p}_2) - \frac{1}{4}\vec{\sigma}_1\cdot\frac{\vec{r}_2}{r_2^3}\vec{\sigma}_2\cdot\frac{\vec{r}_1}{r_1^2}\Bigg)\\ &+ Z\Bigg(\frac{\vec{\sigma}_1\cdot\vec{r}\times(\vec{p}_1 + \vec{p}_2)}{r_1r^3} - \frac{\vec{\sigma}_1\cdot(\vec{r}_1\times\vec{r})}{r_1^3r^3}\vec{r}_1\cdot(\vec{p}_1 + \vec{p}_2)\Bigg)\Bigg].\end{aligned} \tag{7.138}$$

Finally, recoil corrections to the Lewis–Serafino calculation of the second-order contribution must be treated. This is a far less trivial task than including leading QED corrections, and has at present only been carried out by Drake [32], who quotes 0.010 19 MHz for the contribution to $\nu_{12}^{(4r)}$ and -0.010 81 MHz for $\nu_{01}^{(4r)}$. The result given for $\nu^{(4r)}$ in Eq. (7.72) is the sum of the three corrections discussed above.

Table 7.3. Results in kHz for the recoil correction operators, listed with prefactor $2\,\alpha^4 \frac{m}{m_\alpha}\mathrm{Ry}_\infty c$ understood, for the large and small helium fine structure intervals.

Contribution	ν_{01}	ν_{12}
$\frac{iZ}{4} p_1^2 \frac{1}{r_1} \vec{\sigma}_1 \cdot (\vec{p}_1 \times \vec{p}_2)$	0.480	0.959
$-\frac{iZ}{4}\, p_1^2\, \frac{\vec{r}_1}{r_1^3}\, (\vec{\sigma}_1 \cdot \vec{r}_1 \times \vec{p}_1) \cdot (\vec{p}_1 + \vec{p}_2)$	0.165	0.330
$-\frac{3Z}{4}\, p_1^2\, \vec{\sigma}_1 \cdot \frac{\vec{r}_1}{r_1^3} \times (\vec{p}_1 + \vec{p}_2)$	-1.103	-2.206
$Z\, \vec{\sigma}_1 \cdot \frac{\vec{r}}{r_1\, r^3} \times (\vec{p}_1 + \vec{p}_2)$	0.129	0.258
$Z\, \vec{\sigma}_1 \cdot \frac{\vec{r}}{r^3} \times \frac{\vec{r}_1}{r_1^3}\, (\vec{r}_1 \cdot (\vec{p}_1 + \vec{p}_2))$	0.042	0.085
$Z^2\, \vec{\sigma}_1 \cdot \frac{\vec{r}_1}{r_1^3} \times \frac{\vec{r}_2}{r_2^3}\, (\vec{r}_1 \cdot \vec{p}_1)$	-0.236	-0.472
$-\frac{Z^2}{2}\, \vec{\sigma}_1 \cdot \frac{\vec{r}_1}{r_1^4} \times (\vec{p}_1 + \vec{p}_2)$	-0.471	-0.941
$-\frac{Z^2}{4} \vec{\sigma}_1 \cdot \frac{\vec{r}_2}{r_2^3}\, \vec{\sigma}_2 \cdot \frac{\vec{r}_1}{r_1^3}$	-0.354	0.141
Sum	-1.348	-1.847

7.8. Determination of $\nu^{(5)}$

By far the most difficult part of this calculation is the accurate determination of corrections of order $m\alpha^7$. As can be seen from the discussion of the hydrogenic case, even for a one-electron system these calculations are extremely complex. Several recent advances in bound state QED have made the calculation of higher-order corrections to helium fine structure possible. Firstly, Yelkhovsky in Ref. [41] has shown how to use dimensional regularization in the calculation of helium energy levels, and together with Korobov has obtained in [42] numerical values for the α^4 Ry contributions to the ground state. Although their results contain several mistakes [30], the approach in general is a very powerful one. Secondly, in Ref. [16], a Foldy–Wouthuysen (F.W.) transformed QED Langrangian was used to derive all effective α^4 Ry operators for arbitrary states of few electron atoms. It will be demonstrated below that instead of the full QED Lagrangian, one can use an approximate F.W. form when, as is the case here, all electron momenta involved are much smaller than the electron mass. This includes corrections involving the anomalous magnetic moment a_e. More recently, K.P. together with Jentschura and Czarnecki, has obtained in Ref. [43] general formulae for α^5 Ry corrections to hydrogenic energy levels. The calculational approach of these works [4, 42, 44] and the present chapter is based on dimensionally regularized QED. The parameter ϵ, related to the space dimension $d = 3 - 2\,\epsilon$, plays the role of both an infrared and ultraviolet regulator, as some α^5 Ry terms are divergent in $d = 3$ space.

This artificial parameter ϵ is used to regulate various terms, and we will explicitly demonstrate its cancellation in their sum.

The fine structure in order $m\,\alpha^7$ (α^5 Ry) can be written as [45]

$$E^{(7)} = \langle H^{(7)}\rangle + 2\left\langle H^{(4)}\,\frac{1}{(E_0-H_0)'}\,H^{(5)}\right\rangle + E_L, \tag{7.139}$$

where E_L is the Bethe logarithmic correction defined below in Eq. (7.161), and $H^{(i)}$ is an effective Hamiltonian of order $m\,\alpha^i$. We now present complete derivations of $H^{(7)}$, the second-order term (which will be referred to in the following as E_S), and E_L. These contributions have been derived and calculated in a series of works [14, 45]. Important terms of order $m\,\alpha^7\,\ln\alpha$, previously calculated in Ref. [46], have been confirmed in [45].

$H^{(7)}$ consists of exchange terms, where photons are exchanged between the electrons, and radiative corrections, where a photon is emitted and absorbed by the same electron. We consider first the exchange terms. Their derivation in general is quite complicated. We find that only two-photon exchange diagrams contribute and there are no three-body terms, the result of an internal cancellation. A feature of the calculation that leads to considerable simplification is the fact that the order being calculated is non-analytic in α^2. For example, $H^{(5)}$ (see Eq. (7.157)) consists of Dirac delta $\delta^3(r)$ and $1/r^3$ terms only, and they can be derived from the two-photon exchange scattering amplitude. Similar results hold for the spin-dependent $m\,\alpha^7$ terms. If $H^{(7)}$ represents an effective Hamiltonian, it has to give the same scattering amplitude as in full QED. Therefore, we obtain the exchange contribution δH from the spin-dependent part of the two-photon scattering amplitude, which is

$$\begin{aligned}\delta_1 H = {}& \frac{i\,e^4}{(2\,\pi)^D}\int d^D k\,\frac{1}{(k+q/2)^2}\,\frac{1}{(k-q/2)^2}\\ & \left[\bar{u}(p_1')\,\gamma^\mu\,\frac{1}{\not k+(\not p_1+\not p_1')/2-1}\,\gamma^\nu\,u(p_1)\right.\\ & \left.+\bar{u}(p_1')\,\gamma^\nu\,\frac{1}{-\not k+(\not p_1+\not p_1')/2-1}\,\gamma^\mu\,u(p_1)\right]\\ & \times\bar{u}(p_2')\,\gamma^\nu\,\frac{1}{\not k+(\not p_2+\not p_2')/2-1}\,\gamma^\mu\,u(p_2)\end{aligned} \tag{7.140}$$

where q has been defined in Eq. (7.103). There are three scales of the k integral: m, $m\,\alpha$, and $m\,\alpha^2$. Only the first two are included here, with the low energy part $k\sim\alpha^2$ treated separately as a contribution to E_L. From the above scattering amplitude one extracts the spin-dependent terms using

$$\bar{u}(p')\,Q\,u(p) = \operatorname{Tr} Q\,u(p)\otimes\bar{u}(p') \rightarrow \begin{cases} \operatorname{Tr} Q\left(\frac{\not p+m}{2m}\right)\left(\frac{\gamma^0+I}{4}\right)\left(\frac{\not p'+m}{2m}\right) \\ \sigma^{ij}\operatorname{Tr} Q\left(\frac{\not p+m}{2m}\right)\left(\frac{\gamma^0+I}{2}\right)\frac{\sigma^{ij}}{4}\left(\frac{\not p'+m}{2m}\right). \end{cases} \quad (7.141)$$

If one expands this amplitude in small external momenta and assumes $k\sim m$, one obtains

$$\begin{aligned}\delta_{1H}H = \alpha^2\Big\{&\sigma_1(j,q)\,\sigma_2(j,q)\,\left[-\frac{23}{36}+\frac{7}{12\,\epsilon}\right] \\ &+i\left[\sigma_1(p_1',p_1)+\sigma_2(p_2',p_2)\right]\left[-\frac{1}{6}-\frac{1}{4\,\epsilon}\right] \\ &+i\left[\sigma_1(p_2',p_2)+\sigma_2(p_1',p_1)\right]\left[\frac{1}{4}\right] \\ &+\frac{1}{8}\,\sigma_1(j,p_1+p_1')\,\sigma_2(j,p_2+p_2') \\ &-\frac{1}{8}\,\sigma_1(j,p_2+p_2')\,\sigma_2(j,p_1+p_1') \\ &+\frac{17}{72}\,\sigma_1(j,p_1-p_2+p_1'-p_2')\,\sigma_2(j,p_1-p_2+p_1'-p_2')\Big\}. \end{aligned} \quad (7.142)$$

A similar expansion, but now assuming that $k\sim m\,\alpha$ leads to

$$\begin{aligned}\delta_{1M}H = \alpha^2\Big\{&\sigma_1(j,q)\,\sigma_2(j,q)\,\left[-\frac{5}{12}-\frac{1}{4\,\epsilon}+\frac{1}{2}\,\ln(q)\right] \\ &+i\left[\sigma_1(p_1',p_1)+\sigma_2(p_2',p_2)\right]\left[\frac{7}{12}-\frac{1}{12\,\epsilon}+\frac{1}{6}\,\ln(q)\right] \\ &+i\left[\sigma_1(p_2',p_2)+\sigma_2(p_1',p_1)\right]\left[\frac{2}{3}-\frac{2}{3\,\epsilon}+\frac{4}{3}\,\ln(q)\right]\Big\}. \end{aligned} \quad (7.143)$$

The sum of $\delta_{1H}H$ and $\delta_{1M}H$ is

$$\begin{aligned}\delta_1 H = \alpha^2\Big\{&\sigma_1(j,q)\,\sigma_2(j,q)\,\left[-\frac{19}{18}+\frac{1}{3\,\epsilon}+\frac{1}{2}\,\ln(q)\right] \\ &+i\left[\sigma_1(p_1',p_1)+\sigma_2(p_2',p_2)\right]\left[\frac{5}{12}-\frac{1}{3\,\epsilon}+\frac{1}{6}\,\ln(q)\right] \\ &+i\left[\sigma_1(p_2',p_2)+\sigma_2(p_1',p_1)\right]\left[\frac{11}{12}-\frac{2}{3\,\epsilon}+\frac{4}{3}\,\ln(q)\right] \\ &+\frac{1}{8}\,\sigma_1(j,p_1+p_1')\,\sigma_2(j,p_2+p_2') \\ &-\frac{1}{8}\,\sigma_1(j,p_2+p_2')\,\sigma_2(j,p_1+p_1') \\ &+\frac{17}{72}\,\sigma_1(j,p_1-p_2+p_1'-p_2')\,\sigma_2(j,p_1-p_2+p_1'-p_2')\Big\}, \end{aligned} \quad (7.144)$$

where $\sigma^{ij} = -i/2\,[\sigma^i, \sigma^j]$, $\sigma(j,q) = \sigma^{ji}\,q^i$, and $q = \sqrt{\vec{q}^{\,2}}$. The $1/\epsilon$ divergences cancel out with the low energy part where photon momenta are of the order of the binding energy. This low energy contribution is also evaluated in D-dimensions. Specifically, the integral over photon energy is split into two parts $\int_0^\infty d\omega = \int_0^\Lambda d\omega + \int_\Lambda^\infty d\omega$, then an expansion around $D = 3$ is made, followed finally by an expansion in α and taking the limit of large Λ. The first part is finite at $D = 3$, and will be evaluated later together with the self-energy contribution in Eq. (7.161), while the second part is given by

$$\delta E_L = \alpha \int_\Lambda^\infty \frac{d^d k}{(2\,\pi)^d\, 2\,k}\, P_{ij}(k) \times \delta \left\langle \phi \left| p_1^i \, \frac{1}{E - H - k}\, p_2^j \right| \phi \right\rangle + (1 \leftrightarrow 2). \tag{7.145}$$

Here δ indicates both including the first-order correction to ϕ along with the change in H and E due to the spin-dependent part of the Breit-Pauli Hamiltonian $H^{(4)}$, which in D–dimensions is

$$\delta H = \frac{\alpha}{4\,m^2\,q^2}\left[i\,p_1'^{\,i}\,p_1{}^j(\sigma_1^{ij} + 2\,\sigma_2^{ij}) + i\,p_2'^{\,i}\,p_2{}^j(\sigma_2^{ij} + 2\,\sigma_1^{ij}) - \sigma_1^{ik}\,\sigma_2^{jk}\,q^i\,q^j\right]. \tag{7.146}$$

The resulting correction is a sum of two terms. The first one contributes to E_S in Eq. (7.139), and the second term is the effective Hamiltonian

$$\delta_2 H = \alpha^2\left[\frac{5}{9} + \frac{1}{3\,\epsilon} + \frac{2}{3}\,\ln[(Z\,\alpha)^{-2}]\right]\,[i\,\sigma_1(p_1', p_1) + i\,\sigma_2(p_2', p_2) + 2\,i\,\sigma_1(p_2', p_2) + 2\,i\,\sigma_2(p_1', p_1) - \sigma_1(j,q)\,\sigma_2(j,q)], \tag{7.147}$$

where we have omitted $\ln 2\Lambda$. (A canceling $\ln 2\Lambda$ appears in the $\int_0^\Lambda d\omega$ integral, and therefore can be consistently dropped in both parts of the calculation). Together with Eq. (7.144) this gives the complete contribution due to exchange terms. When calculating expectation values between ^3P_J states further simplifications can be performed. One of them is using the fact that the expectation value of a Dirac delta function with both momenta on the right or on the left hand side vanishes. Secondly, as noted earlier, the expectation value of σ_1 is equal to that of σ_2 for triplet states. As a result the total exchange contribution $H_E = \delta_1 H + \delta_2 H$ is

$$H_E = \alpha^2\left[6 + 4\ln[(Z\,\alpha)^{-2}] + 3\,\ln q\right] i\,\sigma_1(p_1', p_1) + \alpha^2\left[-\frac{23}{9} - \frac{2}{3}\,\ln[(Z\,\alpha)^{-2}] + \frac{1}{2}\,\ln q\right] \times \sigma_1(j,q)\,\sigma_2(j,q). \tag{7.148}$$

The treatment of radiative corrections is different. We argue that these corrections can be incorporated by the use of the same electromagnetic form factors introduced in the hydrogen section along with a Uehling correction to the Coulomb potential

$$\begin{aligned} F_1(-\vec{q}^{\,2}) &= 1 + \frac{\alpha}{\pi}\left(\frac{1}{8} + \frac{1}{6\,\epsilon}\right)\vec{q}^{\,2} \\ F_2(-\vec{q}^{\,2}) &= \frac{\alpha}{\pi}\left(\frac{1}{2} - \frac{1}{12}\,\vec{q}^{\,2}\right) \\ F_V(-\vec{q}^{\,2}) &= \frac{\alpha}{\pi}\,\frac{1}{15}\,\vec{q}^{\,2}. \end{aligned} \tag{7.149}$$

In principle there are additional corrections quadratic in the electromagnetic field. However, terms formed out of $\vec{E}, \vec{B}, \vec{p}, \vec{\sigma}$ can be shown to contribute only at higher order, and thus can be neglected. Corrections coming from Eq. (7.149) are obtained by rederiving the Breit–Pauli Hamiltonian H_{SD} using these modified electromagnetic vertices and photon propagator, with the result

$$\begin{aligned} \delta_3 H = {} & \pi\, Z\, \alpha(F_1' + 2\, F_2' + F_V')i\,[\sigma_1(p_1'', p_1) + \sigma_2(p_2'', p_2)] \\ & -\pi\, \alpha(2\, F_1' + 2\, F_2' + F_V')i\,[\sigma_1(p_1', p_1) + \sigma_2(p_2', p_2)] \\ & -2\,\pi\, \alpha(2\, F_1' + F_2' + F_V')i\,[\sigma_1(p_2', p_2) + \sigma_2(p_1', p_1)] \\ & +\pi\, \alpha(2\, F_1' + 2\, F_2' + F_V')\,\sigma_1(j, q)\,\sigma_1(j, q)\,, \end{aligned} \tag{7.150}$$

where by p'' we denote momentum scattered off the Coulomb potential of a nucleus. There is also a low-energy contribution which is calculated in a way similar to this in Eq. (7.145), namely

$$\begin{aligned} \delta E_L = {} & \alpha \int_\Lambda^\infty \frac{d^d k}{(2\,\pi)^d\, 2\, k}\, P_{ij}(k) \\ & \times \delta \left\langle \phi \left| p_1^i\, \frac{1}{E - H - k}\, p_1^j \right| \phi \right\rangle + (1 \to 2). \end{aligned} \tag{7.151}$$

The resulting effective Hamiltonian is

$$\begin{aligned} \delta_4 H = {} & \alpha^2 \left[\frac{5}{9} + \frac{1}{3\epsilon} + \frac{2}{3}\,\ln[(Z\,\alpha)^{-2}]\right] \\ & \times \left[\frac{i\,Z}{2}\,\sigma_1(p_1'', p_1) + \frac{i\,Z}{2}\,\sigma_2(p_2'', p_2) \right. \\ & -i\,\sigma_1(p_1', p_1) - i\,\sigma_2(p_2', p_2) - 2\,i\,\sigma_2(p_1', p_1) \\ & \left. -2\,i\,\sigma_1(p_2', p_2) + \sigma_1(j, q)\,\sigma_2(j, q)\right]. \end{aligned} \tag{7.152}$$

The complete radiative correction is the sum of Eqs. (7.150) and (7.152), $H_R = \delta_3 H + \delta_4 H$. Using symmetry $1 \leftrightarrow 2$ it takes the form

$$H_R = Z\,\alpha^2\left[\frac{91}{180} + \frac{2}{3}\,\ln[(Z\,\alpha)^{-2}]\right]\,i\,\sigma_1(p_1'', p_1)$$
$$+\alpha^2\left[\frac{73}{180} + \frac{2}{3}\,\ln[(Z\,\alpha)^{-2}]\right]\,\sigma_1(j,q)\,\sigma_2(j,q)$$
$$-\alpha^2\left[\frac{21}{10} + 4\,\ln[(Z\,\alpha)^{-2}]\right]\,i\,\sigma_1(p_1', p_1). \tag{7.153}$$

It is convenient to consider the sum of Eqs. (7.148) and (7.153), as several logarithmic terms cancel out. In coordinate space, using atomic units, $H_Q = H_E + H_R$ takes the form

$$H_Q = Z\,\alpha^7\left(\frac{91}{180} + \frac{2}{3}\,\ln[(Z\,\alpha)^{-2}]\right)\,i\,\vec{p}_1 \times \delta^3(r_1)\,\vec{p}_1\cdot\vec{\sigma}_1$$
$$+\alpha^7\left(-\frac{83}{60} + \frac{\ln\alpha}{2}\right)\,(\vec{\sigma}_1\cdot\vec{\nabla})\,(\vec{\sigma}_2\cdot\vec{\nabla})\delta^3(r)$$
$$-\alpha^7\,\frac{15}{8\,\pi}\,\frac{1}{r^7}\,(\vec{\sigma}_1\cdot\vec{r})\,(\vec{\sigma}_2\cdot\vec{r})$$
$$+\alpha^7\left(\frac{69}{10} + 3\,\ln\alpha\right)\,i\,\vec{p}_1\times\delta^3(r)\,\vec{p}_1\cdot\vec{\sigma}_1$$
$$-\alpha^7\,\frac{3}{4\,\pi}\,i\,\vec{p}_1\times\frac{1}{r^3}\,\vec{p}_1\cdot\vec{\sigma}_1, \tag{7.154}$$

where the singular $\int dr/r$ integral is defined with an implicit lower cut off ϵ and a term $\ln\epsilon + \gamma$ is subtracted out. The logarithmic terms in the above equation agree with Refs. [34, 46]. Numerical results for the nonlogarithmic terms are presented in Table 7.4.

Table 7.4. Operators due to exchange diagrams, slope of form factors and the vacuum polarization, in atomic units with a prefactor $m\,\alpha^7/\pi$.

Operator	ν_{01} [kHz]	ν_{12} [kHz]
$Q_1 = \frac{91\,\pi}{180}\,Z\,i\,\vec{p}_1\times\delta^3(r_1)\,\vec{p}_1\cdot\vec{\sigma}_1$	2.854	5.709
$Q_2 = -\frac{83\,\pi}{60}\,\vec{\sigma}_1\cdot\vec{\nabla}\,\vec{\sigma}_2\cdot\vec{\nabla}\delta^3(r)$	10.886	-4.355
$Q_3 = -\frac{15}{8}\,\frac{1}{r^7}\,\vec{r}\cdot\vec{\sigma}_1\,\vec{r}\cdot\vec{\sigma}_2$	4.132	-1.653
$Q_4 = \frac{69\,\pi}{10}\,i\vec{p}_1\times\delta^3(r)\vec{p}_1\cdot\vec{\sigma}_1$	5.186	10.372
$Q_5 = -\frac{3\,i}{4}\,\vec{p}_1\times\frac{1}{r^3}\,\vec{p}_1\cdot\vec{\sigma}_1$	-1.328	-2.656
$E_Q = \sum_{i=1,5} Q_i$	21.731	7.418

The remaining contribution to $H^{(7)}$ involves the anomalous magnetic moment (amm) correction to the spin-dependent operators. They do not lead to any divergences, and therefore can be calculated without dimensional regularization. The amm corrections to the second-order contribution have already been included in $\nu_A^{(4)}$, as discussed in Section 4.1. The amm part of $H^{(7)}$ is calculated with the help of a NRQED Hamiltonian which is obtained by a Foldy–Wouthuysen transformation of the Dirac Hamiltonian including the magnetic moment anomaly [43],

$$\begin{aligned}H_{FW} = &\frac{\vec{\pi}^2}{2} + e\,A^0 - \frac{e}{2}\,(1+a_e)\,\vec{\sigma}\cdot\vec{B} - \frac{\vec{\pi}^4}{8}\\ &-\frac{e}{8}\,(1+2\,a_e)\,[\vec{\nabla}\cdot\vec{E} + \vec{\sigma}\cdot(\vec{E}\times\vec{\pi} - \vec{\pi}\times\vec{E})]\\ &+\frac{e}{8}(\{\vec{\sigma}\cdot\vec{B},\vec{\pi}^2\} + a_e\,\{\vec{\pi}\cdot\vec{B},\vec{\pi}\cdot\vec{\sigma}\})\\ &-\frac{(3+4\,a_e)}{64}\,\{\vec{p}^{\,2}, e\,\vec{E}\times\vec{p}\cdot\vec{\sigma}\}. \end{aligned} \tag{7.155}$$

All the $m\,\alpha^6$ operators obtained by Douglas and Kroll (DK) in [27], derived in this chapter using time-dependent perturbation theory, can also be obtained using this Hamiltonian, as shown in Ref. [16]. The anomalous magnetic moment operators can be derived in a very similar way. They differ from the DK operators only by multiplicative factors, as shown in Table 7.5. There is a one to one correspondence with the listing of DK operators given in Table I of Ref. [38] with three exceptions. The operator H_8 from our Table 7.5 cancels out in the DK calculation, as mentioned after Eq. (7.132). The other two exceptions are related to the different spin structure of the next to last term in Eq. (7.155), which leads to the operators H_{16} and H_{17} in our Table 7.5.

Apart from the H_i and Q_i operators, second-order contributions and low energy Bethe–logarithmic type corrections contribute to the fine structure. These contributions have been obtained in [45]. The second-order contribution, beyond the anomalous magnetic moment terms is the second term in the general formula (7.139)

$$E_S = 2\Big\langle\phi\Big|H_{SD}\,\frac{1}{(E-H)'}\,H^{(5)}\Big|\phi\Big\rangle, \tag{7.156}$$

where H_{SD} is the leading relativistic spin-dependent correction given in Eq. (7.91) with $a_e = 0$ and $H^{(5)}$ is the spin-independent effective Hamiltonian at order $m\,\alpha^5$,

$$\begin{aligned}H^{(5)} = &-\frac{7}{6\,\pi}\,\frac{\alpha^2}{r^3} + \frac{38\,Z\,\alpha^2}{45}\,\left[\delta^3(r_1)+\delta^3(r_2)\right]\\ &+\frac{4\,Z\,\alpha^2}{3}\,\ln(Z\,\alpha)^{-2}\,\left[\delta^3(r_1)+\delta^3(r_2)\right]. \end{aligned} \tag{7.157}$$

Table 7.5. Operators due to magnetic moment anomaly in atomic units with the prefactor $m\,\alpha^7/\pi$.

Operator	ν_{01}[kHz]	ν_{12}[kHz]
$H_1 = -\frac{Z}{4}\,p_1^2\,\frac{\vec{r}_1}{r_1^3}\times\vec{p}_1\cdot\vec{\sigma}_1$	3.239	6.478
$H_2 = -\frac{3\,Z}{4}\,\frac{\vec{r}_1}{r_1^3}\times\frac{\vec{r}}{r^3}\cdot\vec{\sigma}_1\,(\vec{r}\cdot\vec{p}_2)$	0.267	0.534
$H_3 = \frac{3\,Z}{4}\,\frac{\vec{r}}{r^3}\cdot\vec{\sigma}_1\,\frac{\vec{r}_1}{r_1^3}\cdot\vec{\sigma}_2$	0.332	-0.133
$H_4 = \frac{1}{2\,r^4}\,\vec{r}\times\vec{p}_2\cdot\vec{\sigma}_1$	0.749	1.498
$H_5 = -\frac{3}{4\,r^6}\,\vec{r}\cdot\vec{\sigma}_1\,\vec{r}\cdot\vec{\sigma}_2$	2.638	-1.055
$H_6 = \frac{1}{4}\,p_1^2\,\frac{\vec{r}}{r^3}\times\vec{p}_1\cdot\vec{\sigma}_1$	-0.807	-1.614
$H_7 = -\frac{1}{4}\,p_1^2\,\frac{\vec{r}}{r^3}\times\vec{p}_2\cdot\vec{\sigma}_1$	-1.237	-2.474
$H_8 = -\frac{Z}{4\,r}\,\frac{\vec{r}_1}{r_1^3}\times\vec{p}_2\cdot\vec{\sigma}_1$	-0.460	-0.920
$H_9 = -\frac{i}{2}\,p_1^2\,\frac{1}{r^3}\,\vec{r}\cdot\vec{p}_2\,\vec{r}\times\vec{p}_1\cdot\vec{\sigma}_1$	0.093	0.187
$H_{10} = \frac{3\,i}{4\,r^5}\,\vec{r}\times(\vec{r}\cdot\vec{p}_2)\,\vec{p}_1\cdot\vec{\sigma}_1$	-0.376	-0.752
$H_{11} = -\frac{3}{8\,r^5}\,\vec{r}\times(\vec{r}\times\vec{p}_1\cdot\vec{\sigma}_1)\,\vec{p}_2\cdot\vec{\sigma}_2$	-0.193	0.077
$H_{12} = -\frac{1}{8\,r^3}\,\vec{p}_1\cdot\vec{\sigma}_2\,\vec{p}_2\cdot\vec{\sigma}_1$	-0.447	0.179
$H_{13} = \frac{21}{16}\,p_1^2\,\frac{1}{r^5}\,\vec{r}\cdot\vec{\sigma}_1\,\vec{r}\cdot\vec{\sigma}_2$	-14.908	5.963
$H_{14} = -\frac{3\,i}{8}\,p_1^2\,\frac{\vec{r}}{r^3}\cdot\vec{\sigma}_1\,\vec{p}_1\cdot\vec{\sigma}_2$	4.411	-1.764
$H_{15} = \frac{i}{8}\,p_1^2\,\frac{1}{r^3}\,\big(\vec{r}\cdot\vec{\sigma}_2\,\vec{p}_2\cdot\vec{\sigma}_1 + (\vec{r}\cdot\vec{\sigma}_1)$ $\times(\vec{p}_2\cdot\vec{\sigma}_2) - \frac{3}{r^2}\,\vec{r}\cdot\vec{\sigma}_1\,\vec{r}\cdot\vec{\sigma}_2\,\vec{r}\cdot\vec{p}_2\big)$	4.618	-1.847
$H_{16} = -\frac{1}{4}\,\vec{p}_1\cdot\vec{\sigma}_1\,\vec{p}_1\times\frac{\vec{r}}{r^3}\cdot\vec{p}_2$	-0.483	-0.967
$H_{17} = \frac{1}{8}\,\vec{p}_1\cdot\vec{\sigma}_1\,\big(-\vec{p}_1\cdot\vec{\sigma}_2\,\frac{1}{r^3}$ $+3\vec{p}_1\cdot\vec{r}\,\frac{\vec{r}}{r^5}\cdot\vec{\sigma}_2\big)$	-1.643	0.657
$E_H = \sum_{i=1,17}\langle H_i\rangle$	-4.208	4.047

The low energy contribution E_L comes from photon momenta $k < \Lambda$. We derive from the low energy form of the electromagnetic interaction Hamiltonian

$$H_I = \sum_a -e\,\vec{r}_a\cdot\vec{E} - \frac{e}{2\,m}\,\sigma_a^i\,r_a^j\,B^i_{,j} - \frac{e}{4\,m^2}\,\vec{\sigma}_a\cdot\vec{E}_a\times\vec{p}_a \qquad (7.158)$$

instead of the general nonrelativistic Hamiltonian in Eq. (A.39). This is because of the high degree of cancellation between various terms, which is more visible using Eq. (7.158). Specifically, the contribution from the second and the third terms cancel out between themselves and only the first term leads to E_L,

$$E_L = \frac{2}{3}\,\alpha\int_0^\Lambda \frac{d^3k}{(2\,\pi)^3\,2\,k}\,k^2\,\delta\Big\langle\phi\Big|(\vec{r}_1+\vec{r}_2)\,\frac{1}{E-H-k}\,(\vec{r}_1+\vec{r}_2)\Big|\phi\Big\rangle, \qquad (7.159)$$

where $\delta\langle\ldots\rangle$ denotes the correction to the matrix element $\langle\ldots\rangle$ due to H_{SD}. Using the relation

$$[H, \vec{r}_1 + \vec{r}_2] = -\frac{i}{m}(\vec{p}_1 + \vec{p}_2) + \frac{1}{4\,m^2}\left[\vec{p}_1 \times \vec{\sigma}_1 + \vec{p}_2 \times \vec{\sigma}_2, H_0 - E_0\right] \quad (7.160)$$

E_L is transformed to

$$E_L = -\frac{2\,\alpha}{3\,\pi}\,\delta\left\langle\phi\left|(\vec{p}_1 + \vec{p}_2)\,(H - E)\ln\left[\frac{2(H-E)}{(Z\,\alpha)^2}\right]\right.\right.$$
$$\left.\left.(\vec{p}_1 + \vec{p}_2)\right|\phi\right\rangle + \frac{i\,Z^2\,\alpha^3}{3\,\pi}\left\langle\phi\left|\left(\frac{\vec{r}_1}{r_1^3} + \frac{\vec{r}_2}{r_2^3}\right)\right.\right.$$
$$\left.\left.\times\frac{(\vec{\sigma}_1 + \vec{\sigma}_2)}{2}\ln\left[\frac{2(H_0 - E_0)}{(Z\,\alpha)^2}\right]\cdot\left(\frac{\vec{r}_1}{r_1^3} + \frac{\vec{r}_2}{r_2^3}\right)\right|\phi\right\rangle. \quad (7.161)$$

Numerical results for all these contributions are presented in Table 7.6.[a]

Table 7.6. Summary of contributions to helium fine structure, $E^{(4)}$ and $E^{(6)}$ including nuclear recoil corrections and the electron anomalous magnetic moment at the level of the Breit–Pauli Hamiltonian; α^{-1} = 137.03599911(46), m/m_α = 1.37093355575(61)10^{-4}, Ry c = 3.289841960360(22) 10^{15} Hz. Not indicated is the uncertainty due to α, which is 0.20 kHz for ν_{01}. The last row includes the most recent experimental values.

	ν_{01}[kHz]	ν_{12}[kHz]	Ref.
E_Q	21.73	7.42	
E_H	−4.21	4.05	
E_S	11.37(02)	−1.25(01)	[45]
E_L	−29.76(16)	−12.51(27)	[45]
$E^{(7)}$	−0.87(16)	−2.30(27)	
$E^{(7)}_{\log}$	82.59	−10.09	[46]
$E^{(6)}$	−1557.50(06)	−6544.32(12)	[32, 36, 37, 53]
$E^{(4)}$	29 618 418.79(01)	2 297 717.84	[32, 36]
total	29 616 943.01(17)	2 291 161.13(30)	
Drake	29 616 946.42(18)	2 291 154.62(31)	[32]
exp.	29 616 951.66(70)	2 291 175.59(51)	[31, 48–51]

7.9. Conclusions

Since all relevant contributions to helium fine structure splitting have now been treated, we are at a position to present our final theoretical predictions following Ref. [14]. We begin with a discussion of the small interval ν_{12}, as

[a]Please see the *Note added in proof* at the end of this chapter for results of a new calculation that updates the values for E_L.

its treatment is unambiguous. By this we mean that unlike ν_{01}, which can either be used to determine α or else be directly compared with experiment using the $g-2$ value of α, only the latter approach is available for ν_{12}. As shown in Table 7.6, the essentially α-independent theory predictions disagree both with each other and with experiment. A notable feature of the experimental situation is the fact that two independent groups, using very different techniques, published results consistent with one another within experimental error at the same time [49, 50]. We have quoted a later determination [31] of even greater accuracy, which is also in agreement with the earlier measurements, so we regard the experiment determination as non-controversial. The difference in theory comes primarily from $\nu^{(5)}$, so we defer its discussion to the large interval treatment.

We begin the treatment of the large interval by first assuming α is known, using the $g-2$ result of Eq. (7.8). From Table 7.6 we see that our theory is significantly discrepant. Drake's results are discrepant with ours, but in somewhat better agreement with experiment.

However, the main point of this chapter is the determination of α from helium fine structure. As shown in Eq. (7.73), the value of the fine structure constant as determined by our theory is strongly discrepant, which is of course the same discrepancy as noted above. Again, because Drake's result was less discrepant, the value of α coming from his theory is

$$\alpha^{-1}(\text{He fs}) = 137.036987(2), \tag{7.162}$$

with the error corresponding to his estimate of 1 kHz theoretical uncertainty. In terms of $\delta\nu_{01}$, the difference of the two calculations is that Ref. [14], described in this chapter, has -1.475 78 MHz and Ref. [32] has -1.472 37 MHz, with the difference coming solely from ν_{01}^{5}. While the difference of 3.4 kHz is significant, we note that it is smaller than the 8.6 and 5.2 kHz respective differences of the two theories from experiment. This discrepancy is far worse for the small interval for both theories. It is clearly premature to quote either value as a definitive helium fine structure determination. We now turn to a discussion of this quite unsatisfactory situation.

There are a number of possible sources of the discrepancy. A problem that was quite important in the early days of helium fine structure was the quality of the nonrelativistic wave functions, determined with variational techniques, so we begin by discussing the numerical error expected from the wave functions used here. A central parameter used to gauge the quality of a wave function is the nonrelativistic energy. Our wave function, consisting of 1500 explicitly correlated exponential functions reproduces the nonrelativistic $2P$ energy with 21 significant digits, consistent with the result of Drake in [32]. Matrix elements determined with this wave function are not as accurate as the nonrelativistic energy, but they are sufficiently accurate for evaluating the leading fine structure operators, and again our results agree with the independent calculation of Drake [32]. For example, $E^{(4)}$ agrees to 0.01 kHz and $E^{(6)}$ to 0.1 kHz. In fact almost all numerical calculations have been performed by us and by Drake independently with one exception: We have not obtained recoil correction to the second-order matrix element with Breit operators in $E^{(6)}$.

A much more important issue is the complexity of the derivation of the operators that contribute in order $m\alpha^7$, specifically H_i and Q_i. We purposely derived H_i in a way very similar way to the derivation of the D–K operators to avoid error. We note that the Q_i operators were obtained from the one-loop scattering amplitude in an almost automatic way, in contrast to the former very lengthy derivation of Zhang [15, 54, 55], with which we are in disagreement (see the summary of Zhang results in Ref. [32]). In previous papers by the present authors [37, 45] we have pointed out several computational mistakes and inconsistencies in Zhang's calculations, and therefore we consider the result of Drake, given in Table 7.6, to be incomplete. While it is possible that we have made a mistake somewhere, the other probable explanation of the discrepancy with experiments is the neglect of higher order terms, specifically those of order $m\alpha^8$. An indication of their importance is the recoil correction to the second-order contribution, obtained by Drake in [32]. In spite of the small electron-alpha particle mass ratio, this correction is very significant; for example $\delta\nu_{01} = -10.81$ kHz. The mass ratio $m/m_\alpha \approx 0.00014$ is not much different from $\alpha^2 \approx 0.000053$, and for this reason it is possible that iteration of the Breit–Pauli Hamiltonian to third order could give a significant contribution. However, we expect that most $m\alpha^8$ operators should be negligible, as $E^{(7)}$ is already at the few kHz level, so that an additional power of α will make these operators contribute well below the experimental accuracy. However, while it is important to consider higher-order corrections, we feel that it is vital to have an independent calculation of the terms already considered in this chapter carried out.

Note added in proof

In a recent work of V. Yerokhin and (KP) [58] all higher order-corrections to the helium fine structure have been reevaluated. The former result of

Ref. [32] for the recoil correction to the Lewis–Serafino second order contribution has been corrected. The new results are −0.47, 0.21 kHz for ν_{01} and ν_{12} correspondingly. The former result of (KP) for the second term of Eq. (7.161) had a wrong sign, and the overall accuracy has been significantly improved. As a result, new values for E_L in Table 7.6 are −26.90, 4.23 kHz for ν_{01} and ν_{12} correspondingly. The new result for the small interval $\nu_{12} = 2\,291\,177.28 \pm 1.6$ kHz is in disagreement with all former experiments, but it is in excellent agreement with a very recent measurement by Borbely et al. [59] which is $2\,291\,177.53 \pm 0.35$ kHz. The new result for the large interval $\nu_{01} = 29\,616\,946.20 \pm 1.6$ kHz remains in disagreement with all the experiments and we do not have an explanation for this curious situation.

Acknowledgments

The work of KP was supported in part by NIST grant PMG 60NANB7D6153, and that of JS by NSF grant PHY-0757125.

A.1. Dimensionally Regularized QED of Bound States

We take the dimension of space to be $d = 3 - 2\,\epsilon$. The surface area of a d-dimensional unit sphere can be obtained by considering the integral

$$\mathrm{I} = \int d^d k\, e^{-\vec{k}^2}. \tag{A.1}$$

In Cartesian coordinates it is a product of d one-dimensional integrals

$$\mathrm{I} = \left[\int dk\, e^{-k^2}\right]^d = \pi^{d/2}, \tag{A.2}$$

while in spherical coordinates it is

$$\mathrm{I} = \int d\Omega_d \int_0^\infty dk\, k^{d-1}\, e^{-k^2} = \Omega_d\, \frac{1}{2}\, \Gamma(d/2). \tag{A.3}$$

From comparison of Eq. (A.2) and (A.3) one obtains

$$\Omega_d = \frac{2\,\pi^{d/2}}{\Gamma(d/2)}. \tag{A.4}$$

The d-dimensional Laplacian is $\nabla^2 = \partial^i\,\partial^i$. For spherically symmetric functions f and g

$$\begin{aligned}
\int d^d r\, \nabla^2(f)\, g &= -\int d^d r\, \nabla(f)\cdot\nabla(g) = -\Omega_d \int dr\, r^{d-1}\, \partial_r f\, \partial_r g \\
&= \Omega_d \int dr\, \partial_r(r^{d-1}\,\partial_r f)\, g \\
&= \int d^d r\, r^{1-d}\, \partial_r(r^{d-1}\,\partial_r f)\, g\,,
\end{aligned} \tag{A.5}$$

so the Laplacian takes the form

$$\nabla^2 = r^{1-d}\partial_r\, r^{d-1}\partial_r. \tag{A.6}$$

The photon propagator, and thus Coulomb interaction preserves its form in the momentum representation, while in the coordinate representation it is

$$\mathcal{V}(r) = \int \frac{d^d k}{(2\,\pi)^d}\,\frac{4\,\pi}{k^2}\,e^{i\,\vec{k}\cdot\vec{r}} = \pi^{\epsilon-1/2}\,\Gamma(1/2-\epsilon)\,r^{2\,\epsilon-1} \equiv \frac{C_1}{r^{1-2\,\epsilon}}. \tag{A.7}$$

An alternative derivation of $\mathcal{V}(r)$ which does not require a Fourier transform is the following. Consider the equation

$$\nabla^2\mathcal{V}(r) = -4\,\pi\delta^d(r). \tag{A.8}$$

If one assumes that $\mathcal{V}(r)$ is of the form $\mathcal{V}(r) = C\,r^\gamma$, then for $r \neq 0$

$$r^{1-d}\partial_r\, r^{d-1}\partial_r(C\,r^\gamma) = 0, \tag{A.9}$$

and therefore $\gamma = 2 - d = 2\,\epsilon - 1$. The coefficient C is obtained by considering the integral with a trial function f

$$\begin{aligned}
4\,\pi\,f(0) &= -\int d^d r\,\nabla^2(\mathcal{V})\,f = \int d^d r\,\nabla(\mathcal{V})\cdot\nabla(f)\\
&= \lim_{\varepsilon\to 0}\int_\varepsilon dr\,r^{d-1}\,\Omega_d\,\partial_r(\mathcal{V})\,\partial_r(f)\\
&= \lim_{\varepsilon\to 0}\left[r^{d-1}\,\Omega_d\,\partial_r(\mathcal{V})\,f\Big|_{r=\varepsilon} - \int_\varepsilon dr\,\Omega_d\,\partial_r(r^{d-1}\,\partial_r(\mathcal{V}))\,f\right]\\
&= \lim_{\varepsilon\to 0}\,\Omega_d\,\varepsilon^{d-1}\,\partial_\varepsilon(C\,\varepsilon^{2-d})\,f(\varepsilon)\\
&= \Omega_d\,(2-d)\,C\,f(0).
\end{aligned} \tag{A.10}$$

Therefore

$$C \equiv C_1 = \frac{4\,\pi}{(d-2)\,\Omega_d} = \pi^{\epsilon-1/2}\,\Gamma(1/2-\epsilon). \tag{A.11}$$

We are now ready to consider quantum mechanics in d-dimensions. The nonrelativistic Hamiltonian of a hydrogen-like system is

$$H_0 = \frac{\vec{p}^{\,2}}{2} - Z\,\frac{C_1}{r^{1-2\,\varepsilon}}, \tag{A.12}$$

and of a helium-like system

$$H_0 = \frac{\vec{p_1}^{\,2}}{2} + \frac{\vec{p_2}^{\,2}}{2} - Z\,\frac{C_1}{r_1^{1-2\,\epsilon}} - Z\,\frac{C_1}{r_2^{1-2\,\epsilon}} + \frac{C_1}{r^{1-2\,\epsilon}}. \tag{A.13}$$

We denote the solution of the stationary Schrödinger equation $H_0\,\phi = E_0\,\phi$ by ϕ, but do not need to refer to its explicit (and unknown) form in d-dimensions. Instead, we will use only the generalized cusp condition to eliminate various singularities from matrix elements with relativistic operators. Specifically, we expect that for small r

$$\phi(r) \approx \phi(0)\,(1 - C\,r^{\gamma}), \tag{A.14}$$

with the coefficients C and γ to be obtained from the two-electron Schrödinger equation around $r = 0$

$$[-\nabla^2 + \mathcal{V}(r)]\phi(0)\,(1 - C\,r^{\gamma}) \approx E\,\phi(0)\,(1 - C\,r^{\gamma}). \tag{A.15}$$

From the cancellation of small r singularities on the left hand side of above equation, one obtains

$$\gamma = 1 + 2\,\epsilon, \tag{A.16}$$

$$C \equiv C_2 = \frac{1}{4}\,\pi^{\epsilon-1/2}\,\Gamma(-1/2 - \epsilon). \tag{A.17}$$

Therefore the two-electron wave function around $r = 0$ behaves as

$$\phi(\vec{r}_1, \vec{r}_2) \approx \phi(r_{12} = 0)\,\big(1 - C_2\,r_{12}^{1+2\,\epsilon}\big). \tag{A.18}$$

Apart from the Coulomb potential $\mathcal{V}(r)$ in coordinate space, we need additional functions which appear in the calculations of relativistic operators, namely

$$\mathcal{V}_2(r) = \int \frac{d^dk}{(2\,\pi)^d}\,\frac{4\,\pi}{k^4}\,e^{i\,\vec{k}\cdot\vec{r}}, \tag{A.19}$$

$$\mathcal{V}_3(r) = \int \frac{d^dk}{(2\,\pi)^d}\,\frac{4\,\pi}{k^6}\,e^{i\,\vec{k}\cdot\vec{r}}. \tag{A.20}$$

They can be obtained from the differential equations

$$-\nabla^2\mathcal{V}_2(r) = \mathcal{V}(r), \tag{A.21}$$

$$-\nabla^2\mathcal{V}_3(r) = \mathcal{V}_2(r), \tag{A.22}$$

(A.23)

with the results

$$\mathcal{V}_2(r) = C_2\,r^{1+2\,\epsilon}, \tag{A.24}$$

$$\mathcal{V}_2(r) = C_3\,r^{3+2\,\epsilon}, \tag{A.25}$$

(A.26)

with C_2 defined in Eq. (A.17) and

$$C_3 = \frac{1}{32}\,\pi^{\epsilon-1/2}\,\Gamma(-3/2 - \epsilon). \tag{A.27}$$

Using $\mathcal{V}_i$ we calculate various integrals involving the photon propagator in Coulomb gauge, namely

$$\int \frac{d^d k}{(2\,\pi)^d}\,\frac{4\,\pi}{k^4}\,P_{ij}(k)\,e^{i\,\vec{k}\cdot\vec{r}} = \delta^{ij}\,\mathcal{V}_2 + \partial^i\partial^j\,\mathcal{V}_3$$
$$= \pi^{\epsilon-1/2}\,r^{-1+2\,\epsilon}\left[\frac{3}{16}\,\delta^{ij}\,\Gamma(-1/2-\epsilon)\,r^2 + \frac{1}{8}\,\Gamma(1/2-\epsilon)\,r^i\,r^j\right]$$
$$\equiv \left[\frac{1}{8\,r}(r^i\,r^j - 3\,\delta^{ij}\,r^2)\right]_\epsilon, \tag{A.28}$$

$$\int \frac{d^d k}{(2\,\pi)^d}\,\frac{4\,\pi}{k^2}\,P_{ij}(k)e^{i\,\vec{k}\cdot\vec{r}} = \delta^{ij}\mathcal{V} + \partial^i\partial^j\,\mathcal{V}_2$$
$$= \pi^{\epsilon-1/2}\,r^{-3+2\,\epsilon}\left[\frac{1}{2}\,\delta^{ij}\,\Gamma(1/2-\epsilon)\,r^2 + \Gamma(3/2-\epsilon)\,r^i\,r^j\right]$$
$$\equiv \left[\frac{1}{2\,r^3}(\delta^{ij}\,r^2 + r^i\,r^j)\right]_\epsilon, \tag{A.29}$$

and

$$\int \frac{d^d k}{(2\pi)^d}\,4\pi P_{ij}(k)e^{i\vec{k}\cdot\vec{r}} = \frac{(d-1)}{d}\delta^{ij}4\pi\delta^d(r) + \left(\partial^i\partial^j - \frac{\delta^{ij}}{d}\partial^2\right)\mathcal{V}$$
$$= \frac{(d-1)}{d}\delta^{ij}4\pi\delta^d(r) + \pi^{\epsilon-\frac{1}{2}}r^{-5+2\epsilon}\left[-2\delta^{ij}\Gamma(\frac{3}{2}-\epsilon)r^2 + 4\Gamma(\frac{5}{2}-\epsilon)r^ir^j\right]$$
$$= \left[\frac{2}{3}\delta^{ij}4\pi\delta^3(r) + \frac{1}{r^5}(3r^ir^j - \delta^{ij}r^2)\right]_\epsilon \equiv \delta^{ij}_\perp. \tag{A.30}$$

A.2. Foldy–Wouthuysen Transformation in d-Dimensions

The Foldy–Wouthuysen (FW) transformation [57] is the nonrelativistic expansion of the Dirac Hamiltonian in an external electromagnetic field. Following Ref. [30] we extend this transformation to the case where the dimension d of space is arbitrary. The Dirac Hamiltonian in an external electromagnetic field is

$$H = \vec{\alpha}\cdot\vec{\pi} + \beta\,m + e\,A^0\,, \tag{A.31}$$

where $\vec{\pi} = \vec{p} - e\,\vec{A}$,

$$\alpha^i = \begin{pmatrix} 0 & \sigma^i \\ \sigma^i & 0 \end{pmatrix},\quad \beta = \begin{pmatrix} I & 0 \\ 0 & -I \end{pmatrix}, \tag{A.32}$$

and

$$\{\sigma^i, \sigma^j\} = 2\,\delta^{ij}\,I. \tag{A.33}$$

The FW transformation S [57] leads to a new Hamiltonian

$$H_{\rm FW} = e^{i\,S}\,(H - i\,\partial_t)\,e^{-i\,S}\,, \tag{A.34}$$

which decouples the upper and lower components of the Dirac wave function up to a specified order in the $1/m$ expansion. Here we calculate the FW Hamiltonian up to terms which contribute to $m\,\alpha^6$ correction to the energy. We use a convenient form of the F–W operator S, which can be written as

$$\begin{aligned} S = -\frac{i}{2m}\biggl\{&\beta\vec{\alpha}\cdot\vec{\pi} - \frac{1}{3m^2}\beta(\vec{\alpha}\cdot\vec{\pi})^3 + \frac{1}{2m}[\vec{\alpha}\cdot\vec{\pi}, eA^0 - i\partial_t] + \frac{\beta}{5m^4}(\vec{\alpha}\cdot\vec{\pi})^5 \\ &-\frac{\beta e}{4m^2}\vec{\alpha}\cdot\dot{\vec{E}} + \frac{ie}{24m^3}[\vec{\alpha}\cdot\vec{\pi},[\vec{\alpha}\cdot\vec{\pi},\vec{\alpha}\cdot\vec{E}]] \\ &-\frac{ie}{3m^3}\{(\vec{\alpha}\cdot\vec{\pi})^2,\vec{\alpha}\cdot\vec{E}\}\biggr\}. \end{aligned} \tag{A.35}$$

The FW Hamiltonian is expanded in a power series in S

$$H_{\rm FW} = \sum_{j=0}^{6}\mathcal{H}^{(j)} + \ldots \tag{A.36}$$

where

$$\begin{aligned} \mathcal{H}^{(0)} &= H, \\ \mathcal{H}^{(1)} &= [i\,S\,, \mathcal{H}^{(0)} - i\,\partial_t], \\ \mathcal{H}^{(j)} &= \frac{1}{j}\,[i\,S\,, \mathcal{H}^{(j-1)}] \text{ for } j > 1, \end{aligned} \tag{A.37}$$

and higher order terms in this expansion, denoted by dots, are neglected. The calculation of subsequent commutators is rather tedious but the result is simply

$$\begin{aligned} H_{\rm FW} = e\,A^0 &+ \frac{(\vec{\sigma}\cdot\vec{\pi})^2}{2\,m} - \frac{(\vec{\sigma}\cdot\vec{\pi})^4}{8\,m^3} + \frac{(\vec{\sigma}\cdot\vec{\pi})^6}{16\,m^5} - \frac{i\,e}{8\,m^2}\,[\vec{\sigma}\cdot\vec{\pi},\vec{\sigma}\cdot\vec{E}] \\ &-\frac{e}{16\,m^3}\,\{\vec{\pi}\,,\,\partial_t\vec{E}\} - \frac{i\,e}{128\,m^4}\,[\vec{\sigma}\cdot\vec{\pi},[\vec{\sigma}\cdot\vec{\pi},[\vec{\sigma}\cdot\vec{\pi},\vec{\sigma}\cdot\vec{E}]]] \\ &+\frac{i\,e}{16\,m^4}\,\Bigl\{(\vec{\sigma}\cdot\vec{\pi})^2\,,\,[\vec{\sigma}\cdot\vec{\pi},\vec{\sigma}\cdot\vec{E}]\Bigr\}. \end{aligned} \tag{A.38}$$

There is some arbitrariness in the operator S, which means that $H_{\rm FW}$ is not unique. The standard approach [57], which relies on subsequent use of FW-transformations, differs from this one in $d = 3$ by the transformation S with an additional even operator.

We aim to obtain FW Hamiltonian suitable for calculations of QED contributions to energy levels of an arbitrary light atom, up to the order

$m\,\alpha^7$. For this one can neglect the vector potential $\vec{A}$ in all the terms having m^4 and m^5 in the denominator. Moreover, less obviously, one can neglect terms with $\vec{\sigma}\cdot\vec{A}\,\vec{\sigma}\cdot\dot{\vec{E}}$ and $\vec{B}^2$. This is because they are of second order in electromagnetic fields which additionally contain derivatives, and thus contribute only at higher orders. After these simplifications, $H_{\rm FW}$ takes the form

$$
\begin{aligned}
H_{\rm FW} = {} & e\,A^0 + \frac{\pi^2}{2\,m} - \frac{e}{4\,m}\,\sigma^{ij}\,B^{ij} - \frac{\pi^4}{8\,m^3} - \frac{e}{8\,m^2}\Big(\vec{\nabla}\cdot\vec{E} + \sigma^{ij}\,\big\{E^i\,,\,\pi^j\big\}\Big) \\
& + \frac{e}{16\,m^3}\big\{\sigma^{ij}B^{ij}\,,\,p^2\big\} - \frac{e}{16\,m^3}\,\big\{\vec{p},\,\partial_t\vec{E}\big\} + \frac{3\,e}{32\,m^4}\,\big\{\sigma^{ij}\,E^i\,p^j\,,\,p^2\big\} \\
& + \frac{e}{128\,m^4}\,[p^2,[p^2,A^0]] - \frac{3\,e}{64\,m^4}\,\big\{p^2\,,\,\nabla^2A^0\big\} + \frac{p^6}{16\,m^5},
\end{aligned}
\tag{A.39}
$$

where

$$\sigma^{ij} = \frac{1}{2\,i}\,[\sigma^i\,,\,\sigma^j], \tag{A.40}$$

$$B^{ij} = \partial^i\,A^j - \partial^j\,A^i, \tag{A.41}$$

$$E^i = -\nabla^i A^0 - \partial_t A^i. \tag{A.42}$$

References

[1] A. Sommerfeld, Ann. der Phys. **51**, 1-94, 125-167 (1916).
[2] J.C. Baird, J. Brandenberger, K.-I. Gondaira, and H. Metcalf, Phys. Rev. A **5**, 564 (1972).
[3] P.J. Mohr and B.N. Taylor, Rev. Mod. Phys. **77**, 1 (2005).
[4] D. Hanneke, S. Fogwell, and G. Gabrielse, Phys. Rev. Lett. **100**, 120801 (2008).
[5] T. Aoyama, M. Hayakawa, T. Kinoshita and M. Nio, Phys. Rev Lett. **99**, 110406 (2007).
[6] P. Cladé, E. de Mirandes, M. Cadoret, S. Guellati-Khélifa, C. Schwob, F. Nez, L. Julien, and F. Biraben, Phys. Rev. A **74**, 052109 (2006).
[7] A. Wicht, J.M. Hensley, E. Sarajlic, and S. Chu, Phys. Scripta **T102**, 82 (2002).
[8] K. Pachucki, Ann. Phys. (N.Y.) **226**, 1 (1993).
[9] K. Pachucki, Phys. Rev. Lett. **72**, 3154 (1994).
[10] K. Melnikov and T. van Ritbergen, Phys. Rev. Lett. **84**, 1673 (2000).
[11] U.D. Jentschura, P.J. Mohr, and G. Soff, Phys. Rev. Lett. **82**, 53 (1999).
[12] U.D. Jentschura and K. Pachucki, Phys. Rev. Lett. **91**, 113005 (2003).
[13] V.A. Yerokhin, P. Indelicato, and V.M. Shabaev, Phys. Rev. A **71**, 040101(R), (2005).
[14] K. Pachucki, Phys. Rev. Lett. **97**, 013002 (2006).
[15] T. Zhang, Phys, Rev. A **53**, 3896 (1996).
[16] K. Pachucki, Phys. Rev. A **71**, 012503 (2005).

[17] Considerable effort has gone into attempts to generalize the Dirac equation to helium, but the proper treatment of negative energy states requires a field theoretic treatment, as discussed by G.E. Brown and D.G. Ravenhall, Proc. R. Soc. London, Ser. A **208**, 552 (1951).
[18] K. Pachucki, Phys. Rev. A **56**, 297 (1997).
[19] A. Czarnecki, K. Melnikov, and A. Yelkhovksy, Phys. Rev. Lett. **82**, 311 (1999).
[20] J. Sapirstein and D. Yennie, in *Quantum Electrodynamics*, (ed.) T. Kinoshita, World Scientific (Singapore) (1991).
[21] W.A. Barker and F.N. Glover, Phys. Rev. **99**, 317 (1955).
[22] V.M. Shabaev, Teor. Mat.Fiz. **63**, 394 (1985). [Translated in Theor. Math. Phys. **63**, 588 (1985).]
[23] A.N. Artemyev, V.M. Shabaev, and V.A. Yerokhin, Phys. Rev. A **52**, 1884 (1995).
[24] A.N. Artemyev, V.M. Shabaev, and V.A. Yerokhin, J. Phys. B **28**, 5201 (1995).
[25] G.S. Adkins, S. Morrison, and J. Sapirstein, Phys. Rev. A **76**, 042508 (2007).
[26] Baranger, H. Bethe, and R.P. Feynman, Phys. Rev. **92**, 482 (1953).
[27] M. Douglas and N.M. Kroll, Ann. Phys. NY **82**, 89 (1974).
[28] U. Jentschura and K. Pachucki, Phys. Rev. A **54**, 1853 (1996).
[29] C. Schwartz, Phys. Rev. **5**, A1181 (1964).
[30] K. Pachucki, Phys. Rev. A. **74**, 022512 (2006).
[31] T. Zelevinsky, D. Farkas, and G. Gabrielse, Phys. Rev. Lett. **95**, 203001 (2005).
[32] G.W.F. Drake, Can. J. Phys. **80**, 1195 (2002).
[33] V. Korobov, Phys. Rev. A **61**, 064503 (2000).
[34] K. Pachucki, J. Phys. B **32**, 137 (1999).
[35] M.L. Lewis and P.H. Serafino, Phys. Rev. A **18**, 867 (1978).
[36] K. Pachucki and J. Sapirstein, J. Phys. B. **35**, 1783 (2002).
[37] K. Pachucki and J. Sapirstein, J. Phys. B **36**, 803 (2003).
[38] J. Daley, M. Douglas, L. Hambro, and N.M. Kroll, Phys. Rev. Lett. **29**, 12 (1972).
[39] T. Zhang, Phys. Rev. A **56**, 270(1997).
[40] T. Zhang, Phys. Rev. A **54**, 1252 (1996).
[41] A. Yelkhovsky, Phys. Rev. A **64**, 062104 (2001).
[42] V. Korobov and A. Yelkhovsky, Phys. Rev. Lett **87**, 193003 (2001).
[43] U. Jentschura, A. Czarnecki and K. Pachucki, Phys. Rev. A **72**, 062102 (2005).
[44] A. Czarnecki, U. Jentschura, and K. Pachucki, Phys. Rev. Lett. **95**, 180404 (2005).
[45] K. Pachucki and J. Sapirstein, J. Phys. B **33**, 5297 (2000).
[46] T. Zhang, Z-C. Yan and G.W.F. Drake, Phys. Rev. Lett. **77**, 1715 (1996).
[47] T. Zhang, Phys. Rev. A **53**, 3896 (1996).
[48] M.C. George, L.D. Lombardi, and E.A. Hessels, Phys. Rev. Lett. **87**, 173002 (2001).
[49] C.H. Storey, M.C. George, and E.A. Hessels, Phys. Rev. Lett. **84**, 3274

(2000).
[50] J. Castillega, D. Livingston, A. Sandars and D. Shiner, Phys. Rev. Lett. **84**, 4321 (2000).
[51] F. Minardi, G. Bianchini, P. Cancio Pastor, G. Giusfredi, F. S. Pavone, and M. Inguscio, Phys. Rev. Lett. **82**, 1112 (1999).
[52] G. Giusfriedi, P. de Natale, D. Mazzotti. P.C. Pastor, C. de Mauro, L. Fallani, G. Hagel, V. Krachmalnicoff, and M. Inguscio, Can. J. Phys. **83**, 301 (2005).
[53] Z-C. Yan and G.W.F. Drake, Phys. Rev. Lett. **74**, 4791 (1995).
[54] T. Zhang, Phys, Rev. A **54**, 1252 (1996).
[55] T. Zhang and G. W. F. Drake, Phys. Rev. A **54**, 4882 (1996).
[56] P. Clade *et al.*, Phys. Rev. Lett. **96**, 033001 (2006).
[57] C. Itzykson and J. B. Zuber, *Quantum Field Theory*, McGraw-Hill, New York (1990).
[58] Krzysztof Pachucki and Vladimir A. Yerokhin, Phys. Rev. A **79**, 062516 (2009).
[59] J. S. Borbely, et al., Phys. Rev. A **79**, 060503 (2009).

Chapter 8

Hadronic Vacuum Polarization and the Lepton Anomalous Magnetic Moments

Michel Davier

Laboratoire de l'Accélérateur Linéaire, IN2P3-CNRS et Université de Paris Sud, 91898 Orsay, France

davier@lal.in2p3.fr

Contents

8.1. Introduction

Hadronic vacuum polarization (HVP) originates from quantum fluctuations in the photon propagator. It contributes to the running of α and as such it plays an important role in many precision tests of the Standard Model. It turns out to be the case for the anomalous part of the magnetic moments of leptons. At lowest order (α^2) the corresponding Feynman diagram is shown in Fig. 8.1 which involves one hadronic insertion.

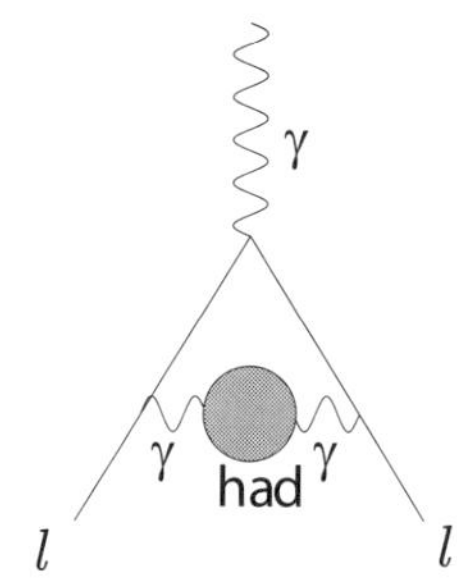

Fig. 8.1. The lowest-order hadronic contribution: The bubble inserted in the photon propagator represents quantum fluctuations involving hadrons.

Unlike the QED part, the contribution from hadronic polarization in the photon propagator cannot currently be computed from theory alone, because most of the contributing hadronic physics occurs in the low-energy nonperturbative QCD regime. However, by virtue of the analyticity of the vacuum polarization correlator, the HVP contribution to the magnetic anomaly a_l of a lepton l can be calculated via the dispersion integral [1, 2]

$$a_l^{\mathrm{had,LO}} = \frac{\alpha^2(0)}{3\pi^2} \int_{4m_\pi^2}^{\infty} ds \, \frac{K_l(s)}{s} R(s) \,, \tag{8.1}$$

where the QED kernel $K_l(s)$ is given by

$$K_l(s) = \int_0^1 dy \, \frac{y^2(1-y)}{y^2 + \frac{s}{m_l^2}(1-y)} \,, \tag{8.2}$$

where m_l is the lepton mass, $R = \sigma(e^+e^- \to \mathrm{hadrons})/\sigma_{\mathrm{pt}}$ with $\sigma_{\mathrm{pt}} = 4\pi\alpha^2/3s$, and s is the square of the e^+e^- center-of-mass energy.

It is immediately clear from Eq. (8.2) that, despite the infinite range of the integral, hadronic contributions at masses $s \gg m_l^2$ will be strongly suppressed. This fact has two important consequences: (1) the contribution to the anomaly will be much smaller for the electron as the hadronic threshold is at $4m_\pi^2$ (or even m_π^2 for the $\pi^0\gamma$ final state), and (2), even in the muon case, most of the contribution will come from relatively low masses. In the following we discuss mostly the muon anomaly where the HVP contribution is relatively large, and is much more sensitive to New Physics at high mass scale and as such more interesting to investigate. While being very much smaller, the hadronic contribution to a_e is now larger than the current error of the direct a_e measurement [3], and it should be included in the theoretical prediction. The kernels $K_\epsilon(s)$ and $K_\mu(s)$ are given in Fig. 8.2: Their ratio scales asymptotically as $(m_\mu/m_e)^2 \sim 4.3 \times 10^4$, but even at the hadronic threshold the ratio is already 2/3 of the asymptotic value. Therefore the hadronic contribution to a_e will approximately be 4×10^4 smaller than the corresponding contribution to a_μ. In the following, we concentrate essentially on a_μ.

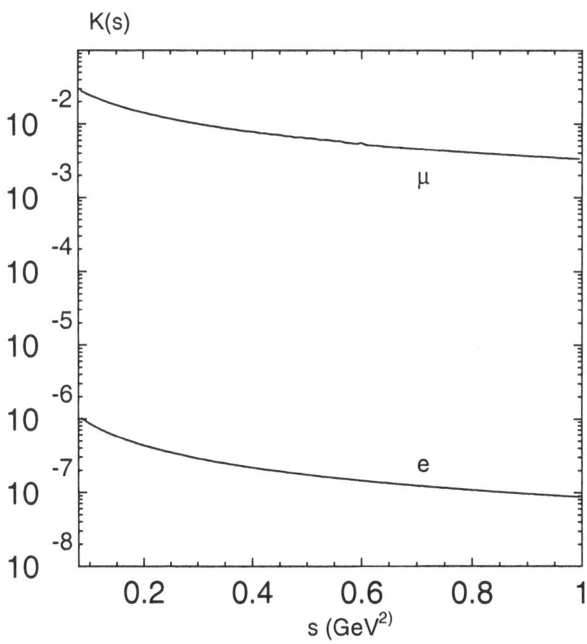

Fig. 8.2. The s dependence of the kernel $K(s)$ of the dispersion integral for the muon and electron magnetic anomalies.

The kernel can be written into a closed analytical form,

$$K_l(s) = x^2\left(1-\frac{x^2}{2}\right)+(1+x)^2\left(1+\frac{1}{x^2}\right)\left(\ln(1+x)-x+\frac{x^2}{2}\right)$$
$$+\frac{(1+x)}{(1-x)}x^2\ln x\ , \tag{8.3}$$

with $x = (1-\beta_l)/(1+\beta_l)$ and $\beta_l = (1-4m_l^2/s)^{1/2}$. In Eq. (8.1), $R(s) \equiv R^{(0)}(s)$ denotes the ratio of the "bare" cross section for e^+e^- annihilation into hadrons to the lowest-order muon-pair cross section. The "bare" cross section is defined as the measured cross section, corrected for initial state radiation, electron-vertex loop contributions and vacuum polarization effects in the photon propagator. The reason for using the "bare" (i.e. lowest order) cross section is that a full treatment of the next order (α^3) is anyhow needed at the level of a_μ, so that the use of "dressed" cross sections would entail the risk of double-counting some of the higher-order contributions, and would still leave anyway other α^3 contributions to be calculated explicitly. However the hadronic insertion can still contain photon propagators, so that final state radiation (FSR) must be included in the input cross section.

The function $K_\mu(s)$ decreases monotonically with increasing s. It gives a strong weight to the low energy part of the integral in Eq. (8.1). About 91% of the total contribution to $a_\mu^{\rm had,LO}$ is accumulated at center-of-mass energies $\sqrt{s}$ below 1.8 GeV and 73% of $a_\mu^{\rm had,LO}$ is covered by the two-pion final state which is dominated by the $\rho(770)$ resonance.

Many calculations of the hadronic vacuum polarization contribution have been carried out in the past taking advantage of the e^+e^- data available at that time. Clearly, the results depend crucially on the quality of the input data which has been improving in time with better detectors and higher luminosity machines. Therefore the later calculations, with more complete and better quality data, supersede the results of the former ones. In addition, some approaches make use of theory constraints not only in the high energy region where perturbative QCD applies [4–6], but even at lower energy [7, 8]. Also, it was proposed [9] to use data on hadronic τ decays to extract the relevant spectral functions, indeed more precisely known than the e^+e^--based results available then.

Calculating a dispersion integral over experimental cross section necessitates a good understanding of the data at hand, as the quality of the input data controls the final result. The technical details of the numerical integration are important,but in principle more straightforward. Therefore the different published results on the HVP contribution to a_μ^{had} are strongly correlated. The only variability comes from using different data sets available at a given time, or from injecting extra information from theory. Here we follow mainly the approach taken by Davier–Eidelman–Höcker–Zhang (DEHZ) [10], which considers both e^+e^- and τ input. Comparison with other analyses will be presented later.

8.2. The Input Data from e^+e^- Annihilation

8.2.1. *The direct measurements*

8.2.1.1. *Overview*

In the past the exclusive low-energy e^+e^- cross sections have been mainly measured by experiments running at e^+e^- colliders in Novosibirsk and Orsay. Due to the higher hadron multiplicity at energies above $\sim$ 2.5 GeV, the exclusive measurement of the many hadronic final states is not practicable. Consequently, the experiments at the high-energy colliders ADONE, SPEAR, DORIS, PETRA, PEP, VEPP-4, CESR and BEPC have measured the total inclusive cross section ratio R. Complete references to published data are given in Ref. [11].

Motivated largely by the necessity to dispose of higher-accuracy data for HVP calculations, new experiments have been performed in the last decade. Precise $e^+e^- \to \pi^+\pi^-$ measurements in the ρ region come from Novosibirsk with the CMD-2 [12] and SND [13] detectors. In both cases revised cross sections have been published after correcting initial problems in the large radiative corrections [14, 15]. In addition, CMD-2 has obtained results below [16] and above [17] the ρ region, as well as a second set of data across the ρ resonance [18]. Both experiments cannot separate reliably enough pions and muons, except near threshold using momentum measurement and kinematics for CMD-2, so that the measured quantity is the ratio $(N_{\pi\pi} + N_{\mu\mu})/N_{ee}$. The pion-pair cross section is obtained after subtracting the muon-pair contribution and normalizing to the BhaBha events, using computed QED cross sections for both, including their respective radiative corrections. The results are corrected for leptonic and hadronic vacuum polarization, and for photon radiation by the pions, so that the deduced cross section corresponds to $\pi^+\pi^-$ including pion-radiated photons and virtual final state QED effects. The overall systematic errors of the final data are quoted to be 0.6% (0.8%) for the two CMD-2 sets in the ρ region and 1.5% for SND, dominated by the uncertainties in the radiative corrections

(0.4%) which should be considered fully correlated for the two experiments as they now use the same programs.

The cross section results from CMD-2 and from previous experiments (corrected for vacuum polarization and FSR, according to the procedure discussed in Section 8.2.1.2) are in agreement within the much larger uncertainties (2–10%) quoted by the older experiments.

Other exclusive channels have important contributions, such as the ω and ϕ resonances and the 4π processes, $e^+e^- \to 2\pi^+2\pi^-$ and $e^+e^- \to \pi^+\pi^-\pi^0\pi^0$ (Fig. 8.3). For the latter process, experiments show rather large discrepancies. In some cases measurements are incomplete, as for example $e^+e^- \to K\overline{K}\pi\pi$ or $e^+e^- \to 6\pi$, and one has to rely on isospin symmetry to estimate or bound the unmeasured cross sections [11].

8.2.1.2. *Radiative corrections*

The evaluation of the integral in Eq. (8.1) requires the use of the "bare" hadronic cross section, so that the input data must be analyzed with care in this respect, especially for the older data where the procedure is often not clear from the publications. While the hadronic cross sections given by the experiments are always corrected for initial state radiation and the effect of loops at the electron vertex, the vacuum polarization correction in the photon propagator is a more delicate point. The cross sections need to be corrected, i.e.

$$\sigma_{\text{bare}} = \sigma_{\text{dressed}} \left(\frac{\alpha(0)}{\alpha(s)} \right)^2 , \tag{8.4}$$

where σ_{dressed} is the measured cross section already corrected for initial state radiation, and $\alpha(s)$ takes into account leptonic and hadronic vacuum polarization. The new data from CMD-2 and SND are explicitly corrected for both leptonic and hadronic vacuum polarization effects (the latter involving in principle an iterative procedure), whereas data from older experiments in general were not.

In Eq. (8.1) $R(s)$ must include the contribution of all hadronic states produced at the energy $\sqrt{s}$, in particular those with FSR. In the $\pi^+\pi^-$ data from CMD-2 and SND most additional photons are experimentally rejected to reduce backgrounds from other channels and the fraction kept is subtracted using the Monte Carlo simulation which includes a model for FSR. Then the full FSR contribution is added back as a correction using scalar QED (point-like pions) [19].

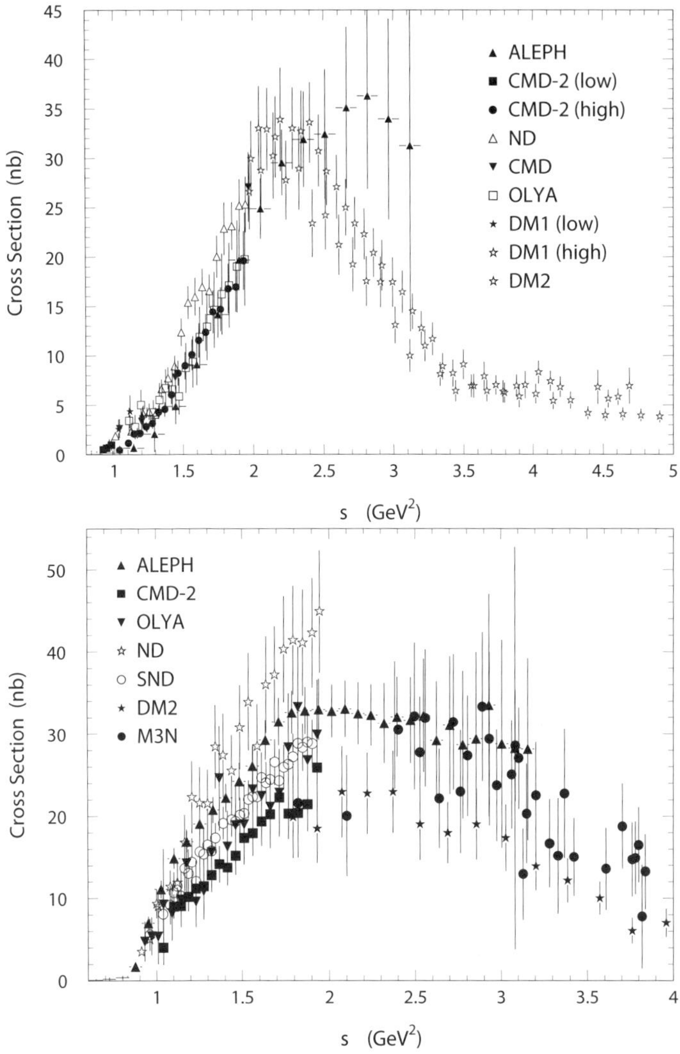

Fig. 8.3. The s dependence of the cross sections $e^+e^- \to 2\pi^+2\pi^-$ (top) and $e^+e^- \to \pi^+\pi^-\pi^0\pi^0$ (bottom), compared to the prediction from the ALEPH τ spectral functions (see Section 8.3.1). References of experiments are given in Ref. [11].

Below 1 GeV the different corrections in the $\pi^+\pi^-$ contribution amount to -2.3% for leptonic vacuum polarization, between -1.0 and $+6.0\%$ for hadronic vacuum polarization, and $+0.8\%$ for FSR. These corrections are preceded by much larger ones, taking into account addition radiation from the incoming electrons and positrons, but which should be well understood from QED.

8.2.2. *Obtaining e^+e^- cross sections from radiative return*

8.2.2.1. *The ISR method*

In recent years the availability of high-luminosity e^+e^- colliders designed as meson factories at fixed energy has opened a new way to measure annihilation cross sections through radiative return. Radiating a photon from the initial state allows one to cover a wide spectrum of energy for e^+e^- processes [20]. This scheme has been implemented for low-energy cross sections by KLOE and BaBar, respectively operating at 1.02 and 10.58 GeV. The very large statistics available in these meson factories more than compensate for the emission probablility of an extra photon (initial state radiation –ISR).

In this way the cross section for $e^+e^- \to X$, where X can be any final state, is deduced from a measurement of the radiative process $e^+e^- \to X\gamma$. The photon energy spectrum (energy E^*_γ in the center of mass) allows one to cover a large range of masses down to threshold for the final state X. Calling $x = 2E^*_\gamma/\sqrt{s}$ the fractional energy of the ISR photon, where s is the square of the e^+e^- center-of-mass energy, the final state mass $\sqrt{s'}$ is derived from the relation $s' = s(1-x)$. The relevant process is shown in Fig. 8.4 for $\mu^+\mu^-$ and $\pi^+\pi^-$ final states.

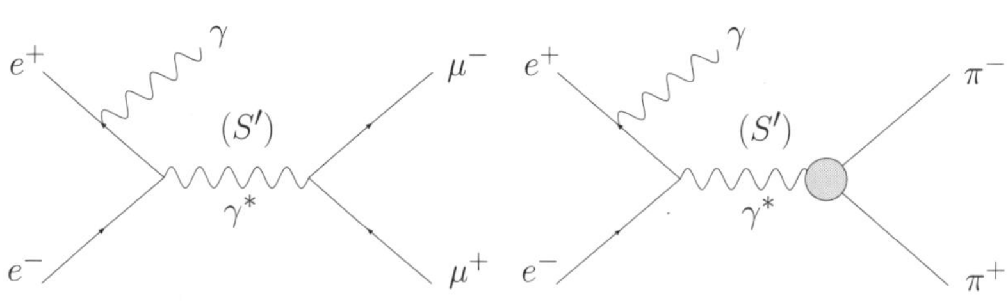

Fig. 8.4. The lowest-order processes for pair production by radiative return (initial state radiation).

8.2.2.2. *The $\pi^+\pi^-$ final state*

Of course, at lowest-order, the radiated photon could come from the initial (ISR) or the final state (FSR), and proper means must be ensured to extract only the $|ISR|^2$ part. Due to the opposite C-parity of the particle pair in the ISR ($C = -1$) and FSR ($C = +1$) amplitudes, the ISR-FSR interference term vanishes for a charge-symmetric event selection. Thus only the $|FSR|^2$ part matters: Its effect is discussed below for KLOE and BaBar.

In addition, an extra photon can be emitted by the final state particle and this contribution must be kept in computing the dispersion integral Eq. (8.1). The observed mass spectrum ($\sqrt{s'} = m_{\pi\pi(\gamma)}$) of $\pi\pi\gamma(\gamma)$ events is given by

$$\frac{dN_{\pi\pi\gamma(\gamma)}}{d\sqrt{s'}} = \frac{dL_{ISR}^{eff}}{d\sqrt{s'}} \varepsilon_{\pi\pi\gamma}(\sqrt{s'})\, \sigma^0_{\pi\pi(\gamma)}(\sqrt{s'}), \tag{8.5}$$

where $\varepsilon_{\pi\pi\gamma}$ is the full acceptance for the event sample, determined by MC with suitable corrections, and $\sigma^0_{\pi\pi(\gamma)}$ is the bare cross section (excluding vacuum polarization) for producing $\pi\pi(\gamma)$ including additional FSR. A similar relation holds for the $\mu\mu(\gamma)$ spectrum with the corresponding $\sigma^0_{\mu\mu(\gamma)}$ cross section.

The effective ISR luminosity function,

$$\frac{dL_{ISR}^{eff}}{d\sqrt{s'}} = L_{ee} \frac{dW}{d\sqrt{s'}} \left(\frac{\alpha(s')}{\alpha(0)}\right)^2 \frac{\varepsilon_{ISR\gamma}(\sqrt{s'})}{\varepsilon_{ISR\gamma}^{MC}(\sqrt{s'})}, \tag{8.6}$$

takes into account the e^+e^- integrated luminosity (L_{ee}), the probability to radiate an ISR photon (with possibly additional ISR photons) ($\frac{dW}{d\sqrt{s'}}$) so that the produced final state (excluding ISR photons) has a mass $\sqrt{s'}$, and the ratio of $\varepsilon_{ISR\gamma}$, the efficiency to detect the main ISR photon, to the same quantity, $\varepsilon_{ISR\gamma}^{MC}$, in simulation. The effective ISR luminosity function can be directly measured from the observed mass spectrum of $\mu\mu\gamma(\gamma)$ events following,

$$\frac{dL_{ISR}^{eff}}{d\sqrt{s'}} = \frac{dN_{\mu\mu\gamma(\gamma)}}{d\sqrt{s'}} \frac{1}{\varepsilon_{\mu\mu\gamma}(\sqrt{s'})\, \sigma^0_{\mu\mu(\gamma)}(\sqrt{s'})} \tag{8.7}$$

inserting for $\sigma^0_{\mu\mu(\gamma)}$ the cross section computed with QED.

The KLOE and BaBar analyses are very different. First, the initial center-of-mass energy is close to the studied energy in the case of KLOE (soft ISR photons), while it is very far in the BaBar case (hard ISR photons). In KLOE the ISR photon is not detected and reconstructed kinematically, assuming no extra photon. Since the cross section strongly peaks along the beams, a large statistics of ISR events is obtained. Pion pairs are separated from muon pairs by the remaining kinematic constraint. In BaBar the ISR photon is detected at a large angle (about 10% efficiency) so that the full event is observed, and an additional photon can be incorporated in the kinematic fit (undetected additional ISR or detected FSR photon). Another big difference concerns the ISR luminosity: In the KLOE analysis it is computed using the next-to-leading order PHOKHARA generator [21], while in BaBar both pion and muon pairs are measured and the ratio $\pi\pi(\gamma)/\mu\mu(\gamma)$ directly provides the $\pi\pi(\gamma)$ cross section. The small-angle ISR photon provides a suppression of the sizeable $|FSR|^2$ contribution in KLOE, and the remaining part is computed from PHOKHARA. In BaBar the $|FSR|^2$ contribution is negligible as it is proportional to the square of the pion form factor $|F_\pi(s)|^2$ at $s = (10.58)$ GeV^2.

KLOE results are available [22], but they are now superseded by new data recently obtained [23]. The new data correct a previous problem affecting an inefficient veto on cosmic events. BaBar results have only been shown in a preliminary form and the published results will appear soon.

8.2.2.3. *Multi-hadronic production at higher energies*

Above 1 GeV many multi-hadronic channels open up and contribute to the dispersion integral. The ISR method has been intensively used with BaBar in order to measure the cross sections for the relevant channels up to 2–2.5 GeV. Most results are published [24] and they are summarized in Fig. 8.5.

8.2.3. *Comparing $e^+e^- \to \pi^+\pi^-$ data from different experiments*

Fig. 8.6 presents a summary of the data up to 1 GeV mass from CMD-2 and SND using the direct measurement, and KLOE with the ISR technique. Older data are much less precise and do not contribute significantly anymore. The important ρ region is well covered, but the region below 600 MeV is less precisely known.

The consistency of the different data sets can be investigated by averaging the results of CMD-2, SND, and KLOE. Each experiment is recast in small energy bins using splines interpolating between the actual data points and the average is performed in each small bin, taking into account the relative weight (statistical + systematic) of the experiments. The error assigned to the average is scaled by $\sqrt{\chi^2/\mathrm{DF}}$ if $\chi^2/\mathrm{DF} > 1$. Finally each set of data is compared to the average in Fig. 8.7. Since the KLOE data have the largest statistics and a relatively small systematic uncertainty,

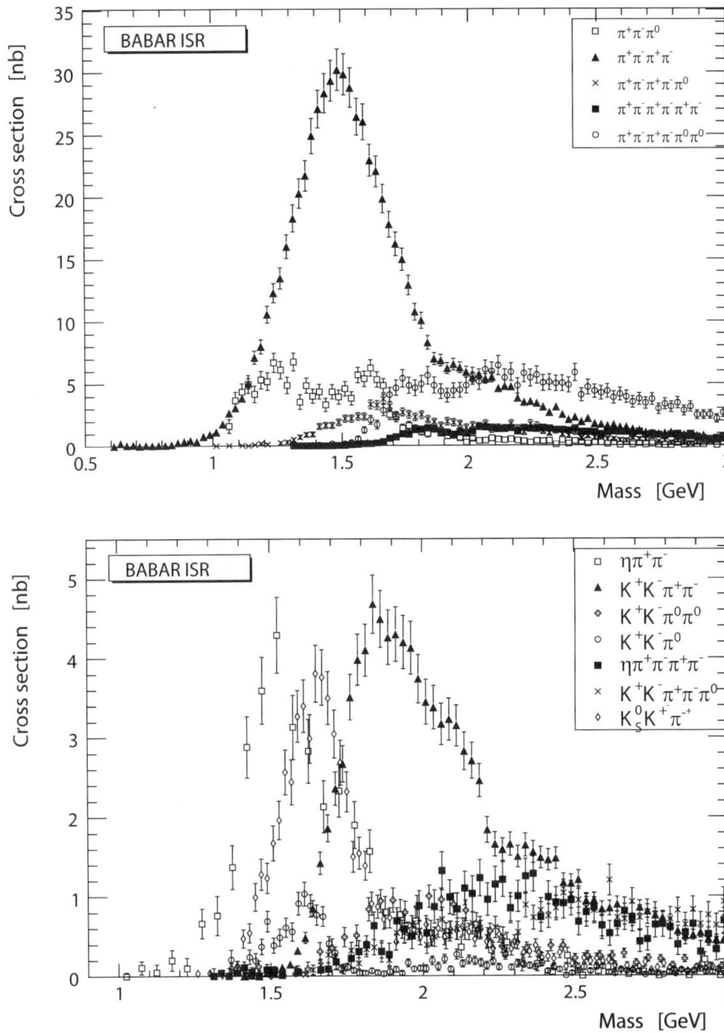

Fig. 8.5. The $\sqrt{s}$ dependence of the cross sections $e^+e^- \to$ hadrons measured by BaBar using the ISR method in the range 1-3 GeV.

they dominate the average. The overall consistency of the three experiments is fair, with some tension between CMD-2 and KLOE at and above the ρ peak. The region between 0.5 and 0.6 GeV is only covered by SND, with relatively large errors, while a similar situation occurs below 0.42 GeV with CMD-2.

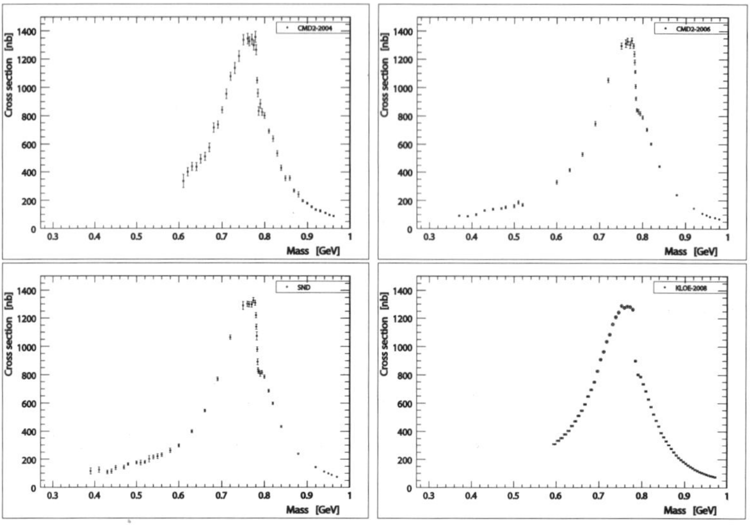

Fig. 8.6. The $\sqrt{s}$ dependence of the cross sections $e^+e^- \to \pi^+\pi^-$ measured (clockwise from upper left) by CMD-2 [14], CMD-2 [16–18], KLOE [23] and SND [15].

8.3. The Input Data from τ Decays

8.3.1. *Spectral functions from τ decays*

Data from τ decays into two- and four-pion final states $\tau^- \to \nu_\tau \pi^- \pi^0$, $\tau^- \to \nu_\tau \pi^- 3\pi^0$ and $\tau^- \to \nu_\tau 2\pi^- \pi^+ \pi^0$, are available from ALEPH [25, 26], CLEO [27, 28], OPAL [29]. High-statistics data have been recently published by Belle [30] for the two-pion mode. A review of the physics of τ hadronic decays can be found in Ref. [31].

It should be pointed out that the experimental conditions at the Z pole (ALEPH, OPAL) and at the $\Upsilon(4S)$ (CLEO, Belle) energies are very different. On the one hand, at LEP, the $\tau^+\tau^-$ events can be selected with high efficiency ($> 90\%$) and small non-τ background ($< 1\%$), thus ensuring little bias in the efficiency determination. Despite higher background and smaller efficiency, CLEO and Belle have the advantage of lower energy for the reconstruction of the decay final state since particles are more separated in space. One can therefore consider ALEPH/OPAL and CLEO/Belle data to be uncorrelated as far as experimental procedures are concerned. The fact that their respective spectral functions for the $\pi^-\pi^0$ and $2\pi^-\pi^+\pi^0$ modes agree is therefore a valuable experimental consistency test.

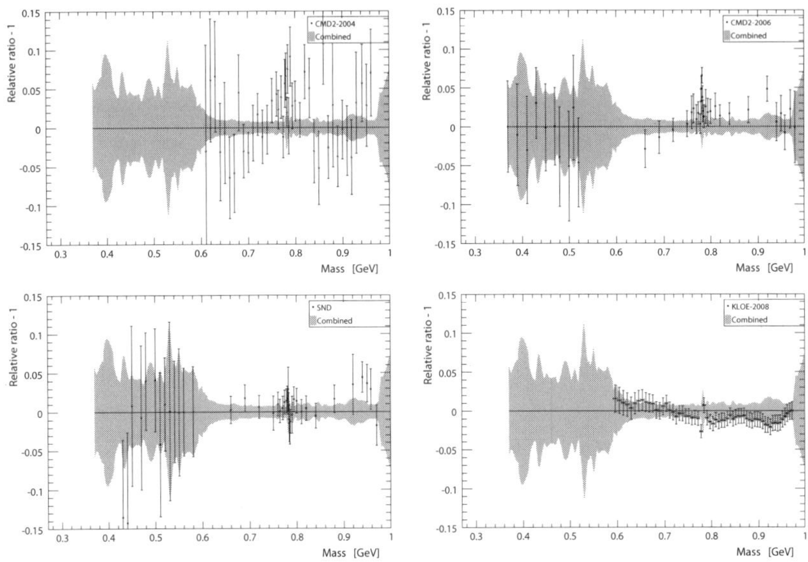

Fig. 8.7. The deviation from unity of the relative ratio of the cross sections for $e^+e^- \to \pi^+\pi^-$ measured by individual experiments to their weighted average as a function of $\sqrt{s}$: (clockwise from upper left) CMD-2 [14], CMD-2 [16–18], KLOE [23] and SND [15].

Assuming (for the moment) isospin invariance to hold, the corresponding e^+e^- isovector cross sections are calculated via the Conserved Vector Current (CVC) relations

$$\sigma^{I=1}_{e^+e^-\to\pi^+\pi^-} = \frac{4\pi\alpha^2}{s}\, v_{\pi^-\pi^0}\ , \tag{8.8}$$

$$\sigma^{I=1}_{e^+e^-\to\pi^+\pi^-\pi^+\pi^-} = 2\cdot\frac{4\pi\alpha^2}{s}\, v_{\pi^-\,3\pi^0}\ , \tag{8.9}$$

$$\sigma^{I=1}_{e^+e^-\to\pi^+\pi^-\pi^0\pi^0} = \frac{4\pi\alpha^2}{s}\left[v_{2\pi^-\pi^+\pi^0} - v_{\pi^-\,3\pi^0}\right]. \tag{8.10}$$

The τ spectral function $v_V(s)$ for a given vector hadronic state V is defined by [32]

$$v_V(s) \equiv$$

$$\frac{m_\tau^2}{6|V_{ud}|^2 S_{\rm EW}}\,\frac{B(\tau^-\to\nu_\tau V^-)}{B(\tau^-\to\nu_\tau e^-\bar{\nu}_e)}\,\frac{dN_V}{N_V ds}\left[\left(1-\frac{s}{m_\tau^2}\right)^2\left(1+\frac{2s}{m_\tau^2}\right)\right]^{-1}, \tag{8.11}$$

where $m_\tau = (1776.84 \pm 0.17)$ MeV [33] and $|V_{ud}| = 0.97418 \pm 0.00019$ obtained from averaging the determinations [34] from nuclear β decays and kaon decays (assuming unitarity of the CKM matrix) and $S_{\rm EW}$ accounts for electroweak radiative corrections as discussed in Section 8.3.3. $(1/N_V)dN_V/ds$ is the normalised invariant mass spectrum of the hadronic final state. The branching fraction of $\tau \to V^-(\gamma)\nu_\tau$ is denoted by BR_V (final state photon radiation is implied for τ branching fractions). The electron branching fraction value $BR_e = (17.818 \pm 0.032)\%$ is obtained [31] assuming lepton universality. The spectral functions are obtained from the corresponding invariant mass distributions, subtracting out the non-τ background and the feedthrough from other τ decay channels, and after a final unfolding from detector response. Again the measured τ spectral functions are inclusive with respect to radiative photons.

8.3.2. *Consistency of τ data from different experiments*

Following the e^+e^- data comparison the τ 2π spectral functions from each experiment are compared to the bin-by-bin combined spectral function. Here the world average value of the $\tau \to \pi^-\pi^0\nu_\tau$ branching fraction is used for each experimental set.

Figure 8.8 shows the relative comparisons. The most precise data is from Belle which dominates the combined spectral function shape. Good

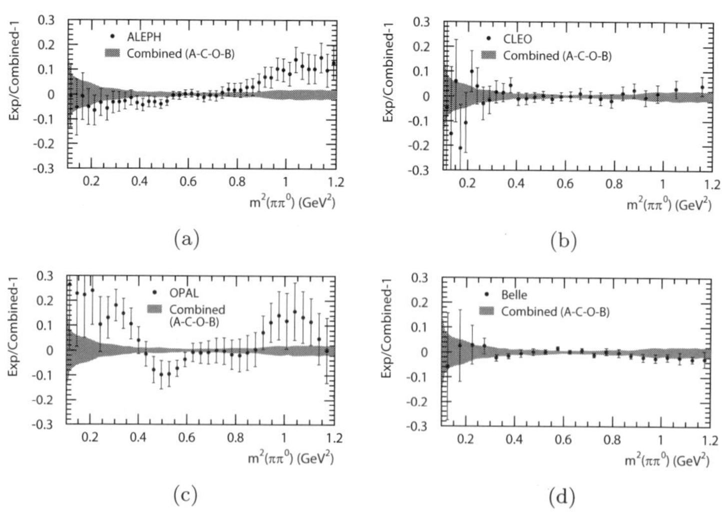

Fig. 8.8. Relative comparison between each individual measurement (data points) from (a) ALEPH; (b) CLEO; (c) OPAL; (d) Belle and the combined mass squared spectrum (A-C-O-B, shaded band).

agreement is observed between Belle and CLEO, whereas some deviations are seen with the LEP experiments, ALEPH and OPAL, in particular for $s > 1$ GeV^2. These deviations are a bit larger than the correlated systematics in this region and they may reflect the more difficult situation at LEP with very collimated tau decays. However, this high-mass region is less relevant for the $(g-2)$ contribution.

8.3.3. *Isospin symmetry breaking*

The relationships given in Eqs. (8.8), (8.9) and (8.10) between e^+e^- and τ spectral functions only hold in the limit of exact isospin invariance. It follows from the factorization of strong interaction physics as produced through the γ and W propagators out of the QCD vacuum. However, symmetry breaking is expected at some level from electromagnetic processes, whereas the small u,d mass splitting leads to negligible effects.

Various identified sources of isospin breaking (IB) are considered for the dominant 2π channel. The IB-corrected τ spectral function reads:

$$v_{\pi^-\pi^0}^{\mathrm{IB-corr}}(s) \; = \; v_{\pi^-\pi^0}(s)\, R_{\mathrm{IB}}(s)\,, \tag{8.12}$$

with

$$R_{\mathrm{IB}}(s) = \frac{(1+\delta_{\mathrm{FSR}})}{G_{\mathrm{EM}}(s)} \frac{\beta_0^3(s)}{\beta_-^3(s)} \left| \frac{F_0(s)}{F_-(s)} \right|^2 , \tag{8.13}$$

$F_\pi^{0,-}(s)$ being the electromagnetic and weak pion form factors, respectively, and $\beta_{0,-} = \beta(s, m_{\pi^-}, m_{\pi^{+,0}})$ is a kinematic factor given in [11], vanishing at threshold. The different contributions are examined in turn.

- Electroweak radiative corrections yield their dominant contribution from the short distance correction to the effective four-fermion coupling $\tau^- \to \nu_\tau(d\bar{u})^-$ enhancing the τ amplitude by the factor $(1 + 3\alpha(m_\tau)/4\pi)(1+2\overline{Q})\ln(M_Z/m_\tau)$, where $\overline{Q}$ is the average charge of the final state partons [35, 36]. While this correction vanishes for leptonic decays, it contributes for quarks. All higher-order logarithms can be re-summed using the renormalization group [35] into an overall multiplicative electroweak factor $S_{\mathrm{EW}}^{\mathrm{had}}$, which is equal to 1.0194. The difference between the re-summed value and the lowest-order estimate (1.0188) can be taken as a conservative estimate of the uncertainty. QCD corrections to $S_{\mathrm{EW}}^{\mathrm{had}}$ have been calculated [35, 36] and found to be small, reducing its value to 1.0189. Subleading non-logarithmic short distance corrections have been calculated to order $O(\alpha)$ for the leptonic width [35], $S_{\rm EW}^{\rm sub,lep} = 1 + \alpha(25/4 - \pi^2)/2\pi \simeq 0.9957$. The total short-distance correction for the 2π channel is thus very well known and amounts to $S_{\rm EW}^{\pi\pi^0} = 1.0233 \pm 0.0006$. The common practice is to include this correction already in the definition of the τ spectral function, Eq. (8.11).

- Long distance radiative corrections are expected to be final-state dependent in general. A consistent calculation of radiative corrections for the $\nu_\tau\pi^-\pi^0$ mode is available at loop level [37, 38], and recently improved [39]. The corresponding correction factor $G_{\rm EM}(s)$, included in Eq. (8.13), corrects the τ spectral function to the bare e^+e^- spectral function. A difficulty here is to know the exact conditions which have been applied to the different measurements regarding the treatment of additional photons in the hadronic final state: exclusive (no radiation allowed, up to a defined energy limit), or inclusive (additional photons kept, within some restrictions). A component of the problem is whether or not the final state $\pi^-\omega \to \pi^-\pi^0\gamma$ is included or not in the $\pi\pi^0(\gamma)$ experimental spectral function. Such a detailed discussion is beyond the scope of this review. So the G_{EM} correction used here will be an average one, taking an uncertainty covering the full range of possible variations.
- The FSR correction $(1 + \delta_{\rm FSR})$ is necessary to include again final state radiation by charged pions, as done for e^+e^- measurements which exclude FSR. It is computed using scalar QED.
- The mass difference between charged and neutral pions, which is essentially of electromagnetic origin introduces some IB as the spectral function has a kinematic factor β^3 which is different in e^+e^- $(\pi^+\pi^-)$ and τ decay $(\pi^-\pi^0)$. This correction is straightforward and very well known.
- Other mass corrections occur in the form factor itself. It is affected by the pion mass difference because the same β^3 factor enters in the $\rho \to \pi\pi$ width. Similarly, mass and width differences between the charged and neutral ρ meson affect the resonance lineshape. Using $m_{\rho^\pm} - m_{\rho^0} = (-0.4 \pm 0.9)$ MeV, a result obtained by KLOE from a fit to the $\phi \to \pi^+\pi^-\pi^0$ Dalitz plot [40], yields the relevant mass difference $m_{\rho^\pm} - m_{\rho^0_{\rm bare}} = (1.0 \pm 0.9)$ MeV. This choice is justified because here $m_{\rho^0_{\rm bare}}$ stands for a bare mass value extracted from bare e^+e^- cross sections with the vacuum polarization effects

removed. Indeed, the expected difference between the dressed and bare mass is $m_{\rho^0} - m_{\rho^0_{\rm bare}} = 3\Gamma(\rho^0 \to e^+e^-)/(2\alpha)$.

- $\rho - \omega$ interference occurs in the $\pi^+\pi^-$ mode only, but its contribution can be readily introduced into the τ spectral function using the parameters determined in e^+e^- data fits.
- Also, electromagnetic decays explicitly break SU(2) symmetry for the ρ width. The major contribution comes from radiative decays $\rho \to \pi\pi\gamma$. A loop-level calculation is now available [41] which moves the value for $\Gamma_{\rho^0} - \Gamma_{\rho^-}$ from -1.1 MeV (no radiative decays, only $\pi^{-,0}$ mass difference) to +0.8 MeV, in contrast to earlier estimates.

The numerical corrections are given in Table 8.3.3. The total correction to $a_\mu^{\rm hadronic}$ from isospin-breaking using the τ 2π data amounts to $(-18.5 \pm 2.9)\ 10^{-10}$, significantly larger than the correction quoted in Ref. [11], $(-13.8 \pm 2.4)\ 10^{-10}$. This difference is due essentially to the ρ width correction from radiative decays [41].

Table 8.1. Contributions to $a_\mu^{\rm had,LO}$ $[\pi\pi, \tau]$ $(\times 10^{-10})$ from the various isospin-breaking corrections.

Source	$\Delta a_\mu^{\rm had,LO}$ $[\pi\pi, \tau]$ (10^{-10})
S_{EW}	-12.2 ± 0.2
G_{EM}	-3.5 ± 2.5
FSR	$+4.6 \pm 0.5$
ρ–ω interference	$+2.4 \pm 1.3$
$m_{\pi^\pm} - m_{\pi^0}$ effect on σ	-7.8
$m_{\pi^\pm} - m_{\pi^0}$ effect on Γ_ρ	$+4.1$
$m_{\rho^\pm} - m_{\rho^0_{\rm bare}}$	-0.1 ± 0.1
$\pi\pi\gamma$, electrom. decays	-6.0 ± 0.6
sum	-17.9 ± 2.9

8.4. Confronting e^+e^- and τ Data

The e^+e^- and the isospin-breaking corrected τ spectral functions, both combining all respective measurements are directly compared for the $\pi\pi$ final state in Fig. 8.9. The corresponding bands include statistical as well as systematic uncertainties (as quoted by the experiments). The e^+e^- and τ data are consistent within 2% below and around the ρ peak, while a discrepancy persists for energies above, reaching 7% at 0.95 GeV. The discrepancy is much reduced if only CMD-2 and SND data are used.

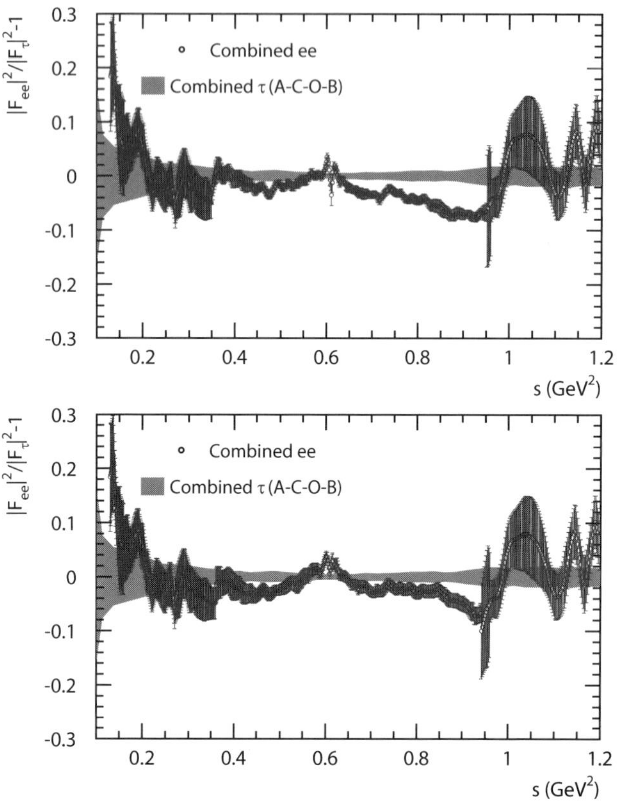

Fig. 8.9. The comparison between the combined spectral functions from $e^+e^- \rightarrow$ hadrons (dark-shaded band) and τ decays (light-shaded band): (top) combination of CMD-2, SND, and KLOE, (bottom) only CMD-2 and SND.

The corresponding integrals yielding $a_\mu^{2\pi,LO}$ are computed for each experiment separately. In the τ case the integral for the 2π channel depends on the normalized spectral function $1/N_{\pi\pi^0} dN_{\pi\pi^0}/ds$ and the corresponding branching fraction $B_{\pi\pi^0}$. Generally, experimental results are quoted by normalizing the spectral function to the world-averaged branching ratio, thus introducing a large correlation between the results from different experiments. Here we prefer to show the values for the integral taking all quantities (shape of spectral function and branching ratio) from the same experiment. In this way, all the results are uncorrelated. However the combined value (which is not exactly the weighted average of the individual results) is obtained by taking the world average value of the branching fraction and combining all spectral functions point-by-point, as shown in Fig. 8.9. This is the optimal combination of the results. For the e^+e^- combination all spectral functions are averaged point-by-point.

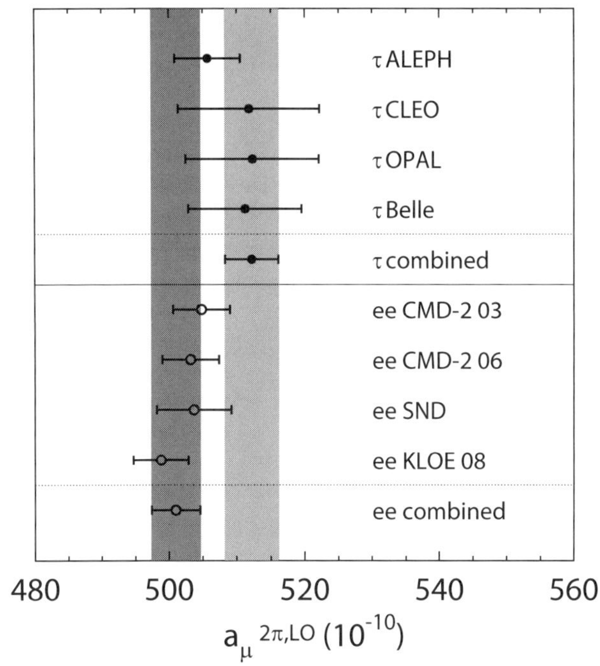

Fig. 8.10. The contributions to $a_\mu^{2\pi,LO}$ obtained at the level of each experiment and their combinations (τ and e^+e^- separately). The combined τ is not the average of the independent results from four experiments, but it is obtained from the average τ spectral function using the world-averaged branching ratio for $\tau \to \pi\pi^0\nu_\tau$.

The results are given in Fig. 8.10. There is a good consistency between the τ individual results, and also for e^+e^-. However, while the τ values are uncorrelated, the e^+e^- results are correlated, as in the mass region below 0.630 and above 0.958 the average is used for all experiments (in fact the truly uncorrelated part corresponds to 71% of the total). The τ and e^+e^- averages are not in good agreement. The τ value is higher by $(11.4 \pm 3.6_{ee} \pm 2.6_\tau \pm 2.9_{IB})\ 10^{-10} = (11.4 \pm 5.3)\ 10^{-10}$ (2.1 σ).

Another way to assess the compatibility between e^+e^- and τ spectral functions is to evaluate the $\tau \to \pi\pi^0\nu_\tau$ decay fractions using the corresponding e^+e^- spectral functions as input. This procedure involves another integral over the spectral function with a weight factor different from $K_\mu(s)$, in fact much more uniform (see the factor explicitly written in Eq. (8.11)). Thus this new kernel will emphasize more the discrepancy observed at larger masses. Using the previously described IB-breaking corrections, the branching ratio is predicted to be

$$\mathcal{B}_{\rm CVC}(\tau^- \to \nu_\tau \pi^- \pi^0) = (24.89 \pm 0.17_{\rm exp} \pm 0.12_{\rm IB})\ \% \ , \tag{8.14}$$

where the errors quoted are split into uncertainties from the experimental input (the e^+e^- annihilation cross sections) and the isospin-breaking corrections when relating τ and e^+e^- spectral functions. The result in Eq. (8.14) is smaller than the direct measurement,

$$\mathcal{B}_{\rm exp}(\tau^- \to \nu_\tau \pi^- \pi^0) = (25.42 \pm 0.10)\ \% \ , \tag{8.15}$$

by $(-0.53 \pm 0.21_{ee} \pm 0.10_\tau)\%$ (2.3 σ). If only CMD-2 and SND are used the discrepancy reduces to $(-0.38 \pm 0.24_{ee} \pm 0.10_\tau)\%$ (1.5 σ). The comparison is shown in Fig. 8.11.

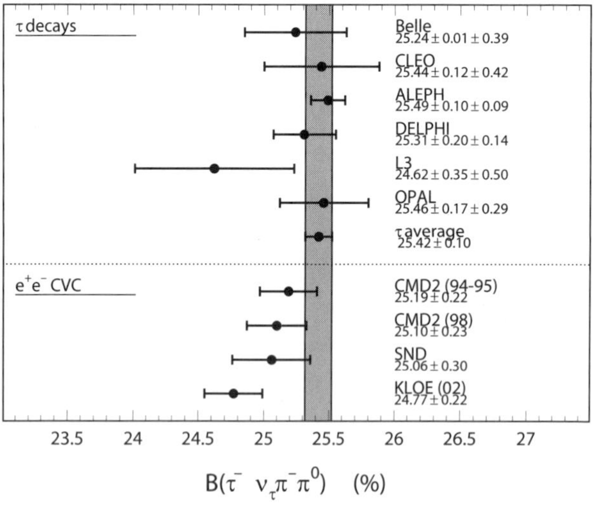

Fig. 8.11. The measured branching fractions for $\tau^- \to \pi^-\pi^0\nu_\tau$ [26, 30, 42–45] compared to the predictions from the $e^+e^- \to \pi^+\pi^-$ spectral functions, applying the isospin-breaking corrections. For the e^+e^- results, the data from the indicated experiments are used in the common $0.630 - 0.958$ GeV range, while the combined e^+e^- data is taken in the remaining energy domains below m_τ.

8.5. Special Cases

8.5.1. *The threshold region*

To overcome the lack of precise data at threshold energies one can benefit from the analyticity property of the pion form factor and use a third order expansion in s:

$$F_\pi^0 = 1 + \frac{1}{6}\langle r^2\rangle_\pi\, s + c_1\, s^2 + c_2\, s^3 + O(s^4). \tag{8.16}$$

Exploiting precise results from space-like data [46], the pion charge radius-squared is constrained to $\langle r^2\rangle_\pi = (0.439 \pm 0.008)\ \mathrm{fm}^2$ and the two parameters $c_{1,2}$ are fitted to the data in the range $[2m_\pi, 0.6\ \mathrm{GeV}]$. Good agreement is observed in the low energy region where the expansion should be reliable. Since the fits incorporate unquestionable constraints from first principles, this parameterization is used for evaluating the integrals in the range up to 0.5 GeV.

8.5.2. *Narrow resonances*

For the ω and ϕ resonances the experimental cross sections can be directly integrated, thus taking into account non-resonant and interference effects. The contributions from the very narrow $c\bar{c}$ resonances are computed using a relativistic Breit–Wigner parametrization for their lineshape, with their known resonance parameters [33].

8.5.3. *QCD for the high energy contributions*

In the asymptotic regime well above quark threshold the experimental spectral function is not known as precisely as its QCD prediction. This observation stems from detailed QCD studies performed with hadronic τ decays [31]. At the τ mass perturbative QCD reproduces within 1% the integral over the τ spectral function with the decay weight factor from Eq. (8.11). The details of the calculation can be found in Ref. [5, 8] and in the references therein.

The perturbative QCD prediction uses a next-to-next-to-leading order ($\mathrm{N^2LO}$) $O(\alpha_s^3)$ expansion of the Adler D-function [47], and recently even to the $\mathrm{N^3LO}$ [48] with second-order quark mass corrections included [49]. $R(s)$ is obtained by evaluating numerically a contour integral in the complex s plane. Nonperturbative effects are considered through the Operator Product Expansion, giving power corrections controlled by gluon and quark condensates. The value $\alpha_s(M_Z^2) = 0.1191 \pm 0.0027$, used for the evaluation of the perturbative part, is taken as the result from the analysis of the Z width in a global electroweak fit [50, 51]. This value has negligible theoretical uncertainties. A test of the QCD prediction can be performed in the energy range between 1.8 and 3.7 GeV. The contribution to $a_\mu^{\rm had,LO}$ in this region is computed to be $(33.9 \pm 0.5)\ 10^{-10}$ using QCD, to be compared with the result, $(34.9 \pm 1.8)\ 10^{-10}$ from the data. The two values agree within the 5% accuracy of the measurements, but the QCD value is more precise.

In Ref. [8] the evaluation of $a_\mu^{\rm had,LO}$ was shown to be improved by applying QCD sum rules. This is no longer the case: The improvement provided by the use of QCD sum rules resulted from a balance between the experimental accuracy of the data and the theoretical uncertainties. The presently achieved precision of e^+e^- and τ data, should they agree, is such that the gain would be now marginal.

8.6. Results for the LO Hadronic Vacuum Polarization Contribution

Figure 8.12 gives a panoramic view of the e^+e^- data in the relevant energy range. The shaded band below 2 GeV represents the sum of the exclusive channels considered in the analysis (it does not yet include the more precise ISR data from BaBar. The QCD prediction is indicated by the cross-hatched band. Note that the QCD band is plotted taking into account the thresholds for open flavour B states, in order to facilitate the comparison with the data in the continuum. However, for the evaluation of the integral, the $b\bar{b}$ threshold is taken at twice the pole mass of the b quark, so that the contribution includes the narrow Υ resonances, according to global quark-hadron duality.

The discrepancy discussed above for the 2π channel is slightly enlarged when including the 4π modes for which τ decays can also be used. Of course the τ spectral functions only provide input for the isovector final states, and the isoscalar contributions have to come solely from e^+e^- data. Summing up all contributions the τ-based estimate is larger by $(14.2 \pm 6.4)\ 10^{-10}$ (2.2 σ).

The results for the lowest order hadronic contribution are

$$\begin{aligned} a_\mu^{\rm had,LO} &= (687.3 \pm 4.2_{\rm exp} \pm 1.9_{\rm rad} \pm 0.7_{QCD})\ 10^{-10} \qquad [{\rm e^+e^-}{-}{\rm based}], \\ a_\mu^{\rm had,LO} &= (701.5 \pm 4.8_{\rm exp+IB} \pm 0.8_{\rm rad} \pm 0.7_{\rm QCD})\ 10^{-10} \qquad [\tau{-}{\rm based}], \end{aligned} \tag{8.17}$$

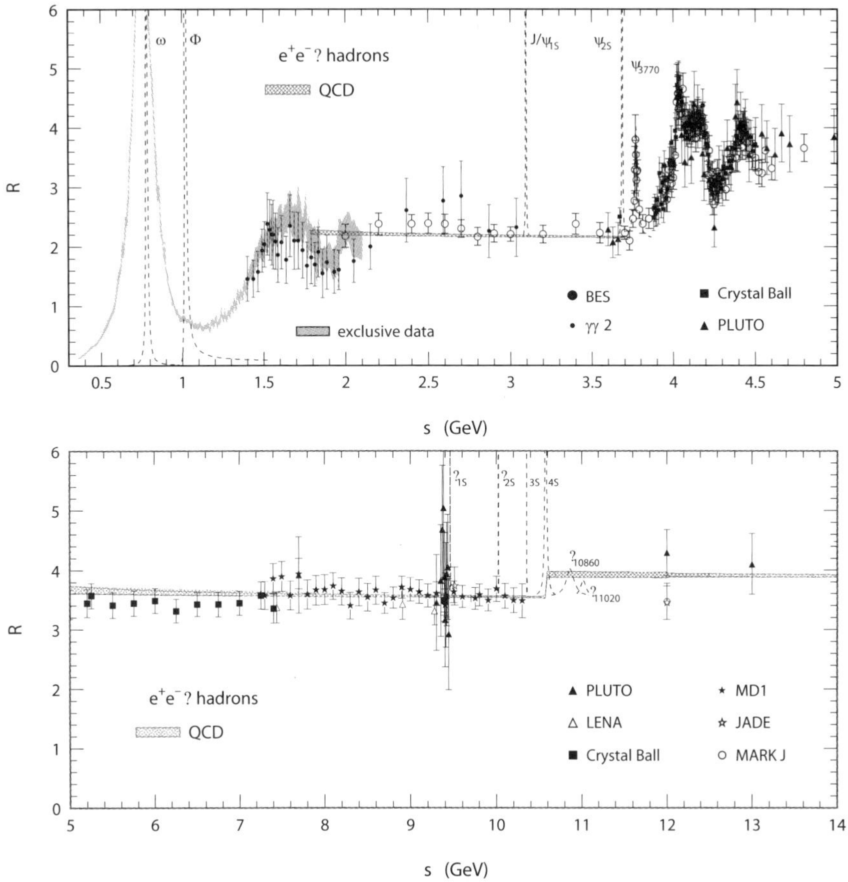

Fig. 8.12. Compilation of the data contributing to $a_\mu^{\rm had,LO}$. Shown is the total hadronic over muonic cross section ratio R. The shaded band below 2 GeV represents the sum of the exclusive channels considered in this analysis, with the exception of the contributions from the narrow resonances which are given as dashed lines. All data points shown correspond to inclusive measurements. The cross-hatched band gives the prediction from (essentially) perturbative QCD (see text).

where the errors labelled 'rad' corresponds to uncertainties in the treatment of radiative corrections in the older e^+e^- experiments.

As discussed in the introduction, the HVP contribution to the electron anomaly is much smaller. An evaluation using e^+e^- data gives the value [8],

$$a_e^{\rm had,LO} = (1.875 \pm 0.017)\ 10^{-12}, \tag{8.18}$$

6 times larger than the best experimental accuracy achieved so far in the a_e measurement [3] of 0.28 10^{-12}. Since this measurement provides by far the most accurate value of $\alpha(0)$, which can in turn be used for computing the QED part of a_μ, it is important to know the hadronic contribution. However the accuracy needed here is much less critical than for the a_μ prediction.

8.7. Comparison Between Different Analyses

Other estimates of $a_\mu^{\rm had,LO}$ have been recently published. As the input data is more or less the same one expects the results to be very close. Differences could occur in the treatment of systematic errors from the experiments, in particular the correlations within the same data set across the masss spectrum. Uncertainties on radiative corrections can also be correlated between different experiments if they use the same programs, as it is the case for CMD-2 and SND.

In the calculation presented by Hagiwara–Martin–Nomura–Teubner (HMNT) [52] the complete set of available exclusive channels up to 1.4 GeV are used except KLOE 2008, and inclusive measurements above. The two main differences between the estimate presented here and HMNT is the treatment of data in the threshold region and the use of inclusive data above 1.4 GeV.

It is more difficult to comment on the determination by Jegerlehner (J) [53] because of limited information about the data used, the way they are handled and the different contributions to the final error.

The values found in these two analyses,

$$\begin{aligned} a_\mu^{\rm had,LO} &= (689.4 \pm 4.2_{\rm exp} \pm 1.8_{\rm rad})\ 10^{-10} \quad &&[\text{HMNT}], \\ a_\mu^{\rm had,LO} &= (690.3 \pm 5.3)\ 10^{-10} &&[\text{J}], \end{aligned} \tag{8.19}$$

are in agreement with the e^+e^--based results in Eq. (8.17). Though the experimental errors should be strongly correlated, differences between the analyses could result from our inclusion of the revised KLOE 2008 data, the treatment of experimental systematic uncertainties, the numerical integration procedure (averaging or not neighboring data points), the treatment of missing radiative corrections, and the use of QCD above 1.8 GeV in our treatment.

8.8. Higher-order Hadronic Contributions

The three-loop hadronic contributions to $a_\mu^{\rm SM}$ involve one hadronic vacuum polarization insertion with an additional loop (either another photon propagator or another leptonic or hadronic vacuum polarization insertion). They can be evaluated [54] using the same $e^+e^- \to$ hadrons data sets described in Section 8.2.1. Calling that subset of $\mathcal{O}(\alpha/\pi)^3$ hadronic contributions $a_\mu^{\rm had,NLO}$, we quote here the result given by HMNT [52],

$$a_\mu^{\rm had,NLO} = -(9.79 \pm 0.08_{\rm exp} \pm 0.03_{\rm rad})\ 10^{-10}\ , \tag{8.20}$$

which is consistent with earlier studies [9, 54].

In the electron case the contribution at α^3 order is very small, but relatively more important than for the muon. Its value is estimated [54] to be $(-0.225\pm0.005)\ 10^{-12})$, comparable to the experimental accuracy of the a_e measurement (see Section 8.6). It reduces the total HVP contribution to a_e to $(1.65 \pm 0.02)\ 10^{-12}$.

Another hadronic contribution originates through a light-by-light scattering process for which a dispersion relation approach using data is not possible. Phenomenological approaches are described in Chapter 9.

8.9. Comparison of Theory and Experiment

It is now possible to collect the different contributions to the muon magnetic anomaly discussed in the other chapters,

$$a_\mu^{\rm QED} = (11\,658\,471.8 \pm 0.1)\ 10^{-10}, \tag{8.21}$$

$$a_\mu^{\rm EW} = (15.4 \pm 0.3)\ 10^{-10}, \tag{8.22}$$

$$a_\mu^{\rm had,LBL} = (10.5 \pm 2.6)\ 10^{-10}, \tag{8.23}$$

and $a_\mu^{\rm had,LO}$ and $a_\mu^{\rm had,NLO}$ from Eqs. (8.17) and (8.20), to obtain the Standard Model prediction for a_μ. Since the situation on $a_\mu^{\rm had,LO}$ is not yet settled finally, it is preferable to quote two values using the e^+e^- and the τ decay data,

$$\begin{aligned} a_\mu^{\rm SM} &= (11\,659\,175.2 \pm 4.7_{\rm had,LO} \pm 2.6_{\rm LBL} \pm 0.3_{\rm QED+EW})\ 10^{-10} \quad [e^+e^-], \\ a_\mu^{\rm SM} &= (11\,659\,189.4 \pm 4.9_{\rm had,LO} \pm 2.6_{\rm LBL} \pm 0.3_{\rm QED+EW})\ 10^{-10} \quad [\tau]\ . \end{aligned} \tag{8.24}$$

The SM values can be compared to the measurement [58],

$$a_\mu^{\rm exp} = (11\,659\,208.0 \pm 6.3)\ 10^{-10}\ . \tag{8.25}$$

Keeping experimental and theoretical errors separate, the differences between measured and predicted values, $\Delta a_\mu = a_\mu^{\rm exp} - a_\mu^{\rm SM}$, are found to be

$$\begin{aligned} \Delta a_\mu &= (32.8 \pm 4.7_{\rm had,LO} \pm 2.6_{\rm other} \pm 6.3_{\rm exp})\ 10^{-10} \quad [e^+e^-], \\ \Delta a_\mu &= (18.6 \pm 4.9_{\rm had,LO} \pm 2.6_{\rm other} \pm 6.3_{\rm exp})\ 10^{-10} \quad [\tau], \end{aligned} \tag{8.26}$$

where the first error quoted is specific to each approach, the second is due to contributions other than hadronic vacuum polarization, and the third is the BNL g-2 experimental error. The last two errors are identical in both evaluations. Adding all errors in quadrature, the differences in Eq. (8.26) correspond to 4.0 and 2.2 standard deviations, respectively. So both approaches yield a Standard Model prediction which deviates from the measurement. A graphical comparison of the results with the experimental value is given in Fig. 8.13. A word of caution is in order about the real meaning of "standard deviations" as the uncertainty in the theoretical prediction is dominated by systematic errors for which a gaussian distribution is questionable.

In principle the e^+e^--based estimate is the most direct one, while the τ approach for the dominant $I = 1$ contribution requires IB-breaking corrections in addition to the measurements. These corrections are better and better understood, so they just introduce one more systematic uncertainty included in the result. Otherwise the purely experimental uncertainties have very different origin in e^+e^- and τ data. Some inconsistencies remain between the two approaches, as seen in the respective spectral functions, however at a level not so different from the deviations observed between individual experiments of each type. The most conservative approach at this point is to average the e^+e^- and τ results, enlarging the final error to take into account the poor χ^2 of the average. In this way the error for the uncommon part is scaled by a factor 2.2 yielding,

$$\Delta a_\mu = (26.0 \pm 7.3_{\rm had,LO} \pm 2.6_{\rm other} \pm 6.3_{\rm exp})\ 10^{-10} \quad [e^+e^- \ \tau], \tag{8.27}$$

corresponding to 2.6 σ away from the Standard Model.

The apparent deviation from the Standard Model prediction is of great interest, even if it is not an overwhelming discrepancy. Ordinarily, one would not necessarily worry about a 2.6 σ effect. In fact, the proper response would be to improve the experimental measurement (which was statistics limited) and to continue to improve theory. With regard to the latter, new $e^+e^- \to$ hadrons data and further study of LBL could potentially reduce the overall theoretical uncertainty. The excitement caused by

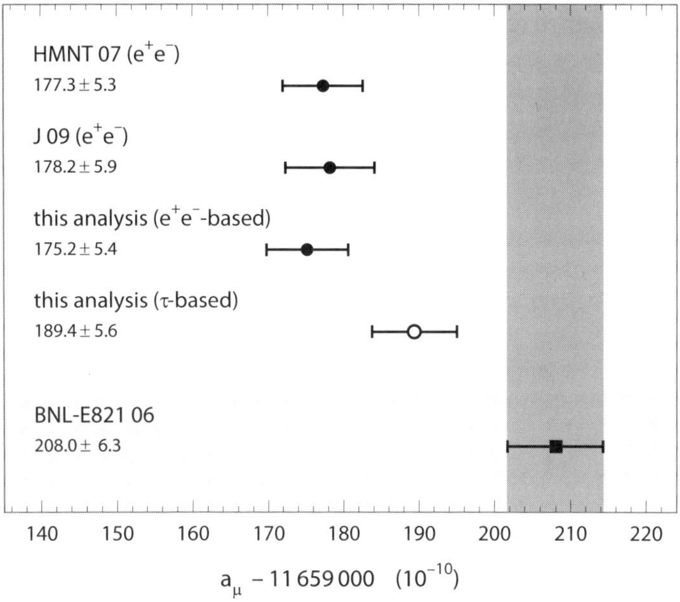

Fig. 8.13. Comparison of the theoretical estimate presented here with other recent analyses [52, 53] with the BNL measurement [58]. The subtracted value is arbitrary.

the deviation stems from the expectation that New Physics could cause a deviation of the magnitude observed in Eq. (8.26), such as extra contributions from supersymmetry.

8.10. Perspectives

It is important to assess the enormous progress accomplished recently in this field. The experiment E821 at Brookhaven has improved the determination of a_μ by about a factor of 14 relative to the classic CERN results of the 1970s [59]. The result is still statistics-limited and could be improved by another factor of 2 or so (to a precision of 30×10^{-11}) before systematics effects become a limitation. The experiment can be improved further and opportunities exist [60]. Pushing the experimental result to a new level of precision seems to be an obvious goal for the long term.

The theoretical prediction within the Standard Model should be concurrently improved. We now have three techniques to measure the $e^+e^- \to$ hadrons cross section: direct measurements, radiative return, and τ decays. More data are expected with all of them from VEPP-2000 in Novosibirsk,

KLOE and BaBar, BaBar and Belle, respectively. Progress is also continuously made in the development of cross-checked Monte Carlo programs, necessary to apply radiative corrections with an increased confidence. Isospin-breaking corrections to τ data should continue to improve. Finally, lattice gauge theories with dynamical fermions can in principle provide a determination from first principles of $a_\mu^{\rm had,LO}$. Some early attempts are underway [61].

Note added in proof

As this book was being finalized for press, a result from the BaBar collaboration became available [62]. It involves a precise measurement of the cross section of the process $e^+e^- \to \pi^+\pi^-(\gamma)$ from threshold to an energy of 3 GeV, obtained with the ISR method and using the measured ratio of the $\pi^+\pi^-\gamma(\gamma)$ to $\mu^+\mu^-\gamma(\gamma)$ yields. The leading-order hadronic contribution to a_μ calculated using the BaBar $\pi\pi(\gamma)$ cross section from threshold to 1.8 GeV is $(514.1 \pm 2.2(\text{stat}) \pm 3.1(\text{syst})) \times 10^{-10}$. This value is larger than the result from the combination of previous e^+e^- data [63] (503.5 ± 3.5), and in better agreement with the updated value from τ decay [63] (515.2 ± 3.4). The BaBar result has a precision comparable to that of the combined value from either e^+e^- or τ data. Analyses are in progress to optimally combine all the available data.

Acknowledgments

It is a pleasure to thank Bogdan Malaescu and Zhiqing Zhang for their help in preparing this review, and the many colleagues who contributed to this exciting field.

References

[1] N. Cabibbo, R. Gatto, Phys. Rev. **124** (1961) 1577.
[2] C. Bouchiat, L. Michel, *J. Phys. Radium* **22** (1961) 121.
[3] D. Hanneke, S. Fogwell, G. Gabrielse, *Phys. Rev. Lett.* **100** (2008) 120801.
[4] A.D. Martin, D. Zeppenfeld, *Phys. Lett.* **B345** (1995) 558.
[5] M. Davier, A. Höcker, *Phys. Lett.* **B419** (1998) 419.
[6] J.H. Kühn, M. Steinhauser, *Phys. Lett.* **B437** (1998) 425.
[7] S. Groote *et al.*, *Phys. Lett.* **B440** (1998) 375.
[8] M. Davier, A. Höcker, *Phys. Lett.* **B435** (1998) 427.
[9] R. Alemany, M. Davier, A. Höcker, *Eur.Phys.J.* **C2** (1998) 123.

[10] M. Davier, S. Eidelman, A. Höcker, Z. Zhang, *Eur.Phys.J.* **C31** (2003) 503.
[11] M. Davier, S. Eidelman, A. Höcker, Z. Zhang, *Eur.Phys.J.* **C27** (2003) 497.
[12] R. Akhmetshin *et al.* (CMD-2 Collaboration), *Phys. Lett.* **B 578** (2002) 161.
[13] M. Achasov *et al.* (SND Collaboration), *Zh. Eksp. Teor. Fiz.* **B 128** (2005) 1201.
[14] R. Akhmetshin *et al.* (CMD-2 Collaboration), *Phys. Lett.* **B 578** (2004) 285.
[15] M. Achasov *et al.* (SND Collaboration), *JETP Lett.* **103** (2006) 380.
[16] V. Aulchenko *et al.* (CMD-2 Collaboration), *JETP Lett.* **82** (2005) 743.
[17] R. Akhmetshin *et al.* (CMD-2 Collaboration), *JETP Lett.* **84** (2006) 413.
[18] R. Akhmetshin *et al.* (CMD-2 Collaboration), *Phys. Lett.* **B 648** (2007) 28.
[19] A. Höfer, J. Gluza, F. Jegerlehner, *Eur. Phys. J.* **C24** (2002) 51.
[20] V.N. Baier and V.S. Fadin, Phys. Lett. B27 (1968) 223; A.B. Arbuzov *et al.*, J. High Energy Phys. 9812 (1998) 009; S. Binner, J.H. Kühn and K. Melnikov, Phys. Lett. B459 (1999) 279; M. Benayoun *et al.*, Mod. Phys. Lett. A14 (1999) 2605.
[21] H. Czyż *et al.*, Eur. Phys. J. C35 (2004) 527; Eur. Phys. J. C39 (2005) 411.
[22] A. Aloisio *et al.*, (KLOE Collaboration), *Phys. Lett.* **B 606** (2005) 12.
[23] F. Ambrosino *et al.*, (KLOE Collaboration), *Phys. Lett.* **B 670** (2009) 285.
[24] B. Aubert *et al.* (BaBar Collaboration), Phys. Rev. **D70** (2004) 072004; Phys. Rev. **D71** (2005) 052001; Phys. Rev. **D73** (2006) 012005; Phys. Rev. **D73** (2006) 052003; Phys. Rev. **D76** (2007) 012008; Phys. Rev. **D76** (2007) 019005; Phys. Rev. **D76** (2008) 092006; Phys. Rev. **D77** (2008) 092002.
[25] R. Barate *et al.*, (ALEPH Collaboration), *Z. Phys.* **C76** (1997) 15.
[26] S. Schael *et al.*, (ALEPH Collaboration), *Phys. Rep.* **421** (2005) 191.
[27] S. Anderson *et al.* (CLEO Collaboration), *Phys.Rev.* **D61** (2000) 112002.
[28] K.W. Edwards *et al.* (CLEO Collaboration), *Phys.Rev.* **D61** (2000) 072003.
[29] K. Ackerstaff *et al.* (OPAL Collaboration), *Eur. Phys. J.* **C7** (1999) 571.
[30] M. Fujikawa *et al.* (Belle Collaboration), *Phys.Rev.* **D78** (2008) 072006.
[31] M. Davier, A. Höcker, and Z. Zhang, *Rev. Mod. Phys.* **78** (2006) 1043.
[32] P. Tsai, *Phys. Rev.* **D4** (1971) 2821.
[33] C. Amsler *et al.* (Particle Data Group), *Phys. Lett.* **B 667** (2008) 1.
[34] J. Charles *et al.* (CKMfitter Group), *Eur. Phys. J.* **C 41** (2005) 1; updates from http://ckmfitter.in2p3.fr.
[35] W.J. Marciano, A. Sirlin, *Phys. Rev. Lett.* **61** (1988) 1815.
[36] A. Sirlin, *Nucl. Phys.* **B196** (1982) 83.
[37] V. Cirigliano, G. Ecker, H. Neufeld, *Phys. Lett.* **B513** (2001) 361.
[38] V. Cirigliano, G. Ecker, H. Neufeld, *JHEP* **0208** (2002) 002.
[39] A. Flores-Tlalpa, F. Flores-Baez, G. Lopez Castro, and G. Toledo Sanchez, *Phys. Rev.* **D 74** (2006) 071301; *Nucl. Phys. Proc. Suppl.* **169** (2007) 250.
[40] A. Aloisio *et al.* (KLOE Collaboration), *Phys. Lett.* **B 561** (2003) 55.
[41] F. Flores-Baez, G. Lopez Castro, and G. Toledo Sanchez, *Phys. Rev.* **D 76** (2007) 096010.
[42] M. Artuso *et al.* (CLEO Collaboration), *Phys. Rev. Lett.* **72** (1994) 3762.
[43] P. Achard *et al.* (L3 Collaboration),
[44] K. Ackerstaff *et al.* (OPAL Collaboration), *Eur. Phys. J.* **C4** (1998) 93.
[45] J. Abdallah *et al.* (DELPHI Collaboration), *Eur. Phys. J.* **C46** (2006) 1.

[46] S.R. Amendolia *et al.* (NA7 Collaboration), *Nucl. Phys.* **B277** (1986) 168.
[47] L.R. Surguladze, M.A. Samuel, Phys. Rev. Lett. **66** (1991) 560;
S.G. Gorishny, K.L. Kataev, S.A. Larin, Phys. Lett. **B259** (1991) 144.
[48] P.A. Baikov, K.G. Chetyrkin, J.H. Kühn, *Nucl. Phys.* **B482** (1996) 213.
[49] K.G. Chetyrkin, J.H. Kühn, M. Steinhauser, Nucl. Phys. **B482** (1996) 213.
[50] M. Davier *et al.*, *Eur. Phys. J.* **C56** (2008) 305.
[51] https://twiki.cern.ch/twiki/bin/view/Gfitter/WebHome
[52] K. Hagiwara, A.D. Martin, D. Nomura, T. Teubner, *Phys. Lett.* **B649** (2007) 173.
[53] F. Jegerlehner, *Nucl. Phys. Proc. Suppl.* **181-182** (2008) 26.
[54] B. Krause, *Phys. Lett.* **B390** (1997) 392.
[55] F. Jegerlehner, A. Nyffeler, arXiv:0902.3360 (Feb. 2009); and references therein.
[56] M. Davier, W.J. Marciano, *Ann. Rev. Nucl. Part. Sc.* **54** (2004) 115; and references therein.
[57] J. Prades, E. de Rafael, A. Vainshtein, arXiv:0901.0306 (Jan. 2009); and references therein.
[58] G.W. Bennett *et al.* (Muon g-2 Collaboration), Phys. Rev. **D73** (2006) 072003.
[59] J. Bailey, *et al.*, *Phys. Lett.* **B68** (1977) 191; F.J.M. Farley, E. Picasso, *The muon (g-2) Experiments, Advanced Series on Directions in High Energy Physics* - Vol. 7 Quantum Electrodynamics, Ed. Kinoshita T, World Scientific (1990).
[60] B.L. Roberts, in *High Intensity Muon Sources*, Eds. Kuno Y, Yokoi T, World Scientific (1999) p69; D. Hertzog, *Nucl. Phys.Proc. Suppl.* **181-182** (2008) 5.
[61] C. Aubin, T. Blum, *Nucl. Phys.Proc. Suppl.* **181** (2006) 251.
[62] B. Aubert *et al.* (BaBar Collaboration), arXiv:0908.3589, submitted to Phys. Rev. Lett.
[63] M. Davier *et al.*, arXiv:0906.5443, sub. to Eur. Phys. J..

Chapter 9

The Hadronic Light-by-Light Scattering Contribution to the Muon and Electron Anomalous Magnetic Moments

Joaquim Prades

CAPFE and Departamento de Física Teórica y del Cosmos,
Universidad de Granada, Campus de Fuente Nueva
E-18002 Granada, Spain

Eduardo de Rafael

Centre de Physique Théorique
CNRS-Luminy Case 907
F-13288 Marseille Cedex 9, France

Arkady Vainshtein

William I. Fine Theoretical Physics Institute
University of Minnesota
Minneapolis, MN 55455, USA

We review the current status of theoretical calculations of the hadronic light-by-light scattering contribution to both the muon and electron anomalous magnetic moments. Different approaches and related issues such as OPE constraints and large breaking of chiral symmetry are discussed. Combining results of different models with educated guesses on the errors we come to the estimate $a_\mu^{\rm HLbL} = (10.5 \pm 2.6) \times 10^{-10}$, and $a_e^{\rm HLbL} = (3.5 \pm 1.0) \times 10^{-14}$.

Contents

9.1. Introduction

From a theoretical point of view the hadronic light-by-light scattering (HLbL) contribution to the muon magnetic moment is described by the vertex function (see Fig. 9.1 below):

$$\Gamma_\mu^{(H)}(p_2,p_1) = ie^6\int\frac{d^4k_1}{(2\pi)^4}\int\frac{d^4k_2}{(2\pi)^4}\frac{\Pi^{(H)}_{\mu\nu\rho\sigma}(q,k_1,k_3,k_2)}{k_1^2k_2^2k_3^2} \times \gamma^\nu(\not p_2+\not k_2-m_\mu)^{-1}\gamma^\rho(\not p_1-\not k_1-m_\mu)^{-1}\gamma^\sigma\,, \quad (9.1)$$

where m_μ is the muon mass and $\Pi^{(H)}_{\mu\nu\rho\sigma}(q,k_1,k_3,k_2)$, with $q = p_2 - p_1 = -k_1 - k_2 - k_3$, denotes the off-shell photon-photon scattering amplitude induced by hadrons,

$$\Pi^{(H)}_{\mu\nu\rho\sigma}(q,k_1,k_3,k_2)=\int d^4x_1\int d^4x_2\int d^4x_3\ \exp[-i(k_1\cdot x_1+k_2\cdot x_2+k_3\cdot x_3)] \times\langle 0|T\{j_\mu(0)\ j_\nu(x_1)\ j_\rho(x_2)\ j_\sigma(x_3)\}|0\rangle\,. \quad (9.2)$$

Here j_μ is the Standard Model electromagnetic current, $j_\mu(x) = \sum_q Q_q\bar q(x)\gamma_\mu q(x)$, where Q_q denotes the electric charge of quark q. The

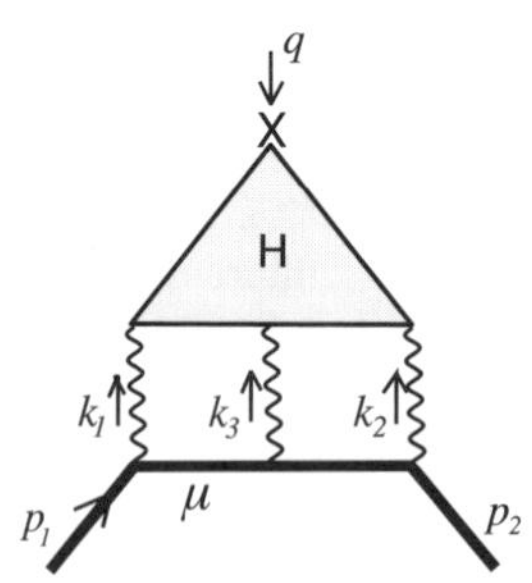

Fig. 9.1. Hadronic light-by-light scattering contribution.

external photon with momentum q represents the magnetic field. We are interested in the limit $q \to 0$ where the current conservation implies that $\Gamma_\mu^{(H)}$ is linear in q,

$$\Gamma_\mu^{(H)} = -\frac{a_\mu^{\rm HLbL}}{4m_\mu}\,[\gamma_\mu\,,\gamma_\nu]\,q^\nu\,. \quad (9.3)$$

The muon anomaly can then be extracted as follows:

$$a_\mu^{\rm HLbL} = \frac{-ie^6}{48m_\mu}\int\frac{d^4k_1}{(2\pi)^4}\int\frac{d^4k_2}{(2\pi)^4}\frac{1}{k_1^2k_2^2k_3^2}\left[\frac{\partial}{\partial q^\mu}\Pi^{(H)}_{\lambda\nu\rho\sigma}(q,k_1,k_3,k_2)\right]_{q=0}$$

$$\times\ {\rm tr}\left\{(\not{p}+m_\mu)[\gamma^\mu,\gamma^\lambda](\not{p}+m_\mu)\gamma^\nu\frac{1}{\not{p}+\not{k}_2-m_\mu}\gamma^\rho\frac{1}{\not{p}-\not{k}_1-m_\mu}\gamma^\sigma\right\}. \quad (9.4)$$

Unlike the case of the hadronic vacuum polarization (HVP) contribution, there is no direct experimental input for the hadronic light-by-light scattering (HLbL) so one has to rely on theoretical approaches. Let us start with the massive quark loop contribution which is known analytically,

$$a_\mu^{\rm HLbL}(\text{quark loop}) =$$
$$\left(\frac{\alpha}{\pi}\right)^3 N_c Q_q^4\left\{\underbrace{\left[\frac{3}{2}\zeta(3)-\frac{19}{16}\right]}_{0.62}\frac{m_\mu^2}{m_q^2}+\mathcal{O}\left[\frac{m_\mu^4}{m_q^4}\log^2\frac{m_\mu^2}{m_q^2}\right]\right\}, \quad (9.5)$$

where N_c is the number of colors and $m_q \gg m_\mu$ is implied. It gives a reliable result for the heavy quarks c, b, t with $m_q \gg \Lambda_{\rm QCD}$. Numerically, however, heavy quarks do not contribute much. For the c quark, with $m_c \approx 1.5$ GeV,

$$a_\mu^{\rm HLbL}(c) = 0.23\times 10^{-10}\,. \quad (9.6)$$

To get a very rough estimate for the light quarks u, d, s let us use a constituent mass of 300 MeV for m_q. This gives $a_\mu^{\rm HLbL}(u,d,s) = 6.4\times 10^{-10}$. QCD tells us that the quark loop should be accurate in describing large virtual momenta, $k_i \gg \Lambda_{\rm QCD}$, i.e. short-distances. What is certainly missing in this constituent quark loop estimate, however, is the low-momenta piece dominated by a neutral pion-exchange in the light-by-light scattering. Adding up this contribution, discussed in more detail below, approximately doubles the estimate to $a_\mu^{\rm HLbL} \approx 12\times 10^{-10}$. While the ballpark of the effect is given by this rough estimate, a more refined analysis is needed to get its magnitude and evaluate the accuracy. Details and comparison of different contributions will be discussed below, but it is already interesting to point out that all existing calculations fall into a range:

$$a_\mu^{\rm HLbL} = (11\pm 4)\times 10^{-10}\,, \quad (9.7)$$

compatible with this rough estimate. The dispersion of the $a_\mu^{\rm HLbL}$ results in the literature is not too bad when compared with the present experimental accuracy of 6.3×10^{-10}. However the proposed new g_μ−2 experiment sets a goal of 1.4×10^{-10} for the error, which calls for a considerable improvement in the theoretical calculations as well. We believe that theory is up to this challenge; a further use of theoretical and experimental constraints could result in reaching such accuracy soon enough.

The history of the evaluation of the hadronic light-by-light scattering contribution is a long one which can be found in the successive review articles on the subject. In fact, but for the sign error in the neutral pion exchange discovered in 2002 [1, 2], the theoretical predictions for $a_\mu^{\rm HLbL}$ have been relatively stable over more than ten years.

Here we are interested in highlighting the generic properties of QCD relevant to the evaluation of Eq. (9.4), as well as their connection with the most recent model-dependent estimates which have been made so far.

9.2. QCD in the Large N_c and Chiral Limits

For the light quark components in the electromagnetic current ($q = u\,, d\,, s$) the integration of the light-by-light scattering over virtual momenta in Eq. (9.4) is convergent at characteristic hadronic scales. We choose the mass of the ρ meson m_ρ to represent that scale. Of course, hadronic physics at such momenta is non-perturbative and the first question to address is what theoretical parameters can be used to define an expansion. Two possibilities are: The large number of colors, $1/N_c \ll 1$, and the smallness of the chiral symmetry breaking, $m_\pi^2/m_\rho^2 \ll 1$. Their relevance can be seen from the expansion of $a_\mu^{\rm HLbL}$ as a power series in these parameters,

$$a_\mu^{\rm HLbL} \sim \left(\frac{\alpha}{\pi}\right)^3 \frac{m_\mu^2}{m_\rho^2}\left[c_1\, N_c + c_2\,\frac{m_\rho^2}{m_\pi^2} + c_3 + \mathcal{O}(1/N_c)\right], \tag{9.8}$$

where $m_\pi > m_\mu$ is implied. Only the power dependencies are shown; possible chiral logarithms, $\ln(m_\rho/m_\pi)$, are included into the coefficients c_i.

9.2.1. *Terms leading in the large N_c limit*

The first term, linear in N_c, comes from the one-particle exchange of a meson M in the HLbL amplitude, see Fig. 9.2(a). In principle, the meson M is any neutral, C-even meson. In particular this includes pseudoscalar mesons π^0, η, η'; scalars f_0, a_0; vectors π_1^0; pseudovectors a_1^0, f_1, f_1^*; spin 2 tensor and pseudotensor mesons f_2, a_2, η_2, π_2 .

The neutral pion exchange is special because of the Goldstone nature of the pion; its mass is much smaller than the hadronic scale m_ρ. In $a_\mu^{\rm HLbL}(\pi^0)$

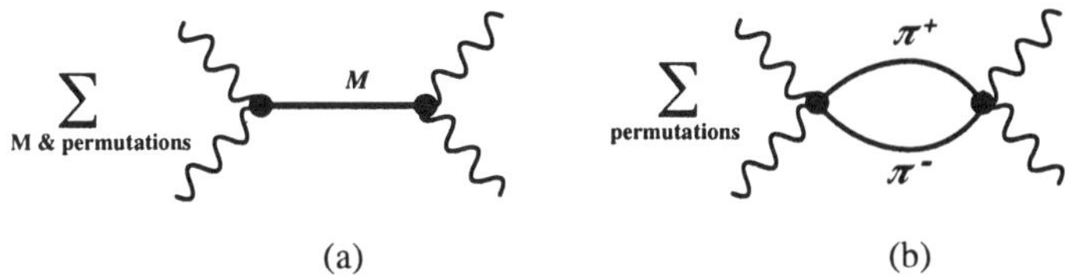

Fig. 9.2. Diagrams for HLbL: (a) meson exchanges, (b) the charged pion loop, the blob denotes the full $\gamma^*\gamma^* \to \pi^+\pi^-$ amplitude.

this leads to an additional enhancement by two powers of a chiral logarithm [2],

$$a_\mu^{\mathrm{HLbL}}(\pi^0) = \left(\frac{\alpha}{\pi}\right)^3 N_c \frac{m_\mu^2 N_c}{48\pi^2 F_\pi^2}\left[\ln^2\frac{m_\rho}{m_\pi} + \mathcal{O}\left(\ln\frac{m_\rho}{m_\pi}\right) + \mathcal{O}(1)\right]. \qquad (9.9)$$

Here the $\pi^0\gamma\gamma$ coupling is fixed by the Adler–Bell–Jackiw anomaly in terms of the pion decay constant $F_\pi \approx 92$ MeV. This constant is $\mathcal{O}\left(\sqrt{N_c}\right)$, therefore N_c/F_π^2 behaves as a constant in the large-N_c limit. The mass of the ρ plays the role of an ultraviolet scale in the integration over k_i in Eq. (9.4) while the pion mass provides the infrared scale. Of course, the muon mass is also important at low momenta but one can keep the ratio m_μ/m_π fixed in the chiral limit.

Equation (9.9) provides the result for a_μ^{HLbL} for the term leading in the $1/N_c$ expansion in the chiral limit where the pion mass is much less than the next hadronic scale. In this limit the dominant neutral pion exchange produces the characteristic universal double logarithmic behavior with the *exact* coefficient given in Eq. (9.9). Testing this limit was particularly useful in fixing the sign of the neutral pion exchange.

Although the coefficient of the $\ln^2(m_\rho/m_\pi)$ term in Eq. (9.9) is unambiguous, the coefficient of the $\ln(m_\rho/m_\pi)$ term depends on low-energy constants which are difficult to extract from experiment [2, 3] (they require a detailed knowledge of the $\pi^0 \to e^+e^-$ decay rate with inclusion of radiative corrections). Model-dependent estimates of the single logarithmic term as well as the constant term show that these terms are not suppressed. It means that we cannot rely on chiral perturbation theory and have to adopt a dynamical framework which takes into account explicitly the heavier meson exchanges as well.

Note that the overall sign of the pion exchange, for physical values of the masses, is much less model-dependent than the previous chiral perturbation theory analysis seems to imply. In fact, if the $\pi^0\gamma^*\gamma^*$ form factor does not change its sign in the Euclidean range of integration over k_i, the overall sign is fixed even without knowledge of the form factor. This implies the same positive sign without use of the chiral limit, i.e. the same sign for exchanges of heavier pseudoscalars, $J^{PC} = 0^{-+}$, where no large logarithms are present. Moreover, one can verify the same positive sign for exchanges by mesons with $J^{PC} = 1^{++}, 2^{-+}$ with an additional assumption about dominance of one of the form factors. Exchanges with $J^{PC} = 0^{++}, 1^{-+}, 2^{++}$ give, however, contributions with a negative sign to $a_\mu^{\rm HLbL}$ under similar assumptions, but they are much smaller.

9.2.2. *Next-to-leading terms in the large N_c limit*

Now let us turn to the next-to-leading terms in $1/N_c$ expansion. Generically these terms are due to two-particle exchanges in the HLbL amplitude, see the diagram in Fig. 9.2(b) with $\pi^+\pi^-$ substituted by any two meson states. What is specific about the charged pion loop is its strong chiral enhancement which is not just logarithmic but power-like in this case. In Eq. (9.8) it is reflected in the term $c_2\, m_\rho^2/m_\pi^2$. The point-like pion loop calculation which gives $a_\mu^{\rm HLbL}(\pi\pi) = -4.6 \times 10^{-10}$ corresponds to $c_2 = -0.065$. The rather small value of c_2 can be contrasted with the one of the coefficient c_1 which is not suppressed: $c_1 \approx 1.7$. As we will see the smallness of c_2 is related to the fact that chiral perturbation theory does not work in this case. To see that this is indeed what happens is sufficient to compare the point-like loop result with the model-dependent calculations where form factors are introduced. Two known results, $a_\mu^{\rm HLbL}(\pi\pi) = -(0.4 \pm 0.8) \times 10^{-10}$ [4, 5] and $a_\mu^{\rm HLbL}(\pi\pi) = -(1.9 \pm 0.5) \times 10^{-10}$ [7, 8], show a 100% deviation from the point-like number. It means that the bulk of the contribution does not come from small virtual momenta k_i and, therefore, chiral perturbation theory should not be applied. In other words, the term c_3 in Eq. (9.8) with no chiral enhancement is comparable with $c_2(m_\rho^2/m_\pi^2)$. It means that loops with heavier mesons should also be included.

Breaking of the chiral perturbation theory looks surprising at first sight. Indeed, the inverse chiral parameter $m_\rho^2/m_\pi^2 \approx 30$ is much larger than $N_c = 3$. What happens is that the leading terms in the chiral expansion are numerically suppressed, which makes chiral corrections governed not by m_π^2/m_ρ^2 but rather by $\approx 40\, m_\pi^2/m_\rho^2$. This can be checked analytically in the case of the HVP contribution to the muon anomaly. The charged pion loop is also enhanced in this case by a factor m_ρ^2/m_π^2 but the relative chiral correction due to the pion electromagnetic radius (evaluated with a cutoff at m_ρ^2 in the $\pi\pi$ spectral function) is $\sim 40\, m_\pi^2/m_\rho^2 \ln(m_\rho/2m_\pi)$. Of course, if the pion mass (together with the muon mass) would be, say, five times smaller than in our real world, the charged pion loop would dominate both in the HVP and the HLbL contributions to the muon anomalous magnetic moment.

In concluding this section, we see that the $1/N_c$ expansion works reasonably well, so one can use one-particle exchanges for the HLbL amplitude. On the other hand, chiral enhancement factors are unreliable, so we cannot limit ourselves to the lightest Goldstone-like states, and this is the case both for the leading and next-to-leading order in the $1/N_c$ expansion.

9.3. Short-Distance QCD Constraints

The most recent calculations of a_μ^{HLbL} in the literature [1, 6, 8, 9] are all compatible with the QCD chiral constraints and large-N_c limit discussed above. They all incorporate the π^0-exchange contribution modulated by $\pi^0\gamma^*\gamma^*$ form factors $\mathcal{F}(k_i^2, k_j^2)$, correctly normalized to the $\pi^0 \to \gamma\gamma$ decay width. They differ, however, in the shape of the form factors, originating in different assumptions: vector meson dominance (VMD) in a specific form of Hidden Gauge Symmetry (HGS) in Refs. [4–6]; a different form of VMD in the extended Nambu–Jona-Lasinio model (ENJL) in Ref. [7, 8]; large-N_c models in Refs. [1, 9]; and on whether or not they satisfy the particular operator product expansion (OPE) constraint discussed in Ref. [9], upon which we next comment.

Let us consider a specific kinematic configuration of the virtual photon momenta k_1, k_2, k_3 in the Euclidean domain. In the limit $q = 0$ these momenta form a triangle, $k_1+k_2+k_3 = 0$, and we consider the configuration where one side of the triangle is much shorter than the others, $k_1^2 \approx k_2^2 \gg k_3^2$. When $k_1^2 \approx k_2^2 \gg m_\rho^2$ we can apply the known operator product expansion for the product of two electromagnetic currents carrying hard moments k_1 and k_2,

$$\int d^4x_1 \int d^4x_2 \mathrm{e}^{-ik_1\cdot x_1 - ik_2\cdot x_2}\, j_\nu(x_1)\, j_\rho(x_2)$$
$$= \frac{2}{\hat{k}^2}\, \epsilon_{\nu\rho\delta\gamma} \hat{k}^\delta \int d^4z\, \mathrm{e}^{-ik_3\cdot z}\, j_5^\gamma(z) + \mathcal{O}\left(\frac{1}{\hat{k}^3}\right). \tag{9.10}$$

Here $j_5^\gamma = \sum_q Q_q^2 \bar{q}\gamma^\gamma\gamma_5 q$ is the axial current where different flavors are weighted by squares of their electric charges and $\hat{k} = (k_1 - k_2)/2 \approx k_1 \approx -k_2$. As illustrated in Fig. 9.3 this OPE reduces the HLbL amplitude, in the special kinematics under consideration, to the AVV triangle amplitude.

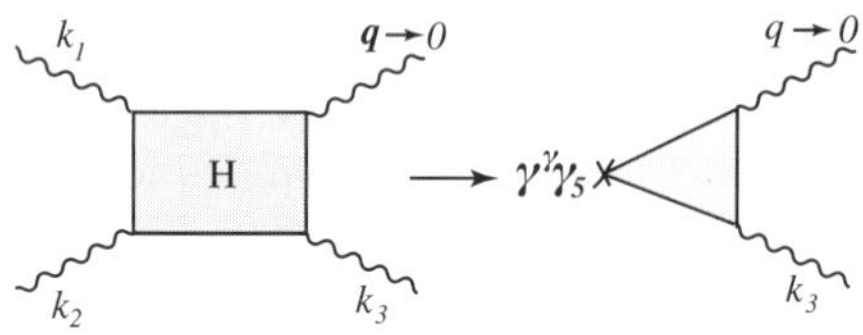

Fig. 9.3. OPE relation between the HLbL scattering and the AVV triangle amplitude.

There are a few things we can learn from the OPE relation in Eq. (9.10). The first one is that the pseudoscalar and pseudovector meson exchanges are dominant at large $k_{1,2}$. Indeed, only 0^- and 1^+ states are coupled to the axial current. It also provides the asymptotic behavior of form factors at large $k_1^2 \approx k_2^2$. In particular, we see that the $\pi^0\gamma^*\gamma^*$ form factor $\mathcal{F}(k^2,k^2)$ goes as $1/k^2$ and similar asymptotics hold for the axial-vector couplings. The relation in Eq. (9.10) does not imply that other mesons, for example scalars, do not contribute to HLbL, it is just that their $\gamma^*\gamma^*$ form factors should fall off faster at large $k_{1,2}^2$.

The AVV triangle amplitude consists of two parts: The anomalous, longitudinal part and the non-anomalous, transverse one; we consider the chiral limit where $m_\pi^2 \to 0$. Because of the absence of both perturbative and non-perturbative corrections to the anomalous AVV triangle graph in the chiral limit, the pion pole description for the isovector part of the axial current works at all values of k_3^2 connecting regions of soft and hard virtual momenta. This, in particular, implies the absence of a form factor $\mathcal{F}(0,k_3^2)$ in the vertex which contains the external magnetic field. At first sight, this conclusion seems somewhat puzzling because for non-vanishing external momentum q the form factor $\mathcal{F}(q^2,k_3^2)$ certainly is attributed to the pion exchange. The answer is provided by the observation that this form factor enters not in the longitudinal anomalous part, but in the transverse part. It is for this reason that the axial anomaly is not corrected by the form factor. In the transverse part the form factor shows up together with the massless pion pole in the form

$$\frac{\mathcal{F}(q^2,k_3^2) - \mathcal{F}(0,0)}{(k_3+q)^2}\,. \tag{9.11}$$

At $q = 0$ this combination contains no pion pole at $k_3^2 = 0$. It means that the discussed piece conspires with the pseudovector exchange to produce the transverse result and in this sense becomes part of what could be called the pseudovector exchange. It provides the leading short-distance constraint for the pseudovector exchange. Contrary to the case of the longitudinal component, the transverse, non-anomalous part of the AVV triangle is, however, corrected non-perturbatively [10, 11].

Additional constraints on subleading terms in the $\mathcal{F}(k_i^2, k_j^2)$ form factor, which were derived in Ref. [12], are also taken into account in the calculation quoted in Ref. [9].

The large momentum behavior which singles out pseudoscalar and pseudovector exchanges is, however, not sufficient to fix *per se* a unique model for the evaluation of a_μ^{HLbL} because the bulk of the integral in Eq. (9.4) comes from momenta k_i of the order of an hadronic scale. However, the faster decreasing of exchanges other than pseudoscalar and pseudovector ones makes these contributions numerically smaller. Moreover, the importance of asymmetric momenta configurations with two momenta much larger than the third one was checked in [9, 13] numerically. This check is related to a question which we next discuss.

There are other short-distance constraints than those associated with the particular kinematic configuration governed by the AVV triangle. At present, none of the light-by-light hadronic parameterizations made so far in the literature can claim to satisfy fully all the QCD short-distance properties of the HLbL amplitude which is needed for the evaluation of Eq. (9.4). In fact, within the large-N_c framework, it has been shown [14] that, in general, for other than two-point functions and two-point functions with soft insertions, this requires the inclusion of an infinite number of narrow states. However, a numerical dominance of certain momenta configuration could help. In particular, in the model of Ref. [9] with a minimal set of pseudoscalar and pseudovector exchanges, the corrections due to additional constraints not satisfied in the model turn out to be quite small numerically. Note that in the frameworks of the ENJL model [7, 8] the QCD short-distance constraints are accounted for by adding up the quark loop with virtual momenta larger than the cutoff scale of the model.

9.4. Hadronic Model Calculations

In the previous section we have mentioned a few models used for the calculations of a_μ^{HLbL}: HGS model in [4–6], ENJL model in [7, 8], the pseudoscalar exchange only in [1], the OPE-based model of pseudoscalar and pseudovector exchanges in [9]. In order to compare different results it is convenient to separate the hadronic light-by-light contributions which are leading in the $1/N_c$-expansion from the non-leading ones [15].

9.4.1. *Contributions leading in the $1/N_c$ expansion*

Among these contributions, the pseudoscalar meson exchanges which incorporate the π^0, and to a lesser degree the η and η' exchanges, are the dominant ones. As discussed above, there are good QCD theoretical reasons for that. In spite of the different definitions of the pseudoscalar meson exchanges and the associated choices of the $\mathcal{F}(k_i^2, k_j^2)$ form factors used in the various model calculations, there is a reasonable agreement among the final results, which we reproduce in Table 9.1.

Table 9.1. Contribution to $a_\mu^{\rm HLbL}$ from π^0, η and η' exchanges.

Result	Reference
$(8.5 \pm 1.3) \times 10^{-10}$	[7, 8]
$(8.3 \pm 0.6) \times 10^{-10}$	[4–6]
$(8.3 \pm 1.2) \times 10^{-10}$	[1]
$(11.4 \pm 1.0) \times 10^{-10}$	[9]

In fact, the agreement is better than this table shows. One should keep in mind that in the ENJL model (the first line) the momenta higher than a certain cutoff are accounted separately via quark loops while in the OPE based model these momenta are already included into the result (the last line in the Table 9.1). Assuming that the bulk of the quark loop contribution is associated with the pseudoscalar exchange channel one gets 10.7×10^{-10} in the ENJL model instead of 8.5×10^{-10}. In the calculations quoted in the two other entries, the higher momenta were suppressed by an extra form factor in the soft photon vertex and no separate contribution was added to compensate for this.

Closely related to pseudoscalar exchanges is the exchange by the pseudovectors. Both enter the axial-vector current implying relations between form factors (see the discussion of the triangle amplitude in the previous section). Again, here the estimates in the literature differ by the shape of the form factors used for the $A\gamma^*\gamma^*$ and $A\gamma^*\gamma$ vertex. Different assumptions

Table 9.2. Contribution to a_μ^{HLbL} from axial-vector exchanges.

Result	**Reference**
$(0.25 \pm 0.10) \times 10^{-10}$	[7, 8]
$(0.17 \pm 0.10) \times 10^{-10}$	[4–6]
$(2.2 \pm 0.5) \times 10^{-10}$	[9]

on hadronic mixing is another source of uncertainty. Although the contribution from axial-vector exchanges is found to be much smaller than the one from the Goldstone-like exchanges by all the authors, the central values, shown in Table 9.2, differ quite a lot. The authors of Ref. [9] attribute this to the influence of the OPE constraint for the non-anomalous part of the AVV triangle amplitude, discussed above. Further study of the discrepancy in this channel is certainly needed.

The scalar exchange contributions have only been taken into account in Refs. [7, 8]. In fact, within the framework of the ENJL model, these contributions are somewhat related to the constituent quark loop contribution. The result is:

$$-(0.7 \pm 0.2) \times 10^{-10}\,. \tag{9.12}$$

It is much smaller than the contribution from the Goldstone-like exchanges and negative. In comparison with the pseudovector exchange, the magnitude for the scalar is a few times smaller than for the pseudovector in the OPE-based model but a few times larger in HGS and ENJL models.

As we discussed in section 9.2, there is some number of other C-even mesonic resonances in the mass interval 1–2 GeV, not accounted for in the ENJL model, which could contribute to a_μ^{HLbL} comparably to the contribution from scalars. These contributions are of both signs depending on quantum numbers. At the moment we can only guess about their total effect. Thus, it seems reasonable to use the scalar exchange result rather as an estimate of error associated with these numerous contributions.

9.4.2. *Contributions subleading in the $1/N_c$ expansion*

As we discussed in section 9.2, the charge pion loop chirally enhanced as m_ρ^2/m_π^2 is a priori the dominant contribution in the subleading $1/N_c$ order. It occurs, however, that the chiral enhancement does not work and

Table 9.3. *Contribution to* $a_\mu^{\rm HLbL}$ *from a dressed pion loop.*

Result	Reference
$-(0.45 \pm 0.85) \times 10^{-10}$	[4, 5]
$-(1.9 \pm 0.5) \times 10^{-10}$	[7, 8]
$(0 \pm 1) \times 10^{-10}$	[9]

loops involving other heavier mesons can compete with the simple pion loop contribution.

The dressed pion loop results are considerably smaller than the one for the point-like pion. They are presented in Table 9.3. The last line from Ref. [9] is not the result of a calculation. Strictly speaking it represents an error estimate of the meson loop contributions subleading in $1/N_c$-expansion. One can probably increase this error to cover the ENJL result in the second line.

9.5. Numerical Conclusions and Outlook

What final result can one give at present for the hadronic light-by-light contribution to the muon anomalous magnetic moment? It seems to us that, from the above considerations, it is fair to proceed as follows:

(1) *Contribution to* $a_\mu^{\rm HLbL}$ *from* π^0, η *and* η' *exchanges*
Because of the effect of the OPE constraint discussed above, we suggest to take as central value the result of Ref. [9] with, however, the largest error quoted in Refs. [7, 8]:

$$a_\mu^{\rm HLbL}(\pi\,,\eta\,,\eta') = (11.4 \pm 1.3) \times 10^{-10}\,. \qquad (9.13)$$

Let us recall this central value is quite close to the one in the ENJL model when the short-distance quark loop contribution is added there.

(2) *Contribution to* $a_\mu^{\rm HLbL}$ *from pseudovector exchanges*
The analysis made in Ref. [9] suggests that the errors in the first and second entries of Table 9.2 are likely to be underestimates. Raising their ± 0.10 errors to ± 1 puts the three numbers in agreement within one sigma. We suggest then as the best estimate at present

$$a_\mu^{\rm HLbL}(\text{pseudovectors}) = (1.5 \pm 1) \times 10^{-10}\,. \qquad (9.14)$$

(3) *Contribution to $a_\mu^{\rm HLbL}$ from scalar exchanges*
The ENJL model should give a good estimate for these contributions. We keep, therefore, the result of Ref. [7, 8] with, however, a larger error which covers the effect of other unaccounted meson exchanges,

$$a_\mu^{\rm HLbL}({\rm scalars}) = -(0.7 \pm 0.7) \times 10^{-10} \,. \tag{9.15}$$

(4) *Contribution to $a_\mu^{\rm HLbL}$ from a dressed pion loop*
Because of the instability of the results for the charged pion loop and unaccounted loops of other mesons, we suggest using the central value of the ENJL result but with a larger error:

$$a_\mu^{\rm HLbL}(\pi\text{–dressed loop}) = -(1.9 \pm 1.9) \times 10^{-10} \,. \tag{9.16}$$

From these considerations, adding the errors in quadrature, as well as the small charm contribution in Eq. (9.6), we get

$$a_\mu^{\rm HLbL} = (10.5 \pm 2.6) \times 10^{-10} \tag{9.17}$$

as our final estimate.

We wish to emphasize, however, that this is only what we consider to be our best estimate at present. In view of the proposed new g_μ–2 experiment, it would be nice to have more independent calculations in order to make this estimate more robust. More experimental information on the decays $\pi^0 \to \gamma\gamma^*$, $\pi^0 \to \gamma^*\gamma^*$ and $\pi^0 \to e^+e^-$ (with radiative corrections included) could also help to confirm the result of the main contribution in Eq. (9.13).

More theoretical work is certainly needed for a better understanding of the other contributions which, although smaller than the one from pseudoscalar exchanges, have nevertheless large uncertainties. This refers, in particular, to pseudovector exchanges in Eq. (9.14) but other C-even exchanges are also important. Experimental data on radiative decays and two-photon production of C-even resonances could be helpful. An evaluation of $1/N_c$-suppressed loop contributions present even a more difficult task. New approaches to the dressed pion loop contribution, in parallel with experimental information on the vertex $\pi^+\pi^-\gamma^*\gamma^*$, would be very welcome. Again, measurement of the two-photon processes like $e^+e^- \to e^+e^-\pi^+\pi^-$ could give some information on that vertex and help to reduce the model dependence and therefore the present uncertainty in Eq. (9.16).

9.6. Hadronic L-B-L Contribution to the Electron Anomaly

In view of the remarkable accuracy in the experimental determination of the anomalous magnetic moment of the electron a_e [16], and prospects for its future improvement, the question about the size of the contribution from the hadronic light-by-light scattering (HLbL) to a_e becomes a relevant issue.[a] The model dependence of the HLbL contribution is the main source of the theoretical uncertainty for both a_e and a_μ. Once the model is fixed, the results for $a_l^{\rm HLbL}$ ($l = \mu, e$) are given by an integral with a well known kernel.

For hadronic exchanges with mass scales much larger than m_μ, the simple scaling: $a_l^{\rm HLbL} \propto m_l^2$ applies. Deviations from this simple scaling are particularly important for the neutral pion exchange where the pion mass is not that different from the muon one. Accounting for the lepton mass leads to the following modification of the double logarithmic term in Eq. (9.9):

$$\ln^2 \frac{m_\rho}{m_\pi} \longrightarrow \ln^2 \frac{m_\rho}{m_\pi} - \frac{m_l^2}{m_\pi^2 - m_l^2} \ln \frac{m_\pi}{m_l} \left(2 \ln \frac{m_\rho}{m_\pi} + \ln \frac{m_\pi}{m_l} \right). \tag{9.18}$$

Numerically, the second term in the case of the muon diminishes the leading $\ln^2$-term by almost 50%. As a result, the neutral pion exchange contribution to a_e becomes enhanced with respect to the simple scaling. This enhancement, however, does not apply to non-logarithmic terms in the pion exchange, neither to the other hadronic contributions, and the simple m_l^2 scaling can be applied there. Altogether we have the two contributions, shown numerically in parentheses,

$$a_e^{\rm HLbL} = a_e^\pi[\ln^2](2.2\times 10^{-14}) + \frac{m_e^2}{m_\mu^2}\,[a_\mu^{\rm HLbL} - a_\mu^\pi(\ln^2 \text{corrected})](1.3\times 10^{-14}), \tag{9.19}$$

and with the same relative error as for $a_\mu^{\rm HLbL}$ we get

$$a_e^{\rm HLbL} = (3.5 \pm 1.0) \times 10^{-14}. \tag{9.20}$$

Acknowledgments

EdeR is grateful to Marc Knecht for very helpful discussions. AV is thankful to H. Leutwyler, K. Melnikov and A. Nyffeler for helpful discussions.

[a]See Chapters 5 and 6.

The work of JP and EdeR has been supported in part by the EU RTN network FLAVIAnet [Contract No. MRTN-CT-2006-035482]. Work by JP has also been supported by MICINN, Spain [Grants No. FPA2006-05294 and Consolider-Ingenio 2010 CSD2007-00042 -CPAN-] and by Junta de Andalucía [Grants No. P05-FQM 101, P05-FQM 467 and P07-FQM 03048]. The work of AV has been supported in part by DOE grant DE-FG02-94ER408.

References

[1] M. Knecht and A. Nyffeler, Phys. Rev. D **65** (2002) 073034.
[2] M. Knecht, A. Nyffeler, M. Perrottet and E. de Rafael, Phys. Rev. Lett. **88** (2002) 071802.
[3] M. Ramsey-Musolf and M. B. Wise, Phys. Rev. Lett. **89** (2002) 041601.
[4] M. Hayakawa, T. Kinoshita and A.I. Sanda, Phys. Rev. Lett. **75** (1995) 790; Phys. Rev. D **54** (1996) 3137.
[5] M. Hayakawa and T. Kinoshita, Phys. Rev. D **57** (1998) 465; Phys. Rev. D **66** (2002) 073034 (Erratum).
[6] M. Hayakawa and T. Kinoshita, Phys. Rev. D **66** (2002) 073034 (Erratum).
[7] J. Bijnens, E. Pallante and J. Prades, Nucl. Phys. B **474** (1996) 379; Phys. Rev. Lett. **75** (1995) 1447; Erratum-*ibid.* **75** (1995) 3781.
[8] J. Bijnens, E. Pallante and J. Prades, Nucl. Phys. B **626** (2002) 410.
[9] K. Melnikov and A. Vainshtein, Phys. Rev. D **70** (2004) 113006.
[10] A. Vainshtein, Phys. Lett. B **569** (2003) 187.
[11] M. Knecht, Ş. Peris, M. Perrottet and E. de Rafael, JHEP **0403** (2004) 035.
[12] V.A. Novikov, M.A. Shifman, A.I. Vainshtein, M.B. Voloshin and V.I. Zakharov, Nucl. Phys. B **237** (1984) 525.
[13] J. Bijnens and J. Prades, Mod. Phys. Lett. A **22** (2007) 767.
[14] J. Bijnens, E. Gamiz, E. Lipartia, and J. Prades, JHEP **0304** (2003) 055.
[15] E. de Rafael, Phys. Lett. B **322** (1994) 239.
[16] D. Hanneke, S. Fogwell, and G. Gabrielse, Phys. Rev. Lett. **100**, 120801, (2008).

Chapter 10

General Prescriptions for One-loop Contributions to $a_{e,\mu}$

Kevin R. Lynch

Department of Physics, Boston University
590 Commonwealth Ave
Boston, MA, 02215, USA
krlynch@bu.edu

We derive general expressions at one-loop order for the anomalous magnetic moments of fundamental, charged Dirac fermions. In particular, we provide the expressions for charged and neutral scalar and charged and neutral gauge boson contributions with general scalar, pseudoscalar, vector and axial couplings to the fermion of interest. Our expressions reproduce the Standard Model electroweak contributions to a_μ yet are flexible enough to allow one to handle many scenarios of New Physics beyond the Standard Model.

Contents

10.1. Introduction

The recent measurement of the anomalous magnetic moment of the muon, a_μ, by the Brookhaven E821 Collaboration [1], shows a possibly significant discrepancy between experimental measurement and the Standard Model theoretical expectation. This looming discrepancy has generated substantial new interest in improved theoretical calculations, as well as renewed pedagogical interest.

In this chapter we present general expressions at one-loop order for contributions to the anomalous magnetic moment of fundamental, charged Dirac fermions. In particular, we derive the contributions made by scalar and vector bosons with arbitrary couplings to the fermion of interest. We have explicitly allowed general couplings, such as the possibility of tree-level flavor changing couplings. We do not, however, allow for non-standard couplings to the photon.

Other authors have presented similar results before; in particular we should cite the first calculations of the weak contributions [2–6], which include the first calculation in general renormalizable gauges [3] and the first calculation for general gauge models [7]. The main difference is that this chapter takes a pedagogical approach, in the hope that it will be helpful to those learning the techniques necessary for these types of calculations.

10.2. The Photon-Fermion Vertex Function

The Feynman diagram for fermion-photon scattering is shown in Fig. 10.1. The amplitude for leptons scattering off a static background field, $\tilde{A}^{\rm cl}_\mu(q)$,

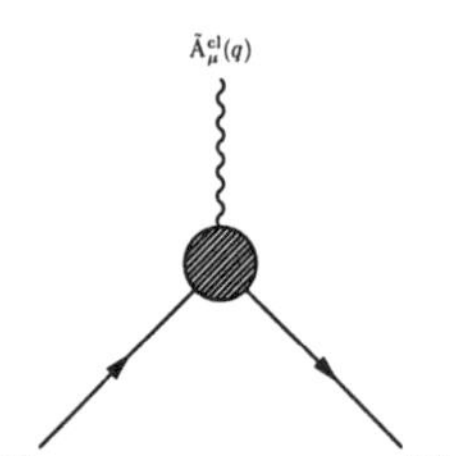

Fig. 10.1. A Feynman diagram cartoon for fermion scattering from a static, classical background of photons, $\tilde{A}^{\rm cl}_\mu(q)$. The shaded circle is a stand-in for the full vertex function.

is given by

$$i\mathcal{M} = -ieQ_\ell \tilde{A}^{\rm cl}_\mu(q)\overline{\rm u}(p')\Gamma^\mu(q^2){\rm u}(p), \tag{10.1}$$

where Lorentz covariance restricts the vertex operator Γ^μ to be a function of q, γ^μ, and γ^5 only. One conventional combination is written

$$\Gamma^\mu(q^2) = F_1(q^2)\gamma^\mu + F_2(q^2)\frac{i\sigma^{\mu\nu}q_\nu}{2m_\ell} + F_3(q^2)\gamma^5\frac{\sigma^{\mu\nu}q_\nu}{2m_\ell}, \tag{10.2}$$

where the last term, which is associated with a permanent electric dipole moment (EDM), will be ignored in the subsequent discussion. The terms are chosen in this way because, in the limit $q^2 \to 0$ (i.e. when external particles are put on shell), the functions F_i correspond to classical definitions of the electric charge, $F_1(0)$, and the anomalous magnetic moment, $F_2(0)$, of the fermion. We are fortunate to find that all of the integrals in the one-loop diagrams that give rise $F_2(0)$ are convergent, so we will not need to involve ourselves at all in the renormalization program; we will say no more here, beyond noting that $F_1(0)$ is constrained to be one by the QED renormalization conditions. At higher loop orders, of course, this is no longer the case. These observations will simplify the pedagogical task.

To get the notation straight, let us calculate the F_i at tree level in QED. The amplitude for an electron scattering from a static photon background is given by

$$i\mathcal{M} = \overline{\rm u}(p')\left(-ieQ_\ell\gamma^\mu\right)\tilde{A}^{\rm cl}_\mu(q){\rm u}(p).$$

Clearly, when we rearrange this as in Eq. (10.1), we find $F_1(q^2) = 1$ and $F_2(q^2) = 0$.

In this notation, the anomalous magnetic moment of the fermion f corresponds to

$$a_{\rm f} = \frac{g_{\rm f} - 2}{2} = F_2(0),$$

which can be compared directly to the results of experiments which measure the anomaly.

To make a prediction for $a_{\rm f}$ in a concrete model, our program is clear: perform a loop expansion of the photon vertex operator in the model, and map the results onto Eq. (10.2) to extract the form factor $F_2(q^2)$. Below, we will perform this expansion at one loop for four generalized models: For neutral and charged scalars with general scalar and pseudoscalar couplings, and for neutral and charged vectors with general vector and axial vector couplings. We will then find the one-loop contributions of the Standard Model electroweak sector to the anomaly of the muon.

10.3. Scalar Boson Contributions

There are two types of one-loop diagrams including scalars that contribute to a_{f}, both shown in Fig. 10.2. The diagram in Fig. 10.2(a), the "neutral diagram", can be populated either by electrically neutral or charged scalars; the diagram in Fig. 10.2(b), the "charged diagram", is only available for electrically charged bosons.

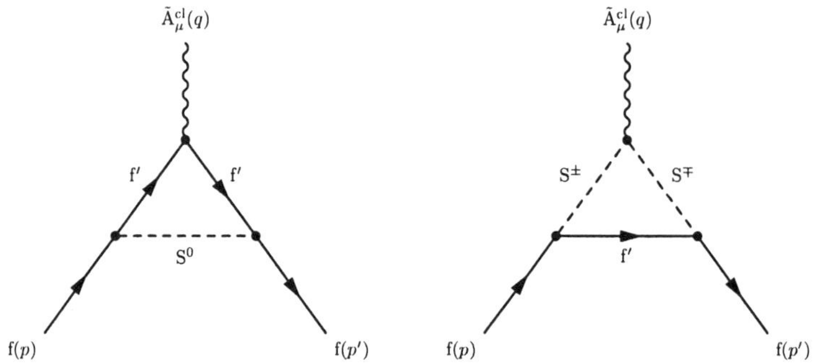

(a) Neutral/Charged Scalars (b) Charged Scalars

Fig. 10.2. One-loop Feynman diagrams for scalar boson contributions to a_{f}. The left-hand diagram can occur for either electrically neutral or charged scalars, while the right-hand diagram can only occur for electrically charged scalars.

For either type of scalar, the most general (potentially flavor changing) coupling between the scalar and two fermions, $\mathrm{f} \to \mathrm{f}'\mathrm{S}$, gives a Feynman Rule, including both scalar and pseudoscalar terms, $ig(\mathfrak{s} + \mathfrak{p}\gamma^5)$.

10.3.1. *Neutral scalar diagram*

Using the kinematic conventions in Fig. 10.3, it is straightforward to derive the amplitude for the "neutral" scalar diagram (which, again, can contain either neutral or charged scalar participants) of Fig. 10.2(a):

$$i\mathcal{M} = \int \frac{\mathrm{d}^4 k}{(2\pi)^4} \bar{\mathrm{u}}(p') ig\left(\mathfrak{s} + \mathfrak{p}\gamma^5\right) \frac{i\left(\not{k}' + m_{\rm int}\right)}{k'^2 - m_{\rm int}^2} ieQ_{\rm int}\gamma^\mu \tilde{A}^{\rm cl}_\mu(q) \times \\ \frac{i\left(\not{k} + m_{\rm int}\right)}{k^2 - m_{\rm int}^2} ig\left(\mathfrak{s}^\dagger - \mathfrak{p}^\dagger\gamma^5\right) \frac{i}{(p-k)^2 - M^2} \mathrm{u}(p),$$

where $m_{\rm int}$ (the "internal" fermion mass) corresponds to fermion f′ while $m_{\rm ext}$ (the "external" fermion mass) corresponds to f in our Feynman rule.

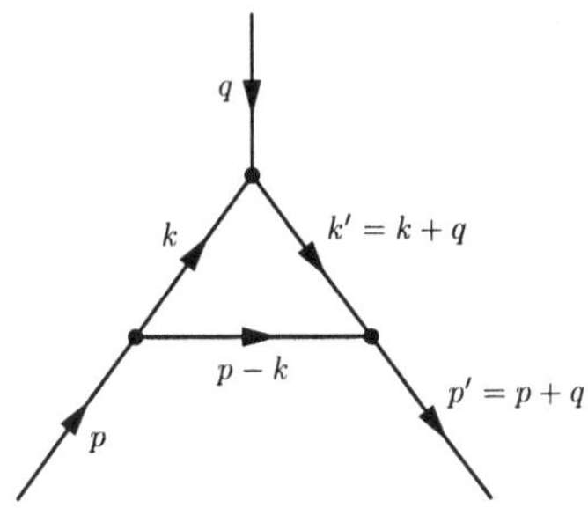

Fig. 10.3. The definitions and directionality of our momentum variables.

We can simplify this greatly in a few steps. For a denominator with three distinct terms, Feynman parametrization gives

$$\mathcal{I} = \frac{1}{ABC} = \int_0^1 \mathrm{d}x\mathrm{d}y\mathrm{d}z\delta(1-x-y-z)\frac{2}{(xA+yB+zC)^3} = \int_0^1 \mathrm{d}x\mathrm{d}y\mathrm{d}z\delta(1-x-y-z)\frac{2}{D^3},$$

where

$$\begin{aligned} D &= x\left(k^2 - m_{\text{int}}^2\right) + y\left(k'^2 - m_{\text{int}}^2\right) + z\left((p-k)^2 - M^2\right) \\ &= (x+y+z)k^2 - (x+y)m_{\text{int}}^2 - zM^2 + zp^2 + yq^2 + 2k(yq - zp) \end{aligned}$$

substituting $p^2 = m_{\text{ext}}^2$, and completing the square, we find

$$= (k + yq - zp)^2 - (yq - zp)^2 + yq^2 - (x+y)m_{\text{int}}^2 - zM^2 + zm_{\text{ext}}^2.$$

If we combine the second and third terms of the last equation, we find

$$\begin{aligned} yq^2 - (yq - zp)^2 &= yq^2 - y^2q^2 + 2yzqp - z^2p^2 \\ &= xyq^2 - z^2m_{\text{ext}}^2 \end{aligned}$$

which gives

$$\begin{aligned} D &= (k + yq - zp)^2 + xyq^2 - z^2m_{\text{ext}}^2 + zm_{\text{ext}}^2 - zM^2 - (x+y)m_{\text{int}}^2 \\ &= \ell^2 - \Delta + xyq^2, \end{aligned}$$

where we have defined

$$\begin{aligned} \ell &= k + yq - zp \\ \Delta &= z(z-1)m_{\text{ext}}^2 + zM^2 + (x+y)m_{\text{int}}^2. \end{aligned}$$

We can choose a different parametrization, $u = x + y$ and $v = x - y$, where Δ simplifies to

$$\Delta = u(u-1)m_{\text{ext}}^2 + (1-u)M^2 + um_{\text{int}}^2. \tag{10.3}$$

With this, the amplitude will be an integral over u alone. Therefore, it makes sense for us to change variables throughout. In particular,

$$\begin{aligned}\mathcal{I} &= \int_0^1 \mathrm{d}x \int_0^1 \mathrm{d}y \int_0^1 \mathrm{d}z \delta(1-x-y-z) f(x+y) \\ &= \int_0^1 \mathrm{d}x \int_0^{1-x} \mathrm{d}y f(x+y) = \int_0^1 \mathrm{d}u \int_{-u}^{u} \mathrm{d}v f(u)|J|,\end{aligned}$$

where the Jacobian determinant, J, is given by

$$J(u,v) = \begin{vmatrix} \frac{\partial u}{\partial x} & \frac{\partial u}{\partial y} \\ \frac{\partial v}{\partial x} & \frac{\partial v}{\partial y} \end{vmatrix} = -\frac{1}{2}.$$

We find

$$\mathcal{I} = \int_0^1 \mathrm{d}u u f(u),$$

which vastly simplifies our situation.

We can now return to extracting the anomaly contribution from the amplitude. We must match the simplified amplitude to Eq. (10.2). Since the anomaly is contained in the coefficients of the $\sigma^{\mu\nu}q_\nu$ terms, we are free to drop any contributions which do not match this pattern, once we have made them manifest. The Gordon identity

$$\overline{\mathrm{u}}(p')\gamma^\mu \mathrm{u}(p) = \frac{1}{2m}\overline{\mathrm{u}}(p')\Big((p'+p)^\mu + i\sigma^{\mu\nu}q_\nu\Big)\mathrm{u}(p), \tag{10.4}$$

allows us to exchange momenta in the numerator for the required gamma matrices. Suppressing the spinors we simplify the numerator

$$\begin{aligned}&\left(\mathfrak{s} + \mathfrak{p}\gamma^5\right)\left(\not{k}' + m_{\text{int}}\right)\gamma^\mu\left(\not{k} + m_{\text{int}}\right)\left(\mathfrak{s}^\dagger - \mathfrak{p}^\dagger\gamma^5\right) = \\ &\qquad \mathfrak{s}\mathfrak{s}^\dagger\left(\not{k}' + m_{\text{int}}\right)\gamma^\mu\left(\not{k} + m_{\text{int}}\right) + \mathfrak{p}\mathfrak{p}^\dagger\left(\not{k}' - m_{\text{int}}\right)\gamma^\mu\left(\not{k} - m_{\text{int}}\right).\end{aligned}$$

Applying the Gordon identity and the equations of motion to

$$\not{k}'\gamma^\mu\not{k} \pm m_{\text{int}}\left(\not{k}'\gamma^\mu + \gamma^\mu\not{k}\right)$$

we find

$$\frac{i\sigma^{\mu\nu}q_\nu}{2m_{\text{ext}}}2m_{\text{ext}}\Big[-m_{\text{ext}}u(u-1) \pm m_{\text{int}}u\Big].$$

Thus, the remaining pieces contain only the following contributions to the numerator:

$$\overline{\mathrm{u}}(p')\frac{i\sigma^{\mu\nu}q_\nu}{2m_{\rm ext}}2m_{\rm ext}\Big[-m_{\rm ext}u(u-1)\left(\mathfrak{s}\mathfrak{s}^\dagger+\mathfrak{p}\mathfrak{p}^\dagger\right)+m_{\rm int}u\left(\mathfrak{s}\mathfrak{s}^\dagger-\mathfrak{p}\mathfrak{p}^\dagger\right)\Big]\mathrm{u}(p). \tag{10.5}$$

Since this numerator is ℓ independent, the loop integration

$$\int\frac{\mathrm{d}^4\ell}{(2\pi)^4}\frac{2}{\left(\ell^2-\Delta+xyq^2\right)^3}$$

is finite, without need of renormalization, as promised. In the limit $q^2\to0$, the integral gives

$$\frac{-i}{16\pi^2\Delta}.$$

Finally, extracting the form factor $F_2(q^2)$, and taking the limit as $q^2\to0$, we find the anomaly contribution

$$\frac{m_{\rm ext}g^2}{8\pi^2}\frac{Q_{\rm int}}{Q_{\rm ext}}\int_0^1 du\,u^2\frac{m_{\rm int}\left(\mathfrak{s}\mathfrak{s}^\dagger-\mathfrak{p}\mathfrak{p}^\dagger\right)-m_{\rm ext}\left(\mathfrak{s}\mathfrak{s}^\dagger+\mathfrak{p}\mathfrak{p}^\dagger\right)(u-1)}{(1-u)M^2+um_{\rm int}^2+u(u-1)m_{\rm ext}^2}. \tag{10.6}$$

10.3.2. *Charged scalar diagram*

The "charged" scalar contribution is given by the Feynman diagram in Fig. 10.2(b). The amplitude for this graph is

$$i\mathcal{M}=\int\frac{\mathrm{d}^4k}{(2\pi)^4}\overline{\mathrm{u}}(p')ig\left(\mathfrak{s}+\mathfrak{p}\gamma^5\right)\frac{i\left(\not p-\not k+m_{\rm int}\right)}{\left((p-k)^2-m_{\rm int}^2\right)}ig\left(\mathfrak{s}^\dagger-\mathfrak{p}^\dagger\gamma^5\right)\times \frac{i}{k^2-M^2}ieQ_{\rm S}\left(k+k'\right)^\mu\tilde{A}_\mu^{\rm cl}(q)\frac{i}{k'^2-M^2}\mathrm{u}(p),$$

where we have used the photon-scalar Feynman rule $ieQ_{\rm S}(p+p')^\mu\tilde{A}_\mu^{\rm cl}(q)$. The Feynman parametrization and denominator simplification follow directly from the neutral scalar case, swapping M for $m_{\rm int}$ in Eq. (10.3). The numerator simplification is similarly straightforward, and results in

$$\overline{\mathrm{u}}(p')\frac{-i\sigma^{\mu\nu}q_\nu}{2m_{\rm ext}}\left\{\left(\mathfrak{s}\mathfrak{s}^\dagger+\mathfrak{p}\mathfrak{p}^\dagger\right)um_{\rm ext}+\left(\mathfrak{s}\mathfrak{s}^\dagger-\mathfrak{p}\mathfrak{p}^\dagger\right)m_{\rm int}\right\}(1-u)2m_{\rm ext}\mathrm{u}(p).$$

Combining these pieces, extracting the form factor $F_2(q^2)$, and taking the limit $q^2\to0$, we find the following anomaly contribution

$$\frac{-m_{\rm ext}g^2}{8\pi^2}\frac{Q_{\rm S}}{Q_{\rm ext}}\int_0^1 du\,u(1-u)\frac{\left(\mathfrak{s}\mathfrak{s}^\dagger+\mathfrak{p}\mathfrak{p}^\dagger\right)um_{\rm ext}+\left(\mathfrak{s}\mathfrak{s}^\dagger-\mathfrak{p}\mathfrak{p}^\dagger\right)m_{\rm int}}{uM^2+u(u-1)m_{\rm ext}^2+(1-u)m_{\rm int}^2}. \tag{10.7}$$

10.4. Vector Boson Contributions

There are, as in the scalar case, two general types of one-loop vector boson diagrams that contribute to a_{f}, shown in Fig. 10.4. As before, one diagram, Fig. 10.4(a), can contain neutral or charged vector exchange, which we'll call the "neutral diagram", while the second diagram, Fig. 10.4(b), is only available in models with electrically charged vector bosons.

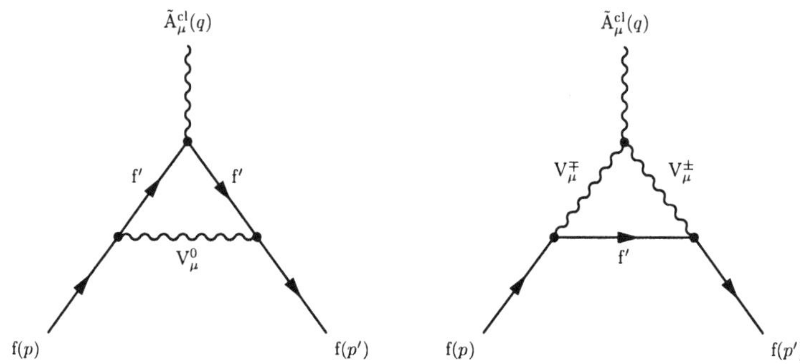

(a) Neutral/Charged Vectors (b) Charged Vectors

Fig. 10.4. One-loop Feynman diagrams for vector boson contributions to a_{f}. The left-hand diagram can occur for either electrically neutral or charged vectors, while the right-hand diagram can only occur for electrically charged vectors.

For either type of vector, the most general Feynman rule containing both vector and axial vector contributions is $ig\gamma^\mu(\mathfrak{v} + \mathfrak{a}\gamma^5)$. While there are many possible gauge choices, we restrict ourselves to Feynman gauge; this is a natural choice for one-loop calculations such as ours, but requires the calculation of additional unphysical scalar diagrams to cancel the extra degrees of freedom present in the Feynman gauge propagator. For the neutral diagram we have already calculated the scalar contribution, Eq. (10.6); however, we need the correct (gauge independent) scalar-fermion coupling derived from the vector-fermion coupling. This can of course be derived directly from the full gauge-fixed Lagrangian. Alternatively, we can examine tree level exchange diagrams, including both the vector and unphysical scalar contributions, and demanding gauge invariance of the full amplitude. The unphysical fermion to scalar plus fermion transition $f_{\mathrm{ext}} \to S f_{\mathrm{int}}$ has the coupling

$$ig(\mathfrak{s} + \mathfrak{p}\gamma^5) = i\frac{g}{M}\left((m_{\mathrm{int}} - m_{\mathrm{ext}})\mathfrak{v} - (m_{\mathrm{int}} + m_{\mathrm{ext}})\mathfrak{a}\right). \tag{10.8}$$

As we shall see, the leading factor of m/M generally suppresses the contributions of the unphysical scalars to the anomaly of light fermions.

10.4.1. *Neutral vector diagram*

Using the same kinematic conventions as in Section 10.3, the amplitude for the diagram in Fig. 10.4(a):

$$i\mathcal{M} = \int \frac{d^4k}{(2\pi)^4} \overline{\mathfrak{u}}(p') ig\gamma^\rho \left(\mathfrak{v} + \mathfrak{a}\gamma^5\right) \frac{i\left(\not{k}' + m_{\text{int}}\right)}{k'^2 - m_{\text{int}}^2} ieQ_{\text{int}}\gamma^\mu \tilde{A}_\mu^{\text{cl}}(q) \times \frac{i\left(\not{k} + m_{\text{int}}\right)}{k^2 - m_{\text{int}}^2} ig\gamma^\sigma \left(\mathfrak{v}^\dagger + \mathfrak{a}^\dagger\gamma^5\right) \frac{-ig_{\rho\sigma}}{(p-k)^2 - M^2} \mathfrak{u}(p).$$

If the vector is massless, we treat M as an infrared regulator, and take the limit $M \to 0$ after completing all manipulations. Regardless, we simplify the numerator, using the same techniques as before, obtaining

$$\overline{\mathfrak{u}}(p') \frac{i\sigma^{\mu\nu} q_\nu}{2m_{\text{ext}}} 2m_{\text{ext}} \Big[2m_{\text{ext}}(u-1)(u-2)\left(\mathfrak{v}\mathfrak{v}^\dagger + \mathfrak{a}\mathfrak{a}^\dagger\right) + 4m_{\text{int}}(u-1)\left(\mathfrak{v}\mathfrak{v}^\dagger - \mathfrak{a}\mathfrak{a}^\dagger\right) \Big] \mathfrak{u}(p).$$

When the vectors are massive, we add the contribution of an unphysical scalar mode. We calculated this contribution in Fig. 10.2(a); we replace the couplings as noted in Eq. (10.8). Extracting the $F_2(q^2)$ values in the limit $q^2 \to 0$ for both contributions and summing them gives the result

$$-\frac{m_{\text{ext}} g^2}{4\pi^2} \frac{Q_{\text{int}}}{Q_{\text{ext}}} \int_0^1 du\, u(u-1) \frac{m_{\text{ext}}(u-2)\left(\mathfrak{v}\mathfrak{v}^\dagger + \mathfrak{a}\mathfrak{a}^\dagger\right) + 2m_{\text{int}}\left(\mathfrak{v}\mathfrak{v}^\dagger - \mathfrak{a}\mathfrak{a}^\dagger\right)}{(1-u)M^2 + um_{\text{int}}^2 + u(u-1)m_{\text{ext}}^2} +$$
$$-\frac{m_{\text{ext}} g^2}{8\pi^2} \frac{Q_{\text{int}}}{Q_{\text{ext}}} \frac{1}{M^2} \int_0^1 du\, u^2 \left\{ m_{\text{int}} \frac{\left(m_{\text{int}} - m_{\text{ext}}\right)^2 \mathfrak{v}\mathfrak{v}^\dagger - \left(m_{\text{int}} + m_{\text{ext}}\right)^2 \mathfrak{a}\mathfrak{a}^\dagger}{(1-u)M^2 + um_{\text{int}}^2 + u(u-1)m_{\text{ext}}^2} - m_{\text{ext}}(u-1) \frac{\left(m_{\text{int}} - m_{\text{ext}}\right)^2 \mathfrak{v}\mathfrak{v}^\dagger + \left(m_{\text{int}} + m_{\text{ext}}\right)^2 \mathfrak{a}\mathfrak{a}^\dagger}{(1-u)M^2 + um_{\text{int}}^2 + u(u-1)m_{\text{ext}}^2} \right\}. \quad (10.9)$$

The first line holds the contribution of the vector itself; the remaining lines hold the contribution of the unphysical scalar. As alluded to before, the unphysical scalar contribution is suppressed by factors of $(m/M)^2$ compared to the vector. For a massless vector in an unbroken theory, where there are no unphysical scalar modes, only the first line is retained, and we take the regulator $M \to 0$.

10.4.2. *Charged vector diagram*

Finally, we consider charged vector boson contributions of the type diagrammed in Fig. 10.4(b). Since we are considering only non-anomalous photon couplings, the vectors here can only arise in electroweak gauge models where at least part of the photon is contained in the weak gauge group.

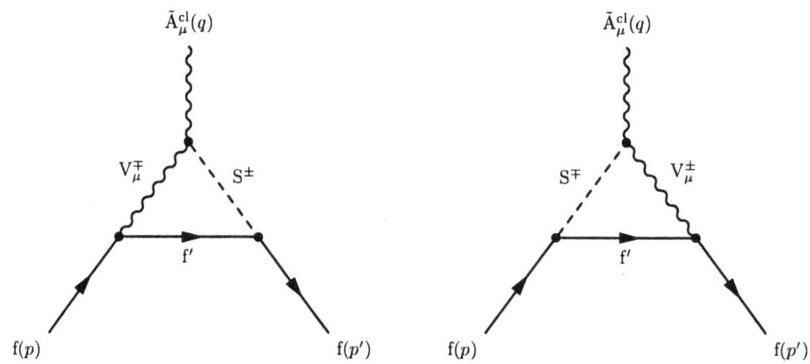

Fig. 10.5. The two additional unphysical scalar diagrams needed to calculate the contributions of the charged vector diagram of Fig. 10.4(b).

This final calculation is the most complicated to organize, but straightforward to calculate using the foregoing techniques. In addition to the primary diagram of Fig. 10.4(b), there are three additional diagrams where vector propagators are replaced with all possible combinations of unphysical scalar modes: a diagram like Fig. 10.2(b), as well as the pair of diagrams in Fig. 10.5. Given the Feynman rule for vector-fermion coupling, we've already shown how to derive the scalar-fermion coupling; we need only the vector-vector-photon (VVP) and scalar-vector-photon (SVP) couplings in order to complete the calculation. In the kinematic notation of Fig. 10.3, the VVP vertex rules have the form

$$G\left[g^{\mu\nu}(q-k)^\rho + g^{\nu\rho}(k+k')^\mu + g^{\rho\mu}(-k'-q)^\nu\right], \tag{10.10}$$

where G is the relevant coupling strength, including the gauge coupling, mixing angles, and the structure constants of the group. The SVP vertex rule must be of the form $i\tilde{G}g^{\mu\nu}$; G and $\tilde{G}$ will be related by the Lagrangian of the theory, but it is simplest here to leave them independent, and derive the appropriate form when applying our results to a particular model.

As the path is now prepared, and the route is familiar, we will move directly to the result

$$\frac{m_{\text{ext}}g^2}{4\pi^2}\frac{G}{eQ_{\text{ext}}}\int_0^1 du\,u^2\frac{m_{\text{ext}}(2u+1)\left(\mathfrak{v}\mathfrak{v}^\dagger+\mathfrak{a}\mathfrak{a}^\dagger\right)-3m_{\text{int}}\left(\mathfrak{v}\mathfrak{v}^\dagger-\mathfrak{a}\mathfrak{a}^\dagger\right)}{uM^2+(1-u)m_{\text{int}}^2+u(u-1)m_{\text{ext}}^2}+$$

$$-\frac{m_{\text{ext}}g^2}{8\pi^2}\frac{Q_V}{Q_{\text{ext}}}\frac{\tilde{G}}{eM}\int_0^1 du\,\frac{u^2\left((m_{\text{int}}-m_{\text{ext}})\mathfrak{v}\mathfrak{v}^\dagger-(m_{\text{int}}+m_{\text{ext}})\mathfrak{a}\mathfrak{a}^\dagger\right)}{uM^2+(1-u)m_{\text{int}}^2+u(u-1)m_{\text{ext}}^2}+$$

$$-\frac{m_{\text{ext}}g^2}{8\pi^2}\frac{Q_V}{Q_{\text{ext}}}\frac{1}{M^2}\int_0^1 du\,u(1-u)\times$$

$$\left\{\frac{m_{\text{ext}}\left\{(m_{\text{int}}-m_{\text{ext}})^2\mathfrak{v}\mathfrak{v}^\dagger+(m_{\text{int}}+m_{\text{ext}})^2\mathfrak{a}\mathfrak{a}^\dagger\right\}u}{uM^2+(1-u)m_{\text{int}}^2+u(u-1)m_{\text{ext}}^2}+\right.$$

$$\left.\frac{m_{\text{int}}\left\{(m_{\text{int}}-m_{\text{ext}})^2\mathfrak{v}\mathfrak{v}^\dagger-(m_{\text{int}}+m_{\text{ext}})^2\mathfrak{a}\mathfrak{a}^\dagger\right\}}{uM^2+(1-u)m_{\text{int}}^2+u(u-1)m_{\text{ext}}^2}\right\}. \tag{10.11}$$

The first line is the pure vector (VVP) contribution, the second line is from the mixed (SVP) diagrams, while the remaining complicated lines are the pure scalar (SSP) contributions. As with the neutral vector interaction, the pure scalar contributions are suppressed by an additional factor of $(m/M)^2$. Interestingly, the SVP contributions are generally unsuppressed as $\tilde{G}$ is usually of order M. In the Standard Model photon-$W^\pm$ interaction, for instance, $G=-e$ and $\tilde{G}=eM_W$.

10.5. Example: The Standard Electroweak Contributions

We now apply the results we obtained in the previous section to the Standard Model electroweak contributions to the a_f of any light fermion (that is, $m_{\text{int}}, m_{\text{ext}} \ll M_{Z^0}$). In this limit, the diagrams containing only unphysical scalars will be negligible compared to those containing vectors, since they are suppressed by additional factors of $m_{\text{ext}}/M_{Z^0,W}$, and will not be displayed. The results we obtain here are valid for all of the charged fermions except the top quark, whose mass is certainly not small compared to the vectors. In that case, we can't ignore the mass of the top quark compared to the vector (and unphysical scalar) masses. We do not deal with that case here.

10.5.1. *The* $\mathbf{Z}^0$ *contribution*

In this small fermion mass limit, the Z^0 contribution of Eq. (10.9), where the unphysical scalar contribution is suppressed

$$a_{\rm ext}^{Z^0}(0) = \frac{m_{\rm ext}}{8\pi^2}\left(\frac{g}{M_{Z^0}\cos\theta_W}\right)^2 \int_0^1 du u \times$$
$$\left\{2m_{\rm ext}(u-2)\left(\mathfrak{v}\mathfrak{v}^\dagger + \mathfrak{a}\mathfrak{a}^\dagger\right) + 4m_{\rm int}\left(\mathfrak{v}\mathfrak{v}^\dagger - \mathfrak{a}\mathfrak{a}^\dagger\right)\right\}.$$

With the conventional definitions $v = (C_R{+}C_L)/2$ and $a = (C_R{-}C_L)/2$, we can replace the vector and axial couplings with the left and right couplings, to place this result in terms more suitable for Standard Model phenomenology,

$$\mathfrak{v}\mathfrak{v}^\dagger + \mathfrak{a}\mathfrak{a}^\dagger = \frac{1}{2}\left(C_L^2 + C_R^2\right)$$
$$\mathfrak{v}\mathfrak{v}^\dagger - \mathfrak{a}\mathfrak{a}^\dagger = C_L C_R.$$

Performing the integrations and substituting for the couplings

$$a_{\rm ext}^{Z^0}(0) = -\frac{m_{\rm ext}^2}{8\pi^2}\frac{G_{\rm F}}{\sqrt{2}}\frac{16}{3}\left(C_L^2 + C_R^2 - 3C_L C_R\right),$$

where $M_{Z^0}\cos\theta_W = M_{\rm W}$, and $g^2/8M_{\rm W}^2 \equiv G_{\rm F}/\sqrt{2}$.

10.5.2. *The* $\mathbf{W}^\pm$ *contribution*

We can similarly simplify the integrals for the charged W contribution in the small fermion mass limit. The SSP diagram is highly suppressed, but the others are not. The VVP diagram contributes

$$F_2(0)^{\rm W(\mathit{vvp})}(0) = \frac{1}{Q_{\rm ext}}\frac{m_{\rm ext}}{8\pi^2}\frac{g^2}{4M_{\rm W}^2}\frac{1}{3}\left(7m_{\rm ext}\left(C_L^2 + C_R^2\right) - 18m_{\rm int}C_L C_R\right),$$

where $Q_{\rm ext}$ is the electric charge of the incoming fermion and $G = -e$. The SVP diagram contributes

$$F_2^{\rm W(\mathit{svp})}(0) = \frac{m_{\rm ext}}{8\pi^2}\frac{1}{Q_{\rm ext}}\frac{g^2}{4M_{\rm W}^2}\left(m_{\rm ext}\left(C_L^2 + C_R^2\right) - 2m_{\rm int}C_L C_R\right),$$

where $\tilde{G} = eM_{\rm W}$, and the charges are defined as above. Summing these two contributions, we obtain

$$a_{\rm ext}^{\rm W} = \frac{m_{\rm ext}}{8\pi^2}\frac{1}{Q_{\rm ext}}\frac{G_{\rm F}}{\sqrt{2}}\frac{2}{3}\left(10m_{\rm ext}\left(C_L^2 + C_R^2\right) - 24m_{\rm int}C_L C_R\right).$$

10.5.3. *The γ contribution*

To calculate the photon contribution, we use the result calculated for Z^0-like gauge bosons, but we drop the unphysical scalar modes and take the gauge mass to zero. QED admits only vector couplings without flavor-changing contributions. Applying these constraints to the results of Section 10.4.1, we obtain

$$a_{\rm ext}^{\gamma} = \frac{e^2 Q_{\rm ext}^2}{8\pi^2} = \frac{\alpha_{\rm em}}{2\pi} Q_{\rm ext}^2.$$

10.5.4. *The Higgs contribution*

Calculating the Higgs boson contribution is somewhat less transparent than for the vectors. The Standard Model Higgs has a pure scalar coupling fermion mass, with a Feynman rule of $im_{\rm f}/v$, with v the electroweak VEV. This simplifies Eq. (10.6),

$$a_{\rm ext}^{\rm H} = \frac{m_{\rm ext}^2}{8\pi^2 v^2} r^2 \int_0^1 {\rm d}u \frac{u^2(2-u)}{1-u+u^2r^2},$$

with $r = m_{\rm ext}/M_{\rm H}$. While nontrivial to simplify further, given the direct search limits on the Higgs mass ($r \ll 1$), the numerical value of this integral is orders of magnitude smaller than the other electroweak contributions to the anomaly. The standard Higgs contribution can safely be ignored.

10.5.5. *Summary for a_μ*

The following table gives the charges and masses of the Standard Model couplings to the weak gauge bosons:

	Z^0	W
$m_{\rm int}$	m_μ	$m_\nu = 0$
$m_{\rm ext}$	m_μ	m_μ
C_L	$-\frac{1}{2} + \sin^2\theta_W$	$\frac{1}{\sqrt{2}}$
C_R	$\sin^2\theta_W$	0

Substituting these values into the expressions in the previous subsections, we find as expected

$$a_\mu^{Z^0} = -\frac{m_\mu^2 G_{\rm F}}{8\pi^2\sqrt{2}} \frac{4}{3}\left(1 + 2\sin^2\theta_W - 4\sin^4\theta_W\right)$$

$$a_\mu^{\rm W} = \frac{m_\mu^2 G_{\rm F}}{8\pi^2\sqrt{2}} \frac{10}{3} \qquad \text{and} \qquad a_\mu^{\gamma} = \frac{\alpha_{\rm em}}{2\pi}.$$

10.6. Conclusions

As promised, we have derived general expressions at one-loop order for the contribution of a wide range of particle types to the anomalous magnetic moment of charged Dirac fermions. Our expressions are both general and flexible, and have allowed us to readily reproduce the well known expressions for the Standard Model electroweak contributions to a_μ.

Acknowledgments

Preparation of this work was supported in part by U.S. National Science Foundation Grant PHY-0758603.

References

[1] G. W. Bennett *et al.*, Final report of the muon E821 anomalous magnetic moment measurement at BNL, *Phys. Rev.* **D73**, 072003, (2006). doi: 10.1103/PhysRevD.73.072003.
[2] R. Jackiw and S. Weinberg, Weak interaction corrections to the muon magnetic moment and to muonic atom energy levels, *Phys. Rev.* **D5**, 2396–2398, (1972).
[3] K. Fujikawa, B. W. Lee, and A. I. Sanda, Generalized renormalizable gauge formulation of spontaneously broken gauge theories, *Phys. Rev.* **D6**, 2923–2943, (1972). doi: 10.1103/PhysRevD.6.2923.
[4] G. Altarelli, N. Cabbibo and L. Maiani, Phys. Lett. **40B**, 415 (1972).
[5] W.A. Bardeen, R. Gastmans and B Lautrup, Nucl. Phys. **B46**, 319 (1972).
[6] I. Bars and M. Yoshimura, Phys. Rev. D **6**, 374 (1972).
[7] J. P. Leveille, The second order weak correction to $(g-2)$ of the muon in arbitrary gauge models, *Nucl. Phys.* **B137**, 63, (1978).

Chapter 11

Measurement of the Muon $(g-2)$ Value

James P. Miller and B. Lee Roberts

Department of Physics, Boston University
Boston, MA 01890 U.S.A.
miller@bu.edu, roberts@bu.edu

Klaus Jungmann

Kernfysisch Versneller Instituut, University of Groningen,
NL-9747 AA, Groningen, The Netherlands
jungmann@kvi.nl

The muon anomalous magnetic moment has now been measured to a precision of 0.54 ppm. This level of sensitivity is adequate to probe the few-hundred GeV mass scale, and to place significant constraints on physics beyond the Standard Model. In this chapter we briefly review the history of such measurements, and then describe the most recent experiment, E821 at the Brookhaven Alternating Gradient Synchrotron.

Contents

11.1. The Discovery of the Muon and Determination of its Spin

In 1933, the first published observation of the muon was reported by Kunze [1] using a Wilson cloud chamber, where it was reported to be "a particle of uncertain nature." In 1936 Anderson and Neddermeyer [2] reported the presence of "particles less massive than protons but more penetrating than electrons" in cosmic rays, which was confirmed in 1937 by Street and Stevenson [3], Nishina, Tekeuchi and Ichimiya [4], and by Crussard and Leprince-Ringuet [5]. This discovery became a topic of great interest, and Tomonaga and Araki published a paper in the Physical Review discussing the effect of the nuclear Coulomb field on the nuclear capture of slow mesons [6].

The Yukawa theory of the nuclear force had predicted such a particle, but this "mesotron" as it was called, interacted too weakly with matter to be the carrier of the strong force. It took ten years for this fact to become clear – it was about the right mass; it had some of the characteristics, but not the strong interaction with nuclear matter that one would expect for the carrier of the nuclear force [7]. By 1941, cosmic ray studies indicated that the spin of the muon was most likely "spin 0, or possibly spin $\frac{1}{2}$" [8].

11.1.1. *Measurements of the muon magnetic moment*

By 1949 evidence had accumulated that the spin of the muon was $\frac{1}{2}$ [9], perhaps meaning that the muon behaved as a heavy electron. With the advent of cyclotrons in the 1950s which had sufficient proton-beam energy to produce pions, it became possible to produce pions, and thus muons, in the laboratory. This development permitted studies of muon decay, and presented the possibility of making "exotic" atoms in the laboratory with a μ^- orbiting about a positive nucleus. Since the Bohr radius is inversely proportional to the orbiting particle's mass, the muon quickly moves well inside of the atomic electron cloud and becomes a hydrogen-like atom with nuclear charge Z.

In 1953 at the Columbia-Nevis Cyclotron, Fitch and Rainwater [10] studied the x rays from muonic atoms for a range of atomic number Z to search for fine-structure splitting in the muonic x-ray spectrum. For a spin $\frac{1}{2}$ Dirac particle bound to a nucleus of charge Z and having quantum numbers (n, ℓ), the fine-structure splitting is

$$\Delta E_{n,\ell} = \frac{(Z\alpha)^4}{2n^3} \frac{m_\mu}{\ell(\ell+1)} \, . \tag{11.1}$$

The splitting is largest for the lowest n and highest Z, so the $2p \rightarrow 1s$ transition in a high-Z element, which has two fine-structure components, would have the largest splitting. While limited by the poor energy resolution of their NaI(Tl) photon detector and the inability to perform non-linear least-square fits to the spectra, Fitch and Rainwater concluded: "...we believe our results for Pb can best be explained in terms of the expected fine structure splitting for spin $\frac{1}{2}$ and the expected Dirac magnetic moment." This experiment represented the first attempt to measure the magnetic moment of the muon.

Upon hearing about the discovery of the muon, I.I. Rabi is reputed to have asked "who ordered that?". Since it took ten years to show conclusively that the muon was not the Yukawa particle, it's not clear when exactly, or whether at all, this statement was made. However, it is a good question and one for which we still have no answer. Nevertheless, we do know that the electron, muon and tauon, along with their neutrinos, are the leptons of the Standard Model, where the negatively charged leptons are "particles" and the positively charged ones are antiparticles.

Unlike the electron which appears to be stable, the muon decays through the weak force, the dominant decay being $\mu^- \rightarrow e^- + \bar{\nu}_e + \nu_\mu$. This three-body decay tells us that the individual lepton numbers, electron and muon, are conserved separately, and that the two flavors (kinds) of neutrinos are distinct particles [13].

In their 1956 paper [11], Lee and Yang proposed several experimental tests of parity non-conservation, including a measurement of the angular correlation between the muon momentum in pion decay, $\pi^+ \rightarrow \mu^+ + \nu_\mu$ and the positron from muon decay. This parity violation was subsequently observed in two different experiments [16, 17], with the experiment of Garwin *et al.* [16], observing the spin rotation of a muon in a magnetic field for the first time. The torque exerted by a magnetic field on the muon's magnetic moment produces a spin precession frequency

$$\vec{\omega}_S = -\frac{gq\vec{B}}{2m} - \frac{q\vec{B}}{\gamma m}(1-\gamma) \, , \tag{11.2}$$

where $\gamma = (1-\beta^2)^{-\frac{1}{2}}$, $\beta = v/c$, and the muon charge is $q = \pm e$. Garwin *et al.* [16] found that the observed rate of spin rotation gave $g_\mu = 2.0 \pm 0.10$, indicating "the very strong probability that the spin of the μ^+ is $\frac{1}{2}$." This experiment provided the first clear indication that the muon behaved like a heavy electron.

A second muon spin rotation experiment by Garwin *et al.* [18], obtained a 12% measurement of the muon anomaly, $a_\mu^+ = 0.001\,13^{+0.00016}_{-0.00012}$ which agreed very well with the expected Schwinger value of $\alpha/2\pi \simeq 0.001161\ldots$. This experiment showed conclusively that the muon did indeed have the characteristics of a heavy electron.

Following the experiments at the Nevis Cyclotron, three experiments were carried out at CERN, the first at the synchrocylotron [19, 20], and the second two at the proton synchrotron [21, 23, 24]. These experiments have been well documented in an earlier volume in this series [26], and we will not discuss them in detail. The relative precision obtained in the final CERN experiment was ±7.3 parts per million (ppm) [24]. This experiment verified the ≃ 60 ppm contribution of virtual hadrons to the muon anomaly. Measurements of the muon magnetic moment are summarized below in Table 11.1.

Table 11.1. Measurements of the muon anomalous magnetic moment. When the uncertainty on the measurement is the size of the next term in the QED expansion, or the hadronic or weak contributions, the term is listed under "sensitivity". The "?" indicates a result that differs by greater than two standard deviations with the Standard Model. For completeness, we include the experiment of Henry *et al.* [22], which is not discussed in the text.

±	Measurement	σ_{a_μ}/a_μ	Sensitivity	Reference
μ^+	$g = 2.00 \pm 0.10$		$g = 2$	Garwin *et al* [16]
μ^+	$0.001\,13^{+0.00016}_{-0.00012}$	12.4%	$\frac{\alpha}{\pi}$	Garwin *et al* [18]
μ^+	0.001 145(22)	1.9%	$\frac{\alpha}{\pi}$	Charpak *et al* [19]
μ^+	0.001 162(5)	0.43%	$(\frac{\alpha}{\pi})^2$	Charpak *et al* [20]
$\mu^\pm$	0.001 166 16(31)	265 ppm	$(\frac{\alpha}{\pi})^3$	Bailey *et al* [21]
μ^+	0.001 060(67)	5.8%	$\frac{\alpha}{\pi}$	Henry*et al* [22]
$\mu^\pm$	0.001 165 895(27)	23 ppm	$(\frac{\alpha}{\pi})^3$ + Hadronic	Bailey *et al* [23]
$\mu^\pm$	0.001 165 911(11)	7.3 ppm	$(\frac{\alpha}{\pi})^3$ + Hadronic	Bailey *et al* [24]
μ^+	0.001 165 919 1(59)	5 ppm	$(\frac{\alpha}{\pi})^3$ + Hadronic	Brown *et al* [31]
μ^+	0.001 165 920 2(16)	1.3 ppm	$(\frac{\alpha}{\pi})^4$ + Weak	Brown *et al* [32]
μ^+	0.001 165 920 3(8)	0.7 ppm	$(\frac{\alpha}{\pi})^4$ + Weak + ?	Bennett *et al* [33]
μ^-	0.001 165 921 4(8)(3)	0.7 ppm	$(\frac{\alpha}{\pi})^4$ + Weak + ?	Bennett *et al* [34]
$\mu^\pm$	0.001 165 920 80(63)	0.54 ppm	$(\frac{\alpha}{\pi})^4$ + Weak + ?	Bennett *et al* [25, 34]

11.2. Experiment E821 at the Brookhaven AGS

Around 1984, Vernon Hughes began to put together a collaboration to improve on the measurement of the muon anomaly. The goal was a relative error of ±0.35 ppm ($\pm 40 \times 10^{-11}$), which was chosen to be one fifth of the lowest-order electroweak contribution of $a_\mu^{EW} = 195 \times 10^{-11}$. The present authors were among the first to join this effort. At that time, well before the Large Electron Positron Collider (LEP) became operational, the motivation was to check the renormalizability of the electroweak theory, and to search for possible physics beyond the Standard Model. Space considerations require that in this discussion of E821, we omit many details that can be found in the final report of the E821 Collaboration [25] and/or in the review [35].

The measurement of the magnetic anomaly uses the time evolution of its spin in a magnetic field. For a muon moving in a magnetic field, spin and momentum rotate with the frequencies:

$$\vec{\omega}_S = -\frac{gq\vec{B}}{2m} - \frac{q\vec{B}}{\gamma m}(1-\gamma) \quad \text{and} \quad \vec{\omega}_C = -\frac{q\vec{B}}{m\gamma}. \tag{11.3}$$

The spin precession *relative to the momentum* occurs at the difference frequency, ω_a, between the spin and cyclotron frequencies,

$$\vec{\omega}_a = \vec{\omega}_S - \vec{\omega}_C = -\left(\frac{g-2}{2}\right)\frac{q\vec{B}}{m} = -a_\mu \frac{q\vec{B}}{m}. \tag{11.4}$$

The precession frequency and magnetic field are averages over the muon ensemble. This technique has been used in all but the first experiments by Garwin *et al.* [16, 18], which used stopped muons to measure their anomaly.

The weak interaction and parity violation play a central role in the measurement of the muon anomaly. Once parity violation was observed [14, 16, 17] it was realized that one could make beams of polarized muons in the pion decay reactions

$$\pi^- \to \mu^- + \bar{\nu}_\mu \quad \text{or} \quad \pi^+ \to \mu^+ + \nu_\mu.$$

The pion has spin zero, the neutrino (antineutrino) has a helicity of -1 (+1), and the weak force in the decay process is very short range, so the orbital angular momentum in the final state is zero. Thus conservation of angular momentum requires that the μ^- (μ^+) helicity be +1 (-1) in the pion rest frame. The muons from pion decay at rest are always polarized.

From a beam of pions traversing a straight beam-channel consisting of focusing and defocusing elements (FODO), a beam of polarized muons can be produced by selecting the "forward" or "backward" decays. The forward muons are those produced, in the pion rest frame, nearly parallel to the pion laboratory momentum and are the decay muons with the highest laboratory momenta. The backward muons are those produced nearly anti-parallel to the pion momentum and have the lowest laboratory momenta. The forward μ^- (μ^+) are polarized along (opposite) their lab momenta respectively; the polarization reverses for backward muons. The most recent muon $(g-2)$ experiment, E821 at the Brookhaven National Laboratory (BNL) Alternating Gradient Synchrotron (AGS) used forward muons produced by a pion beam with an average momentum of $p_\pi \approx 3.15$ GeV/c. Under a Lorentz transformation from the pion rest frame to the laboratory frame, the decay muons have momenta in the range $0 < p_\mu < 3.15$ GeV/c. After momentum selection, forward muons are injected into a circular storage ring possessing a uniform magnetic field. The average momentum of muons stored in the ring is the "magic" value $p_{magic} = 3.094$ GeV/c (which is explained below), with an average polarization in excess of 95%.

Parity violation is also important in measuring the muon spin direction at the time of decay. Polarized muons are confined in a 7.1 m diameter magnetic storage ring. As their spin precesses relative to the momentum according to Eq. (11.4), the muons decay. As discussed below, in the decay of a $\mu^{-(+)}$ the highest-energy electrons (positrons) are emitted preferentially anti-parallel (parallel) to the muon spin, thereby providing the means to determine the spin direction at the time of the decay.

Detectors are placed on the inside of the storage ring, so that the decay electrons spiral inward and are detected. These detectors measure both the arrival time, and the energy of the decay electron. One obtains a time spectrum showing the exponential decay of the muon, modulated by the $(g-2)$ precession of Eq. (11.4) (see Fig. 11.16).

One needs to provide vertical focusing in the storage ring, since the helical path of a muon in a uniform magnetic field would quickly result in the beam being lost. Traditionally the focusing in storage rings is done with magnetic gradients. However in the $(g-2)$ experiments significant magnetic gradients would compromise the the knowledge of the average magnetic field, since the magnetic field $\vec{B}$ that enters into Eq. (11.4) is the magnetic field averaged over the muon distribution. The presence of gradients limits the ability to determine $\langle B \rangle$ by several orders of magnitude less than the necessary part in 10^7. This problem was overcome in the third CERN experiment by using electrostatic quadrupoles. While a laboratory electric field appears as a combination of an electric and a magnetic field in the rest frame of a relativistic particle, its effect on the particle's spin precession cancels at one "magic" value of the Lorentz factor $\gamma_m = 29.3$.

With an electric field present ω_a becomes [27]

$$\vec{\omega}_S = -\frac{q}{m}\left[\left(\frac{g}{2} - 1 + \frac{1}{\gamma}\right)\vec{B} - \left(\frac{g}{2} - 1\right)\frac{\gamma}{\gamma+1}(\vec{\beta}\cdot\vec{B})\vec{\beta}\right] + \frac{q}{m}\left[\left(\frac{g}{2} - \frac{\gamma}{\gamma+1}\right)\left(\frac{\vec{\beta}\times\vec{E}}{c}\right)\right], \tag{11.5}$$

which simplifies to

$$\vec{\omega}_a = -\frac{q}{m}\left[a_\mu\vec{B} - \left(a_\mu - \frac{1}{\gamma^2-1}\right)\frac{\vec{\beta}\times\vec{E}}{c}\right], \tag{11.6}$$

if $\vec{\beta}\cdot\vec{B} = 0$. For $[a_\mu - 1/(\gamma^2-1)] = 0$ (the "magic" $\gamma_m = 29.3$), the electric field does not contribute to the spin motion relative to the momentum. Thus, vertical focusing achieved using electrostatic quadrupoles permits the use of a *uniform* dipole magnetic field, which is the same principle used in a Penning trap. With the storage ring field set such that the central orbit momentum is the "magic" value of 3.09 GeV/c, only a small correction from the electric field is necessary to account for the stored muons that do not have the magic momentum.

Equation (11.4) suggests that the magnetic field be measured in units of the muon magneton, i.e. $\frac{q}{m_\mu}$. For practical reasons the field is determined first with nuclear magnetic resonance (NMR) signals of protons in water samples over the full azimuth of the storage ring, inside of the 44 mm radius circle where the muons are stored. The NMR measurements are tied to a calibration with a spherical water sample that when averaged over the full azimuth and over the muon distribution gives the average Larmor frequency of a *free* proton, which we call ω_p. The link between the muon and proton magnetic moments is closed with a measurement of the Zeeman effect in the muonium atom (μ^+e^-) hyperfine structure [28].

Although the determination of the muon anomaly a_μ from the Lorentz invariant Eq. (11.4) requires in principle the precise knowledge of only one fundamental constant, i.e. the muon magnetic moment, the practical implementation requires a larger set: Muon mass and charge along with the proton and electron masses and charges, and their magnetic moments–which for most practical purposes are interlinked in the fundamental constant adjustment. Assuming the muon and proton have the same charge[a]

[a]For positive muons and electrons the equality of electric charge has been verified at the 2 ppb level by laser spectroscopy in muonium [29].

$|e|$, the uncertainty on each of these parameters is on the order of 85 parts per billion (ppb). Thus the overall uncertainty on a_μ from these fundamental constants would be around 150 ppb, which is significant compared to the experimental errors. Instead, we used the fundamental constant $\lambda^+ = \mu_\mu^+/\mu_p$ which has been determined from muonium atom spectroscopy to a precision of 120 ppm, which with further theory input can be refined to $\lambda^+ = 3.183\,345\,39(10)$ (± 30 ppb) [28]. The "+" subscript is to remind the reader that this value is obtained from measurements on the positive muon, and one must assume CPT invariance to use λ^+ to determine a_{μ^-}.

Ignoring the signs of the muon and proton charges, we have the two equations:

$$\omega_a = \frac{e}{m} a_\mu B \quad \text{and} \quad \omega_p = g_p \left(\frac{eB}{2m_p} \right), \tag{11.7}$$

where ω_p is the Larmor frequency for a *free* proton. Dividing and solving for a_μ we find

$$a_\mu = \frac{\omega_a/\omega_p}{\lambda^+ - \omega_a/\omega_p} = \frac{\mathcal{R}}{\lambda^+ - \mathcal{R}}. \tag{11.8}$$

In the evaluation of the real experiment we use $\tilde{\omega}_a$ in place of ω_a which is the measured muon spin frequency adjusted for two small corrections: For the radial electric field, and and for the vertical pitching motion of the muons, the latter being necessary since $\vec{\beta} \cdot \vec{B} \simeq 0$. Both of these corrections are discussed below.

As previously mentioned, the third CERN experiment [24], introduced the use of the magic γ. A pion beam was brought to the edge of the storage region through a pulsed coaxial line that canceled the storage-ring field. Muons were kicked onto stored orbits by the $\pi \to \mu\nu$ decay resulting in 125 muons stored per million injected pions. The remainder of the pions struck objects in the ring producing a significant *flash* in the electron calorimeters. The CERN magnet was shimmed to an average azimuthal uniformity of ± 10 ppm, with the total systematic error from all issues related to the magnetic field of ± 1.5 ppm [24].

The BNL based collaboration used the general principle of the third CERN experiment, most significantly the use of electrostatic focusing with the magic γ. The goal of a total error of ± 0.35 ppm on a_μ required a number of significant innovations:

(1) A superferric storage ring, with a field uniformity of ± 1 ppm when averaged over azimuth;

(2) A scheme for direct muon injection into the storage ring that did not perturb the magnetic field seen by the stored muons was developed in order to suppress injection background and store an adequate number of muons to reach the statistical design;
(3) A system of nuclear magnetic resonance (NMR) probes to map and monitor the magnetic field to a part in 10^7, which could map the field without having to cycle the magnet power;
(4) A static superconducting inflector magnet, with no leakage field, to bring the beam to the edge of the storage region;
(5) Detectors and timing circuits which could withstand the high instantaneous rates following injection, with rate-dependent timing shifts from early to late counting-times of less that 20 ps on average.

11.2.1. *Muon decay*

Before discussing the experimental details, we first discuss muon decay, which provides the experimental signal. The pure $(V - A)$ three-body weak decay of the muon, $\mu^- \to e^- + \nu_\mu + \bar{\nu}_e$ or $\mu^+ \to e^+ + \bar{\nu}_\mu + \nu_e$, is "self-analyzing", that is, the parity-violating correlation between the directions in the muon rest frame (MRF) of the decay electron and the muon spin can provide information on the muon spin orientation at the time of the decay.[b]

Consider the case when the decay electron has the maximum allowed energy in the MRF, $E'_{\rm max} \approx (m_\mu c^2)/2 = 53$ MeV. The neutrino and anti-neutrino are directed parallel to each other and at 180° relative to the electron direction. The $\nu\bar{\nu}$ pair carry zero total angular momentum, since the neutrino is left-handed and the anti-neutrino is right-handed; the electron spin carries the muon's angular momentum of 1/2. The electron, being a lepton, is preferentially emitted left-handed in a weak decay, and thus has a larger probability to be emitted with its momentum *anti-parallel* rather than parallel to the μ^- spin. By the same line of reasoning, in μ^+ decay, the highest-energy positrons are emitted *parallel* to the muon spin in the MRF.

In the other extreme, when the electron kinetic energy is zero in the MRF, the neutrino and anti-neutrino are emitted back-to-back and carry a total angular momentum of one. In this case, the electron spin is directed opposite to the muon spin in order to conserve angular momentum. Again,

[b] In the following text, we often use "electron" generically for either e^- and e^+ from the decay of the $\mu^\mp$.

the electron is preferentially emitted with helicity -1, however in this case its momentum will be preferentially directed *parallel* to the μ^- spin. The positron, in μ^+ decay, is preferentially emitted with helicity +1, and therefore its momentum will be preferentially directed *anti-parallel* to the μ^+ spin.

With the approximation that the energy of the decay electron $E' >> m_e c^2$, the differential decay distribution in the muon rest frame is given by [15],

$$dP(y', \theta') \propto n'(y') \left[1 \pm \mathcal{A}(y') \cos\theta'\right] dy' d\Omega' \tag{11.9}$$

where y' is the momentum fraction of the electron, $y' = p'_e / p'_{e\ \max}$, $d\Omega'$ is the solid angle, $\theta' = \cos^{-1}(\hat{p}'_e \cdot \hat{s})$ is the angle between the muon spin and $\vec{p}\,'_e$, $p'_{e\ \max} c \approx E'_{max}$, and the $(-)$ sign is for negative muon decay. The number distribution $n(y')$ and the decay asymmetry $\mathcal{A}(y')$ are given by

$$n(y') = 2y'^2(3 - 2y') \quad \text{and} \quad \mathcal{A}(y') = \frac{2y' - 1}{3 - 2y'}. \tag{11.10}$$

Note that both the number and asymmetry reach their maxima at $y' = 1$, and the asymmetry changes sign at $y' = \frac{1}{2}$, as shown in Fig. 11.1(a).

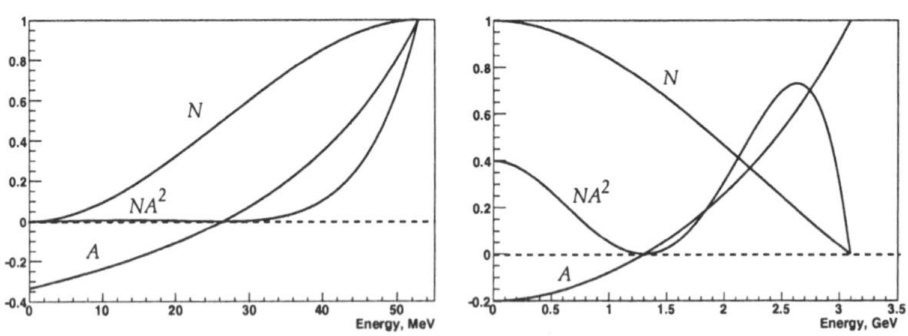

(a) Muon Rest Frame (b) Laboratory Frame

Fig. 11.1. Number of decay electrons per unit energy, N (arbitrary units), value of the asymmetry A, and relative figure of merit NA^2 (arbitrary units) as a function of electron energy. Detector acceptance has not been incorporated, and the polarization is unity. For the third CERN experiment and E821, $E_{max} \approx 3.1$ GeV ($p_\mu = 3.094$ GeV/c) in the laboratory frame.

The CERN and Brookhaven based muon $(g - 2)$ experiments stored relativistic muons in a uniform magnetic field, which resulted in the muon spin precessing with constant frequency $\vec{\omega}_a$, while the muons traveled in circular orbits. If *all* decay electrons were counted, the number detected as a function of time would be a pure exponential; therefore we seek cuts on the laboratory observables to select subsets of decay electrons whose numbers oscillate at the precession frequency. Recalling that the number of decay electrons in the MRF varies with the angle between the electron and spin directions, the electrons in the subset should have a preferred direction in the MRF when weighted according to their asymmetry as given in Eq. (11.9). At $p_\mu \approx 3.094$ GeV/c the directions of the electrons resulting from muon decay in the laboratory frame are very nearly parallel to the muon momentum regardless of their energy or direction in the MRF. Therefore the only practical remaining cut is on the electron's laboratory energy. Typically, selecting an energy subset will have the desired effect: There will be a net component of electron MRF momentum either parallel or anti-parallel to the laboratory muon direction. For example, suppose that we only count electrons with the highest laboratory energy, around 3.1 GeV. Let $\hat{z}$ indicate the direction of the muon laboratory momentum. The highest-energy electrons in the laboratory are those near the maximum MRF energy of 53 MeV, and with MRF directions nearly parallel to $\hat{z}$. There are more of these high-energy electrons when the μ^- spins are in the direction opposite to $\hat{z}$ than when the spins are parallel to $\hat{z}$. Thus the number of decay electrons reaches a maximum when the muon spin direction is opposite to $\hat{z}$, and a minimum when they are parallel. As the spin precesses the number of high-energy electrons will oscillate with frequency ω_a. More generally, at laboratory energies above ~ 1.2 GeV, the electrons have a preferred average MRF direction parallel to $\hat{z}$ (see Fig. 11.1). In this discussion, it is assumed that the spin precession vector, $\vec{\omega}_a$, is independent of time, and therefore the angle between the spin component in the orbit plane and the muon momentum direction is given by $\omega_a t + \phi$, where ϕ is a constant.

Equations (11.9) and (11.10) can be transformed to the laboratory frame to give the electron number oscillation with time as a function of electron energy,

$$N_d(t,E) = N_{d0}(E) e^{-t/\gamma\tau}[1 + A_d(E)\cos(\omega_a t + \phi_d(E))], \tag{11.11}$$

or, taking all electrons above threshold energy E_{th},

$$N(t,E_{th}) = N_0(E_{th}) e^{-t/\gamma\tau}[1 + A(E_{th})\cos(\omega_a t + \phi(E_{th}))]. \tag{11.12}$$

In Eq. (11.11) the differential quantities are,

$$A_d(E) = \mathcal{P}\frac{-8y^2+y+1}{4y^2-5y-5}, \quad N_{d0}(E) \propto (y-1)(4y^2-5y-5), \tag{11.13}$$

and in Eq. (11.12),

$$N(E_{th}) \propto (y_{th}-1)^2(-y_{th}^2+y_{th}+3), \qquad A(E_{th}) = \mathcal{P}\frac{y_{th}(2y_{th}+1)}{-y_{th}^2+y_{th}+3}. \tag{11.14}$$

In the above equations, $y = E/E_{max}$, $y_{th} = E_{th}/E_{max}$, $\mathcal{P}$ is the polarization of the muon beam, and E, E_{th} and $E_{max} = 3.1$ GeV are the electron laboratory energy, threshold energy, and maximum energy, respectively.

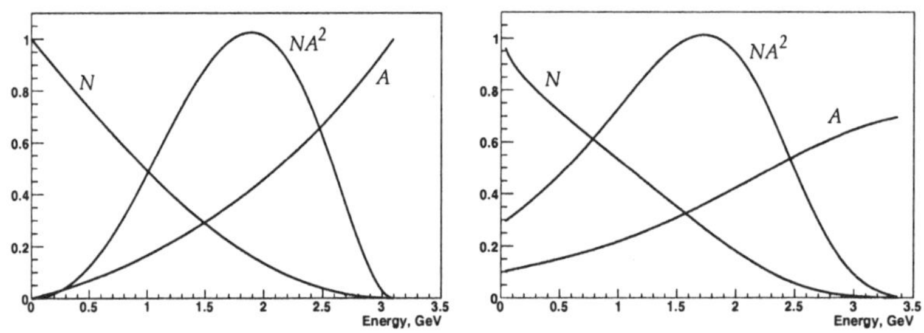

(a) No detector acceptance or energy resolution included (b) Detector acceptance and energy resolution included

Fig. 11.2. The integral N, A, and NA^2 (arbitrary units) for a single energy-threshold as a function of the threshold energy; (a) in the laboratory frame, not including and (b) including the effects of detector acceptance and energy resolution for the E821 calorimeters discussed below. For the third CERN experiment and E821, $E_{max} \approx 3.1$ GeV ($p_\mu = 3.094$ GeV/c) in the laboratory frame.

The fractional statistical error on the precession frequency, when fitting data collected over many muon lifetimes to the five-parameter function (Eq. (11.12)), is given by

$$\delta\epsilon = \frac{\delta\omega_a}{\omega_a} = \frac{\sqrt{2}}{2\pi f_a \tau_\mu N^{\frac{1}{2}} A}, \tag{11.15}$$

where N is the total number of electrons, and A is the asymmetry, in the given data sample. For a fixed magnetic field and muon momentum, the statistical figure of merit is NA^2, the quantity to be maximized in order to minimize the statistical uncertainty.

The energy dependences of the numbers and asymmetries used in Eqs. (11.11) and (11.12), along with the figures of merit NA^2, are plotted in Figs. 11.1 and 11.2 for the case of E821. The statistical power is greatest for electrons at 2.6 GeV (Fig. 11.1). When a fit is made to all electrons above some energy threshold, the optimal threshold energy is about 1.7–1.8 GeV (Fig. 11.2).

11.2.2. *The design and construction of E821*

The technique used in E821 represented a logical extension of the third CERN experiment [24]. While the technology used in E821 was significantly updated, the completely new idea was direct muon injection into the storage ring, which was first suggested by Fred Combly of the University of Sheffield. This new injection scheme required the development of a fast muon kicker which left minimal residual magnetic field behind, the specification being that the contribution of the kicker field to the integral of $\vec{B} \cdot d\vec{\ell}$ for times greater than 20 μs be ≤ 0.1 ppm. A comparison of the basic features of the two experiments is given below in Table 11.2.

Table 11.2. A comparison of the features of the E821 and the third CERN muon $(g-2)$ experiment [24]. Both experiments operated at the "magic" $\gamma = 29.3$, and used electrostatic quadrupoles for vertical focusing. Bailey *et al.* [24], do not quote a systematic error on the muon frequency ω_a.
*Estimated value. The kicker efficiency is being studied in detail for a proposed new experiment.

System	E821	CERN
Magnet	Superconducting	Room Temperature
Yoke Construction	Monolithic Yoke	40 Separate Magnets
Magnetic Field	1.45 T	1.47 T
Magnet Gap	180 mm	140 mm
Stored Energy	6 MJ	
Field mapped in situ?	yes	no
Central Orbit Radius	7112 mm	7000 mm
Averaged Field Uniformity	±1 ppm	±10 ppm
Muon Storage Region	90 mm Diameter Circle	120 × 80 mm^2 Rectangle
Injected Beam	Muon	Pion
Inflector	Static Superconducting	Pulsed Coaxial Line
Kicker	Pulsed Magnetic	$\pi \to \mu\ \nu_\mu$ decay
Kicker Efficiency*	∼ 4%	125 ppm
Muons stored/fill	10^4	350
Ring Symmetry	Four-fold	Two-fold
$\sqrt{\beta_{max}/\beta_{min}}$	1.03	1.15
Detectors	Pb-Scintillating Fiber	Pb-Scintillator "Sandwich"
Electronics	Waveform Digitizers	Discriminators
Systematic Error on B-field	0.17 ppm	1.5 ppm
Systematic Error on ω_a	0.21 ppm	Not given
Total Systematic Error	0.28 ppm	1.5 ppm
Statistical Error on ω_a	0.46 ppm	7.0 ppm
Final Total Error on a_μ	0.54 ppm	7.3 ppm

The Brookhaven AGS was capable of providing a total of 60 to 70 $\times 10^{12}$ protons (Tp) per 2.7 s machine cycle, thereby providing a proton flux 180 times above that available at CERN in the 1970s. These protons were contained in a number of bunches equally spaced around the AGS ring, which is commonly referred to as the harmonic number H. There were three major data collection periods. In 1999, $H = 8$. In 2000 $H = 6$ and in 2001 $H = 12$. The proton intensity in a bunch, and the resulting pile-up (accidental coincidences between two electrons) in the detectors, is minimized by maximizing the number of proton bunches. Since pulse pile-up in the detectors following injection into the storage ring is one of the systematic issues requiring careful study in the data analysis, the best beam conditions were realized in the 2001 run.

Each proton bunch was extracted separately at 33 ms intervals, and transported to a production target. The counting time of the experiment typically terminated after about ten muon lifetimes (640 μs). The 33 ms time between bunches was determined by limitations on some parts of the extraction equipment.

At this juncture, we wish to acknowledge the original papers containing physics results that have been published by the E821 collaboration [30–34] which are summarized in Ref. [25]. Many review articles have been written on the experiments [26, 35–38]. The theory is discussed in detail in several articles in this volume, as well as in Refs. [35, 39].

11.2.2.1. *The proton and muon beamlines*

The primary proton beam from the AGS was brought to a water-cooled nickel production target. Because of mechanical shock considerations, the intensity of a bunch was limited to less than 7 Tp ($= 7\times 10^{12}$ protons/pulse). The beamline, shown in Fig. 11.3, accepts pions produced at 0° at the production target. They are collected by the first two quadrupoles, momentum analyzed, and brought into the decay channel by four dipoles. A pion momentum of 3.15 GeV/c, 1.7 % higher than the magic momentum, was selected.

The beam then enters a straight 80-meter-long focusing-defocusing quadrupole channel, where those muons from pion decays that are emitted approximately parallel to the pion momentum, so-called forward decays, are collected and transported downstream. Muons with momentum 3.094 MeV/c, average polarization of 95%, are separated from the slightly higher momentum pions at the second momentum slit. However, after this

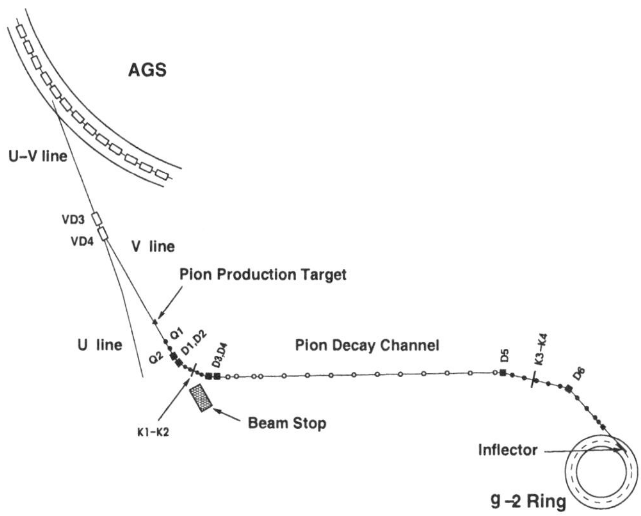

Fig. 11.3. The E821 beamline and storage ring. Pions produced at 0° are collected by the quadrupoles Q1-Q2 and the momentum is selected by the collimators K1-K2. The pion decay channel is 72 m in length. Forward muons at the magic momentum are selected by the collimators K3-K4. (This figure was reprinted with permission from [25]. Copyright 2006 by the American Physical Society.)

momentum selection a rather large pion component, which causes significant injection related background, remains in the beam. The beam composition was measured to be 1:1:1, $e^+ : \mu^+ : \pi^+$. The proton content was calculated to be approximately one-third of the pion flux [31]. The secondary muon beam intensity incident on the storage ring was about 2×10^6 per fill of the ring, which can be compared with 10^8 particles per fill with "pion injection" [30] which was used in the 1997 engineering run.

The injection flash is most severe with the pion injection scheme used at CERN where the $\pi \to \mu\bar{\nu}_\mu$ decay was used to kick muons onto a stored orbit. In E821 when this scheme was tried, the upstream photomultiplier tubes had to be gated off for 120 μs (1.8 muon lifetimes) following injection, to allow the signals to return to the nominal baseline. To reduce this "flash" and to increase significantly the number of muons stored per fill of the storage ring, a fast muon kicker was developed which permitted direct muon injection into the storage ring.

Each proton bunch resulted in a narrow time bunch of pions $\sigma = 25$ ns, which were momentum selected and separated from the incident proton beam at a set of collimators indicated by K1-K2 in Fig. 11.3. The pions then entered a focusing-defocusing (FODO) channel where a forward muon beam was collected, and then separated from the pion beam at the collimators labeled K3-K4. The resulting muon beam was injected into the storage ring, which is shown schematically in Fig. 11.4.

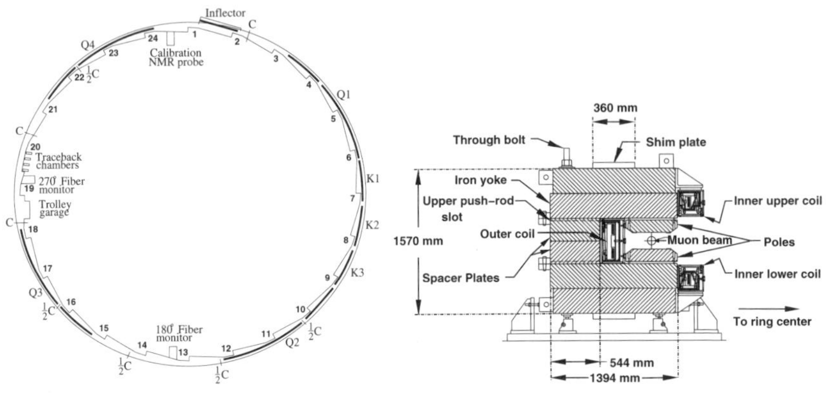

(a) (b)

Fig. 11.4. (a) The layout of the storage ring, as seen from above, showing the location of the inflector, the kicker sections (labeled K1–K3), and the quadrupoles (labeled Q1–Q4). The beam circulates in a clockwise direction. Also shown are the collimators, which are labeled "C", or "$\frac{1}{2}$C" indicating whether the Cu collimator covers the full aperture, or half the aperture. The collimators are rings with inner radius 45 mm; outer radius 55 mm; thickness 3 mm. The scalloped vacuum chamber consists of 12 sections joined by bellows. The chambers containing the inflector, the NMR trolley garage, and the trolley drive mechanism are special chambers. The other chambers are standard, with either quadrupole or kicker assemblies installed inside. An electron calorimeter is placed behind each of the radial windows, at the position indicated by the calorimeter number. (b) The cross-section of the storage-ring magnet. The beam center is at a radius of 7112 mm. The pole pieces are separated from the yoke by an air gap.(This figure was reprinted with permission from [25]. Copyright 2006 by the American Physical Society.)

11.2.2.2. *The Inflector and the fast muon kicker*

In order to get the muons into the storage ring undeflected a superconducting septum magnet called an "inflector" was used to cancel the main storage ring field [40, 41]. The need for this magnet can easily be understood from Fig. 11.5, where it can be seen that the beam would otherwise have to traverse almost 2 m in the magnetic field before arriving at the edge of the storage region. The inflector, along with a calculated field map are shown in Fig. 11.6. The inflector is a truncated, double-cosine theta

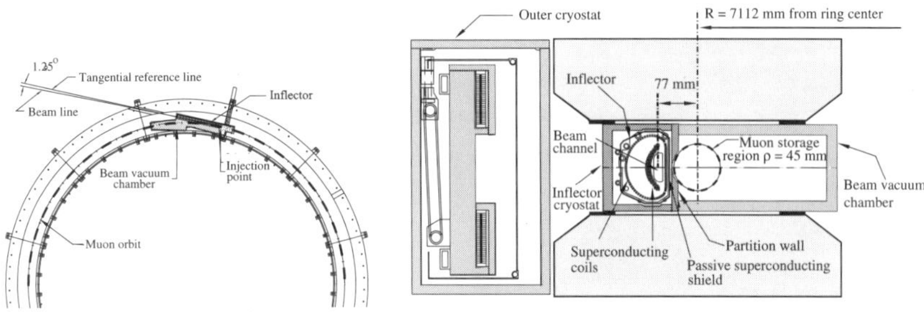

(a) (b)

Fig. 11.5. (a) A plan view of the inflector-storage-ring geometry. The dot-dash line shows the central muon orbit at 7112 mm. The beam enters through a hole in the back of the magnet yoke, then passes into the inflector. The inflector cryostat has a separate vacuum from the beam chamber, as can be seen in the cross-sectional view. The cryogenic services for the inflector are provided through a radial penetration through the yoke at the upstream end of the inflector. (b) A cross-sectional view of the pole pieces, the outer-radius coil-cryostat arrangement, and the downstream end of the superconducting inflector. The muon beam direction at the inflector exit is into the page. The center of the storage ring is to the right. The outer-radius coils which excite the storage-ring magnetic field are shown, but the inner-radius coils are omitted. (This figure was reprinted with permission from [25]. Copyright 2006 by the American Physical Society.)

magnet, shown in Fig. 11.5(b) at its downstream end, with the muon velocity going *into* the page. In the inflector, the current flows into the page down the central "C"-shaped layer of superconductor, then out of the page through the "backward-D"-shaped outer conductor layer. At the inflector exit, the center of the injected beam is 77 mm from the central orbit. For μ^+ stored in the ring, the main field points up, and $q\vec{v} \times \vec{B}$ points to the right in Fig. 11.5(b), toward the ring center.

In the inflector design shown in Fig. 11.5(b) and Fig. 11.6(b), the beam channel aperture is rather small compared to the flux return area. The field is ~ 3 T in the return area (inflector plus central storage-ring field), and the flux density is sufficiently high to lower the critical current in superconductor placed in that region. If the beam channel aperture were to be increased by pushing the coil further into the flux return area, the design would have to be changed, either by employing a superconductor with larger critical current, or by using more conductors in a revised geometry, further complicating the fabrication of this magnet. The result of the small inflector aperture is a rather poor phase-space match between the inflector and the storage ring and, as a consequence, a loss of stored muons.

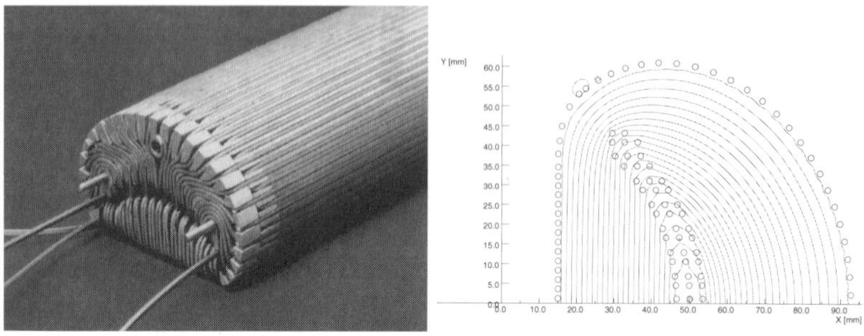

(a) (b)

Fig. 11.6. (a) A photo of the prototype inflector showing the crossover between the two coils. The beam channel is covered by the lower crossover. (b) The magnetic design of the inflector. Note that the magnetic flux is largely contained inside of the inflector volume.

As can be seen from Fig. 11.6(a), the entrance (and exit) to the beam channel is covered with superconductor, as well as by aluminum windows that are not visible in the photograph. This design was chosen to maximize the mechanical stability of the superconductor in the magnetic field, thus reducing the risk of motion which would quench the magnet. However, multiple scattering in the material at both the entrance and exit windows causes about half the incident muon beam to be lost.

The distribution of conductor on the outer surface of the inflector magnet (the "D-shaped" arrangement) prevents most of the magnetic flux from leaking outside of the inflector volume, as seen from Fig. 11.6(b). To prevent flux leakage from entering the beam storage region, the inflector is wrapped with a passive superconducting shield that extends beyond both inflector ends, with a 2 m seam running longitudinally along the inflector side away from the storage region. With the inflector at zero current and the shield warm, the main storage ring magnet is energized. Next the inflector is cooled down, so that the shield goes superconducting and pins the precision field inside the inflector region. When the inflector magnet is powered, the supercurrents in the shield prevent the leakage field from penetrating into the storage region behind the shield.

The beam exited the inflector 77 mm from the central orbit of the storage ring. A fast muon kicker was developed to kick the beam particles onto orbits, which otherwise would make one turn in the ring before hitting the inflector and being lost. As shown schematically in Fig. 11.7, the role

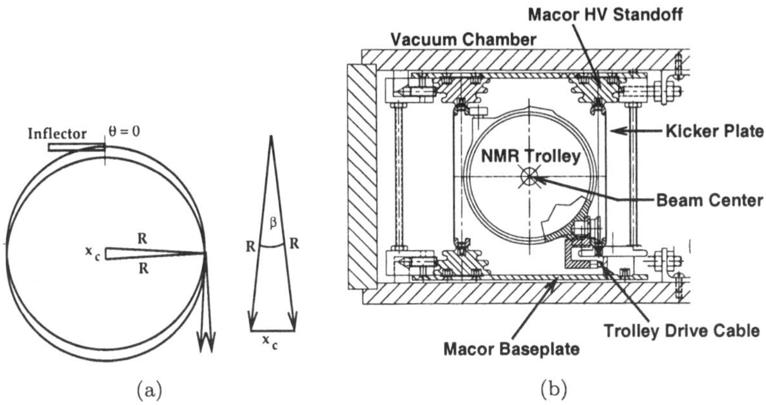

Fig. 11.7. (a) A sketch of the beam geometry. $R = 7112$ mm is the storage ring radius, $x_c = 77$ mm is the distance between the inflector center and the center of the storage region. This is also the distance between the centers of the circular trajectory that a particle entering at the inflector center (at $\theta = 0$ with $x' = 0$) will follow, and the circular trajectory a particle at the center of the storage volume (at $\theta = 0$ with $x' = 0$) will follow. (b) An elevation view of the kicker plates, showing the ceramic cage supporting the kicker plates, and the NMR trolley riding on the kicker plates. (The trolley is removed during data collection).

of the fast muon kicker is to briefly reduce a portion of the main storage field in order to move the center of the muon orbit to the geometric center of the storage ring. The 77-mm offset at the injection point, between the center of the entering beam and the central orbit, requires that the beam be kicked outward by approximately 10 mrad. The kick should be made at about 90 degrees around the ring, plus a few degree correction due to the defocusing effect of the electric quadrupoles between the injection point and the kicker, as shown in Figs. 11.5(b) and 11.7(a).

The requirements on the fast muon kicker are rather stringent. While electric, magnetic, and combination electromagnetic kickers were considered, the collaboration settled on a magnetic kicker design [42] because it was thought to be technically easier and more robust than the other options. Because of the very stringent requirements on the storage ring magnetic field uniformity, no magnetic materials could be used. Thus the kicker field had to be generated and shaped solely with currents, rather than using ferrite cores. Even with the kicker field generated by currents, there existed the potential problem of inducing eddy currents which might affect the magnetic field seen by the stored muons.

The length of the kicker is limited to the ~ 5 m azimuthal space between the electrostatic quadrupoles (see Fig. 11.4), so each of the three sections is 1.76 m long. The cross section of the kicker is shown in Fig. 11.7(b). The two parallel conductors are connected with cross-overs at each end, forming a single current loop. The kicker plates also have to serve as "rails" for the NMR field-mapping trolley (discussed below), and the trolley is shown riding on the kicker rails in Fig. 11.7(b).

The kicker current pulse is formed by an under-damped LCR circuit. A capacitor is charged to 95 kV through a resonant charging circuit. Just before the beam enters the storage ring, the capacitor is shorted to ground by firing a deuterium thyratron. The peak current in an LCR circuit is given by $I_0 = V_0/(\omega_d L)$ making it necessary to keep the system inductance, L, low to maximize the magnetic field for a given voltage V_0. For this reason, the kicker was divided into three sections, each powered by a separate pulse-forming network. The resulting current waveform is shown in Fig. 11.8.

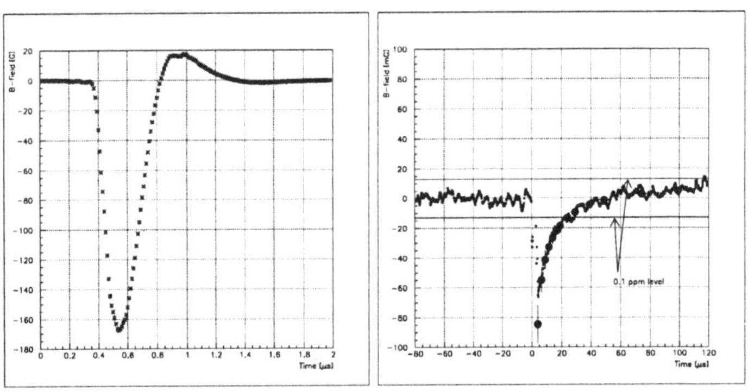

(a) (b)

Fig. 11.8. (a) The main magnetic field of the kicker measured with the Faraday effect. (b) The residual magnetic field measured using the Faraday effect. The solid points are calculations from OPERA [43], and the small $\times$ are the experimental points measured with the Faraday effect. The solid horizontal lines show the ± 0.1 ppm band for affecting $\int \vec{B} \cdot d\vec{\ell}$.

The cyclotron period of the ring is 149 ns, substantially less than the kicker pulse base-width of $\sim$ 400 ns, so that the injected beam is kicked on the first few turns. Nevertheless, approximately 10^4 muons are stored per fill of the ring, corresponding to an injection efficiency of about 3 to 5% (ratio of stored to incident muons). The storage efficiency with muon injection is much greater than that obtained with pion injection for a given proton flux, with only 1% of the hadronic flash.

11.2.2.3. *The electrostatic quadrupoles*

The electric quadrupoles, which are arranged around the ring with four-fold symmetry, provide vertical focusing for the stored muon beam. The quadrupoles cover 43% of the ring in azimuth, as shown in Fig. 11.4. While the ideal vertical profile for a quadrupole electrode would be hyperbolic, beam dynamics calculations determined that the higher multipoles present with flat electrodes, which are much easier to fabricate, would not cause an unacceptable level of beam losses. The flat electrodes are shown in Fig. 11.9 Only certain multipoles are permitted by the four-fold symmetry, and a judicious choice of the electrode width relative to the separation between opposite plates minimizes the lowest of these. With this configuration, the 20-pole is the largest, being 2% of the quadrupole component and an order of magnitude greater than the other allowed multipoles [47].

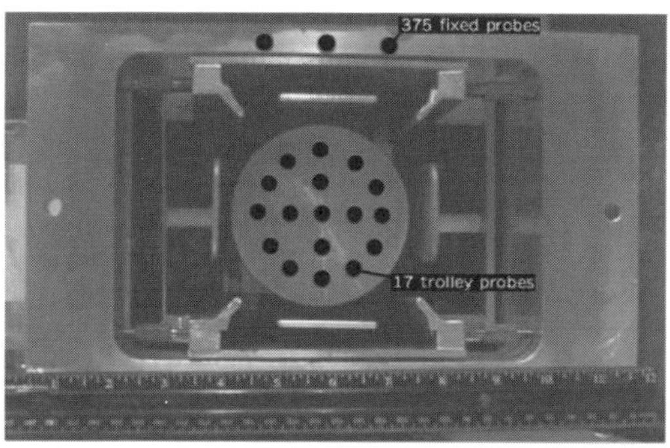

Fig. 11.9. A photograph of an electrostatic quadrupole assembly inside a vacuum chamber. The cage assembly doubles as a rail system for the NMR trolley which is resting on the rails. The location of the NMR probes inside the trolley are shown as black circles. The probes are located just behind the front face. The inner (outer) circle of probes has a diameter of 3.5 cm (7 cm) at the probe centers. The storage region has a diameter of 9 cm. The vertical location of three upper fixed probes is also shown. The fixed probes are located symmetrically above and below the vacuum chamber.

In the quadrupole regions, the combined electric and magnetic fields can lead to electron trapping. The electron orbits run longitudinally along the inside of the electrode, and then return on the outside. Excessive trapping in the relatively modest vacuum of the storage ring can cause sparking.

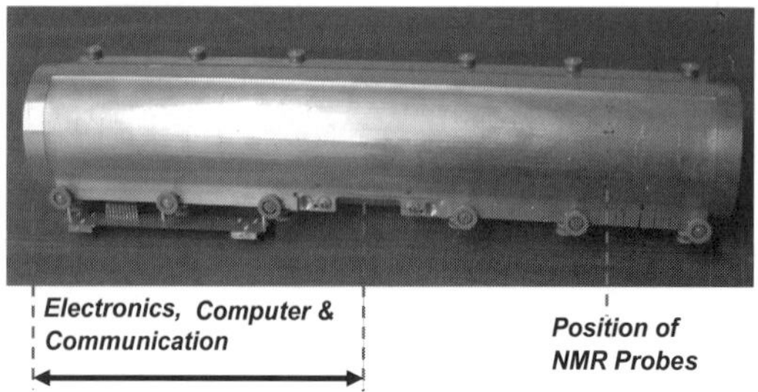

Fig. 11.10. A photograph of the NMR trolley. It carries a full NMR spectrometer which is controlled via an on-board micro computer. It is made from non-magnetic materials, such as the aluminum housing and PEEK wheels with glass ball bearings. The electronics components were selected individually not to contain any spurious ferromagnetic materials. The positions of the centers of the cylindrical NMR probes are indicated. (Photograph by K. Jungmann)

To minimize trapping, the leads were arranged to introduce a dipole field at the end of the quadrupole thus sweeping away trapped electrons. In addition, the quadrupoles were pulsed, so that after each fill of the ring all trapped particles could escape. Since some protons (antiprotons) were stored in each fill of μ^+ (μ^-), they were also released at the end of each storage time. This lead arrangement worked so well in removing trapped electrons that for the μ^+ polarity it would have been possible to operate the quadrupoles in a dc mode. For the storage of μ^-, this was not true; some electrons were trapped by the quadrupole field which caused sparking. This problem was reduced by reducing the vacuum in the storage ring by an order of magnitude, and limiting the storage time to less than 700 μs.

Beam losses during the measurement period, which could distort the expected time spectrum of decay electrons, had to be minimized. Beam scraping is used to remove, just after injection, those muons which would likely be lost later on. To this end, the quadrupoles are initially powered asymmetrically, and then brought to their final symmetric voltage configuration. The asymmetric voltages lower the beam and move it sideways in the storage ring. Particles whose trajectories reach too near the boundaries of the storage volume (defined by collimators placed at the ends of the quadrupole sectors) are lost. The scraping time was 17 μs during all data collection runs except 2001, where 7 μs was used. The muon loss rates without scraping were on the order of 0.6% per lifetime at late times in a fill, which dropped to $\sim 0.2\%$ with scraping.

11.2.2.4. *The storage ring magnet*

The storage ring magnet, along with the electrostatic quadrupoles, forms a weak-focusing betatron [38, 44, 45]. A pure quadrupole electric field provides a linear restoring force in the vertical direction, and the combination of the (defocusing) electric field and the central (dipole) magnetic field (B_0) provides a net linear restoring force in the radial direction. The important parameter is the field index, n, which is defined by

$$n = \frac{\kappa R_0}{\beta B_0}, \tag{11.16}$$

where κ is the electric quadrupole gradient and R_0 is the storage ring radius. For a ring with a uniform vertical dipole magnetic field and a uniform quadrupole field that provides vertical focusing covering the full azimuth, the stored particles undergo simple harmonic motion called betatron oscillations, in both the radial and vertical dimensions.

The horizontal and vertical motion are given by

$$x = x_e + A_x \cos(\nu_x \frac{s}{R_0} + \delta_x) \quad \text{and} \quad y = A_y \cos(\nu_y \frac{s}{R_0} + \delta_y), \tag{11.17}$$

where s is the arc length along the trajectory, and $R_0 = 7112$ mm is the radius of the central orbit in the storage ring. The horizontal and vertical tunes are given by $\nu_x = \sqrt{1-n}$ and $\nu_y = \sqrt{n}$. Several n - values were used in E821 for data acquisition: $n = 0.137$, 0.142 and 0.122. The horizontal and vertical betatron frequencies are given by

$$f_x = f_C\sqrt{1-n} \simeq 0.929 f_C \quad \text{and} \quad f_y = f_C\sqrt{n} \simeq 0.37 f_C, \tag{11.18}$$

where f_C is the cyclotron frequency and the numerical values assume that $n = 0.137$. The corresponding betatron wavelengths are $\lambda_{\beta_x} = 1.08(2\pi R_0)$ and $\lambda_{\beta_y} = 2.7(2\pi R_0)$. It is important that the betatron wavelengths are not simple multiples of the circumference, as this minimizes the ability of ring imperfections and higher multipoles to drive resonances that would result in particle losses from the ring. Since the electrostatic quadrupoles are not continuous, these equations are only approximately correct. We return to the topic of beam dynamics in the ring later.

The use of electrostatic focusing permits the magnetic field to be as uniform as possible and thus measured to excellent precision with NMR techniques. The design goal of ±1 ppm was placed on the field uniformity when averaged over azimuth in the storage ring. A "superferric" design, where the field configuration is largely determined by the shape and magnetic properties of the iron, rather than by the current distribution in the superconducting coils, was chosen. To reach the ppm level of uniformity it was important to minimize discontinuities such as holes in the yoke, spaces between adjacent pole pieces, and especially the spacing between pole pieces across the magnet gap containing the beam vacuum chamber. Every effort was made to minimize penetrations in the yoke, and where they are necessary, such as for the beam entrance channel, additional iron is placed around the hole to minimize the effect of the hole on the magnetic flux circuit.

The storage ring, shown in Fig. 11.11, is designed as a continuous C-magnet [46] with the yoke made up of twelve sectors with minimum gaps where the yoke pieces come together. A cross-section of the magnet is shown in Fig. 11.11(b). The largest gap between adjacent yoke pieces after assembly is 0.5 mm. The pole pieces are built in 36 pieces, with keystone rather than radial boundaries to ensure a close fit. They are electrically isolated from each other with 80 μm kapton to prevent eddy currents from running around the ring, especially during a quench or energy extraction from the magnet. The vertical mismatch from one pole piece to the next when going around the ring in azimuth is held to ± 10 μm, since the field strength depends critically on the pole-piece spacing across the magnet gap.

The field is excited by 14 m-diameter superconducting coils, which in 1996 were the largest-diameter such coils ever fabricated. The coil at the outer radius consists of two identical coils on a common mandrel, above and below the plane of the beam, each with 24 turns. Each of the inner-radius coils, which are housed in separate cryostats, also consist of 24 turns (see Figs. 11.5(b)). The nominal operating current is 5200 A, which is driven by a power supply. The choice of using an extremely stable power supply, further stabilized with feedback from the average reading of some 20 selected representative NMR probes, was chosen over operating in a "persistent mode", for two reasons. The switch required to change from the powering mode to persistent mode was not available because it was beyond state-of-the-art technology. Furthermore, unlike the usual superconducting magnet operated in persistent mode, we anticipated the need to cycle the magnet power a number of times during a three-month running period.

At the design stage, calculations suggested that the field could be made quite uniform, and that when averaged over azimuth, a uniformity of ± 1 ppm could be achieved. It was anticipated that, at the initial turn-on of the magnet, the field would have a uniformity of about 1 part in 10^4, and that an extensive program of shimming would be necessary to reach a uniformity of 1 ppm. A number of tools for shimming the magnet were

Fig. 11.11. The storage-ring magnet. The magnet yoke is covered with thermal insulation. The 24 detector stations are positioned inside the ring next to the vacuum chambers. The three kickers are visible at the top of this picture. The racks in the center are the quadrupole pulsers and the kicker driving electronics. The magnet power supply is in the upper left, above the plane of the ring. (Photograph by K. Jungmann)

therefore built into the design. The air gap between the yoke and pole pieces dominates the reluctance of the magnetic circuit outside of the gap that includes the storage region, and decouples the field in the storage region from possible voids, or other defects in the yoke steel. Iron wedges placed in the air gap were ground to the wedge angle needed to cancel the quadrupole field component inherent in a C-magnet. The dipole can be tuned locally by moving the wedge radially. The edge shims bolted to the pole pieces cancel the sextupole component of the field.

After mechanical shimming, the higher multipoles were found to be quite constant in azimuth. They are shimmed out on average by adjusting currents in conductors placed on printed circuit boards going around the ring in concentric circles spaced by 2.5 mm. These boards are glued to the top and bottom pole faces between the edge shims and connected at the pole ends to form a total of 240 concentric circles of conductor, connected in groups of four, to sixty ± 1 A power supplies. These correction coils are quite effective in shimming multipoles up through the octupole. Multipoles higher than octupole are less than 1 ppm at the edge of the storage aperture, and, with our use of a circular storage aperture, are unimportant in determining the average magnetic field seen by the muon beam.

11.2.2.5. *Monitoring of the magnetic field*

The magnetic field is measured and monitored by pulsed Nuclear Magnetic Resonance of protons in water samples [48]. The free induction decay (FID) is picked up by the coil L_S in Fig. 11.12 after a pulsed excitation rotates the proton spin in the sample by 90° to the magnetic field. The proton response signal at frequency f_{NMR} is measured by counting its zero crossings within a well-measured time period the length of which is automatically adjusted to approximately the decay time (1/e) of the FID. It is mixed with a stable reference frequency and filtered to arrive at the difference frequency f_{FID} chosen to be typically in the 50 kHz region. The reference frequency of $f_{\mathrm{ref}} = 61.74$ MHz is obtained from a frequency synthesizer, which is phase locked to a LORAN C secondary frequency standard [51], and it is chosen such that always $f_{\mathrm{ref}} < f_{\mathrm{NMR}}$. The very same LORAN C device also provides the time base for the ω_a measurement. The relationship between

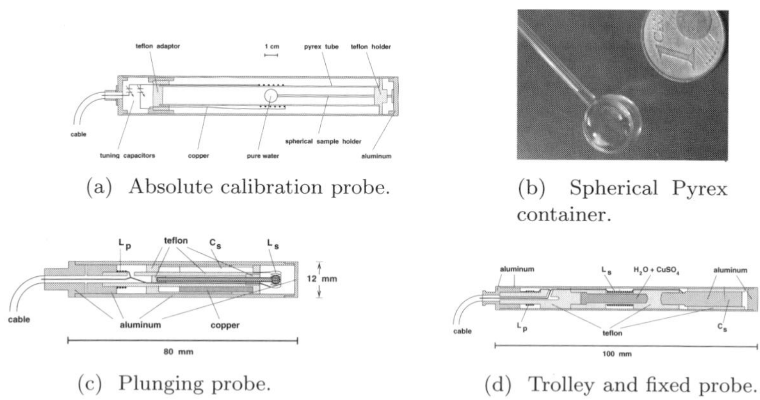

(a) Absolute calibration probe.

(b) Spherical Pyrex container.

(c) Plunging probe.

(d) Trolley and fixed probe.

Fig. 11.12. The different NMR probes. (a) Absolute probe featuring a spherical sample of water. This probe and all its driving and readout electronics are the very same devices employed in reference [28] to determine λ, the muon-to-proton magnetic-moment ratio. (b) The spherical Pyrex container for the absolute probe. (c) Plunging probe, which can be inserted into the vacuum at a specially shimmed region of the storage ring to transfer the calibration to the trolley probes. (d) The standard probes used in the trolley and as fixed probes. The resonant circuit is formed by the two coils with inductances L_s and L_p and a capacitance C_s made by the Al-housing and a metal electrode. (Figures (a,c,d) are reprinted with permission from [25]. Copyright 2006 by the American Physical Society. Photograph in (b) by K. Jungmann.)

the actual field B_{real} and the field corresponding to the reference frequency is given by

$$B_{\text{real}} = B_{\text{ref}} \left(1 + \frac{f_{\text{FID}}}{f_{\text{ref}}} \right). \tag{11.19}$$

The field measurement process has three aspects: calibration, monitoring the field during data collection, and mapping the field. The probes used for these purposes are shown in Fig. 11.12. To map the field, an NMR trolley [50] was built with an array of 17 NMR probes arranged in concentric circles, as shown in Fig. 11.9. While it would be preferable to have information over the full 90-mm aperture, space limitations inside the vacuum chamber, which can be understood by examining Figs. 11.9, 11.7, prevent a larger diameter trolley.

The trolley is built from non-magnetic materials and has a fully functional CPU on-board which controls a full FID excitation and zero crossing counting spectrometer. It is pulled around the storage ring by two cables, one in each direction circling the ring. One of these cables is a thin co-axial cable with only copper conductors and Teflon dielectric and outside protective coating (Suhner 2232-08). It carries simultaneously the dc supply voltage, the reference frequency f_{ref} and two-way communication with the spectrometer via RS232 standard. The other cable is non-conducting nylon (fishing line) to eliminate pickup from the pulsed high voltage on the kicker electrodes.

During muon decay data-collection periods, the trolley is parked in a garage (see Fig. 11.4) in a special vacuum chamber. Every few days, at *random times*, the field is mapped using the trolley. During mapping, the trolley is moved into the storage region and over the course of 2 hours is pulled around the vacuum chamber, measuring the field at some 100,000 points by continuously cycling through the 17 probes while moving. Data were recorded in both possible directions of movement. During the approximately three-month data-collection runs, the storage-ring magnet remains powered continuously for periods lasting from five to twenty days; thus the conditions during mapping are identical to those during the data collection.

To cross calibrate the trolley probes, a two-axis non-magnetic manipulator made from aluminum and titanium only, including titanium bellows, and driven by non-magnetic piezo motors was developed. It was placed at one location in the ring and it permits a special NMR plunging probe, or an absolute calibration probe with a spherical water sample [49], to plunge into the vacuum chamber. In this way the trolley probes can be calibrated by transferring the absolute calibration from the calibration probe shown in Fig. 11.12 to individual probes in the trolley. These measurements of the field at the same spatial point with the plunging, calibration and trolley probes provide both relative and absolute calibration of the trolley probes. During the calibration measurements before, after and occasionally randomly during each running period, the spherical water probe is used to calibrate the plunging probe, and with this then the trolley probes. The absolute calibration probe provides the calibration to the Larmor frequency of the free proton [53], which is called ω_p below.

To monitor the field on a continuous basis during data collection, a total of 378 NMR probes are placed at fixed locations in grooves machined into the outside upper and lower surfaces of the vacuum chamber around the ring. Of these, about half provide useful data for monitoring the field with time. Some of the others are noisy, or have cables damaged over the years or other problems, but a significant number of fixed probes are located in regions near the pole-piece boundaries where the magnetic gradients are sufficiently large to reduce the free-induction decay time in the probe, limiting the precision on the frequency measurement. The number of probes at each azimuthal position around the ring alternates between two and three, at radial positions arranged symmetrically about the magic radius of 7112 mm. Because of this geometry, the fixed probes provide a good monitor of changes in the dipole and quadrupole components of the field around the storage ring.

Initially the trolley and fixed probes contained cylindrical water samples. Over the course of the experiment, the water samples in many of the probes were replaced with petroleum jelly. The jelly has several advantages over water: Low evaporation, favorable relaxation times at room temperature, a proton NMR signal almost comparable to that from water, and a chemical shift (and the accompanying NMR frequency shift) with a temperature coefficient much smaller than that of water, and thus negligible for our experiment.

11.2.2.6. *The detection of the decay electrons*

The detector system consists of a variety of particle detectors: calorimeters, position-sensitive hodoscope detectors, and a set of tracking chambers. There are also horizontal and vertical arrays of scintillating-fiber hodoscopes which could be temporarily inserted into the storage region. A number of custom electronics modules were developed, including event simulators, multi-hit time-to-digital converters(MTDC), and the waveform digitizers (WFD) which are at the heart of the measurement. We refer to the data collected from one muon injection pulse in one detector as a "spill," and we will speak of "early-to-late" effects: namely, the gain or time stability requirements in a given detector at early compared to late decay times in a spill.

The electromagnetic calorimeters [59], together with the custom WFD readout system, are the primary source of data for determining the precession frequency. They provide the energies and arrival times of the electrons, and they also provide signal information immediately before and after the electron pulses, allowing studies of baseline changes and pulse pile-up.

There are 24 lead-scintillating fiber calorimeters [25, 59] placed evenly around the 45-m circumference of the storage region, adjacent to the inside radius of the storage vacuum region as shown in Figs. 11.4 and 11.11. The calorimeters are read out with custom waveform digitizers [25]. Nearly all decay electrons have momenta ($0 < p(lab) < 3.1$ GeV/c, see Fig. 11.1) below the stored muon momentum (3.1 GeV/c $\pm 0.2\%$), and they are swept by the B-field to the inside of the ring where they can be intercepted by the calorimeters. The storage-region vacuum chamber is scalloped so that electrons pass nearly perpendicular to the vacuum wall before entering the calorimeters, minimizing electron pre-showering (see Fig. 11.4). The calorimeters are positioned and sized in order to maximize the acceptance of the highest-energy electrons, which have the largest statistical figure of merit NA^2. The variations of N and A as a function of electron energy are shown in Figs. 11.1 and 11.2.

The electrons with the lowest laboratory energies, while more numerous than high-energy electrons, generally have a lower figure of merit and therefore carry relatively little information on the precession frequency. These electrons have relatively small radii of curvature, and exit the ring vacuum chamber closer to the radial direction than electrons at higher energies, with most of them missing the detectors entirely. Detection of these electrons would require detectors that cover a much larger portion of the circumference than is needed for high-energy electrons, and is not cost effective.

Consequently the detector system is designed to maximize the acceptance of the high-energy decay electrons above approximately 1.8 GeV, with the acceptance falling rapidly below this energy. The detector acceptance reaches a maximum of 87% at 2.3 GeV, decreases to 70% at 1.8 GeV, and continues to decrease roughly linearly to zero as the energy decreases. With increasing energy above 2.3 GeV, the acceptance also decreases because the highest-energy electrons tend to enter the calorimeters at the outer radial edge, increasing the loss of registered energy due to shower leakage, and reducing the acceptance to 80% at 3 GeV.

In a typical analysis, the full data sample consists of all electrons above a threshold energy of about 1.8 GeV, where NA^2 is approximately a maximum, with about 65% of the electrons above that energy detected (Fig. 11.2). The average asymmetry is about 0.35. The loss of efficiency is from the low-energy tail in the detector response characteristic of electromagnetic showers in calorimeters, and from lower energy electrons missing the detectors altogether. The statistical error improves by only 5% if the data sample contains all electrons above 1.8 GeV compared to all above 2.0 GeV. For threshold energies below 1.7 GeV, the decline in the average asymmetry more than cancels the additional number of electrons in NA^2, and the statistical error actually increases. Some of the independent analyses fit time spectra of data formed from electrons in narrow energy bands (about 200 MeV wide). When the results of the separate fits are combined, there was is a 10% reduction in the statistical error on ω_a. However, there is also a slight increase in the systematic error contribution from gain shifts, because the relative number of events moved by a gain shift from one energy band to another increased. One analysis used data weighted by the asymmetry as a function of energy. It can be shown [61] that this produces the same statistical improvement as dividing the data into energy bands.

Gain and timing shift limitations are much more stringent within a single spill than from spill to spill. Shifts at late decay times compared to early times in a given spill, so-called "early-to-late" shifts, can lead directly to serious systematic errors on ω_a. Shifts of gain or the $t = 0$ point from one spill to the next are generally much less serious; they will usually only change the asymmetry, average energy, phase, etc., but to the extent that the measured distribution of particles follows Eq. (11.12), to first order ω_a will be unaffected.

The calorimeters should have pulses with narrow time widths to minimize the probability of two pulses overlapping (pile-up) during the very high electron decay data rates encountered at early decay times, which can reach a MHz in a single detector. The scintillator is chosen to have minimal long-lived components to reduce the afterglow from the intense detector flash associated with beam injection. Laser calibration studies show that the timing stability for a typical detector over any 200 microsecond time interval is better than 15 ps, easily meeting the demands of the measurement of a_μ. For example, a 20 ps timing shift would lead to an uncertainty in a_μ of about 0.1 ppm, which is small compared to the final error. Modest detector energy resolution ($\approx 10-15\%$ at 2 GeV) is required in order to select the desired high-energy electrons for analysis. Better energy resolution also reduces the amount of calibration data needed to monitor the stability of the detector gains.

The stability requirement for the electron energy measurement ("gain") versus time in the spill is largely determined by the energy dependence of the phase of the $(g-2)$ oscillation. In a fit of the data to the 5-parameter function, the oscillation phase is highly correlated to ω_a. Therefore a shift in the gain from early to late decay times, combined with an energy dependence in the (g-2) phase, can lead to a systematic error in the determination of ω_a. There are two main contributing factors to the energy dependence, which appear with opposite signs: 1) The phase ϕ in the 5-parameter function (Eq. (11.12)) depends on the electron drift time. High-energy electrons must travel further, on average, from the point of muon decay to the detector and therefore have longer drift times than low-energy electrons. The change in drift time with energy implies a corresponding energy dependence in the $(g-2)$ phase. 2) For decay electrons at a given energy, those with positive (radially outward) components of momentum at the muon decay point travel further to reach the detectors than electrons with negative (radially inward) components. They spread out more in the vertical direction and may miss the detectors entirely. Consequently, electrons with positive (outward) radial momentum components will have slightly lower acceptance than those with negative components, causing the average spin direction to rotate slightly, leading to a shift in ϕ. Recalling the correlation between electron direction and muon spin, the overall effect is to shift the time at which the number oscillation reaches its maximum, causing a shift in the precession phase in the 5-parameter function. The size of the shift depends on the electron energy. From studies of the data sample and simulations, it is established that the detector gains need to be stable to better than 0.2% over any 200 μs time interval in a spill, in order to keep the systematic error contribution to ω_a less than 0.1 ppm from gain shifts. This requirement is met by all of the calorimeters. The gain from one spill to the next is not coupled to the precession frequency and therefore the requirement on the spill to spill stability is far less stringent than the stability requirement within an individual spill.

In one spill during the time interval from a few tens of microseconds to 640 μs after injection, approximately 20 decay-electrons above 1.8 GeV are recorded on average in each detector. The instantaneous rate of decay electrons above 1 GeV changes from about 300 kHz to almost zero over this period. The gain and timing of photo-multipliers can depend on the data rate. The necessary gain and timing stability is achieved with custom, actively stabilized photo-multiplier bases [60].

To prevent paralysis of the photo-multipliers due to the injection flash, the amplifications in the photo-multipliers are temporarily reduced by a factor of about 1 million during the beam injection. Depending on the intensity of the flash and the duration of the background levels encountered at a particular detector station, the amplifications are restored at times between 2 to 50 μs after injection. The switching of the gain is accomplished in the Hamamatsu R1828 photo-multipliers by swapping the bias voltages on dynodes 4 and 7. With the proper selection of the delay time after injection to let backgrounds die down, the gains typically return to 99.8% of their steady-state value within several microseconds after the tube is turned back on. Other gating schemes, such as switching the photocathode voltage, were found to have either required a much longer time for the gain to recover, or failed to give the necessary reduction in the gain when the tube is gated off.

Several specialized detector systems are employed in the ring to give information complementary to that from the calorimeters: Vertical hodoscopes called front scintillation detectors (FSD); finer-grained $x - y$ hodoscopes called position-sensitive detectors (PSD); two different $x - y$ scintillating fiber "harp-like" arrays located at 180° and 270° that could be placed on demand directly into the beam; and a straw-tube based traceback detector. These different systems provided information on the phase-space parameters of the stored muon beam and their decay electrons. Such measurements are compared to simulation results, and are important, for example, in the study of coherent betatron motion of the stored beam and detector acceptances, and in placing a limit on the electric dipole moment of the muon. A modest knowledge of the beam phase space is necessary in order to calculate the average magnetic and electric fields seen by the stored muons. See Refs. [25, 35] for further details.

For each calorimeter, the arrival times of the signals from the four photo-multipliers are matched to within a nanosecond, and the analog sum is formed. The resulting signal is fed to a custom waveform digitizer (WFD) with 400 MHz equivalent sampling rate, which provides several pulse height samples from each candidate electron. WFD data are added to the data stream only if a trigger is formed, i.e. when the energy associated with a pulse exceeds a pre-assigned threshold, usually taken to be 900 MeV. When a trigger occurs, WFD samples from about 15 ns before the pulse to about

65 ns after the pulse are recorded. There is the possibility of two or more electron pulses being over-threshold in the same 80-ns time window. In that case, the length of the readout period is extended to include both pulses. At the earliest decay times, the detector signals have a large pedestal due to the lingering effects of the injection "flash," and some of the upstream detectors are continuously over-threshold at early times and therefore deliver data continuously to the data acquisition system.

The energy and time of an electron is obtained by fitting a standard pulse shape to the WFD pulse using a conventional χ^2 minimization. The standard pulse shape is established for each calorimeter. It is based on an average of the shapes of a large number of late-time pulses where the problems associated with overlapping pulses and backgrounds are greatly reduced. There are three fitting parameters, time and height of the pulse, and the constant pedestal, with the fits typically spanning 15 samples centered on the pulse. The typical time resolution of an individual electron was about 60 ps.

The period after each pulse is searched for any additional pulses from other electrons. These accidental pulses have the advantage that they do not need to be over the hardware threshold ($\sim$ 900 MeV), but rather over the much lower ($\sim$ 250 MeV) minimum pulse height that can be discriminated from background by the pulse-fitting algorithm. A pile-up spectrum is constructed by combining the triggering pulses and the following accidental pulses [25].

Zero time for a given fill is defined by the trigger pulse to the AGS kicker magnet that extracts the proton bunch and sends it to the pion production target. The resolution of the zero time needs only to be much less than the $(g-2)$ precession period of 4.4 μs in order to minimize loss of the asymmetry amplitude.

A pulsed UV laser signal is fanned out simultaneously by means of an optical fiber system to all elements of the calorimeter stations to monitor the gain and time stabilities. The average timing stability is typically found to be better than 10 ps in any 200 μs-interval when averaged over a number of events, with many stable to 5 ps. This level of timing instability contributes less than a 0.05 ppm systematic error on ω_a.

11.2.3. *Beam dynamics in the storage ring*

The behavior of the beam in the $(g-2)$ storage ring directly affects the measurement of a_μ. Since the detector acceptance for decay electrons depends on the radial coordinate of the muon at the point where it decays, coherent radial motion of the stored beam can produce an amplitude modulation in the observed electron time spectrum. Resonances in the storage ring can cause particle losses, thus distorting the observed time spectrum, and must be avoided when choosing the operating parameters of the ring. Care must be taken in setting the frequency of coherent radial beam motion, the "coherent betatron oscillation" (CBO) frequency, which lies close to the second harmonic of $f_a = \omega_a/(2\pi)$. If $f_{\rm CBO}$ is too close to $2f_a$ the difference frequency $f_- = f_{CBO} - f_a$ complicates the extraction of f_a from the data, and can introduce a significant systematic error.

As mentioned above, the relevant parameter to describe the betatron motion is the field index (Eq. (11.16)) $n = (\kappa R_0)/(\beta B_0)$, where κ is the electric quadrupole gradient. The field index, n, determines the oscillation frequencies as well as the acceptance of the storage ring. The maximum horizontal and vertical angles of the muon momentum are given by

$$\theta^x_{\rm max} = x_{\rm max}\frac{\sqrt{1-n}}{R_0}, \quad \text{and} \quad \theta^y_{\rm max} = y_{\rm max}\frac{\sqrt{n}}{R_0}, \tag{11.20}$$

where $x_{\rm max}, y_{\rm max} = 45$ mm is the radius of the storage aperture. For a betatron amplitude A_x or A_y (see Eqs. (11.17) and (11.18)) less than 45 mm, the maximum angle is reduced, as can be seen from the above equations.

Resonances in the storage ring will occur if $L\nu_x + M\nu_y = N$, where L, M and N are integers, which must be avoided in choosing the operating value of the field index. These resonances form straight lines on the tune plane shown in Fig. 11.13, which shows resonance lines up to fifth order. The operating point lies on the circle $\nu_x^2 + \nu_y^2 = 1$.

For a ring with discrete quadrupoles, the focusing strength changes as a function of azimuth, and the equation of motion looks like an oscillator whose spring constant changes as a function of azimuth s. The motion is described by

$$x(s) = x_e + A\sqrt{\beta(s)}\cos(\psi(s) + \delta), \tag{11.21}$$

where $\beta(s)$ is one of the three Courant–Snyder parameters [45]. The layout of the storage ring is shown in Fig. 11.4. The four-fold symmetry of the quadrupoles was chosen because it provided quadrupole-free regions for the kicker, traceback chambers, fiber monitors, and trolley garage; but the most important benefit of four-fold symmetry over the two-fold used at CERN [24] is that $\sqrt{\beta_{\rm max}/\beta_{\rm min}} = 1.03$. The two-fold symmetry used at

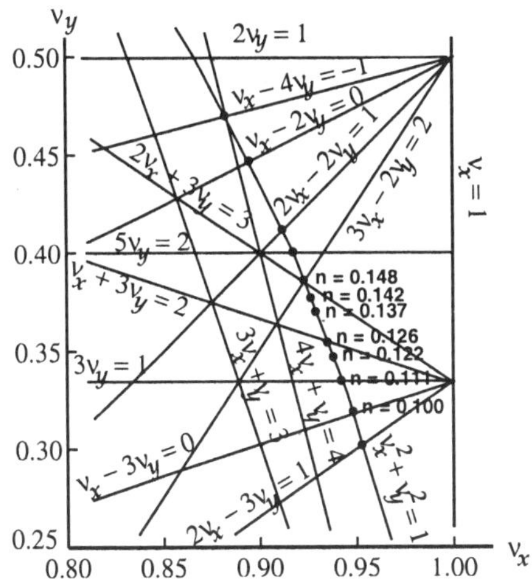

Fig. 11.13. The tune plane, showing the three operating points used during our three years of running. (This figure was reprinted with permission from [25]. Copyright 2006 by the American Physical Society.)

CERN [24] gives $\sqrt{\beta_{\max}/\beta_{\min}} = 1.15$. The CERN magnetic field had significant non-uniformities on the outer portion of the storage region, which when combined with the 15% beam "breathing" from the quadrupole lattice made it much more difficult to determine the average magnetic field weighted by the muon distribution (Eq.(11.6)).

The detector acceptance depends on the radial position of the muon when it decays, so that any *coherent* radial beam motion will amplitude modulate the decay $e^{\pm}$ distribution. The principal frequency will be the "Coherent Betatron Frequency,"

$$f_{\mathrm{CBO}} = f_C - f_x = (1 - \sqrt{1-n}) f_C \simeq 470 \text{ kHZ}, \tag{11.22}$$

which is the frequency at which a single fixed detector sees the beam coherently moving back and forth radially. This CBO frequency is close to the second harmonic of the $(g-2)$ frequency, $f_a = \omega_a/2\pi \simeq 228$ Hz.

An alternative way of thinking about the CBO motion is to view the ring as a spectrometer where the inflector exit is imaged at each successive betatron wavelength, λ_{β_x}. In principle, an inverted image appears at half a betatron wavelength; but the radial image is spoiled by the $\pm 0.3\%$ momentum dispersion of the ring. A given detector will see the beam move radially with the CBO frequency, which is also the frequency at which the horizontal waist precesses around the ring. Since there is no dispersion in the vertical dimension, the vertical waist (VW) is reformed every half wavelength $\lambda_{\beta_y}/2$. A number of frequencies in the ring are tabulated in Table 11.3.

Table 11.3. Frequencies in the $(g-2)$ storage ring, assuming that the quadrupole field is uniform in azimuth and that $n = 0.137$.

Quantity	*Expression*	*Frequency*	*Period*
f_a	$\frac{e}{2\pi m} a_\mu B$	0.228 MHz	4.37 μs
f_c	$\frac{v}{2\pi R_0}$	6.7 MHz	149 ns
f_x	$\sqrt{1-n} f_c$	6.23 MHz	160 ns
f_y	$\sqrt{n} f_c$	2.48 MHz	402 ns
$f_{\rm CBO}$	$f_c - f_x$	0.477 MHz	2.10 μs
$f_{\rm VW}$	$f_c - 2f_y$	1.74 MHz	0.574 μs

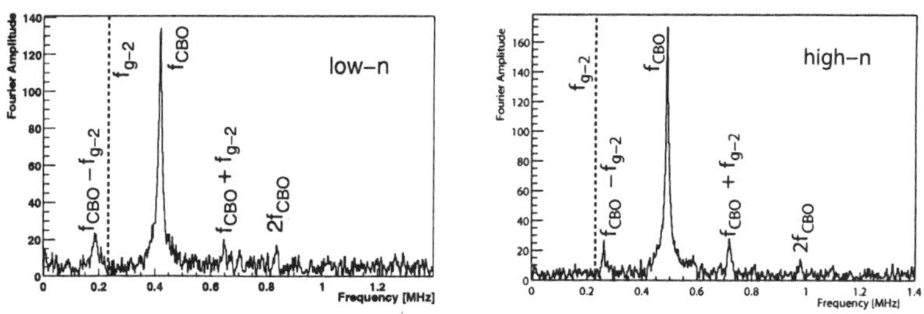

(a) (b)

Fig. 11.14. The Fourier transform to the residuals from a fit to the five-parameter function, showing clearly the coherent beam frequencies. (a) is from 2001 using the low n-value, (b) is from 2001 using the high n-value. (This figure was reprinted with permission from [34]. Copyright 2004 by the American Physical Society.)

The CBO frequency and its sidebands are clearly visible in the Fourier transform to the residuals from a fit to the five-parameter fitting function Eq. (11.12), and are shown in Fig. 11.14. The vertical waist frequency is barely visible. In 2000, the quadrupole voltage was set such that the CBO frequency was uncomfortably close to the second harmonic of f_a, thus placing the difference frequency $f_- = f_{CBO} - f_a$ next to f_a. This nearby sideband forced us to work very hard to understand the CBO and how its related phenomena affect the value of ω_a obtained from fits to the data. In 2001, we carefully set f_{CBO} at two different values, one well above, the other well below $2f_a$, which greatly reduced this problem.

11.2.3.1. *The muon beam profile*

Three tools are available to us to monitor the muon distribution. Study of the beam de-bunching after injection yields information on the distribution of equilibrium radii in the storage ring. The FSDs provide information on the vertical centroid of the beam. The wire chamber system and the fiber beam monitors, described above, also provide valuable information on the properties of the stored beam.

The beam bunch that enters the storage ring has a time spread with $\sigma \simeq 23$ ns, while the cyclotron period is 149 ns. The momentum distribution of stored muons produces a corresponding distribution in radii of curvature. The distributions depend on the phase-space acceptance of the ring, the phase space of the beam at the injection point, and the kick given to the beam at injection. The narrow horizontal dimension of the beam at the injection point, about 18 mm, restricts the stored momentum distribution to about $\pm 0.3\%$. As the muons circle the ring, the muons at smaller radius (lower momentum) eventually pass those at larger radius repeatedly after multiple transits around the ring, and the bunch structure largely disappears after 60 μs.

Only muons with orbits centered at the central radius have the "magic" momentum, so knowledge of the momentum distribution, or equivalently the distribution of equilibrium radii, is important in determining the correction to ω_a caused by the radial electric field used for vertical focusing. Two methods of obtaining the distribution of equilibrium radii from the beam debunching are employed in E821. One method uses a model of the time evolution of the bunch structure. A second, alternative procedure uses modified Fourier techniques [62]. The results from these analyses are shown in Fig. 11.15. The discrete points were obtained using the model, and the dotted curve was obtained with the modified Fourier analysis. The two analyses agree. The measured distribution is used both in determining the average magnetic field seen by the muons and the radial electric field correction discussed below.

11.2.3.2. *Corrections to ω_a: pitch and radial electric field*

If the velocity is not transverse to the magnetic field, or if a muon is not at $\gamma_{\rm magic}$, the difference frequency is modified as indicated in Eq.((11.5). Thus the measured frequency ω_a must be corrected for the effect of a radial electric field (because of the $\vec{\beta} \times \vec{E}$ term), and for the vertical pitching motion of the muons (which enters through the $\vec{\beta} \cdot \vec{B}$ term, see Eq. (11.5).

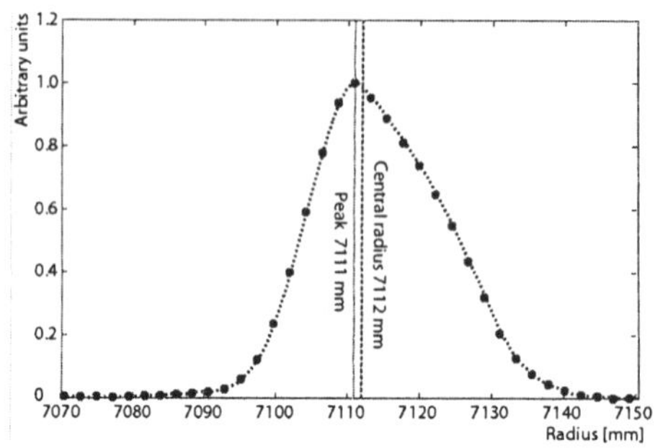

Fig. 11.15. The distribution of equilibrium radii obtained from the beam de-bunching. The solid circles are from a de-bunching model fit to the data, and the dotted curve is obtained from a modified Fourier analysis. (This figure was reprinted with permission from [25]. Copyright 2006 by the American Physical Society.)

These are the *only* corrections made to the ω_a data. The interested reader is referred to Ref. [35] for a derivation of the corrections for E821. For a general derivation the reader is referred to Refs. [26, 63].

The electric field introduces the correction

$$\frac{\Delta\omega}{\omega} = -2n(1-n)\beta^2 \frac{x x_e}{R_0^2 B_y}, \tag{11.23}$$

so clearly the effect of muons in the measurement sample which are not at the magic momentum is to lower the observed frequency. For a quadrupole focusing field plus a uniform magnetic field, the time average of x is just x_e, so the electric field correction is given by

$$C_E = \frac{\Delta\omega}{\omega} = -2n(1-n)\beta^2 \frac{\langle x_e^2 \rangle}{R_0^2 B_y}, \tag{11.24}$$

where $\langle x_e^2 \rangle$ is determined from the fast-rotation analysis. The uncertainty on $\langle x_e^2 \rangle$ is added in quadrature with the uncertainty in the placement of the quadrupoles of $\delta R = \pm 0.5$ mm (± 0.01 ppm), and with the uncertainty in the mean vertical position of the beam, ± 1 mm (± 0.02 ppm). For the low-n 2001 sub-period, $C_E = 0.47 \pm 0.054$ ppm.

The vertical betatron oscillations of the stored muons lead to $\vec{\beta} \cdot \vec{B} \neq 0$. Since the $\vec{\beta} \cdot \vec{B}$ term in Eq. (11.5) is quadratic in the components of $\vec{\beta}$, its contribution to ω_a will not generally average to zero. Thus the spin precession frequency has a small dependence on the betatron motion of the beam. It turns out that the only significant correction comes from the vertical betatron oscillation; therefore it is called the pitch correction (see Eq. (11.5)). As the muons undergo vertical betatron oscillations, the "pitch" angle between the momentum and the horizontal varies harmonically as $\psi = \psi_0 \cos \omega_y t$, where ω_y is the vertical betatron frequency $\omega_y = 2\pi f_y$, given in Eq. (11.18). To derive this correction, we assume that all muons are at the magic γ. The pitch correction is

$$C_p = -\frac{\langle \psi^2 \rangle}{2} = -\frac{\langle \psi_0^2 \rangle}{4} = -\frac{n}{4}\frac{\langle y^2 \rangle}{R_0^2}. \qquad (11.25)$$

The quantity $\langle y_0^2 \rangle$ was both determined experimentally from the traceback detector, and from simulations. For the 2001 period, $C_p = 0.27 \pm 0.036$ ppm, the amount the precession frequency is lowered from that given in Eq. (11.6) because $\vec{\beta} \cdot \vec{B} \neq 0$.

We see that both the radial electric field and the vertical pitching motion *lower* the observed frequency from the simple difference frequency $\omega_a = (e/m)a_\mu B$, which enters into our determination of a_μ using Eq. (11.8). Therefore our observed frequency must be *increased* by these corrections to obtain the measured value of the anomaly. Note that if $\omega_y \simeq \omega_a$ the situation is more complicated, with a resonance behavior that is discussed in References [26, 63].

11.2.4. *The determination of ω_a*

To obtain the muon spin precession frequency ω_a given in Eq. (11.4),

$$\vec{\omega}_a = -a_\mu \frac{q\vec{B}}{m}, \qquad (11.26)$$

which is observed as an oscillation of the number of detected electrons with time

$$N(t, E_{th}) = N_0(E_{th}) e^{-t/\gamma\tau}[1 + A(E_{th}) \cos(\omega_a t + \phi(E_{th}))], \qquad (11.27)$$

it is necessary to:

- Modify the five-parameter function above to include small effects such as the coherent betatron oscillations (CBO), pulse pile-up, muon losses, and gain changes, without adding so many free parameters that the statistical power for determining ω_a is compromised.
- Obtain an acceptable χ_R^2 per degree of freedom in all fits, i.e. consistent with 1, where $\sigma(\chi_R^2) = \sqrt{2/NDF}$.
- Insure that the fit parameters are stable independent of the starting time of the least-square fit. This was found to be a very reliable means of testing the stability of fit parameters as a function of the time after injection.

In general, fits are made to the data out to about 640 μs, about 10 muon lifetimes.

11.2.4.1. *Distribution of decay electrons*

Decay electrons with the highest laboratory energies, typically $E > 1.8$ GeV, are used in the analysis for ω_a, as discussed in Section 11.2.1. In the (excellent) approximations that $\vec{\beta} \cdot \vec{B} = 0$ and the effect of the electric field on the spin is small, the average spin direction of the muon ensemble (i.e. the polarization vector) precesses, relative to the momentum vector, in the plane perpendicular to the magnetic field, $\vec{B} = B_y \hat{y}$, according to

$$\hat{s} = (s_\perp \sin(\omega_a t + \phi)\hat{x} + s_y \hat{y} + s_\perp \cos(\omega_a t + \phi)\hat{z}). \tag{11.28}$$

The unit vectors $\hat{x}$, $\hat{y}$, and $\hat{z}$ are directed along the radial, vertical and azimuthal directions respectively. The (constant) components of the spin parallel and perpendicular to the B-field are $s_y << 1$ and $s_\perp$, with $\sqrt{(s_\perp^2 + s_y^2)} = 1$. The polarization vector precesses in the plane perpendicular to the magnetic field at a rate independent of the ratio $s_y/s_\perp$. Since $a_\mu = (g-2)/2 > 0$, the spin vector rotates in the same plane, but slightly faster than the momentum vector. Note that the present experiment is, apart from small detector acceptance effects, insensitive to whether the spin vector rotates faster or slower than the momentum vector rotation, and therefore it is insensitive to the sign of a_μ. There are small geometric acceptance effects in the detectors which demonstrate that our result is consistent with $a_\mu > 0$.

The value for ω_a is determined from the data using a least-square χ^2 minimization fit to the time spectrum of electron decays, $\chi^2 = \sum_i (N_i - N(t_i))^2/N(t_i)$, where the N_i are the data points, and $N(t_i)$ is the fitting function. The statistical uncertainty, in the limit where data are taken over an infinite number of muon lifetimes, is given by Eq. (11.15). The statistical figure of merit is $F_M = A\sqrt{N}$, which reaches a maximum at about $y = 0.8$, or $E \approx 2.6$ GeV/c (see Fig. 11.1). If all electrons are taken above some minimum energy threshold, $F_M(E_{th})$ reaches a maximum at about $y = 0.6$, or $E_{thresh} = 1.8$ GeV/c (Fig. 11.2).

The spectra to be fit are in the form of histograms of the number of electrons detected versus time (see Fig. 11.16), which in the ideal case follow the five-parameter distribution function, Eq. (11.12). While this is a fairly good approximation for the E821 data sets, small modifications, due mainly to detector acceptance effects, must be made to the five-parameter

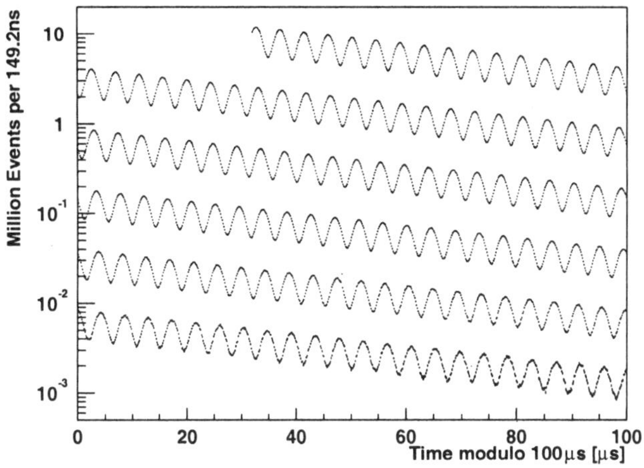

Fig. 11.16. Histogram of the total number of electrons above 1.8 GeV versus time (modulo 100 μ s) from the 2001 μ^- data set. The bin size is the cyclotron period, $\approx$ 149.2 ns, and the total number of electrons is 3.6 billion. (This figure was reprinted with permission from [25]. Copyright 2006 by the American Physical Society.)

function to obtain acceptable fits to the data. The most important of these effects are described in the next section.

The five-parameter function has an important, well-known invariance property. A sum of arbitrary time spectra, each obeying the five-parameter distribution and having the same λ and ω, but different values for N_0, A, and ϕ, also has the five-parameter functional form with the *same* values for λ and ω. That is,

$$\sum_i B_i e^{-\lambda(t_i - t_{i0})}(1 + A_i \cos(\omega(t_i - t_{i0}) + \phi_i)) = Be^{-\lambda t}(1 + A\cos(\omega t + \phi)). \tag{11.29}$$

This invariance property has significant implications for the way in which data are handled in the analysis. The final histogram of electrons versus time is constructed from a sum over the ensemble of the time spectra produced in individual spills. It extends in time from less than a few tens of microseconds after injection out to 640μs, a period of about 10 muon lifetimes. To a very good approximation, the spectrum from each spill follows the five-parameter probability distribution. From the invariance property, the t=0 points and the gains from one spill to the next do not need to be precisely aligned.

Pulse shape and gain stabilities are monitored primarily using the electron data themselves rather than laser pulses, or some other external source

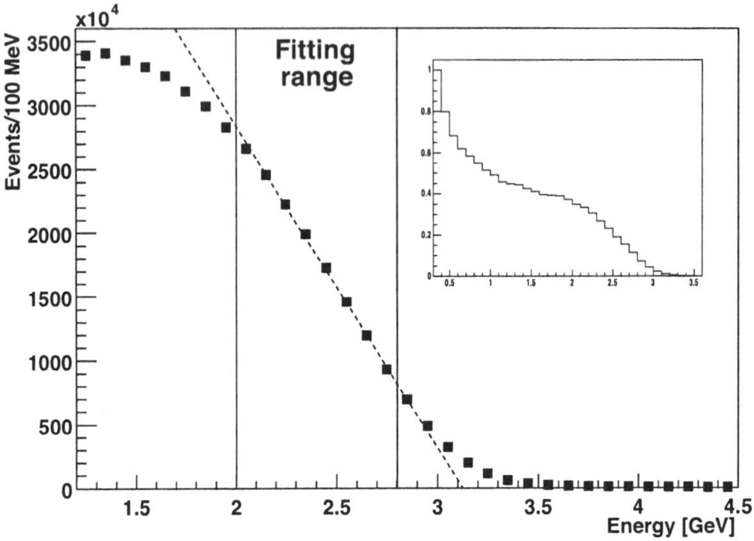

Fig. 11.17. Typical calorimeter energy distribution, with an endpoint fit superimposed. The inset shows the full range of reconstructed energies, from 0.3 to 3.5 GeV. (This figure was reprinted with permission from [25]. Copyright 2006 by the American Physical Society.)

of pulses. The electron times and energies are given by fits to standard pulse shapes, which are are established for each detector by taking an average over many pulses at late times. The variations in pulse shapes in all detectors are found to be small as a function of energy and decay time, and contribute negligibly to the uncertainty in ω_a. In order to monitor the gains, the energy distributions integrated over one spin precession period and corrected for pile-up are collected at various times relative to injection. The high-energy portion of the energy distribution is well-described by a straight line between the energy points at heights of 20% and 80% of the plateau in the spectrum (see Fig. 11.17). The position of the x-axis intercept is taken to be the endpoint energy, 3.1 GeV. It is found that the energy stability of the detectors on the "quiet" side of the ring (furthest from the beam injection point) stabilize earlier in the spill than those on the "noisy" side of the ring. Some of the gain shift is due to the PMT gating operation. Since the noisy detectors are gated on later in the spill than the quiet ones, their gains tend to stabilize later. The starting times for the detectors are chosen so that most of their gains are calibrated to better than 0.2%. On the quiet side of the ring, data fitting can begin as early as a couple of microseconds after injection; however it is necessary to delay at least until the beam-scraping process is completed. For the noisiest detectors, just downstream of the injection point, the start of fitting may be delayed to 30 μs or more. The time histograms are then accumulated after applying the gain correction to the energy of each electron. An uncertainty in the gain stability on average over a fill affects N, λ, ϕ, and A to a small extent. The result is a systematic error on ω_a on the order of 0.1 ppm.

While the five-parameter function gives a qualitative description of the time spectrum of the high-energy electrons, it is necessary make a number of modifications to the fitting function to include small, but statistically important features. These include:

(1) Pulse pile-up,
(2) The coherent betatron motion of the stored beam,
(3) Muon losses from the storage ring other than through decay,

In the interest of brevity, we only discuss the coherent beam oscillations here and refer the reader to Refs. [25, 35] for details.

The coherent betatron oscillations, or oscillations in the average position and width of the stored beam, (see Section 11.2.3) cause unwanted oscillations in the muon decay time spectrum, with the resulting effects generically referred to as CBO. Some of the important CBO frequencies are given in Table 11.3, which necessitate small modifications to the five-parameter functional form of the spectrum, and, like pile-up, can cause a shift in the derived value of ω_a if they are not properly accounted for in the analysis.

The most serious issues come from the horizontal CBO which leads to oscillation in the average radial position of the beam, since the detector acceptance depends on the radial position of the muon decay. Thus the coherent horizontal beam oscillations produce an amplitude modulation of the decay electron arrival-time spectrum, and it causes oscillations in the average detected energy. For a time spectrum constructed from the decay electrons in a given energy band, oscillation in the parameter N is due primarily to the oscillation in the detector acceptance. Oscillations induced in A and ϕ, on the other hand, depend primarily on the oscillation in the average energy. In either case, each of the parameters N, A and ϕ acquire small CBO-induced oscillations of the general form

$$P_i = P_{i0}[1 + B_i e^{-\lambda_{CBO}} \cos(\omega_{CBO} t + \theta_i)], \tag{11.30}$$

which introduce f_{CBO} and its harmonics, along with the sum and difference frequencies associated with beating between f_{CBO} and $f_a = \omega_a/2\pi \approx 229.1$ kHz. If the CBO effects are *not* included in the fitting function, it will pull the value of ω_a in the fit by an amount related to how close f_{CBO} is to the second harmonic of f_a (see Fig. 11.14), introducing a serious systematic error (see Table 11.3). This effect is shown qualitatively in Fig. 11.18.

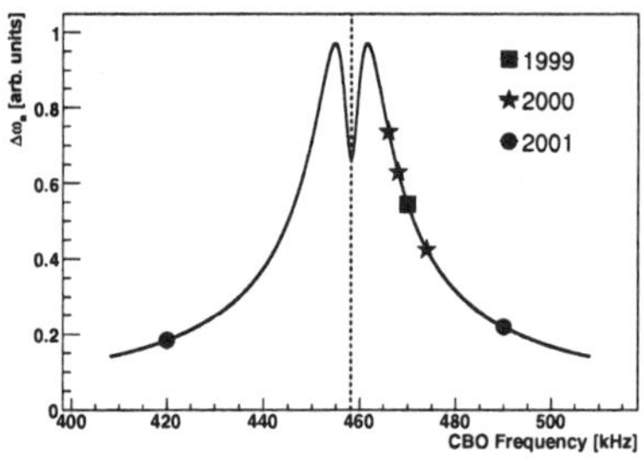

Fig. 11.18. The relative shift in the value obtained for ω_a as a function of the CBO frequency, when the CBO effects are neglected in the fitting function. The vertical line is at $2f_a$, and the operating point for each of the data collection periods is indicated on the curve. (This figure was reprinted with permission from [25]. Copyright 2006 by the American Physical Society.)

The problems posed by the CBO in the fitting procedure were solved in a variety of ways in the many independent analyses. All analyses took advantage of the fact that the CBO phase varied fairly uniformly from 0 to 2π around the ring in going from one detector to the next; the CBO oscillations should tend to cancel when data from all detectors are summed together, and would be perfect if all the detector acceptances were identical, or even if opposite pairs of detectors at 180° in the ring were identical. Imperfect cancellation is due to the reduced performance of some of the detectors and to slight asymmetries in the storage-ring geometry. This was especially true of detector number 20, where there were modifications to the vacuum chamber and whose position was displaced to accommodate the traceback chambers. The 180° symmetry is broken for detectors near the kicker because the electrons pass through the kicker plates. Also the fit start times for detectors near the injection point are inevitably later than for detectors on the other side of the ring because of the presence of the injection flash.

In addition to relying on the partial CBO cancellation around the ring, all of the other analysis approaches use a modified function in which all parameters except ω_a and λ oscillate according to Eq (11.30). In fits to the time spectra, the CBO parameters ω_{CBO} and $\lambda_{CBO} \approx 100\mu s$ (the frequency and lifetime of the CBO oscillations, respectively), are typically held fixed to values determined in separate studies. They were established in fits to time spectra formed with independent data from the FSDs and calorimeters, in which the amplitude of CBO modulation is enhanced by aligning the CBO oscillation phases of the individual detectors and then adding all the spectra together.

An important alternative analysis method to determine ω_a utilizes the so-called "ratio method" which removes effects which vary slowly (compared with $2\pi/\omega_a$). Each electron event is randomly placed with equal probability into one of four time histograms, $N_1 - N_4$, each looking like the usual time spectrum, Fig. 11.16. A spectrum based on the ratio of combinations of the histograms is formed:

$$r(t_i) = \frac{N_1(t_i + \frac{1}{2}\tau_a) + N_2(t_i - \frac{1}{2}\tau_a) - N_3(t_i) - N_4(t_i)}{N_1(t_i + \frac{1}{2}\tau_a) + N_2(t_i - \frac{1}{2}\tau_a) + N_3(t_i) + N_4(t_i)}. \tag{11.31}$$

For a pure five-parameter distribution, keeping only the important large terms, this reduces to

$$r(t) = A\cos(\omega_a t + \phi) + \frac{1}{16}(\frac{\tau_a}{\tau_\mu})^2, \tag{11.32}$$

where $\tau_a = 2\pi/\omega_a$ is an estimate ($\sim$ 10 ppm is easily good enough) of the spin precession period, and the small constant offset produced by the exponential decay is $\frac{1}{16}(\frac{\tau_a}{\tau_\mu})^2 = 0.000287$. Construction of independent histograms N_3 and N_4 simplifies the estimates of the statistical uncertainties in the fitted parameters. This technique removes the the exponential decay of the muon itself, along with muon losses and small shifts in PMT gains due to the high rates encountered at early decay times. There are only three parameters, Eq. (11.32), compared to five, Eq. (11.12), in the regular spectrum. Unfortunately faster-varying effects such as the CBO will not cancel in the ratio and must be handled in ways similar to the standard analyses. One of the ratio analyses of the 2001 data set used a fitting function formed from the ratio of functions h_i, consisting of the five-parameter function modified to include parameters to correct for acceptance effects such as the CBO:

$$r_{fit} = \frac{h_1(t + \frac{1}{2}\tau_a) + h_2(t - \frac{1}{2}\tau_a) - h_3(t) - h_4(t)}{h_1(t + \frac{1}{2}\tau_a) + h_2(t - \frac{1}{2}\tau_a) + h_3(t) + h_4(t)}. \tag{11.33}$$

The systematic errors for three yearly data sets, 1999 and 2000 for μ^+ and 2001 for μ^-, are given in Table 11.4.

Table 11.4. Systematic errors for ω_a in the 1999, 2000 and 2001 data periods. In 2001, systematic errors for the AGS background, timing shifts, E-field and vertical oscillations, beam de-bunching/randomization, binning and fitting procedure together equaled 0.11 ppm and this is indicated by ‡ in the table.

$\sigma_{\text{syst}}\ \omega_a$	1999 (ppm)	2000 (ppm)	2001 (ppm)
Pile-up	0.13	0.13	0.08
AGS Background	0.10	0.01	‡
Lost Muons	0.10	0.10	0.09
Timing Shifts	0.10	0.02	‡
E-field and Pitch	0.08	0.03	‡
Fitting/Binning	0.07	0.06	‡
CBO	0.05	0.21	0.07
Gain Changes	0.02	0.13	0.12
Total for ω_a	0.3	0.31	0.21

11.2.5. *The determination of ω_p*

In the data analysis for E821, great care was taken to insure that the results were not biased by previous measurements or the theoretical value expected from the Standard Model. This was achieved by a blind analysis which guaranteed that no single member of the collaboration could calculate the value of a_μ before the analysis was complete. Two frequencies, ω_p, the Larmor frequency of a free proton which is proportional to the B field, and ω_a, the frequency with which the muon spin precesses relative to its momentum, are measured. The analysis was divided into two separate efforts, ω_a and ω_p, with no collaboration member permitted to work on the determination of both frequencies.

In the first stage of each year's analysis, each independent ω_a (or ω_p) analyzer presented intermediate results with his own concealed offset on ω_a (or ω_p). Once the independent analyses of ω_a appeared to be mutually consistent, an offset common to all independent ω_a analyses was adopted, and a similar step was taken by the independent analyses of ω_p. The ω_a offsets were kept strictly concealed, especially from the ω_p analyzers. Similarly, the ω_p offsets were kept strictly concealed, especially from the ω_a analyzers. The nominal values of ω_a and ω_p were known at best to many ppm error, much larger than the eventual result, and could not be guessed with any precision. No one person was allowed to know both offsets, and it was therefore impossible to calculate the value of a_μ until the offsets were publicly revealed, after all analyses were declared to be complete.

For each of the four yearly data sets, 1998–2001, there were between four and five largely independent analyses of ω_a, and two independent analyses of ω_p. Typically, on ω_a there were one or two physicists conducting independent analyses in two successive years, and one on ω_p, providing continuity between the analysis of the separate data sets. Each of the fit parameters, and each of the potential sources of systematic error were studied in great detail. For the high statistics data sets, 1999, 2000 and 2001, it was necessary in the ω_a analysis to modify the five-parameter function given in Eq. (11.12) to account for a number of small effects. Often different approaches were developed to account for a given effect, although there were common features between some of the analyses.

All intermediate results for ω_a were presented in terms of $\Re$, which is defined by

$$\omega_a = 2\pi \cdot 0.2291 MHz \cdot [1 \pm (\Re \pm \Delta\Re) \times 10^{-6}] \tag{11.34}$$

where $\pm\Delta\Re$ is the concealed offset. Similarly, the ω_p analyzers maintained a constant offset which was strictly concealed from the rest of the collaboration.

To obtain the value of $\mathcal{R} = \omega_a/\omega_p$ to use in Eq. (11.8), the pitch and radial electric-field corrections discussed in section 11.2.3.2 were added to the measured frequency ω_a obtained from the least-squares fit to the time spectrum. Once these two corrections were made, the value of a_μ was obtained from Eq. (11.8) and was published with no other changes.

11.2.6. *The average magnetic field: the ω_p analysis*

The magnetic field data consist of three separate sets of measurements: The calibration data taken before, after, and occasionally during each running period; maps of the magnetic field obtained with the NMR trolley at intervals of a few days at random hours; and the field measured by each of the fixed NMR probes located in the vacuum chamber walls. For the latter measurements groups of 20 probes were connected via one of 20 analog multiplexers to one of 20 readout channels, each consisting of a frequency mixer and a custom-designed FID zero crossing counting device [48]. The plunging probe and the calibration probe [49] were also connected to one of the multiplexer inputs. The probes of each group were sequentially excited and their FID was read in full cycles repeated approximately every 5 seconds all throughout the experimental periods and whenever the magnet was energized. The data taken concurrent with the muon spin-precession data were tied to the field mapped by the trolley, which were used to determine the average magnetic field in the storage ring, and subsequently the value of ω_p to be used in Eq. (11.8).

11.2.6.1. *Calibration of the trolley probes*

The errors arising from the cross-calibration of the trolley probes with the plunging probes are caused both by the uncertainty in the relative positioning of the trolley probe and the plunging probe, and by the local field inhomogeneity. At this point in azimuth, trolley probes are fixed with respect to the frame that holds them, and to the rail system on which the trolley rides. The vertical and radial positions of the trolley probes with respect to the plunging probe are determined by applying a sextupole field and comparing the change of field measured by the two probes. The field shimming at the calibration location minimizes the error caused by the relative-position uncertainty, which in the vertical and radial directions has an inhomogeneity less than 0.2 ppm/cm, as shown in Fig. 11.19(b). The full multipole components at the calibration position are given in Table 11.5, along with the multipole content of the full magnetic field averaged over azimuth. For the estimated rms 1 mm-position uncertainty, the uncertainty on the relative calibration is less than 0.02 ppm.

The absolute calibration utilizes a probe with a spherical water sample (see Figs. 11.12(a), 11.12(b)) [49]. The Larmor frequency of a proton in a spherical water sample is related to that of the free proton through [52, 54]

$$f_{\rm L}({\rm sph}-{\rm H_2O},T) = [1-\sigma({\rm H_2O},T)]\, f_{\rm L}({\rm free}), \qquad (11.35)$$

where $\sigma(H_2O, T)$ is from the diamagnetic shielding of the proton in the water molecule, determined from [53]

$$\sigma({\rm H_2O}, 34.7°C) = 1 - \frac{g_p({\rm H_2O}, 34.7°C)}{g_J(H)}\frac{g_J(H)}{g_p(H)}\frac{g_p(H)}{g_p({\rm free})} \qquad (11.36)$$

$$= 25.790(14)\times 10^{-6}. \qquad (11.37)$$

The g-factor ratio of the proton in a spherical water sample to the electron in the hydrogen ground state ($g_J(H)$) is measured to 10 parts per billion (ppb) [53]. The ratio of electron to proton g-factors in hydrogen is known to 9 ppb [55]. The bound-state correction relating the g-factor of the proton bound in hydrogen to the free proton are calculated in References [56, 57]. The temperature dependence of σ is corrected for using $d\sigma(H_2O,T)/dT = 10.36(30)\times 10^{-9}/°C$ [58]. The free proton frequency is determined to an accuracy of 0.05 ppm.

The fundamental constant $\lambda_+ = \mu_{\mu^+}/\mu_p$ (see Eq.(11.8)) can be computed from the hyperfine structure of muonium (the μ^+e^- atom) [54], or from the Zeeman splitting in muonium [28]. The latter experiment used the very same calibration probe as well as the essential NMR field monitoring and mapping devices and techniques, including all the driving and readout electronics, as we used in our $(g-2)$ experiment. The magnetic environments of the two experiments were slightly different, so that perturbations of the probe materials on the surrounding magnetic field differed by a few ppb between the two experiments, which can be neglected at our level of accuracy. We have therefore a direct robust link of our magnetic field to the muon magneton (proton NMR has only the role of a fly wheel), which is independent of possible future changes in fundamental constants in the regular adjustment procedures [54], unless the muon magneton will be remeasured experimentally.

The errors in the calibration procedure result both from the uncertainties on the positions of the water samples inside the trolley and the calibration probe, and from magnetic field inhomogeneities. The precise location of the trolley in azimuth, and the location of the probes within the trolley, are not known better than a few mm. The uncertainties in the relative calibration resulting from position uncertainties are 0.03 ppm. Temperature and power-supply voltage dependences contribute 0.05 ppm, and the paramagnetism of the O_2 molecules in the air-filled trolley causes an experimentally verified 0.037 ppm shift in the field.

11.2.6.2. *Mapping the magnetic field*

During a trolley run, the value of B is measured by each probe at approximately 6000 locations in azimuth around the ring. The magnitude of the

Table 11.5. Multipoles at the outer edge of the storage volume (radius = 4.5 cm). The left-hand set are for the plunging station where the plunging probe and the calibration are inserted. The right-hand set are the multipoles obtained by averaging over azimuth for a representative trolley run during the 2000 period.

Multipole	Calibration		Azimuthal Averaged	
[ppm]	Normal	Skew	Normal	Skew
Quadrupole	-0.71	-1.04	0.24	0.29
Sextupole	-1.24	-0.29	-0.53	-1.06
Octupole	-0.03	1.06	-0.10	-0.15
Decupole	0.27	0.40	0.82	0.54

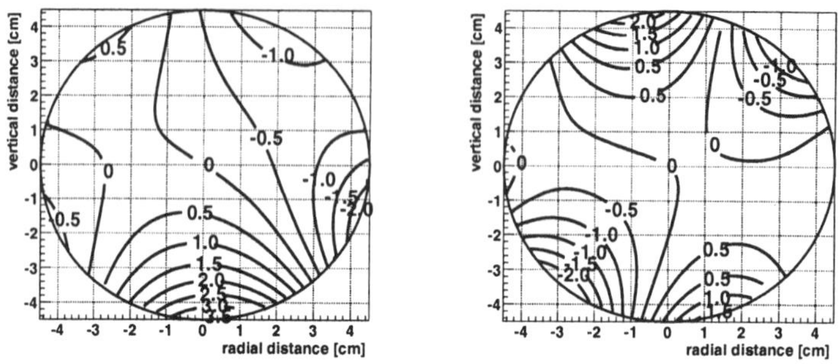

(a) Calibration position.

(b) Azimuthal average.

Fig. 11.19. Homogeneity of the field (a) at the calibration position and (b) for the azimuthal average for one trolley run during the 2000 period. In both figures, the contours correspond to 0.5 ppm field differences between adjacent lines. (This figure was reprinted with permission from [25]. Copyright 2006 by the American Physical Society.)

field measured by the central probe is shown as a function of azimuth in Fig. 11.20 for one of the trolley runs. The insert shows that the fluctuations in this map that appear quite sharp are in fact quite smooth, and are not noise. The field maps from the trolley are used to construct the field profile averaged over azimuth. This contour plot for one of the field maps is shown in Fig. 11.19(b). Since the storage ring has weak focusing, the average over azimuth is the important quantity in the analysis. Because the recorded NMR frequency is only sensitive to the magnitude of B and not to its direction, the multipole distributions must be determined from azimuthal magnetic field averages, where the field can be written as

$$B(r, \theta) = \sum_{n=0}^{n=\infty} r^n \left(c_n \cos n\theta + s_n \sin n\theta\right), \tag{11.38}$$

where in practice the series is limited to 5 terms.

11.2.6.3. *Tracking the magnetic field in time*

During data-collection periods the field is monitored with the fixed probes. To determine how well the fixed probes permitted us to monitor the field

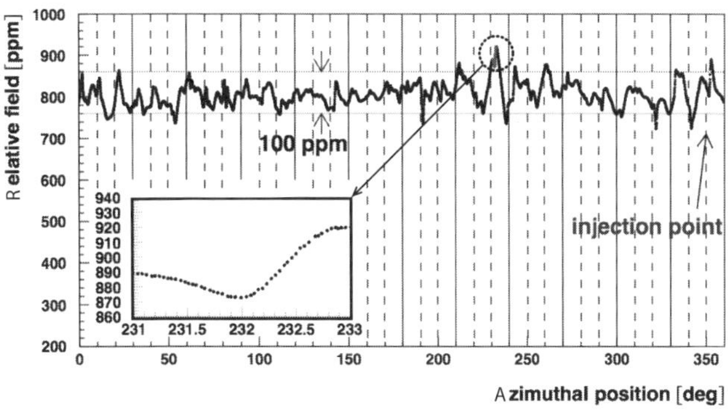

Fig. 11.20. The magnetic field measured at the center of the storage region vs. azimuthal position. Note that while the sharp fluctuations appear to be noise, when the scale is expanded the variations are quite smooth and represent true variations in the field. (This figure was reprinted with permission from [25]. Copyright 2006 by the American Physical Society.)

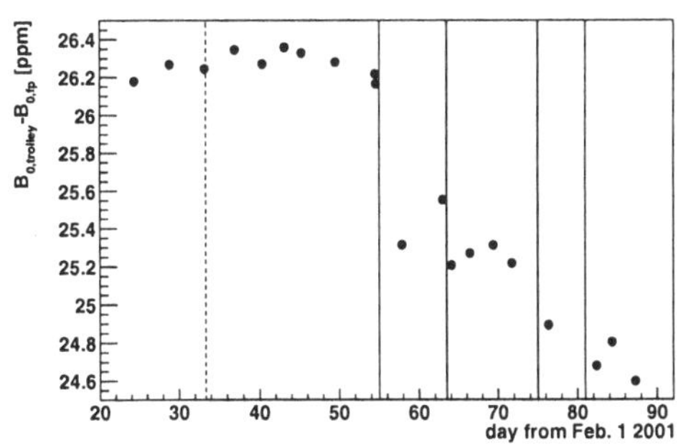

Fig. 11.21. The difference between the average magnetic field measured by the trolley and that inferred from tracking the magnetic field with the fixed probes between trolley maps. The vertical lines show when the magnet was powered down and then back up. After each powering of the magnet, the field does not exactly come back to its previous value, so that only trolley runs taken between magnet powerings can be compared directly. (This figure was reprinted with permission from [25]. Copyright 2006 by the American Physical Society.)

felt by the muons, the measured field, and that predicted by the fixed probes is compared for each trolley run. The results of this analysis for the 2001 running period is shown in Fig. 11.21. The rms distribution of these differences is 0.10 ppm.

11.2.6.4. *Determination of the average magnetic field:* ω_p

The value of ω_p entering into the determination of a_μ is the field profile weighted by the muon distribution. The multipoles of the field, Eq. (11.38), are folded with the muon distribution,

$$M(r,\theta) = \sum [\gamma_m(r)\cos m\theta + \sigma_m(r)\sin m\theta], \tag{11.39}$$

to produce the average field,

$$\langle B \rangle_{\mu-\mathrm{dist}} = \int M(r,\theta)B(r,\theta) r dr d\theta, \tag{11.40}$$

where the moments in the muon distribution couple moment-by-moment to the multipoles of $\vec{B}$. Computing $\langle B \rangle$ is greatly simplified if the field is quite uniform (with small higher multipoles), and the muons are stored in a circular aperture, thus reducing the higher moments of $M(r,\theta)$. This worked quite well in E821, and the uncertainty on $\langle B \rangle$ weighted by the muon distribution was ± 0.03 ppm.

The weighted average was determined both by a tracking calculation that used a field map and calculated the field seen by each muon, and also by using the quadrupole component of the field and the beam center determined from a fast-rotation analysis to determine the average field. These two agreed extremely well, vindicating the choice of a circular aperture and the ± 1 ppm specification on the field uniformity, that were set in the design stage of the experiment. [25]

11.2.6.5. *Summary of the magnetic field analysis*

The limitations on our knowledge of the magnetic field come from measurement issues, i.e. systematics and not statistics, so in E821 the systematic errors from each of these sources had to be evaluated and understood. The results and errors are summarized in Table 11.6.

11.2.7. *The determination of* a_μ *from E821*

The values obtained in E821 for a_μ are given in Table 11.1 However, the experiment measures $\mathcal{R} = \tilde{\omega}_a/\omega_p$, *not* a_μ directly, where the tilde over ω_a means that the pitch and radial electric field corrections have been included (see section 11.2.3.2). The fundamental constant $\lambda_+ = \mu_{\mu^+}/\mu_p$ (see Eq. (11.8)) connects the two quantities. The values obtained for $\mathcal{R}$ are given in Table 11.7.

Table 11.6. Systematic errors for the magnetic field for the different run periods. †Higher multipoles, trolley temperature and its power-supply voltage response, and eddy currents from the kicker.

Source of errors	1999 [ppm]	2000 [ppm]	2001 [ppm]
Absolute calibration of standard probe	0.05	0.05	0.05
Calibration of trolley probes	0.20	0.15	0.09
Trolley measurements of B_0	0.10	0.10	0.05
Interpolation with fixed probes	0.15	0.10	0.07
Uncertainty from muon distribution	0.12	0.03	0.03
Inflector fringe field uncertainty	0.20	–	–
Other†	0.15	0.10	0.10
Total systematic error on ω_p	0.4	0.24	0.17
Muon-averaged field [Hz]: $\omega_p/2\pi$	61 791 256	61 791 595	61 791 400

Table 11.7. The frequencies ω_a and ω_p obtained from the three major data-collection periods. The radial electric field and pitch corrections applied to the ω_a values are given in the second column. Total uncertainties for each quantity are shown. The right-hand column gives the values of $\mathcal{R}$, where the tilde indicates the muon spin precession frequency corrected for the radial electric field and the pitching motion. The error on the average includes correlations between the systematic uncertainties of the three measurement periods.

Period	$\omega_a/(2\pi)$ [Hz]	E–pitch [ppm]	$\omega_p/(2\pi)$ [Hz]	$\mathcal{R} = \tilde{\omega}_a/\omega_p$
1999 (μ^+)	229 072.8(3)	+0.81(8)	61 791 256(25)	0.003 707 204 1(5 1)
2000 (μ^+)	229 074.11(16)	+0.76(3)	61 791 595(15)	0.003 707 205 0(2 5)
2001 (μ^-)	229 073.59(16)	+0.77(6)	61 791 400(11)	0.003 707 208 3(2 6)
Average	–	–	–	0.003 707 206 3(2 0)

The results are

$$\mathcal{R}_{\mu^-} = 0.003\,707\,208\,3(2\,6) \tag{11.41}$$

and

$$\mathcal{R}_{\mu^+} = 0.003\,707\,204\,7(2\,5). \tag{11.42}$$

The *CPT* theorem predicts that the magnitudes of $\mathcal{R}_{\mu^+}$ and $\mathcal{R}_{\mu^-}$ should be equal. The difference is

$$\Delta\mathcal{R} = \mathcal{R}_{\mu^-} - \mathcal{R}_{\mu^+} = (3.5 \pm 3.4) \times 10^{-9}. \tag{11.43}$$

Note that it is the quantity $\mathcal{R}$ that must be compared for a *CPT* test, rather than a_μ, since the quantity λ_+ which connects them is derived from measurements of the hyperfine structure of the μ^+e^- atom (see Eq. (11.8)).

Using the latest value $\lambda_+ = \mu_{\mu^+}/\mu_p = 3.183\,345\,39(10)$ [28, 54] gives the values for a_μ shown in Table 11.1 Assuming *CPT* invariance, we combine all measurements of a_μ^+ and a_μ^- to obtain the "world average" [25]

$$a_\mu(\text{Expt}) = 11\,659\,208.0(6.3) \times 10^{-10} \quad (0.54\,\text{ppm}). \tag{11.44}$$

The values of a_μ obtained in E821 are shown in Fig. 11.22. The final combined value of a_μ represents an improvement in precision of a factor of 13.5 over the CERN experiments. The final error of 0.54 ppm consists of a 0.46 ppm statistical component and a 0.28 systematic component.

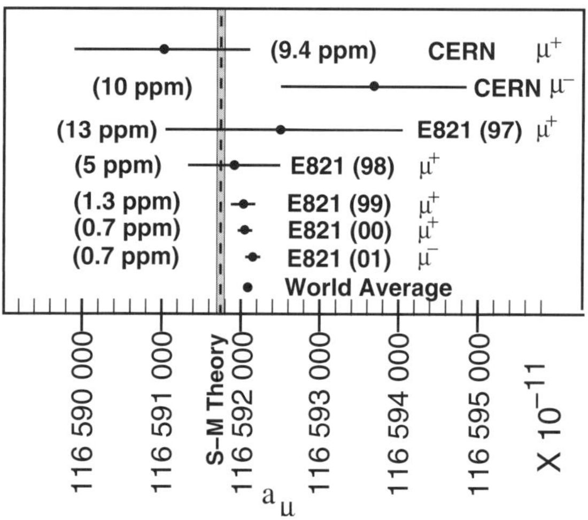

Fig. 11.22. Results for the E821 individual measurements of a_μ by running year, together with the final average. The theory value is taken from Ref. [64]. The accuracy of the theoretical value depends mostly on the knowledge of hadronic vacuum polarization contributions, which is, as discussed elsewhere in this volume, is a subject of much ongoing research.

11.2.8. *Other results*

In addition to the CPT test shown in Eq. (11.43), we were able to place limits on other CPT- and Lorentz-violating effects in muon spin precession. The data-collection periods covered a several-month period in each of the three years, by searching for a sidereal variation in $\mathcal{R}$. The recorded data over a several-month period permitted them to be analyzed for effects of Lorentz invariance breaking [65], through searches for differences in the anomaly frequencies ω_a^+ and ω_a^- of μ^+ and μ^- respectively and for sidereal variations in both ω_a^+ and ω_a^-. No significant effect is found, and several limits on Lorentz- and CPT-violating parameters for positive and negative muons are set at the level of $10^{-23} - 10^{-24}$ GeV [66]. These bounds are well below the ratio of the muon mass to the Planck mass, which would set a natural scale for the occurrence of such effects.

A permanent electric dipole moment of the muon, d_μ, would cause the muon spin precession to move out of the orbital plane, producing an up-down oscillation in the decay electrons that is out of phase by $\pi/4$ with the main $(g-2)$ precession. This has been searched for using the FSD and PSD hodoscopes, and the traceback array. An upper limit of $|d_\mu| < 1.8 \times 10^{-19}$ (e cm) at 95% confidence level was established [67], a factor of 5 smaller than the limit placed in the last CERN muon $(g-2)$ experiment.

11.2.9. *Future issues and prospects*

The potential difference between the measured and Standard-Model values of a_μ provides significant motivation for an improved experiment. It appears that with improvements to the basic technology employed in BNL E821, one could improve by about one order of magnitude before one starts to encounter the systematic limitations of the "magic-γ" technique. Such an improvement would require a more intense source of muons, along with adequate running time. The traditional measurement technique of determining a time and energy for each decay electron, along with a reduced systematic error on the pile-up contribution to the time spectrum, requires a lower instantaneous rate, rather than a higher one. A modified technique, which integrates the energy in a calorimeter as a function of time, could permit much higher instantaneous rates, and certainly any new experiment would want to collect data in both the traditional and this integral way.

Two laboratories could host this experiment: Fermilab and the Japan Proton Accelerator Research Complex (J-PARC). The beam structure that could be available at Fermilab appears to be ideal, with many more pulses of muons per hour than were available at Brookhaven. At J-PARC, which will have a proton intensity per machine cycle that is five times greater than the Brookhaven AGS could also provide enough muons, but the intensity of each muon bunch could be significantly larger than was obtained at BNL. Exploratory discussions have occurred at both laboratories, and early in 2009 a proposal was submitted to Fermilab. The goal is equal statistical and systematic errors of 0.1 ppm, for a total error of 0.14 ppm. An improvement in our knowledge of the hadronic contribution to the muon anomaly will also be important for the comparison with the Standard Model. For a discussion of the New Physics implications of a new experiment, see Ref. [68] and references therein.

11.3. Conclusions

The E821 collaboration successfully completed a program of muon $(g-2)$ experiments at BNL which resulted in a new and significantly improved value for the muon magnetic anomaly, thus providing an important new calibration point for all theoretical models. In particular, theoretical approaches that try to expand the Standard Model in order to give a deeper explanation of its shortcomings will have to satisfy this new bound. If a significant deviation of the measured magnetic anomaly from the calculated value within the Standard Model can be unambiguously established in the future, then this difference will not only indicate – like the decisive fermion g-factor measurements of the past century – the existence of New Physics beyond present standard theory. Furthermore, it will constructively limit the parameter space of all new models trying to explain this New Physics. This provides a very strong incentive to improve on both experiment and theory.

Acknowledgments

We wish to thank our many colleagues on E821 for numerous conversations on various parts of the experiment. Preparation of this work was supported in part by U.S. National Science Foundation Grant PHY-0758603 and the Dutch FOM under program 48 (TRμP).

We gratefully acknowledge permission from the American Physical Society to use Figures 3–5, 12, 13, 15, 16, 18–21 from the published E821 papers, Refs. [25, 34]; from IOP Publishing to use Figures 1, 2, and Tables 1 and 2 taken from the review [35] (Copyright (2007) by IOP Publishing); and from Elsevier for Figs. 7, 8 from Ref. [42], Fig. 9 from Ref. [47], and Fig. 6(b) from Ref. [41] (Copyright (2003,2002) by Elsevier).

References

[1] See Figure 5 in Paul Kunze, Z. Phys. **83**, 1 (1933).
[2] Carl D. Anderson and Seth H. Neddermeyer, Phys. Rev. **50** (1936) 263, and Seth H. Neddermeyer and Carl D. Anderson, Phys. Rev. **51** (1937) 844.
[3] J. C. Street, E.C. Stevenson, Phys. Rev. **52** (1937) 1003.

[4] Y. Nishina, M. Tekeuchi and T. Ichimiya, Phys. Rev. **52** (1937) 1198.
[5] M.M. Jean Crussard and L. Leprince-Ringuet, Compt. Rend. **204** (1937) 240.
[6] S.Tomonaga, G. Araki, Phys. Rev. 58, 90 (1940)
[7] C.M.G. Lattes, H. Muirhead, G.P.S. Occhianlini and C.F. Powell, Nature **159** (1947) 694; C.M.G. Lattes, G.P.S. Occhianlini, and C.F. Powell, Nature **160** (1947) 453 and 486; E. Gardner and C.M.G. Lattes, Science **107** (1948) 270; W.Y. Chang, Rev. Mod. Phys. **21** (1949) 166.
[8] R. F. Christy and S. Kusaka, Phys. Rev. **59** (1941) 414.
[9] John A. Wheeler, Rev. Mod. Phys. **21** (1949) 133.
[10] Val L. Fitch and James Rainwater, Phys. Rev. **92** (1953) 789.
[11] T.D. Lee and C.N. Yang, Phys. Rev. **104** (1956) 254.
[12] T. Kinoshita and M. Nio, Phys. Rev. **D73**, 053007 (2006).
[13] G. Danby, *et al.*, Phys. Rev. Lett. 9, 36 (1962).
[14] C.S. Wu, E. Ambler, R.W. Hayward, D.D. Hoppes, R.P. Hudson, Phys. Rev. **105**, 1413 (1957).
[15] E.J. Konopinski, Ann. Rev. Nucl. Sci. **9** 99, (1959).
[16] R.L. Garwin, L.M. Lederman, M. Weinrich, Phys. Rev. **105**, 1415, (1957).
[17] J.I. Friedman and V.L. Telegdi, Phys. Rev. **105**, 1681 (1957).
[18] R.L. Garwin, D.P. Hutchinson, S. Penman and G. Shapiro, Phys. Rev. **118**, 271 (1960).
[19] G. Charpak *et al.*, Phys. Rev. Lett. **6**, 128 (1961), Nuovo Cimento **22**, 1043 (1961), Phys. Lett. **1**, 16 (1962), and Nuovo Cimento **37** 1241 (1965).
[20] G. Charpak, et al, Phys. Lett. **1**, 16 (1962),
[21] J. Bailey, *et al.*, Phys. Lett. **28B**, 287 (1968). Additional details can be found in J. Bailey, *et al.*, Nuovo Cimento **A9**, 369 (1972) and references therein.
[22] G.R. Henry, G. Schrank and R.A. Swanson, Nuovo Cim. **A63**, 995 (1969).
[23] J. Bailey, *et al.*, Phys. Lett. **B55**, 420 (1975).
[24] J. Bailey, *et al.*, Nucl. Phys. **B150**, 1 (1979).
[25] G. Bennett, *et al.*, (Muon $(g-2)$ Collaboration), Phys. Rev. **D73**, 072003 (2006).
[26] F.J.M. Farley and E. Picasso, in Quantum Electrodynamics, Adv. Series on Dir. in H.E.P., V7, T. Kinoshita, ed., World Scientific, 479, (1990).
[27] L. H. Thomas, Phil. Mag. **3**, 1 (1927); V. Bargmann, L. Michel, and V. L. Telegdi, Phys. Rev. Lett. **2**, 435 (1959).
[28] W. Liu et. al., Phys. Rev. Lett. **82**, 711 (1999); D.E. Groom, et al, (Particle Data Group), Eur. Phys. J. **C15**, 1 (2000).
[29] V. Meyer *et al.*, Phys. Rev. Lett. **84**, 1137 (2000).
[30] R.M. Carey *et al.*, Phys. Rev. Lett. **82**, 1632 (1999).
[31] H.N. Brown *et al.*, (Muon $(g-2)$ Collaboration), Phys. Rev. **D62**, 091101 (2000).
[32] H.N. Brown, *et al.*, (Muon $(g-2)$ Collaboration), Phys. Rev. Lett. **86** 2227 (2001).
[33] G.W. Bennett, *et al.*, (Muon $(g-2)$ Collaboration), Phys. Rev. Lett. **89**, 101804 (2002).
[34] G.W. Bennett, *et al.*, (Muon $(g-2)$ Collaboration), Phys. Rev. Lett. **92**,

161802 (2004).
[35] J.P. Miller, E. de Rafael and B.L. Roberts, hep-ph/0703049, and Rep. Prog. Phys. **70**, 795-881 (2007).
[36] F.J.M. Farley and Y.K. Semertzidis, Prog. in Nucl. and Part. Phys. **52**, 1 (2004);
[37] D.W. Hertzog and W.M. Morse, Ann. Rev. Nucl. and Part. Phys. **54**, 141 (2004).
[38] F. Combley and E. Picasso, Phys. Rept. **14**, 1 (1974).
[39] M. Davier and W.J. Marciano, Ann. Rev. Nucl. and Part. Phys. **54**, 115 (2004);
[40] F. Krienen, D. Loomba and W. Meng, Nucl. Inst. and Methods Phys. Res. **A 283**, 5 (1989).
[41] A. Yamamoto, *et al.*, Nucl. Instrum. and Methods Phys. Res. **A491** 23-40 (2002).
[42] Efstratios Efstathiadis, *et al.*, Nucl. Inst. and Methods Phys. Res. **A496**, 8-25 (2002).
[43] OPERA, Electromagnetic Fields Analysis Program,
[44] H. Wiedemann, *Particle Accelerator Physics* Vol. 1, Springer-Verlag, (1993) p. 54.
[45] D.A. Edwards and M.J. Syphers, *An Introduction to the Physics of High Energy Accelerators*, John Wiley & Sons, (1993) p. 75.
[46] G.T. Danby, *et al.*, Nucl. Instr. and Methods Phys. Res. **A 457**, 151–174 (2001).
[47] Y.K. Semertzidis, Nucl. Instrum. Methods Phys. Res. **A503** 458–484 (2003)
[48] R. Prigl, *et al.*, Nucl. Inst. Methods Phys. Res. **A374** 118 (1996).
[49] X. Fei, V. Hughes and R. Prigl, Nucl. Inst. Methods Phys. Res. **A394**, 349 (1997).
[50] A. Grossmann, doctoral thesis, University of Heidelberg (1998)
[51] LORAN-C User Handbook, OMDTPUB P16562.6, available at http://www.navcen.uscg.gov/loran/handbook/h-book.htm
[52] see A. Abragam, "Principles of Nuclear Magnetism", Oxford U. Press, (1961), pps. 173-178.
[53] W.D. Phillips *et al.*, Metrologia **13**, 179 (1979).
[54] P.J. Mohr and B.H. Taylor, Rev. Mod. Phys. **77**, 1 (2005).
[55] P.F. Winkler, D. Kleppner, T. Myint, and F.G. Walther, Phys. Rev. **A5**, 83 (1972).
[56] W.E. Lamb Jr., Phys. Rev. **60**, 817 (1941).
[57] H. Grotch and R.A. Hegstrom, Phys. Rev. **A4**, 59 (1971).
[58] B.W. Petley *et al.*, Metrologia **20**, 81 (1984).
[59] S.A. Sedyk *et al.*, Nucl. Inst. and Methods Phys. Res. **A455**, 346 (2000).
[60] J. Ouyang *et al.*, Nucl. Inst. and Methods Phys. Res. **A374**, 215 (1996).
[61] G.W. Bennett, *et al.* (Muon $(g-2)$ Collaboration), Nucl. Inst. and Methods Phys. Res. **A579** 1096-1116 (2007).
[62] Y. Orlov, *et al.*, Nucl. Instrum. Meth. Phys. Res. **A482**, 767 (2002).
[63] F.J.M. Farley, Phys.Lett. **B 42**, 66 (1972), and J.H. Field, and G. Fiorentini, Nuovo Cimento, **21 A**, 297 (1974).

[64] Eduardo de Rafael *Present Status of the Muon Anomalous Magnetic Moment*; To appear in the proceedings of 14th High-Energy Physics International Conference in Quantum Chromodynamics (QCD 08), Montpellier, France, 7-12 Jul 2008. e-Print: arXiv:0809.3085.
[65] V.A. Kostelecky and N. Russell, arXiv:0801.0287v1 (2008); R. Bluhm, V.A. Kostelecký and C.D. Lane, Phys. Rev. Lett. **84**, 1098 (2000); and V.A. Kostelecký and C.D. Lane, Phys. Rev. **D 60** 116010 (1999); J. Math. Phys. **40**, 6245 (1999).
[66] G.W. Bennett, *et al.*, Phys.Rev.Lett. **100**, 091602(2008)
[67] G.W. Bennett, *et al.*, submitted to Phys. Rev. **D**, and arXiv:0811.1207v1[hep-ex], Nov. 2008.
[68] D.W. Hertzog, *et al.*, arXiv:0705.4617v1[hep-ph], 2007.

Chapter 12

Muon $(g-2)$ and Physics Beyond the Standard Model

Dominik Stöckinger

Technische Universität Dresden
Institut für Kern- und Teilchenphysik
D-01062 Dresden, Germany
stoeckinger@physik.tu-dresden.de

An overview of the role of the muon magnetic moment in the study of physics beyond the Standard Model is given. Contributions of very different types of physics beyond the Standard Model are compared on a general level. Supersymmetry as a particularly promising scenario is discussed in detail, and examples of the impact and the importance of $(g-2)_\mu$ for New Physics are reviewed.

Contents

12.1. Introduction

The magnetic moment of the muon is one of the most precisely measured and calculated quantities in elementary particle physics. The precision of the experimentally measured value and of the theory evaluation within the Standard Model has reached the level of 0.5 parts per million (ppm). Thanks to this fantastic precision the comparison of theory and experiment is not only a sensitive test of all SM interactions but also of possible New Physics at the electroweak scale. For many years, the muon magnetic moment has played an important role in constraining physics beyond the Standard Model.

In recent years, a 3σ deviation between experiment and Standard Model theory has been established. Progress has been achieved in particular on the determination of the hadronic light-by-light contributions and the hadronic vacuum polarization contributions, based on better experimental determinations of the hadronic e^+e^- cross section. For recent reviews we refer to the other chapters of the present book and to [1–4]. The 3σ deviation is tantalizing – it is too large to be ignored but it is not sufficient to prove the existence of physics beyond the Standard Model.

The 3σ deviation is one of the strongest indications for physics beyond the Standard Model at the electroweak scale. It is different from other arguments based on fine-tuning, naturalness or grand unification because it is purely based on observations and independent of theoretical prejudice. It is also different from evidence for dark matter or the baryon/antibaryon asymmetry of the universe, which prove the existence of physics beyond the Standard Model but not necessarily at the electroweak scale.

In the following we review the role of the muon magnetic moment $(g-2)_\mu$ for physics beyond the Standard Model. After a brief account of the current experimental and Standard Model status we first discuss very different types of New Physics and show how valuable $(g-2)_\mu$ can be to discriminate between them. Then we focus on supersymmetry as a particularly well-motivated scenario, which could explain the observed deviation very nicely. We describe the contributions from supersymmetry to $(g-2)_\mu$ in detail. Finally, we review examples which illustrate the impact of $(g-2)_\mu$ on New Physics. Parts of this chapter are based on updates of the review [5].

12.1.1. *Comparison of experiment and Standard-Model theory*

The anomalous magnetic moment of the muon $a_\mu = \frac{1}{2}(g-2)_\mu$ has been measured by the E821 experiment at Brookhaven to a precision of 0.54

ppm [6],

$$a_\mu^{\rm E821} = (116\,592\,080 \pm 63) \times 10^{-11} \quad (0.54\,{\rm ppm}). \tag{12.1}$$

The result represents a 14-fold improvement of the earlier CERN experiment [7]. The success of the Brookhaven experiment has inspired tremendous progress also on the SM theory evaluation of a_μ, particularly on the hadronic vacuum polarization contributions and hadronic light-by-light contributions, the two contributions with the by far largest theory uncertainties, see e.g. Refs. [1, 2].

For the purpose of the present article we use the Standard Model (SM) evaluation from Ref. [2],

$$a_\mu^{\rm SM} = (116\,591\,785 \pm 51) \times 10^{-11}\,(0.44\,{\rm ppm}). \tag{12.2}$$

The deviation is

$$\Delta a_\mu({\rm E821} - {\rm SM}) = (295 \pm 81) \times 10^{-11}\,. \tag{12.3}$$

It is noteworthy that the SM theory prediction is already more precise than the experiment and can be expected to further improve in the future. The experimental precision is mainly statistics limited and could be improved by a new experiment [8, 9].

12.1.2. *Varieties of physics beyond the Standard Model*

The role of $(g-2)$ as a discriminator between very different models of physics beyond the Standard Model is well illustrated by a general relation due to Czarnecki and Marciano [10]: If a New Physics model with a mass scale Λ contributes to the muon mass δm_μ(N.P.), it will also lead to a loop contribution to the magnetic moment, a_μ(N.P.). Clearly, the structure of loop diagrams contributing to m_μ or a_μ is the same, even including the necessary chirality flip, except for the additional external photon in the case of a_μ. Only due to the different dimensionality of the two corresponding mass/magnetic moment operators, the contributions to a_μ are suppressed by $1/M^2$. Hence the two contributions are related as

$$a_\mu({\rm N.P.}) = \mathcal{O}(1) \times \left(\frac{m_\mu}{\Lambda}\right)^2 \times \left(\frac{\delta m_\mu({\rm N.P.})}{m_\mu}\right). \tag{12.4}$$

The ratio $C({\rm N.P.}) \equiv \delta m_\mu({\rm N.P.})/m_\mu$ is typically between $\mathcal{O}(\alpha/4\pi)$ (perturbative contributions to the muon mass) and $\mathcal{O}(1)$ (the muon mass is essentially due to radiative corrections).

Therefore, the value of the constant C is highly model-dependent. It is important that the $\mathcal{O}(1)$ factors do not contain any coupling constants or $1/(16\pi^2)$ factors – those are all contained in the constant C. Several explicit examples for this relation can be found in [10].

This relation explains why New Physics at the electroweak/TeV scale naturally contributes to a_μ at the ppm-level, the current level of precision. Conversely, if perturbativity is assumed, $|C| = |\delta m_\mu(\text{N.P.})/m_\mu| < 1$, a general, model-independent bound is obtained:

$$|a_\mu(\text{N.P.})| < \mathcal{O}(1) \times \left(\frac{m_\mu}{M}\right)^2 . \tag{12.5}$$

This bound clearly shows that if the deviation Eq. (12.3) is due to New Physics, the New Physics scale must be at or below the TeV scale.

The relation Eq. (12.4) also allows a classification of types of New Physics and their respective contributions to a_μ:

- For models with radiative muon mass generation at some scale Λ ($C(\text{N.P.}) \simeq 1$) the New Physics contribution to a_μ can be very large [10],

$$a_\mu(\Lambda) \simeq \frac{m_\mu^2}{\Lambda^2} \simeq 1100 \times 10^{-11} \left(\frac{1\ \text{TeV}}{\Lambda}\right)^2 , \tag{12.6}$$

  and the difference Eq. (12.3) can be used to place a lower limit on the New Physics mass scale, a limit which is in the few TeV range [11].
- In models with extra weakly interacting gauge bosons Z', W', $C(\text{N.P.}) = \mathcal{O}(\alpha/4\pi)$, see e.g.Ref. [1, 4] and references therein. A difference as large as Eq. (12.3) is then very hard to accommodate unless the mass scale is very small, of the order of M_Z. The situation is similar in certain models with universal extra dimensions [12, 13]. In a model with $\delta = 1$ (or 2) universal extra dimensions, other measurements already imply a lower bound of 300 (or 500) GeV on the masses of the extra states, and the one-loop contributions to a_μ are correspondingly small,

$$a_\mu(\text{UED}) \simeq -5.8 \times 10^{-11}(1 + 1.2\delta)S_{\text{KK}} \tag{12.7}$$

  with $|S_{\text{KK}}| \lesssim 1$ [14].
- Supersymmetric models lie in between these two extremes. Compared to generic perturbative models, supersymmetry provides an enhancement to $C(\text{SUSY}) = \mathcal{O}(\tan\beta\alpha/4\pi)$ and $a_\mu(\text{SUSY})$ by a factor $\tan\beta$ (the ratio of the vacuum expectation values of the two Higgs fields). As discussed below in more detail, in a model with

SUSY masses equal to Λ the supersymmetric contribution to a_μ is given by [10]

$$a_\mu(\text{SUSY}) \simeq (\text{sgn}\mu)\ 130 \times 10^{-11}\ \tan\beta \left(\frac{100\ \text{GeV}}{\Lambda}\right)^2 \tag{12.8}$$

which indicates the dependence on $\tan\beta$ and the SUSY mass scale. Thus muon $(g-2)$ is sensitive to any SUSY model with large $\tan\beta$. Conversely, SUSY models with Λ in the few hundred GeV range would provide the most natural explanation of the deviation Eq. (12.3).

For many more examples for beyond-the-Standard-Model contributions to a_μ see the review [1] and references therein. The examples confirm the important role of $(g-2)$ in investigations of New Physics, and they confirm that among the well-known scenarios supersymmetry could provide the most natural explanation of the observed deviation Eq. (12.3).

12.1.3. *Brief introduction to supersymmetry and the MSSM*

In order to prepare the subsequent sections and to fix notation, we present a brief introduction to the minimal supersymmetric standard model (MSSM). The MSSM is the appropriate framework for a general discussion of a_μ and SUSY. The unknown supersymmetry breaking mechanism is parametrized in terms of a set of in principle arbitrary soft SUSY breaking parameters. Specific models of supersymmetry breaking can be accommodated within the MSSM by suitable restrictions on these parameters.

The MSSM as the minimal supersymmetric extension of the SM contains all SM particles and corresponding SUSY partners, see Table 12.1.

Table 12.1. The field/particle content of the MSSM. Only 2nd generation (s)leptons and 3rd generation (s)quarks are listed explicitly. The mass eigenstates corresponding to the electroweak gauge and Higgs bosons and their superpartners are indicated.

	(s)leptons	(s)quarks	Higgs Higgsinos	gauge bosons gauginos
SM/ THDM	$\begin{pmatrix}\nu_\mu\\ \mu_L\end{pmatrix}\cdots$	$\begin{pmatrix}t_L\\ b_L\end{pmatrix}\cdots$	$\mathcal{H}_1, \mathcal{H}_2$	$B^\mu, W^{a\mu};\quad G^{a\mu}$
	$\mu_R\cdots$	$t_R, b_R\cdots$	$\gamma, Z, W^\pm, G^{0,\pm}$, $h^0, H^0, A^0, H^\pm$	
SUSY partners	$\begin{pmatrix}\tilde\nu_\mu\\ \tilde\mu_L\end{pmatrix}\cdots$	$\begin{pmatrix}\tilde t_L\\ \tilde b_L\end{pmatrix}\cdots$	$\tilde H_1, \tilde H_2$	$\tilde B, \tilde W^a;\quad \tilde g^a$
	$\tilde\mu_R\cdots$	$\tilde t_R, \tilde b_R\cdots$	$\chi^0_{1,2,3,4}, \chi^\pm_{1,2}$	

In addition it also contains a second Higgs doublet with associated SUSY partner; hence the MSSM can actually be regarded as the SUSY version of the two-Higgs-doublet model (THDM). Two Higgs doublets $\mathcal{H}_{1,2}$ of opposite hypercharge ∓ 1 are required for the cancellation of chiral gauge anomalies caused by the corresponding Higgsinos. Thus the field content of the MSSM comprises the THDM fields, including five physical Higgs bosons, see Eqs. (12.29)–(12.31) below, scalar SUSY partners of each chiral SM fermion, called sfermions $\tilde{f}_{L,R}$, Higgsino doublets $\tilde{H}_{1,2}$, and U(1), SU(2) and SU(3) gauginos (called bino, winos and gluinos) $\tilde{B}$, $\tilde{W}^{\pm,3}$, $\tilde{g}$. Right-handed (s)neutrinos as well as non-vanishing neutrino masses are not relevant for this review and are ignored.

Two central MSSM parameters that are of particular importance for a_μ are related to the two Higgs doublets. The first of these is the ratio of the two vacuum expectation values,

$$\tan\beta = \frac{v_2}{v_1}. \tag{12.9}$$

SUSY and gauge invariance require that the doublet $\mathcal{H}_1$ gives masses to down-type fermions, while $\mathcal{H}_2$ gives masses to up-type fermions. As a result, the top- and bottom-Yukawa couplings in the MSSM are enhanced by factors $1/\sin\beta$ and $1/\cos\beta$, respectively. In order to avoid non-perturbative values of these Yukawa couplings, $\tan\beta$ is commonly restricted to the range between about 1 and 50. High values $\tan\beta = \mathcal{O}(50)$ lead to similar top and bottom Yukawa couplings and are therefore favored by the idea of top-bottom Yukawa coupling unification [15].

The second important parameter relating the two Higgs doublets is the μ-parameter, which appears in the MSSM Lagrangian in the terms

$$\mu\tilde{H}_1\tilde{H}_2 - \mu F_{H_1}\mathcal{H}_2 - \mu F_{H_2}\mathcal{H}_1 + h.c. \tag{12.10}$$

The first term describes a Higgsino mass term, while in the other terms $F_{H_{1,2}}$ are auxiliary fields whose elimination gives rise to interactions of $\mathcal{H}_{1,2}$ with sfermions of the opposite type compared to the Yukawa couplings, e.g. to $\mathcal{H}_1^0\tilde{t}_L\tilde{t}_R^\dagger$ and $\mathcal{H}_2^0\tilde{\mu}_L\tilde{\mu}_R^\dagger$.

In addition the MSSM contains a large number of parameters that parametrize soft SUSY breaking. Except where explicitly stated we will restrict the number of these parameters by neglecting generation mixing in the sfermion sectors. Furthermore, we restrict ourselves to the case of R-parity conservation, since R-parity violating interactions have not much impact on a_μ.

The SUSY particles of particular importance for a_μ are the smuons, muon-sneutrino, gauginos and Higgsinos since they appear in the SUSY one-loop contributions. At higher order, also other sectors of the MSSM become relevant, most notably the third generation squarks and the Higgs sector. Since in the MSSM all particles of equal quantum numbers can mix and these mixings have an important influence on the a_μ-prediction, we briefly discuss the mixing of the individual sectors in the following.

The sfermions $\tilde{f}_{L,R}$ for each flavor can mix, and the mass matrices corresponding to the $\tilde{f}_L, \tilde{f}_R$ basis read

$$M_{\tilde{f}}^2 = \begin{pmatrix} M_{LL}^2 & m_f X_f^* \\ m_f X_f & M_{RR}^2 \end{pmatrix}, \tag{12.11}$$

where

$$M_{LL}^2 = m_f^2 + m_{L,\tilde{f}}^2 + M_Z^2 \cos 2\beta (I_3^f - Q^f s_W^2), \tag{12.12}$$

$$M_{RR}^2 = m_f^2 + m_{R,\tilde{f}}^2 + M_Z^2 \cos 2\beta \; Q^f s_W^2, \tag{12.13}$$

$$X_f = A_f - \mu^* \{\cot\beta, \tan\beta\} \tag{12.14}$$

with $\{\cot\beta, \tan\beta\}$ for up- and down-type sfermions, respectively. m_f, I_3^f and Q^f denote the mass, weak isospin and electric charge of the corresponding fermion; $s_W^2 \equiv \sin^2\theta_W = 1 - M_W^2/M_Z^2$, where θ_W denotes the weak mixing angle and $M_{W,Z}$ the W and Z boson masses. The quantities A_f are soft SUSY breaking parameters for trilinear interactions of sfermions with Higgs bosons of the form $\tilde{f}_L$–$\tilde{f}_R$–Higgs. The remaining entries $m_{L,R}$ of the diagonal elements are governed by the five independent soft SUSY-breaking parameters for each generation:

$$m_{L,\tilde{t}} = m_{L,\tilde{b}} \equiv M_{Q3}, \qquad m_{R,\tilde{b}} \equiv M_{D3}, \qquad m_{R,\tilde{t}} \equiv M_{U3}, \tag{12.15}$$

$$m_{L,\tilde{\mu}} = m_{L,\tilde{\nu}_\mu} \equiv M_{L2}, \qquad m_{R,\tilde{\mu}} \equiv M_{R2}. \tag{12.16}$$

We have given these relations for third generation squarks and second generation sleptons as these are most important for our purposes. Analogous formulas hold for the other generations. The mass matrices can be diagonalized by unitary matrices $U^{\tilde{f}}$ in the form

$$U^{\tilde{f}} M_{\tilde{f}}^2 U^{\tilde{f}\dagger} = \mathrm{diag}(m_{\tilde{f}_1}^2, m_{\tilde{f}_2}^2), \tag{12.17}$$

and sfermion mass eigenstates can be defined by

$$\begin{pmatrix} \tilde{f}_1 \\ \tilde{f}_2 \end{pmatrix} = U^{\tilde{f}} \begin{pmatrix} \tilde{f}_L \\ \tilde{f}_R \end{pmatrix}. \tag{12.18}$$

The mass of the sneutrino $\tilde{\nu}_\mu$ is given by

$$m^2_{\tilde{\nu}_\mu} = m^2_{L,\tilde{\mu}} + \frac{1}{2} M_Z^2 \cos 2\beta \tag{12.19}$$

and similar for the other generations.

The superpartners of the charged gauge and Higgs bosons also mix, and the mass and mixing terms can be easily expressed in terms of the Weyl spinor combinations $\psi^- = (\tilde{W}^-, \tilde{H}_1^-)$, $\psi^+ = (\tilde{W}^+, \tilde{H}_2^+)$. The mass term for these fields is given by $\psi^- X \psi^+ + h.c.$ with the mass matrix

$$X = \begin{pmatrix} M_2 & M_W\sqrt{2}\sin\beta \\ M_W\sqrt{2}\cos\beta & \mu \end{pmatrix}, \tag{12.20}$$

where M_2 is the soft SUSY breaking parameter corresponding to the SU(2) gaugino mass. This matrix can be diagonalized using two unitary matrices U, V in the form

$$U^* X V^{-1} = \mathrm{diag}(m_{\chi_1^\pm}, m_{\chi_2^\pm}). \tag{12.21}$$

Similarly, the mass matrix Y corresponding to the superpartners of the neutral gauge and Higgs bosons in the basis $\psi^0 = (\tilde{B}, \tilde{W}^3, \tilde{H}_1^0, \tilde{H}_2^0)$ is given by

$$Y = \begin{pmatrix} M_1 & 0 & -M_Z s_W \cos\beta & M_Z s_W \sin\beta \\ 0 & M_2 & M_Z c_W \cos\beta & -M_Z c_W \sin\beta \\ -M_Z s_W \cos\beta & M_Z c_W \cos\beta & 0 & -\mu \\ M_Z s_W \sin\beta & -M_Z c_W \sin\beta & -\mu & 0 \end{pmatrix}, \tag{12.22}$$

where M_1 is a U(1) gaugino (bino) soft SUSY breaking parameter and $c_W = M_W/M_Z$. It can be diagonalized with the help of one unitary matrix N in the form

$$N^* Y N^{-1} = \mathrm{diag}(m_{\chi_1^0}, \ldots, m_{\chi_4^0}). \tag{12.23}$$

The mass eigenstates corresponding to the charged and neutral gauginos and Higgsinos are called charginos and neutralinos, and they are related to the interaction eigenstates by

$$\chi_i^+ = V_{ij}\psi_j^+, \tag{12.24}$$

$$\chi_i^- = U_{ij}\psi_j^-, \tag{12.25}$$

$$\chi_i^0 = N_{ij}\psi_j^0. \tag{12.26}$$

The gluinos do not mix, and their tree-level mass is given by $|M_3|$ in terms of the SU(3) gaugino mass parameter M_3.

The SUSY parameters μ, A_f, $M_{1,2,3}$ can be complex. However, not all complex phases can appear in observables. The only physical phases of the MSSM (beyond the phase in the CKM matrix) are the ones of the combinations

$$\mu A_f, \quad \mu M_{1,2,3}. \tag{12.27}$$

Hence the frequently adopted convention that M_2 is real and positive constitutes no restriction.

After spontaneous symmetry breaking the two MSSM Higgs doublets lead to five physical Higgs bosons and three unphysical Goldstone bosons. Parametrizing the two doublets in the form

$$\mathcal{H}_1 = \begin{pmatrix} v_1 + \frac{1}{\sqrt{2}}(\phi_1^0 - i\chi_1^0) \\ -\phi_1^- \end{pmatrix}, \qquad \mathcal{H}_2 = \begin{pmatrix} \phi_2^+ \\ v_2 + \frac{1}{\sqrt{2}}(\phi_2^0 + i\chi_2^0) \end{pmatrix}, \tag{12.28}$$

the mass eigenstates are given by

$$\begin{pmatrix} H^0 \\ h^0 \end{pmatrix} = \begin{pmatrix} \cos\alpha & \sin\alpha \\ -\sin\alpha & \cos\alpha \end{pmatrix} \begin{pmatrix} \phi_1^0 \\ \phi_2^0 \end{pmatrix}, \tag{12.29}$$

$$\begin{pmatrix} G^0 \\ A^0 \end{pmatrix} = \begin{pmatrix} \cos\beta & \sin\beta \\ -\sin\beta & \cos\beta \end{pmatrix} \begin{pmatrix} \chi_1^0 \\ \chi_2^0 \end{pmatrix}, \tag{12.30}$$

$$\begin{pmatrix} G^\pm \\ H^\pm \end{pmatrix} = \begin{pmatrix} \cos\beta & \sin\beta \\ -\sin\beta & \cos\beta \end{pmatrix} \begin{pmatrix} \phi_1^\pm \\ \phi_2^\pm \end{pmatrix}, \tag{12.31}$$

where the mixing angle α is related to β and the mass M_A of the CP-odd scalar A^0 by

$$\tan 2\alpha = \tan 2\beta \, \frac{M_A^2 + M_Z^2}{M_A^2 - M_Z^2}, \qquad -\frac{\pi}{2} < \alpha < 0. \tag{12.32}$$

The physical Higgs degrees of freedom are the light and heavy CP-even scalars h^0, H^0, the CP-odd scalar A^0 and the charged Higgs bosons $H^\pm$. For $M_A \gg M_Z$ the masses of H^0 and $H^\pm$ are of the order M_A. The three unphysical Goldstone bosons $G^{0,\pm}$ are eaten to give masses to the Z and $W^\pm$ bosons.

For the discussion of two-loop contributions to a_μ with Higgs exchange it is important to note that the muon receives its mass from the doublet $\mathcal{H}_1$. In the case that $M_A \gg M_Z$ and $\tan\beta$ is large the heavy Higgs bosons H^0, A^0 and $H^\pm$ are predominantly composed of $\mathcal{H}_1$-components. In this case they have larger couplings to the muon than the light Higgs boson h^0.

12.2. The Muon Magnetic Moment and Supersymmetry

The deviation between experimental and theoretical value of a_μ could be due to contributions from supersymmetry (SUSY). Independent of a_μ, SUSY at the electroweak scale is one of the most compelling ideas of physics beyond the SM (see Refs. [16–20] for reviews). SUSY is the unique symmetry that relates fermions and bosons in relativistic quantum field theories. It eliminates the quadratic divergences associated with the Higgs boson mass and thus stabilizes the weak scale against quantum corrections from ultra-high scales. SUSY at the weak scale also automatically leads to gauge coupling unification, and the lightest SUSY particle (LSP) can be neutral and stable and constitutes a natural candidate for cold dark matter. Moreover, in contrast to many other scenarios for physics beyond the SM, the MSSM is a weakly coupled, renormalizable gauge theory [21], such that quantum effects are computable and well-defined, and it has survived many non-trivial electroweak precision tests [22]. In the present section we give an overview of the known SUSY contributions to a_μ, discuss their qualitative behavior and provide the most important analytical formulas.

12.2.1. *Relevant symmetries*

The anomalous magnetic moment of the muon is a property of the muon in presence of an electromagnetic field. In quantum field theory it is related to the muon–photon vertex function, in particular to the magnetic form factor $F_M(q^2)$. The gyromagnetic ratio is given by $g = 2[1 - 2m_\mu F_M(0)]$, or equivalently

$$a_\mu = -2m_\mu F_M(0). \tag{12.33}$$

It has been known for a long time that in a theory with unbroken SUSY, the form factor $F_M(0)$ has to vanish and thus $a_\mu = 0$ [23, 24], that is the SUSY contribution exactly cancels the SM contribution. In realistic SUSY models such as the MSSM, soft SUSY breaking prevents this cancellation, and the genuine SUSY contributions are suppressed by a factor $m_\mu^2/M_{\rm SUSY}^2$ as already mentioned in Section 12.1.2.

This behavior, as well as the relation Eq. (12.4), is due to chiral symmetry. The form factor F_M corresponds to a chirality-flipping interaction between the left- and right-handed muon. If the MSSM were invariant under the discrete chiral transformation

$$\left.\begin{matrix} \begin{pmatrix} \nu_\mu \\ \mu_L \end{pmatrix}, & \begin{pmatrix} \tilde{\nu}_\mu \\ \tilde{\mu}_L \end{pmatrix} \\ \mu_R, & \tilde{\mu}_R \end{matrix}\right\} \to \left\{\begin{matrix} +\begin{pmatrix} \nu_\mu \\ \mu_L \end{pmatrix}, & +\begin{pmatrix} \tilde{\nu}_\mu \\ \tilde{\mu}_L \end{pmatrix}, \\ -\ \mu_R, & -\ \tilde{\mu}_R \end{matrix}\right. \tag{12.34}$$

of the left-handed doublets and right-handed singlets, F_M and thus a_μ would vanish in the MSSM. F_M is proportional to m_μ because the invariance of the MSSM under (12.34) is broken by the muon mass, or more precisely by all terms in the MSSM Lagrangian that are proportional to the muon Yukawa coupling. In each Feynman diagram that contributes to a_μ, the μ-chirality has to be flipped by one of these terms. The main possibilities for the chirality flip are illustrated in Fig. ds-fig:chiralityflips and are the following:

- at a μ-line through a muon mass term, contributing a factor m_μ,
- at a Yukawa coupling of $\mathcal{H}_1$ to μ_R and μ_L or ν_μ, contributing a factor y_μ,
- at a $\tilde{\mu}$-line, corresponding to the transition $\tilde{\mu}_L$–$\tilde{\mu}_R$, contributing a factor $m_\mu X_\mu \approx m_\mu \tan\beta\, \mu$ for large $\tan\beta$ from the smuon mass matrix,
- at a SUSY Yukawa coupling of a Higgsino to μ and $\tilde{\mu}$ or $\tilde{\nu}_\mu$, contributing a factor y_μ.

The muon Yukawa coupling y_μ is given by

$$y_\mu = \frac{m_\mu}{v_1} = \frac{m_\mu g_2}{\sqrt{2} M_W \cos\beta}, \tag{12.35}$$

where $g_2 = e/s_W$, and is thus enhanced by the factor $1/\cos\beta \approx \tan\beta$ for large $\tan\beta$ compared to its SM value. Hence, while all four chirality flips are proportional to the muon mass m_μ, the last three are enhanced by a factor $\tan\beta$ for large $\tan\beta$. SUSY contributions to a_μ that make use of these enhanced chirality flips are themselves enhanced compared to the generic estimate $m_\mu^2/M_{\rm SUSY}^2$.

12.2.2. *How large can the SUSY contributions be?*

Before presenting the full SUSY one-loop contributions to a_μ [25–36], it is instructive to discuss the dominant parameter dependence on an intuitive level and to obtain useful estimates. As mentioned above the contributions from SUSY particles of a generic mass $M_{\rm SUSY}$ are of the order $m_\mu^2/M_{\rm SUSY}^2$, and hence suppressed by a factor $M_W^2/M_{\rm SUSY}^2$ compared to the SM electroweak contributions.

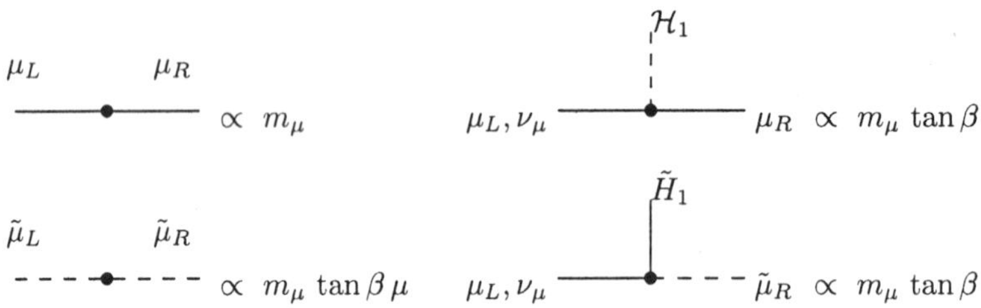

Fig. 12.1. Possibilities for chirality-flips along the line carrying the μ-lepton number.

However, it has been observed in [29] and further stressed and discussed in [34–36] that the SUSY contributions can be significantly enhanced if $\tan\beta$ is large. Moreover, for large $\tan\beta$ the sign of the one-loop contributions is mainly determined by the sign of the μ-parameter introduced in Eq. (12.10). We will see here that not only the one-loop but also the leading two-loop contributions behave in this way.

This $\tan\beta$-enhancement is due to chirality flips governed by the $\tan\beta$-enhanced muon Yukawa coupling in Eq. (12.35) as discussed in the previous subsection. It is instructive to analyze the leading behavior with the help of diagrams that are written in terms of interaction eigenstates, where the insertions of mass and mixing terms and chirality flips are explicitly shown [36]. The five diagrams in Figs. 12.2, 12.3 exemplify the main enhancement mechanisms. The basic reason for the $\tan\beta$-enhancement is the fact that the muon Yukawa coupling in the MSSM is larger by a factor $1/\cos\beta \approx \tan\beta$ for large $\tan\beta$ than its SM counterpart. This Yukawa coupling enters the diagrams in Figs. 12.2, 12.3 in the vertices where the muon chirality is flipped, i.e. in the couplings of the muon to the Higgsino or Higgs boson in cases (C,N2,C_{2L},$\tilde{t}_{2L}$), and in the $\tilde{\mu}_L$–$\tilde{\mu}_R$ transition, given by $(M^2_{\tilde{\mu}})_{12}$, in case (N1).

The second important parameter entering all five diagrams is the μ-parameter, which governs the mixing between the two Higgs doublets. In all cases, the enhancement due to this mixing can be traced back to the fact that $\mathcal{H}_2$ has the larger vacuum expectation value and strongly couples to top quarks, while only $\mathcal{H}_1$ couples to muons.

In diagrams (C,N2,C_{2L}) the μ-parameter enters via the Higgsino $\tilde{H}_1$–$\tilde{H}_2$ transitions. These transitions enhance the diagrams because the following $\tilde{H}_2$–gaugino transitions are by a factor $v_2 : v_1 = \tan\beta$ larger than $\tilde{H}_1$–gaugino transitions. In diagram (N1) μ enters via the dominant part of the

(C)

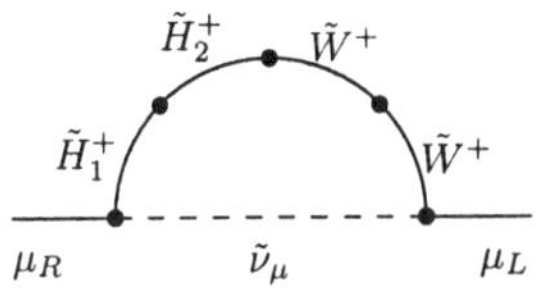

$\propto m_\mu^2 \tan\beta \, \mu \, M_2 \, F(\mu, M_2, m_{\tilde{\mu}_L})$

(N1)

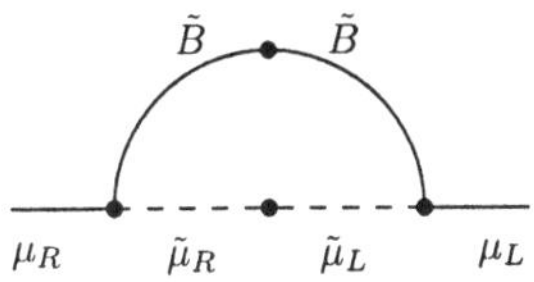

$\propto m_\mu^2 \tan\beta \, \mu \, M_1 \, F(M_1, m_{\tilde{\mu}_{L,R}})$

(N2)

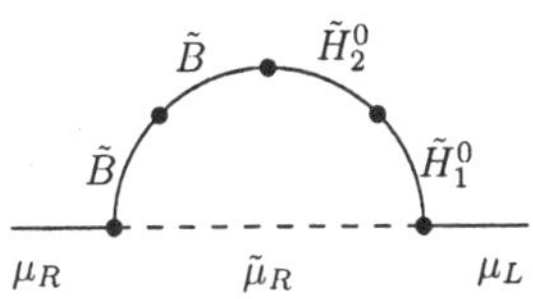

$\propto m_\mu^2 \tan\beta \, \mu \, M_1 \, F(\mu, M_1, m_{\tilde{\mu}_R})$

Fig. 12.2. Three sample one-loop mass-insertion diagrams. Vertices and mass insertions are denoted by dots, and the interaction eigenstates corresponding to each line are displayed explicitly. The external photon has to be attached in all possible ways to the charged internal lines. The diagrams (C), (N1), (N2) have been discussed also in Ref. [36]. The loop functions F in the results are all different and depend on different masses.

smuon mixing. This mass insertion is obtained from the F-term $F_{H_1}\mathcal{H}_2$, see (12.10), by replacing $\mathcal{H}_2$ by its large vacuum expectation value. Finally, in diagram ($\tilde{t}_{2\mathrm{L}}$) the dominant part of the Higgs–stop coupling originates from $F_{H_2}\mathcal{H}_1$ and thus enables $\mathcal{H}_1$ to couple with the top-Yukawa coupling.

The remaining mass insertions in the diagrams provide additional factors of the gaugino mass $M_{1,2}$ and stop mixing parameter X_t. They are necessary in order to obtain an even number of γ-matrices in the fermion line and in order to connect $\tilde{t}_L$ and $\tilde{t}_R$, respectively. As an illustration, the relevant factors of diagram (C) are given by

$$y_\mu \, X_{22} \, X_{12} \, X_{22} = \frac{m_\mu}{v_1} \, \mu \, (g_2 v_2) \, M_2 = g_2 \, m_\mu \, \tan\beta \, \mu \, M_2. \tag{12.36}$$

Combining the enhancement factors of all diagrams leads to the estimates

(C_{2L})

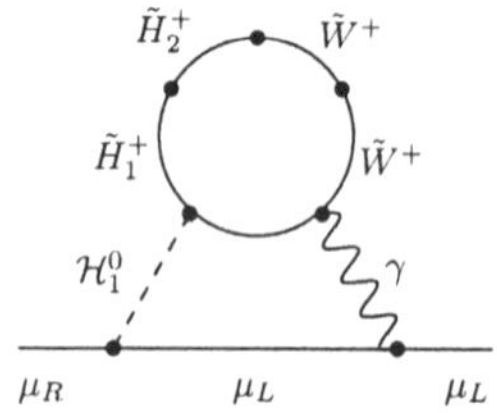

$\propto m_\mu^2 \tan\beta\, \mu M_2\, F(\mu, M_2, M_{\mathcal{H}_1})$

($\tilde{t}_{2L}$)

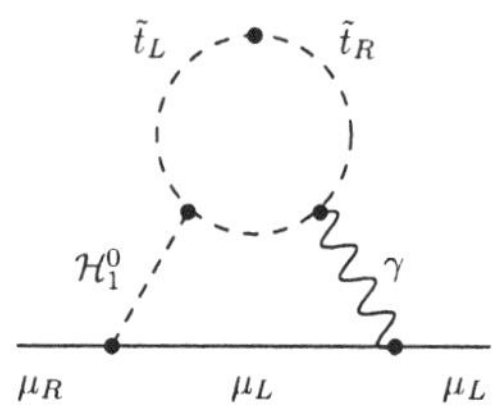

$\propto m_\mu^2 \tan\beta\, \frac{\mu m_t}{M_W^2}\, (m_t X_t) F(m_{\tilde{t}_{L,R}}, M_{\mathcal{H}_1^0})$

Fig. 12.3. Two sample two-loop mass-insertion diagrams. Description see Fig. 12.2.

given in Figs. 12.2, 12.3. They all have a similar form,

$$a_\mu^{(\mathrm{C,C_{2L};N1,N2})} \propto m_\mu^2 \tan\beta\, \mu\, M_{2;1}\, F, \tag{12.37}$$

$$a_\mu^{(\tilde{t}_{2L})} \propto m_\mu^2 \tan\beta\, \frac{\mu\, m_t}{M_W^2}\, (m_t X_t)\, F, \tag{12.38}$$

where $M_{2;1}$ corresponds to $(\mathrm{C, C_{2L}})$ and $(\mathrm{N1,N2})$, respectively. The loop functions F are different in the five cases, depend on the masses appearing in the respective diagrams and generally behave as $F \propto M_{\mathrm{SUSY}}^{-4}$ for large SUSY masses. In these formulas one power of m_μ is due to the a_μ–F_M relation (12.33) and gauge couplings have been suppressed.

Therefore all leading one- and two-loop contributions are approximately linear in $\tan\beta$, and their sign is given by the sign of μ, together with the sign of $M_{1,2}$ or X_t. Generally, all diagrams are suppressed by two powers of the SUSY mass scale. Hence the basic behavior of diagrams ($\mathrm{C,C_{2L},N1,N2}$) is given by

$$a_\mu^{(\mathrm{C,C_{2L};N1,N2})} \propto \frac{m_\mu^2}{M_{\mathrm{SUSY}}^2}\, \tan\beta\, \mathrm{sign}(\mu M_{2;1}) \tag{12.39}$$

if all SUSY masses are set equal to a common scale M_{SUSY}. However, it is important to keep in mind that the relevant SUSY masses are different in the five diagrams.

In particular, diagrams (N1) and ($\tilde{t}_{2L}$) are special because they increase linearly with μ, while the all other diagrams are suppressed for large μ by their μ-dependent loop functions F. Likewise, only the one-loop diagrams are sensitive to the smuon and sneutrino masses. If these are large, the one-loop diagrams can be suppressed and the two-loop diagrams can become dominant. The chargino-loop diagram (C_{2L}) can be large if the chargino and Higgs masses are small.

The discussion of the stop-loop diagram ($\tilde{t}_{2L}$) is complicated by the fact that the seemingly linear dependence on $(m_t X_t)$ is cut off by the requirement that both stop mass eigenvalues are positive. This diagram is largest for maximal stop mixing, i.e. if $(m_t X_t)$ is large but both eigenvalues are positive and $m_{\tilde{t}_1} \ll m_{\tilde{t}_2}$, and if $m_{\tilde{t}_1}$ and the Higgs boson mass are small. In this case, diagram ($\tilde{t}_{2L}$) has the behavior

$$a_\mu^{(\tilde{t}_{2L})} \propto \frac{m_\mu^2}{M_{\text{SUSY}}^2} \frac{\mu\, m_t}{M_W^2} \tan\beta\, \text{sign}(X_t), \tag{12.40}$$

where M_{SUSY} denotes here the common mass scale of the appearing Higgs boson and the lightest stop. Thus the diagram is linearly enhanced by large μ, and its sign is determined by $\text{sign}(X_t)$. In the following subsections we will provide the exact analytical formulas for all these diagrams and also derive the numerical prefactors in the proportionalities (12.39) and (12.40).

12.2.3. *One-loop contributions*

Each diagram that contributes to a_μ contains one line carrying the μ-lepton number. This fact allows to divide the MSSM one-loop diagrams into two classes:

(1) SM-like diagrams, where the μ-lepton number is carried only by μ and/or ν_μ.
(2) SUSY diagrams, where the μ-lepton number is carried also by $\tilde{\mu}$ and/or $\tilde{\nu}_\mu$.

The diagrams of the first class involve only SM-particles, and they are essentially identical in the SM and the MSSM. The only non-identical diagrams involve two couplings of physical SM or MSSM Higgs bosons to the muon line. Owing to the additional suppression factor m_μ^2/M_W^2 such diagrams are entirely negligible both in the SM and the MSSM.

Therefore the SUSY one-loop contribution, i.e. the difference between a_μ in the MSSM and the SM, is given entirely by the diagrams of the second class. They are displayed in Fig. 12.4 and involve either a chargino–sneutrino or a neutralino–smuon loop. In contrast to the diagrams in Figs. 12.2, 12.3 they are written in terms of interaction eigenstates, which is more appropriate for an exact evaluation. The diagrams have been evaluated in Refs. [25–28] with various restrictions on the masses and mixings. These restrictions have been dropped in Refs. [29–32], and exact results have been derived. Later, more comprehensive and general evaluations of these diagrams have been presented in the context of particular supersymmetric models [33–35] and the unconstrained MSSM [36] (see also Refs. [37, 38] for related results on weak dipole moments in the MSSM). We present the general result in the form given in Ref. [39]:

$$a_\mu^{\rm SUSY,1L} = a_\mu^{\chi^0} + a_\mu^{\chi^\pm}, \tag{12.41}$$

with

$$\begin{aligned} a_\mu^{\chi^0} &= \frac{m_\mu}{16\pi^2}\sum_{i,m}\Big\{-\frac{m_\mu}{12m^2_{\tilde\mu_m}}(|n^L_{im}|^2+|n^R_{im}|^2)F_1^N(x_{im}) \\ &\qquad + \frac{m_{\chi^0_i}}{3m^2_{\tilde\mu_m}}\mathrm{Re}[n^L_{im}n^R_{im}]F_2^N(x_{im})\Big\}, \end{aligned} \tag{12.42}$$

$$\begin{aligned} a_\mu^{\chi^\pm} &= \frac{m_\mu}{16\pi^2}\sum_{k}\Big\{\frac{m_\mu}{12m^2_{\tilde\nu_\mu}}(|c^L_k|^2+|c^R_k|^2)F_1^C(x_k) \\ &\qquad + \frac{2m_{\chi^\pm_k}}{3m^2_{\tilde\nu_\mu}}\mathrm{Re}[c^L_k c^R_k]F_2^C(x_k)\Big\}, \end{aligned} \tag{12.43}$$

where $i = 1\ldots 4$ and $k = 1,2$ denote the neutralino and chargino indices, $m = 1,2$ denotes the smuon index, and the couplings are given by

$$n^L_{im} = \frac{1}{\sqrt{2}}(g_1 N_{i1} + g_2 N_{i2})U^{\tilde\mu}{}_{m1}{}^* - y_\mu N_{i3} U^{\tilde\mu}{}_{m2}{}^*, \tag{12.44}$$

$$n^R_{im} = \sqrt{2} g_1 N_{i1} U^{\tilde\mu}{}_{m2} + y_\mu N_{i3} U^{\tilde\mu}{}_{m1}, \tag{12.45}$$

$$c^L_k = -g_2 V_{k1}, \tag{12.46}$$

$$c^R_k = y_\mu U_{k2}. \tag{12.47}$$

Fig. 12.4. The two SUSY one-loop diagrams, written in terms of mass eigenstates. The external photon line has to be attached to the charged internal lines.

The kinematic variables are defined as the mass ratios $x_{im} = m^2_{\chi^0_i}/m^2_{\tilde{\mu}_m}$, $x_k = m^2_{\chi^\pm_k}/m^2_{\tilde{\nu}_\mu}$, and the loop functions are given by

$$F_1^N(x) = \frac{2}{(1-x)^4}\big[1 - 6x + 3x^2 + 2x^3 - 6x^2\log x\big], \tag{12.48}$$

$$F_2^N(x) = \frac{3}{(1-x)^3}\big[1 - x^2 + 2x\log x\big], \tag{12.49}$$

$$F_1^C(x) = \frac{2}{(1-x)^4}\big[2 + 3x - 6x^2 + x^3 + 6x\log x\big], \tag{12.50}$$

$$F_2^C(x) = \frac{3}{2(1-x)^3}\big[-3 + 4x - x^2 - 2\log x\big], \tag{12.51}$$

normalized such that $F_i^j(1) = 1$. The U(1) and SU(2) gauge couplings are given by $g_{1,2} = e/\{c_W, s_W\}$, such that the one-loop contributions are of the order $\alpha = e^2/(4\pi)$. A class of large two-loop logarithms can be taken into account by the replacement $\alpha \to \alpha(M_{\rm SUSY})$ (see Section 12.2.4.2 for more details).

For discussing the one-loop contributions $a_\mu^{\chi^{0,\pm}}$ it is noteworthy that the terms linear in $m_{\chi^{0,\pm}}$ are not enhanced by a factor $m_{\chi^{0,\pm}}/m_\mu$ compared to the other terms. Rather, these terms involve either an explicit factor of the muon Yukawa coupling y_μ or of the combination $U^{\tilde{\mu}}{}_{m1}U^{\tilde{\mu}}{}_{m2}/m^2_{\tilde{\mu}_m}$, which in turn is proportional to $(M^2_\mu)_{12}$ and thus to y_μ. Hence, all terms are of the same basic order $m_\mu^2/M^2_{\rm SUSY}$, and the terms linear in $m_{\chi^{0,\pm}}$ are enhanced merely by a factor $\tan\beta$ from the muon Yukawa coupling.

It is instructive to close this subsection by deriving a simple approximation of Eqs. (12.42), (12.43) for large $\tan\beta$ and the case that all SUSY mass parameters in the smuon, chargino and neutralino mass matrices are equal to a common scale $M_{\rm SUSY}$. In this case only the terms linear in $m_{\chi^{0,\pm}}$ have to be considered, and the loop functions $F_i^j(x)$ can be approximated by a Taylor series around $x = 1$. For example, using $F_2^C(x) \approx 1 - \frac{3}{4}(x-1)$, the factors $m_{\chi_k^\pm} c_k^L c_k^R F_2^C(x_k)$ appearing in $a_\mu^{\chi^\pm}$ can be approximated as[a]

$$-g_2 y_\mu \sum_k U_{k2} V_{k1} m_{\chi_k^\pm} \left(\frac{7}{4} - \frac{3}{4} \frac{m^2_{\chi_k^\pm}}{m^2_{\tilde{\nu}_\mu}} \right)$$
$$\approx \frac{3 g_2 y_\mu}{4} \frac{X_{22}(X^\dagger)_{21} X_{11}}{m^2_{\tilde{\nu}_\mu}} \approx \frac{3 g_2 y_\mu}{4} \text{sign}(\mu M_2) X_{12}. \qquad (12.52)$$

Here terms that are suppressed by $1/\tan\beta$ or M_W/M_{SUSY} have been neglected. Note that the factors on the right-hand side correspond directly to the mass-insertion diagram (C) in Fig. 12.2 and the approximation (12.39). The factors appearing in $a_\mu^{\chi^0}$ can be similarly approximated. Inserting these approximations, one obtains [36]

$$a_\mu^{\chi^0} = \frac{g_1^2 - g_2^2}{192\pi^2} \frac{m_\mu^2}{M^2_{\text{SUSY}}} \text{ sign}(\mu M_2) \tan\beta \left[1 + \mathcal{O}\left(\frac{1}{\tan\beta}, \frac{M_W}{M_{\text{SUSY}}} \right) \right], \qquad (12.53)$$

$$a_\mu^{\chi^\pm} = \frac{g_2^2}{32\pi^2} \frac{m_\mu^2}{M^2_{\text{SUSY}}} \text{ sign}(\mu M_2) \tan\beta \left[1 + \mathcal{O}\left(\frac{1}{\tan\beta}, \frac{M_W}{M_{\text{SUSY}}} \right) \right], \qquad (12.54)$$

where real parameters and equal signs of M_1 and M_2 have been assumed.

12.2.4. *Two-loop contributions*

It is useful to classify the MSSM two-loop diagrams similar to the one-loop diagrams, into

(1) two-loop corrections to SM one-loop diagrams, where the μ-lepton number is carried only by μ and/or ν_μ.
(2) two-loop corrections to SUSY one-loop diagrams, where the μ-lepton number is carried also by $\tilde{\mu}$ and/or $\tilde{\nu}_\mu$.

The first class contains in particular SM-like diagrams with an insertion of a loop of SUSY particles, e.g. of $\tilde{t}$, $\tilde{b}$ or $\chi^\pm$. Such diagrams are particularly interesting since they constitute SUSY two-loop contributions that involve other particles and have a completely different parameter dependence than the SUSY one-loop contributions. Most importantly, these two-loop contributions can be large even if $a_\mu^{\text{SUSY,1L}}$ is suppressed. The contributions of this class are exactly known [41, 42].

[a] Note the subtle but important difference between setting the SUSY mass parameters equal and setting the mass eigenvalues equal. Only in the former case one obtains the $\tan\beta$-enhanced terms [40].

The SUSY two-loop contributions of the second class involve the same particles as the SUSY one-loop contributions (possibly plus additional ones). Hence they can be expected to have a similar parameter dependence as $a_\mu^{\rm SUSY,1L}$. Only parts of the contributions of this class are known. Leading QED-logarithms have been evaluated in Ref. [43], contributions enhanced by two powers of $\tan\beta$ in Ref. [44], and a subclass of the remaining contributions has been investigated in Ref. [45].

12.2.4.1. *Two-loop corrections to SM one-loop diagrams*

The MSSM two-loop contributions of the first class can be decomposed into a SM- and SUSY-part,

$$a_\mu^{\rm SM,2L} + a_\mu^{\rm SUSY,2L(a)}, \tag{12.55}$$

where $a_\mu^{\rm SM,2L}$ denotes the SM two-loop contributions. The genuine SUSY contributions of this class can be split into four parts:

$$a_\mu^{\rm SUSY,2L(a)} = a_\mu^{\chi,\rm 2L} + a_\mu^{\tilde f,\rm 2L} + a_\mu^{\rm SUSY,ferm,2L} + a_\mu^{\rm SUSY,bos,2L}. \tag{12.56}$$

The first two terms correspond to diagrams involving a closed chargino/neutralino or sfermion loop, respectively. These diagrams are further categorized according to the particles coupling to the muon line,

$$a_\mu^{X,\rm 2L} = a_\mu^{(XVH)} + a_\mu^{(XVV)} + a_\mu^{(XVG)}, \qquad X = \chi, \tilde f. \tag{12.57}$$

Diagrams where one gauge boson and one physical Higgs boson couple to the muon line are denoted as (XVH) with $V = \gamma, W, Z$ and $H = h^0, H^0, A^0, H^\pm$. Diagrams where only gauge bosons or unphysical Goldstone bosons couple to the muon are denoted as (XVV), (XVG).[b] Sample diagrams are shown in Fig. 12.5.

The remaining two terms in (12.56) correspond to diagrams involving only SM- or two-Higgs-doublet model particles and no SUSY particles. These diagrams are different in the MSSM and the SM due to the additional Higgs bosons and the modified Higgs boson couplings. $a_\mu^{\rm SUSY,ferm,2L}$ denotes the difference between the MSSM- and SM-evaluation of the diagrams involving a SM fermion (i.e. quark or lepton) loop; likewise, $a_\mu^{\rm SUSY,bos,2L}$ denotes the corresponding difference of the diagrams without fermion loop, the so-called bosonic contributions. Sample diagrams are shown in Fig. 12.6.

[b]Diagrams of the form (XHH), (XHG) etc. in which two Higgs or Goldstone bosons couple to the muon line are suppressed by an additional muon Yukawa coupling and can be neglected.

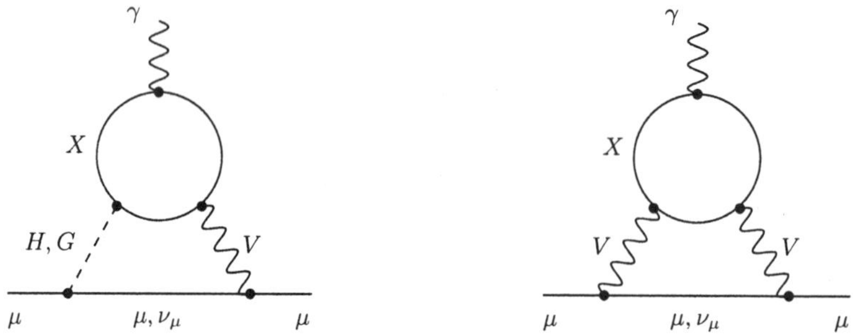

Fig. 12.5. Sample two-loop diagrams with closed chargino/neutralino or sfermion loop, contributing to $a_\mu^{\chi,2L}$ and $a_\mu^{\tilde{f},2L}$. The diagrams are categorized into classes (XVH), (XVG) and (XVV), where $X = \chi^{\pm,0}, \tilde{f}$. $V = \gamma, Z, W^\pm$ denotes gauge bosons, $H = h^0, H^0, A^0, H^\pm$ denotes physical Higgs bosons, and $G = G^{\pm,0}$ denotes Goldstone bosons. See [41, 42] for more details on the possible diagram topologies.

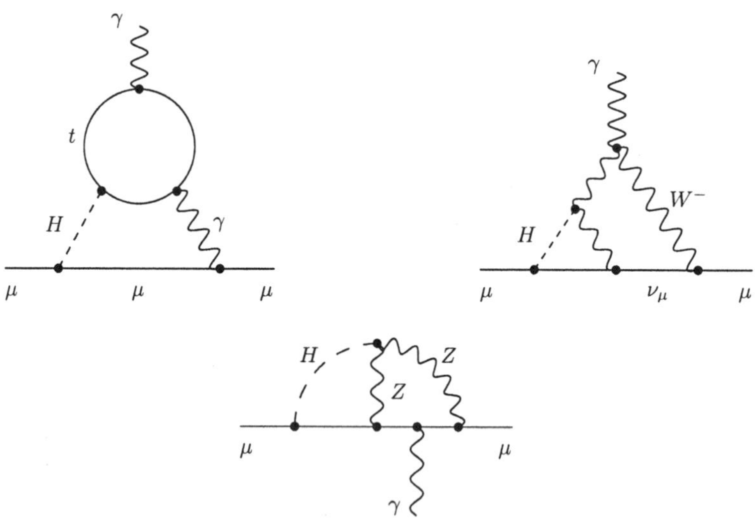

Fig. 12.6. Sample two-loop diagrams involving only SM- or two-Higgs-doublet model particles and either with or without fermion loop. These diagrams are different in the MSSM and the SM due to the modified Higgs sector, and this difference constitutes the SUSY contributions $a_\mu^{\mathrm{SUSY,ferm,2L}}$ and $a_\mu^{\mathrm{SUSY,bos,2L}}$.

The SUSY two-loop diagrams can be conveniently evaluated by first applying a large mass expansion [46], where the muon mass is treated as small and all other masses as large. This results in a separation of scales, and all remaining integrals are of one of two types. One type are one-scale two-point integrals with external momentum $p^2 = m_\mu^2$ and all internal masses being either zero or equal to m_μ. The other type are integrals where all internal masses are heavy but the external momentum can be neglected. All these integrals and the corresponding prefactors can be evaluated analytically [41, 42]. In addition to the genuine two-loop diagrams, one-loop counterterm diagrams have to be evaluated. These contain renormalization constants corresponding to charge, mass, and tadpole renormalization, which are defined in the on-shell renormalization scheme [21, 47, 48].

The diagrams of classes (XVH), $X = \chi, \tilde{f}, f$ can be calculated in an alternative way. In these so-called Barr–Zee diagrams [49], a closed loop generates an effective γ–V–H vertex, and this vertex can be evaluated first by a one-loop computation. By inserting the result and performing the second loop integral one obtains a simple integral representation for the full two-loop diagram. Barr–Zee diagrams were first considered because they give rise to important contributions to electric dipole moments in extensions of the SM (see Refs. [49–51]). The contributions from particular Barr–Zee diagrams of the classes $(f\gamma A^0)$, $(f\gamma H)$, $(\tilde{f}\gamma H)$, $(\tilde{f}W^\pm H^\mp)$ to a_μ were considered in Refs. [52–55], respectively.

In a series of papers [56], a similar method was developed for a more general class of heavy fermion-loop diagrams. The method was successfully applied to evaluate e.g. $a_\mu^{\chi,2\text{L}}$ and to cast the results of these contributions into a rather compact form.

In Refs. [41, 42] the numerical results of all two-loop contributions in (12.56) were compared and analyzed in detail, taking into account that the SUSY parameters are constrained by experimental bounds on b-decays, M_h and other quantities. It turned out that the numerical values of the various contributions are very different:

- The largest contributions by far are the ones from the photonic Barr–Zee diagrams $(\chi\gamma H)$ and $(\tilde{f}\gamma H)$, where H denotes the neutral physical Higgs bosons h^0, H^0, A^0. As explained in Section 12.2.2 they are enhanced by a factor $\tan\beta$ and, in the case of the sfermion loop diagrams, by the potentially large Higgs–sfermion coupling. They can have values up to
$$a_\mu^{(\chi\gamma H)}, a_\mu^{(\tilde{f}\gamma H)} \sim \mathcal{O}(10) \times 10^{-10}. \tag{12.58}$$
Barr–Zee diagrams with Z or $W^\pm$ exchange have a similar parameter dependence but are typically smaller by a factor of about 3–5.
- The diagrams of classes (χVV) and $(\tilde{f}VV)$ and the corresponding Goldstone diagrams (χVG) and $(\tilde{f}VG)$ involve no enhanced muon–

Higgs Yukawa coupling and thus no $\tan\beta$-enhancement, and they do not involve any other enhancement factors. Their numerical impact is tiny. For SUSY masses larger than 100 GeV these contributions are smaller than 0.1×10^{-10}.

- The genuine SUSY contributions to the SM-like diagrams $a_\mu^{\rm SUSY,ferm,2L}$ and $a_\mu^{\rm SUSY,bos,2L}$ depend only on $\tan\beta$ and M_A and are small. Only for $M_A < 200$ GeV they can reach 10^{-10}, but for larger M_A they are typically below 0.5×10^{-10}.

Hence, for the purpose of the present review, we only present the analytical result for the dominant contributions from the photonic Barr–Zee diagrams with physical Higgs bosons. They can be written as

$$a_\mu^{(\chi\gamma H)} = \frac{\alpha^2 m_\mu^2}{8\pi^2 M_W^2 s_W^2} \sum_{k=1,2} \Big[{\rm Re}[\lambda_\mu^{A^0} \lambda_{\chi_k^+}^{A^0}] \, f_{PS}(m_{\chi_k^+}^2/M_{A^0}^2) + \sum_{S=h^0,H^0} {\rm Re}[\lambda_\mu^S \lambda_{\chi_k^+}^S] \, f_S(m_{\chi_k^+}^2/M_S^2) \Big], \quad (12.59)$$

$$a_\mu^{(\tilde f\gamma H)} = \frac{\alpha^2 m_\mu^2}{8\pi^2 M_W^2 s_W^2} \sum_{\tilde f=\tilde t,\tilde b,\tilde\tau} \sum_{i=1,2} \Big[(N_c Q^2)_{\tilde f} \times \sum_{S=h^0,H^0} {\rm Re}[\lambda_\mu^S \lambda_{\tilde f_i}^S] \, f_{\tilde f}(m_{\tilde f_i}^2/M_S^2) \Big]. \quad (12.60)$$

The Higgs–muon and Higgs–chargino coupling factors are given by

$$\lambda_\mu^{\{h^0,H^0,A^0\}} = \left\{ -\frac{s_\alpha}{c_\beta}, \frac{c_\alpha}{c_\beta}, t_\beta \right\}, \quad (12.61)$$

$$\lambda_{\chi_k^+}^{\{h^0,H^0,A^0\}} = \frac{\sqrt{2} M_W}{m_{\chi_k^+}} \big(U_{k1} V_{k2} \{c_\alpha, s_\alpha, -c_\beta\} + U_{k2} V_{k1} \{ -s_\alpha, c_\alpha, -s_\beta\} \big). \quad (12.62)$$

In the Higgs–sfermion couplings we neglect terms that are subleading in $\tan\beta$ and that give rise to negligible contributions to a_μ:

$$\lambda_{\tilde t_i}^{\{h^0,H^0\}} = \frac{2m_t}{m_{\tilde t_i}^2 s_\beta} \big(+\mu^* \{s_\alpha, -c_\alpha\} + A_t \{c_\alpha, s_\alpha\} \big) \, (U_{i1}^{\tilde t})^* \, U_{i2}^{\tilde t}, \quad (12.63)$$

$$\lambda_{\tilde b_i}^{\{h^0,H^0\}} = \frac{2m_b}{m_{\tilde b_i}^2 c_\beta} \big(-\mu^* \{c_\alpha, s_\alpha\} + A_b \{ -s_\alpha, c_\alpha\} \big) \, (U_{i1}^{\tilde b})^* \, U_{i2}^{\tilde b}, \quad (12.64)$$

$$\lambda_{\tilde \tau_i}^{\{h^0,H^0\}} = \frac{2m_\tau}{m_{\tilde \tau_i}^2 c_\beta} \big(-\mu^* \{c_\alpha, s_\alpha\} + A_\tau \{ -s_\alpha, c_\alpha\} \big) \, (U_{i1}^{\tilde \tau})^* \, U_{i2}^{\tilde \tau}. \quad (12.65)$$

The loop integral function f_{PS} can be given either as a one-dimensional integral or in terms of dilogarithms:

$$f_{PS}(z) = z\int_0^1 \frac{dx \log\frac{x(1-x)}{z}}{x(1-x)-z} = \frac{2z}{y}\Big[\mathrm{Li}_2\Big(1-\frac{1-y}{2z}\Big) - \mathrm{Li}_2\Big(1-\frac{1+y}{2z}\Big)\Big] \tag{12.66}$$

with $y = \sqrt{1-4z}$. Note that $f_{PS}(z)$ is real and analytic even for $z \geq 1/4$. The other loop functions are related to f_{PS} as

$$f_S(z) = (2z-1)f_{PS}(z) - 2z(2+\log z), \tag{12.67}$$

$$f_{\tilde{f}}(z) = \frac{z}{2}\Big[2+\log z - f_{PS}(z)\Big]. \tag{12.68}$$

We remark that useful numerical estimates for the leading two-loop contributions can be obtained by taking into account only the $\tan\beta$-enhanced terms in the couplings and by approximating the loop functions. In the case of the sfermion-loop contributions, simple approximations for the most important Barr–Zee diagrams $(\tilde{t}HV)$, $(\tilde{b}HV)$ with stop or sbottom loops can be derived [41]. The approximation for $(\tilde{t}HV)$ agrees with the estimate (12.40) discussed in Section 12.2.2, and the approximation for $(\tilde{b}HV)$ has a similar form. In the case of the chargino/neutralino-loop diagrams the parameter dependence is simpler, and the diagrams with W- and Z-exchange can be included in the approximation [42]. We will collect all these approximations below in Eqs. (12.83)–(12.85).

12.2.4.2. *Two-loop corrections to SUSY one-loop diagrams*

Sample diagrams for the two-loop corrections to SUSY one-loop diagrams are shown in Fig. ds-fig:SUTLb. The diagrams all involve the same particles and the same couplings as the SUSY one-loop diagrams (possibly plus additional ones). Hence the overall parameter dependence of $a_\mu^{\mathrm{SUSY,2L(b)}}$ and of $a_\mu^{\mathrm{SUSY,1L}}$ can be expected to be similar, up to the additional two-loop suppression of $a_\mu^{\mathrm{SUSY,2L(b)}}$.

So far, this class of contributions is not completely known, but two dominant parts have been identified and computed: large QED-logarithms [43] and $(\tan\beta)^2$-enhanced corrections [44]. In the following we will first describe these two dominant contributions and then discuss a partial evaluation of the remaining two-loop diagrams [45].

The $(\tan\beta)^2$-enhanced corrections are particularly interesting since, as explained at length in the previous sections, all contributions discussed so far are enhanced by at most one factor of $\tan\beta$. The physical origin of

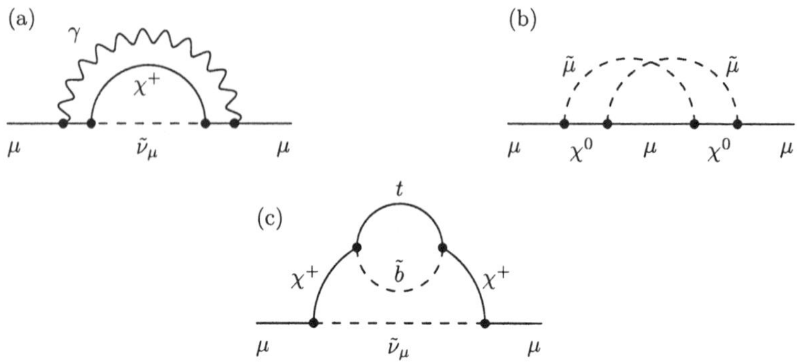

Fig. 12.7. Sample two-loop diagrams contributing to $a_\mu^{\rm SUSY,2L(b)}$, i.e. involving a SUSY one-loop diagram. The external photon can be attached to all charged internal lines. (a) shows a diagram with additional photon loop, giving rise to large QED-logarithms. (b) shows a diagram of the class computed in [45]. (c) shows a diagram with an additional fermion/sfermion loop.

these $(\tan\beta)^2$-corrections is a shift in the muon Yukawa coupling y_μ due to $\tan\beta$-enhanced one-loop effects. In the computation of $a_\mu^{\rm SUSY}$ beyond the one-loop level this shift appears in the muon mass renormalization constant δm_μ, defined in the on-shell scheme:

$$m_\mu + \delta m_\mu = \frac{m_\mu}{1+\Delta_\mu} + \text{non-}\tan\beta\text{-enhanced terms},$$
$$y_\mu = \frac{m_\mu}{v_1(1+\Delta_\mu)}\left(1+\mathcal{O}(\cot\beta)\right), \tag{12.69}$$

where m_μ is the physical, pole-mass of the muon and where the shift $\Delta_\mu \propto \alpha\tan\beta$ will be given below. This type of $\tan\beta$-enhanced corrections has been studied intensely in the down-quark sector [57, 58]. In Ref. [44] the method of Ref. [58] was applied to explicitly identify $\tan\beta$-enhanced loop diagrams and re-sum them to all orders in perturbation theory. Equation (12.69) contains the desired effect to all orders $\alpha^l\tan^l\beta$, $l=1,2,\ldots$. For the phenomenology of a_μ only the term with $l=1$, contributing to $a_\mu^{\rm SUSY}$ at the two-loop level, is relevant.

The shift Δ_μ is given by the $\tan\beta$-enhanced terms of the muon self energy. In terms of the loop function

$$I(a,b,c) = \frac{a^2b^2\log\frac{a^2}{b^2} + b^2c^2\log\frac{b^2}{c^2} + c^2a^2\log\frac{c^2}{a^2}}{(a^2-b^2)(b^2-c^2)(a^2-c^2)}, \tag{12.70}$$

which satisfies $I(a,a,a) = 1/(2a^2)$, it can be written as

$$\begin{aligned}\Delta_\mu = &- \mu \ \tan\beta \ \frac{g_2^2 \ M_2}{16\pi^2} \ I(m_1, m_2, m_{\tilde{\nu}_\mu}^{(\infty)}) \\ &- \mu \ \tan\beta \ \frac{g_2^2 \ M_2}{16\pi^2} \ \frac{1}{2} I(m_1, m_2, m_{\tilde{\mu}_L}^{(\infty)}) \\ &- \mu \ \tan\beta \ \frac{g_1^2 \ M_1}{16\pi^2} \ \Big[I(\mu, M_1, m_{\tilde{\mu}_R}^{(\infty)}) \\ &- \frac{1}{2} I(\mu, M_1, m_{\tilde{\mu}_L}^{(\infty)}) - I(M_1, m_{\tilde{\mu}_L}^{(\infty)}, m_{\tilde{\mu}_R}^{(\infty)}) \Big]. \end{aligned} \quad (12.71)$$

Here we have defined

$$\begin{aligned} m_{1,2}^2 &= \frac{1}{2} \Big[(M_2^2 + \mu^2 + 2M_W^2) \\ &\qquad \mp \sqrt{(M_2^2 + \mu^2 + 2M_W^2)^2 - 4M_2^2\mu^2} \Big], \\ m_{\tilde{\nu}_\mu}^{(\infty)2} &= m_{L,\tilde{\mu}}^2 - \frac{M_Z^2}{2}, \quad m_{\tilde{\mu}_L}^{(\infty)2} = m_{L,\tilde{\mu}}^2 - M_Z^2(s_W^2 - \frac{1}{2}), \\ m_{\tilde{\mu}_R}^{(\infty)2} &= m_{R,\tilde{\mu}}^2 + M_Z^2 s_W^2. \end{aligned} \quad (12.72)$$

While the chargino contributions are exact in the large-$\tan\beta$ limit, the neutralino contributions in Eq. (12.71) have been simplified using the approximation $M_Z \ll \mu, M_1, M_2$.

The contribution $a_\mu^{\mathrm{SUSY},\Delta_\mu}$ of the $(\tan\beta)^2$-enhanced contributions to $a_\mu = -2m_\mu F_M(0)$ can be easily obtained by noting that the magnetic form factor $F_M(0)$ is proportional to y_μ, apart from numerically irrelevant terms with three or more powers of y_μ. In all $\tan\beta$-enhanced one-loop contributions, i.e. the $\mathrm{Re}[c^L c^R]$ and $\mathrm{Re}[n^L n^R]$ terms in Eqs. (12.42), (12.43), y_μ appears explicitly in the higgsino-muon couplings or in the Higgs-smuon coupling triggering the left-right mixing in the smuon mass matrix. At higher orders the shift (12.69) has to be taken into account, and therefore

$$a_\mu^{\mathrm{SUSY,1L}} + a_\mu^{\mathrm{SUSY},\Delta_\mu} = a_\mu^{\mathrm{SUSY,1L}} \left(\frac{1}{1+\Delta_\mu} \right). \quad (12.73)$$

Note that this formula is only correct for the enhanced terms of order $\alpha^l \tan^l \beta$, but this is the desired order. An alternative way to derive this result is to eliminate y_μ using Eq. (12.35) in favor of m_μ, to evaluate the two-loop counterterm contributions involving the renormalization constant δm_μ, and then to use Eq. (12.69).

Numerically, the $(\tan\beta)^2$-enhanced contributions can be quite sizable. In the case where all SUSY masses are equal and much larger than M_W,

$$\Delta_\mu = -0.0018 \tan\beta \, \mathrm{sign}\, \mu. \quad (12.74)$$

Hence, in the region with $\tan\beta \sim 50$ the corrections can shift the one-loop contributions by roughly 10%. Importantly, being a dimensionless quantity, Δ_μ does not decouple for arbitrarily large SUSY masses. For slight mass splittings, Δ_μ can be even larger than in Eq. (12.74).

In order to discuss the QED-logarithmic contributions we start with the decomposition

$$a_\mu^{\rm SUSY,2L(b)} = c_L^{\rm SUSY,2L(b)} \log\frac{M_{\rm SUSY}}{m_\mu} + c_0^{\rm SUSY,2L(b)}. \tag{12.75}$$

The first piece contains the large logarithm of the ratio $M_{\rm SUSY}/m_\mu$, where $M_{\rm SUSY}$ is the generic SUSY mass scale, and the second piece contains at most small logarithms of ratios of different SUSY masses.

The decomposition Eq. (12.75) is analogous to the one of the bosonic two-loop contributions in the SM, which have been evaluated in [42, 59, 60]. In the case of the SM, the term enhanced by $\log M_Z/m_\mu$ is roughly a factor 10 larger than the non-logarithmic piece. In the case of the MSSM, the non-logarithmic piece is enhanced by the $(\tan\beta)^2$-terms, but nevertheless it is clear that the QED-logarithms constitute an important class of the two-loop contributions.

The large QED-logarithms in Eq. (12.75) arise from two-loop diagrams that involve a SUSY one-loop diagram and an additional photon loop. The loop integrals of such diagrams have an infrared singularity in the limit $m_\mu \to 0$ and therefore give rise to terms $\propto \log m_\mu$. As shown by [43], the appropriate framework to evaluate these logarithms efficiently is the framework of effective field theories.

The relevant effective field theory is obtained from the MSSM by integrating out all fields of mass $\geq M_{\rm SUSY}$ and retaining only the muon and photon. All further light or heavy SM fields are irrelevant in this analysis and can be ignored. The resulting theory is QED with additional higher-dimensional terms, described by

$$\mathcal{L}_{\rm eff} = -2\sqrt{2}G_\mu \sum_i C_i(\mu)\mathcal{O}_i \tag{12.76}$$

in the notation of Ref. [43]. The $\mathcal{O}_i$ are higher-dimensional operators. The analysis of Ref. [43] shows that in the MSSM, like in many New Physics models, only one higher-dimensional operator has to be considered, namely the one corresponding to the muon anomalous magnetic moment,

$$H_\mu = -\frac{e}{16\pi^2}\, m_\mu\, \bar\mu\, \sigma^{\nu\rho}\, \mu\, F_{\nu\rho}. \tag{12.77}$$

The prefactors $C_i(\mu)$ are renormalization-scale dependent Wilson coefficients, which can be determined by matching the effective theory to the full MSSM at the high scale $M_{\rm SUSY}$. Determining the Wilson coefficient $C_{H_\mu}(M_{\rm SUSY})$ thus corresponds to the one-loop computation of $a_\mu^{\rm SUSY}$.

By construction, the large logarithms are identical in the full MSSM and the effective theory. However, in the effective theory the logarithms can be obtained simply from the one-loop renormalization-group running of the Wilson coefficient $C_{H_\mu}(\mu)$ from $\mu = M_{\rm SUSY}$ down to $\mu = m_\mu$. This running is described by

$$C_{H_\mu}(\mu) = C_{H_\mu}(M_{\rm SUSY}) - \gamma(H_\mu, H_\mu)\,\frac{\alpha(\mu)}{4\pi}\,\log\frac{M_{\rm SUSY}}{\mu}\,C_{H_\mu}(M_{\rm SUSY}) \tag{12.78}$$

where $\gamma(H_\mu, H_\mu)$ is the anomalous dimension of H_μ. On a diagrammatic level, the correspondence of this formula to the two-loop computation in the full MSSM is easy to see. In the MSSM, the logarithms arise from diagrams like the one in Fig. 12.7(a). Corresponding diagrams in the effective theory are obtained by contracting the insertion of the SUSY one-loop diagram to a point. The resulting diagrams are one-loop contributions to H_μ, involving the effective vertex H_μ. Their UV-divergence, and thus their $\log\mu$-terms, determine the anomalous dimension $\gamma(H_\mu, H_\mu)$.

The value of the anomalous dimension is

$$\gamma(H_\mu, H_\mu) = 16. \tag{12.79}$$

As a result, the QED-logarithms in the two-loop contributions to $a_\mu^{\rm SUSY}$ are given by

$$a_\mu^{\rm SUSY,2L(b)} = -\frac{4\alpha}{\pi}\log\frac{M_{\rm SUSY}}{m_\mu}\,a_\mu^{\rm SUSY,1L} + c_0^{\rm SUSY,2L(b)}. \tag{12.80}$$

This logarithmic correction is negative, and it amounts to $-7\%\ldots-9\%$ of the SUSY one-loop contributions for $M_{\rm SUSY}$ between 100 and 1000 GeV. This result can be compared to the case of the bosonic SM two-loop contributions, where the logarithms amount to -19% of the SM electroweak one-loop result.

Depending on the value of $\tan\beta$, either the QED-logarithms or the $(\tan\beta)^2$-enhanced corrections can be the dominant two-loop effect.

As mentioned before, the non-logarithmic terms in $a_\mu^{\rm SUSY,2L(b)}$, $c_0^{\rm SUSY,2L(b)}$, are not known so far. A first evaluation of a subclass of diagrams has been carried out in Ref. [45]. The considered diagrams involve only sleptons, charginos and neutralinos, and only topologies as in Fig. 12.7(b) are taken into account that contain no self-energy subdiagrams. These diagrams constitute a finite contribution to $c_0^{\rm SUSY,2L(b)}$. However, the result of Ref. [45] can only be viewed as intermediate because there are more diagrams, e.g. containing self-energy subdiagrams or W- or Z-exchange, that would involve exactly the same coupling constants, such that non-trivial cancellations could be possible.

Nevertheless, the investigation of Ref. [45] provides a first insight to the possible values of the remaining two-loop contributions. The numerical values are surprisingly large. In a range of SUSY parameters with SUSY masses of the order 300...500 GeV, the values are mostly below 10^{-10} but can become up to 2×10^{-10}, which is significantly larger than the corresponding non-logarithmic terms of the bosonic SM two-loop contributions.

12.2.5. *Summary of known contributions and error estimate*

To summarize, the SUSY contributions to a_μ up to the two-loop level, i.e. the difference of a_μ in the MSSM and the SM, are given by

$$\begin{aligned} a_\mu^{\rm SUSY} = a_\mu^{\rm SUSY,1L} & \left(1 - \frac{4\alpha}{\pi}\log\frac{M_{\rm SUSY}}{m_\mu}\right)\left(\frac{1}{1+\Delta_\mu}\right) \\ & + a_\mu^{(\chi\gamma H)} + a_\mu^{(\tilde{f}\gamma H)} \\ & + a_\mu^{(\chi\{W,Z\}H)} + a_\mu^{(\tilde{f}\{W,Z\}H)} + a_\mu^{\rm SUSY,ferm,2L} + a_\mu^{\rm SUSY,bos,2L} + \ldots, \end{aligned} \tag{12.81}$$

where the terms in the first and second line have been given analytically in Eqs. (12.41), (12.59), (12.60), (12.71), (12.80). Note that the discussion of the two-loop QED-logarithms and Eq. (12.78) also show that the one-loop result should be parametrized in terms of the running $\alpha(M_{\rm SUSY})$.

For many applications it should be sufficient to take into account the terms in the first, possibly the second line. The explicitly written terms in the third line are known, and they can be up to $\mathcal{O}(1) \times 10^{-10}$, but in the largest part of the MSSM parameter space they are much smaller. The dots denote the known but negligible contributions of the type (χVV), $(\tilde{f}VV)$, the contributions evaluated in Ref. [45], and the remaining unknown two-loop contributions.

Useful approximations for the dominant terms are given by

$$a_\mu^{\rm SUSY,1L} \approx \; 13 \times 10^{-10} \left(\frac{100\,{\rm GeV}}{M_{\rm SUSY}}\right)^2 \tan\beta \; {\rm sign}(\mu M_2), \tag{12.82}$$

$$a_\mu^{(\chi VH)} \approx \quad 11 \times 10^{-10} \left(\frac{\tan\beta}{50}\right)\left(\frac{100\ \text{GeV}}{M_{\text{SUSY}}}\right)^2 \text{sign}(\mu M_2), \tag{12.83}$$

$$a_\mu^{(\tilde{t}\gamma H)} \approx -13 \times 10^{-10} \left(\frac{\tan\beta}{50}\right)\left(\frac{m_t}{m_{\tilde{t}}}\right)\left(\frac{\mu}{20 M_H}\right) \text{sign}(X_t), \tag{12.84}$$

$$a_\mu^{(\tilde{b}\gamma H)} \approx -3.2 \times 10^{-10} \left(\frac{\tan\beta}{50}\right)\left(\frac{m_b \tan\beta}{m_{\tilde{b}}}\right)\left(\frac{A_b}{20 M_H}\right) \text{sign}(\mu). \tag{12.85}$$

The first two are valid if all SUSY masses are approximately equal (note that the relevant masses are different in the two cases), and the third and fourth are valid if the stop/sbottom mixing is large and the relevant stop/sbottom and Higgs masses are of similar size. The result for the $(\tilde{b}\gamma H)$ contribution has not been discussed so far, but it can be understood in the same way as the $(\tilde{t}\gamma H)$ result [41].

In the following we list the missing contributions and estimate the theory error of the SUSY prediction (12.81).

- Two-loop QED-corrections beyond the leading logarithm (12.80). The leading-log approximation does not exactly fix the scale M_{SUSY} in the logarithm and in $\alpha(M_{\text{SUSY}})$ (the latter appears in the one-loop result). The exact form of the logarithms and of the non-logarithmic terms can only be found by a complete computation of the two-loop diagrams with a SUSY one-loop diagram and additional photon exchange. The error of the approximation (12.80) can be estimated by varying M_{SUSY} in the range 100...1000 GeV to about 2% of the SUSY one-loop contributions. If the SUSY contributions to a_μ are the origin of the observed deviation Eq. (12.3), they are certainly smaller than roughly 50×10^{-10}, and then this error is below 1×10^{-10}.
- Further electroweak and SUSY two-loop corrections to SUSY one-loop diagrams. These corrections include two-loop diagrams similar to Fig. 12.7(a) but with W-, Z-, Higgs- instead of photon-exchange, and like in Fig. 12.7(b) with purely SUSY particles in the loops. Given the result for the subclass evaluated in Ref. [45], we assign an error of $\pm 2 \times 10^{-10}$ to these diagrams. Note that this is a factor of 10 larger than the known result of the corresponding non-logarithmic bosonic two-loop contributions in the SM [42, 59, 60].
- Two-loop corrections to SUSY one-loop diagrams with fermion/sfermion-loops (see Fig. ds-fig:SUTLb(c) for an example). This class of diagrams involves in particular top/stop- and

bottom/sbottom-loops, which are enhanced by the large 3rd generation Yukawa couplings. We estimate the numerical value of these diagrams to be less than $\pm 0.5 \times 10^{-10}$ for not too light SUSY masses for the following reasons. SUSY relates these diagrams to SM diagrams with top/bottom loops, which amount to about 0.6×10^{-10} but are not suppressed by possibly heavy SUSY masses. SUSY also relates the fermion/sfermion-loop diagrams to pure sfermion-loop diagrams such as $a_\mu^{(\tilde{f}\gamma H)}$, see Eqs. (12.84), (12.85). However, this relation should be most accurate for rather small $A_{t,b}$ and μ, since the fermion/sfermion-loop diagrams are not enhanced by making these parameters large. In that case, the approximations Eqs. (12.84), (12.85) lead to values below 0.5×10^{-10}.

- In general, three-loop contributions can be expected to be significantly smaller than the two-loop contributions. Two potential exceptions are three-loop diagrams that correspond to the two-loop contributions of the types $(\chi\gamma H)$, $(\tilde{f}\gamma H)$ with subloop-corrections to the Higgs-boson masses or the b-quark Yukawa coupling. It is well-known that the one-loop corrections to the Higgs-boson masses, in particular to M_h, and to y_b can be very large. Hence, in cases where the diagrams with h-exchange and/or sbottom loop are very large, the missing three-loop contributions could amount to $\mathcal{O}(1) \times 10^{-10}$. Fortunately, however, the influence of the lightest Higgs boson mass and y_b on the $(\chi\gamma H)$, $(\tilde{f}\gamma H)$ diagrams is small in the largest part of the MSSM parameter space. Hence we neglect the theory error associated with the missing three-loop contributions.

To summarize, we estimate the theory error associated with (12.81) to

$$\delta a_\mu^{\rm SUSY}(\text{unknown}) = 0.02\ a_\mu^{\rm SUSY,1L} + 2.5 \times 10^{-10}, \tag{12.86}$$

where the errors associated with the individual classes of missing diagrams have been added linearly. If $a_\mu^{\rm SUSY}$ is approximated by only the first line of (12.81), the error increases by the neglected contributions in the second line. An upper limit of these can be well approximated by [41, 42]

$$\delta a_\mu^{\rm SUSY}(\text{2nd line}) = 0.3\ \left(a_\mu^{(\chi\gamma H)} + a_\mu^{(\tilde{f}\gamma H)}\right) + 0.3 \times 10^{-10}. \tag{12.87}$$

It should be noted that the error estimate is deliberately conservative. The later numerical analysis shows that often already the known two-loop contributions are much smaller than 10^{-10} (see Table 12.2). In these cases, it is reasonable to assume that the theory error due to the unknown higher-order corrections is also much smaller than the estimate Eq. (12.86). In order to further improve the accuracy of the $a_\mu^{\rm SUSY}$, the full SUSY two-loop contributions will need to be evaluated.

12.3. Impact of a_μ on New Physics Phenomenology

The muon magnetic moment has been an important constraint on physics beyond the Standard Model for many years. With the recent experimental and theoretical progress it has become one of the most precisely known quantities in particle physics, and it is now sensitive to physics beyond the Standard Model at the TeV scale.

The TeV scale appears to be a crucial scale in particle physics. It is linked to electroweak symmetry breaking, and many arguments indicate that radically new concepts such as supersymmetry, extra dimensions, or new interactions, could be realized at this scale. Furthermore, cold dark matter particles could have weak-scale/TeV-scale masses, and Grand Unification prefers the existence of supersymmetry at the TeV scale.

As discussed in Section 12.1.2, different scenarios such as supersymmetry, extra dimensions, or a muon substructure, lead to very different predictions for a_μ. With the precise determination of a_μ and the deviation Eq. (12.3), a_μ constitutes an excellent discriminator between these different scenarios.

In the present section we discuss in detail the impact of a_μ on the phenomenology of physics beyond the Standard Model. We focus again on SUSY as a particularly well-motivated case. We first give general examples how a_μ can be used to constrain SUSY; then we focus on the role a_μ can play in the LHC era as a complementary observable.

12.3.1. *Constraints from a_μ*

Already before the Brookhaven $g-2$ experiment, the muon magnetic moment was an important observable in SUSY phenomenology. Since the SM theory prediction agreed with the older CERN measurement of a_μ [7], it was possible to derive lower bounds on SUSY masses [25–35]. Even after many of these bounds were superseded by LEP bounds, taking into account a_μ remained complementary in corners of parameter space where light SUSY particles could escape LEP detection [61].

Today, after the more than 10-fold improvement of the experimental precision reached at the E821 experiment, and after corresponding improvements on the theory side, much tighter constraints on SUSY parameters can be derived. Since the first result of the Brookhaven experiment was published in 2001, a large number of studies analyzed the impact of a_μ on supersymmetry. Examples are Refs. [39, 62–90]. In particular, upper and lower limits on SUSY masses could be derived under the assumption that SUSY is responsible for the deviation Δa_μ. In general, parameter restrictions can be derived not only in the MSSM but also in modified or extended models such as the next-to minimal-supersymmetric Standard Model. Furthermore, correlations between a_μ and other observables from b-physics or dark matter were found in restricted models, such as minimal supergravity.

We focus here on a very conservative approach that was taken in the "superconservative" analysis of Ref. [91]. A general MSSM with very few restrictions on the parameters was assumed and it was only required that $a_\mu^{\rm SUSY}$ falls into a more-than-5 σ region, a region that nobody could seriously argue with. Interestingly, even in this approach significant bounds on SUSY parameters could be obtained. These bounds can be regarded as definite, robust against any future theoretical or experimental developments. As an example, for $\tan\beta = 30$ and a lighter chargino mass of 100 GeV, an absolute lower bound on the heavier smuon mass of 200 GeV (for $\mu > 0$) or 300 GeV (for $\mu < 0$) has been derived.

The existence of such robust bounds, which cannot be obtained from any other experiments, shows that a_μ is an important and independent probe of New Physics. This role of a_μ does not even depend on the magnitude of the deviation Eq. (12.3), only on its precision.

The value of a_μ for New Physics phenomenology can also be seen from Table 12.2 and Fig. 12.8, where the numerical results for $a_\mu^{\rm SUSY}$ are shown in several different SUSY scenarios. Evidently, a_μ excludes many SUSY parameter points and prefers others.

Table 12.2 shows the results for $a_\mu^{\rm SUSY}$ in the Snowmass Points and Slopes (SPS) benchmark points [92]. These results provide an overview and reference of the SUSY contributions that can be expected in various well-motivated and often considered parameter scenarios.

The SPS points 1a, 1b, 3, 6, correspond to various minimal supergravity, (mSUGRA)-scenarios with $\tan\beta$ between 10 and 30 and SUSY masses in the range 100 ... 1000 GeV. They lead to predictions of a_μ very close to the observed value Eq. (12.3). The same is true for points 7, 8, which correspond to gauge-mediated SUSY breaking. The point SPS 2 does not

Table 12.2. Results of the SUSY contributions to a_μ in units of 10^{-10} for the SPS benchmark parameter points. The two-loop QED-logarithms (12.80) are already included in the improved one-loop result. The remaining known SUSY two-loop corrections have been split up into the $(\tan\beta)^2$-enhanced terms and the full two-loop corrections to SM one-loop diagrams.

SPS Point	$a_\mu^{\text{SUSY,1L}}$(improved)	$a_\mu^{\text{SUSY},\Delta_\mu}$	$a_\mu^{\text{SUSY,2L(a)}}$
SPS 1a	29.29	0.54	0.52
SPS 1b	31.84	1.87	0.65
SPS 2	1.65	0	0.13
SPS 3	13.55	0.26	0.30
SPS 4	49.04	3.55	1.51
SPS 5	8.59	0.07	0.27
SPS 6	16.87	0.30	0.41
SPS 7	23.71	0.75	0.61
SPS 8	17.33	0.45	0.44
SPS 9	-8.98	0.06	0.05

fit so well to Eq. (12.3) because it corresponds to the focus-point region in which sfermions are very heavy. SPS 4, 5, involve very large/small $\tan\beta$, respectively, and therefore yield too high/low values for a_μ^{SUSY}, and SPS 9, which corresponds to anomaly-mediated SUSY breaking, involves negative $(\mu M_{1,2})$ and thus leads to negative a_μ^{SUSY}.

Figure 12.8 shows the results of a general, model-independent MSSM parameter scan that summarizes the current status of a_μ in SUSY. This scan shows the maximum results for a_μ in the MSSM if all parameters are independently varied and the range of SUSY masses for which the MSSM can accommodate the current experimental result. The full set of one- and two-loop contributions in (12.81) are taken into account, and the results are compared with the current deviation between the experimental and SM result Eq. (12.3).

The MSSM parameters have been varied in the ranges

$$|\mu|, M_2, m_{L,\tilde{f}}, m_{R,\tilde{f}}, |A_f| \leq 3000 \text{ GeV}, \quad M_A = 90\dots3000 \text{ GeV}, \tag{12.88}$$

with $\tan\beta = 50$, where the upper mass limit is motivated by naturalness arguments and the lower limit on M_A corresponds to the experimental limit. The parameter $\tan\beta$ has not been varied because of the essentially linear $\tan\beta$ dependence of a_μ^{SUSY}. All other parameters have been varied independently, except that M_1 has been fixed via the GUT relation $M_1/M_2 = 5g_1^2/3g_2^2$, $A_\mu = 0$, $m_{L,\tilde{b}} = m_{L,\tilde{\tau}}$, $m_{R,\tilde{b}} = m_{R,\tilde{\tau}}$ and $A_b = A_\tau$.

These constraints do not have much impact on the maximum contributions [62]. The third generation sfermion parameters are significantly restricted by experimental constraints on M_h, $\Delta\rho$ and b-decays. Only parameter points have been considered that are in agreement with these constraints, according to the same criteria as in the "weak bounds" of Ref. [41]. Note that the precise values used in these constraints have not much influence on the maximum results in Fig. 12.8 as they affect essentially only the two-loop sfermion contributions.

The light region of Fig. 12.8 corresponds to all data points and thus to all possible values of $a_\mu^{\rm SUSY}$ compatible with the parameter range (12.88) and the experimental constraints from b-decays, M_h and $\Delta\rho$. The SUSY contributions can accommodate the observed result Eq. (12.3) within 1σ for LOSP masses below about 540 GeV. For LOSP masses below about 400 GeV, the SUSY contributions can be even too large, and thus the a_μ-measurement significantly restricts the MSSM parameter space in this low-mass region.

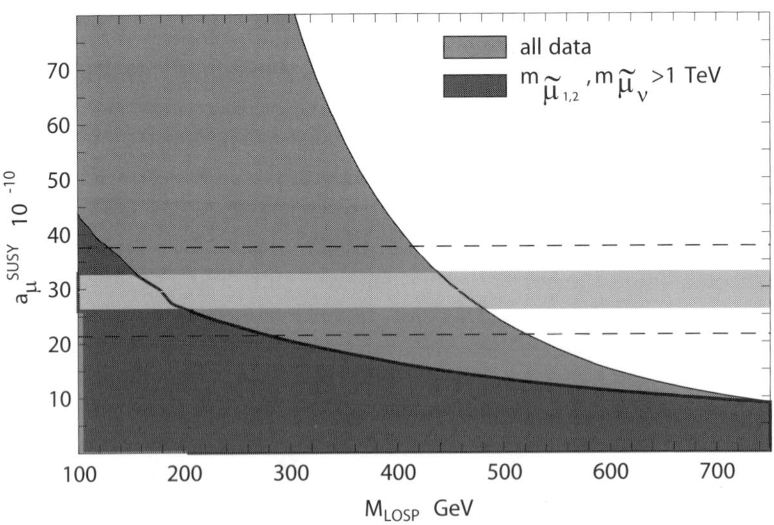

Fig. 12.8. Allowed values of $a_\mu^{\rm SUSY}$ as a function of the mass of the lightest observable supersymmetric particle $M_{\rm LOSP} = \min(m_{\tilde\chi_1^\pm}, m_{\tilde\chi_2^0}, m_{\tilde f_i})$, from a scan of the MSSM parameter space in the range (12.88) and for $\tan\beta = 50$. The 1σ regions corresponding to the current deviation between experimental and SM values Eq. (12.3) and the proposed future precision are indicated. The light gray region corresponds to all data points that satisfy the experimental constraints from b-decays, M_h and $\Delta\rho$. In the dark gray region, the smuons and sneutrinos are heavier than 1 TeV.

The dark region corresponds to the situation that the smuons and sneutrinos are heavy,

$$m_{\tilde{\mu}_{1,2}},\ m_{\tilde{\nu}_\mu} > 1000\ \text{GeV}, \tag{12.89}$$

while charginos, neutralinos, stops and sbottoms can still be light. A dedicated analysis of this situation is of interest since heavy first and second generation sfermions are sometimes considered as a possible explanation of the absence of observable SUSY contributions to flavor-changing neutral currents and CP-violating observables. Heavy smuons and sneutrinos significantly suppress the maximum SUSY contributions to a_μ. Nevertheless, the contributions can be in the 2σ region Eq. (12.3) for M_{LOSP} smaller than about 560 GeV.

12.3.2. *Complementarity to LHC measurements*

In the next decade, LHC experiments will for the first time directly probe physics at the TeV scale. As discussed above, TeV-scale physics can be expected to be very rich, and it is likely that the LHC will discover physics beyond the Standard Model.

In the quest to identify the nature of TeV-scale physics and to answer questions related to, for example, electroweak symmetry breaking and Grand Unification, information from the LHC needs to be combined and cross-checked with information from as many complementary experiments as possible. This need is highlighted by the unprecedented complexity of the LHC accelerator and experiments, the involved initial and final states, and the huge backgrounds at the LHC. In all these respects, $(g-2)$ provides an indispensable complement.

In fact, a new, improved $(g-2)$ measurement with 2.5-fold improvement in precision is feasible and has been proposed [8, 9]. Similar improvements on the precision of the SM theory prediction seem likely. Taken together, the combined experimental and theoretical uncertainty of $\Delta a_\mu(\text{Exp} - \text{SM})$ could go from 8.1×10^{-10} in Eq. (12.3) down to 3.9×10^{-10} or less [8, 9]. These improvements can be achieved already rather early in the LHC era, if the new $(g-2)$ measurement is carried out.

In the following we discuss in more detail how a_μ will be useful in understanding TeV-scale physics even after the LHC has established the existence of physics beyond the Standard Model, following the discussion in Ref. [8].

It has been established that the LHC is sensitive to virtually all proposed weak-scale extensions of the Standard Model, ranging from supersymmetry to extra dimensions, little Higgs models and others. However, even if the existence of physics beyond the Standard Model is established, it will be far from easy for the LHC alone to identify which of the possible alternatives is realized. a_μ, in particular with an improved measurement will be highly valuable in this respect since it will provide a benchmark and stringent selection criterion that can be imposed on any model that is tested at the LHC.

For example, a situation is possible where the LHC finds many new mass states, which are compatible with both minimal-supersymmetric and universal-extra-dimension model predictions [93]. The muon $(g-2)$ would especially aid in the selection. Universal extra dimension models predict a tiny effect to a_μ, see Eq. (12.7), while the SUSY contributions can be much larger.

Likewise, within SUSY itself there are many different well-motivated scenarios that are not always easy to distinguish at the LHC. Figure 12.9 shows a graphical distribution of the 10 Snowmass Points and Slopes model benchmark predictions [92] for a_μ(SUSY), see Table 12.2, and the UED prediction Eq. (12.7). The predictions range considerably and can be positive and negative, due to the factor sgn(μ) in Eq. (12.8), where this sign is particularly difficult to determine at the LHC.

The discriminating power of the current and an improved $(g-2)$ measurement – even if the actual value of Δa_μ turned out to be smaller than the current value – is evident from the figure.

In a much more general approach, Ref. [94] discussed "supersymmetry without prejudice". First a large set of supersymmetry parameter points ("models") in a 19-dimensional parameter space was identified which was in agreement with many important existing experimental and theoretical constraints. Then the postdictions for observables such as $(g-2)$ were studied. The result for $(g-2)$ was rather similar to Fig. 12.9. The entire range $a_\mu^{\rm SUSY} \sim (-100 \ldots + 300) \times 10^{-11}$ was populated by a large number of "models". Therefore, $(g-2)$ constitutes a crucial tool to rule out a large fraction of models and thus determine supersymmetry parameters.

In a complementary strategy, often special, highly constrained models of New Physics are considered. A popular example is the constrained MSSM (CMSSM) [95]. In this scenario supersymmetry breaking is assumed to take place in a hidden sector and to be transmitted to the observable sector via gravitational interactions. The Kähler potential of the underlying supergravity theory is assumed to be minimal, i.e. in particular to involve no generation-dependent couplings. Furthermore, at the GUT scale

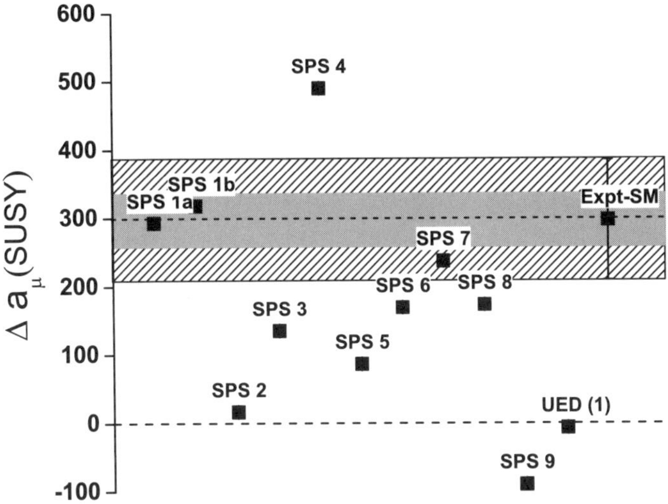

Fig. 12.9. The Snowmass Points and Slopes predictions for a_μ(SUSY) for various scenarios [92], and the UED prediction for one extra dimension [14]. The wide band is the present 1σ difference between experiment and theory. The narrow band represents the proposed improved precision, given the same central value. Figure courtesy of D. W. Hertzog.

$M_{\rm GUT} \approx 2 \times 10^{16}$ GeV the SM gauge interactions unify. The free parameters of this model are

$$m_0, m_{1/2}, A_0, \tan\beta, \mathrm{sign}(\mu), \tag{12.90}$$

where m_0, $m_{1/2}$ and A_0 are the universal values of all scalar mass, gaugino mass, and A parameters of the MSSM at the GUT scale. The low-energy values of the MSSM soft-breaking parameters are determined by renormalization-group running, and the value of $|\mu|$ is determined by the requirement that electroweak symmetry breaking leads to the correct value of M_Z.

Although the CMSSM has only four free continuous parameters it fits rather well to existing data. One precise measurement such as the future determination of Δa_μ effectively fixes one parameter as a function of the others and thus reduces the number of free parameters by one. A large number of recent analyses have made use of this, see Refs. [96–100]. In fact, the CMSSM is very sensitive not only to a_μ but also to the dark matter (assumed to consist of neutralinos) relic density and the b-decay $b \to s\gamma$. As shown in Fig. 12.10, the three observables lead to complementary constraints in CMSSM parameter space. Already now it is non-trivial that

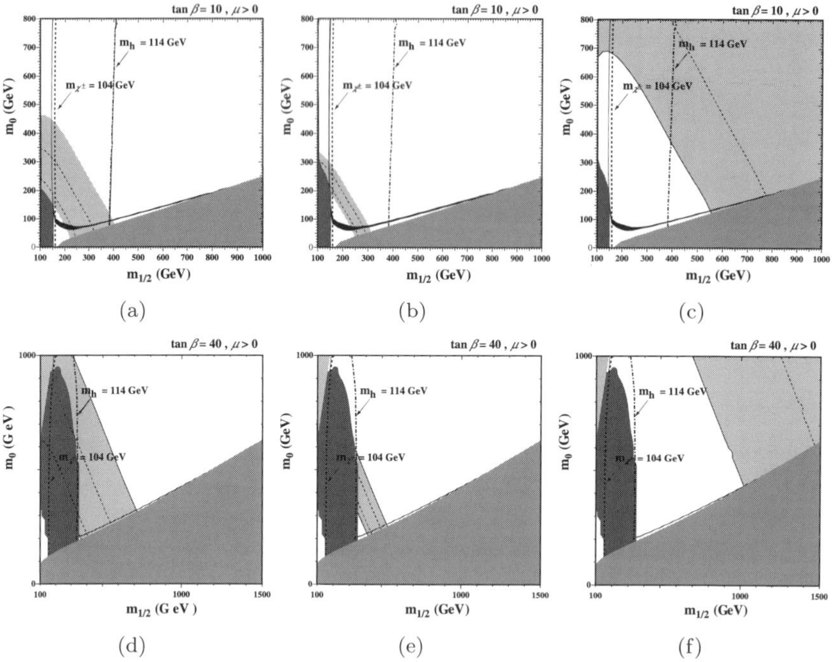

Fig. 12.10. The m_0(scalar mass)–$m_{1/2}$(gaugino mass) plane of the CMSSM parameter space for $\tan\beta = (10; 40)$, $A_0 = 0$, $\mathrm{sgn}(\mu) = +$:
(a;d) The $\Delta a_\mu^{(\mathrm{today})} = 295(81) \times 10^{-11}$ between experiment and standard-model theory is from Eq. (12.3). The medium gray wedge on the lower right is excluded by the requirement the dark matter be neutral. Direct limits on the Higgs and chargino $\chi^\pm$ masses are indicated by vertical lines, with the region to the left excluded. Restrictions from the WMAP satellite data are shown as a dark gray hyperbola. The $(g-2)$ 1 and 2-standard deviation boundaries are shown in light gray. The dark region on the left is excluded by $b \to s\gamma$. (b;e) The plot with $\Delta a_\mu = 295(39) \times 10^{-11}$. (c;f) The same errors as (b), but $\Delta a_\mu = 0$. (Figures courtesy of K. Olive, following Ref. [100]).

points exist which are in agreement with all current constraints. Statistical surveys of the CMSSM parameter space in Ref. [99] indicate a certain tension between a_μ and $b \to s\gamma$, i.e. the majority of CMSSM parameter points are not compatible with at least one of these observables. Clearly, any future improvement on the determination of these observables will be highly welcome. Combining these constraints with forthcoming LHC data will allow for very stringent tests of the CMSSM.

For unraveling the mysteries of TeV-scale physics it is not sufficient to determine which type of New Physics is realized, but it is necessary to determine model parameters as precisely as possible. Here the complementarity between the LHC and a_μ becomes particularly important.

A difficulty at the LHC is the very indirect relation between LHC observables (cross sections, mass spectra, edges, etc.) and model parameters such as masses and couplings, let alone more underlying parameters such as supersymmetry-breaking parameters or the μ-parameter in the MSSM. Generally, the LHC Inverse problem [101] states that several different points in the supersymmetry parameter space can give rise to indistinguishable LHC signatures. It has been shown that a promising strategy is to determine the model parameters by performing a global fit of a model such as the MSSM to all available LHC data. Several tools for such global fits in scenarios beyond the Standard Model have been developed [102, 103]. However, one shortcoming of global fits is that there can be several almost degenerate local minima of χ^2 as a function of the model parameters. Independent observables such as the ones available at the proposed International Linear Collider [104] or a_μ will be highly valuable to break such degeneracies and in this way to unambiguously determine the model parameters.

In the following we provide further examples for the complementarity of LHC and a_μ for the well-studied case of the MSSM. The LHC has only a weak sensitivity to two central parameters: the sign of the μ-parameter and $\tan\beta$, the ratio of the two Higgs vacuum expectation values. According to Eq. (12.8) the MSSM contributions to a_μ are highly sensitive to both of these parameters. Therefore, a future improved a_μ measurement has the potential to establish a definite positive or negative sign of the μ-parameter in the MSSM, which would be a crucial piece of information. Statistical investigations of sign(μ) in the CMSSM have already confirmed the importance of a_μ in this respect [97].

In order to give a concrete illustration of a $\tan\beta$ measurement, we reconsider the case discussed in Ref. [102] and assume that the MSSM reference point SPS1a [92] is realized. Using the comprehensive LHC-analysis of [102], $\tan\beta$ can be determined only rather poorly to $\tan\beta^{\text{LHC fit}} = 10.0 \pm 4.5$. In such a situation one can study the MSSM prediction for a_μ as a function of $\tan\beta$ (all other parameters are known from the global fit to LHC data) and compare it to the measured value, in particular after an improved measurement. The result is shown in Fig. 12.11. As can be seen from the plot, already using today's value for a_μ would strongly improve the determination of $\tan\beta$, but the improvement will be even more impressive after a future more precise a_μ measurement. A similar but more comprehensive study in Ref. [105], where a_μ has been incorporated into the

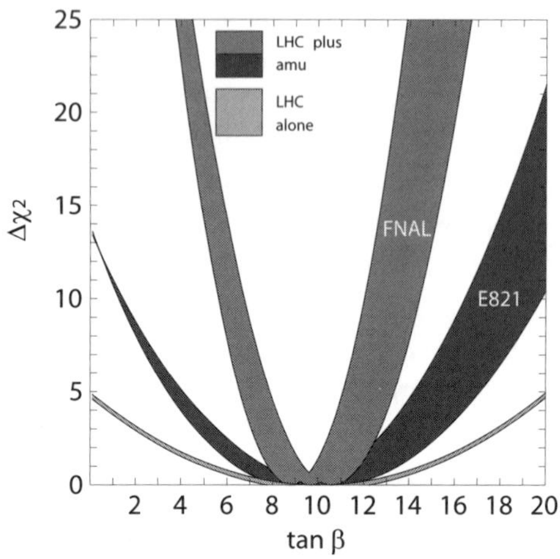

Fig. 12.11. Possible future $\tan\beta$ determination from the measurement of a_μ, assuming that the MSSM reference point SPS1a is realized. The lowest, thin parabola indicates the precision of the LHC determination $\tan\beta^{\mathrm{LHC\ fit}} = 10.0 \pm 4.5$ from Ref. [102, 105]. The darker gray band corresponds to the present value for a_μ, the lighter gray band corresponds to $\Delta a_\mu^{\mathrm{future}} = 295(34) \times 10^{-11}$, which could be achieved with an improved $(g-2)$ measurement at Fermilab. The gray bands show $\Delta\chi^2 = \left(\frac{a_\mu^{MSSM}(\tan\beta) - a_\mu^{\exp}}{\{81;34\}\times 10^{-11}}\right)^2$ as a function of $\tan\beta$, where in $a_\mu^{\mathrm{MSSM}}(\tan\beta)$ all parameters except $\tan\beta$ have been set to the values determined at the LHC. The width of the gray bands results from the expected LHC-uncertainty of the parameters (mainly smuon masses and M_2, μ) [105]. The plot shows that the precision for $\tan\beta$ that can be obtained using a_μ is limited by the precision of the other input parameters but is still much better than the determination using LHC data alone.

global fit, confirms this role of a_μ as an excellent observable to measure $\tan\beta$.

One should note that even if better ways to determine $\tan\beta$ at the LHC alone might be found in the future, an independent determination using a_μ will still be highly valuable. $\tan\beta$ is one of the central MSSM parameters, and it appears in all sectors and in almost all observables. Therefore, measuring $\tan\beta$ in two different ways, e.g. using certain Higgs- or b-decays at the LHC and using a_μ, would constitute a non-trivial and indispensable test of the universality of $\tan\beta$ and thus of the structure of the MSSM.

The anomalous magnetic moment of the muon is sensitive to contributions from a wide range of models beyond the Standard Model, and will continue to place stringent restrictions on them. In the LHC era it will constitute an indispensable tool to discriminate between very different types of new physics, and it is highly sensitive to parameters which are difficult to measure at the LHC. It will play an essential complementary role in the quest to understand physics beyond the Standard Model at the TeV scale.

12.4. Conclusions

The final result of the Brookhaven $a_\mu = (g_\mu - 2)/2$ measurement shows a deviation of $29.5\,(8.1) \times 10^{-10}$, corresponding to more than 3σ, from the corresponding current SM prediction. This observed deviation is one of the strongest direct indications for physics beyond the Standard Model. Independent of theoretical arguments based for example on fine-tuning, it singles out the electroweak or TeV-scale as the required New Physics scale. Different types of New Physics can give very different contributions to a_μ, depending on whether the respective contributions to the muon mass are of $\mathcal{O}(1)$, of the order of a loop factor $\alpha/(4\pi)$ or in between. Hence a precise determination of a_μ will always be a valuable discriminator between different types of New Physics, independent of the actual result.

Supersymmetry below the TeV-scale fits particularly well to the observed deviation. Within supersymmetry, a_μ provides strong constraints on the possible values of supersymmetry parameters. The SUSY contributions to a_μ are essentially proportional to $\tan\beta\ \mathrm{sign}(\mu)/M^2_{\mathrm{SUSY}}$. Hence, positive μ and moderate to large $\tan\beta$ are preferred. But even more detailed, upper and lower mass limits can be derived. Bounds derived from a_μ are different from and therefore complementary to bounds derived from all other observables, e.g. collider, dark matter, or b-physics observables.

The qualitative behavior of a_μ^{SUSY} can be well understood using the mass insertion technique. The $\tan\beta$-enhancement arises in diagrams where the necessary chirality flip occurs at a muon Yukawa coupling, either to a Higgsino or Higgs boson, because this coupling is enhanced by $1/\cos\beta \approx \tan\beta$ compared to its SM value. The μ-parameter mediates the transition between the two Higgs/Higgsino doublets $H_{1,2}$, and this transition enhances diagrams because only H_1 couples to muons while H_2 has the larger vacuum expectation value. For a precise quantitative analysis, the MSSM prediction of a_μ should be known with a precision that matches the one of the SM prediction. Currently the full one-loop and leading two-loop contributions are known, and the precision of the SUSY contributions is better than the one of the SM hadronic contributions but worse than the one of the electroweak SM contributions. In order to further improve the precision the full SUSY two-loop contributions need to be evaluated.

In spite of the impressive current status, progress on a_μ is important and has to come both from the experimental and the theoretical side. Already a small improvement of the precision could be sufficient to increase the discrepancy between the experimental and SM values of a_μ to 4–5σ and thus to establish the existence of non-SM contributions.

After more than 40 years of experimental and theoretical progress, the observable a_μ has become a sensitive probe of physics at the electroweak scale. Already today it is one of the most important constraints of physics beyond the SM. In the near future, particle physics will enter a new era where the detailed structure of physics at the electroweak scale and above will be unraveled by LHC experiments. The magnetic moment of the muon will provide a crucial cross-check and complement of the forthcoming experiments. It will be very important for constraining, studying and identifying physics beyond the Standard Model.

Acknowledgments

Parts of the present chapter are based on the Topical Review [5]. Permission from IOP Publishing to use this material (Copyright (2007) by IOP Publishing) is gratefully acknowledged. It is a pleasure to acknowledge collaboration and discussions with B. L. Roberts and D. W. Hertzog during the preparation of this chapter. This work has been supported in part by the DFG grant STO 876/1-1.

References

[1] F. Jegerlehner and A. Nyffeler, arXiv:0902.3360 [hep-ph].

[2] Eduardo de Rafael *Present Status of the Muon Anomalous Magnetic Moment*; To appear in the proceedings of 14th High-Energy Physics International Conference in Quantum Chromodynamics (QCD 08), Montpellier, France, 7-12 Jul 2008. e-Print: arXiv:0809.3085.

[3] Kirill Melnikov, Arkady Vainshtein, *Theory of the Muon Anomalous Magnetic Moment* (Springer Tracts in Modern Physics).

[4] Friedrich Jegerlehner, *The Anomalous Magnetic Moment of the Muon* (Springer Tracts in Modern Physics Vol. 226).

[5] D. Stöckinger, "The muon magnetic moment and supersymmetry," J. Phys. G **34** (2007) R45.

[6] G.W. Bennett, *et al.*, (Muon $(g-2)$ Collaboration), Phys. Rev. D **73**, 072003 (2006).

[7] J. Bailey *et al.* [CERN-Mainz-Daresbury Collaboration], Nucl. Phys. B **150** (1979) 1.
[8] D. W. Hertzog, J. P. Miller, E. de Rafael, B. Lee Roberts and D. Stöckinger, arXiv:0705.4617 [hep-ph].
[9] D. W. Hertzog, B. Lee Roberts *et al.*, Fermilab Proposal P-989, March 2009, http://lss.fnal.gov/archive/test-proposal/0000/fermilab-proposal-0989.shtml
[10] Andrzej Czarnecki and William J. Marciano, Phys. Rev. **D64** 013014 (2001).
[11] E. Eichten, *et al.*, Phys. Rev. Lett. **45**, 225 (1980), K. Lane, arXiv hep-ph/0102131, 2001.
[12] T. Appelquist, H. C. Cheng and B. A. Dobrescu, Phys. Rev. D **64** (2001) 035002.
[13] T. G. Rizzo, Phys. Rev. D **64** (2001) 095010.
[14] T. Appelquist and B. A. Dobrescu, Phys. Lett. B **516** (2001) 85.
[15] B. Ananthanarayan, G. Lazarides and Q. Shafi, Phys. Rev. D **44**, 1613 (1991).
[16] H. P. Nilles, Phys. Rept. **110** (1984) 1.
[17] H. E. Haber and G. L. Kane, Phys. Rept. **117** (1985) 75.
[18] S. P. Martin, arXiv:hep-ph/9709356.
[19] D. J. H. Chung, L. L. Everett, G. L. Kane, S. F. King, J. D. Lykken and L. T. Wang, Phys. Rept. **407** (2005) 1 [arXiv:hep-ph/0312378].
[20] I. J. R. Aitchison, arXiv:hep-ph/0505105.
[21] W. Hollik, E. Kraus, M. Roth, C. Rupp, K. Sibold and D. Stöckinger, Nucl. Phys. B **639**, 3 (2002) [arXiv:hep-ph/0204350].
[22] S. Heinemeyer, W. Hollik and G. Weiglein, Phys. Rept. **425** (2006) 265 [arXiv:hep-ph/0412214].
[23] S. Ferrara and E. Remiddi, Phys. Lett. B **53** (1974) 347.
[24] S. Ferrara and M. Porrati, Phys. Lett. B **288** (1992) 85.
[25] P. Fayet, in *Unification of the Fundamental Particles Interactions*, ed. S. Ferrara, J. Ellis, and P. van Nieuwenhuizen (Plenum, New York, 1980), p. 587.
[26] J. A. Grifols and A. Mendez, Phys. Rev. D **26**, 1809 (1982).
[27] J. R. Ellis, J. S. Hagelin and D. V. Nanopoulos, 'Phys. Lett. B **116** (1982) 283.
[28] R. Barbieri and L. Maiani, Phys. Lett. B **117**, 203 (1982).
[29] D. A. Kosower, L. M. Krauss and N. Sakai, Phys. Lett. B **133**, 305 (1983).
[30] T. C. Yuan, R. Arnowitt, A. H. Chamseddine and P. Nath, Z. Phys. C **26**, 407 (1984).
[31] J. C. Romao, A. Barroso, M. C. Bento and G. C. Branco, Nucl. Phys. B **250**, 295 (1985).
[32] I. Vendramin, Nuovo Cim. A **101** (1989) 731.
[33] S. A. Abel, W. N. Cottingham and I. B. Whittingham, Phys. Lett. B **259** (1991) 307.
[34] J. L. Lopez, D. V. Nanopoulos and X. Wang, Phys. Rev. D **49**, 366 (1994).

[35] U. Chattopadhyay and P. Nath, Phys. Rev. D **53**, 1648 (1996).
[36] T. Moroi, Phys. Rev. D **53** (1996) 6565 [Erratum-*ibid.* **56** (1997) 4424].
[37] W. Hollik, J. I. Illana, S. Rigolin and D. Stöckinger, Phys. Lett. B **416** (1998) 345.
[38] W. Hollik, J. I. Illana, S. Rigolin and D. Stöckinger, Phys. Lett. B **425** (1998) 322.
[39] S. Martin and J. Wells, Phys. Rev. D **64** (2001) 035003.
[40] W. S. Hou, F. F. Lee and C. Y. Ma, arXiv:0812.0064 [hep-ph].
[41] S. Heinemeyer, D. Stöckinger and G. Weiglein, Nucl. Phys. B **690** (2004) 62.
[42] S. Heinemeyer, D. Stöckinger and G. Weiglein, Nucl. Phys. B **699** (2004) 103.
[43] G. Degrassi and G. F. Giudice, Phys. Rev. D **58** (1998) 053007.
[44] S. Marchetti, S. Mertens, U. Nierste and D. Stockinger, Phys. Rev. D **79**, 013010 (2009).
[45] T. F. Feng, X. Q. Li, L. Lin, J. Maalampi and H. S. Song, Phys. Rev. D **73** (2006) 116001.
[46] V. Smirnov, *Applied Asymptotic Expansions in Momenta and Masses*, Springer Verlag, Berlin (2002).
[47] K. I. Aoki, Z. Hioki, M. Konuma, R. Kawabe and T. Muta, Prog. Theor. Phys. Suppl. **73**, 1 (1982).
[48] M. Böhm, H. Spiesberger and W. Hollik, Fortsch. Phys. **34**, 687 (1986).
[49] S. M. Barr and A. Zee, Phys. Rev. Lett. **65** (1990) 21.
[50] D. Chang, W. Y. Keung and A. Pilaftsis, Phys. Rev. Lett. **82** (1999) 900 [Erratum-ibid. **83** (1999) 3972].
[51] A. Pilaftsis, Phys. Lett. B **471** (1999) 174.
[52] D. Chang, W. F. Chang, C. H. Chou and W. Y. Keung, Phys. Rev. D **63** (2001) 091301.
[53] K. m. Cheung, C. H. Chou and O. C. W. Kong, Phys. Rev. D **64** (2001) 111301.
[54] A. Arhrib and S. Baek, Phys. Rev. D **65** (2002) 075002.
[55] C. H. Chen and C. Q. Geng, Phys. Lett. B **511** (2001) 77.
[56] T. F. Feng, L. Sun and X. Y. Yang, Phys. Rev. D **77** (2008) 116008; Nucl. Phys. B **800** (2008) 221; T. F. Feng and X. Y. Yang, Nucl. Phys. B **814** (2009) 101.
[57] M. Carena, M. Olechowski, S. Pokorski and C.E.M. Wagner, Nucl. Phys. **B426** (1994) 269; L.J. Hall, R. Rattazzi and U. Sarid, Phys. Rev. **D50** (1994) 7048; C. Hamzaoui, M. Pospelov and M. Toharia, Phys. Rev. D **59** (1999) 095005; K. S. Babu and C. F. Kolda, Phys. Rev. Lett. **84** (2000) 228; G. Isidori and A. Retico, JHEP **0111** (2001) 001; A. J. Buras, P. H. Chankowski, J. Rosiek and L. Slawianowska, Nucl. Phys. B **659** (2003) 3; T. Blazek, S. Raby and S. Pokorski, Phys. Rev. D **52** (1995) 4151.
[58] M. S. Carena, D. Garcia, U. Nierste and C. E. M. Wagner, Nucl. Phys. B **577** (2000) 88.
[59] A. Czarnecki, B. Krause and W. Marciano, Phys. Rev. Lett. **76** (1996) 3267.

[60] T. Gribouk and A. Czarnecki, Phys. Rev. D **72** (2005) 053016.
[61] M. Carena, G. F. Giudice and C. E. M. Wagner, Phys. Lett. B **390** (1997) 234.
[62] J. L. Feng and K. T. Matchev, Phys. Rev. Lett. **86**, 3480 (2001).
[63] L. L. Everett, G. L. Kane, S. Rigolin and L. T. Wang, Phys. Rev. Lett. **86** (2001) 3484.
[64] U. Chattopadhyay and P. Nath, Phys. Rev. Lett. **86** (2001) 5854.
[65] S. Komine, T. Moroi and M. Yamaguchi, Phys. Lett. B **506** (2001) 93.
[66] T. Ibrahim, U. Chattopadhyay and P. Nath, Phys. Rev. D **64** (2001) 016010.
[67] J. R. Ellis, D. V. Nanopoulos and K. A. Olive, Phys. Lett. B **508** (2001) 65.
[68] R. Arnowitt, B. Dutta, B. Hu and Y. Santoso, Phys. Lett. B **505** (2001) 177.
[69] K. Choi, K. Hwang, S. K. Kang, K. Y. Lee and W. Y. Song, Phys. Rev. D **64** (2001) 055001.
[70] H. Baer, C. Balazs, J. Ferrandis and X. Tata, Phys. Rev. D **64** (2001) 035004.
[71] A. Djouadi, M. Drees and J. L. Kneur, JHEP **0108** (2001) 055.
[72] E. A. Baltz and P. Gondolo, Phys. Rev. Lett. **86** (2001) 5004.
[73] J. Hisano and K. Tobe, Phys. Lett. B **510**, 197 (2001).
[74] Y. G. Kim and M. M. Nojiri, Prog. Theor. Phys. **106** (2001) 561.
[75] G. C. Cho and K. Hagiwara, Phys. Lett. B **514** (2001) 123.
[76] R. Arnowitt, B. Dutta and Y. Santoso, Phys. Rev. D **64** (2001) 113010.
[77] G. Belanger, F. Boudjema, A. Cottrant, R. M. Godbole and A. Semenov, Phys. Lett. B **519** (2001) 93.
[78] A. Dedes, H. K. Dreiner and U. Nierste, Phys. Rev. Lett. **87** (2001) 251804.
[79] S. Komine, T. Moroi and M. Yamaguchi, Phys. Lett. B **507** (2001) 224.
[80] S. Baek, P. Ko and H. S. Lee, Phys. Rev. D **65**, 035004 (2002).
[81] S. Baek, T. Goto, Y. Okada and K. i. Okumura, Phys. Rev. D **64**, 095001 (2001).
[82] K. Enqvist, E. Gabrielli and K. Huitu, Phys. Lett. B **512** (2001) 107.
[83] D. G. Cerdeno, E. Gabrielli, S. Khalil, C. Munoz and E. Torrente-Lujan, Phys. Rev. D **64** (2001) 093012.
[84] M. Graesser and S. D. Thomas, Phys. Rev. D **65**, 075012 (2002).
[85] Z. Chacko and G. D. Kribs, Phys. Rev. D **64** (2001) 075015.
[86] T. Blazek and S. F. King, Phys. Lett. B **518**, 109 (2001).
[87] F. Domingo and U. Ellwanger, JHEP **0807** (2008) 079.
[88] J. Cao and J. M. Yang, Phys. Rev. D **78** (2008) 115001.
[89] J. Cao, H. E. Logan and J. M. Yang, arXiv:0901.1437 [hep-ph].
[90] M. Byrne, C. Kolda and J. E. Lennon, Phys. Rev. D **67** (2003) 075004.
[91] S. P. Martin and J. D. Wells, Phys. Rev. D **67**, 015002 (2003).
[92] B. C. Allanach *et al.*, in *Proc. of the APS/DPF/DPB Summer Study on the Future of Particle Physics (Snowmass 2001)* ed. N. Graf, Eur. Phys. J. C **25** (2002) 113.
[93] J. M. Smillie and B. R. Webber, JHEP **0510** (2005) 069.

[94] C. F. Berger, J. S. Gainer, J. L. Hewett and T. G. Rizzo, arXiv:0812.0980 [hep-ph].
[95] G. L. Kane, C. F. Kolda, L. Roszkowski and J. D. Wells, Phys. Rev. D **49**, 6173 (1994).
[96] S. Komine and M. Yamaguchi, Phys. Rev. D **65** (2002) 075013; U. Chattopadhyay and P. Nath, Phys. Rev. D **65** (2002) 075009; Phys. Rev. D **66** (2002) 093001; H. Baer and C. Balazs, JCAP **0305** (2003) 006; U. Chattopadhyay, A. Corsetti and P. Nath, Phys. Rev. D **68** (2003) 035005; J. R. Ellis, K. A. Olive, Y. Santoso and V. C. Spanos, Phys. Rev. D **69** (2004) 095004; H. Baer, A. Belyaev, T. Krupovnickas and A. Mustafayev, JHEP **0406** (2004) 044; A. Djouadi, M. Drees and J. L. Kneur, JHEP **0603** (2006) 033.
[97] B. C. Allanach and C. G. Lester, Phys. Rev. D **73** (2006) 015013; B. C. Allanach, Phys. Lett. B **635** (2006) 123; B. C. Allanach, C. G. Lester and A. M. Weber, JHEP **0612** (2006) 065; B. C. Allanach, K. Cranmer, C. G. Lester and A. M. Weber, JHEP **0708**, 023 (2007); F. Feroz, B. C. Allanach, M. Hobson, S. S. AbdusSalam, R. Trotta and A. M. Weber, JHEP **0810** (2008) 064.
[98] J. R. Ellis, S. Heinemeyer, K. A. Olive, A. M. Weber and G. Weiglein, JHEP **0708** (2007) 083; S. Heinemeyer, X. Miao, S. Su and G. Weiglein, JHEP **0808**, 087 (2008). O. Buchmueller *et al.*, JHEP **0809** (2008) 117.
[99] R. Ruiz de Austri, R. Trotta and L. Roszkowski, JHEP **0605** (2006) 002; JHEP **0704** (2007) 084; JHEP **0707** (2007) 075; F. Feroz, M. P. Hobson, L. Roszkowski, R. R. de Austri and R. Trotta, arXiv:0903.2487 [hep-ph].
[100] J. R. Ellis, K. A. Olive, Y. Santoso and V. C. Spanos, Phys. Lett. **B 565** 176 (2003); John Ellis, Keith A. Olive, Yudi Santoso, and Vassilis C. Spanos, Phys. Rev. **D71** 095007 (2005), and references therein.
[101] N. Arkani-Hamed, G. L. Kane, J. Thaler and L. T. Wang, JHEP **0608**, 070 (2006).
[102] R. Lafaye, T. Plehn, M. Rauch and D. Zerwas, Eur. Phys. J. C **54**, 617 (2008).
[103] P. Bechtle, K. Desch and P. Wienemann, Comput. Phys. Commun. **174** (2006) 47.
[104] C. F. Berger, J. S. Gainer, J. L. Hewett, B. Lillie and T. G. Rizzo, arXiv:0712.2965 [hep-ph].
[105] M. Alexander, S. Kreiss, R. Lafaye, T. Plehn, M. Rauch, and D. Zerwas, Chapter 9 in M. M. Nojiri *et al.*, arXiv:0802.3672 [hep-ph].

Chapter 13

Probing CP Violation with Electric Dipole Moments

Maxim Pospelov

Department of Physics and Astronomy, University of Victoria, Victoria, BC, V8P 1A1 Canada
and
Perimeter Institute for Theoretical Physics, Waterloo, Ontario N2J 2W9, Canada
pospelov@uvic.ca

Adam Ritz

Department of Physics and Astronomy, University of Victoria, Victoria, BC, V8P 1A1 Canada
aritz@uvic.ca

We review several aspects of CP violation beyond the Kobayashi-Maskawa phase, focusing on the flavor-diagonal sector. In particular, we discuss the manifestation of CP violation within QCD, from the origin of the strong CP problem, through some tentative resolutions, to the most stringent experimental constraints due to searches for electric dipole moments (EDMs) of leptons, nucleons, atoms and molecules. We dwell on the calculational aspects of applying the EDM constraints, and we also discuss the current status of the ensuing constraints on the underlying sources of CP-violation in physics beyond the Standard Model, focusing on weak-scale supersymmetry.

Contents

13.1. Introduction

The search for violations of fundamental symmetries has played a central role in the development of particle physics in the twentieth century. In particular, tests of the discrete symmetries, charge conjugation C, parity P, and time-reversal T, have been of paramount importance in establishing the Standard Model (SM). Perhaps the most famous example was the discovery of parity violation in the weak interactions [1], which led to the realization that matter fields should be combined into asymmetric left- and right-handed chiral multiplets, one of the cornerstones of the Standard Model. The observation of CP violation via the mixing of Kaons [2], also subsequently provided strong evidence for the presence of three quark and lepton generations, via the Kobayashi–Maskawa mechanism [3], prior to direct experimental evidence for the third family.

One of the first tests of this kind, actually pre-dating the discovery of parity violation in the weak interactions, was a probe of parity violation within the – at the time unknown – theory of the strong interactions. This was the observation of Purcell and Ramsey [4] in 1949 that parity violation in the strong interactions would allow for an electric dipole moment (EDM) of the neutron. Using existing data they were able to exclude such a possibility at the level of a few orders of magnitude below the characteristic nucleon size. Intrinsic electric dipole moments also violate T in addition to P, and it was only some 25 years later with the emergence of QCD that the possibility of T-violation (or CP-violation, on assuming the CPT theorem) in the strong interactions had some theoretical underpinning. Indeed, QCD allows for the addition of a dimension-four term, known as the θ-term, with a dimensionless coefficient θ which, if nonzero, would signify the violation of both P and T. This term is somewhat unusual, being purely topological, i.e. a boundary term, but its presence is intrinsically tied to another elegant feature of QCD, namely the mechanism via which the mass of the η' meson is lifted well above the scale one might naturally expect given its apparent status as a Goldstone boson. 't Hooft's resolution of this $U(1)$ *problem* [5] naturally implies the physical relevance of the θ-term. Moreover, were $\theta \sim \mathcal{O}(1)$, one would predict a neutron EDM of sufficient size to ensure that the original analysis of Purcell and Ramsey would have detected it. In fact θ is now known to be tuned to zero, or at least to cancel, to better than one part in 10^9! This tuning is the well-known *strong CP problem* of the Standard Model, which has been with us for more than 25 years, i.e. since the early days of QCD. A number of possible dynamical mechanisms have been proposed to explain this tuning, and some of these have important consequences and predictions for other aspects of particle physics and cosmology, including the possibility of an ultra-light pseudoscalar, the axion.

The required tuning of this CP-odd parameter in QCD comes into sharp focus when we put QCD into its rightful place within the Standard Model, which necessarily means coupling it to the electroweak sector and the quarks in particular. In this case the physical value of θ acquires a contribution from the overall phase of the quark mass matrix. In this sense the strong CP problem can be phrased as the absence, to high precision, of *flavor-diagonal* CP-violation within the Standard Model. This situation could not contrast more strongly with the situation in the flavor-changing sector, which is where all currently observed CP-violating effects reside. Indeed, the original discovery of CP-violation in the system of neutral Kaons [2], can be explained within this sector through the elegant and indeed rather minimal model of Kobayashi and Maskawa, which links CP-violation to the single physical phase in the unitary mixing matrix (known as the CKM matrix) describing transitions between the three generations of quarks [3]. This picture has recently received significant support – indeed essential confirmation – through experiments using neutral B mesons [6]. In contrast to θ, the phase in the CKM mixing matrix requires no tuning at all – its effects are nicely masked in the appropriate channels by the flavor structure of the Standard Model. Indeed, it turns out that the predictions for any CP-violating effect in the flavor-conserving channel induced by CKM mixing are minuscule, thus denying any hopes of detecting the experimental manifestation of CKM physics in these channels in the foreseeable future.

Searches for flavor-diagonal CP-violation, while insensitive to the CKM phase, thus inherit on the flip-side the status as one of the unique, essentially "background" free, probes of new physics. Electric dipole moments, through continuous experimental development since the work of Purcell and Ramsey, remain our most sensitive probes of this sector. All existing searches have failed to detect any intrinsic EDM, and indeed the precision to which EDMs are now known to vanish is remarkable, and sufficient to render them some of the most important precision tests of the Standard Model. In this more general context, the strong CP problem, associated with the tuning of θ, becomes just the most highly tuned example among many possible CP-odd operators which could contribute to the observable EDMs of nucleons, leptons, atoms and molecules. Anticipating the presence of such CP-odd sources is not without motivation. Indeed, one of the strongest motivations comes from cosmology, where the dynamical generation of a baryon asymmetry in the universe apparently requires additional sources of CP-violation [7]. Moreover, much theoretical and experimental work is currently focused on the elucidation of the physics of the Fermi scale, i.e. the mechanism for electroweak symmetry breaking. There are several theoretical motivations to believe that new physics, beyond the SM Higgs, should become apparent at, or just above, this scale, with weak-scale supersymmetry (SUSY) being a prominent example. Flavor-diagonal CP-violation constitutes a powerful probe of these scales, since any new physics need not provide the same flavor-dependent suppression factors as in the SM, while the SM itself constitutes a negligible background. These precision tests are thus highly complementary to direct searches at colliders. A rough estimate, based on the decoupling of new physics as the inverse square of its characteristic energy scale Λ, currently gives us the possibility to probe an order one CP-violating source at up to $\Lambda \sim 10^6$ GeV. In many weakly coupled theories, such as SUSY, this scale is somewhat lower, but often is still beyond the reach of existing and/or projected colliders. As with the link between the Kobayashi–Maskawa mechanism and the three-generation structure, one might hope that flavor-diagonal CP-violation, or perhaps the lack thereof, will tell us something profound about the matter sector.

This review is split into two main parts. The first deals with the origin of the strong CP problem, and some of the suggested resolutions. We being by recalling the motivations for introducing the concept of θ-vacua into gauge theories, and we then spend some time discussing the interplay between this concept and the resolution of the U(1) problem, namely the large mass of the η' meson, which naively appears to be the Goldstone boson associated with the spontaneous breaking of the (anomalous) U(1) axial current. We emphasize how the anomaly in this current provides on the one hand the basis for a mechanism via which the η' obtains its large mass, and on the other aids in the calculation of the θ-dependence of physical quantities in low energy QCD. This makes the strong CP problem manifest and we then turn to some suggested resolutions, splitting them into two basic categories. The first corresponds to dynamical relaxation through the spontaneous breaking of an additional axial U(1) (or Peccei–Quinn [8]) symmetry, leading to the prediction of an axion. The second corresponds to various models which enforce exact P and/or CP at high scales, and then allow for it to be spontaneously broken.

The second part of the review generalizes this discussion to the broader concept of flavor-diagonal CP violation, of which the strong CP problem is one (important) part, and the use of the extremely precise constraints thereon arising from the experimental bounds on EDMs of nucleons, leptons, atoms and molecules. We discuss the current status of the experimental constraints within three generic classes, namely the neutron EDM, and the EDMs of paramagnetic and diamagnetic atoms, and then review the calculational aspects, starting with the observable EDMs and working upwards in energy scale through the use of several effective CP-odd Lagrangians defined at the appropriate threshold scales. We concentrate in particular on the QCD sector which currently is the source of some of the more significant uncertainties in the application of EDM bounds (for a detailed discussion of many other aspects we refer the reader to [9, 10]). We then consider the generation of these observables within specific models of CP-violation, reviewing first the significant sources of suppression within the Standard Model, covering the interplay with models of baryogenesis and then focusing on weak scale supersymmetry, and the MSSM in particular, as the source of new physics at the electroweak scale. We discuss the generic constraints that EDMs impose on combinations of CP-violating parameters in the SUSY-breaking sector, and also explore some additional effects which may arise in special parameter regimes. Finally, we conclude with an outlook on future experimental and theoretical developments.

13.2. θ-Vacua and the Strong CP Problem

13.2.1. *Introduction*

The strong CP problem, or in other words the observed CP invariance of the strong interactions to a high degree of precision, can be argued to arise primarily from the interplay of two crucial facts. One is experimental, namely the discovery of P and CP violation in the weak interactions. This implies that these are not symmetries of Nature, and moreover, since this sector is coupled to QCD through the quarks, one must anticipate that CP

violation could manifest itself purely within QCD. The second is partly theoretical, and concerns the interplay between CP violation in QCD and the resolution of the so-called *U(1) problem*, namely the surprisingly large mass of the η' meson given its apparent status as a Goldstone boson for the breaking of the $U(1)_A$ flavor symmetry. It is precisely this interplay that we will discuss in some detail in this section. However, let us first consider, purely from a symmetry point of view, how CP violation may arise in QCD, or in fact in pure Yang–Mills theories.

13.2.1.1. θ-vacua

Since P and CP are violated in the electroweak sector, one is entitled to ask whether QCD is also P- and CP-violating? From an effective field theory point of view, we can simply assume that these symmetries are violated and write down a set of operators allowed under the residual symmetry, and which may be added to the QCD Lagrangian. Setting aside the quarks for the moment, there is a unique operator in this list which has minimal dimension, namely dimension four, and thus is unsuppressed by any heavy scale. This is the θ-term,

$$\Delta\mathcal{L}^{\theta} = \frac{\theta g^2}{32\pi^2} G^a_{\mu\nu}\tilde{G}^{a\mu\nu}, \tag{13.1}$$

where $\tilde{G}^a_{\mu\nu} = \frac{1}{2}\epsilon_{\mu\nu\rho\sigma}G^{a\,\rho\sigma}$ is the dual of the gluon field strength.

At first sight this operator looks irrelevant, as it is a total derivative,

$$G^a_{\mu\nu}\tilde{G}^{a\,\mu\nu} = \partial^{\mu}\mathcal{K}^{\rm CS}_{\mu}, \qquad \mathcal{K}^{\rm CS}_{\mu} = 2\epsilon_{\mu\nu\rho\sigma}\left(A^a_{\nu}\partial_{\rho}A^a_{\sigma} + \frac{1}{3}gf^{abc}A^a_{\nu}A^b_{\rho}A^c_{\sigma}\right), \tag{13.2}$$

and thus plays no role in perturbation theory. The fact that it cannot be ignored has to do in general with the resolution of the U(1) problem alluded to above. However, we can first provide some more direct motivation for its relevance by noting that classical field configurations do exist for which it is nonvanishing, namely *instantons*.

In particular, in Euclidean space, one can consider field configurations that tend to the vacuum, or a gauge transform thereof, at infinity and so have finite action. These instanton configurations [11] are classified by a topological charge,

$$\begin{aligned}\nu &= \frac{g^2}{32\pi^2}\int d^4x_E G^a_{\mu\nu}\tilde{G}^{a\,\mu\nu} \\ &= \frac{g^2}{32\pi^2}\int d^3x\, \mathcal{K}^{\rm CS}_0(x,t)\Big|^{t=+\infty}_{t=-\infty} \equiv n_+ - n_- \in \mathbb{Z}.\end{aligned} \tag{13.3}$$

The first line here involves an integral in Euclidean space. However, to avoid having to change signature in the ensuing discussion, it is convenient to go to the $A_0 = 0$ gauge where, restricting our attention to an SU(2) gauge group, we can specify such a configuration as one which provides the following interpolation:

$$A_i(x, t \to -\infty) = 0 \quad \longrightarrow \quad A_i(x, t \to +\infty) = \frac{1}{g} U^\dagger \partial_i U, \tag{13.4}$$

where $U(x)$ tends to unity at spatial infinity, and for winding number $n = 1$ can be taken as

$$U(x_i) = \exp\left(-\frac{i\pi x_i \sigma_i}{\sqrt{x_i^2 + \rho^2}}\right), \tag{13.5}$$

where ρ is a parameter, referred to as the instanton size.

The map U between the two configurations in Eq. (13.4) is a residual gauge transformation within the $A_0 = 0$ gauge, known as a *large* gauge transformation, as it changes the topology as specified by n, i.e. $U|n\rangle = |n+1\rangle$. Consequently, we observe that eigenstates of fixed winding number are not gauge invariant and thus not physical. One can nonetheless define a gauge invariant vacuum state by superposing the eigenstates of fixed winding number,

$$|\theta\rangle = \sum_n e^{-in\theta}|n\rangle. \tag{13.6}$$

The values of θ label superselection sectors of the theory [12, 13] as one may verify by considering the off-diagonal expectation value, $\langle\theta|\mathcal{O}|\theta'\rangle$, or a gauge invariant operator $\mathcal{O}$. Expanding in the basis Eq. (13.6) we observe, from the fact that $\mathcal{O}$ is gauge invariant, that $\langle n_-|\mathcal{O}|n_+\rangle = \langle\mathcal{O}\rangle_\nu$ can depend only on the topological charge, $\nu = n_+ - n_-$. We thus find as advertised,

$$\langle\theta|\mathcal{O}|\theta'\rangle = \sum_n e^{in(\theta-\theta')} \sum_\nu e^{i\frac{\nu}{2}(\theta+\theta')}\langle\mathcal{O}\rangle_\nu = 2\pi\delta(\theta-\theta')\sum_\nu e^{i\nu\theta}\langle\mathcal{O}\rangle_\nu. \tag{13.7}$$

Returning to a covariant path integral expression for this expectation value, and recalling the explicit form of the topological charge in Eq. (13.3), we can write

$$\langle\theta|\mathcal{O}|\theta\rangle = \sum_\nu \int [d\mu] e^{i(S+\Delta S^\theta)}\delta\left(\nu - \frac{g^2}{32\pi^2}\int G\tilde{G}\right), \tag{13.8}$$

which takes us back to the original CP-odd addition to the QCD action that we motivated above purely on symmetry grounds. What this admittedly formal argument demonstrates is that, in a semi-classical approximation where in the path integral we sum over a set of saddle points – i.e. instantons – the θ-term is necessarily present by virtue of the lack of gauge invariance of the eigenstates of fixed winding number.

One might well question the validity of this semi-classical picture in a confining gauge theory such as QCD. However, as we will discuss, this qualitative picture is required on more general grounds, namely the second fact alluded to above, which is the intricate interplay between the topological structure we have discussed here at a somewhat formal level, leading to the theta vacua, and the observation that the "ninth" (SU(3) singlet) Goldstone boson, the η', is abnormally heavy. As we will review below, if we accept the topological solution of the U(1) problem, we are forced to accept that the θ-term has physical consequences, and indeed it would have been detected were $|\theta| \geq 10^{-9}$. This required tuning of θ, all the more puzzling these days with recent experimental evidence that the Kobayashi–Maskawa phase in the quark sector is apparently of $\mathcal{O}(1)$, is then a precise statement of the *strong CP problem.*

To present these interconnections more clearly, we will need to delve a little deeper into QCD, and its chiral structure in particular.

13.2.2. *Anomalies, the U(1) problem and the η'-mass*

To motivate in more detail why one should take the topological structure seriously, we will first review 't Hooft's solution of the U(1) problem [5].

13.2.2.1. *The U(1) anomaly*

We first recall that, in contrast to the non-singlet $\mathrm{SU}(N_f)_L \times \mathrm{SU}(N_f)_R$ flavor currents, the singlet axial-vector U(1) current,

$$J_A^\mu = \sum_{i=1}^{N_f} \bar{q}_i \gamma^\mu \gamma_5 q_i, \tag{13.9}$$

has an Adler–Bell–Jackiw anomaly,

$$\partial_\mu J_A^\mu = N_f \left(\frac{g^2}{16\pi^2} \right) G^a_{\mu\nu} \tilde{G}^{a\,\mu\nu}, \tag{13.10}$$

where, for the moment, we have taken the quarks to be massless.

In perturbation theory, a non-anomalous U(1) current can still be defined by virtue of the fact that $G\tilde{G}$ is a total derivative. In particular, from Eq. (13.2) and Eq. (13.9), we observe that

$$\tilde{J}_A^\mu = J_A^\mu - N_f \left(\frac{g^2}{16\pi^2} \right) \mathcal{K}^\mu_{\mathrm{CS}} \tag{13.11}$$

is conserved: $\partial_\mu \tilde{J}^\mu_A = 0$. However, in a manner now familiar from our earlier discussion, this redefined current $\tilde{J}_\mu$ is not gauge invariant under large gauge transformations. This is clearly not a concern in perturbation theory, but would render $\tilde{J}_\mu$ unphysical if QCD were to admit nonperturbative processes changing the gauge field topology. In particular, we observe that,

$$U\tilde{Q}_A U^{-1} = \tilde{Q}_A + 2N_f, \tag{13.12}$$

where $\tilde{Q}_A = \int d^3x \tilde{J}^0_A$, from which one can deduce that the θ-vacua are not invariant under chiral rotations [5, 12, 13],

$$e^{i\alpha\tilde{Q}_A}|\theta\rangle = |\theta + 2N_f\alpha\rangle, \tag{13.13}$$

as one may verify by applying U to the left hand side and using Eq. (13.12).

Before we explore the consequences of this non-gauge invariant current more closely, let us note that Eq. (13.13) immediately implies the important fact that, if there are (any) massless quarks, θ is unphysical and may be rotated to zero. Moreover, if we now turn on a generic mass matrix,

$$\mathcal{L}_q = -\bar{q}_R M q_L - \bar{q}_L M^\dagger q_R, \tag{13.14}$$

the phase of $\det(M)$ acts as a spurion for the axial U(1) symmetry, and thus contributes to the physical value of θ,

$$\bar{\theta} \equiv \theta + \mathrm{ArgDet}(M), \tag{13.15}$$

which, assuming all quarks are massive, is now necessarily invariant under chiral rotations.

13.2.2.2. *The Veneziano ghost*

The observation above that the formally conserved axial U(1) current is actually not gauge invariant opens up the possibility of escaping the naive conclusion that after chiral symmetry breaking there would necessarily be a massless Goldstone, the η'. To explore this in more detail, we need to study correlation functions of the axial current. To formally verify that there is no massless state coupling to this current, it is enough to show that the zero momentum residue of correlators of this current vanishes, i.e. $k^\mu\langle T(J^A_\mu, \mathcal{O})\rangle|_{k=0} = 0$ for some gauge invariant operator $\mathcal{O}$, which for simplicity we take to be an isosinglet. Note that we are now necessarily referring to the gauge invariant, but anomalous, current J^A_μ.

Using the anomaly relation, we can write (with $N_f = 2$ in the present discussion)

$$\begin{aligned}&\int d^4x \partial^\mu \langle 0|T\{J^A_\mu(x), \mathcal{O}(0)\}|0\rangle \\ &\qquad = -\eta_{\mathcal{O}}\langle\mathcal{O}\rangle + \frac{\alpha_s}{2\pi}\int d^4x \langle 0|T\{G^a_{\mu\nu}\tilde{G}^{a\mu\nu}(x), \mathcal{O}(0)\}|0\rangle,\end{aligned} \tag{13.16}$$

where the first term on the right-hand side involves the "chirality" $\eta_{\mathcal{O}}$ of the operator $\mathcal{O}$, which is defined using the conserved axial charge $\tilde{Q}^A$, via the equal-time commutator,

$$\left[\mathcal{O}(x), \tilde{Q}^A(0)\right] \equiv \eta_{\mathcal{O}} \mathcal{O} \delta^{(3)}(x). \tag{13.17}$$

It is clear that the residue on the right-hand side of Eq. (13.16) can only vanish if there is a conspiracy between the two terms. Given the normalization of the θ-term, this constraint can be compactly written in the form $(2N_f \partial_\theta - i\eta_{\mathcal{O}}) \langle \mathcal{O} \rangle$.

Recalling that $G\tilde{G} = \partial^\mu \mathcal{K}_\mu$, we see that the second term on the right-hand side of Eq. (13.16) has the form $k^\mu \langle \mathcal{K}_\mu, \mathcal{O} \rangle|_{k=0}$. Since this must be nonzero in order for the total residue to vanish, we come to the surprising conclusion that the current $\mathcal{K}_\mu$ couples to a massless state. This state is certainly unphysical – a "ghost" – since $\mathcal{K}_\mu$ is not gauge invariant, but it turns out to provide a very useful physical picture of the resolution of the U(1) problem. This structure was first discussed within the Schwinger model by Kogut and Susskind [14], noted in the present context by Crewther [15], and its role made transparent by Veneziano [16]. We will follow in part a detailed discussion by Diakonov and Eides [17]

In fact it is more convenient to consider the diagonal correlator $\langle \mathcal{K}_\mu, \mathcal{K}_\nu \rangle$, so that the presence of the pole directly implies,

$$k^\mu k^\nu i \int d^x e^{ik\cdot x} \langle 0|T\{\mathcal{K}_\mu(x), \mathcal{K}_\nu(0)\}|0\rangle \,|_{k=0} = -\chi(0) \neq 0, \tag{13.18}$$

where we have introduced $\chi(0) = \lim_{k\to 0} \chi(k)$, with

$$\chi(k) = -i \int d^4x e^{ik\cdot x} \left\langle 0 \left| T\left\{ \frac{\alpha_s}{2\pi} G^a_{\mu\nu} \tilde{G}^{a\mu\nu}(x), \frac{\alpha_s}{2\pi} G^a_{\mu\nu} \tilde{G}^{a\mu\nu}(x) \right\} \right| 0 \right\rangle, \tag{13.19}$$

(we have dropped a commutator term, associated with the T-product, which is not relevant for the present discussion). $\chi(0)$ is known as the topological susceptibility and the the condition Eq. (13.18) then implies that it is nonvanishing, which is now a perfectly gauge-invariant statement. In this representation, we see that the lifting of $m_{\eta'}$ is tied in general to quantum fluctuations which generate $\chi(0) \neq 0$.

If we now recall the earlier discussion, we see that in principle instantons provide an "existence proof" that configurations exist for which $\chi(0) \neq 0$. Indeed, if we approximate the path integral by a sum over these saddle points then the cancellation of the residue in Eq. (13.16) necessarily occurs. The reason is that the operators that can be generated by an instanton vertex have a chirality specified by the number of instanton zero modes. This in turn is determined by the topological instanton number ν which by virtue of Eq. (13.8) precisely matches their θ-dependence. In other words, as first noted by 't Hooft [5], the operator corresponding to the instanton generated vertex has precisely the quantum numbers to be a mass term for η' (or rather η for $N_f = 2$). However, while this logic has more recently been applied with notable success in supersymmetric gauge theories [18], this is limited to phases in which one can control the semi-classical approximation, namely when the theory is conformal or in a weakly-coupled Higgs phase. Thus one may question its quantitative reliability in a confining theory such as QCD.

The picture outlined above, however, is quite general and is not dependent on the precise mechanism via which the η' obtains its mass. Before turning to another consistency check on the latter, let us conclude this subsection by spelling out the interpretation of the Veneziano ghost in a little more detail [17].

If we return once again to the Hamiltonian formalism, in the $A_0 = 0$ gauge, we can consider the following coordinate in field space,

$$K_0 = \int d^3x \mathcal{K}_0(x), \tag{13.20}$$

which is invariant under small gauge transformations, but shifts under large gauge transformations, e.g. $U(x)$, by an integer,

$$K_0 \to K_0 + n. \tag{13.21}$$

The crucial feature is that, since the potential for the theory is gauge invariant, it must be periodic in K_0. The "ghost" is then the gap-less excitation along this periodic coordinate. This picture has a precise parallel with the motion of an electron in a periodic crystal. Near the bottom of the first band, the electron behaves like a free particle, $E \propto p^2$, where p is the quasi-momentum. However, the state energy is actually periodic in p. In fact the analogy becomes precise if one identifies $\theta = p$, and the θ-vacuum Eq. (13.6) then corresponds to the Bloch superposition for the wave function in the crystal.

This picture also leads to an interpretation of how the η' gains its mass which has certain parallels with the Higgs mechanism. Namely, one can regard it as arising through the mixing of the "ghost" (corresponding in this analogy to the longitudinal mode of the gauge field), with the isosinglet Goldstone mode $\sim \sum_{i=u,d,s} \bar{q}_i \gamma_5 q_i$.

Having motivated this heuristic picture, rather than pursue it in more gauge invariant quantitative terms, we will now consider another approach which provides a very satisfying consistency check of this framework within the large N expansion, as discussed by Witten [19] and Veneziano [16].

13.2.2.3. *Large N and the η' mass*

As we have discussed in some detail, the solution of the U(1) problem can be formulated in several different ways. To emphasize the relation to θ we will recall here the argument of Witten [19], which makes use of 't Hooft's large N_c limit, where we send $N_c \to \infty$ and $g \to 0$ with $\lambda = g^2 N_c$ held fixed. This provides a very succinct consistency check relating the solution of the U(1) problem to the physical relevance of θ.

In the large N limit, the chiral anomaly takes the form,

$$\partial_\mu J_A^\mu = \frac{N_f}{N_c} \frac{\lambda}{16\pi^2} G^a_{\mu\nu} \tilde{G}^{a\,\mu\nu}, \tag{13.22}$$

and thus vanishes at leading order in $1/N$. We conclude that in this limit, the η' indeed descends to become a genuine Goldstone boson. We may also anticipate that, as for the quark mass dependence of pions, at next-to-leading order in $1/N$ the mass of η' should be linear in the deformation parameter, i.e. $m^2_{\eta'} \sim 1/N$.

This expectation can be justified by studying the dependence of the vacuum energy (or in principle some other observable) on θ. By considering the formal path integral expression for the partition function, we find

$$\left.\frac{d^2 E}{d\theta^2}\right|_{\theta=0} = \frac{1}{N^2} \frac{1}{4\pi^2} \lim_{k\to 0} \chi_\lambda(k), \tag{13.23}$$

where $\chi_\lambda(k) \equiv N^2 \chi(k)$ remains finite in the 't Hooft limit.

At large N, we can make use of the important simplification that interactions between hadrons are subleading in $1/N$. Thus we may write any two-point function in terms of a dispersion relation as a sum over hadron poles. In the present case, the quantum numbers correspond to pseudoscalar glueballs and mesons,

$$\chi_\lambda(k) = \sum_{\text{glueballs}} \frac{iN^2 a_n^2}{k^2 - m_g^2} + \sum_{\text{mesons}} \frac{iN b_n^2}{k^2 - m_m^2}, \tag{13.24}$$

where the residues are determined by large-N counting,

$$N a_n = \langle 0|G\tilde{G}|\mathrm{n}^{th} \text{ glueball}\rangle, \tag{13.25}$$

$$\sqrt{N} b_n = \langle 0|G\tilde{G}|\mathrm{n}^{th} \text{ meson}\rangle, \tag{13.26}$$

with a_n and b_n order one constants.

The crucial point is that, if there are any massless quarks, one can rotate θ away via the chiral anomaly, i.e. it is unphysical and so Eq. (13.23) must vanish. Thus, at $k = 0$, we must find that the two terms in Eq. (13.24) conspire to cancel. The effect of quark loops is subleading at large N and thus their presence or absence cannot change the N-dependence. We conclude that cancellation can only occur if there is one pseudoscalar meson, namely the lightest η', whose mass-squared scales as $1/N$ [19].

Although we will not need the details here, one can actually follow this argument through to obtain a formula for the η' mass in terms of the topological susceptibility (in pure Yang–Mills),

$$m_{\eta'}^2 = \frac{4N_f}{f_\pi^2}\left(\frac{d^2E}{d\theta^2}\right)_{\theta=0}^{\mathrm{YM}}, \tag{13.27}$$

where $f_\pi^2 \sim \mathcal{O}(N)$. This is the Witten–Veneziano formula [16, 19], which can be argued to form part of the effective Lagrangian for η' at leading order in $1/N$. It provides a clear illustration of the interplay between θ and the solution of the U(1) problem within the context of the large N expansion. However, one may again note that the $1/N$ scaling is apparently not born out precisely in the QCD spectrum (i.e. in the physical case $N = 3$, where we know that $m_{\eta'}$ is numerically not suppressed).

13.2.3. *The strong CP problem*

The conclusion we would like to emphasize from the above review of the axial U(1) structure in QCD is, simply put, that since QCD at a practical level does solve the U(1) problem, i.e. $m_{\eta'} \gg m_\pi$, one must take the θ-term seriously. The strong CP problem then automatically arises from the severe constraints that experimental bounds place on the value of θ, or more precisely $\bar{\theta}$. However, such bounds are not straightforward to derive since a dependence of a given observable on θ can necessarily arise only at the nonperturbative level, a regime that in QCD is generally far from tractable. However, one can make progress based once again on the use of anomalous Ward identities. In particular, we will now effectively turn the logic around. We will assume $m_{\eta'} \gg m_\pi$, and deduce some consequences for the θ-dependence.

13.2.3.1. *θ and observables*

Although one can consider various observables, and we will focus specifically on EDMs later on, let us consider an example closely related to the vacuum energy discussed above. From the preceding subsection, it is apparent that, for small θ, $E(\theta) \sim \mathcal{O}(\theta^2)$, while we expect that generic observables will have a linear dependence on θ at leading order. The simplest example is $\langle G\tilde{G}\rangle$, which to leading order in θ is again given in terms of the topological susceptibility,

$$\left\langle 0 \left| \frac{\alpha_s}{\pi} G^a_{\mu\nu} \tilde{G}^{a\mu\nu} \right| 0 \right\rangle_\theta = -\frac{1}{2}\theta \lim_{k\to 0} \chi(k) + \mathcal{O}(\theta^2), \tag{13.28}$$

where $\chi(k)$ was defined in Eq. (13.19).

Thus we are required to compute the topological susceptibility in QCD, which might seem a tall order, but in fact (anomalous) symmetries are very constraining and one can obtain a precise result – we will follow the elegant derivation by Shifman, Vainshtein and Zakharov [20], that will also be useful later on. Note that a related analysis was also performed somewhat earlier by Crewther [15] in the case $m_u = m_d$.

We consider for simplicity the case of two light quark flavors and, noting that $\chi(0) = 0$ for massless quarks since θ can be rotated away, in considering the chiral anomaly we now restore the dependence on the quark mass which provides a classical term violating U(1) charge conservation. A useful calculational simplification follows if we take as the anomaly relation a linear combination of the singlet equations for the u and d quarks. In particular, we use

$$\partial_\mu J^\mu_A = 2m_*(\overline{u}i\gamma_5 u + \overline{d}i\gamma_5 d) + \frac{\alpha_s}{2\pi} G^a_{\mu\nu}\tilde{G}^{a\mu\nu}, \tag{13.29}$$

where

$$J^A_\mu = \frac{m_*}{m_u}\overline{u}\gamma_\mu\gamma_5 u + \frac{m_*}{m_d}\overline{d}\gamma_\mu\gamma_5 d \tag{13.30}$$

is the conveniently normalized anomalous current, and we have introduced the reduced mass,

$$m_* \equiv \frac{m_u m_d}{m_u + m_d}, \tag{13.31}$$

that plays an important role in the θ-dependence of QCD observables. The reason being that, since there can be no physical θ-dependence if any of the quark masses vanish, the leading dependence will be through m_* (or its appropriate generalization for three light flavors).

The idea, as announced above, is now simply to determine the consequences of the anomaly equation under the assumption that the U(1) problem is resolved, i.e. that $m_{\eta'} \gg m_\pi$. To make this explicit, we introduce the additional correlators,

$$\Pi^{(1)}_{\mu\nu}(k) = i \int d^4x e^{ik\cdot x} \left\langle 0 \left| T\{J^A_\mu(x), J^A_\nu(0)\} \right| 0 \right\rangle, \tag{13.32}$$

$$\Pi^{(2)}_{\mu}(k) = i \int d^4x e^{ik\cdot x} \left\langle 0 \left| T\{J^A_\mu(x), \frac{\alpha_s}{2\pi} G^a_{\mu\nu}\tilde{G}^{a\mu\nu}(0)\} \right| 0 \right\rangle. \tag{13.33}$$

Since there are no massless particles coupling to these currents, and thus no poles in the correlators, we have the following constraints:

$$k^\mu k^\nu \Pi^{(1)}_{\mu\nu}(k)\Big|_{k\to 0} = 0, \qquad k^\mu \Pi^{(2)}_{\mu}(k)\Big|_{k\to 0} = 0. \tag{13.34}$$

In evaluating these equations, we integrate by parts to obtain T-products of $\partial^\mu J^A_\mu$ and $G\tilde{G}$. There is also a boundary (or contact) term which is determined by the equal time commutator,

$$\left[2m_* \sum_{i=u,d} \bar{q}_i(x)\gamma_5 q_i(x), J^A_0(0)\right] \delta(x_0) = 4m_* \sum_{i=u,d} \bar{q}_i q_i \delta^{(4)}(x). \tag{13.35}$$

This clearly leads to a local contribution (independent of x).

By substituting the anomaly relation Eq. (13.29), we can manipulate the resulting two equations Eq. (13.34) to eliminate one unknown T-product, and obtain the result,

$$\chi(0) = -16m_* \langle 0|\bar{q}q|0\rangle$$
$$-i \int d^4x \left\langle 0 \left| T\left\{ m_* \sum_{i=u,d} \bar{q}_i\gamma_5 q_i(x)\, ,\, m_* \sum_{i=u,d} \bar{q}_i\gamma_5 q_i(0) \right\} \right| 0 \right\rangle. \tag{13.36}$$

The nonlocal contribution to this correlator, the second term above, is $O(m^2)$ in light quark masses. Nonetheless, this term would cancel the local contribution were there an intermediate state with mass squared of $O(m)$ – for example the Goldstone boson in the singlet channel. Inverting the logic used earlier in this section, we now make use of the fact that the lightest intermediate state η (or η' in the three-flavor case) has $m_\eta \gg m_\pi$ and thus the second term can be neglected at leading order in $(m_\pi/m_\eta)^2$ [20].

Thus, we find

$$\left\langle 0 \left| \frac{\alpha_s}{\pi} G^a_{\mu\nu}\tilde{G}^{a\mu\nu} \right| 0 \right\rangle_\theta = 8\theta m_* \langle 0|\bar{q}q|0\rangle = -8\theta f_\pi^2 m_\pi^2 \frac{m_*}{m_u + m_d}, \tag{13.37}$$

where in the last equality we have made use of the Gell-Mann–Oakes–Renner relation to eliminate the dependence on $\langle\bar{q}q\rangle$.

The general structure of the θ-dependence of the $\langle G\tilde{G}\rangle$ condensate allows us to make some simple estimates of other CP-odd quantities. In particular, we observe that the dependence essentially follows the dictates of symmetry. Indeed on general grounds, we could write $\langle \alpha_s G\tilde{G}\rangle \sim \theta m_* \Lambda_{\rm had}^3$ with Λ_{had} the relevant dimensionful scale, that the calculation above identifies with the quark condensate. For example, let us consider the EDM of a nucleon that one can use to directly bound θ. We will discuss a more systematic calculation of this quantity in following sections, but for now a simple estimate will suffice. The EDM d_n (see Section 13.3 for definitions) has the dimensions of length, and so we may expect,

$$d_n(\theta) \sim e\frac{\theta m_*}{\Lambda_{\rm had}^2} \sim \theta \cdot (6\times 10^{-17})\, e\,{\rm cm}, \tag{13.38}$$

where we identified $\Lambda_{\rm had} = m_n$ and used conventional values for the light quark masses. The current experimental bound on the neutron EDM,

$$|d_n| < 6\times 10^{-26} e\,{\rm cm}, \tag{13.39}$$

then translates into the constraint,

$$|\bar{\theta}| < 10^{-9}. \tag{13.40}$$

Although the arguments above amount to no more than a rough estimate, we will see in later sections that this bound is indeed quite reliable. Of course one can also consider other physical quantities which are sensitive to θ in a similar way, e.g. CP-odd decay rates such as $A(\eta \to \pi^+\pi^-)$, but the current constraints are not competitive with EDMs.

It is important to emphasize that, in this derivation, we have used very little input besides the solution of the U(1) problem and the consequent decoupling of η' from the chiral sector. This illustrates rather clearly how the topological solution of the U(1) problem necessitates the strong bound on θ.

13.2.3.2. *Renormalization of θ*

Before turning to the question of how one might resolve the apparent tuning of θ necessitated by the EDM bounds, we should first clarify the physical status of θ with regard to renormalization and scale dependence. Since θ multiplies an operator of dimension four, we might expect that it could be subject to logarithmic renormalization. This might seem odd in a term that we argued above could have no physical consequences in perturbation theory, but one must distinguish the physical dependence on θ from its effective value at a given scale.

It turns out that after coupling to the electroweak sector θ is indeed renormalized, but the leading order effect arises at no less than 14th order in the weak coupling [21, 22]. This may sound rather esoteric, but in fact it is not too difficult to understand. We can write the divergent term schematically as [23]

$$\delta\theta \propto g_W^2 \mathcal{J} \ln \Lambda_{\rm uv}, \tag{13.41}$$

where $\mathcal{J}$ is the full CP-violating invariant introduced by Jarlskog [24], built from the Yukawa coupling matrices,

$$\mathcal{J} = \text{Im Det}\left[Y_u Y_u^\dagger, Y_d Y_d^\dagger\right]. \tag{13.42}$$

More explicitly, on inserting factors of the electroweak v.e.v. $(1/v) = (g_W/2m_W)$,

$$\mathcal{J} = \left(\frac{g_W}{2m_W}\right)^{12} \prod_{i>j=u,c,t} (m_i^2 - m_j^2) \prod_{k>l=d,s,b} (m_k^2 - m_l^2)\, J_{CP}, \tag{13.43}$$

where J_{CP} is the basis invariant form of the KM phase,

$$J_{CP} = \text{Im}(V_{cs}V_{us}^* V_{ud} V_{cd}^*) \simeq 3 \times 10^{-5}, \tag{13.44}$$

with V the CKM mixing matrix.

The dependence on twelve powers of the weak coupling then follows directly from the constraint that the result be proportional to a CP-odd invariant of the Yukawa couplings. The additional two powers of g_W (or in fact the U(1) coupling g') in Eq. (13.41) are required by a further symmetry which needs to be broken through some dependence on hypercharge. This UV sensitivity may seem problematic, but in fact it is numerically completely negligible if we restrict the cutoff to the Planck scale.

Besides the logarithmic piece, the θ-term also receives finite additive contributions at the mass threshold for fermions charged under $SU(3)_c$. In the SM such finite contributions to θ arise at three-loop order, specifically $\mathcal{O}(g_W^4 g_s^2)$ [25]. This contribution also turns out to be many orders of magnitude below the current EDM bounds,

$$\delta\theta^{\rm SM} \sim 10^{-19}, \tag{13.45}$$

but is of conceptual importance in demonstrating that, even within the Standard Model, there are nonzero contributions to $\bar{\theta}$.

13.2.4. *Resolving the strong CP problem*

Having spent some time introducing the strong CP problem, we now turn to some of its proposed resolutions, of which there are several generic classes. One approach is simply to argue that in fact there is no problem at all, namely to accept that for some accidental reasons the θ-term is fine-tuned. An accidental cancellation of the different contributions to $\bar{\theta}$ at the 10^{-10} level or below is certainly possible in principle but looks entirely unmotivated from a theoretical perspective. If we discard this possibility, and instead search for an explanation for why $\bar{\theta}$ is very small, then we find that the existing theoretical attempts to solve the strong CP problem can be divided into those that are based either on continuous symmetries or on spontaneously broken discrete symmetries.

One can also distinguish the motivations for these two approaches by looking at two extreme reference points, namely when $\bar{\theta}$ is either fully rotated to sit in front of $G\tilde{G}$, or to manifest itself as an overall phase of the quark mass matrix. Although inherently basis-dependent, the former viewpoint suggests that θ is essentially tied to the gluonic structure of QCD, while the latter emphasizes instead its links to the flavor sector.

13.2.4.1. *Dynamical relaxation of $\bar{\theta}$: Axions*

The energy of the QCD vacuum as a function of $\bar{\theta}$ Eq. (13.23) has a minimum at $\bar{\theta} = 0$. Thus the relaxation of the θ-parameter to zero is possible if one promotes it to a dynamical field, called the *axion* [8, 21, 26]. This is motivated by the assumption that the Standard Model, augmented by appropriate additional fields, admits a chiral symmetry $\mathrm{U}(1)_{\mathrm{PQ}}$,

$$\psi_L \to e^{i\alpha}\psi_L, \qquad \psi_R \to e^{-i\alpha}\psi_R, \tag{13.46}$$

where ψ is a fermion charged under $\mathrm{SU}(3)_c$, not necessarily a quark. When this symmetry is spontaneously broken at a necessarily high scale f_a, a pseudoscalar Goldstone boson – the axion – survives as the only low energy manifestation. We can therefore ignore the precise details of how this symmetry is broken, and note that the coupling to $\mathrm{SU}(3)_c$, via the axial anomaly, implies that the essential components of the axion Lagrangian are very simple,

$$\mathcal{L}_a = \frac{1}{2}\partial_\mu a \partial^\mu a + \frac{a(x)}{f_a}\frac{\alpha_s}{8\pi}G\tilde{G}, \tag{13.47}$$

leading to a field-dependent shift of θ,

$$\bar{\theta} \to \bar{\theta} + \frac{a}{f_a}. \tag{13.48}$$

If the effects of non-perturbative QCD are ignored, this Lagrangian possesses a symmetry, $a \to a+\text{const}$, and a is a massless field with derivative couplings to the SM fields, i.e. $\partial_\mu a \bar{\psi}\gamma_\mu\gamma_5\psi$, that are not important for the solution of the strong CP problem.

Below the QCD scale, one finds that $U(1)_{PQ}$ is explicitly broken by the chiral anomaly, and thus the axion is in reality a *pseudo*-Goldstone boson and acquires a potential. The form of this potential can be read directly from our earlier discussion of the θ-dependence of the vacuum energy, namely $E(\theta) \sim \chi(0)\theta^2/2 + \cdots$. Accounting for the shift in Eq. (13.48), the effective axion Lagrangian becomes,

$$\mathcal{L}_a^{\text{eff}} = \frac{1}{2}\partial_\mu a \partial^\mu a - \frac{1}{2}\chi(0)\left(\bar{\theta} + \frac{a}{f_a}\right)^2 + \cdots, \qquad (13.49)$$

where $\chi(0)$ is the topological susceptibility calculated before in Eq. (13.36), and we have corrections of quartic order and higher in the potential. We see from Eq. (13.48) that the vacuum expectation value of the axion field $\langle a \rangle$ renormalizes the value of $\bar{\theta}$ so that all observables depend on the $(\bar{\theta}+\langle a\rangle/f_a)$ combination. At the same time, such a combination must vanish in the vacuum as it minimizes the value of the axion potential in Eq. (13.49). This dynamical relaxation then solves the strong CP problem. This cancellation mechanism works independently of the "initial" value of $\bar{\theta}$, which is why it is very appealing. However, the excitations around $\langle a \rangle$ correspond to a massive axion particle with

$$m_a \sim \frac{1}{f_a}|\chi(0)|^{1/2}, \qquad (13.50)$$

a formula analogous to that discussed earlier for η' Eq. (13.27). For large f_a the axion is very light and thus has significant phenomenological consequences that we will briefly come to shortly.

An aspect of the axion mechanism that is perhaps not stressed as often as it should be is that there can be other contributions to the axion potential which shift its minimum away from $(\bar{\theta} + a/f_a) = 0$ [27]. In particular, in the analysis above, we included only the leading term corresponding to the vacuum energy. However, if there are other CP-odd operators $\mathcal{O}_{\text{CP}}$ present at low scales, QCD effects may also generate terms linear in θ via nonzero mixed correlators of the form

$$\chi_{\mathcal{O}_{\text{CP}}}(0) = -i \lim_{k\to 0} \int d^4x e^{ik\cdot x}\langle 0|T(G\tilde{G}(x), \mathcal{O}_{\text{CP}}(0))|0\rangle. \qquad (13.51)$$

An example of this type is the quark chromoelectric dipole moment, $\mathcal{O}_{\text{CP}} = \tilde{d}_q \bar{q} G\sigma\gamma_5 q$, that we will discuss in more detail in Section 13.3. The axion potential is then modified,

$$\mathcal{L}_a^{\text{eff}} = \frac{1}{2}\partial_\mu a \partial^\mu a - \chi_{\mathcal{O}_{\text{CP}}}(0)\left(\bar{\theta} - \frac{a}{f_a}\right) + \frac{1}{2}\chi(0)\left(\bar{\theta} + \frac{a}{f_a}\right)^2 + \cdots, \quad (13.52)$$

and exhibits a minimum shifted from zero. The size of this *induced* contribution to θ, i.e. $\theta_{\text{ind}} = -\chi_{\mathcal{O}_{\text{CP}}}(0)/\chi(0)$, is linearly related to the coefficient of the CP-odd operator $\mathcal{O}_{\text{CP}}$ generating $\chi_{\mathcal{O}_{\text{CP}}}(0)$. These effects therefore need to be taken into account in computing the observable consequences of CP-odd sources in axion scenarios.

The presence of a non-renormalizable term in Eq. (13.47) points to the presence of new physics at the scale f_a, where the symmetry $a \to a+\text{const}$ is realized linearly. In the initial proposal of Peccei and Quinn [8], and elucidated in the work of Weinberg [26] and Wilczek [21], f_a was suggested to be at the weak scale with the colored fermions ψ introduced above identified with the quarks. However, the ensuing negative results of direct and indirect searches for axions has restricted f_a to significantly larger scales, $f_a > 10^{10}$ GeV. Models which decouple the $\text{U}(1)_{\text{PQ}}$ and electroweak breaking scales were subsequently introduced, differing in how this decoupling is achieved. The KSVZ model [28] uses additional colored fermions as above, while the ZDFS [29] approach retains the quarks as the colored fermions but enlarges the Higgs sector. Searches for these so-called "invisible axions" have thus far proved unsuccessful, but axion-related physics has since created an intriguing sub-field in particle physics, cosmology and astrophysics.

Before moving on, it is worth recalling that, were it realized, the simplest solution to the strong CP problem would fall into the class we are discussing, namely the possibility that $m_u = 0$ in the Standard-Model Lagrangian normalized at a high scale M, or more generically, $\det Y_u(M) = 0$. In this situation, the Lagrangian already possesses the appropriate chiral symmetry without the addition of extra fields and, as we have discussed, $\theta(M)$ then becomes unphysical. Notice that such a condition does not allow for removal of the Kobayashi–Maskawa phase. However, it sets $m_*(M)$ to zero and removes the dependence of observable quantities on $\bar{\theta}$. (Alternatively, one could simply require a factor of 10^{11} suppression of m_u relative to its commonly accepted value.) We have emphasized the dependence on the scale M here, because once one runs down to the regime where chiral symmetry is broken, the identification of the light quark masses becomes less straightforward. Indeed, since our information on these masses comes precisely from this regime, indirectly via meson and baryon spectra and chiral perturbation theory, the possibility that $m_u(M) = 0$, and its relevance, has been debated at length in the literature. Indeed, even if $m_u(M) = 0$, it has been argued that higher-order corrections in chiral perturbation theory, quadratic in the nonzero quark masses, could in principle mimic the presence of a nonzero m_u [30], although such effects would need to be exceedingly large to accomodate $m_u \sim 4$ MeV. In this context, the constraint $\det Y_u(M) = 0$ could be rendered natural through the imposition of an accidental U(1) symmetry [31].

It should be emphasized, however, that the possibility of $m_u(M) = 0$ is strongly disfavored by the conventional chiral perturbation theory analysis, with recent results implying $m_u/m_d = 0.553 \pm 0.043$ [32], and this conclusion is beginning to be backed up by unquenched (but chirally extrapolated) lattice simulations which suggest similar values, $m_u/m_d = 0.43 \pm 0.1$ [33].

13.2.4.2. *Engineering $\bar{\theta} \simeq 0$: Spontaneously broken P or CP*

Another way to approach the strong CP problem is to assume that either P or CP or both are exact symmetries of Nature at some high-energy scale. Then one can declare that $\theta G\tilde{G}$ be zero at this high scale as a result of symmetry. Of course, to account for the parity- and CP-violation observed in the SM, one has to assume that these symmetries are spontaneous broken at a particular scale $\Lambda_{P(CP)}$.

The model building problem that this sets up – one which has been made particularly manifest by the consistency of the recent B-physics CP-violation with the KM mechanism – is that one needs to ensure that the subsequent corrections to θ are small, while still allowing for an order one KM phase. Symmetry breaking at $\Lambda_{P(CP)}$ may generate the θ-term at tree level through, for example imaginary corrections to the quark mass matrices M_u and M_d,

$$\begin{aligned} \bar{\theta} &\sim \text{Arg Det}(M_u M_d) + \cdots && (13.53) \\ &= \text{Arg Det}(Y_u Y_d) + \text{Arg Det}(v_u v_d) + \cdots . && (13.54) \end{aligned}$$

Here v_u and v_d are the Higgs expectation values, and in the SM $v_u = v_d^*$. The dots stand for the mass-matrix phases of other colored fermions which may be considerably heavier than the SM quarks but still have to be included in the calculation of the residual $\bar{\theta}$.

For comparison, the SM CKM-type phase (in basis-invariant form) is (see Eq. (13.42))

$$\theta_{EW} \sim \text{Arg Det}\left[Y_u Y_u^\dagger, Y_d Y_d^\dagger\right], \qquad (13.55)$$

and one is then led to consider models for flavor in which the second phase Eq. (13.55) can be large, as is required, while the first Eq. (13.54) vanishes, or is at least highly suppressed. Let us now review some of the proposals put forward in this regard.

Exact parity at some high-energy scale would imply $L \leftrightarrow R$ reflection symmetry in the Yukawa sector, and as a consequence,

$$Y_u = Y_u^\dagger; \; Y_d = Y_d^\dagger. \tag{13.56}$$

Hermitian Yukawa matrices give no contribution to $\bar{\theta}$, and therefore such a symmetry can be considered a first step towards the solution of the strong CP problem [34]. Note that both Yukawa matrices can be complex thus easily accommodating the Kobayashi–Maskawa phase. The use of exact parity necessitates the extension of the SM gauge group by the right-handed group $SU(2)_R$ and "unification" of the u_R, and d_R fields in a single multiplet. The reality of $v_{u(d)}$ comes as an additional constraint on the model and can be achieved for example in its supersymmetric versions [35, 36].

Models attempting to solve the strong CP problem via spontaneous breaking of CP, in contrast, do not require an extension of the gauge group. In such models, the Yukawa couplings are real and CP violation typically comes via complex vacuum expectation values of additional scalar fields.

For example, one may introduce a heavy vector-like quark T [37, 38] with mass M which couples to the $SU(2)$-singlet down-type quarks of the SM via an additional scalar field S, $h_i d_{Ri} T S$, where i is the generation index and h_i are the corresponding Yukawa couplings. The resulting 4×4 mass matrix in the d-sector takes the following form [37, 38],

$$M_d = \begin{pmatrix} Y_d v_d & h_i S \\ 0 & M \end{pmatrix}, \tag{13.57}$$

where S now stands for a complex v.e.v. of the S scalar. Such a mass-matrix has complex entries, yet the phase of its determinant is zero, thus providing no contributions to $\bar{\theta}$ at tree level.

The real challenge for this type of model is to create a plausible CKM-type phase, or in other words the CP-odd combination of the CKM mixing angles $\mathrm{Im}(V_{cs} V_{us}^* V_{ud} V_{cd}^*)$. Typically, this invariant comes out too small, prompting the prediction of a super-weak type of CP violation for K and B mesons. In view of the recent discoveries of CP violation in the B-meson sector, which at the time of writing are in very good accord with the CKM predictions, such models have become disfavored.

A interesting possibility to use models with low-scale supersymmetry breaking for the solution of the strong CP problem has been proposed in [39] (see also earlier ideas [40, 41]). The model postulates the spontaneous breaking of CP at some high-energy scale Λ_{CP} where SUSY is exact. The sector of the model that breaks CP spontaneously cannot lead to renormalization of the Yukawa interactions in the superpotential as they are protected by SUSY. On the other hand, kinetic terms in the quark sector can be renormalized due to CP-odd interactions at Λ_{CP} in a flavor-dependent way. These wave function renormalization factors Z_i are in general complex, but are hermitian and positive definite due to the reality of the Kähler potential, and thus the phases contained therein cannot contribute to θ. To see this, one notes that such Z_i, for $i = Q, u, d$, can be written in the form $Z_i = (T_i)^2$ with T_i also hermitian and positive definite. Thus the rescaling required to go to a canonical normalization implies,

$$Y_u \to T_Q Y_u T_u, \qquad Y_d \to T_Q Y_d T_d, \tag{13.58}$$

and thus the rescaled Yukawa couplings continue to satisfy ArgDet$(Y_{u,d})$=0 by the hermiticity of T. Consequently, this renormalization does not induce $\bar{\theta}$ while SUSY is unbroken. At the same time, the renormalization of the CKM matrix can in principle be substantial allowing for a sizable δ_{KM}. In practice, it turns out that this is possible only if there is an additional source of strong dynamics at Λ_{CP} such that Z_i deviate significantly from the unit matrix through threshold effects.

All models of the type discussed above that attempt to solve the strong CP problem by postulating exact parity or CP at high scales, have to cope with the very tight bound on $\bar{\theta}$. Indeed, it is not enough to obtain $\bar{\theta} = 0$ at tree level, as loop effects at and below $\Lambda_{P(CP)}$ can lead to a substantial renormalization of the θ-term (see, e.g. [42, 43]). If the effective theory reduces to the SM below the scale $\Lambda_{P(CP)}$, the residual *low-scale* corrections to the θ-term can only come via the Kobayashi–Maskawa phase and the resulting value for $\bar{\theta}(\delta_{KM})$ is small. However, this does not guarantee that the threshold corrections at $\Lambda_{P(CP)}$ are also small, as they will depend on different sources of CP-violation and do not have to decouple in the limit of large $\Lambda_{P(CP)}$. Such corrections are necessarily model-dependent. However, if the underlying theory is supersymmetric at the scale $\Lambda_{P(CP)}$ and the breaking of supersymmetry occurs at a lower scale Λ_{SUSY}, one expects the corrections to $\bar{\theta}$ to be suppressed by power(s) of the small ratio $\Lambda_{SUSY}/\Lambda_{P(CP)}$ [39].

To summarize this section, we comment that the way the strong CP problem is resolved affects the issue of how large additional non-CKM CP-violating sources can be. The axion solution, as well as $m_u = 0$, generically allows for the presence of *arbitrarily* large CP-violating sources above a certain energy scale. This scale is determined by comparison of higher-dimension CP-odd operators (i.e. dim$\geq$ 5) induced by these sources with the current EDM constraints. On the contrary, models using a discrete symmetry solution to the strong CP problem usually have tight restrictions on the amount of additional CP-violation even at higher scales in order to avoid potentially dangerous contributions to the θ-term.

13.3. Electric Dipole Moments as Probes of New Physics

The idea to use the electric dipole moments of particles as high-precision probes of symmetry properties of the strong interactions is due to Purcell and Ramsey [4]. Remarkably, it precedes not only the discovery of CP violation in K mesons, but also the discovery of parity violation in weak interactions. The main motivation behind the initial idea was the suggestion that the (at the time unknown) theory of the strong interactions may not be parity symmetric. As we saw in the previous section, it was only 25 years later that the establishment of QCD as the theory of strong interactions led to the possibility of P and CP violation by the θ-term.

Towards the end of Section 13.2, we emphasized that EDMs of nucleons, atoms, and molecules play a dominant role in the experimental constraints on $\bar{\theta}$, and in probes of flavor-diagonal CP-violation more generally. Although they are clearly not the only observables sensitive to non-CKM sources of CP-violation, the remarkable degree of precision to which they can currently be measured endows them with a privileged status. In this section, we will explore in some detail the theoretical techniques required to exploit these constraints, which are somewhat involved as the physics scales relevant to the discussion range from the TeV scale down to the atomic scale.

To begin, let us recall that when placed in a magnetic and an electric field, a neutral nonrelativistic particle of spin S can be described by the following Hamiltonian, containing electric (d) and magnetic (μ) dipole moments,

$$H = -\mu \mathbf{B} \cdot \frac{\mathbf{S}}{S} - d\mathbf{E} \cdot \frac{\mathbf{S}}{S} \, . \tag{13.59}$$

Under the reflection of space coordinates, $P(\mathbf{B} \cdot \mathbf{S}) = \mathbf{B} \cdot \mathbf{S}$, whereas $P(\mathbf{E} \cdot \mathbf{S}) = -\mathbf{E} \cdot \mathbf{S}$. The presence of a nonzero d signifies the existence of parity and time-reversal violation. Indeed, under time reflection, $T(\mathbf{B} \cdot \mathbf{S}) = \mathbf{B} \cdot \mathbf{S}$ and $T(\mathbf{E} \cdot \mathbf{S}) = -\mathbf{E} \cdot \mathbf{S}$. Therefore a nonzero d may

Table 13.1. Current constraints within three representatve classes of EDMs.

Class	EDM	Current Bound
Paramagnetic	^{205}Tl	$\lvert d_{\mathrm{Tl}}\rvert < 9 \times 10^{-25} e\,\mathrm{cm}$ (90% C.L.) [45]
Diamagnetic	^{199}Hg	$\lvert d_{\mathrm{Hg}}\rvert < 3 \times 10^{-29} e\,\mathrm{cm}$ (95% C.L.) [46, 47]
Nucleon	n	$\lvert d_n\rvert < 3 \times 10^{-26} e\,\mathrm{cm}$ (90% C.L.) [44]

exist if and only if both parity and time reveral invariance are broken. In the initial work of Purcell and Ramsey, analysis of the existing experimental data on neutron scattering from spin zero nuclei led to the conclusion, $|d_n| < 3 \times 10^{-18} e\mathrm{cm}$ [4]. Such a result probes physics at distances much shorter than the typical scale of nuclear froces $\sim 1\mathrm{fm}$, or the Compton wavelength of the neutron. This initial limit on the neutron EDM implied that P and T were good symmetries of the strong interactions at percent-level precision.

On applying the CPT theorem, one concludes that the breaking of T also requires the breaking of CP. Following the discovery of CP violation in the mixing of neutral kaons [2], the EDM search intensified, and the level of experimental precision has improved steadily ever since. Indeed, following significant progress throughout the past decade, the EDMs of the neutron [44], and of several heavy atoms and molecules [45, 46, 48–51] have been measured to vanish to remarkably high precision. From the present standpoint, it is convenient to classify the EDM searches into three main categories, distinguished by the dominant physics which would induce the EDM, at least within a generic class of models. These categories are: the EDMs of paramagnetic atoms and molecules; the EDMs of diamagnetic atoms; and the EDMs of hadrons, and nucleons in particular. For these three categories, the experiments that currently champion the best bounds on CP-violating parameters are the atomic EDMs of thallium and mercury and that of the neutron, as listed in Table 13.1.[a]

The upper limits on EDMs obtained in these experiments can be translated into tight constraints on the CP-violating physics at and above the

[a]Since this chapter was written, a new limit on the ^{199}Hg EDM has been published by the Seattle group [47]. This new result is included in Table 13.1, but the older limit [46] was used when calculating the constraints on various parameters that are described in the subsequent text. See Chapter 16 for more details.

electroweak scale, with each category of EDM primarily sensitive to different CP-odd sources. For example, the neutron EDM can be induced by CP violation in the quark sector, while paramagnetic EDMs generally result from CP-violating sources that induce the electron EDM. Despite the apparent difference in the actual numbers in Table 13.1, all three limits on d_n, $d_{\rm Tl}$, and $d_{\rm Hg}$ actually have comparable sensitivity to fundamental CP violation, e.g. superpartner masses and CP-violating phases, and thus play complementary roles in constraining fundamental CP-odd sources. This fact can be explained by the way the so-called Schiff screening theorem [52] is violated in paramagnetic and diamagnetic atoms. The Schiff theorem essentially amounts to the statement that, in the nonrelativistic limit and treating the nucleus as point-like, the atomic EDMs will vanish due to screening of the applied electric field within a neutral atom. The paramagnetic and diamagnetic EDMs result from violations of this theorem due respectively to relativistic and finite-size effects, and in heavy atoms such violation is maximized. For heavy paramagnetic atoms, i.e. atoms with nonzero electron angular momentum, relativistic effects actually result in a net enhancement of the atomic EDM over the electron EDM. For diamagnetic species, the Schiff screening is violated due to the finite size of the nucleus, but this is a weaker effect and the induced EDM of the atom is suppressed relative to the EDM of the nucleus itself. These factors equilibrate the sensitivities of the various experimental constraints in Table 13.1 to more fundamental sources of CP violation.

In this section, we will review this role of EDMs in some detail (see Refs. [9, 10] for further details).

13.3.1. *EDMs as probes of CP violation*

The majority of EDM experiments are performed with matter as opposed to anti-matter. Therefore, the conclusion about the relation between d and CP violation relies on the validity of the CPT theorem. The interaction $d\mathbf{E}\cdot\mathbf{S}$ for a spin 1/2 particle then has the following relativisitic generalization

$$H_{\rm T,P-odd} = -d\mathbf{E}\cdot\frac{\mathbf{S}}{S} \qquad \longrightarrow \qquad \mathcal{L} = -d\frac{i}{2}\overline{\psi}\sigma^{\mu\nu}\gamma_5\psi F_{\mu\nu}. \qquad (13.60)$$

Parenthetically, it is worth remarking that the precision of EDM experiments has now reached a level sufficient to provide competetive tests of CPT invariance, since one can also consider a CP-even, but CPT-odd, relativisitic form of $d\mathbf{E}\cdot\mathbf{S}$, namely $\mathcal{L} = d\overline{\psi}\gamma^\mu\gamma_5\psi F_{\mu\nu}n^\nu$, with a preferred frame $n^\nu = (1,0,0,0)$, which spontaneously breaks Lorentz invariance and CPT.

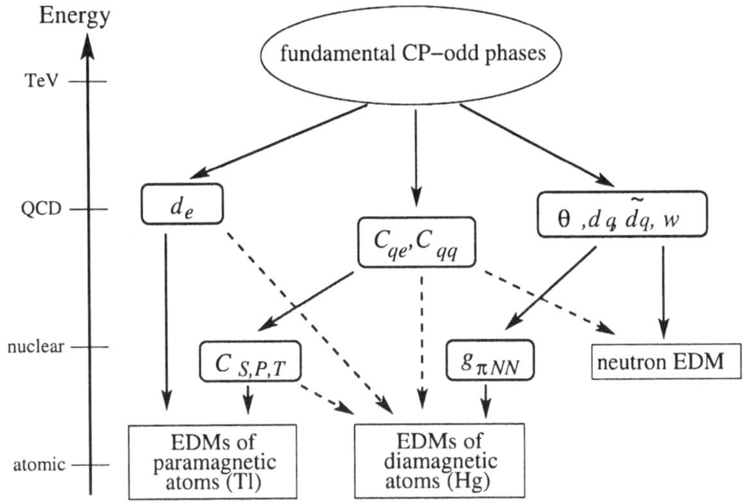

Fig. 13.1. A schematic plot of the hierarchy of scales between the CP-odd sources and three generic classes of observable EDMs. The dashed lines indicate generically weaker dependencies.

The problem of calculating an observable EDM from the underlying CP violation in a given particle physics model can be conveniently separated into different stages, depending on the characteristic energy/momentum scales. At each step the result can be expressed as an effective Lagrangian in terms of light degrees of freedom with Wilson coefficients that encode information about CP violation at higher energy scales. As usual in effective field theory, it is very convenient to classify all possible effective CP violating operators in terms of their dimension, with the operators of lowest dimension usually leading to the largest contributions. This logic may need to be refined if symmetry requirements imply that certain operators are effectively of higher dimension than naive counting would suggest. This is actually the case for certain EDM operators due to gauge invariance, as discussed in more detail below.

We will present this analysis systematically in order of increasing energy scale, working our way upwards in the dependency tree outlined in Fig. 13.1, which allows us to remain entirely model-independent until the final step where some high-scale model of CP violation can be imposed and then subjected to EDM constraints.

13.3.1.1. *Observable EDMs*

Let us begin by reviewing the lowest level in this construction, namely the precise relations between observable EDMs and the relevant CP-odd operators at the nuclear scale. At leading order, such effects may be quantified in terms of EDMs of the constituent nucleons, d_n and d_p (where the neutron EDM is already an observable), the EDM of the electron d_e, and CP-odd electron-nucleon and nucleon-nucleon interactions. In the relevant channels these latter interactions are dominated by pion exchange, and thus we must also consider the CP-odd pion-nucleon couplings $\bar{g}_{\pi NN}$ which can be induced by CP-odd interactions between quarks and gluons. To be more explicit, we write down the relevant CP-odd terms at the nuclear scale,

$$\mathcal{L}_{eff}^{\rm nuclear} = \mathcal{L}_{\rm edm} + \mathcal{L}_{\pi NN} + \mathcal{L}_{eN}, \tag{13.61}$$

which can be split into terms for the nucleon (and electron) EDMs,

$$\mathcal{L}_{edm} = -\frac{i}{2} \sum_{i=e,p,n} d_i \, \overline{\psi}_i (F\sigma)\gamma_5 \psi, \tag{13.62}$$

the CP-odd pion nucleon intercations,

$$\begin{aligned} \mathcal{L}_{\pi NN} = \bar{g}^{(0)}_{\pi NN} \bar{N}\tau^a N \pi^a + \bar{g}^{(1)}_{\pi NN} \bar{N} N \pi^0 \\ + \bar{g}^{(2)}_{\pi NN} (\bar{N}\tau^a N\pi^a - 3\bar{N}\tau^3 N\pi^0), \end{aligned} \tag{13.63}$$

and finally CP-odd electron-nucleon couplings,

$$\begin{aligned} \mathcal{L}_{eN} = C_S^{(0)} \bar{e} i\gamma_5 e \bar{N} N + C_P^{(0)} \bar{e} e \bar{N} i\gamma_5 N + C_T^{(0)} \epsilon_{\mu\nu\alpha\beta} \bar{e}\sigma^{\mu\nu} e \bar{N}\sigma^{\alpha\beta} N \\ + C_S^{(1)} \bar{e} i\gamma_5 e \bar{N}\tau^3 N + C_P^{(1)} \bar{e} e \bar{N} i\gamma_5 \tau^3 N + C_T^{(1)} \epsilon_{\mu\nu\alpha\beta} \bar{e}\sigma^{\mu\nu} e \bar{N}\sigma^{\alpha\beta}\tau^3 N. \end{aligned} \tag{13.64}$$

In certain rare cases, CP-odd nucleon-nucleon forces are not mediated by pions, in which case the effective Lagrangian must be extended by a variety of *contact* terms, e.g. $\bar{N}N\bar{N}i\gamma_5 N$, and the like.

The dependence of the observable EDMs on the corresponding Wilson coefficients relies on atomic and nuclear many-body calculations which would go beyond the scope of this review to cover here (see the reviews [9, 53] for further details). However, we will briefly summarize the current status of these calculations, before turning to our major focus which is the calculation of these coefficients in terms of higher scale CP-odd sources.

As alluded to earlier on, it is convenient to split the discussion into three parts, corresponding roughly to the three classes of observable EDMs which currently provide constraints at a similar level of precision: EDMs of paramagnetic atoms and molecules, EDMs of diamagnetic atoms, and the neutron EDM.

• *EDMs of paramagnetic atoms – thallium EDM*

Paramagnetic systems, namely those with one unpaired electron, are primarily sensitive to the EDM of this electron. At the nonrelativistic level, this is far from obvious due to the Schiff shielding theorem which implies, since the atom is neutral, that any applied electric field will be shielded and so an EDM of the unpaired electron will not induce an atomic EDM. Fortunately, this theorem is violated by relativistic effects. In fact, it is violated strongly for atoms with a large atomic number, and even more strongly in molecules which can be polarized by the applied field. For atoms, the parameteric enhancement of electron EDM is given by [53–55]

$$d_{\text{para}}(d_e) \sim 10 \frac{Z^3\alpha^2}{J(J+1/2)(J+1)^2} d_e, \tag{13.65}$$

up to numerical $\mathcal{O}(1)$ factors, with J the angular momentum and Z the atomic number. This enhancement is significant, and for large Z, the applied field can be enhanced by a factor of a few hundred within the atom. This feature explains why atomic systems provide such a powerful probe of the electron EDM, since the "effective" electric field can be much larger than one could actually produce in the lab.

Although the electron EDM is the predominant contributor to any paramagnetic EDM in most models, one should bear in mind that other contributions may also be significant in certain regimes. In particular, significant CP-odd electron-nucleon couplings may also be generated, due for example to CP violation in the Higgs sector. Among these couplings, C_S plays by far the most important role for paramagnetic EDMs because it couples to the spin of the electron and is enhanced by the large nucleon number in heavy atoms.

Among various paramagnetic systems, the EDM of the thallium atom currently provides the best constraints on fundamental CP violation. A number of atomic calculations [55–57] (see also Ref. [9] for a more complete list) have established the relation between the EDM of thallium, d_e, and the coefficients of the CP-odd electron-nucleon interactions C_S:

$$d_{\text{Tl}} = -585 d_e - e\ 43\ \text{GeV} \times (C_S^{(0)} - 0.2 C_S^{(1)}), \tag{13.66}$$

with C_S expressed in isospin components. The relevant atomic matrix elements are known to within 10–20% [53].

As we discuss later on, current experimental work is focusing on the use of paramagnetic molecules, e.g. YbF and PbO [51, 58], which can provide an even larger enhancement of the applied field due to polarization effects, have better systematics, and may bring significant progress in measuring/constraining d_e and C_S.

- *EDMs of diamagnetic atoms – mercury EDM*

EDMs of diamagnetic atoms, i.e. atoms with total electron angular momentum equal to zero, also provide an important test of CP violation [9]. In such systems the Schiff shielding argument again holds to leading order. However, in this case it is violated not by relativistic effects but by finite size effects, namely a net misalignment between the distribution of charge and EDM (i.e. first and second moments) in the nucleus of a large atom (see Ref. [53] for a review). However, in contrast to the paramagnetic case, this is a rather subtle effect and the induced atomic EDM is considerably suppressed relative to the underlying EDM of the nucleus.

To leading order in an expansion around the approximation of a point-like nucleus, the contributions arise from an octopole moment (which is only relevant for states with large spin, and will not be relevant for the cases considered here), and the Schiff moment $\vec{S}$, which contributes to the electrostatic potential,

$$V_E = 4\pi \vec{S} \cdot \vec{\nabla} \delta(\vec{r}). \tag{13.67}$$

CP-odd nuclear moments, such as $\vec{S}$, can arise from intrinsic EDMs of the constituent nucleons and also CP-odd nucelon interactions. It turns out that the latter source tends to dominate in diamagnetic atoms and thus, since such interactions are predominantly due to pion exchange, we can ascribe the leading contribution to CP-odd pion nucleon couplings $\bar{g}^{(i)}_{\pi NN}$ for $i = 0, 1, 2$ corresponding to the isospin.

There are of course various additional contributions, which are generically subleading, but may become important in certain models. Schematically, we can represent the EDM in the form

$$d_{\text{dia}} = d_{\text{dia}}(S[\bar{g}_{\pi NN}, d_N], C_S, C_P, C_T, d_e), \tag{13.68}$$

where we note that electron-nucleon interactions may also be significant, as is the electron EDM itself [9] (although in practice the electron EDM tends to be more strongly constrained by limits from paramagentic systems and thus is often neglected).

Currently, the strongest constraint in the diamagnetic sector comes from the bound on the EDM of mercury – at the atomic level, this is in fact the most precise EDM bound in existence. As should be apparent from the above discussion, computing the dependence of $d_{\rm Hg}$ on the underlying CP-odd sources is a nontrivial problem requiring input from QCD and nuclear and atomic physics. In particular, the computation of $S(\bar{g}_{\pi NN})$ is a nontrivial nuclear many-body problem, and has recently been reanalyzed. We quote the results of Dmitriev and Sen'kov [59],

$$S(^{199}\text{Hg}) = -0.0004g\bar{g}^{(0)} - 0.055g\bar{g}^{(1)} + 0.009g\bar{g}^{(2)}\ e\ \text{fm}^3, \qquad (13.69)$$

and also a more recent analysis of de Jesus and Engel [60],

$$S(^{199}\text{Hg}) = -0.010g\bar{g}^{(0)} - 0.074g\bar{g}^{(1)} + 0.018g\bar{g}^{(2)}\ e\ \text{fm}^3, \qquad (13.70)$$

where $g = g_{\pi NN}$ is the CP-even pion-nucleon coupling, and $\bar{g}^{(i)} = \bar{g}^{(i)}_{\pi NN}$ denote the CP-odd couplings. The isoscalar and isotensor couplings vary significantly between the two calculations, and the suppression of the overall coefficient in front of $g\bar{g}^{(0)}$ in the result Eq. (13.69) below $O(0.01)$ is the result of mutual cancellation between several contributions of comparable size, and therefore is in some sense accidental. Nonetheless, these differences do provide some indication of the difficulties inherent in the calculation. Fortunately, the isovector coupling – which generically turns out to be most important for EDMs – has remained relatively stable in most calculations (to within a factor 2). For numerical estimates, we take $S(^{199}\text{Hg}) = -0.06g\bar{g}^{(1)}$ for this coupling.

Putting the pieces together, we can write the mercury EDM in the form,

$$\begin{aligned} d_{\rm Hg} &= (1.8 \times 10^{-3}\,\text{GeV}^{-1})e\,\bar{g}^{(1)}_{\pi NN} + 10^{-2}d_e \\ &\quad + (3.5 \times 10^{-3}\,\text{GeV})e\,C_S^{(0)}, \end{aligned} \qquad (13.71)$$

where we have limited attention to the isovector pion-nucleon coupling and C_S which turns out to the most important for CP violation in supersymmetric models.

- *Neutron EDM*

The final class to consider is that of the neutron itself, whose EDM can be searched for directly with ultracold neutron bottles, and currently provides one of the strongest constraints on new CP-violating physics. In this case, there is clearly no additonal atomic or nuclear physics to deal with, and we must turn directly to the next level in energy scale, namely the use of QCD to compute the dependence of d_n on CP-odd sources at the quark-gluon level. This statement also applies to many of the other quantities we have introduced thus far, including in particular the CP-odd pion-nucleon coupling. Indeed, it is only paramagnetic systems that are partially immune to QCD effects, although even there we have noted the possible relevance of electron-nucleon interactions.

13.3.1.2. *The structure of the low energy Lagrangian at 1GeV*

The effective CP-odd flavor-diagonal Lagrangian normalized at 1 GeV, which is taken to be the lowest perturbative quark/gluon scale, plays a special role in EDM calculations. At this scale, all particles other than the u, d and s quark fields, gluons, photons, muons and electrons can be considered heavy, and thus integrated out. As a result, one can construct an effective Lagrangian by listing all possible CP-odd operators in order of increasing dimension,

$$\mathcal{L}_{\rm eff} = \mathcal{L}_{\rm dim=4} + \mathcal{L}_{\rm dim=5} + \mathcal{L}_{\rm dim=6} + \cdots . \tag{13.72}$$

Accounting for the chiral anomaly, there is only one operator at dimension 4, the QCD theta term,

$$\mathcal{L}_{\rm dim=4} = \frac{g_s^2}{32\pi^2}\,\bar{\theta} G^a_{\mu\nu}\widetilde{G}^{\mu\nu,a}. \tag{13.73}$$

At the dimension 5 level, there are (naively) several operators: EDMs of light quarks and leptons and color electric dipole moments of the light quarks,

$$\mathcal{L}_{\rm dim=5} = -\frac{i}{2}\sum_{i=u,d,s,e,\mu} d_i\,\overline{\psi}_i(F\sigma)\gamma_5\psi_i - \frac{i}{2}\sum_{i=u,d,s}\widetilde{d}_i\,\overline{\psi}_i g_s(G\sigma)\gamma_5\psi_i, \tag{13.74}$$

where $(F\sigma)$ and $(G\sigma)$ are shorthand notations for $F_{\mu\nu}\sigma^{\mu\nu}$ and $G^a_{\mu\nu}t^a\sigma^{\mu\nu}$.

In fact, in most models these operators are really dimension-six operators in disguise. The reason is that, if we proceed in energy above the electroweak scale and assume the system restores SU(2)$\times U(1)$ as in the Standard Model, gauge invariance ensures that these operators must include an insertion of the Higgs field H [61]. Indeed, were we to write the basis of down quark EDMs and CEDMs above the electroweak scale, we should specify the following list of dimension six operators [61],

$$\mathcal{L}^{\rm EW}_{\text{``dim=5''}} = \frac{i}{2\sqrt{2}}\bar{Q}_L\left[2d_1^{\rm EW}(B\sigma) + d_2^{\rm EW}\tau^i(W^i\sigma) \right.$$
$$\left. + d_2^{\rm EW}\lambda^a(G^a\sigma)\right](H/v)D_R + \text{h.c.}, \tag{13.75}$$

which are defined in terms of left-handed doublets $Q_L = (U,D)_L$ and right-handed singlets D_R and in terms of the U(1), SU(2), and SU(3) field strengths $B_{\mu\nu}$, $W^i_{\mu\nu}$ and $G^a_{\mu\nu}$. This representation also points to a number of other classes of dimension six CP-odd operators, e.g. dipole operators purely defined in terms of fermion doublets [61], which do not flip chirality, but it would take us too far afield to consider a general parametrization.

The lesson we draw from Eq. (13.75) with regard to EDMs is that, if generated, these operators must be proportional to the Higgs v.e.v. below the electroweak scale, and consequently must scale at least as $1/M^2$ for $M \gg M_W$. In practice, this feature can also be understood in most models by going to a chiral basis, where we see that these operators connect left- and right-handed fermions, and thus require a chirality flip. This is usually supplied by an insertion of the fermion mass, i.e. $d_f \sim m_f/M^2$, again implying that the operators are effectively of dimension six.

This implies that, for consistency, we should also proceed at least to dimension six where we encounter the CP-odd three-gluon Weinberg operator and host of possible four-fermion interactions, $(\bar{\psi}_i \Gamma \psi_i)(\bar{\psi}_j i\Gamma\gamma_5\psi_j)$, where Γ denotes several possible scalar or tensor Lorentz structures and/or gauge structures, which are contracted between the two bilinears. Limiting our attention to a small subset of the latter that will be relevant later on,

$$\mathcal{L}_{\text{dim}=6} = \frac{1}{3} w\, f^{abc} G^a_{\mu\nu} \widetilde{G}^{\nu\beta,b} G_{\beta}{}^{\mu,c} + \sum_{i,j} C_{ij}\, (\bar{\psi}_i\psi_i)(\bar{\psi}_j i\gamma_5\psi_j) + \cdots\,. \qquad (13.76)$$

In this formula, the operators with C_{ij} are summed over all light fermions. Going once again to a chiral basis, we can argue as above that the four-fermion operators, which require two chirality flips, are in most models effectively of dimension eight. Nonetheless, in certain cases they may be non-negligible.

To proceed to the next level in energy scale in Fig. 13.1, we need to determine the dependence of the nucleon EDMs, pion-nucleon couplings, etc., on these quark-gluon Wilson coefficients normalized at 1 GeV, i.e.

$$\begin{aligned} d_n &= d_n(\bar{\theta}, d_i, \tilde{d}_i, w, C_{ij}), \\ \bar{g}_{\pi NN} &= \bar{g}_{\pi NN}(\bar{\theta}, \tilde{d}_q, w, C_{ij}). \end{aligned} \qquad (13.77)$$

The systematic project of deducing this dependence was first initiated some 20 years ago by Khriplovich and his collaborators, and is clearly a nontrivial task as it involves nonperturbative QCD physics. It is nonetheless crucial in terms of extracting constraints, and in particular one would like to do much better than order of magnitude estimates so that the different dependencies of the observable EDMs may best be utilized in constrained models for new physics.

It is this problem that we will turn to next. In order to be concrete, we will limit our discussion to the nucleon EDMs and pion-nucleon couplings. The electron nucleon couplings, of which C_S plays the most important role for the EDM of paramagnetic atoms, receive contributions from the semi-leptonic four-fermion couplings C_{qe} in Eq. (13.76), which may be determined straightforwardly using low-energy theorems for the matrix elements of quark bilinears in the nucleon. (See Ref. [62].)

13.3.2. *QCD calculation of EDMs*

Since this is a nonperturbative QCD problem, the tools at our disposal are limited. Ultimately, the lattice may provide the most systematic treatment, but for the moment we are limited to various approximate methods. While, one can make use of various models of the infrared regime of QCD, we prefer here to limit our discussion to three (essentially) model-independent approaches, which vary both in their level of QCD input, and in genericity as regards the calculations to which they may be applied.

However, we will first recall what is perhaps the most widely used approach for estimating the contribution of quark EDMs to the EDM of the neutron. This is the use of the SU(6) quark model, wherein one associates a nonrelativistic wave function to the neutron which includes three constituent quarks and allows for the two spin states of each. The contribution of quark EDMs to d_n then amounts to evaluating the relevant Clebsch–Gordan coefficients and one finds,

$$d_n(d_q)^{\mathrm{QM}} = \frac{1}{3}(4d_d - d_u). \tag{13.78}$$

Although one may raise many questions regarding the reliability, and expected precision, of this result, we will emphasize here only the significant disadvantage that this approach cannot be used for a wider class of CP-odd sources, relevant to the generation of d_n.

13.3.2.1. *Naive dimensional analysis*

Although historically not the first, conceptually the simplest approach is a form of QCD power-counting which goes under the rather unassuming name of "naive dimensional analysis" (NDA) [63]. This is a scheme for estimating the size of some induced operator by matching loop corrections to the tree level term at the specific scale where the interactions become strong. In practice, one uses a dimensionful scale $\Lambda_{\mathrm{had}} \sim 4\pi f_\pi$ characteristic of chiral symmetry breaking, and a dimensionless coupling $\Lambda_{\mathrm{had}}/f_\pi$ to parametrize the coefficients. The claim is that, to within an order of magnitude, the dimensionless "reduced coupling" of an operator below the scale Λ_{had} is given by the product of the reduced couplings of all operators in the effective Lagrangian above Λ_{had} which may generate it. The reduced couplings are determined by demanding that loop corrections match the tree level terms, and for the coefficient $c_{\mathcal{O}}$ of an operator $\mathcal{O}$ of dimension D, containing N fields, is given by $(4\pi)^{2-N}\Lambda_{\rm had}^{D-4}c_{\mathcal{O}}$. A crucial, and often rather delicate, point is the precise scale at which one should perform this matching. Within the quark sector, the identification of this scale with $\Lambda_{\rm had}$ often seems to work quite well. However, for gluonic operators, the implied matching occurs at a very low scale where g_s is very large, up to $g_s \sim 4\pi$, and NDA has proved more problematic in this sector.

To illustrate this approach, let us consider the neutron EDM induced by θ, in this case realized as an overall phase θ_q of the quark mass matrix, and also the EDM and CEDM of a light quark. The dimension five neutron EDM operator has reduced coupling $d_n\Lambda_{\rm had}/(4\pi)$. Above the scale $\Lambda_{\rm had}$ we need the reduced couplings of the electromagnetic coupling of the quark, $e/(4\pi)$, and the CP-odd quark mass term, $\theta_q m_q/\Lambda_{\rm had}$. Thus we find,

$$d_n(\theta_q,\mu) \sim e\theta_q(\mu)\frac{m_q(\mu)}{\Lambda_{\rm had}^2}, \tag{13.79}$$

where the μ-dependence reflects the choice of matching scale. To obtain a similar estimate for the contribution of a light quark EDM, we note simply that it has a reduced coupling given by $d_d\Lambda_{\rm had}/(4\pi)$ and thus

$$d_n(d_q,\mu) \sim d_q(\mu), \tag{13.80}$$

which can be contrasted with the quark model estimate above. The contribution of the quark CEDM is similar, but one needs in addition the reduced electromagnetic coupling of the quark, $e/(4\pi)$, so that

$$d_n(\tilde{d}_q,\mu) = \frac{eg_s(\mu)}{4\pi}\tilde{d}'_q(\mu), \tag{13.81}$$

where we have redefined the CEDM operator so that $\tilde{d}_q = g_s\tilde{d}'_q$. This makes the factor g_s explicit, which seems crucial to the success of NDA for gluonic operators as the matching needs to be performed at a large value of g_s, e.g. $g_s \sim 4\pi$ as noted above.

These examples indicate, on one hand the simplicity of this approach and also its general applicability, but also the fact that it does not easily allow one to combine different contributions into a single result for the neutron EDM. In particular, these estimates have uncertain signs and thus can only be used independently with an assumption that the physics which generates them does not introduce any correlations. This will not generically be the case.

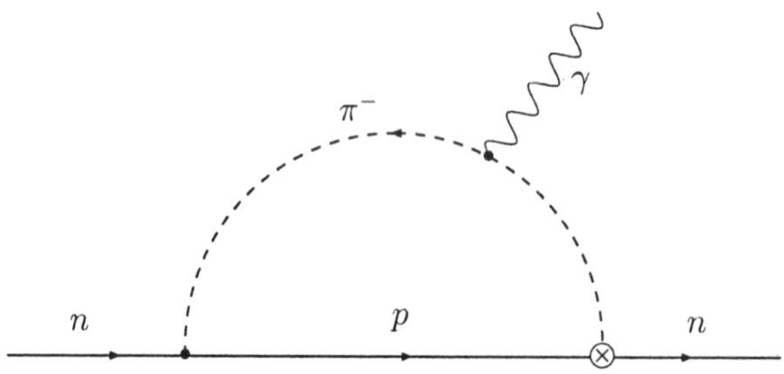

Fig. 13.2. Chirally enhanced contribution to the neutron EDM.

13.3.2.2. *Chiral techniques*

Historically, the first model-independent calculation of the neutron EDM [64] made use of chiral techniques to isolate an infrared log-divergent contribution in the chiral limit. This was one of the landmark calculations which made the strong CP problem, and indeed the magnitude of the required tuning of θ, quite manifest.

The basic observation was that, given a CP-odd pion-nucleon coupling $\bar{g}_{\pi NN}$, one could generate a contribution to the neutron EDM via a π^--loop (see Fig. 13.2) which was infrared divergent in the chiral limit. In reality this log-divergence is cutoff by the finite pion mass, and one obtains,

$$d_n^{\chi\log} = \frac{e}{4\pi^2 M_n} g_{\pi NN} \bar{g}_{\pi NN}^{(0)} \ln \frac{\Lambda}{m_\pi}, \tag{13.82}$$

where Λ is the relevant UV cutoff, i.e. $\Lambda = m_\rho$ or M_n. One can argue that such a contribution cannot be systematically canceled by other, infrared finite, pieces and thus the bound one obtains on $\bar{g}_{\pi NN}^{(0)}$ in this way is reliable in real-world QCD.

This reduces the problem to one of computing the relevant CP-odd pion-nucleon couplings. For a given CP-odd source $\mathcal{O}_{CP}$, we have

$$\langle N\pi^a|\mathcal{O}_{CP}|N'\rangle = \frac{i}{f_\pi}\langle N|[\mathcal{O}_{CP}, J_{05}^a]|N'\rangle + \text{rescattering}, \tag{13.83}$$

justified by the small t-channel pion momentum. The possible rescattering corrections will be discussed below. If we now specialize to the θ-term, as in [64], with $\mathcal{O}_{CP} = -\theta_q m_* \sum_f \bar{q}_f i\gamma_5 q_f$ then the commutator reduces to the triplet nucleon sigma term, and we find

$$\bar{g}_{\pi NN}^{(0)}(\theta_q) = \frac{\theta_q m_*}{f_\pi}\langle p|\bar{q}\tau^3 q|p\rangle \left(1 - \frac{m_\pi^2}{m_\eta^2}\right). \tag{13.84}$$

One can then determine $\langle N|\bar{q}\tau^a q|N\rangle$ from lattice calculations or, as was done in [64], by using chiral symmetry to relate it to measured splittings in the baryon octet.

The final factor on the right-hand side of Eq. (13.84) reflects the vanishing of the result in the limit that the chiral anomaly switches off and η (or η' in the three-flavor case) is a genuine Goldstone mode. This factor is numerically close to one and was ignored in [64]. It arises because in Eq. (13.83) we should also take into account the fact that the CP-odd mass term can produce η from the vacuum and thus, in addition to the PCAC commutator, there are rescattering graphs with η produced from the vacuum and then coupling to the nucleon, and the soft pion radiated via the CP-even pion-nucleon coupling [65].

Although this technique is not universally applicable, one can also contemplate computing the contribution of certain other sources, e.g. the quark CEDMs. Following the same approach, the induced CP-odd pion-nucleon (isovector) coupling depends on specific quark-gluon condensates over the nucleon, i.e. $\langle N|\bar{q}G\sigma q|N\rangle$, which are difficult to estimate. Moreover, as we will discuss below in more detail, the rescattering graphs are now very significant and not suppressed by m_π^2/m_η^2. They tend to reduce the relevant matrix elements making the estimates of $\bar{g}^{(1)}_{\pi NN}$ highly uncertain.

This limited applicability is one problem that currently afflicts the chiral approach. A more profound issue is that the terms enhanced by the chiral log, while conceptually distinct, are not necessarily numerically dominant. Indeed there are infrared finite corrections to Eq. (13.82) which, while clearly subleading for $m_\pi \to 0$, are not obviously so in the physical regime. This dependence on threshold corrections has been observed to provide a considerable source of uncertainty [66] .

13.3.2.3. *QCD sum-rules techniques*

An alternative to considering the chiral regime directly, is to first start at high energies, making use of the operator product expansion, and attempt to construct QCD sum rules [67] for the nucleon EDMs, or the CP-odd pion-nucleon couplings. This approach in principle allows for a systematic treatment of all the sources, and is motivated in part by the success of such approaches to the calculation of baryon masses [68] and magnetic moments [69]. For a recent review of some aspects of the application of QCD sum rules to nucleons, see Ref. [70].

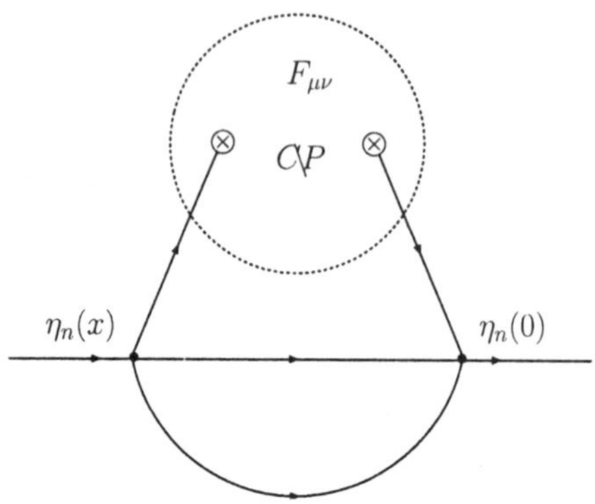

Fig. 13.3. A leading contribution to the neutron EDM within QCD sum rules. Sensitivity to the CP-violating source enters through the two soft quark lines which lead to a dependence on the chiral condensate.

- *Nucleon EDM calculations*

The basic idea is familiar from other sum-rules applications. One considers the two-point correlator of currents $\eta_N(x)$, with quantum numbers of the nucleon in question (e.g. a possible choice for the neutron is $2\epsilon_{abc}(d_a^T C\gamma_5 u_b)d_c$), in a background with nonzero CP-odd sources and an electromagnetic field $F_{\mu\nu}$,

$$\Pi(Q^2) = i\int d^4x e^{ip\cdot x}\langle 0|T\{\eta_N(x)\overline{\eta}_N(0)\}|0\rangle_{\not{CP},F}, \qquad (13.85)$$

where $Q^2 = -p^2$, with p the current momentum. One then computes the correlator at large Q^2 using the operator product expansion (OPE), generalized to incorporate condensates of the fields, and then matches this to a phenomenological parametrization corresponding to an expansion of the nucleon propagator to linear order in the background field and CP-odd sources, and corresponding higher excited states in the relevant channel. In practice, one makes use of a Borel transform to suppress the contribution of excited states, and then checks for a stability domain in Q^2, or rather the corresponding Borel mass M, where the two asymptotics may be matched. To isolate the EDM, it turns out that there is a unique Lorentz structure reducing to the nonrelativistic EDM which is chirally invariant. This structure is $\{F\sigma\gamma_5, \not{p}\}$ which is therefore the natural quantity on which to focus in constructing a sum rule for the EDM.

Although it would take us too far afield to describe this procedure in detail (see Ref. [10] for a more detailed review), we can exhibit some of the dominant physics by looking at just one class of diagrams which arise in evaluating the OPE for Eq. (13.85). In particular, in Fig. 13.3, two of the quarks in the nucleon current propagate without interference, carrying the large current momentum, while the third is taken to be soft and so induces a dependence on the chiral quark condensate. We may then make use of similar arguments to those of Section 13.2 to determine the dependence of this condensate on the CP-odd source. In particular, for the leading source of dimension four, namely the θ-term, we have

$$\begin{aligned} m_q\langle 0|\bar{q}\sigma_{\mu\nu}\gamma_5 q|0\rangle_{\theta,F} &= im_*\theta\langle 0|\bar{q}\sigma_{\mu\nu}q|0\rangle_F + \mathcal{O}(m_*^2) \\ &= i\chi e_q\theta m_* F_{\mu\nu}\langle\bar{q}q\rangle + \mathcal{O}(m_*^2), \end{aligned} \tag{13.86}$$

where in the first equality the dependence on θ has been determined as in the computation of $\langle G\tilde{G}\rangle_\theta$ in Section 13.2; while in the second we have introduced the so-called electromagnetic susceptibility of the vacuum, χ, defined via

$$\langle\bar{q}\sigma_{\mu\nu}q\rangle_F \equiv \chi e_q F_{\mu\nu}\langle\bar{q}q\rangle, \tag{13.87}$$

which numerically is rather large, $\chi = -(5-9)\ \mathrm{GeV}^{-2}$ [71, 72], and results in the diagram in Fig. 13.3 being numerically very important. We refer the reader to Refs. [65, 73–76] for more of the details involved in these calculations.

This contribution, among others, leads to a leading order sum-rules estimate for the the dependence on the θ-term of

$$d_n(\bar{\theta}) = -(0.8 \pm 0.4)e\,\chi m_*\bar{\theta}. \tag{13.88}$$

This result for d_n is numerically consistent with the determinations quoted above, although slightly smaller, and implies a bound $|\bar{\theta}| < 3\times 10^{-10}$.

To proceed to sources of higher dimension, one first needs to invoke some mechanism for supressing $\bar{\theta}$. If one assumes Peccei–Quinn (PQ) symmetry and uses a generic form of the invisible axion, it is important to recall, as noted in Section 13.2, that linear terms in the axion potential are generated in the presence of CP-odd sources which couple to $G\tilde{G}$. This is the case for CEDM sources, and these then imply an induced correction to $\bar{\theta}$ in the vacuum, given by [77]:

$$\theta_{\mathrm{ind}} = \frac{m_0^2}{2}\sum_{q=u,d,s}\frac{\tilde{d}_q}{m_q}, \tag{13.89}$$

where m_0^2 denotes the following condensate,

$$g_s\langle\bar{q}G\sigma q\rangle \equiv -m_0^2\langle\bar{q}q\rangle, \tag{13.90}$$

independent of the specific details of the axion mechanism. Consequently, we find additional vacuum contributions to the EDM.

If one now considers the dimension five CP-odd sources, the quark EDMs and CEDMs, this shift has the significant effect of precisely canceling the direct contribution from the s-quark CEDM at leading order in the OPE. The remaining contributions associated with the light quark EDMs and CEDMs, after PQ rotation, take the numerical form,

$$d_n^{\rm PQ}(d_q \tilde{d}_q) = (1 \pm 0.5)\frac{|\langle \overline{q}q \rangle|}{(225\text{MeV})^3}\left[1.1e(\tilde{d}_d + 0.5\tilde{d}_u) + 1.4(d_d - 0.25d_u)\right]. \quad (13.91)$$

Note also that an overall factor of $\langle \overline{q}q \rangle$ combines with the light quark masses from short-distance expressions for d_i and $\tilde{d}_i$ to give a result $\sim f_\pi^2 m_\pi^2 (1 + O(m_u/m_d))$, thus reducing the uncertainty due to poor knowledge of the absolute values of quark masses and condensates.

Compared to the techniques outlined previously, this approach has the significant advantage that all of the sources up to dimension five can be handled systematically and thus relative signs and magnitudes can be consistently tracked. As indicated in Eq. (13.91), one can also make a systematic estimate of the precision of the result, where the errors are due to contribution of excited states, neglected higher dimensional operators in the OPE, and also an ambiguity in the nucleon current.

Comparing the numerical result with those obtained using NDA and chiral techniques one finds, as is to be expected, that the results agree in terms of order of magnitude (in fact to within a factor of two in most cases). Although consistent within errors, it is worth noting that this suppression seems in accord with more recent lattice computations of nucleon tensor charges [78], which also indicate results somewhat below quark model estimates.

An important conceptual aspect of this approach is that it must combine the OPE with chiral determinations of the dependence of condensates on CP-odd sources. Although the relevant scales can be consistently separated, the use of nucleon currents with only valence quarks leads one to suspect that gluon and sea-quark contributions may be underestimated, since they enter only at higher orders. In this regard, it is worth noting that the contribution of the strange quark CEDM before imposing PQ symmetry is given by,

$$d_n(\tilde{d}_s) = (0.4 \pm 0.2) e \chi m_0^2 m_* \frac{\tilde{d}_s}{m_s}, \quad (13.92)$$

which, assuming $\tilde{d}_s \propto m_s$, is comparable to the contribution of the light quark CEDMs. This term is removed by PQ rotation at leading order, but one suspects that related contributions could re-enter at higher orders in the OPE. It is possible that the question of $d_n(\tilde{d}_s)$ may be resolved in future lattice simulations, given an appropriate lattice implementation of chiral symmetry.

In progressing to consider the contribution from sources of higher dimension, problems arise through the appearance of certain infrared divergences at low orders in the OPE, while a number of unknown condensates also enter and render a corresponding calculation for dimension six sources intractable. One can nonetheless estimate the contribution of these operators by utilizing a trick which involves relating the EDM contribution to the measured anomalous magnetic moment μ_n via the γ_5-rotation of the nucleon wave function induced by the CP-odd source [27],

$$d_n \sim \mu_n \frac{\langle N|\delta\mathcal{L}_{CP}|N\rangle}{m_n \bar{N} i\gamma_5 N}. \qquad (13.93)$$

One may analyze the γ_5-rotation using conventional sum-rules techniques, and for the Weinberg operator, one can obtain the following estimate [79]

$$|d_n(w)| \sim |\mu_n| \frac{3g_s m_0^2}{32\pi^2} w \ln(M^2/\mu_{IR}^2) \simeq e\ 22\ \text{MeV}\ w(1\ \text{GeV}), \qquad (13.94)$$

taking $M/\mu_{IR} = 2$, where M is the Borel mass and μ_{IR} is an infrared cutoff, and $g_s = 2.1$.

We can also apply this technique for the contribution of four-fermion operators. For SUSY models with generic parameters CP-odd four-fermion operators are negligible due to double helicity-flip requiring an m_q^2 dependence and rendering these operators effectively of dimension eight. However, for large $\tan\beta$, there are enhancements for operators proportional to C_{ij} with $i, j = d, s, b$ which can partially overcome this suppression thus altering the conventional picture of EDM sources (see Fig. 13.1). An important class of contributions in this case involves the four-ferimon operators with a b-quark. The contribution of these sources to d_n can again be estimated using the same technique as above [80],

$$|d_n(C_{ij})| \sim e\ 2.6 \times 10^{-3}\text{GeV}^2 \left(\frac{C_{bd}(m_b)}{m_b} + 0.75\frac{C_{db}(m_b)}{m_b}\right). \qquad (13.95)$$

We should emphasize that both the dimension six estimates above, necessarily have a precision not better than $\mathcal{O}(100\%)$, and one cannot reliably extract the sign. Fortunately, the numerical size of these dimension six

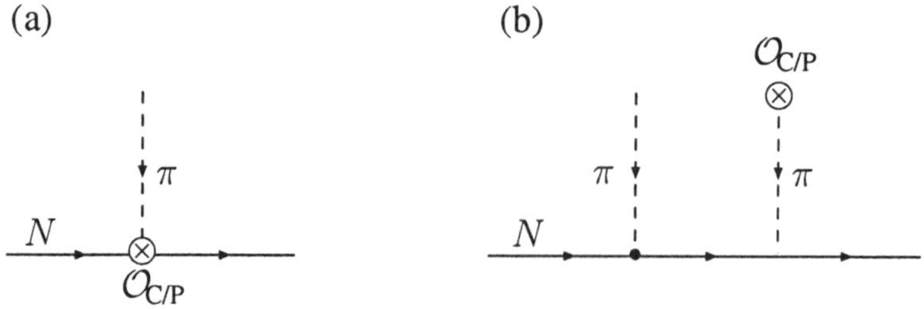

Fig. 13.4. Two classes of diagrams contributing to the CP-odd pion-nucleon coupling constant.

contributions is often negligible, and thus does not significantly impact the phenomenological application of EDM constraints.

- *Calculation of* $\bar{g}_{\pi NN}$

The other primary source of CP-odd nuclear moments, leading to the observable EDMs in diamagnetic atoms, arises through nucleon interactions mediated by pion-exchange with CP violation in the pion-nucleon vertex. As discussed in the preceding subsection, the calculation of these couplings involves essentially two steps: The first is a PCAC-type reduction of the pion in $\langle N\pi^a|\mathcal{O}_{CP}|N'\rangle$ as in Eq. (13.83), and the second is an evaluation of the resulting matrix elements over the nucleon. It is this second part for which QCD sum-rules may usefully be employed, and here we review its application to the computation of the dependence of $\bar{g}_{\pi NN}$ on dimension five CP-odd sources in Eq. (13.74) [81].

Assuming PQ symmetry, the dominant sources are the quark CEDMs which predominantly generate the isovector coupling $\bar{g}^{(1)}_{\pi NN}$ in Eq. (13.63) [59, 82]. However, as in our discussion above for θ, one needs to be aware of a subtlety for CEDM sources, first pointed out in [65, 75, 83], namely that a second class of contributions, the pion-pole diagrams (Fig. 13.4b), now contribute at the same order in chiral perturbation theory. In an alternative but physically equivalent approach, one can perform a chiral rotation in the Lagrangian to set $\langle 0|\mathcal{L}_{\not{C}P}|\pi^0\rangle = 0$, thus making this additional source of CP-violation explicit at the level of the Lagrangian and leading to the same result [81].

This subtlety leads to extra contributions in the sum-rule, which are straightforwardly incorporated, but more importantly leads to a precise cancellation of the result at leading order. This can easily be understood as the limit of vacuum saturation, since the relevant sources enter in the combination,

$$\langle N|\bar{q}g_sG\sigma q - m_0^2\bar{q}q|N\rangle. \tag{13.96}$$

This is a rather fundamental problem which limits the precision of this, and other chiral estimates [84, 85], for the dependence of $\bar{g}_{\pi NN}$ on the CEDMs; a point that we alluded to earlier on.

Within the sum-rule analysis, one can nonetheless trace the cancellations, and determine the residual contributions which are subleading numerically, but still enter at leading order in the OPE. Limiting our attention to the relevant isospin-one coupling, one finds the following result at next-to-next-to-leading order [81],

$$\bar{g}_{\pi NN}^{(1)} = 2_{-1}^{+4} \times 10^{-12}\frac{\tilde{d}_u - \tilde{d}_d}{10^{-26}\text{cm}}\frac{|\langle\bar{q}q\rangle|}{(225\text{MeV})^3}, \tag{13.97}$$

where the poor precision is essentially due to the cancellation of the leading terms. We emphasize that a more precise calculation of the matrix element Eq. (13.96) would significantly enhance the quality of constraints one could draw from the experimental bounds on diamagnetic EDMs, and thus constitutes a significant outstanding problem. Again, it seems this progress may have to wait for further developments in lattice QCD.

In considering the contribution of dimension six sources to $\bar{g}_{\pi NN}$, we note firstly that the three-gluon Weinberg operator $GG\tilde{G}$ is additionally suppressed by m_q and can be neglected. However, as for d_n, for SUSY models with large $\tan\beta$, certain four-fermion operators $C_{q_1q_2}$ may be relevant, and can be obtained via vacuum factorization, as the two diagrams in Fig. 13.4 now fail to cancel,

$$\bar{g}_{\pi NN}^{(1)} = -8\times 10^{-3}\text{GeV}^3\left(\frac{0.5C_{dd}}{m_d} + 3.3\kappa\frac{C_{sd}}{m_s} + \frac{C_{bd}}{m_b}(1-0.25\kappa)\right), \tag{13.98}$$

where $\kappa \equiv \langle N|m_s\bar{s}s|N\rangle/$ 220 MeV, with the preferred value $\kappa \simeq 0.5$ [62].

13.3.3. *EDMs in models of CP violation*

We have now moved to the highest level in Fig. 13.1, which is where the EDM constraints can be applied to directly constrain new sources of CP violation. Using the experimental upper limits, we obtain the following set of constraints on the CP-odd sources at 1 GeV (assuming an axion removes the dependence on $\bar{\theta}$):

$$\begin{aligned}
&\left| d_e + e(26\,\text{MeV})^2 \left(3\frac{C_{ed}}{m_d} + 11\frac{C_{es}}{m_s} + 5\frac{C_{eb}}{m_b} \right) \right| \\
&\qquad\qquad < 1.6 \times 10^{-27}\, e\text{cm} \quad \text{from } d_{Tl}, \\
&\left| (\tilde{d}_d - \tilde{d}_u) + \mathcal{O}(\tilde{d}_s, d_e, C_{qq}, C_{qe}) \right| \\
&\qquad\qquad < 2 \times 10^{-26}\,\text{cm} \quad \text{from } d_{Hg}, \\
&\left| e(\tilde{d}_d + 0.56\tilde{d}_u) + 1.3(d_d - 0.25 d_u) + \mathcal{O}(\tilde{d}_s, w, C_{qq}) \right| \\
&\qquad\qquad < 2 \times 10^{-26}\, e\text{cm} \quad \text{from } d_n,
\end{aligned} \tag{13.99}$$

where the additional $\mathcal{O}(\cdots)$ dependencies are known less precisely, but may not always be subleading in particular models. The precision of these results varies from 10–15% for the Tl bound, to around 50–100% for the neutron bound, and to a factor of a few for Hg. It is remarkable to note that, accounting for the naive mass-dependence $d_f \propto m_f$, all these constraints are of essentially the same order of magnitude and thus highly complementary.

In this section, we will briefly discuss these constraints, firstly looking at why the Standard Model itself provides such a small background, then discussing the motivation for new CP-odd sources from baryogenesis, and finally showing why most models of new physics, and supersymmetry in particular, tend to overproduce EDMs and are thus subject to stringent constraints.

13.3.3.1. *EDMs in the Standard Model*

The recent discovery and exploration of CP violation in the neutral B-meson system [6] is, along with existing data from CP violation observed in K-mesons, (within current precision) in perfect accord with the minimal model of CP violation known as the Kobayashi–Maskawa (KM) mechanism [3]. This introduces a 3×3 unitary quark mixing matrix V in the charged current sector of up- and down-type quarks taken in the mass eigenstate basis,

$$\mathcal{L}_{cc} = \frac{g}{\sqrt{2}} \left(\bar{U}_L W\!\!\!/^{+} V D_L + (\text{H.c.}) \right). \tag{13.100}$$

This model possesses a single CP-violating invariant in the quark sector, $J_{CP} = \text{Im}(V_{tb}V_{td}^{*}V_{cd}V_{cb}^{*}) \simeq 3 \times 10^{-5}$. This combination, as well as θ_{QCD} that we explored at length earlier on, are the only allowed sources of CP violation in the Standard Model (treating "Standard Model neutrinos" as

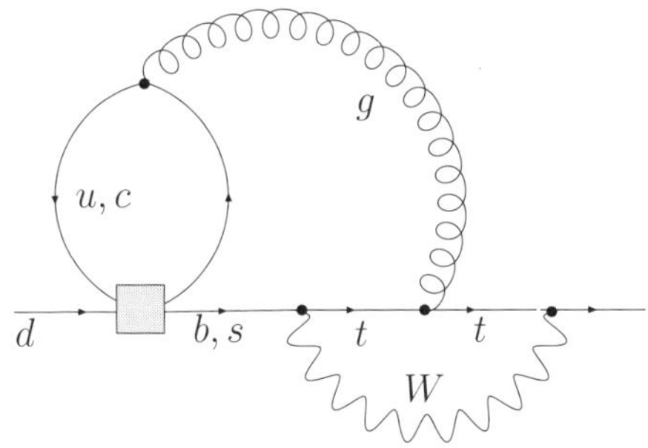

Fig. 13.5. A particular three-loop contribution [86] to the d-quark EDM induced by the KM phase in the Standard Model. The box vertex denotes a contacted W-boson line connected to the light quarks, while it is implicit that the external photon line is to be attached as appropriate to any charged lines.

massless). In addition to this, CP violation in the SM vanishes in the limit of an equal mass for any pair of two quarks of the same isospin, e.g. d and s, u and c, etc. These two conditions are extremely powerful in suppressing any KM-induced CP-odd flavor-conserving amplitude.

- *Quark and nucleon EDMs*

The necessity of four electroweak vertices requires that any diagram capable of inducing a quark EDM have at least two loops. Moreover, it turns out that all EDMs and color EDMs of quarks vanish exactly at the two-loop level [87], and only three-loop diagrams survive [86, 88], as in Fig. 13.5. A leading-log calculation of the three-loop amplitude for an EDM of the d-quark produces the following result [86],

$$d_d = e\frac{m_d m_c^2 \alpha_s G_F^2 J_{CP}}{108\pi^5} \ln^2(m_b^2/m_c^2) \ln(M_W^2/m_b^2). \tag{13.101}$$

Upon the inclusion of the other contributions, it produces a numerical estimate

$$d_d^{\text{KM}} \simeq 10^{-34} e \text{ cm}. \tag{13.102}$$

The only relevant operator that is not zero at two-loop order is the Weinberg operator [89], but its numerical value also turns out to be extremely small. Indeed the largest Standard Model contributon to d_n comes not from quark EDMs and CEDMs, but instead from a four-quark operator generated by a so-called "strong penguin" diagram shown in Fig. 13.6. This is enhanced by long-distance effects, namely the pion loop, and it has been

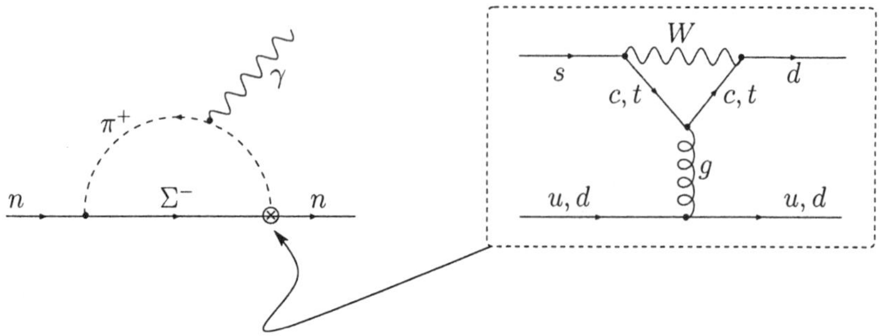

Fig. 13.6. A leading contribution to the neutron EDM in the Standard Model, arising via a four-quark operator generated by a strong penguin, and then a subsequent enhancement via a chiral π^+ loop.

estimated that this mechanism could lead to a KM-generated EDM of the neutron of order [90],

$$d_n^{\mathrm{KM}} \simeq 10^{-32} e \text{ cm}. \tag{13.103}$$

However, this is still six to seven orders of magnitude smaller than the current experimental limit.

- *Lepton EDMs*

The KM phase in the quark sector can induce a lepton via a diagram with a closed quark loop, but a non-vanishing result appears first at the four-loop level [91] and therefore is even more suppressed, below the level of

$$d_e^{\mathrm{KM}} \leq 10^{-38} e \text{ cm}, \tag{13.104}$$

and so small that the EDMs of paramagnetic atoms and molecules would be induced more efficiently by, for example, Schiff moments and other CP-odd nuclear moments.

In this regard, we note that recent data on neutrino oscillations points toward the existence of neutrino masses, mixing angles, and possibly of new CKM-like phase(s) in the lepton sector. Under the assumption that neutrinos are Majorana particles, the presence of these new CP-odd phases in the lepton sector allows for non-vanishing two-loop contributions to d_e [92], without any further additions to the Standard Model. However, recent calculations [93] show that a typical seesaw pattern for neutrino masses and mixings only induces a tiny contribution to the EDMs in this way, of $O(m_e m_\nu^2 G_F^2)$, unless a fine-tuning of the light neutrino masses is tolerated in which case d_e could reach $10^{-33}e$ cm. Therefore, within this minimal extension of the Standard Model allowing for massive neutrinos, the electron EDM is *not* the best way to probe CP violation in the lepton sector.

- *Probing the scale of new physics*

The Standard Model predictions for EDMs described above are well beyond the reach of even the most daring experimental proposals. This implies in turn that the Kobayashi–Maskawa phase provides a negligible background and thus any positive detection of an EDM would necessarily imply the presence of a non-KM CP-violating source. Before we consider some of the models which provide motivations for anticipating such a discovery, it will first be useful to consider in more general terms how high an energy scale one could indirectly probe with EDM meaurements. Indeed, we are led to ask firstly, what energy scale of new CP-violating physics is probed with the current experimental sensitivity to EDMs? Secondly, given the small KM background, we might also ask about the largest energy scale that could be probed *in principle* before reaching the level where the Standard Model KM contibutions would become significant.

To try and answer these questions in a systematic way, let us consider a toy model containing a scalar field ϕ (which is Higgs-like, but needn't be the SM Higgs) coupled to the SM fermions,

$$\mathcal{L}_\phi = \frac{1}{2}\partial^\mu\phi\partial_\mu\phi - \frac{1}{2}M^2\phi^2 - \sum_i \phi\bar{\psi}_i(y_i + iz_i\gamma_5)\psi_i. \tag{13.105}$$

Here we disregard possible flavor-changing effects but assume the presence of both scalar and pseudoscalar couplings y_i and z_i, the simultaneous presence of which breaks CP invariance. In a more realistic framework, this Lagrangian would have to be extended by other scalars so that CP violation could co-exist with $SU(2) \times U(1)$ gauge invariance. However, this simplified setting will be sufficient to consider the generic scales for CP violation probed by EDMs.

Assuming that the scalar mass M is large, we integrate this field out and match the resulting coefficients with the Wilson coefficients listed in Eq. (13.74) and Eq. (13.76). In particular, at tree level, ϕ-exchange generates the following dimension-six four-fermion operators,

$$C_{ij} = \frac{y_i z_j}{M^2}. \tag{13.106}$$

Running down to nuclear scales, these operators will among other things induce the electron-nucleon interaction C_S. In particular, taking the matrix element of the quark bilinear in the operators C_{ie}, and using $(m_u + m_d)\langle N|\bar{u}u + \bar{d}d|N\rangle/2 \simeq 45$ MeV in addition to the definition $\kappa \equiv \langle N|m_s\bar{s}s|N\rangle/$ 220 MeV, we obtain

$$C_S^{(0)} \simeq \frac{1}{M^2} z_e(3(y_d + y_u) + \kappa y_s + \cdots) \qquad (13.107)$$

where the dots stands for the contributions of heavier quark flavors, and the couplings are normalized at 1 GeV. If we now make the assumption that there is no correlation with other sources of CP violation, e.g. the electron EDM d_e, then with the use of the experimental constraint on $d_{\rm Tl}$ and the results of atomic claculations Eq. (13.66), we arrive at the following limit on the CP-odd combination of couplings and the mass M,

$$\frac{1}{M^2} \tilde{z}_e(y_d + y_u + 0.3\kappa y_s) \leq \frac{1}{(1.5 \times 10^6 \text{GeV})^2}. \qquad (13.108)$$

Given the most optimistic assumption about the possible size of these couplings, i.e. $z_e y_q \sim O(1)$, we can conclude that the current experimental EDM sensitivity translates to a bound on M as high as $M_{\rm CP} \sim 10^6$ GeV. If instead we insert the largest Kobayashi Maskawa predictions for the Tl EDM of order $\sim 10^{-35} e$ cm in place of the current sensitivity, we obtain $M_{\rm CP}^{\rm max} \sim 10^{11}$ GeV, as the ultimate scale which can be probed via these dimension six operators before the onset of the "KM background".

For comparison, allowing for arbitrarily large flavor-violating couplings of ϕ to fermions, we can also deduce the sensitivity level to New Flavor Physics (NFP) in a similar way. For example, requiring that four-fermion operators which change flavor by two units, e.g. $\bar{s}\gamma_5 d\bar{s}\gamma_5 d$ and the like, do not introduce new contributions to Δm_B, Δm_K and ϵ_K that are larger than the SM contribution, one typically finds that $M_{\rm NFP} > 10^7 - 10^8$ GeV in this scenario. Thus, we see that the sensitivity of EDMs is already approaching this benchmark and, unlike the contraints from the $\Delta F = 2$ sector, can be significantly improved in the near future.

At this point, we should emphasize that in this example, we are relaxing all constraints on the flavor structure by allowing order-one couplings of the scalar field ϕ to the light fermions. These couplings violate chirality maximally, and if Eq. (13.105) were part of a more realistic construction, for example a two-Higgs doublet model, one would expect that y_i and z_i will have to scale according to the fermion mass m_i. In this case, the sensitivity to M clearly drops dramatically, and the tree-level interactions Eq. (13.106) are not necessarily the dominant contributions to EDMs, as heavy flavors may contribute in a more substantial way via loop effects [94, 95]. Indeed, if new physics above the electroweak scale preserves chirality, as is often assumed, one expects that for the light flavors $d_i \sim e \times (1-10)\,\mathrm{MeV}/M^2$. Taking the electron EDM, and the Tl EDM bound, as a concrete example we find under this more restrictive assumption that

$$d_e \sim e \times \frac{m_e}{M^2} = 10^{-23} e\,\mathrm{cm} \times \left(\frac{1\,\mathrm{TeV}}{M}\right)^2 \quad \Longrightarrow \quad M_{\mathrm{CP}} \sim 70\,\mathrm{TeV}, \tag{13.109}$$

and consequently the current level of sensitivity to new CP-violating chirality-preserving physics drops somewhat, but for reference this scale is still well beyond the centre-of-mass energy of the LHC.

If we put the current EDM bounds into the broader context of precision tests of the Standard Model, we see that the present bounds in Table 13.2 imply that EDMs occupy an intermediate position in sensitivity to mass scales for new physics, between the electroweak precision tests (EWPTs) and very close to flavor violation in $\Delta F = 2$ processes noted above. The EWPTs from LEP impose a bound, $M_{\mathrm{EWPT}} >$ few TeV, through constraints on various dimension six operators, e.g. oblique corrections to gauge boson propagators, in combination with direct exclusion limits on the Higgs mass. Since $M_{\mathrm{EWPT}} \gg M_Z$, this has been dubbed the "little hierarchy problem" [99]. Indeed, while there are general expectations that the Standard Model is an effective theory, and will be corrected at scales of about a TeV, it is clear from this discussion that precision constraints in many sectors do not contain any hints on new physics beyond the Higgs at the weak scale, and in this sense EDMs are no exception. The remarkably large scale M_{CP} implied by EDM limits requires, at least within our level of understanding, a tuning in the CP-odd sector of physics beyond the Standard Model that we currently lack a coherent explanation for. The recent data from BaBar and Belle on CP violation in the neutral B-meson sector, which thus far is consistent with the KM model, within which CP violation is maximal within the confines of the flavor structure, only makes this tuning more pronounced, since we lack a strong motivation to enforce any additional CP-violating phases to be small. Moreover, further experimental progress in the near future could, given null results, push the value of M_{CP} close to and perhaps above M_{NFP}. From this viewpoint EDMs provide our most powerful tool in probing the question of whether CP-violation and flavor physics are intrinsically linked, as indeed they are within the electroweak Standard Model. This issue stands out as one of the most important ways in which EDMs may assist in demystifying some of the less constrained parts of the Standard Model.

13.3.3.2. *Baryogenesis and EDMs*

The search for EDMs provides an important test of the link between particle physics and cosmology. In hot Big Bang cosmology, the baryon asymmetry of the Universe, commonly parametrized by the baryon number density to entropy ratio η_b, is an input parameter. However, recent breakthroughs in observational cosmology point to the existence of an inflationary stage preceeding the hot Big Bang, during which the Universe is essentially in a baryo-symmetric state. This calls for an additional *dynamical* mechanism for generating the baryon asymmetry, i.e. baryogenesis, which has to occur at some point in the history of the Universe between inflation and Big Bang nucleosynthesis. The general criteria which allow for the dynamical generation of a baryon asymmetry from an initial baryo-symmetric state were formulated by Sakharov [7]. They include:

(1) Violation of baryon number,
(2) Departure from thermal equilibrium,
(3) Breaking of C and CP symmetries.

Remarkably, over the years it was realized that the Standard Model does contain all three ingredients. Baryon number fails to be conserved through a combination of nonperturbative thermal processes in the $SU(2)$ gauge sector and an anomaly in the baryon current, thus fulfilling condition (1). This allows for fluctuations of baryon number in the early Universe at $T \gtrsim 100$ GeV, while a combination of (2) and (3) provides a preferred direction for these fluctuations, which can favor baryons over antibaryons. A significant departure from thermal equilibrium in the Standard Model could occur during a first-order electroweak phase transition, as the expansion and cooling of the Universe implies the transition from a hot phase with the unbroken $SU(2) \times U(1)$ symmetry and vanishing Higgs v.e.v. to a broken phase with $\langle H \rangle \neq 0$. Finally, as discussed in the previous chapters, the SM has two sources of CP-violation, $\bar{\theta}$ and the KM phase in the quark mixing matrix.

Despite the existence of all three Sakharov ingredients within the SM, the resulting baryon asymmetry, η_b, that could be dynamically generated, falls more than ten orders of magnitude short of the baryon asymmetry that is observed experimentally. This is because existing Higgs searches point towards a heavier Higgs, $m_h > 100$ GeV, which is inconsistent with the requirement of a strongly first-order phase transition at $T \sim 100$ GeV. On top of that, neither of the two sources of CP violation present in the

SM are adequate for generating a sizable η_b. The KM phase is inefficient largely for the same reason that the SM EDMs are small, as in Eq. (13.101): High loop order, and a high degree of cancellation between different flavors, which becomes even more pronounced when the temperature exceeds the quark mass scale. The effect of $\bar{\theta}$, which is allowed to be order-one at the electroweak scale and later relaxed to zero by the axion mechanism, is proportional to the product of all quark masses normalized by the appropriate power of temperature, and thus is tiny.

The impossibility of having successful baryogenesis within the SM is a very strong motivation for anticipating new degrees of freedom that could enhance the departure from thermal equilibrium and for new sources of CP violation that could be probed with EDMs. Among a multitude of scenarios beyond the SM that allow for the successful generation of the observed baryon asymmetry, we would like to mention leptogenesis and electroweak baryogenesis. Leptogenesis is perhaps the most natural explanation for η_b: It utilizes new heavy degrees of freedom at the energy scale $\Lambda_\nu \sim v^2 m_\nu^{-1}$ suggested by the seesaw mechanism for light neutrino masses m_ν, where v is the electroweak v.e.v. Although there is possibly an interesting imprint of leptogenesis on the neutrino mixing matrix, there are no immediate consequences for EDMs because the energy scale Λ_ν of leptogenesis appears to be too high, although there could in principle be model-dependent indirect contributions. One possible exception is the $\bar{\theta}$ operator, which does not necessitate a power-like suppression by the new scale. In this case, however, the new CP-violating sources would simply renormalize the existing value of $\bar{\theta}$, and any positive detection of EDMs due to the θ-term would not directly relate to the parameters of leptogenesis.

Electroweak baryogenesis, on the other hand, must in its modern guise augment the shortfall of η_b in the SM by postulating new physics, i.e. new degrees of freedom and new sizable couplings, right at the electroweak scale. Such a framework is perfectly testable, both at colliders and by searching for EDMs. Below, we examine the relation between the cosmological quantity η_b, the particle physics parameters m_h and the scale of new physics below TeV, and the low-energy EDM observables within the framework of minimal electroweak baryogenesis.

13.3.3.3. *EDM constraints on minimal electroweak baryogenesis*

The existence of a mechanism for electroweak baryogenesis (EWBG) is an elegant feature of the SM gauge sector. The failure of the Standard Model to explain the precise baryon asymmetry via EWBG could then be viewed as a hint toward the presence of new physics, and indeed EWBG may still be realized in suitably tuned regions of supersymmetric extensions of the SM. The inherent testability of EWBG makes such modifications worthy of further study. In this regard, relatively simple extensions of the Standard Model (SM) Higgs sector, via the introduction of dimension-six operators [96, 97], have been argued to provide a rather minimal realization of consistent EWBG. The new threshold required is admittedly rather low, $\Lambda \sim 500 - 1000$ GeV [97], and this is therefore a scenario for which EDMs may be the most sensitive probes, and we will now briefly review the existing EDM constraints on this class of EWBG models [98].

We will focus on the SM augmented with the following dimension-six operators in the Higgs sector,

$$\mathcal{L}_{\text{dim }6} = \frac{1}{\Lambda^2}(H^\dagger H)^3 + \frac{Z^u_{ij}}{\Lambda^2_{\text{CP}}}(H^\dagger H)U^c_i HQ_j + \frac{Z^d_{ij}}{\Lambda^2_{\text{CP}}}(H^\dagger H)D^c_i HQ_j + \frac{Z^e_{ij}}{\Lambda^2_{\text{CP}}}(H^\dagger H)E^c_i HL_j. \quad (13.110)$$

The first term is required to induce a sufficiently strong first-order transition, while the remaining operators provide the additional source (or sources) of CP-violation. Although only a coupling to the top is strictly needed for EWBG, it is natural at this low scale to follow a framework such as minimal flavor violation and avoid the introduction of any new flavor structure, requiring $Z^{(u,d,e)}_{ij} = Z^{(u,d,e)}Y^{(u,d,e)}_{ij}$. We have introduced two threshold scales for the CP-even and CP-odd sectors, since they are distinguished according to the preserved symmetries.

The primary contributions to the fermion EDMs in this scenario arise at two-loops, in a manner very similar to the Barr–Zee diagrams in the 2HDM [95]. The contributions to d_f can be summarized as those arising from an effective pseudoscalar $hF\tilde{F}$ vertex and those arising from the scalar hFF vertex. The generalization to consider $hZ\tilde{F}$ and hZF vertices is then straightforward, although in fact the Z-mediated contributions are highly suppressed for d_e.

If we assume that the new CP-even and CP-odd physics lies at around the same threshold scale, we can set $\Lambda = \Lambda_{\text{CP}}$, the required baryon asymmetry corresponds to a precise contour in the remaining two-dimensional (Λ, m_h) parameter space. Within this two-dimensional parameter space, the EDM constraints then carve out excluded regions which generally favor lower Higgs mass values in the sense that the threshold Λ may be somewhat larger. This is presented in Fig. 13.7, where three η_b contours are contrasted with bounds from the Tl, Hg and neutron EDMs (the constraint from d_{Hg} is weaker in this case and does not appear on the plot). The contours of the baryon to entropy ratio η_b are labeled in units of the experimental value, taken to be $\eta_b = 8.9 \times 10^{-11}$. The EDM contours are set to *twice* the existing 1σ experimental bound, reflecting estimates for the theoretical precision, and we can interpret these contours as 1σ exclusions in parameter space. Note also that the plots refer to the situation with minimal flavor violation with the up-type phase equal to the down-type phase, which slightly reduces the EDM constraints due to partial cancellations.

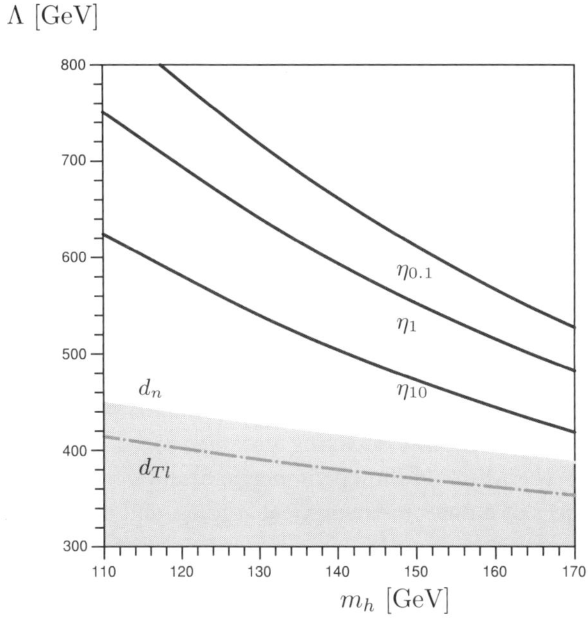

Fig. 13.7. Contours of η_b – labeled as η_x where $\eta_b/\eta_{\text{exp}} = x$ – and the EDMs over the Λ vs m_h plane, with correlated thresholds, $\Lambda_{\text{CP}} = \Lambda$. The shaded region is excluded by the EDMs, primarily the neutron EDM bound in this case. (This figure was reprinted with permission from Ref. [98]. Copyright 2007 by the American Physical Society.)

On general grounds, it is more natural to decouple the two thresholds, and further investigation [98] leads to quite a precisely defined viable region:

$$400 \text{ GeV} < \Lambda, \Lambda_{CP} < 800 \text{ GeV}. \tag{13.111}$$

As a consequence, the predictions for the level of sensitivity attainable in the next-generation EDM experiments have profound implications for these scenarios. If, for the moment, we lock $\Lambda_{\rm CP} = \Lambda$, then the predicted sensitivity attainable in next-generation searches for the electron and neutron EDMs would correspond to a threshold sensitivity of $\Lambda_{\rm CP} \sim 3$ TeV, over the relevant Higgs mass range, which is well beyond the viable region of parameter space for this mechanism of EWBG. Thus, even with a conservative treatment of the EDM precision, it seems clear that EWBG as realized in the form considered here, will be put to the ultimate test with the next generation of experiments.

13.3.3.4. *EDMs in supersymmetric models*

Having demonstrated the generic importance of EDM constraints for TeV-scale physics and for probing the mechanism of baryogenesis, we would now like to make this analysis more concrete by focusing on models with electroweak scale supersymmetry and reviewing their predictions for EDMs.

Supersymmetric extensions of the Standard Model provide perhaps the most natural solution to the gauge hierearchy problem by automatically cancelling the quadratically divergent contributions to the Higgs mass. Supersymmetry is thought of here as a symmetry of Nature at high energies, whereas at the electroweak scale and below it is obviously broken. Ensuring that supersymmetry breaking does not re-introduce quadratic divergences, and is compatible with the observed low energy spectrum, still allows for a large number of new dimensionful parameters, unfixed by any symmetry, that are usually called the soft breaking parameters. The minimal realization, known as the Minimal Supersymmetric Standard Model (MSSM), has been the subject of numerous theoretical studies, and also experimental searches, for over two decades. While no experimental evidence for SUSY exists, the MSSM retains a pre-eminent status among models of TeV-scale physics in part through several indirect virtues, e.g. gauge coupling unification and a "natural" dark matter candidate. For full details of the MSSM spectrum and the parametrization of the soft-breaking terms, we address the reader to any of the comprehensive reviews on MSSM phenomenology [100].

The unbroken sector of the MSSM contains, besides the gauge interactions, the Yukawa couplings parametrized by 3×3 Yukawa matrices in flavor space, $\mathbf{Y}_u$, $\mathbf{Y}_d$ and $\mathbf{Y}_e$. These matrices source the tree-level masses of matter fermions,

$$\mathbf{M}_u = \mathbf{Y}_u\langle H_2\rangle,\ \mathbf{M}_d = \mathbf{Y}_d\langle H_1\rangle,\ \mathbf{M}_e = \mathbf{Y}_e\langle H_1\rangle, \tag{13.112}$$

where $\langle H_1\rangle$ and $\langle H_2\rangle$ are two Higgs vacuum expectation values related to the SM Higgs v.e.v. via $\langle H_2\rangle^2 + \langle H_1\rangle^2 = v^2/2$. In SUSY models, anomaly cancellation in the Higgsino sector requires the introduction of at least two Higgs superfields as above. In addition to Yukawa couplings, the supersymmetry-preserving sector contains the so-called μ-term that provides a Dirac mass to the the higgsinos (the superpartners of the Higgs bosons) and contributes to the mass term of the Higgs potential,

$$V_{\text{Higgs}} = m_1^2|H_1|^2 + m_2^2|H_2|^2 + m_{12}^2 H_1 H_2 + |\mu|^2(|H_1|^2 + |H_2|^2) + \cdots, \tag{13.113}$$

where the dots denotes quartic terms fixed by supersymmetry and gauge invariance [100]. m_1^2, m_2^2 and m_{12}^2 are soft-breaking parameters that may attain negative values thus driving electroweak symmetry breaking. By suitable phase redifinitions of H_1 and H_2, one can restrict to real Higgs v.e.v.s and introduce the parameter, $\tan\beta = \langle H_2\rangle/\langle H_1\rangle$.

Among the remaining soft-breaking parameters one has to distinguish the gaugino mass terms and the squark and slepton masses,

$$-\mathcal{L}_{\text{mass}} = \frac{1}{2}\sum_{i=1,2,3} M_i\bar{\lambda}_i\lambda_i + \sum_{S=Q,U,D,L,E} \tilde{S}^\dagger \mathbf{M}_{\tilde{S}}^2 \tilde{S}, \tag{13.114}$$

where λ_i are the gaugino (Majorana) spinors, with i labeling the corresponding gauge group, U(1), SU(2) or SU(3). Each gaugino mass M_i can be complex. The second sum spans all the squarks and sleptons and contains five Hermitian 3×3 mass matrices in flavor space. Finally, the soft-breaking terms also include three-boson couplings allowed by gauge invariance, such as $\tilde{Q}H_2\mathbf{A}_u\tilde{U}$, that are called A-terms and are parametrized by three arbitrary complex matrices $\mathbf{A}_u$, $\mathbf{A}_d$ and $\mathbf{A}_e$. In the construction above we have limited the discussion to the R-parity conserving case, which only allows an *even* number of superpartners in each physical vertex, and is imposed to reduce problems with baryon number violation. Even with this restriction, if we count all the free parameters in this model we find a huge number, of $\mathcal{O}(100)$, with a few dozen new CP-violating phases! Truncating this number is fully justified only within the context of a fully specified supersymmetry breaking mechanism, which may then enforce additional symmetries and relations among parameters.

Without going into the details of the dynamics behind SUSY breaking, it will be enough for our purposes to simply assume that the following, very restrictive, conditions are fulfilled:

$$\begin{aligned} \mathbf{M}_S^2 &= m_S^2\mathbf{1}; \text{ for } S = Q, U, D, L, E, \quad \text{“degeneracy”} \\ \mathbf{A}_i &= A_i\mathbf{Y}_i; \text{ for } i = u, d, e, \quad \text{“proportionality”}. \end{aligned} \tag{13.115}$$

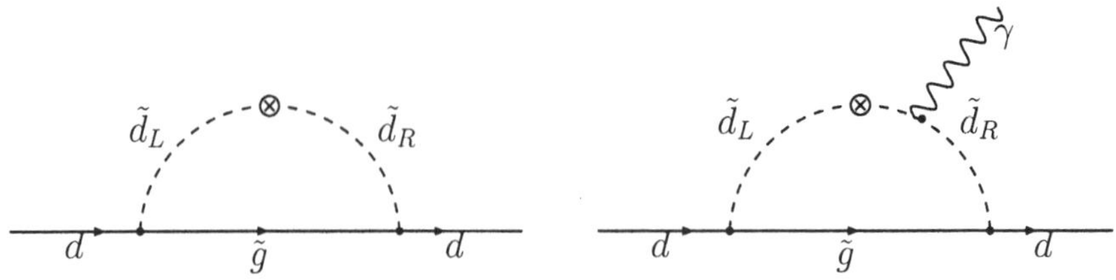

Fig. 13.8. One-loop SUSY threshold corrections in the down quark sector induced by a gluino-squark loop. On the left, a threshold correction generating $\text{Im}(m_d)$, while on the right the analogous diagram for the EDM. The CP-violating source enters via the highlighted vertex, squark-mixing in the present case.

Strictly speaking, such conditions can only be imposed (to a limited accuracy, due to threshold effects) at a specific normalization point above the weak scale, as the renormalization group evolution of the MSSM parameters will modify these relations. Moreover, these conditions can only be imposed with limited precision at this scale due to threshold effects. Nonetheless, such a restrictive flavor universality ansatz in the scalar mass sector, and proportionality of the trilinear soft breaking terms to the Yukawa matrices has the utility that it greatly reduces the number of independent soft-breaking parameters. Even so, a significant number of CP-violating phases remain,

$$\text{Arg}(m_{12}^2 M_i);\ \text{Arg}(\mu M_i);\ \text{Arg}(M_i A_j). \qquad (13.116)$$

Going to an even more restrictive framework, by assuming a common phase for the gaugino masses and another common phase for A_i reduces the number of independent CP-violating parameters to two. Using phase redefinitions, one can choose the phase of the gaugino mass to be zero, and use $\theta_A = \text{Arg}(A)$ and $\theta_\mu = \text{Arg}(\mu)$ as the basis for parametrizing CP violation.

It has been known for over twenty years that even in the absence of New Flavor Physics, large EDMs can be induced within one generation and at the one loop level [101, 102]. Thus, one should generically expect large EDMs as both of the reasons that made $d_i(\delta_{KM})$ small, namely high-loop order and also mixing angle/Yukawa coupling suppression, are not present for EDMs induced by the phases of the soft-breaking parameters.

Fig. 13.8 exhibits examples of one-loop diagrams at the supersymmetric threshold that generate nonzero contributions to the CP-odd Lagrangians Eq. (13.73) and Eq. (13.74). If we leave aside the problematic s-quark CEDM, then at one-loop we can concentrate on diagrams involving just the first generation of quarks and leptons. Within the parametrization described above, the phases residing in μ and A permeate the squark, selectron, chargino and neutralino spectrum, which in the mass eigenstate basis translates into complex phases in the quark-squark-gluino and fermion-sfermion-chargino(neutralino) vertices. To make this explicit, for a moment let us truncate the flavor space to one generation and write down the expression for the 2×2 d-squark mass matrix at the electroweak scale in the basis of $\tilde{d}_L$ and $\tilde{d}_R$,

$$M_{\tilde{d}}^2 = \begin{pmatrix} m_Q^2 + O(v^2) & -m_d(\mu \tan\beta + A_d^*) \\ -m_d(\mu^* \tan\beta + A_d) & m_D^2 + O(v^2) \end{pmatrix}, \tag{13.117}$$

where we further assume that the soft masses m_Q^2 and m_D^2 are large relative to the weak scale, and thus we can ignore subleading $O(v^2)$ corrections to the diagonal entries. Similar expressions can be written for the selectron mass matrix with the obvious substitutions in Eq. (13.117), and for the u squark, where in addition one has to exchange $\tan\beta$ by $\cot\beta$. In the generic case of three generations, M^2 becomes a 6×6 matrix with 3×3 blocks which are traditionally called M_{LL}^2, M_{LR}^2, M_{RL}^2 and M_{RR}^2. For our purposes, the crucial terms in Eq. (13.117) are the off-diagonal components,

$$(M_{\tilde{d}}^2)_{LR} = -m_d(\mu \tan\beta + A_d^*) \tag{13.118}$$

which contain the CP-odd phases. By virtue of being proportional to the small mass m_d, such a term can be treated as a perturbation and accounted for by an explicit mass insertion on the squark line, as in Fig. 13.7. Note that the natural range for $\tan\beta$, in the interval between 1 and 60, allows for a significant enhancement of the μ-dependent term in Eq. (13.117). The phase of μ also modifies the spectrum of charginos and neutralinos. Charginos, a common name for the mass-eigenstates comprising charged Winos (the superpartners of W-bosons) and Higgsinos, have a mass matrix in the gauge eigenstate basis given by

$$M_{\text{chargino}} = \begin{pmatrix} M_2 & \sqrt{2}M_W \sin\beta \\ \sqrt{2}M_W \cos\beta & \mu \end{pmatrix}. \tag{13.119}$$

In the limit $\mu, M_2 \gg M_W$, the off-diagonal terms can again be treated as a perturbation and accounted for by mass insertions. However, this is a much poorer approximation than for $(M_{\tilde{d}}^2)_{LR}$, and explicit diagonalization of Eq. (13.119) may be warranted in general. For this diagonalization, as well as for the explicit form of the neutralino mass matrix, we refer the reader to the existing MSSM reviews [100].

With this notation in hand we see that, for example, the squark-gluino loop diagram generates an imaginary d-quark mass correction that contributes to $\Delta\bar{\theta}_{\rm rad}$,

$$\operatorname{Im} m_d = -m_d \frac{\alpha_s}{3\pi} \frac{M_3(\mu \tan\beta \sin\theta_\mu - A_d \sin\theta_A)}{M_Q^2} I(M_3, m_Q, m_D). \quad (13.120)$$

The loop function I is normalized in such a way that $I(m,m,m) = 1$; its exact form (see Refs. [80, 103]) is not important for our discussion. The ratio $\operatorname{Im}(m_d)/m_d$, along with contributions from other quark flavors, represent a one-loop renormalization of the θ–term. It is important to observe that it only depends on the SUSY mass ratio and thus does not decouple if $A, \mu, M_3, m_{Q(D)}$ are pushed far above the electroweak scale. Applying the bound on the θ–term to the combined tree level and one-loop results Eq. (13.120), with degenerate SUSY mass parameters as above, we find $|\bar{\theta}_{\rm tree} + 10^{-2}\delta_{\rm CP}| < 10^{-9}$, where $\delta_{\rm CP}$ is a linear combination of $\sin\theta_\mu$ and $\sin\theta_A$ with $\mathcal{O}(1)$ coefficients. If there is no axion and $\bar{\theta}_{\rm tree}$ vanishes instead by symmetry arguments, it follows that the phases of the soft-breaking parameters must be tuned to within a factor of 10^{-7} in order to satisfy the EDM bounds. Therefore, an incredibly tight constraint on the phases of the SUSY soft-breaking parameters can be obtained in models which invoke high-scale symmetries to resolve the strong CP problem.

However, if the PQ symmetry removes the θ–term, such radiative corrections to $\bar{\theta}$ have no physical consequences, and the residual EDMs are determined by higher-dimensional operators. The relevant expressions for the one-loop-induced d_i and $\tilde{d}_i$ contributions can be found in Ref. [103]. Here we would just like to demonstrate the main point implied by these SUSY EDM calculations in a simplistic model in which *all* soft-breaking parameters are taken to the same value $M_{\rm SUSY}$ at the electroweak scale, i.e. $M_i = m_Q = m_D = \cdots = |\mu| = |A_i| = M_{\rm SUSY}$. Working at leading order in $v^2/M_{\rm SUSY}^2$, we can then present the following compact results for all dimension five operators (with $q = d, u$),

$$\begin{aligned}
\frac{d_e}{e\kappa_e} &= \frac{g_1^2}{12}\sin\theta_A + \left(\frac{5g_2^2}{24} + \frac{g_1^2}{24}\right)\sin\theta_\mu \tan\beta, \\
\frac{d_q}{e_q\kappa_q} &= \frac{2g_3^2}{9}\Big(\sin\theta_\mu[\tan\beta]^{\pm 1} - \sin\theta_A\Big) + O(g_2^2, g_1^2), \qquad (13.121) \\
\frac{\tilde{d}_q}{\kappa_q} &= \frac{5g_3^2}{18}\Big(\sin\theta_\mu[\tan\beta]^{\pm 1} - \sin\theta_A\Big) + O(g_2^2, g_1^2).
\end{aligned}$$

The notation $[\tan\beta]^{\pm 1}$ implies that one uses the plus(minus) sign for $d(u)$ quarks, g_i are the gauge couplings, and $e_u = 2e/3$, $e_d = -e/3$. For the quarks we quoted the explicit result only for the gluino-squark diagram that dominates in this limit. All these contributions to d_i are proportional to κ_i, a universal combination corresponding to the generic dipole size,

$$\kappa_i = \frac{m_i}{16\pi^2 M_{\rm SUSY}^2} = 1.3 \times 10^{-25}{\rm cm} \times \frac{m_i}{1{\rm MeV}} \left(\frac{1{\rm TeV}}{M_{\rm SUSY}}\right)^2, \qquad (13.122)$$

which varies by a factor of a few for $i = e, d, u$ depending on the value of the fermion mass. The perturbative nature of the MSSM provides a loop suppression factor in Eq. (13.122) so that κ_i is about two orders of magnitude smaller than the estimate Eq. (13.109). Correspondingly, the reach of the current EDM constraints in SUSY models cannot exceed the scale of a few TeV.

In Eq. (13.122) the quark masses should be normalized at the high scale, $M_{\rm SUSY}$. To make the explicit connection with the dipole operators in Eq. (13.74), the results of Eq. (13.121) should be evolved down to the low-energy normalization point of 1 GeV using the relevant anomalous dimensions (see Ref. [80]). Plugging these results into the expressions for d_n, $d_{\rm Tl}$ and $d_{\rm Hg}$ and comparing them to the current experimental bounds, we arrive at a set of constraints on θ_A and θ_μ depending on $M_{\rm SUSY}$ and $\tan\beta$. In Fig. 13.9, we plot these constraints in the (θ_μ,θ_A)-plane for $M_{\rm SUSY} = 500$ GeV and $\tan\beta = 3$. The region allowed by the EDM constraints is at the intersection of all three bands around $\theta_A = \theta_\mu = 0$. One can observe that the combination of all three constraints strengthens the bounds on the phases, and protects against the accidental cancellation of large phases that can occur within one particular observable. Note that the uncertainties in the QCD calculation of $\bar{g}^{(1)}_{\pi NN}$, and the nuclear calculation of $S(g^{(1)}_{\pi NN})$, discussed earlier may affect the width of the $d_{\rm Hg}$ constraint band, but do not significantly change its slope on the (θ_μ, θ_A) plane.

Before we briefly review the most common approaches to address the "overproduction" of EDMs in supersymmetric models, for completeness we will briefly discuss some of the additional contributions which become important in certain parts of the parameter space, e.g. when $\tan\beta$ is large, a regime favored for consistency of the MSSM Higgs sector with the final LEP results [104]. One simple observation is that the EDMs of down quarks and electrons, induced by θ_μ at one-loop, grow linearly with $\tan\beta$ Eq. (13.121). However, at the two-loop level, there are additional contributions from the phase of the A-parameter which may also be $\tan\beta$–enhanced [105]. A typical representative of the two-loop family is presented in Fig. 13.10. At large $\tan\beta$ the additional loop factor can be overcome, and these two-loop

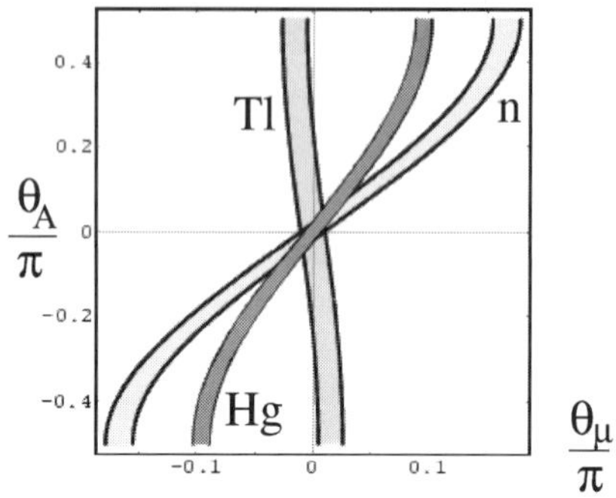

Fig. 13.9. The combination of three most sensitive EDM constraints, d_n, $d_{\rm Tl}$ and $d_{\rm Hg}$, for $M_{\rm SUSY} = 500$ GeV, and $\tan\beta = 3$. The region allowed by EDM constraints is at the intersection of all three bands around $\theta_A = \theta_\mu = 0$.

effects have to be taken into account alongside the one-loop contributions in Eq. (13.121). For example, the stop-loop contribution to the electron EDM in the same limit of a large universal SUSY mass is given by

$$d_e^{\rm two\ loop} = -e\kappa_e \frac{\alpha Y_t^2}{9\pi} \ln\left[\frac{M_{\rm SUSY}^2}{m_A^2}\right] \sin(\theta_A + \theta_\mu)\tan\beta, \tag{13.123}$$

where m_A is the mass of the pseudoscalar Higgs boson, that we took to be smaller than $M_{\rm SUSY}$, $Y_t \simeq 1$ is the top quark Yukawa coupling in the SM, and $\kappa_e \simeq 0.6 \times 10^{-25}$cm. For very large values of $\tan\beta$ additional contributions from sbottom and stau loops, which are enhanced by higher powers of $\tan\beta$, also have to be taken into account [80, 105].

Finally, the second, and in some sense more profound change is that at large $\tan\beta$, the observable EDMs of neutrons and heavy atoms receive contributions not only from the EDMs of the constituent particles, e.g. d_e and d_q, but also from CP-odd four-fermion operators [106]. The relevant Higgs-exchange diagram is given in Fig. 13.10. The CP violation in the Higgs-fermion vertex originates from the CP-odd correction to the fermion mass operator in Fig. 13.8. These diagrams, since they are induced by Higgs exchange, receive an even more significant enhancement by $(\tan\beta)^3$. In the same approximation as before, the value of the thallium EDM induced by this Higgs-exhange mechanism, and normalized to the current experimental limit, is given by

$$\frac{d_{\rm Tl}}{[d_{\rm Tl}]_{\rm exp}} \simeq \frac{\tan^3\beta}{330}\left(\frac{100{\rm GeV}}{m_A}\right)^2 \Big[\sin\theta_\mu + 0.04\sin(\theta_\mu + \theta_A)\Big]. \tag{13.124}$$

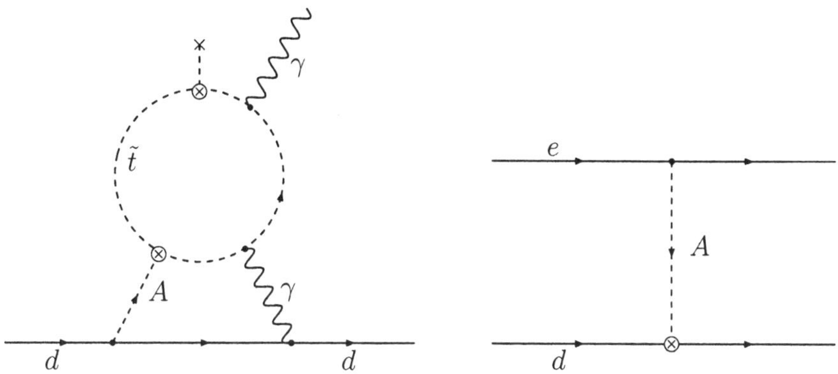

Fig. 13.10. Additional corrections to the EDMs. On the left, two-loop Barr–Zee type graphs mediated by a stop-loop and a pseudoscalar Higgs, while on the right we have a Higgs-mediated electron-quark interaction C_{de} with CP violation at the Higgs-quark vertex. There is a second diagram with CP violation at the Higgs-electron vertex mediated by H.

Notice that this result *does not* scale to zero as $M_{\rm SUSY} \to \infty$. Although just an $O(10^{-3} - 10^{-2})$ correction for $\tan\beta \sim O(1)$, these Higgs-exchange contributions become very large for $\tan\beta \sim O(50)$ [80, 106].

13.3.3.5. *The SUSY CP problem*

Figure 13.9 exemplifies the so-called SUSY CP problem: Either the CP-violating phases are small, or the scale of the soft-breaking masses is significantly larger than 1TeV, or schematically,

$$\delta_{CP} \times \left(\frac{1\text{TeV}}{M_{\rm SUSY}}\right)^2 < 1. \tag{13.125}$$

The need to provide a plausible explanation to the SUSY CP problem has spawned a sizable literature, and the following modifications to the SUSY spectrum have been discussed.

- *Heavy superpartners.* If the masses of the supersymmetric partners exhibit certain hierarchy patterns the SUSY CP problem can be alleviated. One of the more actively discussed possibilities is an inverted hierarchy among the slepton and squark masses, i.e. with the squarks of the first two generations being much heavier than the stops, sbottoms and staus, i.e. $(\mathbf{M}_S^2)_{ij} \gg (\mathbf{M}_S^2)_{i3}$, $(\mathbf{M}_S^2)_{33}$, where $i, j = 1, 2$ is the generation index [107]. It is preferable to have masses of the third generation sfermions under the TeV scale because they enter into radiative corrections to the Higgs potential, and making them too heavy would re-introduce the fine-tuning of the Higgs mass whose resolution was one of the primary motivations for weak-scale SUSY. Such a framework suppresses the one-loop EDMs which become immeasurably small if the scale of the u and d squarks is pushed all the way to ~ 50TeV, as suggested by the absence of SUSY contributions in $\Delta m_{K(B)}$. This does not mean, however, that the EDMs in such models become comparable to $d_i(\delta_{\rm KM})$. Indeed the two-loop contributions to d_i and w involving the third generation sfermions are not small in this framework, and indeed are at (or sometimes above) the level of current experimental sensitivity. Also, this means of suppressing the EDMs would not necessarily work in the large $\tan\beta$ regime where Higgs exchange may induce a large value for C_S that is not as sensitive to $M_{\rm SUSY}$ as the EDM operators. We note that future improvements in experimental precision will allow a stringent probe of such scenarios.

- *Small phases.* A rather obvious possibility for suppressing EDMs is the assumption of an exact (or approximate) CP symmetry of the soft-breaking sector. This is essentially a "model-building" option, and various ways of avoiding the SUSY flavor and CP problem in this way have been suggested in the past fifteen years [108–110]. The idea of using low-energy supersymmetry breaking looks especially appealing, as it can also help in constructing an axionless solution to the strong CP problem [39]. If the CP-odd phases in the soft-breaking sector are exactly zero and the conditions Eq. (13.115) are imposed *exactly* at the unification scale as a constraint on the high scale model, what is the scale of EDMs induced by SUSY diagrams due purely to $\delta_{\rm KM}$? Since such an MSSM framework would possess the same flavor properties as the SM, one expects proportionality to the same CP-odd invariant combination of mixing angles, namely $J_{\rm CP}$, and suppression by differences of Yukawa couplings [102]. Then it is easy to understand that the superpartner contributions to the down quark (chromo-)EDM will necessarily be suppressed by the equivalent of $\delta_{\rm CP} \sim JY_c^2 \sim 10^{-9}$, which is again six to seven orders of magnitude below current experimental capabilities, and thus not significantly larger than the EDMs induced in the SM [111].

- *Accidental cancellations.* Another possibility entertained in recent years [112] is the partial or complete cancellation between the contributions of several CP-odd sources to physical observables, thus allowing for $\delta_{\rm CP} \sim O(1)$ with $M_{\rm SUSY} < 1$ TeV. Since the number of potential CP-odd phases is large, and the superpartner mass spectrum is clearly unknown, one cannot exclude this possibility in principle. However, as we illustrated in Fig. PRfig8, d_n, $d_{\rm Tl}$ and $d_{\rm Hg}$ depend on different combinations of phases, and the possibility of such a cancellation looks improbable. A more thorough exploration of the MSSM parameter space in search of acceptable solutions that pass the EDM constraints was performed in [113–115].
- *No electroweak scale supersymmetry.* Of course, there is always the possibility that other mechanisms (or no easily identifiable mechansisms at all) lie behind the gauge hierarchy problem and the SM is a good effective theory valid up to energy scales *much larger* than 1 TeV. In this case there is no SUSY CP problem by definition. One of the recently suggested scenarios [116] exploits the possibility of a large number of electroweak vacua to invoke anthropic reasoning for selecting the "right" vacuum, thus side-stepping naturalness arguments for expecting new physics at the weak scale. Ref. [116] assumes that all the scalar superpartners are very heavy, but leaves gauginos and Higgsinos under a TeV, in order to preserve gauge-coupling unification and a dark matter candidate. This eliminates the one-loop induced EDMs, but leaves room for two-loop contributions [105, 117] generated by chargino loops via a diagram similar to that shown in Fig. 13.9 with A replaced by the light Higgs. This scenario can also be probed with the predicted sensitivity of future EDM experiments.

Given the large parameter space of supersymmetric models, and the fact that many of the leading contributions to EDMs do depend on details of the SUSY particle spectrum, the current situation might be better phrased as follows. Namely, to what extent can the MSSM spectrum be manipulated to avoid these leading order contributions, and at what level do the secondary constraints from numerous, and more robust, two-loop contributions and four-fermion sources limit ones ability to avoid the EDM constraints? Indeed, if we consider two extreme cases: (i) The 2HDM, where all SUSY fermions and sfermions are very heavy; and (ii) split SUSY, where all SUSY scalars are very heavy; one finds that while one-loop EDMs are suppressed, two-loop contributions are already very close to the current bounds. This bodes well for the ability of next-generation experiments to provide a comprehensive test of large SUSY phases at the electroweak scale, regardless of the detailed form of the SUSY spectrum.

13.3.3.6. *EDM constraints on new CP-odd supersymmetric thresholds*

Given the existing constraints on new CP violation in the soft-breaking sector reviewed above, and if SUSY is indeed discovered (e.g. at the LHC) but with no sign of phases in the soft sector, we may then ask about the ability of EDMs to detect new supersymmetric CP-odd thresholds. At low energies, such thresholds manifest themselves through various higher-dimensional operators, the most significant being of dimension five. At this order in supersymmetric theories, there are several well-known R-parity conserving operators associated with neutrino masses, H_uLH_uL, and baryon and lepton number violation, $UUDE$, $QQQL$ [121]. We will discuss the remaining operators at dimension five with regard to their impact on CP- (and flavor-) violating observables. We write the superpotential as

$$\begin{aligned}\mathcal{W} = \mathcal{W}_{\text{MSSM}} &+ \frac{y_h}{\Lambda_h} H_d H_u H_d H_u \\ &+ \frac{Y^{qe}_{ijkl}}{\Lambda_{qe}} (U_i Q_j) E_k L_l + \frac{Y^{qq}_{ijkl}}{\Lambda_{qq}} (U_i Q_j)(D_k Q_l) \\ &+ \frac{\tilde{Y}^{qq}_{ijkl}}{\Lambda_{qq}} (U_i t^A Q_j)(D_k t^A Q_l),\end{aligned} \tag{13.126}$$

where y_h, Y_{qe}, Y_{qq} and $\tilde{Y}_{qq}$ are dimensionless coefficients, the latter three being tensors in flavor space. A renormalizable realization of Eq. (13.126) can easily be obtained, e.g. the MSSM extended by a singlet N (the NMSSM) or an extra pair of heavy Higgses.

The phenomenological consequences of these dimension-five terms arise primarily from the dimension-three and -six operators obtained by integrating out the superpartners at the SUSY threshold, and we will now briefly discuss the resulting sensitivity to Λ_{qe} and Λ_{qq} in various experimental channels [122]. Of course, one of the most important issues is the flavor structure of the new couplings, Y^{qe}, Y^{qq} and $\tilde{Y}^{qq}$. Assuming that these coefficients are of order one, and do *not* factorize: $Y^{qe} \neq Y_u Y_e$, we should first determine the natural scale for Λ such that the corrections to SM fermion masses do not exceed their measured values.

Particle masses and θ-term: With the soft scale of $\mathcal{O}(300)$ GeV, one-loop corrections to fermion masses imply a naturalness scale of $\Lambda_{qe} > 10^7$ GeV by requiring that $\Delta m_e < m_e$. However, a strikingly high naturalness scale emerges from consideration of the corresponding shift of $\bar{\theta}$ and the existing bound on the neutron EDM,

$$\Delta\bar{\theta} \sim \frac{\text{Im } m_d}{m_d} \sim 10^{-10} \times \frac{10^{17} \text{ GeV}}{\Lambda^{qq}}, \tag{13.127}$$

which translates directly to an extremely strong bound on $\Lambda^{qq} \sim 10^{17}$ GeV in scenarios where $\bar{\theta} \simeq 0$ is engineered by hand using high scale symmetries. n.b.: This conclusion does not apply for the axion scenario.

Electric dipole moment constraints: At one-loop, one can also generate various CP-odd four-fermion operators at the SUSY threshold, e.g.

$$\mathcal{L}_{CP} = -\frac{\alpha_s \text{Im} Y^{qe}_{1111}}{6\pi\Lambda_{qe} m_{\text{susy}}} \left[(\bar{u}u)\bar{e}i\gamma_5 e + (\bar{u}i\gamma_5 u)\bar{e}e\right], \tag{13.128}$$

which in turn induces the CP-odd electron-nucleon interactions, $\mathcal{L} = C_S \bar{N}N\bar{e}i\gamma_5 e + C_P \bar{N}i\gamma_5 N\bar{e}e$, and one finds $C_S \sim 2 \times 10^{-4}/(1\text{GeV} \times \Lambda^{qe})$. The limits on C_S (and C_P) deduced from the Tl and Hg EDM bounds discussed in Section 13.3.13.1 then imply [122],

$$\Lambda_{qe} \geq 3 \times 10^8 \text{ GeV} \qquad \text{from Tl EDM} \tag{13.129}$$

$$\Lambda_{qe} \geq 1.5 \times 10^8 \text{ GeV} \qquad \text{from Hg EDM} \tag{13.130}$$

$$\Lambda_{qq} \geq 3 \times 10^7 \text{ GeV} \qquad \text{from Hg EDM.} \tag{13.131}$$

The last relation results from sensitivity to the CP-violating operators $(\bar{d}i\gamma_5 d)(\bar{u}u)$ leading to the Schiff nuclear moment and the Hg EDM. These are remarkably large scales, and indeed not far below the scales suggested by neutrino physics. In fact, the next generation of atomic/molecular EDM experiments may reach sensitivities sufficient to push Λ_{qe} into regions close to the suggested scale of right-handed neutrinos.

Semileptonic operators involving heavy quark superfields are in turn strongly constrained by two-loop corrections to the dipole amplitudes; the bound on d_{Tl} then implies $\Lambda_{qe} \geq 1.3 \times 10^8$GeV. It is important to note that the level of these constraints is quite comparable to the sensitivity achieved from constraints on lepton flavor violation, e.g. the bounds on $\mu \to e\gamma$ decay and and $\mu \to e$ conversion in nuclei, which also imply a sensitivity to $\Lambda^{qe} \sim 10^8$ GeV [122].

The high sensitivity to $QULE$ and $QUQD$ arises primarily because they can flip the light fermion chirality without Yukawa suppression. It would

Table 13.2. Sensitivity to the threshold scale. The naturalness bound on $\mathrm{Im}(Y^{qq})$ doesn't apply to the axionic solution of the strong CP problem, the best sensitivity to $\mathrm{Im}(y_h)$ is achieved at maximal $\tan\beta$, and the Hg EDM constraint on $\mathrm{Im}(Y^{qq})$ applies when at least one pair of quarks belongs to the 1st generation. This table was reprinted with permission from [122]. Copyright 2006 by the American Physical Society.

operator	sensitivity to Λ (GeV)	source
Y^{qe}_{3311}	$\sim 10^7$	naturalness of m_e
$\mathrm{Im}(Y^{qq}_{3311})$	$\sim 10^{17}$	naturalness of $\bar{\theta}$, d_n
$\mathrm{Im}(Y^{qe}_{ii11})$	$10^7 - 10^9$	Tl, Hg EDMs
Y^{qe}_{1112}, Y^{qe}_{1121}	$10^7 - 10^8$	$\mu \to e$ conversion
$\mathrm{Im}(Y^{qq})$	$10^7 - 10^8$	Hg EDM
$\mathrm{Im}(y_h)$	$10^3 - 10^8$	d_e from Tl EDM

then come as no surprise if $H_uH_dH_uH_d$ were to have little implication for CP and flavor-violating observables. Remarkably enough, it turns out that EDMs do exhibit a high sensitivity to $H_uH_dH_uH_d$ at large $\tan\beta$ through corrections to the Higgs potential, and in particular the effective shift of the m_{12}^2 parameter which enters the one-loop diagrams contributing to EDMs. The effect scales as $(\tan\beta)^2$ and provides significant sensitivity to Λ_h at large $\tan\beta$.

The full set of constraints is summarized in Table 13.2. Note that, since these effects decouple linearly, an increase in sensitivity by just two orders of magnitude would already start probing the scales relevant for neutrino physics.

13.3.3.7. *EDMs from flavor physics in SUSY models*

In addition to the examples of new thresholds above, EDMs can also serve as a sensitive probe of non-minimal flavor physics more generally, e.g. within the soft-breaking sector. Indeed, the assumptions of proportionality and universality in the soft-breaking sector Eq. (13.115) at a given high-energy scale are highly idealized, and are not expected to hold with arbitrary precision. In this subsection, we would like to show that EDMs are sensitive to flavor-changing terms in the soft-breaking sector, and provide significant constraints on SUSY models with non-minimal flavor structure (see for example Ref. [10]).

For concreteness, let us assume that Eq. (13.115) holds approximately, and the perturbations are small. Around the electroweak scale, and in the basis of diagonal quark mass matrices, the soft-breaking mass matrices can

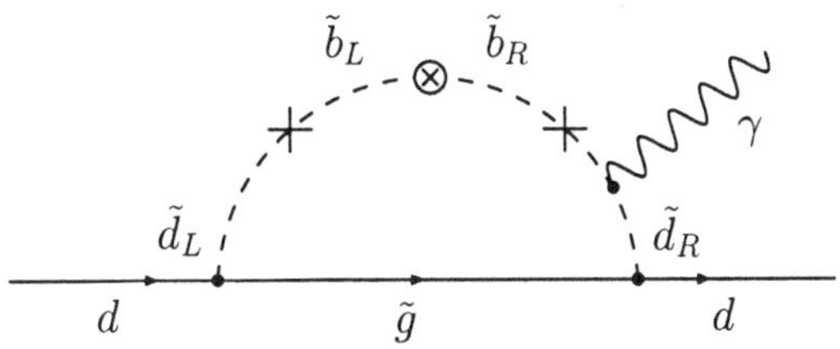

Fig. 13.11. Contribution of flavor changing processes to the d-quark EDM. The middle insertion on the sfermion line corresponds to LR mixing proportional to m_b; the insertions on the left and on the right correspond to flavor transitions in LL and RR squark mass sector.

be approximated as

$$\mathbf{M}_S^2 = \mathrm{diag}(m_{S11}^2, m_{S22}^2, m_{S33}^2) + \delta M_{Sij}^2, \tag{13.132}$$

where, as before, S labels the different squarks and sleptons, and $i \neq j$. Using this approximation, we can calculate the contributions to the relevant observables using δM_{Sij}^2 as a perturbation via insertions along the squark line, as in Fig. 13.11.

Calculating the gluino one-loop diagram in the approximation of equal SUSY masses, $(\mathbf{M}_S^2)_{ii} = M_i^2 = |\mu|^2 = M_{\rm SUSY}^2$, we arrive at the following result for the d-quark EDM, and the imaginary correction to the d-quark mass,

$$\begin{aligned} \mathrm{Im}\, m_d &= -\delta_{131}^d \times m_b \,\frac{\alpha_s \tan\beta}{18\pi} \\ d_d &= \delta_{131}^d \times e_d m_b \,\frac{\alpha_s \tan\beta}{45\pi M_{\rm SUSY}^2} \end{aligned} \tag{13.133}$$

where δ_{131}^d denotes the following CP-odd dimensionless combination,

$$\delta_{131}^d = \frac{\mathrm{Im}(\delta M_{Q13}^2 e^{i\theta_\mu} \delta M_{D31}^2)}{M_{\rm SUSY}^4}. \tag{13.134}$$

In Eq. (13.133), for simplicity, we neglected the contributions from the A parameters, and retained only the mixing coefficients between the first and the third generations. There are two important points about Eq. (13.133) that we should emphasize here: δ_{131} can be nonzero even if $\theta_\mu = 0$, and both $\mathrm{Im}(m_d)$ and d_d are enhanced relative to Eq. (13.120) and Eq. (13.121) by the large ratio $(m_b/m_d) \sim 10^3$, which can compensate the suppression associated with flavor violation. In the case of u quark operators, this enhancement factor is even larger, $m_t/m_u \sim 10^5$.

As we have seen in the previous subsection, renormalization of $\bar{\theta}$ ~Im$(m_q)/m_q$ can be very large, capable of producing bounds on δ^d_{131} and δ^u_{131} at the 10^{-9} level or better unless $\bar{\theta}$ is removed via PQ symmetry. In the latter case, using Eq. (13.133) and similar results in the lepton sector, one obtains the following sensitivity of EDMs to the above combination of flavor-changing transitions on electron, u, and d quark lines for $M_{\rm SUSY} = 1{\rm TeV}$,

$$\delta^e_{131} \sim 10^{-4} - 10^{-3}; \;\; \delta^u_{131} \sim 10^{-6} - 10^{-5}; \;\; \delta^d_{131} \sim 10^{-4} - 10^{-3}. \quad (13.135)$$

Thus, EDMs independently provide very stringent constraints on the combined sources of flavor- and CP violation in the soft-breaking sector. These constraints are complementary to those coming from K and B meson physics and searches for lepton flavor violation.

It is important to realize that the apparent enhancement of EDMs in Eq. (13.133) by the ratios of heavy to light quark and lepton masses occurred because of the presence of flavor-changing terms in *both* LL and RR sectors of the squark/slepton mass matrices. Indeed, to make this point transparent we can write

$$\delta^d_{131} = {\rm Arg}[(\delta^d_{13})_{LL}(\delta^d_{33})_{LR}(\delta^d_{31})_{RR}], \quad (13.136)$$

in terms of "mass insertions" [118] $\delta^f_{ij} = M^2_{\tilde{f}ij}/m^2_{\tilde{f}}$, which, although the distinction is not crucial here, are usually defined on a slightly different basis to the one we have been using. $m^2_{\tilde{f}}$ denotes here the average sfermion mass-squared. The status of the LL and RR insertions is in general rather different, and particularly so within the MSSM where the latter are essentially absent.

To see this in more detail, we recall that flavor-changing terms in the LL sector are natural, as they are induced by renormalization group evolution of the soft-breaking parameters (see Ref. [119]) even if one assumes the conditions Eq. (13.115) at the unification scale. Starting from the universal boundary conditions Eq. (13.115) for all scalar masses, equal to m_0^2, and A parameters at some high-energy scale Λ_{UV}, one can obtain the expression for $\mathbf{M}^2_Q$ at a lower energy scale Λ, which at one-loop is given by

$$\mathbf{M}^2_Q = m_0^2\,\mathbf{1} - \frac{3m_0^2 + A^2}{16\pi^2}\ln\left(\frac{\Lambda^2_{UV}}{\Lambda^2}\right)\left(\mathbf{Y}^\dagger_u\mathbf{Y}_u + \mathbf{Y}^\dagger_d\mathbf{Y}_d\right) + \cdots . \quad (13.137)$$

The dots denotes "flavor-blind" contributions and also higher-order terms. Depending on the particular model of SUSY breaking Λ_{UV} can be anywhere between a few tens of TeV and the Planck scale. The presence of *both* up and down Yukawa matrices in Eq. (13.137) guarantees the appearence of flavor-changing contributions in the LL entries of the squark mass matrices. At the superpartner threshold, $\lambda = M_{\rm SUSY}$, the flavor-changing terms in the down squark sector will evidently be

$$\delta M^2_{\tilde{d}ij} \simeq -\frac{3m_0^2 + A^2}{16\pi^2} \ln\left(\frac{\Lambda^2_{UV}}{M^2_{\rm SUSY}}\right) Y_t^2 V^*_{ti} V_{tj}, \tag{13.138}$$

where V is the CKM matrix, and Y_t is the top quark Yukawa coupling. If the scale Λ_{UV} is very high, i.e. comparable to the Planck or GUT scales, the logarithm is large and can entirely offset the loop factor. Therefore, the natural size of the 13 entry in the down squark LL sector is $\sim M^2_{\rm SUSY} V_{td} \simeq 0.01 M^2_{\rm SUSY}$.

The situation in the RR sector is completely different. There the absence of any $\mathbf{Y}_u$-dependence in the RG equations for $\mathbf{M}^2_D$ forbids the generation of substantial flavor-changing transitions, unless the MSSM spectrum is modified above certain energies so that the RG equations for the right-handed squark masses acquire flavor dependence.

A number of SUSY scenarios have been proposed which describe plausible patterns of small deviations from Eq. (13.115), allowing for significant RR contributions, and we would like to mention a few:

- *SO(10) unification with* $\Lambda_{UV} > \Lambda_{\rm GUT}$. If the running of the soft-breaking parameters extends *above* the unification scale, the RG equations are modified by the presence of new field degrees of freedom. For SO(10) GUTs this modification introduces significant flavor dependence in the RR sector of squark and slepton mass matrices [120], even if the restrictions Eq. (13.115) are imposed at the Planck scale. The resulting flavor-changing terms for down squarks δM^2_{Dij} are of the same order of magnitude as in the LL sector, leading to the prediction $\delta^d_{131} \sim 10^{-4}$, which is right at the borderline of current experimental sensitivity Eq. (13.135), [123, 124].
- *Heavy sterile neutrinos.* The light neutrino mass scale might, via the seesaw mechanism, be pointing to the existence of a new energy scale, $M_R \sim Y_\nu^2 v^2/(0.1 - 0.001 \text{ eV})$ related to heavy sterile (or "right-handed") neutrinos. If Λ_{UV} is larger than M_R, the RG equations for sleptons will be modified above M_R with an effect similar to that above, namely a non-trivial flavor dependence will be imprinted on the slepton mass matrices. The importance of such an effect will depend on the size of the neutrino Yukawa couplings Y_ν and, with certain Yukawa patterns, an observable or nearly-observable electron EDM might be induced [125]. Of course, if

the scale of SUSY breaking is lower than M_R (or Λ_{GUT}) there are no significant consequences for EDMs unless one allows for other "diagonal" phases in this sector.

- *Horizontal symmetries and alignment.* It might be that the hierarchy of quark and lepton masses and mixing angles *and* the suppressed flavor-changing effects in the sfermion sector find a common explanation. A candidate for such an explanation, namely SUSY-adapted "horizontal" flavor symmetries [126], might predict an approximate allignment of squark and quark mass matrices with the off-diagonal terms being suppressed by powers of Wolfenstein's parameter $\lambda \simeq 0.22$. In such models, EDMs can provide additional constraints that could help to assign the horizontal charges to different fields.
- *String-inspired models.* Some ideas for how to get an approximate flavor symmetry in string-derived models [127] have resulted in $O(0.01)$ predictions for the off-diagonal squark mass entries, which should have observable effects in EDMs at today's accuracy level. In a separate development, deviations from the proportionality condition $\mathbf{A} = A\mathbf{Y}$ have been investigated in certain string scenarios, with the conclusion that EDMs are often over-produced unless additional flavor symmetries are imposed [128].

13.4. Conclusions and Future Directions

Recent years have seen a dramatic improvement in our understanding of the origins of the CP violation observed in Nature. Experimental verification of direct CP violation in Kaon decay, but most of all the spectacular measurments of CP asymmetries for neutral B mesons at BaBar and Belle, has provided solid confirmation of the correctness of the Kobayashi–Maskawa mechanism. The current status of CP violation in flavor changing processes is such that (within errors) it does not necessitate the introduction of any additional CP-violating sources. At the same time, there is ample (experimental) room for the existence of new CP-violating physics for which the K and B meson data is not sensitive. This concerns, most of all, CP violation in flavor-conserving channels. The existence of such new sources is hinted at, albeit indirectly, by the baryon asymmetry in the Universe. The search for CP violation in flavor-conserving channels, and the search of EDMs in particular, should thus remain high on the priority list for particle physics. The strong suppression of EDMs that are induced purely by the Kobayashi–Maskawa phase, combined with prospects for improving the experimental sensitivity, places EDM searches at the forefront in probing CP-violating physics beyond the Standard Model.

We began our review by discussing the origin of the strong CP problem and some of its proposed resolutions. This issue has been with us for nearly thirty years and, while it can be resolved in some of the ways we discussed – most of the basic mechanisms for which have also been with us for much of that time – we have as yet no experimental verification of any of these ideas. From our subsequent discussion of EDMs and the manner in which they can be generated it becomes clear that the θ-term, in an effective field theory sense, is just one – albeit the most constrained – among a number of possible flavor-diagonal sources of CP violation beyond the Standard Model. In this sense, and beyond their direct sensitivity, current (and future) null results for EDM searches also provide very powerful constraints on models for new physics. Indeed, as we have discussed, the sensitivity for example to CP violation in the soft-breaking sector of SUSY models, allows us to probe soft-breaking masses as large as a few TeV. In this indirect sense, EDMs are often sensitive to energy scales beyond the reach of future collider experiments, and play a central role in the full suite of precision tests of the Standard Model. As discussed earlier on, the scales probed by EDMs and also by the constraints on flavor-changing neutral currents are not too dissimilar, and may come even closer with future progress on EDM searches. This only heightens the tension between the observed CP violation in the flavor-changing sector and the lack thereof in flavor-diagonal channels, of which the strong CP problem is the most manifest example. We seem compelled to question whether CP and flavor are as intrinisically linked in general as they are within the Kobayashi–Maskawa model? This is one aspect of what we might hope would be answered by a general "theory of flavor". EDMs will clearly continue to provide a crucial probe in tackling this question.

In the remaining pages of this review, we would like to emphasize some directions on the experimental and theoretical side that are likely to bring future progress in establishing the nature of CP violation at and above the electroweak scale.

- *Experimental Developments*

There are a number of experimental developments in techniques to search for EDMs which promise to narrow the gap between the current

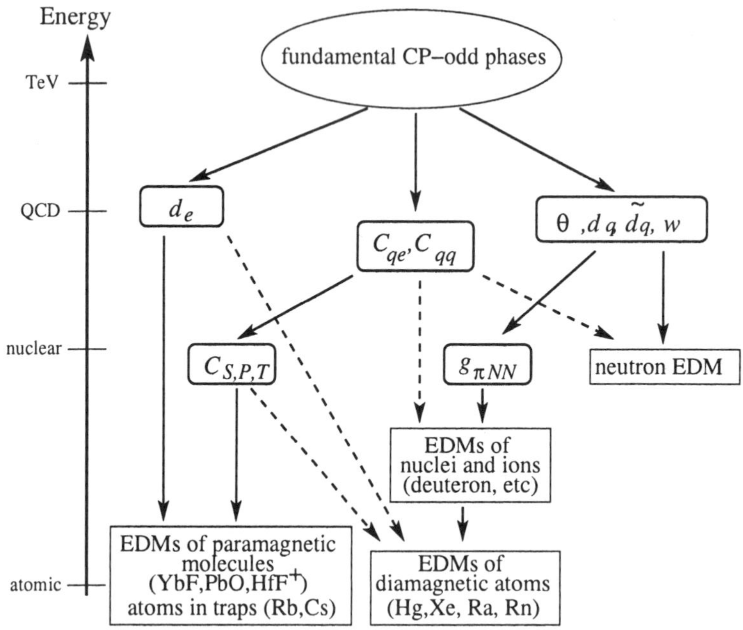

Fig. 13.12. A schematic (and futuristic) plot of the hierarchy of scales between the CP-odd sources and three generic classes of observable EDMs, as may apply in a few years when several new experiments come online.

limits and the KM background in all of the classes of EDMs discussed in this review. We will only briefly mention a few of them here as many will be covered in detail elsewhere in this volume. A forecast of how the generic sensitivities to CP-odd sources may be probed in a few years is shown in Fig. 13.12, in analogy with the current situation shown earlier in Fig. 13.1. This activity is occurring on many fronts, which means importantly that it covers each of the three primary EDM classes required to provide complementary information on the underlying sources of CP violation. Several new experiments aiming to probe the neutron EDM are in development using cryogenic techniques [130–132]. New proposals to search for CP-odd nuclear moments, falling into the diamagnetic class in terms of underlying sensitivity to, e.g. $\bar{g}_{\pi NN}$, include the study of exotic nuclei [134] with enhanced octopole moments [133] and also EDMs of charged nuclei such as the deuteron using storage rings [137]. Next-generation experiments probing paramagnetic EDMs are making use of polarizable molecules [51, 58], paramagnetic atoms in atoms traps, and also solid-state devices [135].

To place this activity in context, we should bear in mind probably the single most important question for particle physics – the origin of electroweak symmetry breaking – will be subjected to serious experimental scrutiny with the Large Hadron Collider coming online this year. Besides the discovery of the Higgs boson(s), it may provide an answer to the gauge hierarchy problem, and indeed uncover a plethora of new particles or resonances above the electroweak scale. EDM experiments, which might of course discover new physics inaccessible to the LHC, can also play a complementary role in providing constraints on (or signatures of) CP-violating couplings (e.g. in the Higgs sector of the MSSM). The projected level of sensitivity in coming years will be more than competitive in this regard with collider probes. Moreover, strangely enough, the absence of new physics (beyond the Higgs – or whatever plays this role) at the TeV scale would not remove motivations for EDM searches. Indeed, as we argued in this review, EDMs are sensitive to CP violation at multi-TeV scales, and thus represent one of the few classes of low-energy precision measurements that are sensitive to such high-energy scales. In fact, the future discovery of EDMs at new levels of sensitivity could, given an appropriate theoretical framework, point to the existence of new physics beyond the reach of the LHC, thus providing further motivation for the development of the next generation of colliders.

Another important experimental direction relevant to CP-violating physics is the search for axions. As we reviewed, one of the more natural resolutions of the strong CP problem predicts the existence of a light pseudo-scalar particle, the axion. The developments of recent years in cosmology have lent considerable weight to the presence of a non-baryonic cold dark matter component of the energy density in the Universe. Although the popularity of supersymmetric models continues to focus attention on the lightest supersymmetric particle (or LSP) as a natural dark matter candidate at the weak scale, axions with a coupling f_a^{-1} below its astrophysical bound in fact still represent a viable alternative, thus providing additional motivation for the continuation of axion searches.

- *Theoretical Developments*

On the theoretical side, beyond questions of the precise generation mechanisms of CP-odd sources in specific new physics models, it is clear that the primary limitation on the full application of the observational bounds arises through the limited precision of QCD and nuclear calculations. Perhaps the most afflicted quantity at present is the CP-odd pion-nucleon constant, as induced in particular by the CEDMs of light quarks. As we have discussed this is a fundamental parameter controlling the level of the constraints imposed by diamagnetic atoms, but can currently be calculated only to limited precision due to large cancellations in the relevant nucleon matrix elements. Another important issue concerns the strange quark CEDM contribution both to $\bar{g}_{\pi NN}$ and the neutron EDM d_n, and whether or not it is underestimated in the leading-order sum-rules analysis [83]. It would clearly be worthwhile to revisit these aspects. However, it seems likely that significant quantitative progress will come only from *ab initio* lattice calculations. This is a very challenging task, since a successful lattice calculation would necessarily have to respect chiral symmetry both at the level of quarks and gluons and also among the observable matrix elements between the hadronic states, since this is the underlying reason for the suppression of $d_n(\bar{\theta})$ by m_* and the partial suppression of $d_n(\tilde{d}_q)$. To that end, it will be important to implement a calculation displaying all the required symmetries, and in this sense $d_n(\bar{\theta})$ is a good starting point (see Ref. [138]), as many features of the answer, such as the dependence on m_* and on $\bar{\theta} = \theta + \arg \det M_q$, are enforced by symmetry allowing for independent checks of the calculation. On the nuclear side, we noted that recent re-analyses of the Schiff moment indicate that various many-body effects, e.g. polarization, can be significant and thus further progress in this area would assist significantly in improving the quality of constraints on $\bar{g}_{\pi NN}$ in different isospin channels. It will also be important, in guiding future experimental ideas, to clarify the size of the enhancement of CP violation in exotic nuclei with octupole deformations.

In conclusion, the limits on flavor-diagonal CP violation produced by the null results of existing EDM searches already provide strong constraints on new physics at and above the electroweak scale. New CP-violating physics is strongly motivated by the need for a viable mechanism for baryogenesis, and also by the genericity of phases in models of high-scale physics. Thus, developments in coming years promise to provide us with a wealth of new information about the nature of CP violation and TeV-scale physics, complementary to studies of electroweak symmetry breaking at colliders and flavor studies with K and B mesons.

Note added in proof

After the completion of this review, the Mercury EDM group published an updated limit [47] on the EDM of ^{199}Hg, $|d_{\rm Hg}| < 3 \times 10^{-29} e$ cm, significantly improving the bound by a factor of 7. Following the discussion above, this remarkable sensitivity now imposes stringent constraints on CP violation mediated via the Schiff moment, and pion-nucleon couplings, and has important implications for models of electroweak baryogenesis and supersymmetric scenarios. Please see Chapter 16 for further details.

Acknowledgments

We thank D. Demir, S. Huber, O. Lebedev, K. Olive, Y. Santoso, M. Shifman and A. Vainshtein for numerous helpful discussions/collaboration on some of the subjects discussed in this review. This work was supported in part by NSERC of Canada, and research at the Perimeter Institute is also supported in part by NSERC and by the Government of Ontario through MEDT. Figures 1-6 and 9-11 were reprinted from Ref. [10] with permission from Elsevier, Copyright 2005.

References

[1] T. D. Lee and C. N. Yang, Phys. Rev. **104**, 254 (1956); C. S. Wu, E. Ambler, R. W. Hayward, D. D. Hoppes and R. P. Hudson, Phys. Rev. **105**, 1413 (1957).
[2] J. H. Christensen *et al*, Phys. Rev. Lett. **13** 138 (1964).
[3] M. Kobayashi and T. Maskawa, Prog. Theor. Phys. **49**, 652 (1973).
[4] E. M. Purcell and N. F. Ramsey, Phys. Rev. **78**, 807 (1950).
[5] G. 't Hooft, Phys. Rev. Lett. **37**, 8 (1976); Phys. Rev. D **14**, 3432 (1976) [Erratum-*ibid.* D **18**, 2199 (1978)]; Phys. Rept. **142**, 357 (1986).
[6] B. Aubert *et al.* [BABAR Collaboration], Phys. Rev. Lett. **87**, 091801 (2001); K. Abe *et al.* [Belle Collaboration], Phys. Rev. Lett. **87**, 091802 (2001).
[7] A. D. Sakharov, Pisma Zh. Eksp. Teor. Fiz. **5**, 32 (1967) [JETP Lett. **5** 24 (1967 SOPUA,34,392-393.1991 UFNAA,161,61-64.1991)].
[8] R. D. Peccei and H. R. Quinn, Phys. Rev. Lett. **38**, 1440 (1977).
[9] I. B. Khriplovich and S. K. Lamoreaux, *CP Violation Without Strangeness*, Springer, 1997.
[10] M. Pospelov and A. Ritz, Annals Phys. **318**, 119 (2005) [arXiv:hep-ph/0504231].
[11] A. A. Belavin, A. M. Polyakov, A. S. Shvarts and Y. S. Tyupkin, Phys. Lett. B **59**, 85 (1975).

[12] R. Jackiw and C. Rebbi, Phys. Rev. Lett. **37**, 172 (1976).

[13] C. G. . Callan, R. F. Dashen and D. J. Gross, Phys. Lett. B **63**, 334 (1976).

[14] J. B. Kogut and L. Susskind, Phys. Rev. D **11**, 3594 (1975).

[15] R. J. Crewther, Phys. Lett. B **70**, 349 (1977); Riv. Nuovo Cim. **2N8**, 63 (1979).

[16] G. Veneziano, Nucl. Phys. B **159**, 213 (1979).

[17] D. Diakonov and M. I. Eides, Sov. Phys. JETP **54**, 232 (1981) [Zh. Eksp. Teor. Fiz. **81**, 434 (1981)].

[18] N. Dorey, T. J. Hollowood, V. V. Khoze and M. P. Mattis, Phys. Rept. **371**, 231 (2002) [arXiv:hep-th/0206063].

[19] E. Witten, Nucl. Phys. B **156**, 269 (1979).

[20] M. A. Shifman, A. I. Vainshtein and V. I. Zakharov, Nucl. Phys. B **166**, 493 (1980).

[21] F. Wilczek, Phys. Rev. Lett. **40**, 279 (1978).

[22] J. R. Ellis and M. K. Gaillard, Nucl. Phys. B **150**, 141 (1979).

[23] I. B. Khriplovich and A. I. Vainshtein, Nucl. Phys. B **414**, 27 (1994) [arXiv:hep-ph/9308334].

[24] C. Jarlskog, Phys. Rev. Lett. **55**, 1039 (1985).

[25] I. B. Khriplovich, Phys. Lett. B **173**, 193 (1986) [Sov. J. Nucl. Phys. **44**, 659 (1986) (1986 YAFIA,44,1019-1028.1986)].

[26] S. Weinberg, Phys. Rev. Lett. **40**, 223 (1978).

[27] I. I. Bigi and N. G. Uraltsev, Nucl. Phys. B **353**, 321 (1991).

[28] J. E. Kim, Phys. Rev. Lett. **43**, 103 (1979); M. A. Shifman, A. I. Vainshtein and V. I. Zakharov, Nucl. Phys. B **166**, 493 (1980).

[29] A. R. Zhitnitsky, Sov. J. Nucl. Phys. **31**, 260 (1980) [Yad. Fiz. **31**, 497 (1980)]; M. Dine, W. Fischler and M. Srednicki, Phys. Lett. B **104**, 199 (1981).

[30] D. B. Kaplan and A. V. Manohar, Phys. Rev. Lett. **56**, 2004 (1986).

[31] T. Banks, Y. Nir and N. Seiberg, [arXiv:hep-ph/9403203].

[32] J. Gasser and H. Leutwyler, Phys. Rept. **87**, 77 (1982); H. Leutwyler, Nucl. Phys. Proc. Suppl. **94**, 108 (2001).

[33] C. Aubin *et al.* [MILC Collaboration], Phys. Rev. D **70**, 114501 (2004) [arXiv:hep-lat/0407028].

[34] R. N. Mohapatra and G. Senjanovic, Phys. Lett. B **79**, 283 (1978).

[35] R. Kuchimanchi, Phys. Rev. Lett. **76**, 3486 (1996).

[36] R. N. Mohapatra and A. Rasin, Phys. Rev. Lett. **76**, 3490 (1996).

[37] A. E. Nelson, Phys. Lett. B **136**, 387 (1984).

[38] S. M. Barr, Phys. Rev. Lett. **53**, 329 (1984).

[39] G. Hiller and M. Schmaltz, Phys. Lett. B **514**, 263 (2001).

[40] B. Holdom, Phys. Rev. D **61**, 011702 (2000).

[41] G. F. Giudice and R. Rattazzi, Phys. Rept. **322**, 419 (1999).

[42] M. Dine, R. G. Leigh and A. Kagan, Phys. Rev. D **48**, 2214 (1993).

[43] M. E. Pospelov, Phys. Lett. B **391**, 324 (1997).

[44] P. G. Harris *et al.*, Phys. Rev. Lett. **82**, 904 (1999); C. A. Baker *et al.*, Phys. Rev. Lett. **97**, 131801 (2006) [arXiv:hep-ex/0602020].

[45] B. C. Regan *et al.*, Phys. Rev. Lett. **88**, 071805 (2002).

[46] M. V. Romalis, W. C. Griffith, J. P. Jacobs and E. N. Fortson, Phys. Rev. Lett. **86**, 2505 (2001).
[47] W. C. Griffith, M. D. Swallows, T. H. Loftus, M. V. Romalis, B. R. Heckel, E. N. Fortson, Phys. Rev. Lett. **102**, 101601 (2009).
[48] D. Cho, K. Sangster, E.A. Hinds, Phys. Rev. Lett. **63**, 2559 (1989).
[49] M. A. Rosenberry and T. E. Chupp, Phys. Rev. Lett. **86**, 22 (2001).
[50] S. A. Murthy *et al.*, Phys. Rev. Lett. **63**, 965 (1989).
[51] J. Hudson, B. E. Sauer, M. R. Tarbutt, and E. A. Hinds, Phys. Rev. Lett. **89**, 023003 (2002).
[52] L. I. Schiff, Phys. Rev. **132**, 2194 (1963).
[53] J. S. M. Ginges and V. V. Flambaum, Phys. Rept. **397**, 63 (2004).
[54] P. G. H. Sandars, Phys. Lett. B **14**, 194 (1965); **22**, 290 (1966).
[55] Z. W. Liu and H. P. Kelly, Phys. Rev. A **45**, R4210 (1992).
[56] A.-M. Martensson-Pendrill, *Methods in Computational Chemistry, Volume 5: Atomic, Molecular Properties*, ed. S. Wilson (Plenum Press, New York 1992).
[57] E. Lindroth, A.-M. Martensson-Pendrill, Europhys. Lett. **15**, 155 (1991).
[58] D. DeMille *et al.*, Phys. Rev. A **61**, 052507 (2000).
[59] V. F. Dmitriev and R. A. Sen'kov, [arXiv:nucl-th/0304048].
[60] J. H. de Jesus and J. Engel, Phys. Rev. C **72**, 045503 (2005) [arXiv:nucl-th/0507031].
[61] A. De Rujula, M. B. Gavela, O. Pene and F. J. Vegas, Nucl. Phys. B **357**, 311 (1991).
[62] H.-Y. Cheng, Phys. Lett. **B219**, 347 (1989); J. Gasser, H. Leutwyler, and M. E. Sainio, Phys. Lett. **B253**, 252 (1991); H. Leutwyler, [arXiv:hep-ph/9609465]; B. Borasoy and U. G. Meissner, Annals Phys. **254**, 192 (1997); M. Knecht, PiN Newslett. **15**, 108 (1999) [arXiv:hep-ph/9912443].
[63] A. Manohar and H. Georgi, Nucl. Phys. B **234**, 189 (1984).
[64] R. J. Crewther, P. Di Vecchia, G. Veneziano and E. Witten, Phys. Lett. B **88**, 123 (1979) [Erratum-ibid. B **91**, 487 (1980)].
[65] M. Pospelov and A. Ritz, Nucl. Phys. B **573**, 177 (2000) [arXiv:hep-ph/9908508].
[66] A. Pich and E. de Rafael, Nucl. Phys. B **367**, 313 (1991).
[67] M. A. Shifman, A. I. Vainshtein and V. I. Zakharov, Nucl. Phys. B **147**, 385 (1979).
[68] B. L. Ioffe, Nucl. Phys. **B188**, 317 (1981).
[69] B. L. Ioffe and A. V. Smilga, Nucl. Phys. **B232**, 109 (1984); I. I. Balitsky and A. V. Yung, Phys. Lett. **B129**, 328 (1983).
[70] D. B. Leinweber, Ann. Phys. **254**, 328 (1997).
[71] V. M. Belyaev and I. B. Ioffe, Sov. Phys. JETP **100**, 493 (1982); V. M. Belyaev and Ya. I. Kogan, Sov. J. Nucl. Phys. **40**, 659 (1984).
[72] A. Vainshtein, Phys. Lett. B **569**, 187 (2003).
[73] M. Pospelov and A. Ritz, Nucl. Phys. B **558**, 243 (1999) [hep-ph/9903553].
[74] M. Pospelov and A. Ritz, Phys. Rev. Lett. **83**, 2526 (1999) [arXiv:hep-ph/9904483].
[75] M. Pospelov and A. Ritz, Phys. Lett. B **471**, 388 (2000) [arXiv:hep-

ph/9910273].
[76] M. Pospelov and A. Ritz, Phys. Rev. D **63**, 073015 (2001) [hep-ph/0010037].
[77] I. I. Bigi and N. G. Uraltsev, Sov. Phys. JETP **100**, 198 (1991); M. Pospelov, Phys. Rev. **D58**, 097703 (1998).
[78] S. Aoki *et al.*, Phys. Rev. **D56** 433 (1997).
[79] D. A. Demir, M. Pospelov and A. Ritz, Phys. Rev. D **67**, 015007 (2003).
[80] D. A. Demir, O. Lebedev, K. A. Olive, M. Pospelov and A. Ritz, Nucl. Phys. B **680**, 339 (2004).
[81] M. Pospelov, Phys. Lett. B **530**, 123 (2002).
[82] M. A. B. Beg and H. S. Tsao, Phys. Rev. Lett. **41**, 278 (1978);
R. N. Mohapatra and G. Senjanovic, Phys. Lett. **B79**, 278 (1978);
R. N. Mohapatra and A. Rasin, Phys. Rev. **D54**, 5835 (1996);
M. E. Pospelov, Phys. Lett. **B391**, 324 (1997).
[83] T. Falk, K. A. Olive, M. Pospelov and R. Roiban, Nucl. Phys. B **560**, 3 (1999).
[84] V. M. Khatsimovsky, I. B. Khriplovich and A. S. Yelkhovsky, Annals Phys. **186**, 1 (1988).
[85] V. M. Khatsimovsky, I.B. Khriplovich and A.R. Zhitnitsky, Z. Phys. **C36**, 455 (1987).
[86] A. Czarnecki and B. Krause, Phys. Rev. Lett. **78**, 4339 (1997).
[87] E. P. Shabalin, Sov. J. Nucl. Phys. **28**, 75 (1978) [Yad. Fiz. **28**, 151 (1978)].
[88] I. B. Khriplovich, Phys. Lett. B **173**, 193 (1986).
[89] M. E. Pospelov, Phys. Lett. B **328**, 441 (1994).
[90] M. B. Gavela *et al.*, Phys. Lett. B **109**, 215 (1982);
I. B. Khriplovich and A. R. Zhitnitsky, Phys. Lett. B **109**, 490 (1982);
B. H. J. McKellar, S. R. Choudhury, X. G. He and S. Pakvasa, Phys. Lett. B **197**, 556 (1987).
[91] I. B. Khriplovich and M. E. Pospelov, Sov. J. Nucl. Phys. **53**, 638 (1991).
[92] D. Ng and J. N. Ng, Mod. Phys. Lett. A **11**, 211 (1996).
[93] J. P. Archambault, A. Czarnecki and M. Pospelov, Phys. Rev. D **70**, 073006 (2004).
[94] S. Weinberg, Phys. Rev. Lett. **63**, 2333 (1989).
[95] S. M. Barr and A. Zee, Phys. Rev. Lett. **65**, 21 (1990) [Erratum-ibid. **65**, 2920 (1990)].
[96] C. Grojean, G. Servant and J.D. Wells, *Phys. Rev.* **D71** (2005) 036001.
[97] D. Bödeker, L. Fromme, S. J. Huber and M. Seniuch, *JHEP* **0502** (2005) 026.
[98] S. J. Huber, M. Pospelov and A. Ritz, Phys. Rev. D **75**, 036006 (2007)
[99] R. Barbieri and A. Strumia, Phys. Lett. B **462**, 144 (1999); R. Barbieri, talk given at "Frontiers beyond the Standard Model", Minneapolis, October 10-12 (2002).
[100] H. E. Haber and G. L. Kane, Phys. Rept. **117**, 75 (1985); J. F. Gunion and H. E. Haber, Nucl. Phys. B **272**, 1 (1986); S. P. Martin, arXiv:hep-ph/9709356.
[101] J. R. Ellis, S. Ferrara and D. V. Nanopoulos, Phys. Lett. B **114**, 231 (1982);

W. Buchmuller and D. Wyler, Phys. Lett. B **121**, 321 (1983); J. Polchinski and M. B. Wise, Phys. Lett. B **125**, 393 (1983).
[102] Phys. Lett. B **125** (1983) 393; M. Dugan, B. Grinstein and L. Hall, Nucl. Phys. **B255** (1985) 413.
[103] T. Ibrahim and P. Nath, Phys. Lett. B **418**, 98 (1998); Phys. Rev. D **57**, 478 (1998); [Erratum-ibid. D **58**, 019901 (1998)]; Phys. Rev. D **58**, 111301 (1998).
[104] LEP Higgs Working Group for Higgs boson searches, OPAL Collaboration, ALEPH Collaboration, DELPHI Collaboration and L3 Collaboration, Phys. Lett. B **565** (2003) 61 [arXiv:hep-ex/0306033]. *Searches for the neutral Higgs bosons of the MSSM: Preliminary combined results using LEP data collected at energies up to 209 GeV*, LHWG-NOTE-2001-04, ALEPH-2001-057, DELPHI-2001-114, L3-NOTE-2700, OPAL-TN-699, arXiv:hep-ex/0107030; LHWG Note/2002-01, `http://lephiggs.web.cern.ch/LEPHIGGS/papers/July2002_SM/index.html`; J. R. Ellis, G. Ganis, D. V. Nanopoulos and K. A. Olive, Phys. Lett. B **502** (2001) 171; See also H. E. Haber, "Higgs theory and phenomenology in the Standard Model and MSSM," arXiv:hep-ph/0212136.
[105] D. Chang, W. Y. Keung and A. Pilaftsis, Phys. Rev. Lett. **82**, 900 (1999) [Erratum-ibid. **83**, 3972 (1999)]; A. Pilaftsis, Phys. Lett. B **471**, 174 (1999); D. Chang, W. F. Chang and W. Y. Keung, Phys. Rev. D **66**, 116008 (2002).
[106] O. Lebedev and M. Pospelov, Phys. Rev. Lett. **89**, 101801 (2002).
[107] A. G. Cohen, D. B. Kaplan and A. E. Nelson, Phys. Lett. B **388** (1996) 588.
[108] M. Dine, A. E. Nelson and Y. Shirman, Phys. Rev. D **51** (1995) 1362.
[109] L. Randall and R. Sundrum, Nucl. Phys. B **557**, 79 (1999).
[110] Z. Chacko, M. A. Luty, A. E. Nelson and E. Ponton, JHEP **0001** (2000) 003.
[111] A. Romanino and A. Strumia, Nucl. Phys. B **490** (1997) 3; C. Hamzaoui, M. Pospelov and R. Roiban, Phys. Rev. D **56**, 4295 (1997).
[112] T. Falk and K. A. Olive, Phys. Lett. B **375** (1996) 196; M. Brhlik, G. J. Good and G. L. Kane, Phys. Rev. D **59** (1999) 115004; M. Brhlik, L. L. Everett, G. L. Kane and J. Lykken, Phys. Rev. Lett. **83** (1999) 2124.
[113] V. D. Barger, T. Falk, T. Han, J. Jiang, T. Li and T. Plehn, Phys. Rev. D **64** (2001) 056007.
[114] S. Abel, S. Khalil and O. Lebedev, Phys. Rev. Lett. **86** (2001) 5850; Nucl. Phys. B **606** (2001) 151.
[115] M. Argyrou, A. B. Lahanas and V. C. Spanos, JHEP **0805**, 026 (2008) [arXiv:0804.2613 [hep-ph]].
[116] N. Arkani-Hamed and S. Dimopoulos, arXiv:hep-th/0405159.
[117] N. Arkani-Hamed, S. Dimopoulos, G. F. Giudice and A. Romanino, Nucl. Phys. B **709**, 3 (2005).
[118] L. J. Hall, V. A. Kostelecky and S. Raby, Nucl. Phys. B **267**, 415 (1986); F. Gabbiani and A. Masiero, Nucl. Phys. B **322**, 235 (1989).
[119] S. P. Martin and M. T. Vaughn, Phys. Rev. D **50**, 2282 (1994).
[120] S. Dimopoulos and L. J. Hall, Phys. Lett. B **344**, 185 (1995).

[121] S. Weinberg, Phys. Rev. D **26** (1982) 287; N. Sakai and T. Yanagida, Nucl. Phys. B **197** (1982) 533.

[122] M. Pospelov, A. Ritz and Y. Santoso, Phys. Rev. Lett. **96**, 091801 (2006); Phys. Rev. D **74**, 075006 (2006).

[123] R. Barbieri, L. J. Hall and A. Strumia, Nucl. Phys. B **445**, 219 (1995).

[124] I. B. Khriplovich and K. N. Zyablyuk, Phys. Lett. B **383**, 429 (1996).

[125] A. Romanino and A. Strumia, Nucl. Phys. B **622**, 73 (2002); J. R. Ellis, J. Hisano, M. Raidal and Y. Shimizu, Phys. Lett. B **528**, 86 (2002); A. Masiero, S. K. Vempati and O. Vives, Nucl. Phys. B **649**, 189 (2003); I. Masina, Nucl. Phys. B **671**, 432 (2003); Y. Farzan and M. E. Peskin, Phys. Rev. D **70**, 095001 (2004).

[126] Y. Nir and N. Seiberg, Phys. Lett. B **309**, 337 (1993).

[127] A. E. Faraggi and J. C. Pati, Nucl. Phys. B **526**, 21 (1998).

[128] S. Abel, S. Khalil and O. Lebedev, Nucl. Phys. B **606**, 151 (2001).

[129] *CP Violation*, ed. C. Jarlskog, Adv.Ser.Direct.High Energy Phys.3 (1989).

[130] S. N. Balashov *et al.*, arXiv:0709.2428 [hep-ex].

[131] R. Golub and K. Lamoreaux, Phys. Rept. **237**, 1 (1994); S. Lamoreaux *et al.*, talk at: *6th Conference on the Intersections of Particle and Nuclear Physics (CIPANP 97)*, Big Sky, MT, 27 May - 2 Jun. (1997).

[132] see e.g. `http://nedm.web.psi.ch/index.htm`.

[133] N. Auerbach, V. V. Flambaum and V. Spevak, Phys. Rev. Lett. **76**, 4316 (1996); V. V. Flambaum and V. G. Zelevinsky, Phys. Rev. C **68**, 035502 (2003).

[134] T. E. Chupp, talk at ITAMP workshop: *Tests of Fundamental Symmetries in Atoms and Molecules*, 29 Nov. – 1 Dec. (2001).

[135] S. K. Lamoreaux, [arXiv:nucl-ex/0109014].

[136] Y. K. Semertzidis *et al.*, arXiv:hep-ph/0012087; Y. K. Semertzidis *et al.*, Int. J. Mod. Phys. A **16S1B**, 690 (2001).

[137] Y. K. Semertzidis *et al.* [EDM Collaboration], AIP Conf. Proc. **698**, 200 (2004); F. J. M. Farley *et al.*, Phys. Rev. Lett. **93**, 052001 (2004).

[138] F. Berruto, T. Blum, K. Orginos and A. Soni, Phys. Rev. D **73**, 054509 (2006) [arXiv:hep-lat/0512004]; E. Shintani *et al.*, Phys. Rev. D **75**, 034507 (2007) [arXiv:hep-lat/0611032].

Chapter 14

The Electric Dipole Moment of the Electron

Eugene D. Commins

Physics Department
University of California at Berkeley
Berkeley, CA 94720

David DeMille

Department of Physics
Yale University
New Haven, CT 06520

Contents

14.1. Introduction

14.1.1. *Overview of relevant particle theory*

No experimental evidence exists for the electron electric dipole moment (EDM), despite nearly a half-century of search. However, laboratory attempts to find the electron EDM d_e, and EDMs of other particles and nuclei, attract more interest now than ever before. Indeed, in this chapter we shall mention more than a dozen present-day searches or proposed searches for d_e. This extraordinary effort is invested for a very good reason: observation of a non-zero d_e would give definite evidence for physics beyond the Standard Model, and might well illuminate the path taken by that New Physics. Let us explain why this is so.

First of all, no EDM can exist unless both parity (P) and time reversal (T) invariance are violated [1]. To see this we consider a particle of spin 1/2, for example an electron, and assume that it possesses an electric dipole moment $\mathbf{d}$ as well as a magnetic dipole moment $\boldsymbol{\mu}$. Both moments must lie along the spin direction because the spin is the only vector available to orient the particle.[a] The Hamiltonians H_M, H_E that describe the interactions of $\boldsymbol{\mu}$ with a magnetic field $\boldsymbol{B}$, and of $\mathbf{d}$ with an electric field $\boldsymbol{\mathcal{E}}$, respectively, take the following forms in the non-relativistic limit:

$$H_M = -\boldsymbol{\mu} \cdot \boldsymbol{B} = -\mu\boldsymbol{\sigma} \cdot \boldsymbol{B} \tag{14.1}$$

$$H_E = -\boldsymbol{d} \cdot \boldsymbol{\mathcal{E}} = -d\boldsymbol{\sigma} \cdot \boldsymbol{\mathcal{E}} \tag{14.2}$$

where $\boldsymbol{\sigma}$ is the Pauli spin operator. Under a parity transformation the axial vectors $\boldsymbol{\sigma}$ and $\boldsymbol{B}$ remain invariant, but the polar vector $\boldsymbol{\mathcal{E}}$ changes sign. Under a time reversal transformation $\boldsymbol{\mathcal{E}}$ remains invariant, but $\boldsymbol{\sigma}$ and $\boldsymbol{B}$ change sign. Hence, while H_M is invariant under P and T transformations, H_E is invariant under neither transformation.

Now P *is* violated in weak interactions (as is charge conjugation (C) symmetry). Also, the combined symmetry CP is violated in the decays of neutral K and B mesons [2, 3], and this CP violation is equivalent to T violation, assuming CPT invariance. Hence the existence of a P,T-violating

[a]For a particle with internal structure (e.g. an atom or molecule) one can define an electric dipole moment operator $\mathbf{d}$. If the particle is in a state of definite J, J_z, the Wigner–Eckhart theorem ensures that the expectation value of $\mathbf{d}$ must lie along $\langle \boldsymbol{J} \rangle$.

EDM appears quite possible: CP violation and the weak interaction can act jointly to generate an EDM by means of P,T-odd radiative corrections to the P, C, T-conserving electromagnetic interaction. At present the experimental upper limit on the electron EDM d_e is [4]:

$$|d_e| \leq 1.6 \cdot 10^{-27} e \text{ cm} \tag{14.3}$$

where $e = 4.8 \cdot 10^{-10}$ esu is the unit of electronic charge.

14.1.1.1. *Electron EDM in the Standard Model*

Let us compare this limit with what might be expected for d_e from those P,T-odd radiative corrections we have just mentioned.[b] We start with the Standard Model. It is well known that the quark mass eigenstates d, s, b are not identical with the corresponding weak interaction eigenstates. In the Standard Model, this is described by writing the Hermitian conjugate charged weak current of quarks as:

$$J_\lambda^\dagger = \bar{P}_L \gamma_\lambda U N_L \tag{14.4}$$

where P_L, N_L are separate column vectors of left-handed quark fields with electric charges $+2e/3, -e/3$, respectively:

$$P_L = \begin{pmatrix} u \\ c \\ t \end{pmatrix}_L \qquad N_L = \begin{pmatrix} d \\ s \\ b \end{pmatrix}_L \tag{14.5}$$

and U is the 3×3 unitary Cabibbo–Kobayashi–Maskawa (CKM) matrix [5]. Most generally, a complex 3×3 matrix contains $3 \times 3 = 9$ complex numbers or 18 real parameters. The unitary condition $U^\dagger U = I$ imposes 9 constraints and one overall phase is arbitrary, so the number of independent real parameters in U would appear to be 8. However the relative phases of u, c, t and of d, s, b are completely arbitrary. Thus, 4 degrees of freedom remain in U, and most generally it cannot be a real orthogonal 3×3 matrix, which is characterized by only 3 independent real angles. Instead we need 3 angles $\theta_{12}, \theta_{23}, \theta_{13}$ and an additional real parameter δ which is interpreted

[b]The brief and superficial summary of theoretical models of d_e given here is intended mainly for readers who like ourselves, are experimentally inclined. A detailed and authoritative account, with many references to the literature, will be found in Chapter 13 by M. Pospelov and A. Ritz.

as a CP-violating phase. In a standard notation, U is written as:

$$U = \begin{pmatrix} V_{ud} & V_{us} & V_{ub} \\ V_{cd} & V_{cs} & V_{cb} \\ V_{td} & V_{ts} & V_{tb} \end{pmatrix}$$
$$= \begin{pmatrix} c_{12}c_{13} & s_{12}c_{13} & s_{13}e^{-i\delta} \\ -s_{12}c_{23} - c_{12}s_{23}s_{13}e^{i\delta} & c_{12}c_{23} - s_{12}s_{23}s_{13}e^{i\delta} & s_{23}c_{13} \\ s_{12}s_{23} - c_{12}c_{23}s_{13}e^{i\delta} & -c_{12}s_{23} - s_{12}c_{23}s_{13}e^{i\delta} & c_{23}c_{13} \end{pmatrix}, \quad (14.6)$$

where $c_{ij} = \cos\theta_{ij}, s_{ij} = \sin\theta_{ij}$, and $i, j = 1, 2, 3$ are generation labels. In the Standard Model, it can be shown [6] that all CP-violating amplitudes in neutral K and B meson decays are proportional to:

$$J = s_{12}s_{13}s_{23}c_{12}c_{13}^2c_{23}\sin\delta. \quad (14.7)$$

The proportionality of J to the sines of all three mixing angles as well as to $\sin\delta$ appears natural, since the CP-violating phase appears only when three generations are included in the mixing matrix. Various observations of CP violation in K- and B meson decay yield the value [5]:

$$\delta = 1.05 \pm 0.24\,\text{radians}. \quad (14.8)$$

Thus δ is a large phase, but $J \approx 3 \cdot 10^{-5}$ is a very small quantity, because of the small values [5] of $s_{12}, s_{13},$ and s_{23}. Finally, it is notable that in the Standard Model, CP-violating effects vanish in the limit where any two quarks with the same isospin (e.g. u and c, or d and s) have the same mass; this is due to cancellations in the sum over diagrams containing all quark generations.

In the Standard Model with massless neutrinos, there is no analog of the CKM matrix in the lepton sector, and thus no analogous way to generate CP violation. For the electron EDM to arise here we require coupling to virtual quarks via virtual $W^{\pm}$. A priori this requires at least two loops, and naively one might expect a contribution from the two-loop diagram of Fig. 14.1. However, for each contribution V_{ij} from the CKM matrix at one vertex v, there is a contribution V_{ij}^* at the other vertex v'; hence the overall amplitude cannot contain a CP-violating phase. Next, one can consider contributions to the electron EDM at the three-loop level. Here it was shown by Pospelov and Khriplovich [7] that the various three-loop diagrams cancel, yielding a net contribution of zero in the absence of gluonic corrections to the quark lines. (See Fig. 14.2). Hence, four-loop diagrams are required for the electron EDM in the Standard Model, and there is

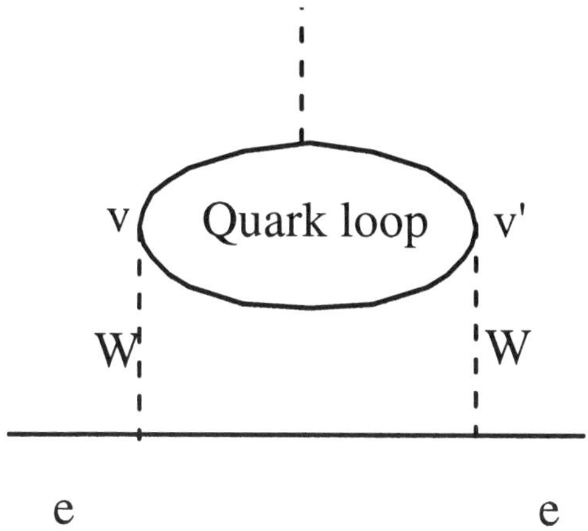

Fig. 14.1. This two-loop diagram cannot contribute to the electron EDM. Although a factor V_{ij} from the CKM matrix appears at vertex v, a factor V_{ij}^* appears at vertex v'; thus there is no net CP-violating phase.

additional suppression because of the smallness of J. This is why the electron EDM is predicted to be so extremely small in the Standard Model: $d_e < 10^{-38} e$ cm.

It is now known from neutrino oscillation experiments that at least two neutrino species have distinct non-zero masses [8]. One can incorporate this into the Standard Model and construct a CKM-like matrix for the lepton sector. Here two of the mixing angles are known to be quite large; however, just as for quarks, the sum over diagrams from all generations gives a result proportional to the mass differences between generations. The neutrino masses are so small that unless very special assumptions are made, the possible values of d_e that result are even less than that arising from the CKM matrix in the quark sector [9].

Although the Standard Model prediction for the lepton EDM is proportional to the lepton mass, and therefore two or three orders of magnitude larger for μ or τ leptons, respectively, than for the electron, the experimental sensitivities for μ, τ at present are seven to nine orders of magnitude poorer than that of the electron [10, 11, 13, 14]. Thus, if the Standard Model is the only source of CP violation, the EDMs of all leptons are far too small to be observed by any practical experiment, now or in the foreseeable future. Conversely, any observation of an EDM implies that the CP-violating effects giving rise to it are not described by the Standard Model.

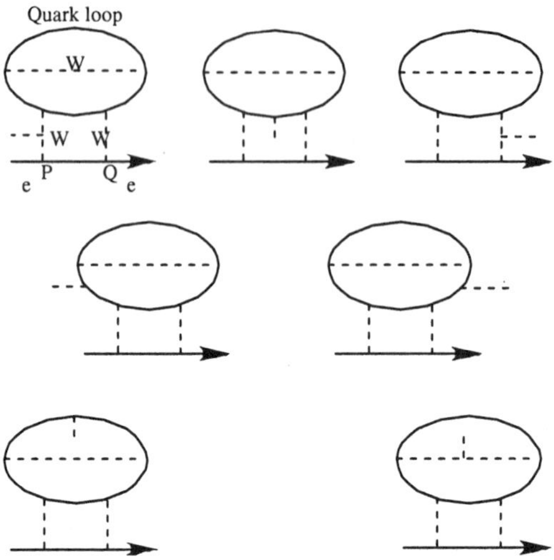

Fig. 14.2. Pospelov and Khriplovich [7] proved that the sum of the contributions to d_e from all three-loop diagrams (shown here) is zero, according to the Standard Model. If each diagram is disconnected from the lepton line at P,Q, one is left with the two-loop contributions to the EDM of an (on-mass-shell) W boson. Thus according to the Standard Model, the EDM of a W boson vanishes in the two-loop approximation.

14.1.1.2. *Electron EDM in extensions of the Standard Model*

Virtually every conceived extension of the Standard Model includes additional scalar fields that allow new complex phases–and thus new sources of CP violation. These hypothetical new particles can induce a non-zero d_e at the two-loop or even one-loop level of perturbation theory, leading to a dramatically enhanced effect. It is difficult to justify any significant suppression of these phases, for it is known that T invariance is not even an approximate symmetry of nature: As we have already noted, the Standard Model T-violating phase $\delta \approx 1$. Furthermore it is generally accepted that the dominance of matter over anti-matter in the observed universe requires additional sources of CP violation beyond that provided by the Standard Model [2].

Supersymmetric (SUSY) models are motivated by the desire to give a natural explanation for the "gauge hierarchy problem". In the standard model, electroweak symmetry breaking is induced by the Higgs mechanism, which imparts masses to $W^{\pm}$ and Z^0 of the order of 100 GeV/c^2 (the "weak scale"). The mass m_H of the Higgs boson itself is still unknown, but direct searches and radiative corrections to electroweak processes constrain it to the range $114 < m_H < 200$ GeV [15] and it must be less than ≈ 1 TeV/c^2 if unitarity is to be preserved in Standard Model perturbation calculations [16]. However, radiative corrections to the Higgs mass itself are quadratically divergent [16] and cannot be controlled within the range $m_H < 1$ TeV/c^2 unless a "fine tuning" is imposed that appears artificial and contrived to many authors. SUSY models attempt to avoid this problem in a natural way by linking physics at the weak scale to physics at the Planck scale. In all SUSY models many new hypothetical particles appear. For each fermion (lepton or quark), one introduces a supersymmetric bosonic partner (slepton, squark); for each Standard Model gauge boson (gluons, Z^0, $W^\pm$, photon) a supersymmetric fermionic partner called a "gaugino" is invoked (gluinos, zino, winos, photino). In addition, even the simplest SUSY models require at least two Higgs supermultiplets as well as their fermionic "higgsino" partners. (Note that none of the particles just mentioned have yet been observed). The wealth of hypothetical new particles and their couplings yields new T-violating phases in addition to Standard Model phase δ, and it becomes possible to generate an electron EDM at the one-loop level. We refer the reader to a detailed discussion of the connection between EDMs and supersymmetry in the chapter by Pospelov and Ritz. In particular see their Fig. 13.9, which shows the constraints already imposed on the "minimal supersymmetric Standard Model" (MSSM) by combining the present experimental limit on d_e with those for the ^{199}Hg EDM [17] and the neutron EDM [18].

In the simplest form of the Standard Model there is only one Higgs boson. However, even in various non-supersymmetric extensions two or more Higgs bosons could exist, and CP violation could then arise in a variety of new ways [19]. Specifically, it could appear directly in the coupling of one Higgs field to another. An electron EDM close to the present experimental limit could be generated from two-loop diagrams. Here the lepton chirality change occurs at the Higgs-lepton-lepton vertex, and the lepton EDM is proportional to the lepton mass. Models of this type might also give rise to an appreciable scalar P,T-odd eN interaction. In one possible class of multi-Higgs models, the lepton EDM is proportional to the cube of the lepton mass, and might be quite substantial for the tau lepton.

Left-right symmetric models, based on the gauge group $SU(2)_L \otimes SU(2)_R \otimes U(1)$, are motivated by the desire to find a natural explanation for the very striking phenomenon of parity violation in weak interactions [20–22]. Here space inversion symmetry is assumed to be valid before spontaneous symmetry breaking. In the simplest left-right symmetric model two Higgs multiplets appear, the first of which is a triplet χ_R that transforms like the (1,3) representation under $SU(2)_L \otimes SU(2)_R$. It gives rise to a very large mass of the right-handed intermediate vector boson W_R and thus breaks parity symmetry. A complex doublet ϕ transforming like (2,2) under $SU(2)_L \otimes SU(2)_R$ contributes to the mass of both W_R and W_L and causes mixing between them. These models also contain a right-handed neutrino N_R for each left-handed neutrino ν_L. The N_R acquires a large Majorana mass from χ_R and mixes with ν_L by means of ϕ. CP violation can occur at the one-loop level from the phases associated with W_L-W_R and N_R-ν_L mixing.

It is instructive to make a crude estimate of the value of d_e that might be expected in almost *any* extension of the Standard Model (see Fig. 14.3(a). The generic one-loop diagram is similar to that responsible for the lowest order radiative correction to the electron g-value: $g - 2 \approx \alpha/\pi$ (see Fig. 14.3(b), and we make use of this similarity to estimate d_e. The new features in Fig. 14.3(a) are (a) the heavy mass m_X of the unknown virtual particle X; (b) the inclusion of a CP-violating phase ϕ; and (c) different couplings (f versus e) at the vertices. Since the electron mass provides the only other energy scale in Fig. 14.3(a), we expect on dimensional grounds that:

$$\frac{d_e}{(g-2)\mu_B} \propto \left(\frac{m_e}{m_x}\right)^2 .$$

Thus we expect

$$d_e \approx \sin\phi \left(\frac{f}{e}\right)^2 \left(\frac{m_e}{m_x}\right)^2 \left(\frac{\alpha}{\pi}\right) \mu_B . \tag{14.9}$$

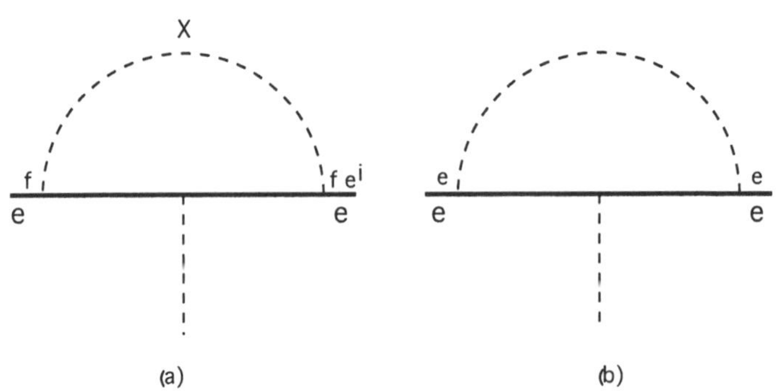

Fig. 14.3. (a) One-loop diagram for electron EDM; (b) Analogous diagram for lowest order correction to g-2.

We shall assume $\sin\phi = 1$ (justifying this assumption with the knowledge that $\delta \approx 1$). Also we shall assume that $f/e = 1$ (basing this assumption on the ground that dimensionless coupling constants should all have the same order of magnitude). This yields $d_e \approx 10^{-24}\ (\frac{100\text{GeV}}{m_X})^2 e$ cm. It is widely expected that new particles like X should have mass m_X in the range 100 GeV–1 TeV. This expectation arises from consideration of the hierarchy problem mentioned earlier: New Physics should yield particles with mass near m_H in order to stabilize the latter. Thus, given our assumptions, one-loop diagrams might be expected to yield $10^{-26} e$ cm $< d_e < 10^{-24} e$ cm. We might also expect that in theories where d_e appears at higher order, each additional loop should introduce a factor of order $f^2/\pi \approx \alpha/\pi \approx 3 \cdot 10^{-3}$. Of course, the assumptions made in this estimate can only be justified in the context of a specific theoretical model. Nevertheless, the crude analysis just presented suggests why the present experimental limit on d_e ($1.6 \cdot 10^{-27} e$ cm) already provides significant constraints on theories that generate d_e with one loop (e.g SUSY theories). We hope that the foregoing paragraphs have made clear why present-day EDM searches attract such great interest.

14.1.2. *Introduction to experimental basis for electron EDM searches*

To detect an electric dipole moment it would seem that one should simply place the particle of interest in an external electric field $\mathcal{E}_{\text{ext}}$ and observe the change in energy that is proportional to $\mathcal{E}_{\text{ext}}$. Obviously this is impractical for a free electron, which would quickly be accelerated out of the region of observation. However, there are alternative approaches. One of the earliest attempts to observe d_e utilized the spin precession of a relativistic free electron in a magnetic field, by means of a $g-2$ experiment [23]. If the electron possesses an EDM, the precession angular velocity is slightly modified because in the electron's rest frame there exists not only a magnetic field, but also a motional electric field to which the EDM is coupled. (This is still the only practical method available to search for the muon EDM [10–12]. Also see Chapter 17 in this volume.)

A far more sensitive method exists for d_e, in which one searches for the EDM of a paramagnetic atom or molecule, and interprets the result in terms of the EDM of the unpaired electron. At first sight this appears to be impossible, because even if $d_e \neq 0$, the atom or molecule cannot exhibit a linear Stark effect to first order in d_e in the limit of non-relativistic quantum mechanics. This is Schiff's theorem [24], which we shall derive and discuss further in Section 14.2.2. A popular qualitative explanation of Schiff's theorem can be stated as follows: *A neutral atom is not accelerated in a homogeneous external electric field. Therefore the average force on each charged particle in the atom must be zero. In the non-relativistic limit, the only forces are electrostatic; hence the average electric field at each charged particle must vanish. The externally applied electric field is canceled, on average, by the internal polarizing field.*

However, as was first demonstrated by P.G.H. Sandars [25, 26], Schiff's theorem fails when relativistic effects are taken into account. Sandars' important result may be expressed in terms of the ratio d_a/d_e (where d_a is the effective EDM of the atom or molecule), or equivalently in terms of the effective electric field $\mathcal{E}_{\text{eff}}$ experienced by d_e. It is convenient to write $\mathcal{E}_{\text{eff}} = QP$ where Q is a factor that includes the relativistic effects as well as details of atomic (or molecular) structure, while P is the degree of polarization of the atom or molecule by the external field. For typical paramagnetic atoms with valence electrons in $s_{1/2}$ or $p_{1/2}$ orbitals, such as Cs and Tl in their ground states,

$$Q \approx 4 \cdot 10^{10}\text{V/cm} \times \left[\frac{Z}{80}\right]^3, \tag{14.10}$$

where Z is the atomic number. Also, for such atoms $P \approx 10^{-3}\left[\frac{\mathcal{E}_{\text{ext}}}{100\ \text{kV/cm}}\right]$, which is only $\approx 10^{-3}$ for the maximum attainable laboratory fields $\mathcal{E}_{\text{ext}} \approx 100$ kV/cm.

Since for paramagnetic atoms in all practical situations, P is proportional to $\mathcal{E}_{\text{ext}}$, the ratio $\mathcal{E}_{\text{eff}}/\mathcal{E}_{\text{ext}}$ is a constant, and is usually called the enhancement factor $R \equiv d_a/d_e$. For the ground states of alkali atoms and for thallium, one finds that $|R| \approx 10Z^3\alpha^2$, where α is the fine structure constant. Although in these cases $P \ll 1$, when Z is sufficiently large the magnitude of R can greatly exceed unity. For example, for thallium ($Z = 81$), one calculates [27] $R = -585$.

The approximate formula (14.10) also applies for a wide range of heavy polar diatomic paramagnetic molecules with valence electrons in σ or π orbitals, such as YbF in the ground $^2\Sigma_{1/2}$ state, or PbO in the metastable $a(1)\ ^3\Sigma_1$ state; in these cases Z is the atomic number of the heavy nucleus. The main difference between atoms and molecules occurs in the factor P. In a typical polar diatomic molecule, nearly complete polarization ($P \approx 1$) can be achieved with relatively modest external fields: ($\mathcal{E}_{\text{ext}} \approx 10 - 10^4$ V/cm). Thus, when $P \approx 1$, $\mathcal{E}_{\text{eff}}$ for a typical paramagnetic molecule such as YbF or

PbO* is approximately three orders of magnitude larger than the maximum attainable with atoms [28].

Experimental searches for d_e using free paramagnetic atoms or molecules employ the standard methods of atomic, molecular, and optical physics: laser and RF spectroscopy, optical pumping, atomic and molecular beams, and so forth. Another way to search for d_e is to apply a large electric field to a suitable paramagnetic solid. In principle, the interaction of the EDMs of the unpaired electrons with the electric field at sufficiently low temperature can yield a net magnetization of the sample, which can be detected by a superconducting quantum interference device (SQUID) magnetometer [29–31]. Alternatively, application of an external magnetic field to a suitable ferrimagnetic solid can yield an EDM-induced electric polarization of the sample, which is detectable in principle by ultra-sensitive charge measurement techniques [32]. Yet another approach has been proposed, in which a sufficiently large external electric field applied to a gaseous sample of *diamagnetic* diatomic molecules could generate an observable P,T-odd magnetization [33]. Details of the various experimental searches for d_e will be discussed in later sections.

14.1.3. *Other sources of atomic and molecular EDMs*

It is important to note that an atomic or molecular EDM can arise from sources other than an electron EDM. A nuclear EDM can be generated by an intrinsic EDM of an unpaired nucleon, and/or by P,T-odd nucleon-nucleon (NN) interactions [34–39]. If the nuclear EDM distribution and the nuclear charge distribution are not the same, the cancellation caused by Schiff's theorem in the atom is incomplete, and a small residual "Schiff moment" remains [24]. Very sensitive searches [17] have been carried out for the Schiff moment in the diamagnetic atom ^{199}Hg, discussed in detail in Chapter 16. A nucleus with nuclear spin $I \geq 1$ could possess a magnetic quadrupole moment M originating from nucleonic EDMs and/or P,T-odd NN interactions, and in a paramagnetic atom or molecule this could couple to the magnetic field resulting from the spin and spatial distribution of the unpaired electron [40–42]. Because this interaction is magnetic, it would not be constrained by Schiff's theorem. P,T-odd electron-nucleon (eN) interactions might also exist [43–48]. These, as well as the P,T-odd NN interactions, could appear in one or several non-derivative coupling forms: "scalar", "tensor", and "pseudoscalar". (P,T-odd electron-electron interactions are also possible but these are likely to yield an extremely small contribution.) Finally, C, T-odd (P-even) eN and NN interactions, and possible T-odd beta decay couplings could cause a P,T-odd atomic or molecular EDM through radiative corrections involving the usual weak interactions of the Standard Model [49]. Of all the possibilities we have just mentioned, the most important for a paramagnetic atomic or molecular EDM, in addition to the electron EDM itself, is the scalar form of the P,T-odd eN interaction. The other contributions are very small by comparison [28].

14.2. Theoretical Basis of Electron EDM Experiments

14.2.1. *Proper-Lorentz-invariant EDM Lagrangian density*

We shall now formulate a gauge-invariant, proper-Lorentz-invariant effective Lagrangian density for the interaction of the EDM of a spin-1/2 fermion with an electromagnetic field. First let us recall the analogous formulation for an anomalous magnetic moment ("Pauli moment") [50]. It is given by the well-known expression:

$$\mathcal{L}_{\text{Pauli}} = -\kappa \frac{\mu_B}{2} \overline{\Psi} \sigma^{\mu\nu} \Psi F_{\mu\nu}. \tag{14.11}$$

Here Ψ is the Dirac field for the fermion, $\overline{\Psi}$ is the Dirac conjugate field, $\sigma^{\mu\nu} = \frac{i}{2}(\gamma^\mu \gamma^\nu - \gamma^\nu \gamma^\mu)$ where $\gamma^{\mu,\nu}$ are the usual 4×4 Dirac matrices,

$$F_{\mu\nu} = \partial_\mu A_\nu - \partial_\nu A_\mu = \begin{pmatrix} 0 & \mathcal{E}_x & \mathcal{E}_y & \mathcal{E}_z \\ -\mathcal{E}_x & 0 & -B_z & B_y \\ -\mathcal{E}_y & B_z & 0 & -B_x \\ -\mathcal{E}_z & -B_y & B_x & 0 \end{pmatrix} \tag{14.12}$$

is the electromagnetic field tensor, μ_B is the Bohr magneton, and κ is a suitable constant. (Here and throughout this chapter we use the notational conventions of Bjorken and Drell [51] for Dirac matrices and algebra, but we define the electromagnetic field tensor conventionally as in Jackson [52].) Rewriting (14.11) in terms of $\boldsymbol{\mathcal{E}}$ and $\boldsymbol{B}$ fields, we obtain:

$$\mathcal{L}_{\text{Pauli}} = \kappa \mu_B \overline{\Psi} [\boldsymbol{\Sigma} \cdot \boldsymbol{B} - i \boldsymbol{\alpha} \cdot \boldsymbol{\mathcal{E}}] \Psi \tag{14.13}$$

where as usual, $\boldsymbol{\Sigma} = \begin{pmatrix} \boldsymbol{\sigma} & 0 \\ 0 & \boldsymbol{\sigma} \end{pmatrix}$, $\boldsymbol{\alpha} = \begin{pmatrix} 0 & \boldsymbol{\sigma} \\ \boldsymbol{\sigma} & 0 \end{pmatrix}$. This Lagrangian density results in the single particle Hamiltonian:

$$H_{\text{Pauli}} = -\kappa \mu_B (\gamma^0 \boldsymbol{\Sigma} \cdot \boldsymbol{B} - i \boldsymbol{\gamma} \cdot \boldsymbol{\mathcal{E}}), \tag{14.14}$$

which reduces in the non-relativistic limit to the Hamiltonian in (14.1). Of course, $\mathcal{L}_{\text{Pauli}}$ of (14.11) or (14.13) and H_{Pauli} of (14.14) are each P- and

T- invariant. We can render them P,T odd by replacing $\boldsymbol{\mathcal{E}}$ by $-\boldsymbol{B}$ and $\boldsymbol{B}$ by $\boldsymbol{\mathcal{E}}$, which is equivalent to the replacement of $F_{\mu\nu}$ by the tensor $-F^*_{\mu\nu}$, where:

$$F^*_{\mu\nu} = \frac{1}{2}\varepsilon_{\mu\nu\alpha\beta}F^{\alpha\beta} = \begin{pmatrix} 0 & B_x & B_y & B_z \\ -B_x & 0 & \mathcal{E}_z & -\mathcal{E}_y \\ -B_y & -\mathcal{E}_z & 0 & \mathcal{E}_x \\ -B_z & \mathcal{E}_y & -\mathcal{E}_x & 0 \end{pmatrix}.$$

Alternatively one obtains the same Lagrangian density by replacing $\sigma^{\mu\nu}$ in (14.11) with $i\sigma^{\mu\nu}\gamma^5$ (where, as usual, $\gamma^5 = i\gamma^0\gamma^1\gamma^2\gamma^3$), with no change in $F_{\mu\nu}$. Making this latter transformation and replacing $\kappa\mu_B$ by d we obtain the EDM Lagrangian density:

$$\mathcal{L}_{\text{EDM}} = -i\frac{d}{2}\overline{\Psi}\sigma^{\mu\nu}\gamma^5\Psi F_{\mu\nu} = d\overline{\Psi}[\boldsymbol{\Sigma}\cdot\boldsymbol{\mathcal{E}} + i\boldsymbol{\alpha}\cdot\boldsymbol{B}]\Psi, \tag{14.15}$$

which was first described by Salpeter [53]. This in turn yields the single-particle Hamiltonian:

$$H_{\text{EDM}} = -d(\gamma^0\boldsymbol{\Sigma}\cdot\boldsymbol{\mathcal{E}} + i\boldsymbol{\gamma}\cdot\boldsymbol{B}). \tag{14.16}$$

As we shall see in Sec. 14.2.2, the appearance of γ^0 in the first term on the right-hand side of (14.16) is of crucial significance: It leads to the failure of Schiff's theorem and the resulting enhancement discovered by Sandars. In the non-relativistic limit, the first term on the right-hand side of (14.16) reduces to the right-hand side of (14.2), while the second term on the right-hand side of (14.16) gives no contribution.

14.2.2. *Schiff's theorem*

We now derive Schiff's theorem for a paramagnetic atom, and show how it fails for the unpaired electron when relativistic motion is taken into account. For the purposes of this discussion we assume the central field approximation and begin by writing the one-electron Dirac Hamiltonian for an atom in an external electric field $\boldsymbol{\mathcal{E}}_{\text{ext}}$, in the absence of an electron EDM:

$$H = c\boldsymbol{\alpha}\cdot\boldsymbol{p} + mc^2\gamma^0 - e(\Phi_i + \Phi_e). \tag{14.17}$$

Here $e > 0$ (the electron charge is $-e$), while Φ_i is the atomic electrostatic central potential and $\Phi_e = -\boldsymbol{\mathcal{E}}_{\textbf{ext}}\cdot\boldsymbol{r}$ is the external electrostatic potential. An eigenstate of H will be denoted by $|\psi\rangle$. Now we introduce the EDM Hamiltonian as a perturbation. From (14.16) with $\boldsymbol{B} = 0$ it is:

$$H_{\text{EDM}} = -d_e\gamma^0\boldsymbol{\Sigma}\cdot\boldsymbol{\mathcal{E}}, \tag{14.18}$$

where $\boldsymbol{\mathcal{E}} = -\nabla\Phi = -\nabla(\Phi_i + \Phi_e)$ is the total electric field. It is convenient to separate the right-hand side of (14.18) into two parts as follows:

$$H_{\mathrm{EDM}} = -d_e\boldsymbol{\Sigma}\cdot\boldsymbol{\mathcal{E}} - d_e(\gamma^0 - 1)\boldsymbol{\Sigma}\cdot\boldsymbol{\mathcal{E}}. \tag{14.19}$$

The first term on the right-hand side of (14.19) is the only portion that survives in the non-relativistic limit, but as we shall now demonstrate, it contributes nothing to the first-order energy shift arising from H_{EDM}. (This is Schiff's theorem). We write:

$$-d_e\boldsymbol{\Sigma}\cdot\boldsymbol{\mathcal{E}} = \frac{d_e}{e}\boldsymbol{\Sigma}\cdot\boldsymbol{\nabla}e\Phi = \frac{id_e}{e}[\boldsymbol{\Sigma}\cdot\boldsymbol{p}, e\Phi].$$

Making use of (14.17) we obtain:

$$-d_e\boldsymbol{\Sigma}\cdot\boldsymbol{\mathcal{E}} = -\frac{id_e}{e}\left[\boldsymbol{\Sigma}\cdot\boldsymbol{p}, (H - c\boldsymbol{\alpha}\cdot\boldsymbol{p} - mc^2\gamma^0)\right].$$

However, $[\boldsymbol{\Sigma}\cdot\boldsymbol{p}, \boldsymbol{\alpha}\cdot\boldsymbol{p}] = 0, [\boldsymbol{\Sigma}\cdot\boldsymbol{p}, \gamma^0] = 0$; hence:

$$\langle\psi| - d_e\boldsymbol{\Sigma}\cdot\boldsymbol{\mathcal{E}}|\psi\rangle = -\frac{id_e}{e}\langle\psi|[\boldsymbol{\Sigma}\cdot\boldsymbol{p}, H]|\psi\rangle = 0, \tag{14.20}$$

which vanishes because $|\psi\rangle$ is an eigenstate of H. Therefore only the second term on the right-hand side of (14.19) can contribute to the first-order energy shift ΔE:

$$\Delta E = \langle\psi| - d_e(\gamma^0 - 1)\boldsymbol{\Sigma}\cdot\boldsymbol{\mathcal{E}}|\psi\rangle. \tag{14.21}$$

A very similar argument shows that the average value of $\boldsymbol{\mathcal{E}}$ is zero. We write:

$$\boldsymbol{\mathcal{E}} = -\frac{i}{e}[\boldsymbol{p}, e\Phi] = \frac{i}{e}[\boldsymbol{p}, H - c\boldsymbol{\alpha}\cdot\boldsymbol{p} - mc^2\gamma^0]$$

Since $\boldsymbol{p}$ commutes with $\boldsymbol{\alpha}\cdot\boldsymbol{p}$ and with γ^0,

$$\langle\psi|\boldsymbol{\mathcal{E}}|\Psi\rangle = \frac{i}{e}\langle\psi|\boldsymbol{p}, H|\psi\rangle = 0.$$

Incidentally, although the right-hand side of (14.18) was separated into two parts in (14.19), it is sometimes convenient to deal directly with (14.18) by writing:

$$\Delta E = \langle\psi| - d_e\gamma^0\boldsymbol{\Sigma}\cdot\boldsymbol{\mathcal{E}}|\psi\rangle = -\frac{id_e}{e}\langle\psi|[\gamma^0\boldsymbol{\Sigma}\cdot\boldsymbol{p}, (H - c\boldsymbol{\alpha}\cdot\boldsymbol{p} - mc^2\gamma^0)]|\psi\rangle$$
$$= \frac{icd_e}{e}\langle\psi|[\gamma^0\boldsymbol{\Sigma}\cdot\boldsymbol{p}, \boldsymbol{\alpha}\cdot\boldsymbol{p}]|\psi\rangle.$$

Taking into account the identities $\boldsymbol{\alpha} = \gamma^5\boldsymbol{\Sigma} = \boldsymbol{\Sigma}\gamma^5, \gamma^0\gamma^5 = -\gamma^5\gamma^0$, and $\boldsymbol{\Sigma}\cdot\boldsymbol{p}\boldsymbol{\Sigma}\cdot\boldsymbol{p} = \mathrm{p}^2$ we arrive at the alternative expression:

$$\Delta E = \frac{2icd_e}{e}\langle\psi|\gamma^0\gamma^5 p^2|\psi\rangle. \tag{14.22}$$

14.2.3. *Enhancement factors for paramagnetic atoms*

Recalling that $|\psi\rangle$ is an eigenstate of Hamiltonian H, which includes the term $-e\Phi_e = e\boldsymbol{\mathcal{E}}_{\mathbf{ext}} \cdot \boldsymbol{r}$, we treat the latter term as a perturbation on the atomic Hamiltonian H_0 with no external field:

$$H_0 = c\boldsymbol{\alpha} \cdot \boldsymbol{p} + mc^2\gamma^0 - e\Phi_i, \tag{14.23}$$

and express $|\psi\rangle$ in terms of the eigenstates $|\psi_n\rangle$ of H_0 to first order in $\mathcal{E}_{\text{ext}}$:

$$|\psi\rangle = |\psi_0\rangle + e\mathcal{E}_{\text{ext}} \sum_{n\neq 0} \frac{|\psi_n\rangle\langle\psi_n|z|\psi_0\rangle}{E_0 - E_n} = |\psi_0\rangle + e\mathcal{E}_{\text{ext}}|\eta\rangle. \tag{14.24}$$

Here $|\psi_0\rangle$ is that state to which $|\psi\rangle$ reduces when $\boldsymbol{\mathcal{E}}_{\text{ext}} = 0$, we have assumed that $\boldsymbol{\mathcal{E}}_{\text{ext}}$ is in the z direction, the E_n are the energy eigenvalues of H_0 corresponding to the $|\psi_n\rangle$, and:

$$|\eta\rangle = \sum_{n\neq 0} \frac{|\psi_n\rangle\langle\psi_n|z|\psi_0\rangle}{E_0 - E_n}. \tag{14.25}$$

Now substituting (14.24) in (14.21) and retaining only terms of first order in $\mathcal{E}_{\text{ext}}$ we obtain:

$$\begin{aligned}\Delta E = {} & -d_e\mathcal{E}_{\text{ext}}\langle\psi_0|(\gamma^0 - 1)\Sigma_z|\psi_0\rangle \\ & -ed_e\mathcal{E}_{\text{ext}}[\langle\eta|(\gamma^0 - 1)\boldsymbol{\Sigma} \cdot \boldsymbol{\mathcal{E}_i}|\psi_0\rangle + \langle\psi_0|(\gamma^0 - 1)\boldsymbol{\Sigma} \cdot \boldsymbol{\mathcal{E}_i}|\eta\rangle].\end{aligned} \tag{14.26}$$

Thus the enhancement factor $R = d_a/d_e$ is given by:

$$\begin{aligned}R = {} & \langle\psi_0|(\gamma^0 - 1)\Sigma_z|\psi_0\rangle \\ & +e\left[\langle\eta|(\gamma^0 - 1)\boldsymbol{\Sigma} \cdot \boldsymbol{\mathcal{E}_i}|\psi_0\rangle + \langle\psi_0|(\gamma^0 - 1)\boldsymbol{\Sigma} \cdot \boldsymbol{\mathcal{E}_i}|\eta\rangle\right],\end{aligned} \tag{14.27}$$

where $\boldsymbol{\mathcal{E}_i} = -\nabla\Phi_i$ is the internal electric field. The operator $\gamma^0 - 1$ connects only small components of Dirac wave functions; matrix elements containing $\gamma^0 - 1$ thus receive a contribution from it of order $Z^2\alpha^2$ and are dominated by the region very close to the nucleus where $\mathcal{E}_i \approx Ze/r^2$. Hence the second and third terms in (14.26) are roughly proportional to Z^3. Meanwhile the first term varies as Z^2 and its coefficient is relatively small; thus when $Z \gg 1$ we may ignore the first term in (14.27), in which case we obtain:

$$\begin{aligned}R &= 2e\sum_{n\neq 0} \frac{\langle\psi_0\left|(\gamma^0 - 1)\boldsymbol{\Sigma} \cdot \boldsymbol{\mathcal{E}_i}\right|\psi_n\rangle\langle\psi_n|z|\psi_0\rangle}{E_0 - E_n} \\ &= 2e\left\langle\psi_o\left|(\gamma^0 - 1)\boldsymbol{\Sigma} \cdot \boldsymbol{\mathcal{E}_i}\right|\eta\right\rangle.\end{aligned} \tag{14.28}$$

Assuming that a single term dominates this sum, we make a crude first estimate of the various factors as follows: $\langle\psi_n|z|\psi_0\rangle \approx a_0, E_0 - E_n \approx 0.2e^2/a_0, \langle\psi_0|(\gamma^0 - 1)\boldsymbol{\Sigma} \cdot \boldsymbol{\mathcal{E}_i}|\psi_n\rangle \approx Z^2\alpha^2 \cdot Ze/a_0^2$; which yields $R \approx 10\ Z^3\alpha^2$.

We now sketch a genuine calculation of R. Here we restrict ourselves to the one-electron central field approximation and to the case where $|\psi_0\rangle$ is a state with $J = 1/2, m_J = 1/2$, which includes the ground states of alkali atoms and thallium. In this case the four-component Dirac wave function ψ_0 can be written:

$$\psi^{\ell}_{J=1/2,m=1/2} = \begin{pmatrix} \frac{iG_{\ell,J=1/2}(r)}{r}\phi^{\ell}_{1/2,1/2} \\ \frac{F_{\ell,1/2}(r)}{r}\boldsymbol{\sigma}\cdot\hat{\boldsymbol{r}}\phi^{\ell}_{1/2,1/2} \end{pmatrix}, \tag{14.29}$$

where we employ the notation:

$$\phi^{\ell=0}_{1/2,1/2} = \begin{pmatrix} Y^0_0 \\ 0 \end{pmatrix}, \phi^{\ell=1}_{1/2,1/2} = \begin{pmatrix} \sqrt{\frac{1}{3}}Y^0_1 \\ -\sqrt{\frac{2}{3}}Y^1_1 \end{pmatrix} \tag{14.30}$$

and where

$$\boldsymbol{\sigma}\cdot\hat{\boldsymbol{r}}\phi^0_{1/2,1/2} = \phi^1_{1/2,1/2}, \quad \boldsymbol{\sigma}\cdot\hat{\boldsymbol{r}}\phi^1_{1/2,1/2} = \phi^0_{1/2,1/2}.$$

Inserting (14.29) in Dirac's equation $(H_0 - E_0|\psi_0\rangle = 0$, defining $W_0 = E_0 - mc^2$, and choosing atomic units where $\hbar = e = m = 1, c = \alpha^{-1} = 137.036$, we obtain the well-known coupled radial equations:

$$\frac{\partial G_{\ell,1/2}}{\partial r} + \frac{\kappa}{r}G_{\ell,1/2} = \alpha\left(W_0 + \frac{2}{\alpha^2} + \Phi_i\right)F_{\ell,1/2} \tag{14.31}$$

$$\frac{\partial F_{\ell,1/2}}{\partial r} - \frac{\kappa}{r}F_{\ell,1/2} = -\alpha(W_0 + \Phi_i)G_{\ell,1/2}, \tag{14.32}$$

where $\kappa = +1$ for $\ell = 1$, $\kappa = -1$ for $\ell = 0$. These equations can be solved analytically or numerically for specified Φ_i, subject to the condition that $\psi^{\ell}_{1/2,1/2}$ is normalized to unity.

To find a useful expression for $|\eta\rangle$ we apply the operator $H_0 - E_0$ to both sides of (14.25) and make use of the completeness relation $\boldsymbol{\Sigma}|\psi_n\rangle\langle\psi_n| = 1$ to obtain:

$$(H_0 - E_0)|\eta\rangle = -z|\psi_0\rangle, \tag{14.33}$$

which is known as the Sternheimer equation [26, 54–56]. It can be seen from (14.25) that $|\eta\rangle$ and $|\psi_0\rangle$ must be of opposite parity, and that, a priori, $|\eta\rangle$ could contain $J = 1/2$ and $J = 3/2$ components. However, because $\boldsymbol{\Sigma}\cdot\boldsymbol{\mathcal{E}}_i$ is a pseudoscalar operator, only the $J = 1/2$ component of $|\eta\rangle$ can contribute to (14.28). Writing this component as:

$$\eta^L_{1/2,1/2} = \begin{pmatrix} \frac{iG^S_{L,1/2}}{r}\phi^L_{1/2,1/2} \\ \frac{F^S_{L,1/2}}{r}\boldsymbol{\sigma}\cdot\hat{\boldsymbol{r}}\phi^L_{1/2,1/2} \end{pmatrix} \tag{14.34}$$

with $L = 0$ or 1, (and where the superscript S stands for Sternheimer), we manipulate (14.33) to obtain the coupled radial equations:

$$r\frac{\partial G^S_{1,1/2}}{\partial r} + G^S_{1,1/2} - \alpha r\left(W_0 + \Phi_i + \frac{2}{\alpha^2}\right)F^S_{1,1/2} = -\frac{\alpha r^2}{3}F_{0,1/2}$$

$$r\frac{\partial F^S_{1,1/2}}{\partial r} - F^S_{1,1/2} + \alpha r(W_0 + \Phi_i)G^S_{1,1/2} = \frac{\alpha r^2}{3}G_{0,1/2} \qquad (14.35)$$

for $^2S_{1/2}$ enhancement factors (as in the ground states of alkali atoms), or:

$$r\frac{\partial G^S_{0,1/2}}{\partial r} - G^S_{0,1/2} - \alpha r\left(W_0 + \Phi_i + \frac{2}{\alpha^2}\right)F^S_{0,1/2} = -\frac{\alpha r^2}{3}F_{1,1/2}$$

$$r\frac{\partial F^S_{0,1/2}}{\partial r} + F^S_{0,1/2} + \alpha r(W_0 + \Phi_i)G^S_{0,1/2} = \frac{\alpha r^2}{3}G_{1,1/2} \qquad (14.36)$$

for $^2P_{1/2}$ enhancement factors (as in the ground state of thallium). Note that $|\eta\rangle$ is not normalized to unity. Instead the magnitudes of $G^S_{L,1/2}, F^S_{L,1/2}$ are determined as solutions to the inhomogeneous equations (14.35) or (14.36) and the requirement that $G^S_{L,1/2}, F^S_{L,1/2}$ vanish as $r \to \infty$.

From the solutions to (14.31), (14.32) and (14.35) or (14.36) we write R as follows (where all quantities are in atomic units):

$$R = 2\left\langle \psi_0 \left|(\gamma^0 - 1)\boldsymbol{\Sigma} \cdot \boldsymbol{\mathcal{E}_i}\right| \eta\right\rangle$$

$$= 4\int \begin{pmatrix} \frac{iG_{\ell,1/2}}{r}\phi^\ell_{1/2,1/2} \\ \frac{F_{\ell,1/2}}{r}\boldsymbol{\sigma}\cdot\hat{\boldsymbol{r}}\phi^\ell_{1/2,1/2} \end{pmatrix}^\dagger \begin{pmatrix} 0 & 0 \\ 0 & 1 \end{pmatrix} \frac{\partial \Phi_i}{\partial r}\boldsymbol{\sigma}\cdot\hat{\boldsymbol{r}} \begin{pmatrix} \frac{iG^S_{L,1/2}}{r}\phi^L_{1/2,1/2} \\ \frac{F^S_{L,1/2}}{r}\vec{\boldsymbol{\sigma}}\cdot\hat{\boldsymbol{r}}\phi^\ell_{1/2,1/2} \end{pmatrix} d^3\boldsymbol{r},$$

which yields:

$$R = 4\int_0^\infty F_{\ell,1/2}F^S_{L,1/2}\frac{\partial \Phi_i}{\partial r}dr. \qquad (14.37)$$

So far, we have employed the Dirac equation with the one-electron central field approximation. When dealing with a many-electron atom, a more careful treatment usually starts with the Hamiltonian $H_{\text{Total}} = H + H_{\text{EDM}}$ where:

$$H = \sum_i \left[c\boldsymbol{\alpha_i}\cdot\boldsymbol{p_i} + \gamma_i^0 mc^2 - e\Phi_{nuc}(\boldsymbol{r}_i) - e\Phi_{\text{ext}}(\boldsymbol{r}_i)\right] + \frac{1}{2}\sum_{\boldsymbol{j\neq 1}}\frac{e^2}{r_{ij}}, \qquad (14.38)$$

$$H_{\text{EDM}} = -d_e\sum_{\boldsymbol{i}}\gamma_i^0\boldsymbol{\Sigma_i}\cdot\boldsymbol{\mathcal{E}_i}, \qquad (14.39)$$

and where

$$\boldsymbol{\mathcal{E}_i} = -\boldsymbol{\nabla}\left[\Phi_{nuc}(\mathbf{R}_i) + \Phi_{\text{ext}}(r_i) - \sum_{j\neq 1}\frac{e}{r_{ij}}\right]. \qquad (14.40)$$

A number of issues arise at this stage. First, the atomic Hamiltonian in (14.38) contains the instantaneous Coulomb interaction between electrons, but it is missing the Breit interaction. Second, $H_{\rm EDM}$ in (14.39) is missing the magnetic field term $-id_e \sum_i \gamma_i \cdot \boldsymbol{B}(\boldsymbol{r}_i)$, where $\boldsymbol{B}(\boldsymbol{r}_i)$ arises from the motion of the other electrons. (This contribution is somewhat analogous to the Breit interaction.) Third, to avoid the appearance of degenerate but unphysical states that are composed of one positive and one negative energy state, the two-particle operators in (14.38) and (14.39) must be surrounded by positive energy projection operators [57, 58]. This last problem always arises when one employs a Dirac Hamiltonian with two or more electrons.

However, if one makes the usual separation

$$H_{\rm EDM} = -d_e \sum_i \boldsymbol{\Sigma}_i \cdot \boldsymbol{\mathcal{E}}_i - d_e \sum_i (\gamma_i^0 - 1)\boldsymbol{\Sigma}_i \cdot \boldsymbol{\mathcal{E}}_i \tag{14.41}$$

it can be shown that, just as before, only $H_{\rm EDM,eff} = -d_e \sum_i (\gamma_i^0 - 1)\boldsymbol{\Sigma}_i \cdot \boldsymbol{\mathcal{E}}_i$ plays any role in generating an enhancement factor. Then, using $H_{\rm EDM,eff}$, Lindroth, Lynn, and Sandars [59] have shown that to order α^2, the contributions of the Breit interaction and the magnetic terms in the EDM Hamiltonian are very insignificant for R(Cs) and R(Tl) compared to the central field contribution (which originates from the nuclear potential plus the central part of the electron-electron interaction). Finally they have shown that to order α^2 the third difficulty involving degenerate positive and negative energy states is irrelevant for calculation of enhancement factors.

Numerical calculations of R for the alkali and thallium atoms, using (14.37) and based on various semi-empirical potentials Φ_i, have been carried out by a number of authors. *Ab initio* calculations of R have also been done employing a variety of sophisticated many-body techniques. One always finds for $s_{1/2}$ and $p_{1/2}$ orbitals that the integrand on the right-hand side of (14.37) is sharply peaked at the nuclear radius and drops rapidly to zero as r approaches $1/Z (= a_0/Z$ in ordinary units). In all calculations, semi-empirical or *ab initio*, it is important to correct for screening of the external electric field by the electron core. In Table 14.1 we summarize various calculations of R for paramagnetic atoms.

14.2.4. *Is there a simple intuitive explanation for the Sandars effect?*

The discovery by Sandars that Schiff's theorem fails when special relativity is taken into account is so fundamental for our subject that one would like

Table 14.1. Calculated Enhancement Factors $R = d_a/d_e$ for Paramagnetic Atoms. Preferred values are in **boldface**.

Atom	Z	State	Enhancement factor R Semi-empirical	*Ab initio*
Li	3	$2^2S_{1/2}$	**.0043**[a]	
Na	11	$3^2S_{1/2}$	**.32**[a]	
K	19	$4^2S_{1/2}$	**2.42**[a]	
Rb	37	$5^2S_{1/2}$	24[b]	
			16 to 22[b]	**24.6**[b], **25.7**[c]
Cs	55	$6^2S_{1/2}$	119[a]	
			80.3 to 106[b]	**114.9**[b]
Fr	87	$7^2S_{1/2}$	**1150**[a]	
Tl	81	$6^2P_{1/2}$	−700[d]	
			−502 to −607[b]	
			−500[e]	**−585**[f]
Gd^{3+}	64	$^8S_{7/2}$	**−3.3**[gh]	

[a]Ref. 26
[b]Ref. 60
[c]Ref. 61
[d]Ref. 56
[e]Ref. 62
[f]Ref. 27
[g]Ref. 63
[h]Although Z=64, $R(Gd^{3+})$ is small because parity mixing here is mainly between $5d$ and $4f$ orbitals.

to have a simple intuitive explanation for it. Unfortunately, several such explanations that have appeared in the literature are wrong, or at the least misleading. These explanations typically rely on a claim that the presence of relativistic forces can give rise to a net electric field at the electron. However, as we have already shown [see the discussion following (14.21)], the expectation value of the *total* electric field experienced by the electron is zero, even in the presence of all relativistic effects (e.g. spin-orbit and Darwin).

In order to present a legitimate intuitive argument [64], we consider a hydrogen atom exposed to a uniform external electric field. Let the Dirac wave function be $\psi = \begin{pmatrix} \psi_A \\ \psi_B \end{pmatrix}$, where $\psi_{A,B}$ are "large" and "small" two-component wave functions, respectively. The first-order energy shift due to d_e is:

$$\Delta E = -d_e \langle \psi | \gamma^0 \mathbf{\Sigma} \cdot \boldsymbol{\mathcal{E}} | \psi \rangle = -d_e \left[\langle \psi_A | \boldsymbol{\sigma} \cdot \boldsymbol{\mathcal{E}} | \psi_A \rangle - \langle \psi_B | \boldsymbol{\sigma} \cdot \boldsymbol{\mathcal{E}} | \psi_B \rangle \right]. \quad (14.42)$$

For present purposes, we can approximate ψ_B by $\psi_B \cong \frac{\boldsymbol{\sigma} \cdot \boldsymbol{p}}{2mc} \psi_A$, since

$v^2/c^2 \ll 1$. Then it can be shown that:

$$\Delta E \cong -d_e \left\langle \psi_A \left| \boldsymbol{\sigma} \cdot \boldsymbol{\mathcal{E}} - \frac{1}{4m^2c^2} \left[\boldsymbol{p} \cdot \boldsymbol{\mathcal{E}} \boldsymbol{\sigma} \cdot \boldsymbol{p} + \boldsymbol{\mathcal{E}} \cdot \boldsymbol{p} \boldsymbol{\sigma} \cdot \boldsymbol{p} \right] \right| \psi_A \right\rangle . \qquad (14.43)$$

Now we present an intuitive explanation for the operator appearing in the matrix element on the right side of (14.43). Suppose that the EDM, considered classically, is $\boldsymbol{d}_e$ in the electron rest frame, and suppose that in the proton-electron center-of-mass frame the electron moves with velocity $c\boldsymbol{\beta}$. Then since the dipole has dimensions of charge · length, and charge is a Lorentz invariant but length suffers a Lorentz contraction, the dipole moment in the CM frame is:

$$\boldsymbol{d}_e^c = \boldsymbol{d}_e - \frac{\gamma}{1+\gamma} (\boldsymbol{\beta} \cdot \boldsymbol{d}_e) \boldsymbol{\beta}, \qquad (14.44)$$

where as usual, $\gamma = (1-\beta^2)^{-1/2}$. Let $\boldsymbol{\mathcal{E}} = \boldsymbol{\mathcal{E}}_{\mathbf{ext}} + \boldsymbol{\mathcal{E}}_{\boldsymbol{int}}$ be the total electric field in the CM frame, consisting of the external field plus the internal (= atomic) field. The energy of the dipole in the CM frame is:

$$E = -\boldsymbol{d}_e^c \cdot \boldsymbol{\mathcal{E}} = -\boldsymbol{d}_e \cdot [\boldsymbol{\mathcal{E}} - \frac{\gamma}{1+\gamma} (\boldsymbol{\beta} \cdot \boldsymbol{\mathcal{E}}) \boldsymbol{\beta}]. \qquad (14.45)$$

Note that for small v^2/c^2 [which is assumed in the derivation of (14.42)],

$$\frac{\gamma}{1+\gamma} (\boldsymbol{\beta} \cdot \boldsymbol{\mathcal{E}}) \, \boldsymbol{d}_e \cdot \boldsymbol{\beta} \rightarrow -\frac{1}{2m^2c^2} (\boldsymbol{p}^c \cdot \boldsymbol{\mathcal{E}}) \, \boldsymbol{d}_e \cdot \boldsymbol{p}^c, \qquad (14.46)$$

where $\boldsymbol{p}^c$ is the classical momentum. The expression in (14.46) closely resembles the operator in the matrix element on the right-hand side of (14.43), and appears to represent the essential physical content of that quantum mechanical statement.

To summarize, the single essential feature in the foregoing intuitive explanation is the Lorentz contraction of the electric dipole moment. Contrary to a number of previously published "intuitive explanations" for the Sandars effect, magnetic (i.e. spin-orbit and Darwin) interactions play no role, and the average electric field at the electron is zero even in the presence of these interactions.

14.2.5. *P, T-odd electron-nucleon interaction*

As previously mentioned, the EDM d_a of an atom or molecule can include a substantial contribution from a P,T-odd electron-nucleon interaction. If we limit ourselves to non-derivative terms, the various possibilities for P-odd eN couplings are easily written by analogy from the theory of nuclear beta decay as follows:

$\bar{N}N \cdot \bar{e}\gamma^5 e$	S-PS (scalar-pseudoscalar)
$\bar{N}\gamma^\mu N \cdot \bar{e}\gamma_\mu\gamma^5 e$	V-A (vector-axial vector)
$\bar{N}\sigma^{\mu\nu} N \cdot \bar{e}\sigma_{\mu\nu}\gamma^5 e$	T-PT (tensor-pesudotensor)
$\bar{N}\gamma^\mu\gamma^5 N \cdot \bar{e}\gamma_\mu e$	A-V (axial vector-vector)
$\bar{N}\gamma^5 N \cdot \bar{e}e$	PS-S (pseudoscalar-scalar).

It is easy to show that under a time reversal transformation, the first, third, and fifth of these forms are odd, while the second and fourth are even. Thus we may write an effective P,T-odd Hamiltonian density as follows:

$$\mathcal{H}_{eN} = \frac{iG_F}{\sqrt{2}}\left[C_S\sum_{i=1}^{A}\bar{N}_i N_i \cdot \bar{e}\gamma^5 e + C_T\sum_{i=1}^{A}\bar{N}_i\sigma^{\mu\nu}N_i \cdot \bar{e}\sigma_{\mu\nu}\gamma^5 e \right.$$
$$\left. + C_P\sum_{i=1}^{A}\bar{N}_i\gamma^5 N_i \cdot \bar{e}e\right]. \qquad (14.47)$$

Here e and N_i are field operators for the electron and the ith nucleon respectively. Also, for convenience we have expressed the coupling strengths in terms of Fermi's constant G_F, and C_S, C_T, and C_P are real coupling constants. Taking into account the complex conjugation properties of the various bilinear forms, a factor of i is included so that $\mathcal{H}_{eN}$ is Hermitian, and also the sums are taken over all nucleons in the nucleus. (For simplicity we do not distinguish between neutrons and protons, although this could be done). In the non-relativistic limit for the nucleons, the term in C_P vanishes, so we neglect it henceforth. In that same limit, the scalar and tensor terms yield the following effective one-particle Hamiltonian:

$$H_{eN} = \frac{iG_F}{\sqrt{2}}\left[AC_S\gamma_e^0\gamma_e^5 + 2C_T\boldsymbol{\gamma}_e \cdot \boldsymbol{\sigma}_N\right]n(\boldsymbol{r}), \qquad (14.48)$$

where $\boldsymbol{\sigma}_N$ is the Pauli spin operator of the last unpaired nucleon, and $n(\boldsymbol{r})$ is the nucleon density. The nucleon number A in the C_S term appears because the nucleons add coherently in that term. The matrix element of each term in (14.48) receives a factor $\approx Z\alpha$ from the Dirac matrices $\gamma_e^0\gamma_e^5$ or $\boldsymbol{\gamma}_e$, which couple large and small components, and another factor of Z because of the zero range nature of the interaction (the nucleon density is very sharply peaked at the origin). Thus matrix elements of the scalar term vary roughly as $AZ^2 \approx Z^3$. Consequently if non-zero atomic EDMs were to be observed in paramagnetic atoms of various atomic numbers Z, it would be difficult from the Z dependence alone to disentangle the electron EDM and the scalar P,T-odd eN contributions. The scalar term in (14.48) is analogous to the dominant contribution to ordinary atomic parity nonconservation (PNC), which arises from the coupling of the axial electronic neutral weak current to the vector nucleonic neutral weak current via Z^0 exchange [65]. Matrix elements of the tensor term vary roughly as Z^2, because only a single valence nucleon with unpaired spin contributes; this term is analogous to the much smaller nuclear spin-dependent PNC contribution arising from vector electronic-axial nucleonic coupling. The tensor term is most strongly bounded by EDM experiments sensitive to the nuclear Schiff moment, rather than electron EDM experiments; thus we ignore it in the following discussion.

Assuming for simplicity a uniform nucleon density within the nuclear volume $V = (4\pi/3)R_{nuc}^3$, we obtain from (14.48) the effective short-range scalar P,T-odd $e - N$ Hamiltonian:

$$H_{eN}^S = \frac{iG_F}{\sqrt{2}} \frac{3A}{4\pi R_{nuc}^3} C_S \gamma_e^0 \gamma_e^5 \qquad (r \leq R_{nuc}).$$

In the presence of $\boldsymbol{\mathcal{E}}_{\text{ext}}$, the first order energy shift of a paramagnetic atom due to H_{eN}^S is then:

$$\begin{aligned} \Delta E_S &= \langle \psi | H_{eN}^S | \psi \rangle = 2e\mathcal{E}_{\text{ext}} \langle \Psi_0 | H_{eN}^S | \eta \rangle \\ &= 2i \frac{G_F}{\sqrt{2}} \frac{3AC_S}{4\pi R_{nuc}^3} e\mathcal{E}_{\text{ext}} \int_{r=0}^{R_{nuc}} \Psi_0^\dagger \gamma_e^0 \gamma_e^5 \eta d^3\, \boldsymbol{r}. \end{aligned} \tag{14.49}$$

Making use of Eqs. (14.29) and (14.34), we thus obtain:

$$H_{eN}^S = \frac{iG_F}{\sqrt{2}} AC_S \gamma_e^0 \gamma_e^5 n(\mathbf{R}). \tag{14.50}$$

If we employ the present upper limit on d_a for atomic thallium, and assume that $d_e = 0$, then calculations similar to that just outlined, with corrections for screening, yield the following bound on C_S:

$$|C_S| \leq 1.6 \cdot 10^{-7}. \tag{14.51}$$

It is interesting to note that if the atomic nucleus has spin and a nuclear magnetic moment, several related effects can generate a small but non-zero d_a from d_e and/or the P,T-odd eN interaction, even if the atom has closed shells and is thus diamagnetic [66–68]. The first and larger effect arises from the first term in (14.16) (and/or the P,T-odd eN interaction) and the hyperfine interaction, which together generate an atomic EDM d_a in third order of perturbation. The second (smaller) effect stems from the second term in (14.16), hitherto ignored, where the magnetic field at the electron is due to the nuclear magnetic moment.

14.2.6. *Paramagnetic molecules*

Certain paramagnetic polar diatomic molecules, listed in Table 14.2, are attractive candidates for experimental electron EDM searches, because, as previously mentioned, the attainable electric field $\mathcal{E}_{\text{eff}}$ can be larger by several orders of magnitude than is possible with heavy atoms. Two of the most promising molecules, YbF and PbO, are currently employed in separate experiments that will be discussed in detail in later sections of this chapter. A proposed experiment on the molecular ion HfF^+ will also be described. Here we confine ourselves to a brief explanation of the basic molecular physics.

In each heavy polar molecule of interest, there is strong hybridization of atomic orbitals, which leads to extremely large internal electric fields $\mathcal{E}_{int}(\approx 10^9\text{–}10^{11}$ V/cm) directed along the internuclear axis $\hat{n}$. For example consider a molecule MF, where M is a heavy metal such as Ba, Yb, or Hg. The M atom in its normal state has two $6s$ electrons, while the ground fluorine configuration is $1s^2 \ldots 2p^5$. In the MF molecule one of the $6s$ electrons is transferred to the fluorine, thereby completing its p shell and creating an ionic bond, with a corresponding molecular dipole moment that is typically 3–5 Debye units. The remaining $6s$ electron moves in a highly polarized orbit in $\mathcal{E}_{\text{int}}$. This unpaired electron is the analog of the valence electron in atomic cesium or thallium.

In the absence of an externally applied electric field, $\hat{n}$ precesses about the molecular angular momentum $\boldsymbol{J}$, and $\mathcal{E}_{int}$ is thus oriented randomly in space. However, as was noted earlier, application of a relatively modest external electric field $\mathcal{E}_{\text{ext}}(\approx 10^2\text{–}10^4$ V/cm) causes $\hat{n}$ to be polarized along the direction of $\mathcal{E}_{\text{ext}}$. The reason is as follows. We recall that the wave function of an *atomic* valence electron in the presence of an external electric field $\mathcal{E}_{\text{ext}}$ is given accurately by the first-order formula:

$$|\psi\rangle = |\psi_0\rangle + \Sigma \frac{|\psi_n\rangle\langle\psi_n|e\boldsymbol{\mathcal{E}}_{\textbf{ext}} \cdot \boldsymbol{r}|\psi_0\rangle}{E_0 - E_n}. \tag{14.52}$$

In an atom $E_0 - E_n$ is the energy difference between electronic levels of opposite parity, which is typically of order $0.1e^2/a_0$, and much larger than $\langle\psi_0|e\mathcal{E}_{\text{ext}}\cdot r|\psi_n\rangle$ for all attainable laboratory electric fields. Thus for an atom the sum on the right-hand side of (14.52) is typically much smaller than $|\psi_0\rangle$ (which justifies the use of first order perturbation theory). However, in a diatomic molecule, the relevant levels of opposite parity are adjacent spin- rotational states, which are sometimes separated by extremely small Ω-doubling splittings, or at most by energies of the order of the rotational constant B. Such splittings are 10^3–10^4 times smaller than in the atomic case; hence, readily attainable external fields can cause nearly complete polarization of $\hat{n}$ along $\boldsymbol{\mathcal{E}}_{\text{ext}}$.

Ignoring P,T-odd effects for the moment, the spin-rotation structure of a molecule in a given electronic and vibrational state is conveniently described by an effective spin-rotational Hamiltonian. As an example we consider YbF. Ytterbium has 7 stable isotopes, of which 5 have zero nuclear spin. For YbF with a spin-zero Yb isotope the spin-rotational Hamiltonian is [79]:

$$H_{\text{spin-rot}} = BN^2 + \gamma \boldsymbol{S}\cdot\boldsymbol{N} + b\boldsymbol{I}\cdot\boldsymbol{S} + c(\boldsymbol{I}\cdot\hat{\boldsymbol{N}})(\boldsymbol{S}\cdot\hat{\boldsymbol{N}}) + C\boldsymbol{I}\cdot\boldsymbol{N}, \quad (14.53)$$

where $\boldsymbol{N}, \boldsymbol{S}$, and $\boldsymbol{I}$ are the molecular rotational angular momentum, electron spin, and fluorine nuclear spin ($I = 1/2$), respectively, and B, γ, b, c, and C are coefficients. The contributions of d_e and the scalar P,T-odd eN interaction are included [80] by adding the following effective P,T-odd Hamiltonian H' to $H_{\text{spin-rot}}$:

$$H' = (W_1^{P,T} C_S + W^d d_e)\boldsymbol{S}\cdot\hat{\boldsymbol{N}}. \quad (14.54)$$

Here, $W^d = \frac{1}{\Omega d_e}\langle\psi|\sum_i H_{edm}(i)|\psi\rangle$, where $H_{edm}(i) = 2d_e\begin{pmatrix}0 & 0\\ 0 & \boldsymbol{\sigma_i}\cdot\boldsymbol{\mathcal{E}}_{\text{int}}\end{pmatrix}$ and ψ is the molecular electronic wave function, while i refers to the ith electron, and Ω is the projection of the electronic angular momentum on the internuclear axis. It is customary to define an effective molecular field by $\mathcal{E}_{\text{eff}} = W^d\Omega$.

The coefficients $W_1^{P,T}$ and W^d, and hence $\mathcal{E}_{\text{eff}}$, have been calculated for a number of the molecules of interest by means of both semi-empirical and *ab initio* methods (see Table 14.2). A useful semi-empirical approach was developed by Kozlov [70], who showed that there is a close connection between the matrix elements of the P,T-odd operators and those of magnetic hyperfine structure operators for coupling of electron spin to non-zero M nuclear spin. In many of the molecules of interest, the hyperfine structure constants are known from experiment, and these data can be used to construct quantitative estimates of the electron spin density near the M nucleus, without direct knowledge of the electronic wave function. However the method cannot always be relied on for very accurate values of $W_1^{P,T}$ and W^d, primarily because of spin-correlations between core electrons and the valence electron(s) of interest. To achieve accurate estimates in any but the simplest molecules (i.e. those with a single valence electron in a σ orbital), until recently it was necessary to resort to *ab initio* calculations that employ sophisticated many-body techniques. While this remains the most reliable method, recently Meyer and Bohn have developed an interesting alternative technique for calculating $\mathcal{E}_{\text{eff}}$ [75, 77]. Using a standard, widely available software package for calculating nonrelativistic molecular electronic structure, they determine the molecular wave functions and write them in the basis of atomic orbitals on M. Relativistic effects are then accounted for using semi-empricial formulae developed for atoms [26]. Comparisons to every case where a full-scale *ab initio* many-body calculation has been performed show that this much simpler method is reliable at the $\sim 20\%$ level.

Table 14.2. Calculated P,T-odd coefficients of polar diatomic paramagnetic molecules.

Molecule	Electronic State	W^d 10^{24} Hz e^{-1}cm^{-1}	$\mathcal{E}_{\text{eff}}$ GV/cm	W_1^{PT} kHz
BaF	$X^2\Sigma^+_{1/2}$	-3.6^{a}	7.5^{a}	-12^{b}
YbF	$X^2\Sigma^+_{1/2}$	-12.1^{c}	26^{c}	-33^{d}
HgF	$X^2\Sigma^+_{1/2}$	-48^{e}	99^{e}	-185^{e}
PbF	$X^2\Pi_{1/2}$	14^{e}	-29^{e}	55^{e}
PbO	$a(1)^3\Sigma_1$	$-(6.1^{+1.8}_{-0.6})^{f}$	26^{f}	
	$B(1)^3\Pi_1$	$-(8.0 \pm 1.6)^{f}$	34^{f}	
ThO	$H^3\Delta_1$		104^{g}	
HI^+	$X^2\Pi_{3/2}$	0.22^{h}	0.34^{h}	
PtH^+	$X^3\Delta_3{}^{i}$		73^{ij}	
HfH^+	$(X?)^3\Delta_1{}^{i}$		-17^{i}	
HfF^+	${}^3\Delta_1{}^{k}$		24^{k}	
ThF^+	${}^3\Delta_1{}^{g}$		90^{g}	

[a] Ref. 69
[b] Ref. 70
[c] Ref. 71
[d] Ref. 72
[e] Ref. 73
[f] Ref. 74
[g] Ref. 75
[h] Ref. 76
[i] Ref. 77
[j] Note that $d_e\mathcal{E}_{\text{eff}}$ is the energy shift for the extreme m sublevel ($m = J$) for the molecular eigenstate with total angular momentum $J = \Omega$. Hence, taking full advantage of the large value of $\mathcal{E}_{\text{eff}}$ in PtH^+ would require measuring the energy difference between $m = \pm 3$ sublevels.
[k] Ref. 78

14.3. Electron EDM Experiments

14.3.1. *General overview*

14.3.1.1. *A simple model experiment*

Experimental searches for d_e differ in their details, but they share many broad features. Virtually every experimental configuration for free atoms or molecules is analogous to an optical interferometer. Each consists of a state selector, where the initial quantum state ψ_0 of the system is prepared; an interaction region or interval in which the system evolves for a time τ in an electric field $\boldsymbol{\mathcal{E}}$ (and often but not always a magnetic field $\boldsymbol{B}$ as well); an analyzer where the resulting quantum state is prepared for detection; and a detector. At least some of these components are separated spatially in beam experiments, but in cell experiments they are not. Also, "analysis" and "detection" are sometimes amalgamated into a single process. Time τ may be the transit time of an atom or molecule in a beam through the interaction region, or the relaxation time of spins in a vapor due to collisions, or the natural lifetime of a metastable state.

To understand the essential features and some of the most important problems encountered, it is helpful to consider a simple model "atom" of spin 1/2 with enhancement factor R, containing an unpaired electron with spin magnetic moment $-g\mu_B/2$ and EDM d_e. Suppose the spin is initially prepared to lie along $\hat{x}$: $\psi_0 = \frac{1}{\sqrt{2}}\begin{pmatrix}1\\1\end{pmatrix}$, while $\boldsymbol{\mathcal{E}}$ and $\boldsymbol{B}$ are parallel to the $\hat{z}$ axis. Then, during the interaction interval the spin rotates in the xy plane by angle $2\phi = -(d_eRE - g\mu_BB/2)\tau/\hbar$, so that at time τ the quantum state has evolved to:

$$\psi = \frac{1}{\sqrt{2}}\begin{pmatrix}e^{-i\phi}\\ e^{i\phi}\end{pmatrix}. \tag{14.55}$$

We choose the analyzer of our simple model to be represented by the unitary matrix $A = \frac{1}{\sqrt{2}}\begin{pmatrix}1 & 1\\ -i & i\end{pmatrix}$, so that when A is applied to ψ one obtains the state $\psi' = \begin{pmatrix}\cos\phi\\ \sin\phi\end{pmatrix}$. Finally, in this model the detector measures the probability of finding the system in the upper component of ψ'. Thus assuming 100% detection efficiency, the signal from a group of N atoms observed in time τ is:

$$S = N\cos^2\phi. \tag{14.56}$$

The angle ϕ is the sum of a large term $\phi_1 = g\mu_B B\tau/4\hbar$ and an extremely small term $\phi_2 = -d_e R\mathcal{E}\tau/2\hbar$. To isolate ϕ_2 one observes the signal $S = N\cos^2(\phi_1 + \phi_2)$ for $\boldsymbol{\mathcal{E}}$ and $\boldsymbol{B}$ both parallel and anti-parallel. Reversing $\boldsymbol{\mathcal{E}} \cdot \boldsymbol{B}$ changes the relative sign of ϕ_1 and ϕ_2 and thus changes S; for given N the largest change in S occurs when $\phi_1 = \pm\pi/4$. Thus, choosing $\phi_1 = -\pi/4$ and taking into account that $|\phi_2| \ll 1$, we have:

$$S_+ \equiv S(\boldsymbol{\mathcal{E}} \cdot \boldsymbol{B} > 0) = \frac{N}{2}(1 + 2\phi_2)$$

$$S_- \equiv S(\boldsymbol{\mathcal{E}} \cdot \boldsymbol{B} < 0) = \frac{N}{2}(1 - 2\phi_2).$$

The simple model we have just described is readily adapted to describe most realistic experimental conditions without radical change of its principal ideas. For example, in the Berkeley Tl experiment [4], where ^{205}Tl atoms in the $6^2P_{1/2}F = 1$ hyperfine state were employed, the angle ϕ_2 corresponds to a phase shift between the $m_F = \pm 1$ components of this state.

14.3.1.2. *Noise*

The uncertainty in a measurement of ϕ_2 is usually caused by shot noise and by fluctuations in various experimental parameters (most significantly the magnetic field) that contribute "phase noise". Let N_0 atoms be observed in a time τ_0 (which is not necessarily equal to τ). The shot noise uncertainty in time τ_0 is $\delta\phi_2^{\text{shot}} = \sqrt{\frac{1}{N_0}}$. If the N_0 atoms are exposed to a common, time-dependent magnetic field $B_z(t)$, the Zeeman effect adds the phase: $g\mu_B \int_0^{\tau_0} B_z(t)dt/4\hbar$. Assuming this contribution fluctuates randomly about zero, the standard deviation in this portion of the phase is: $\delta\phi_2^{\text{mag}} = GD$, where $D = \left(\int_0^{\tau_0} B_z(t)dt\right)_{rms}$ and $G = eg/4mc$. The total uncertainty in the phase for time τ_0 is obtained by adding $\delta\phi_2^{\text{shot}}$ and $\delta\phi_2^{\text{mag}}$ in quadrature. Hence if the experiment is repeated t/τ_0 times for a total time of observation t, the uncertainty in d_e is:

$$\delta d_e = \sqrt{\frac{(GD)^2 + \frac{1}{N_0}}{t\tau_0}} \left|\frac{\hbar}{\mathcal{E}_{\text{eff}}}\right|, \qquad (14.57)$$

where we have replaced $R\mathcal{E}$ by $\mathcal{E}_{\text{eff}}$. As (14.57) reveals, increasing N_0 past the point where magnetic phase noise begins to dominate does not help to improve the precision of a measurement of d_e.

Magnetic phase noise can arise from external sources such as laboratory equipment, building elevators, nearby electric railways, etc., and careful shielding can reduce such noise by orders of magnitude. However, magnetic phase noise can also be generated by thermal fluctuations in the electric current density in conducting parts of experimental apparatus (including the shields themselves). Such fluctuations occur even in the absence of applied voltage and go by the name "magnetic Johnson noise" (MJN). The effects of MJN on electron EDM experiments have been analyzed in detail by Munger [81], following earlier work by Lamoreaux [82] and by Nenonen, Montonen, and Katila [83]. It has been shown that at a point at distance z from the surface of an infinite slab of thickness d, resistivity ρ, magnetic permeability μ, and absolute temperature T_0, the RMS value of the magnetic field in the z direction is $\left(\int_0^\infty B_{n,z}^2(\nu)d\nu\right)^{1/2}$, where:

$$B_{n,z}(\nu) = \mu_0\sqrt{\frac{k_B T_0 d}{8\pi\rho z(z+d)}}\theta, \tag{14.58}$$

and where we here employ S.I. units, μ_0 is the permeability of vacuum, k_B is Boltzmann's constant, and θ is a dimensionless integral. For all frequencies $0 \leq \nu \leq \nu_c$, with $\nu_C = \rho/(2\pi\mu z^2)$, one finds $\theta \approx 1$, but for $\nu > \nu_C$, θ decreases to zero. One can show that the quantity D appearing in Eq. (14.57) is related to $B_{n,z}(\nu)$ by:

$$D = \sqrt{\frac{\tau_0}{2}}\left[\int_0^\infty B_{n,z}^2(\nu)\frac{\sin^2\pi\nu\tau_0}{(\pi\nu\tau_0)^2}\tau_0 d\nu\right]^{1/2}. \tag{14.59}$$

However, for almost all practical situations except those in which very high permeability conductors are present, it is sufficient to employ the zero frequency limit:

$$D \approx B_{n,z}(0)\sqrt{\frac{\tau_0}{2}}. \tag{14.60}$$

In any real experiment, atoms or molecules occupy a finite volume V near conductors. How well correlated are the fluctuations at different points within V? Roughly speaking, the correlation length is $\approx z$ parallel or perpendicular to the slab.

Using this result (which together with particle velocity is relevant for the choice of τ_0), as well as knowledge of the geometry, materials, and other parameters in various electron EDM experiments, Munger has estimated the limits on precision due to MJN, and these estimates have the following implications. MJN was about an order-of-magnitude less significant than other sources of noise in the Berkeley Tl experiment, a conclusion in agreement with unpublished estimates made by Regan *et al.* [4]. MJN is unlikely to be a problem for paramagnetic molecule experiments until they reach a precision of $\approx 2\text{–}5 \cdot 10^{-30}$ e cm. However, for atomic systems with $R \approx 100$ (e.g. Cs) or less, if one wishes to improve the existing limit on d_e substantially and remain limited by shot noise, then unless special magnetic field cancellation techniques are employed, metallic electric field plates are likely to generate too much noise. Also, MJN from vacuum system walls must be screened, or else the walls must be made of non-conducting material. Standard high permeability metallic magnetic shields generate roughly as much MJN as ordinary metals (copper, stainless steel, titanium, etc.). Thus it is advisable that the innermost of a nested set of magnetic shields be made of ferrite or some other material with high resistivity as well as high permeability and low coercivity.

14.3.1.3. *Systematic errors*

In any EDM experiment P,T violation is revealed by a term in the signal proportional to a P,T-odd pseudoscalar such as $\boldsymbol{\mathcal{E}} \cdot \boldsymbol{B}$. However, a false term of the form $\boldsymbol{\mathcal{E}} \cdot \boldsymbol{B}$ will appear even without P,T violation if $\boldsymbol{B}$ depends on the sign of $\boldsymbol{\mathcal{E}}$. Such dependence can occur in various ways. First, a component of $\boldsymbol{B}$ in the direction of $\boldsymbol{\mathcal{E}}$ can be generated by leakage currents flowing through the insulator(s) separating the electric field electrodes. Careful design can minimize this possibility, but it is almost impossible to eliminate it completely. Thus to a greater or lesser extent, this is a common problem for almost all EDM experiments.

Second, in beam experiments, where the atoms or molecules have a well-defined velocity $\boldsymbol{v}$ through the interaction region, a motional magnetic field $\boldsymbol{B}_{\text{mot}} = \frac{1}{c}\boldsymbol{\mathcal{E}} \times \boldsymbol{v}$ exists in addition to the applied magnetic field $\boldsymbol{B}$. Let $\boldsymbol{v} = v\hat{x}$ and $\boldsymbol{\mathcal{E}} = \mathcal{E}\hat{z}$, and assume that the applied field $\boldsymbol{B}$, which is nominally in the z direction, has a small unintended component in the y direction: $\boldsymbol{B} = B_y\hat{y} + B_z\hat{z}$ with $B_y \ll B_z$. Then the total magnetic field (applied plus motional) is $\boldsymbol{B}_{\text{total}} = \left(B_y + \frac{v\mathcal{E}}{c}\right)\hat{y} + B_z\hat{z}$. The impact of the resulting systematic effect depends dramatically on the presence or absence of a quadratic Stark effect in the system [84]. We can see the behavior in both cases simultaneously by analyzing as an example an atom or molecule in a state of total angular momentum $F = 1$.

The Zeeman shifts of the $F = 1, m_F = \pm 1$ levels in B_z are $E_\pm = \pm g_F \mu_B B_z$. Let B_z be sufficiently weak that the Zeeman shift of $F = 1, m_F = 0$ (proportional to B^2) is quite negligible. In addition suppose that the quadratic Stark effect causes $m_F = 0$ to be shifted relative to the average energy of $m_F = \pm 1$ by an amount $-\Delta = a\mathcal{E}^2$, where a is a constant. Then, ignoring a possible scalar Stark shift that affects all three Zeeman components by the same amount, the Hamiltonian matrix for $F = 1$ is

$$H = \begin{pmatrix} k_1 & -ik_2 & 0 \\ ik_2 & k_3 & -ik_2 \\ 0 & ik_2 & -k_1 \end{pmatrix}, \tag{14.61}$$

where the rows (and columns) are labeled by $m_F = +1, 0, -1$, respectively, while $k_1 = g_F \mu_B B_z, k_2 = g_F \frac{\mu_B}{\sqrt{2}} \left(B_y + \frac{\mathcal{E}v}{c}\right)$, and $k_3 = -\Delta$. Assuming $|k_2| \ll |k_1|$, we diagonalize the matrix to obtain the energy eigenvalues:

$$\begin{aligned} \lambda_+ &= k_1 + \frac{k_2^2}{k_1 - k_3} + \text{ higher order terms} \\ \lambda_- &= -k_1 - \frac{k_2^2}{k_1 + k_3} + \cdots \\ \lambda_0 &= k_3 + \frac{k_2^2 k_3}{k_3^2 - k_1^2} + \cdots . \end{aligned}$$

The quantity of interest is the energy difference $\lambda_+ - \lambda_-$, given with sufficient accuracy by:

$$\delta E = \lambda_+ - \lambda_- = 2k_1 + 2\frac{k_2^2 k_1}{k_1^2 - k_3^2}. \tag{14.62}$$

If $k_1^2 \gg k_3^2$, (that is if $\Delta \ll g_F \mu_B B_z$), then:

$$\delta E \cong 2k_1 \left(1 + \frac{k_2^2}{k_1^2}\right) = 2g_F \mu_B \left(B_z + \frac{B_y^2 + \frac{\mathcal{E}^2 v^2}{c^2} + 2\frac{\mathcal{E}v}{c} B_y}{2|B_z|}\right). \tag{14.63}$$

In this expression the troublesome term responsible for the $\boldsymbol{\mathcal{E}} \times \boldsymbol{v}$ effect is the last one on the right-hand side. It is odd in $\boldsymbol{\mathcal{E}}$ and in $\boldsymbol{B}$ (as well as in $\boldsymbol{v}$). If $k_3^2 \gg k_1^2$, that is if $\Delta \gg g_F \mu_B B_z$, this troublesome term also appears, but its coefficient is smaller by the factor $\left(\frac{g\mu_B B_z}{\Delta}\right)^2$:

$$\begin{aligned} \delta E &\cong 2k_1 \left(1 - \frac{k_2^2}{k_3^2}\right) \\ &= 2g_F \mu_B \left(B_z - \frac{1}{2}\left(\frac{g\mu_B B_z}{\Delta}\right)^2 \frac{B_y^2 + \frac{\mathcal{E}^2 v^2}{c^2} + 2\frac{\mathcal{E}v}{c} B_y}{|B_z|}\right). \end{aligned} \tag{14.64}$$

In the most troublesome case [eq. (14.63)], the $\boldsymbol{\mathcal{E}} \times \boldsymbol{v}$ effect cannot be distinguished from a genuine EDM signal by varying the magnitude of the applied magnetic field, since it is proportional to $B_y/|B_z|$. However, since it is odd in $\boldsymbol{v}$ it can in principle be canceled by using opposing atomic (molecular) beams.

Another systematic effect, often related to the one just described, involves the geometric phase (sometimes called "Berry's phase"), and appears if the direction of the quantization axis varies in a certain way between the state selector and the analyzer [85]. For example, consider a spin with expectation value aligned along a magnetic field at time $t = 0$. Now imagine that as time elapses the direction of the field slowly changes in the particle rest frame, so that the tip of the field vector traces out a closed curve, coming back to its starting point after time τ. If the change is slow, the spin follows the field vector adiabatically, and the spin expectation value also returns to its original orientation at time τ. Nevertheless the spin wave function accrues a (geometric) phase proportional to the solid angle traced out by the tip of the magnetic field vector. If a portion of the magnetic field is motional, the solid angle is altered by the reversal of $\boldsymbol{\mathcal{E}}$ as well as of $\boldsymbol{B}$. It is not even necessary for the tip of the magnetic field vector to describe a closed curve, for open curves also result in $\boldsymbol{\mathcal{E}}$-odd, $\boldsymbol{B}$-odd geometric phases. The geometric phase effect was significant in the Berkeley Tl experiment, and even played a role in neutron EDM experiments utilizing neutrons trapped in a cell [86]. Further analysis of the geometric phase effect with application to neutron EDM experiments has been presented by Lamoreaux and Golub [87]. In beam experiments with polar molecules, the quantization axis can be determined by $\boldsymbol{\mathcal{E}}$ if it is sufficiently strong that the internuclear axis is significantly polarized along $\boldsymbol{\mathcal{E}}$. In this case, if $\boldsymbol{\mathcal{E}}$ varies in direction, geometric-phase-related systematic errors might arise. Geometric phase effects for molecules confined in a Stark-gravitational trap have also been studied theoretically [88].

Systematic errors from light shifts [89] can affect EDM experiments whenever intense beams of laser light interact with atoms during the time interval τ, as would be the case in an optical dipole trap. One such effect is the "optical Zeeman shift" which appears if there is a residual component of circular polarization in the trapping light. It results in a "vector" shift of Zeeman levels (that is, a shift linear in the m_F value of a Zeeman sublevel of a hyperfine component F). Hyperfine interactions in the excited state of the atom of interest, together with the trapping light and the applied $\boldsymbol{\mathcal{E}}$ field, cause tensor shifts (quadratic in m_F) in the ground state F level. Finally, in the presence of $\boldsymbol{\mathcal{E}}$ each atomic level acquires a small admixture of states of the opposite parity (Stark mixing), which causes interference between the primary E1 optical transition amplitude and the Stark-induced

M1 and E2 transition amplitudes. The resulting light shifts (calculated in third-order perturbation theory) are proportional to $\boldsymbol{\mathcal{E}}$, and this last effect has the potential to cause serious systematic error.

A variety of approaches are employed to deal with systematic effects. Some experiments utilize diamagnetic atoms or paramagnetic atoms with low Z as co-magnetometers, in addition to the atoms of interest. The co-magnetometers have negligible or small enhancement factors, but are sensitive to leakage currents, and/or the $\boldsymbol{\mathcal{E}} \times \boldsymbol{v}$ and geometric phase effects. In cell experiments where velocities are randomized by multiple collisions with buffer gas and/or cell walls, the $\boldsymbol{\mathcal{E}} \times \boldsymbol{v}$ effect and the geometric phase effect are strongly suppressed. Light shifts can be mitigated by minimizing residual circular polarization of trapping light (Zeeman shift), and/or by appropriate choice of relative orientations of light linear polarization, light propagation direction, and $\boldsymbol{\mathcal{E}}$ (third-order light shift). In the paramagnetic molecule experiments, the ratio $\boldsymbol{\mathcal{E}}_{\text{eff}}/\boldsymbol{\mathcal{E}}_{\text{ext}}$ is very large, and sensitivity to some systematics is correspondingly reduced. In addition, in molecular states with $F \geq 1$ (e.g., YbF and PbO*) the $\boldsymbol{\mathcal{E}} \times \boldsymbol{v}$ effect is mitigated by large tensor polarization due to quadratic Stark effect [recall (14.64)]. In the PbO cell experiment and the proposed HfF^+ experiment, use is made of both components of an Ω-doublet, which respond with opposite signs to d_e but respond virtually identically to systematics associated with magnetic fields. The case of PbF is of some interest because it has been shown that the electric field-dependent g factor of its $^2\Pi_{1/2}$ ground state should vanish when a suitable external electric field $\boldsymbol{\mathcal{E}}_0$ is applied [90]. An experiment on PbF has been suggested [91]. If it were performed at $\boldsymbol{\mathcal{E}}_0$ (calculated to be ≈ 67 kV/cm) several potential magnetic field-related systematic errors might be avoided. In all molecular experiments, the saturation of the molecular polarization (and hence $\mathcal{E}_{\text{eff}}$) at a finite value of $\mathcal{E}_{\text{ext}}$ leads to a well-understood non-linear dependence of the EDM signal on $\mathcal{E}_{\text{ext}}$ that can be employed in principle to discriminate against certain systematic effects.

14.3.2. *The Berkeley thallium atomic beam experiment*

In this experiment [4], two pairs of vertical counter-propagating atomic beams, separated by 2.54 cm, and each consisting of Tl ($Z = 81$) and Na ($Z = 11$), were employed (See Fig. 14.4). State-selection and analysis of the $6^2P_{1/2}(F = 1)$ state of Tl and the $3^2S_{1/2}(F = 2, F = 1)$ states of Na were accomplished by laser optical pumping and atomic beam magnetic resonance with separated oscillating RF fields of the Ramsey type.

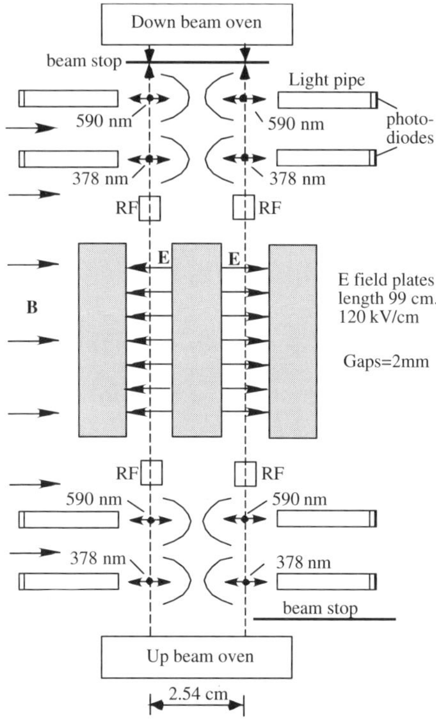

Fig. 14.4. Schematic diagram of the Berkeley thallium experiment [4], not to scale. Laser beams for state selection and analysis at 590 nm (for Na) and 378 nm (for Tl) are perpendicular to the page, with indicated linear polarizations. The diagram shows the up-going atomic beams active.

Detection was achieved by observation of laser-induced fluorescence in the analyzer region. Between the 2 RF fields was a region of length ≈ 1 meter where the spatially separated atomic beams were exposed to nominally identical $\boldsymbol{B}$ fields, and opposite $\boldsymbol{\mathcal{E}}$ fields of ≈ 120 kV/cm. This provided common-mode noise rejection and control of some systematic effects. Use of counter-propagating atomic beams served to cancel all but a very small remnant of the $\boldsymbol{\mathcal{E}} \times \boldsymbol{v}$ effect, and various auxiliary measurements, including use of Na as a co-magnetometer, further reduced this remnant and isolated the geometric phase effect. Leakage currents were monitored by observing the decay of $\boldsymbol{\mathcal{E}}$ after disconnecting the high voltage power supply from the electric field plates. $\boldsymbol{\mathcal{E}}$ could be measured very precisely by using the quadratic Stark effect. A number of other small systematic effects were also dealt with effectively by auxiliary measurements. About $5.2 \cdot 10^{13}$

photo-electrons of signal per up/down beam pair were collected by the fluorescence detectors. Assuming the enhancement factor $R = -585$, the final result is:

$$d_e = (6.9 \pm 7.4) \cdot 10^{-28} e \text{ cm}, \tag{14.65}$$

which yields the limit:

$$|d_e| \leq 1.6 \cdot 10^{-27} e \text{ cm} \qquad (90\% \text{ conf.}) \tag{14.66}$$

already referred to in (14.3).

14.3.3. *Cesium optical pumping experiments*

An optical pumping experiment to search for d_e in Cs was carried out at Amherst by L. Hunter and co-workers [92] and reported in 1989. In its time it achieved the best limit on d_e, and although that result has now been surpassed, we describe it here because the method is interesting and has been resuscitated in a present-day search by a Princeton group led by M. Romalis [93]. The Amherst experiment was carried out with two glass cells, one stacked on the other in the z direction. The plane surfaces parallel to the xy plane were coated with tin oxide, and ±4kV was applied to the center electrode, while the outer ones were grounded. Thus nominally equal and opposite $\boldsymbol{\mathcal{E}}$ fields were applied in the two cells. The cells were filled with cesium (number density $n(\text{Cs}) \approx 4 \cdot 10^{10}$ cm^{-3}) and nitrogen ($n(\text{N}_2) \approx 9 \cdot 10^{18}$ cm^{-3}), the latter employed to minimize Cs ground state spin relaxation. A circularly polarized laser beam for each cell directed along x and tuned to the cesium $6S_{1/2}F = 3 \rightarrow 6P_{1/2}$ transition was utilized for optical pumping, which resulted in initial polarization along x of the $6S_{1/2}F = 4$ state of $\approx 70\%$. The polarization of the pump beam was periodically reversed by means of a Pockels cell. The cells were mounted inside a multi-layer magnetic shield, and with the aid of compensation coils, the magnetic field components in all three directions were reduced to less than 10^{-7} G, except for a small field applied along the x axis to compensate for the Zeeman light shift produced by the pump laser beam. Thus precession of the atomic polarization in the xy plane was nominally due to $\boldsymbol{\mathcal{E}}$ alone, and was monitored by a probe laser beam, directed along y, and tuned to the $6S_{1/2}F = 4 \rightarrow 6P_{1/2}$ transition. The effective time interval for polarization precession was the ground state spin relaxation time $\tau \approx$ 15 ms. The signals were the intensities of the probe beams transmitted through each cell, and a non-zero EDM would have been indicated by a dependence of these signals on the pump and probe circular polarizations

$\boldsymbol{\sigma}, \boldsymbol{J}$ respectively and the sign of $\boldsymbol{\mathcal{E}}$, manifested in a component of each signal proportional to the rotational invariant $\boldsymbol{J} \cdot (\boldsymbol{\sigma} \times \boldsymbol{\mathcal{E}})\,\tau$. The most important sources of possible systematic error were leakage currents, which could not be monitored adequately, and imperfect reversal of the electric field, which could be monitored by observing the tensor polarization of the ground state of Cs arising from quadratic Stark effect. The result was:

$$d_e = (-1.5 \pm 5.5 \pm 1.5) \cdot 10^{-26} e \text{ cm}. \tag{14.67}$$

In the current version of this experiment at Princeton, there are a number of fundamental modifications. In addition to cesium, and nitrogen gas at a fraction of atmospheric pressure (now utilized to quench spontaneous emission from excited Cs atoms), each cell also contains ^{129}Xe at several atmospheres pressure. Optical pumping of Cs results in polarization of the valence electron spins, and polarization is transferred to the ^{129}Xe nuclei by spin-exchange collisions. Here the principal mechanism is the "contact" hyperfine interaction between the Cs valence electron and the ^{129}Xe nucleus. This interaction causes relatively large frequency shifts in the Cs electron spin resonance and the ^{129}Xe nuclear magnetic resonance. In the case of extremely small applied magnetic fields, such frequency shifts can be larger than the Larmor frequencies themselves. This results in novel and rather complex "hybrid" resonance behavior that has been studied in detail experimentally by the Princeton group [94, 95] in an analogous alkali-rare gas system: potassium and ^{3}He. A phenomenological theoretical description of the hybrid resonances, worked out in terms of Bloch's equations, yields predictions in good agreement with experiment. One feature is of special relevance for an electron EDM search. This is a self-compensation mechanism, predicted by the Bloch equation formalism and observed experimentally, where slow changes in components of magnetic field transverse to the initial polarization axis are nearly canceled by interaction between the alkali electron spin and the noble gas nuclear spin, leaving only a signal proportional to an anomalous interaction (e.g. interaction of an EDM with $\boldsymbol{\mathcal{E}}_{\text{eff}}$) that does not scale with the magnetic moments. This mechanism is important because it has the potential to reduce magnetic Johnson noise, as well as systematic error from leakage currents. The Princeton group has succeeded in applying electric fields of 15 kV/cm to their Cs-Xe cells with the aid of electrodes external to the cells, but a special surface coating on the inner cell walls is necessary to prevent disappearance of Cs, which would otherwise occur when high voltage is applied. Such disappearance may have been due to the build-up of stray charges on the inner walls.

A separate cesium experiment that employs optical pumping and a slow "fountain" atomic beam has been proposed and developed by H. Gould and co-workers at the Lawrence Berkeley National Laboratory [96].

14.3.4. *Cesium optical trap experiments*

Trapped and cooled paramagnetic atoms offer some advantages for electron EDM searches, and experiments of this type with cesium have been proposed by a number of investigators, including S. Chu *et al.* [97], D. Heinzen [98], and D. S. Weiss [99]. Cooling and trapping make possible long coherence times, which can compensate for the fact that smaller numbers of atoms may be available for use compared to the numbers in conventional cell or beam experiments. Trapping randomizes atomic velocities and cooling reduces them by orders of magnitude. Thus linewidths are greatly narrowed, and the $\boldsymbol{\mathcal{E}} \times \boldsymbol{v}$ effect is essentially eliminated as a source of systematic error. Also, different atomic species (e.g. Cs and Rb) can be loaded simultaneously into the same far-detuned optical lattice, so that co-magnetometry can be employed for further reduction in systematic error.

However, several potentially serious problems confront optical trap experiments. We have already referred to problems caused by light shifts and the fact that in all Cs experiments where a substantial improvement in the present limit on d_e is desired, magnetic Johnson noise is a problem that must be overcome.

Two electron EDM searches with trapped Cs are being developed: One by D. S. Weiss and co-workers at Pennsylvania State University [99] and another by D. Heinzen and co-workers at the University of Texas [98]. The Texas apparatus consists of two side-by-side far-off-resonance optical dipole traps with trapping wavelength $\lambda = 1.3$ microns, far to the red of the Cs 894 nm resonance line. These traps are placed between three parallel electric field plates which generate nominally equal and opposite $\boldsymbol{\mathcal{E}}$ fields in the two traps. There is also a $\boldsymbol{B}$ field of several mG. The traps are housed in a Ti vacuum chamber, which is inside a five-layer magnetic shield. The optical trap is in a vertical one-dimensional optical lattice configuration; standing waves are sustained in two-mirror optical resonators interior to the vacuum chamber. To load the Cs atoms into the optical lattice, Heinzen and co-workers plan to create a cold atomic beam with a 2D magneto-optical trap exterior to the shields and vacuum system, and to capture those Cs atoms with optical molasses between the $\boldsymbol{\mathcal{E}}$ plates. The concept and design of the Pennsylvania State experiment is very similar.

14.3.5. *The francium optical trap experiment*

Francium ($Z = 87$), discovered in 1939, is the heaviest alkali atom, with an enhancement factor $R = 1150$ (10 times that of Cs). Unfortunately, there are no stable isotopes of Fr, the longest lifetime (22 min) being that of ^{223}Fr. An experiment to search for the EDM of ^{210}Fr($\tau = 3.2$ min) has been proposed and is being developed by a group at the Research Center of Nuclear Physics (RCNP), Osaka University, Japan [100]. Radioactive francium is produced in the heavy-ion fusion reaction:

$$^{197}\mathrm{Au}(^{18}\mathrm{O}, xn)^{209-211}\mathrm{Fr} \tag{14.68}$$

using ^{18}O ions formed in an ECR source and accelerated at the RCNP cyclotron. The ^{18}O beam thus generated is incident on a gold target, where at ≈ 100 MeV beam energy, one can produce $\approx 1.3 \cdot 10^5$ ^{210,211}Fr ions per second. The francium ions are transported to an yttrium target, which acts as a neutralizer. The resulting neutral francium atoms are to be trapped and cooled in a magneto-optic trap apparatus equipped with electric field plates. At the time of writing, this experiment is in an early stage of development.

14.3.6. *The YbF experiment*

E. A. Hinds and co-workers [101] at Imperial College, London have developed a molecular beam experiment for investigation of d_e in the $X^2\Sigma^+_{1/2}(v = 0, N = 0)$ ground state of YbF. Fig. 14.5 is a schematic diagram of the apparatus. YbF molecules are generated by pulsed laser beam ablation of a

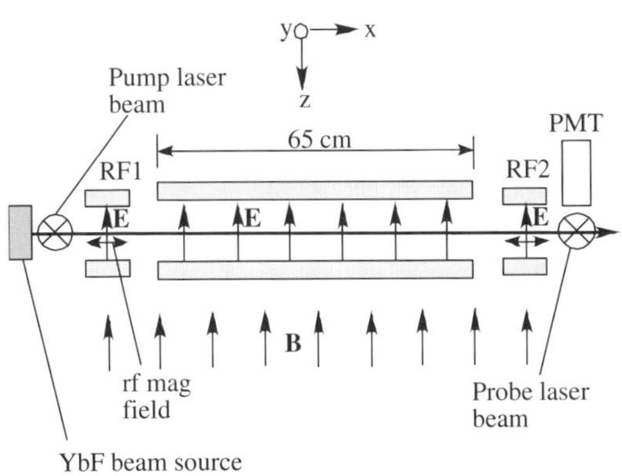

Fig. 14.5. Schematic diagram of the Imperial College YbF experiment [99], not to scale.

solid disk of Yb; the resulting Yb atoms and ions are entrained in a supersonic carrier gas of Ar or Xe with a few percent admixture of SF_6, which is admitted to the vacuum chamber by a pulsed valve that is synchronized with the ablation laser [102]. Reactions between Yb and SF_6 then yield a substantial amount of YbF. The translational and rotational temperatures of the YbF beam molecules (which are essentially the same) can be as low as 1.4K, and the vibrational temperature is sufficiently low that 98% of the molecules are in the ground vibrational state.

Hinds and his group have investigated the relevant spectroscopic parameters of YbF [79]. Fig. 14.6 shows the hyperfine structure of the $X\ ^2\Sigma^+_{1/2}(v = 0, N = 0)\ J = 1/2$ state of a ^{174}YbF molecule. ^{174}Yb has zero nuclear spin and natural abundance 31.8%. Since the nuclear spin of fluorine is 1/2, there are two hyperfine components: $F = 1$ and $F = 0$, separated by 170 MHz. In the absence of magnetic and electric fields, the 3 sublevels $F = 1, (m_F = \pm 1, m_F = 0)$ are degenerate. However, in the central region of the apparatus of length 65 cm, an external electric field $\mathcal{E} = 8.3$ kV/cm is applied in the z direction (it corresponds to an effective internal field $\boldsymbol{\mathcal{E}}_{\text{eff}} = 13$ GV/cm; see Fig. 14.7). In this field the $m_F = \pm 1$ levels would still be degenerate if d_e were zero (and if we ignore a possible non-zero W_1^{PT}), but otherwise they are split by $2d_e\mathcal{E}_{\text{eff}}$. Also, in this applied field the level $F = 1, m_F = 0$ is shifted downward relative to

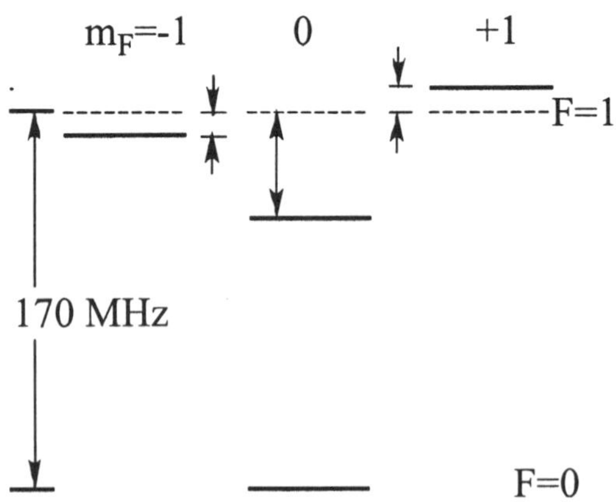

Fig. 14.6. Schematic diagram, not to scale, of the hyperfine structure of the $X^2\Sigma$ electronic state of YbF in the lowest vibrational and rotational level, for the case of an Yb nucleus with zero nuclear spin. Δ is the tensor Stark shift of $m_F = 0$ with respect to the average energies of $m_F = \pm 1$ levels; δ is the shift caused by the combination of the Zeeman effect and the effect of d_e in $\boldsymbol{\mathcal{E}}_{\text{eff}}$.

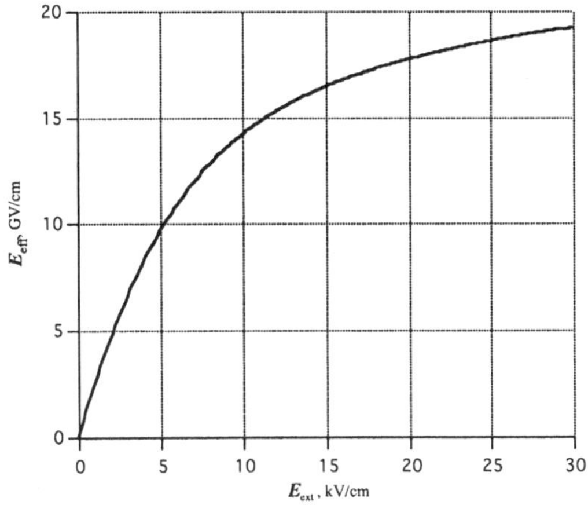

Fig. 14.7. $\mathcal{E}_{\text{eff}}$ versus $\mathcal{E}_{\text{ext}}$ for YbF.

$m_F = \pm 1$ by an amount $\Delta = 6.7$ MHz $\cdot\, h$ (a large tensor quadratic Stark shift). A magnetic field $\boldsymbol{B}$, nominally in the z direction and of the order of 0.1 mGauss, is also applied in the central region. As noted previously, this causes an additional splitting

$$2\mu_B B_z \left[1 - \frac{1}{2}\left(\frac{\mu_B B_\perp}{\Delta}\right)^2\right] + \text{higher order terms} \tag{14.69}$$

between the $m_F = \pm 1$ components. Here $B_\perp$ is the vector sum of an inadvertent component of $\boldsymbol{B}$ in the x-y plane and a motional contribution $\boldsymbol{\mathcal{E}} \times \boldsymbol{v}/c$. The residual systematic term linear in $\boldsymbol{\mathcal{E}} \times \boldsymbol{v}$ is negligible because Δ is so large. The splittings due to d_e and the magnetic interaction are separated, as usual, by making observations with applied $\boldsymbol{\mathcal{E}}$ and $\boldsymbol{B}$ fields both parallel and anti-parallel.

We now follow the YbF molecules as they pass through various components of the apparatus. Laser excitation in the pump region removes all $F = 1$ ground state molecules, leaving only $F = 0$ molecules remaining as they depart from the pump region and enter the region RF1. There, with a 3.3 kV/cm electric field imposed along the z direction, a 170 MHz RF magnetic field along x excites each molecule from $F = 0$ to the coherent superposition $|\psi\rangle = \frac{1}{\sqrt{2}}|F = 1, m_F = 1\rangle + \frac{1}{\sqrt{2}}|1, -1\rangle$. In the central region, the beam is exposed to the fields $(\pm\mathcal{E}, \pm B)\hat{z}$, and the two parts of this wave function develop a relative phase shift:

$$2\phi = 2(\pm d_e \mathcal{E}_{\text{eff}} \mp \mu_B B)\tau/\hbar, \tag{14.70}$$

where τ is the time of transit of the beam through the central region.

In region RF2, similar to RF1, an RF field drives each $F = 1$ molecule back to $F = 0$. Because of the phase shift 2ϕ developed in the central region, the final population of $F = 0$ molecules is proportional to $\cos^2 \phi$. Finally, the YbF molecular beam is detected by laser-induced fluorescence in the probe region, where the laser is tuned to the $F = 0$ component in the Q(0) line of the $A^2\Pi_{1/2} - X^2\Sigma^+$ transition at 553 nm.

The two RF fields do not form a "Ramsey pair": they are not necessarily coherent. Instead, the phase ϕ, and thus the signal, proportional to $\cos^2 \phi$, are varied by changing the magnitude of $\boldsymbol{B}$ in the central region. Data are acquired by setting $\boldsymbol{B}$ near the points of steepest slope of $\cos^2 \phi$ (where $\phi = \pm\pi/4$) on either side of the central interference fringe (where $\phi = 0$).

Possible systematic error arising from the "$\boldsymbol{\mathcal{E}} \times \boldsymbol{v}$" effect is greatly diminished in this experiment because of the large tensor Stark shift Δ. However, significant systematic error might arise from variation in the direction and magnitude of $\boldsymbol{\mathcal{E}}$ on the beam axis from the RF1 region, through the central region, and into the RF2 region. In particular, if the direction of $\boldsymbol{\mathcal{E}}$ changes in an absolute sense, a geometric phase could be generated, and if $\boldsymbol{B}$ changes relative to $\boldsymbol{\mathcal{E}}$, the magnetic precession phase, proportional to $\boldsymbol{\mathcal{E}} \cdot \boldsymbol{B}/|\boldsymbol{\mathcal{E}}|$, could be affected [103].

A preliminary result of the YbF experiment [101], published in 2002, is:

$$\left[d_e = (-0.2 \pm 3.2) \cdot 10^{-26} e \text{ cm}\right]. \tag{14.71}$$

Many significant improvements have been made since 2002. The current statistical sensitivity δd_e for one day of integration has been reported [104] as $\delta d_e = 0.9 \cdot 10^{-27} e$ cm, and detailed techniques have been developed for mapping (and subsequently minimizing) field inhomogeneities in the apparatus that could give rise to systematic errors [103]. It appears likely that this experiment will yield a much more precise result in the near future.

14.3.7. *The PbO experiment*

Table 14.3 lists the ground electronic state $X^1\Sigma_0^+$ and the first few excited electronic states of PbO. The $a(1)^3\Sigma_1$ state has a relatively long natural lifetime: $\tau\,[a(1)] = 82(2)\ \mu s$, and is a very good candidate for an EDM search. The $B(1)^3\Pi_1$ state has also been considered as an EDM candidate [105] but it has a much shorter lifetime than $a(1)$. We shall now discuss the principal properties of the $a(1)$ state, and then describe the experimental search for d_e in $a(1)$ being carried out by one of us (D. DeMille) and co-workers at Yale [106–108].

Table 14.3. Low-lying electronic states of PbO.

State($\|\Omega\|$)	T_e, cm^{-1a}
X(0)	0
a(1)	16024
A(0)	19862
B(1)	22285
C(0)	23820
$C'(1)$	24947
D(1)	30199
E(0)	34454
F	51153
G	51661

[a]Note: T_e is the molecular potential energy minimum relative to that of ground state X.

The $a(1)$ state is an example of Hund's case (c) [109], in which the orbital and spin angular momenta $\boldsymbol{\ell}_i$ and $\boldsymbol{s}_i$ of the ith molecular electron couple to form $\boldsymbol{j}_i$, and the $\boldsymbol{j}_i$ couple together to form the total electronic angular momentum $\boldsymbol{J}_e$. $\boldsymbol{J}_e$ precesses about the internuclear axis, forming the projection $\boldsymbol{\Omega}$ on that axis (which is directed along unit vector $\hat{n}$). $\boldsymbol{\Omega}$ and the rotational angular momentum $\boldsymbol{N}$ then couple to form the total molecular angular momentum $\boldsymbol{J}$. Possible values of J are $|\Omega|, |\Omega|+1, |\Omega|+2, \ldots$. In the case of $a(1)$, $|\Omega| = 1$.

In lowest approximation any two states with $\Omega = \pm 1$ and all other quantum numbers the same are degenerate. However, Coriolis coupling of $\boldsymbol{J}_e$ to rotational motion causes a splitting ("Ω doubling") into two states of opposite parity, called e (with parity $(-1)^J$) and f (with parity $(-1)^{J+1}$), which are separated by interval Δ_Ω. Fig. 14.8 is a schematic diagram, not to scale, showing the lowest ($J = 1$) Ω doublet of $a(1)$ with vibrational quantum number $v' = 5$. Also shown are the $J = 2$ levels of $a(1)(v' = 5)$, the lowest $J = 0$ state of $X^1\Sigma_0^+(v = 1)$, and the state $X^1\Sigma_0^+(v = 0)$.

Ignoring P,T-odd effects for the moment, let us consider the effect of an external electric field $\boldsymbol{\mathcal{E}}_{\text{ext}} = \mathcal{E}_{\text{ext}}\hat{z}$ on the $a(1)[J = 1, v' = 5]\Omega$ doublet. It causes mixing of the e^- and f^+ states with the same value of M_J (denoted henceforth by M), yielding states of mixed parity separated in energy by $\Delta E = \sqrt{\Delta_\Omega^2 + (\mu_{JM}\mathcal{E}_{\text{ext}})^2}$. Here $\mu_{J,M} = \frac{\mu_a M}{J(J+1)} = \mu_a \frac{M}{2}$, where $\mu_a = 1.64(3)$ MHz V^{-1}cm is the molecular electric dipole moment in the $a(1)$ state. (Note that states with $M = 0$ do not mix.) As $\boldsymbol{\mathcal{E}}_{\text{ext}}$

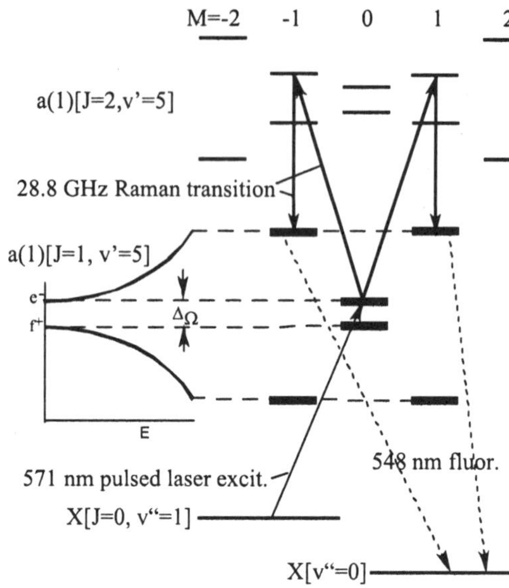

Fig. 14.8. Schematic diagram, not to scale, of energy levels of PbO relevant to the Yale experiment. A pulsed laser at 571 nm excites PbO molecules from the $X[J=0, v"=1]$ state to the $a(1)[J=1^-, M=0, v'=5]$ state. These are transferred to a coherent superposition of $M=\pm 1$ levels in either the upper or lower portion of the Ω-doublet by a Raman transition at 28.8 GHz using the $a(1)[J=2, v'=5]$ intermediate state. The $a(1)[J=1, M=\pm 1, v'=5]$ molecules are detected by fluorescence at 548 nm that accompanies their decay to $X[v"=0]$.

is increased, $\hat{n}$ becomes more polarized along $\boldsymbol{\mathcal{E}}_{\text{ext}}$, the polarization P depending on $\boldsymbol{\mathcal{E}}_{\text{ext}}$ as $P = 2\alpha\beta/(\alpha^2+\beta^2)$, where $\alpha = \mu_{J,M}\mathcal{E}_{\text{ext}}$ and $\beta = \Delta_\Omega/2 + \sqrt{(\Delta_\Omega/2)^2 + (\mu_{JM}\mathcal{E}_{\text{ext}})^2}$. Because $\Delta_\Omega = 11.2$ MHz is so small, $|P|$ rapidly increases toward unity as $\boldsymbol{\mathcal{E}}_{\text{ext}}$ approaches 100 V/cm, already reaching .975 at 30 V/cm, .99 at 50 V/cm, and .997 at 90 V/cm. As previously noted in Table 14.2, when $|P| \approx 1$ the effective molecular field is $\mathcal{E}_{\text{eff}} \cong 26$ GV/cm. In this limit of full polarization, the $M=\pm 1$ levels can be characterized by the quantum numbers M and $N \equiv \text{sign}(\hat{n}\cdot\boldsymbol{\mathcal{E}}_{\text{ext}})$, where $N = -1(+1)$ for the upper (lower) energy pair. The eigenstates can be written as $|N,M\rangle = \frac{1}{\sqrt{2}}\left(|f\rangle + (-1)^M N|e\rangle\right)$. Application of a static magnetic field $\boldsymbol{B} = B\hat{z}$ shifts these components by an additional $\delta_B = g_N\mu_B B\frac{M}{J(J+1)} = \left[g_a + N\delta g\left(\mathcal{E}_{\text{ext}}\right)\right]\mu_B\frac{M}{2}$, where $g_a = 1.86$ is the Lande g factor for the $a(1)$ state and $\delta g\left(\mathcal{E}_{\text{ext}}\right)$ is a small, $\mathcal{E}$-field dependent difference between the g-factors for the $N=\pm 1$ states. Under typical conditions $\delta g/g_a \approx 1.5\cdot 10^{-3}\left[\mathcal{E}_{\text{ext}}/\left(100\text{V/cm}\right)\right]$.

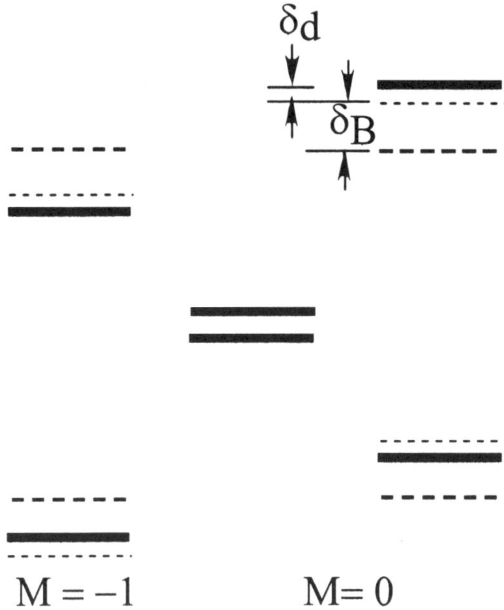

Fig. 14.9. Schematic diagram, not to scale, illustrating the shifts in $M = \pm 1$ levels of the upper ($N = -1$) and lower ($N = +1$) electrically polarized Ω-doublet components of the $a(1)[J = 1, v' = 5]$ state of PbO. Heavy dashed lines: level positions in the absence of external magnetic field and assuming $d_e = 0$. Light dashed lines: Zeeman shifts in presence of $B_z \neq 0$ but $d_e = 0$ still assumed. Solid lines: Shifts for $B_z \neq 0$ and $d_e \neq 0$ assumed.

If we now allow for the possibility $d_e \neq 0$, there is an additional contribution δ_d to the energy of each $|N, M\rangle$ level: $\delta_d = NMd_e\mathcal{E}_{\rm eff}$. We summarize the effects of applied electric and magnetic fields on the $a(1)[J = 1, v' = 5]$-doublet in Fig. 14.9. The difference in sign for the $M = \pm 1$ splitting between upper and lower doublet components arises from the opposite electrical polarizations of these states; it is very significant because it provides an excellent opportunity for effective control of systematic errors. In particular, taking the difference between the $M = \pm 1$ splitting for the $N = -1$ states and that for the $N = +1$ states is equivalent to use of a co-magnetometer, in which the g-factors and internal structure are nearly identical.

The Yale experiment is carried out in a cell containing PbO, which consists of an alumina body with top and bottom end caps supporting flat gold foil electrodes 6 cm in diameter plus surrounding guard ring electrodes, and large flat YAG (yttrium aluminum garnet) windows on all four sides that are sealed to the body with gold foil as a bonding agent. The electric field $\boldsymbol{\mathcal{E}}_{\text{ext}} = \mathcal{E}_{\text{ext}}\hat{z}$ is quite uniform over a cylindrical volume of diameter 5 cm and height 3.8 cm, and is chosen in the range 30–90 V/cm. The magnetic field is controlled by a set of 3 mutually perpendicular Helmholtz coil pairs; B_z is chosen in the range 50–200 mG. The cell is enclosed in an oven which is mounted in a vacuum chamber. At the operating temperature 700 C, the useful PbO density is $\approx 3 \cdot 10^{12}$ cm^{-3}, but the total vapor density, dominated by species such as $Pb_2 0_2$ and $Pb_4 0_4$, is roughly an order of magnitude larger. Collisions with this background gas determine the effective lifetime $\tau_{\text{eff}} \approx 40$ μs.

The relevant states for the experiment are the $J = 1, M = \pm 1$ levels of either the upper or lower part of the polarized $a(1)$ Ω doublet (see Fig. 14.8). A coherent superposition of these states with a particular value of N can be populated with a few techniques; one is described here. A pulsed laser beam with z linear polarization, directed along y and with wavelength 571 nm, excites the transition:

$$X[J = 0^+; v" = 1] \to a(1)[J = 1^-, M = 0; v' = 5].$$

Following the laser pulse a Raman transition is driven by two microwave beams propagating in the y direction. The first, with x linear polarization, excites the upward 28.2 GHz transition:

$$a(1)[J = 1^-, M = 0, v' = 5] \to a(1)[J = 2^+, M = \pm 1; v' = 5].$$

The second, with z linear polarization and detuned to the red or blue with respect to the first by 20–60 MHz, drives the downward transition:

$$a(1)[J = 2^+, M = \pm 1; v' = 5] \to a(1)[J = 1, M = \pm 1; v' = 5].$$

The net result is that about 50% of the $J = 1^-, M = 0$ molecules are transferred to a coherent superposition of $M = \pm 1$ levels in a single desired Ω-doublet component with definite value of N. Because of the shifts δ_B and δ_d separating the $M = \pm 1$ states, their relative phase evolves with time (the coherent state "precesses" in the xy plane). This is detected by observing the frequency of quantum beats in the fluorescence at 548 nm, emitted along the x-direction, that accompanies spontaneous decay to the $X[v" = 0]$ state. The signature of a non-zero EDM is a term in the quantum beat frequency that is proportional to $\boldsymbol{\mathcal{E}}_{\text{ext}} \cdot \boldsymbol{B}$ and that changes sign when one switches from one Ω-doublet component to the other. Given the calculated value $\mathcal{E}_{\text{eff}} \cong 26$ GV/cm one estimates the sensitivity of the quantum beat frequency to an EDM to be:

$$\Delta\nu = 12^{+4}_{-1} \left(\frac{d_e}{10^{-27} e \text{ cm}} \right) \text{ mHz.} \tag{14.72}$$

In the spring of 2008, 41 hours of data were taken and yielded the result $d_e = -19 \pm 20(\text{stat.}) \pm 0.9(\text{syst.}) \cdot 10^{-27} e$ cm. The statistical uncertainty was within a factor of 1.2 of the shot noise limit. During this run the counting rate was dominated by a background due to blackbody radiation from the high-temperature oven surrounding the vapor cell; in addition the contrast of the quantum beats was small ($\sim 4\%$), due to background from off-resonant excitation of higher rotational lines by the broadband laser source. The small value of the systematic uncertainty relative to the statistical error reflects the power of the Ω-doublet states as a co-magnetometer. In particular, all known systematics arise from the combined effect of two or even three imperfections in the system, e.g. non-reversing electric or magnetic field components, magnetic fields due to leakage currents, etc. The size of each imperfection could be extracted from the data by constructing asymmetries odd under reversal of one or two, but not all three, of the primary experimental parameters $\boldsymbol{\mathcal{E}}_{\text{ext}}$, $\boldsymbol{B}$, and N. In all cases the imperfections were found to be consistent with zero, leading to no net systematic correction to the data.

Subsequent improvements to the experiment have decreased the blackbody background (through improved heat shielding) and increased the signal size (by excitation from the $X[v" = 0]$ vibrational level, which has ~ 3 times larger thermal population than the $X[v" = 1]$ level). The present statistical sensitivity is $\delta d_e \approx 7 \cdot 10^{-27} e \text{ cm}/\sqrt{T}$, where T is the integration time in days. A few improvements now underway (such as use of a narrow-band laser source) are anticipated to reduce this by a factor of ~ 5, which should make it possible to improve on the current limit by a significant factor.

14.3.8. *The ThO experiment*

A new method to search for d_e in the $H^3\Delta_1$ electronic state of ThO is being pursued as a collaboration between J. Doyle and G. Gabrielse (Harvard) and one of us (D. DeMille, Yale). This experiment (dubbed ACME, the Advanced Cold Molecule EDM experiment), now under construction at Harvard, combines several features of the YbF and PbO experiments with some attractive new ideas and techniques.

The H state of ThO exhibits Ω-doublet structure similar to that in PbO, and hence an analogous "internal co-magnetometer" method can be used to reject systematic errors in ThO. However, the calculated value of $\mathcal{E}_{\text{eff}} \cong 104$ GV/cm for ThO is four times larger than in PbO (and is in fact the largest value calculated for any molecule to date). The H state of ThO has several other notable features. Its radiative lifetime τ_H is significantly longer than that of the $a(1)$ state of PbO: A preliminary measurement from the ACME collaboration has yielded $\tau_H \geq 1.8$ ms. In addition, the magnetic g-factor g_H of the H state is expected to be unusually small, $g_H \sim 0.01 - 0.1$.[c] This suppresses both noise and systematic effects associated with magnetic fields.

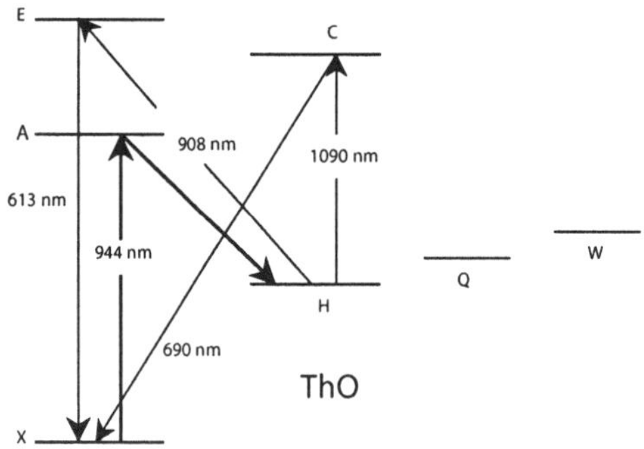

Fig. 14.10. Relevant energy level structure of ThO. The EDM measurement will take place in the metastable $H^3\Delta_1$ state. The H state can be populated by optical pumping from the ground state $X^1\Sigma^+$, via the intermediate state $A^3\Pi_0$, at the wavelength $\lambda_{XA} =$ 944 nm. The H state can be probed in several ways, e.g. by excitation of the $H - E^1\Sigma+$ transition at $\lambda_{HE} = 908$ nm and detection of the subsequent fluorescence accompanying the $E - X$ decay at wavelength $\lambda_{EX} = 613$ nm.

The energy level structure of ThO has been thoroughly measured [110, 111], and interpreted in detail with associated electronic structure calculations [112]. This information has made it possible to identify efficient pathways for laser population and probing of the H state (see Fig. 14.10). Population is achieved by optical pumping, in which laser excitation of the weakly allowed $X^1\Sigma^+ \to A^3\Pi_0$ transition (with transition dipole moment $\sim 0.05ea_0$) is followed by the fully allowed spontaneous decay $A \to H$

[c]This is easily understood [109]. In a $^3\Delta_1$ state the electronic orbital angular momentum projection $\Lambda = 2$. In order to form the total electronic angular momentum projection $\Omega = 1$, the electronic spin angular momentum projection Σ must be $\Sigma = -1$, i.e. the total spin $S = 1$ is oriented opposite to Λ. Since the g-factors associated with electronic orbital and spin angular momenta are $g_L = 1$ and $g_S = 2$ respectively, the magnetic moments associated with the two nominally cancel in the $^3\Delta_1$ state, where $g_H \cong g_L\Lambda + g_S\Sigma = 0$. Spin-orbit effects lead to a small admixture of electronic states with different values of Λ and Σ, leading to the non-zero expected value of g_H.

(transition dipole $\sim 1ea_0$). The ACME collaboration has experimentally verified that this process leads to near-unit efficiency of pumping from the ground state X. The H state population can be probed, e.g. by exciting the weak $H \to E^1\Sigma^+$ transition and monitoring fluorescence on the strong $E - X$ decay. In all cases, the required wavelengths are accessible with convenient and robust diode lasers.

The ThO experiment will be performed in a molecular beam. A key feature of ACME is the use of a new type of cryogenic molecular beam source, which provides a cold and slow beam of molecules that has orders of magnitude higher brightness than available otherwise [113, 114]. In this source, initially hot molecules are injected (by laser ablation from a solid target) into a cell filled with helium buffer gas held at 4 K by contact with a cryostat. The molecules and buffer gas exit the cell via a small aperture to form a beam. In a certain range of conditions, the molecules can be actively swept out of the cell by the hydrodynamic flow of buffer gas; this leads to near-unit efficiency for extraction of molecules into the beam. Once in the beam, the molecules have an average forward velocity v_f characteristic of He atoms at 4 K ($v_f \sim 150$ m/s), but a typical transverse velocity $v_\perp$ determined by the 4 K Boltzmann distribution for the (much heavier) molecules: $v_\perp \sim 15$ m/s for ThO. Hence the beam intensity is strongly peaked in the forward direction.

The combination of high beam brightness, a large value of $\mathcal{E}_{\text{eff}}$, and a long state lifetime leads to very promising projections for the statistical sensitivity δd_e attainable with this approach. With one day of integration, the ACME team anticipates reaching $\delta d_e \sim 1 \cdot 10^{-29} e$ cm with a simple first-generation apparatus, and ultimately $\delta d_e \sim 2 \cdot 10^{-32} e$ cm. Preliminary estimates of all known systematics have revealed none that are anticipated to be a problem even at this level.

14.3.9. *The proposed* HfF^+ *experiment*

E. Cornell and co-workers at the Joint Institute for Laboratory Astrophysics, Boulder, have proposed an experiment [115] to search for d_e in the $^3\Delta_1$ electronic state of HfF^+. The advantage of using a molecular ion is that such ions can be stored in an RF trap, thus making observation times very long. Preliminary calculations [76] suggest that the ground state of HfF^+ is $^1\Sigma$, that there exist relatively high-lying $^1\Pi$ and $^3\Pi$ states, and that $^3\Delta_1$ is a low-lying metastable state with very small Ω-doublet splittings, which would allow it, like the $a(1)$ state of PbO, to be polarized by small

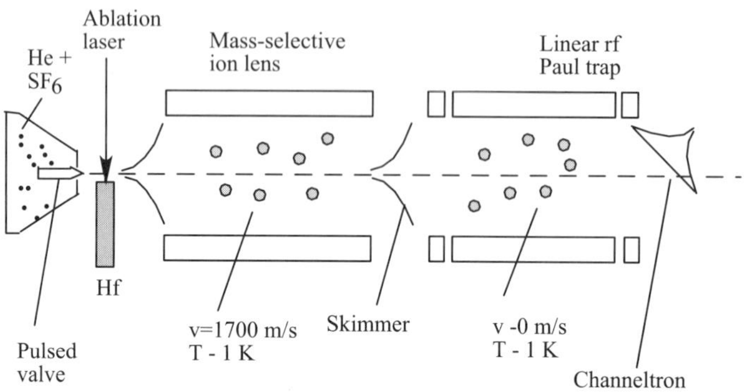

Fig. 14.11. Schematic diagram of HfF^+ experiment, not to scale. Laser ablation of a metal Hf target creates Hf^+ ions that react with SF_6 gas to produce HfF^+ molecular ions. The ions are cooled in a supersonic expansion with a He buffer gas. The mass selective ion lens focuses only $^{180}HfF^+$ into the ion trap where the electron spin resonance spectroscopy is performed. The ions are counted with a channeltron.

external electric fields (≤ 100 V/cm). Also, these preliminary calculations suggest that when $^3\Delta_1$ is fully polarized, its effective molecular field would be quite large ($\mathcal{E}_{\text{eff}} \approx 18$ GV/cm, see Table 14.2).[d]

The present proposal is to produce HfF^+ by laser ablation of a metal Hf target in the presence of SF_6 gas, using the exoergic reaction

$$\mathrm{Hf}^+ + \mathrm{SF}_6 \rightarrow \mathrm{HfF}^+ + \mathrm{SF}_5.$$

The target is to be placed near the opening of a pulsed valve, which allows a mixture of He + 1% SF_6 to expand into vacuum as this gas entrains HfF^+. (See Fig. 14.11 for a schematic diagram of the apparatus). It is expected that supersonic expansion and collisions with He will cool the translational, rotational, and vibrational temperatures of the HfF^+ ions. It is then proposed to filter the masses of the pulsed ion beam using a mass-selective ion lens, so that only $^{180}HfF^+$ ions are focused, decelerated, and confined in a linear RF Paul trap. To search for d_e, electron-spin-resonance spectroscopy, using the Ramsey method, is to be performed in the presence of rotating electric and magnetic fields (see Fig. 14.12). The electric field polarizes the ions and its rotation prevents them from being accelerated out of the trap. The co-rotating magnetic field lifts the degeneracy between $M = +1$ and $M = -1$ spin states. One M level of a single $J = 1$ Ω-doublet of $^3\Delta_1$ is to be populated with a two-photon Raman transition using the well-mixed

[d]We are not aware of any detailed published results concerning the relevant spectroscopic parameters of HfF^+. However, Cornell and co-workers have recently observed the first evidence of laser-induced fluorescence signals from HfF^+ [115].

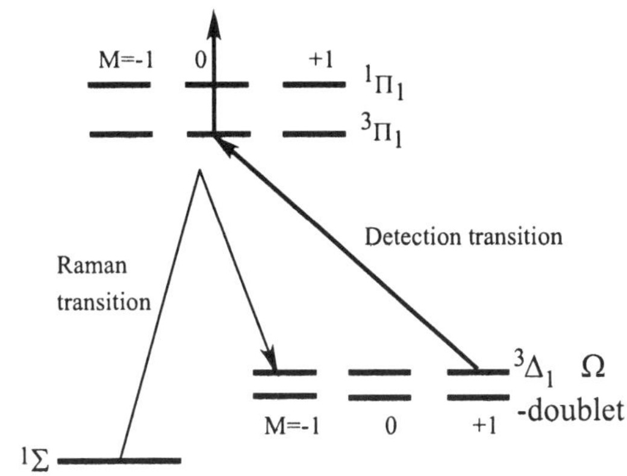

Fig. 14.12. Energy level structure of HfF^+ (not to scale). One M-level of one e/f level of the $^3\Delta_1$ Ω-doublet manifold is to be populated by a two-photon Raman transition from the $^1\Sigma$ ground state. Ramsey spectroscopy will be performed on the $M = +1$ and $M = -1$ levels of the $^3\Delta_1$ state. The final spin composition will be read out by exciting only the $^3\Pi_1(J = 1, M = 0) \leftarrow {}^3\Delta_1(J = 1, M = +1)$ transition using a circularly polarized laser beam and photodissociating the population in the $^3\Pi_1$ state.

$^{1,3}\Pi$ levels as intermediate states. Ramsey spectroscopy of the $M = +1$ and -1 levels is to be performed using two separated RF two-photon $\pi/2$ pulses. The final spin state composition is to be detected by driving a narrow spin-sensitive transition $^3\Pi_1(J = 1, M = 0) \leftarrow {}^3\Delta_1(J = 1, M = +1)$ using a circularly polarized laser beam, followed by photodissociation of the HfF^+ population in $^3\Pi_1$ into $Hf^+ + F$. Thus one spin state will be dissociated into Hf^+, while the other will remain as HfF^+. These ions may be separated by means of their different masses. As in PbO, utilization of both upper and lower Ω-doublet components should yield opposite signs of the EDM signal, but nearly identical signals due to systematic effects. However, this experiment is unique in that it is impossible to reverse the electric field: In the laboratory frame it must always point inward toward the trap center.

14.3.10. *Electron EDM solid-state experiments*

14.3.10.1. *Basic ideas*

Nearly 40 years ago, F. Shapiro [116] suggested that an electron EDM search could be carried out by applying a strong electric field $\mathcal{E}_{\text{ext}}$ to a solid sample with unpaired electron spins. If $d_e \neq 0$, the sample might acquire significant spin-polarization at sufficiently low temperature, and thus a detectable magnetization along the axis of $\mathcal{E}_{\text{ext}}$ could be generated. Following Shapiro's suggestion, an experiment of this kind was performed [117] in 1978 on a nickel-zinc ferrite, in which the ion of interest with unpaired electrons is Fe^{3+}. However, various factors combined to limit the effectiveness of this experiment. First, for iron, where $Z = 26$, the enhancement factor is small. Next, the sample could only support an electric field of 2kV/cm, the temperature (4 K) was not particularly low, and the SQUID magnetometer employed to detect the magnetization was not very sensitive. Thus the result:

$$d_e = -(.81 \pm 1.16) \cdot 10^{-22} e \text{ cm} \tag{14.73}$$

was not very impressive. However, the idea was revived in recent years by S. Lamoreaux [29], who pointed out that a better choice of material, together with larger applied electric field, lower temperatures, and better SQUID magnetometry could improve the sensitivity by more than eight orders of magnitude. The material proposed by Lamoreaux: $Gd_3Ga_5O_{12}$ (gadolinum gallium garnet, or GGG), has a number of attractive properties. Its resistivity is so high ($> 10^{16}$ Ohm-cm for $T < 77$ K) that it can support large applied electric fields ($\mathcal{E}_{\text{ext}} \approx 10$ kV/cm) with very small leakage currents. The ion of interest in GGG, Gd^{3+}, has seven unpaired electrons in the 4f shell, and atomic number $Z = 64$ which implies a non-negligible enhancement factor [63]: $R \approx -3.3$. Furthermore the symmetry of the GGG crystal is such that several magneto-electric effects (e.g. terms in the free energy of the form $H\mathcal{E}, H^2\mathcal{E}$) are ruled out that otherwise could cause systematic error [118]. An experiment of this type on GGG is being carried out by C.-Y. Liu of Indiana University [30, 119], and it will be discussed in some detail below.

A complementary experiment has also been proposed, and is being done by L. Hunter and co-workers [32] at Amherst College. Here, a strong external magnetic field is applied to the ferrimagnetic solid $Gd_3Fe_5O_{12}$ (gadolinum iron garnet, or GdIG), thus causing substantial polarization of the gadolinum electron spins. If $d_e \neq 0$, this must result in electric charge polarization of the sample, and thus a voltage developed across the sample that reverses with applied magnetic field. This experiment will also be described in some detail below.

First, however, we sketch some basic theoretical considerations that must be taken into account to estimate the expected signals [120]. As already stated, in each of these experiments, the ion of interest is Gd^{3+}. (In GdIG, there are also unpaired electron spins in the Fe ion, but the enhancement factor for iron is so small that it is legitimate to neglect their contribution). Considering Gd^{3+} with ground configuration $1s^2 \ldots 5d^{10}4f^7$ as isolated for the moment, one can employ a semi-empirical potential to calculate the electron orbitals with reasonably good accuracy, by solving the Dirac equation numerically. Of course, in a GGG or GdIG crystal the Gd^{3+} ion is not isolated, but rather it is surrounded by 8 O^{2-} ions that form a dodecahedron (distorted cube) structure, with the distance between a Gd and an O being $r_0 = 4.53a_0$ (where a_0 is the Bohr radius). To carry out the relevant estimates it is essential to know the wave functions of the O^{2-} electrons inside the Gd^{3+} ion. Now, the electronic configuration of O^{2-} is $1s^2 2s^2 2p^6$, but it can be shown that $2p_\pi$ orbitals do not penetrate the Gd ion significantly, and one need only consider $2p_\sigma$ orbitals. The effect of the latter on the Gd core can be calculated with sufficient accuracy by approximating the potential generated by the eight oxygen ions as a spherically symmetric attractive shell potential centered on the Gd nucleus:

$$V_0(r) = -A_0 e^{-\left[\frac{r-r_0}{D}\right]^2}, \tag{14.74}$$

and by combining this with the previously mentioned semi-empirical potential for the isolated Gd^{3+} ion. The parameters A_0 and d are determined to $\approx 15\%$ accuracy by matching the resulting orbitals at large distances from the Gd nucleus with previously known orbitals of O^{2-}. One finds $d \approx 0.5$ and $A_0 \approx 1.2$ in atomic units to be the most likely values, although for conservative estimates of the signals to be expected, one can employ $d = 1.0$ and $A_0 = 0.9$.

Once the Gd^{3+} orbitals have been determined, the energy shift due to an EDM d_e can be calculated. To be specific we consider here what happens in a GdIG sample when a strong external magnetic field is applied, causing electron spin polarization along a specific axis. The resulting energy shift ΔE is calculated in third order of perturbation theory, in which 3 perturbations enter, each linearly. The first, of course, is the EDM interaction itself: $V_d = -d_e(\gamma^0 - 1)\mathbf{\Sigma} \cdot \boldsymbol{\mathcal{E}}$. The second is a perturbation related to deformation of the crystal (displacement of the Gd ion with respect to the surrounding O ions by amount x in the direction of spin polarization), and the third is the residual electron-electron Coulomb interaction. As it turns out, there are 15 diagrams corresponding to terms linear in each of these 3 perturbations, of which 4 are direct while 11 are exchange; the latter making relatively large contributions which are not all of the same sign. Taking into account the calculated enhancement factor of Gd, one obtains as a result $\Delta E(x)$ expressed in atomic units in terms of the deformation x

as follows:

$$\Delta E(x) = -Ad_e x, \tag{14.75}$$

where $A = .095$. Now, the crystal lattice has elasticity, and for the extremely small displacements x considered here we may certainly assume that the restoring forces are simple-harmonic. Thus the total energy of the crystal, including elastic restoring forces, is:

$$\Delta E_{\text{Total}} = \frac{1}{2}Kx^2 - Ad_e x. \tag{14.76}$$

The force constant K can be determined from analysis of infrared spectroscopy data on garnets. Equilibrium occurs where ΔE_{Total} is minimized, thus where $x = Ad_e/K$ in atomic units. It turns out coincidentally that $A \approx K$, so one obtains: $x \approx d_e/e$ (in any system of units).

Now we are in a position to calculate the observable effect. When all Gd spins are polarized in the GdIG sample, the resulting macroscopic electric polarization is $P = 3exn_{\text{Gd}}$, where $n_{\text{Gd}} = 1.235 \cdot 10^{22}\ \text{cm}^{-3}$ is the number density of Gd ions in GdIG. From this one obtains the induced electric field:

$$\mathcal{E} = -4\pi P = 12\pi n_{\text{Gd}} d_e,$$

or in practical units:

$$\mathcal{E} = 0.7 \cdot 10^{-10} \left(\frac{d_e}{10^{-27}\text{e cm}}\right) \text{V/cm}. \tag{14.77}$$

A similar calculation can be used to determine the degree of spin polarization of GGG upon application of an external electric field. An electric field of 10 kV/cm acting on a unit cell (corresponding to a macroscopic, externally applied field 3× larger) yields an energy shift:

$$\Delta E = 3.6 \cdot 10^{-22} \left(\frac{d_e}{10^{-27}\text{e cm}}\right) \text{eV}, \tag{14.78}$$

which defines an effective electric field $\mathcal{E}^* = -\Delta E/d_e = 3.6 \cdot 10^5$ V/cm acting on the EDM. The resulting sample magnetization depends on the temperature and internal magnetic interactions of the sample [29]. In a simple model where the Gd^{3+} ions act as free spins, the standard theory of paramagnetism is applicable. This yields an expression for the sample magnetization M:

$$M = n_{\text{Gd}} g_{\text{Gd}} \sqrt{J(J+1)} \frac{d_e \mathcal{E}^*}{k_B T}.$$

Here $g_{\text{Gd}} \cong 2$ and $J = 7/2$ are the g-factor and spin of the Gd^{3+} ion; k_B is the Boltzmann constant; and T is the sample temperature. This yields a magnetic flux $\Phi = 4\pi MS$ over an area S of an infinite flat sheet.

14.3.10.2. *The Indiana GGG experiment*

C. Y. Liu of Indiana University and S. Lamoreaux of Yale have devised a prototype experiment in which two GGG disks 4 cm in diameter and of thickness $\approx$ 1 cm are sandwiched between three planar electrodes. The high voltages are applied so that the electric fields in the top and bottom samples are in the same direction. If $d_e \neq 0$, a magnetic field similar to a dipole field should be generated, and this is to be detected by a flux pick-up coil located in the central ground plane. The latter is designed as a planar gradiometer with three concentric loops, arranged to sum up the returning flux and to reject common-mode magnetic fluctuations. As the electric field polarization is modulated, the gradiometer detects the changing flux and feeds it to a SQUID sensor. The rate of electric field reversals must be small enough to minimize displacement current effects, but large enough to avoid the worst of 1/f noise in the SQUID. The electrodes are made of machinable ceramic coated with graphite to minimize magnetic Johnson noise. The entire assembly is surrounded by magnetic shielding, and is immersed in a liquid helium bath.

The parameters of this prototype experiment are more modest than was suggested in the original proposal of Lamoreaux. The samples are about 10 times smaller in volume, the operating temperatures (1.5–4 K) are $\approx$ 150–400 times higher, and the commercially available SQUID magnetometer noise is about 10 times greater, than the corresponding quantities in the initial proposal. With all these factors taken into account the EDM sensitivity of the prototype experiment is estimated to be $\approx 4 \cdot 10^{-26}$ e-cm. Although this falls short of the ultimate desired sensitivity of 10^{-30} e-cm, the prototype experiment is useful as a learning tool for solving some basic technical problems. These include stable, low-noise SQUID magnetometer operation in a high voltage environment with periodic field reversals and displacement currents, and the necessity to reduce leakage currents to a level less than 10^{-14} A, which is a very stringent requirement.

At Indiana, a second-generation experiment is also being planned, which will operate at much lower temperatures ($\approx$ 10–15 mK), and will employ lower-noise SQUID magnetometers. Here, a major challenge is the magnetic susceptibility χ of GGG at low temperatures: Does it remain sufficiently large? It is known that Gd ions have anti-ferromagnetic interactions, and thus χ obeys a Curie–Weiss relation with a negative Curie-Weiss temperature of -2.3K. In addition, GGG is a geometrically frustrated spin system with a spin-glass phase transition at $\approx$ 200 mK or lower. To prevent the spins from freezing out, it is possible that the exchange field strength could be reduced by substituting Y^{3+} for some of the Gd^{3+} ions. This should bring the Curie-Weiss temperature closer to zero, and should also reduce the spin glass phase transition temperature. However, to understand all this in detail it will be necessary to experiment with samples containing a wide range of yttrium/gadolinium ratios.

Finally, although crystals with inversion symmetry such as GGG and GdIG should not exhibit a linear magneto-electric effect, crystal defects and substitutional impurities can spoil this ideal and thus introduce systematic errors. Furthermore the quadratic magneto-electric effect does exist, and to avoid systematic errors arising from it, good control of field reversal symmetry is required.

14.3.10.3. *The Amherst GdIG experiment*

GdIG is ferrimagnetic, and three different lattices contribute to its magnetization. At $T \approx 0$ K, two iron lattices produce a net magnetic moment per unit cell of $5\mu_B$, and the Gd^{3+} ions generate a magnetic moment per unit cell of $21\mu_B$ in the opposite direction. While the Gd^{3+} magnetization drops rapidly in magnitude with temperature, the iron magnetization falls off more slowly. Hence there exists a "compensation" temperature $T_C = 290$ K where the net magnetization M vanishes. For $T > T_C (< T_C)$, M is dominated by Fe (Gd). As in the proposed second-generation GGG experiment, the gadolinium contribution to M can be reduced by replacing some Gd^{3+} ions with non-magnetic Y^{3+}. Let x be the average number of Gd ions per unit cell, (so that 3-x is the average number of Y ions per unit cell). Then the compensation temperature becomes:

$$T_C = [290 - 115(3 - x)]\,\text{K}. \tag{14.79}$$

This dependence of T_C on x is exploited in the Amherst GdIG experiment. A toroidal sample is employed, consisting of two half-toroids, each in the shape of the letter C. (See Fig. 14.13.) One "C" has $x = 1.35$ with a corresponding $T_C = 103$ K. The other "C" has $x = 1.8$ with a corresponding $T_C = 154$ K. These are joined together, with .0025 cm thick copper foil electrodes bonded to both C's by conductive epoxy. At $T = 127$ K, the magnetizations of the 2 C's are identical, but their Gd magnetizations are nominally opposite. When a magnetic field H is applied to the sample with a toroidal current coil, all Gd spins are nominally oriented toward the same copper electrode. Thus EDM signals from C_1 and C_2 add constructively. However, below 103 K (above 154 K) the Gd magnetization is parallel

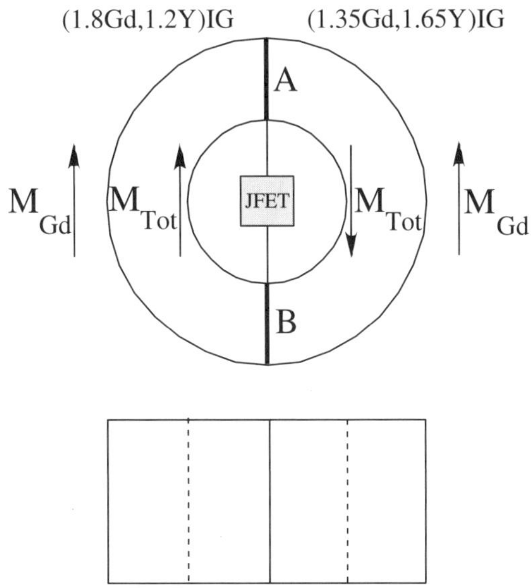

Fig. 14.13. Sketch of the split toroid employed in the Amherst GdIG experiment, (not to scale). At $T = 127$ K, the total magnetization M_{Tot} in each half-toroid "C" is parallel to the applied H field. However, the gadolinium magnetization M_{Gd} in the left "C" is parallel to M_{Tot}, while in the right "C" M_{Gd} and M_{Tot} are anti-parallel. Thus the EDM signals (voltage differences between A and B) contributed by the left and right half-toroids should add constructively.

(antiparallel) to M in both C's, which results in cancellation of one EDM signal by the other.

Data are acquired by observing the voltage difference A (B) between the two foil electrodes for positive (negative) polarity of the applied magnetic field H. An EDM should be revealed by the appearance of an asymmetry $d = A - B$ that has a specific temperature dependence, as is shown by the curve in Fig. 14.14.

However, a large spurious effect is seen that mimics an EDM signal when $T < 180$ K, but which deviates grossly from expectations for $T >$ 180 K. The Amherst group has demonstrated by auxiliary measurements that this effect is associated with a component of magnetization that does not reverse with H (it is thus called the "M-even" effect). They have also demonstrated that it is somehow associated with surface conditions at the interfaces between the two C's, where the copper electrodes are bonded with epoxy: Changing the epoxy changes the size of the effect. At the time of writing, the M-even effect has frustrated efforts to realize the full potential

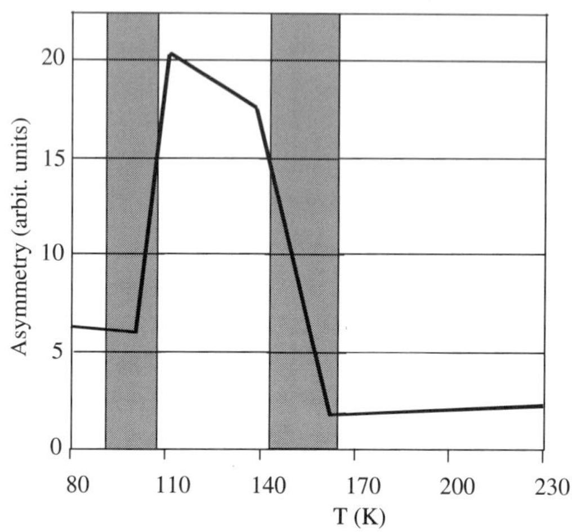

Fig. 14.14. Expected shape of the EDM asymmetry as a function of temperature, Amherst GdIG experiment. The gray rectangles indicate regions where the temperature is so close to the transition temperatures of one of the two "C's" that measurements are not expected to be reliable. Here the magnetization is so small that domain creep follows reversal of applied magnetic field.

of the GdIG experiment, and the best limit that has been achieved so far is that given in the 2005 publication [32]:

$$d_e < 5 \cdot 10^{-24} \text{ e cm}. \tag{14.80}$$

14.3.11. *Atomic T,P-odd polarizability. Molecular T,P-odd magnetic moment*

As we have seen in the previous section, an experiment has been proposed and initiated to search for P,T-violating magnetization of the solid GGG induced by application of an external electric field. B. Ravaine, M. Kozlov, and A. Derevianko [121] have discussed the possibility that a diamagnetic rare gas atom might also acquire a P,T-odd magnetic moment induced by an electric field:

$$\mu^{CP} = \beta^{CP} \boldsymbol{\mathcal{E}}, \tag{14.81}$$

where β^{CP} is called the CP-violating polarizability. To understand how β^{CP} is estimated, we note first of all that two perturbations act on the zeroth order atomic ground state $|\psi_0^0\rangle$: The external electric field Stark effect, and the EDM-plus-P,T-odd eN interaction. Developing the ground state to second order in these perturbations we have:

$$|\psi_0\rangle \cong |\psi_0^0\rangle + |\psi_0^1\rangle + \psi_0^2\rangle, \tag{14.82}$$

where $|\psi_0^1\rangle$ and $|\psi_0^2\rangle$ have opposite parity and the same parity, respectively, compared to $|\psi_0^0\rangle$. Since the magnetic moment operator has even parity, its expectation value to lowest non-vanishing order is then:

$$\begin{aligned}\langle \boldsymbol{\mu}^{CP}\rangle &= \langle \boldsymbol{\mu}^{CP}\rangle_1 + \langle \boldsymbol{\mu}^{CP}\rangle_2 + \langle \boldsymbol{\mu}^{CP}\rangle_3 \\ &= \langle \psi_0^1|\boldsymbol{\mu}|\psi_0^1\rangle + \langle \psi_0^0|\boldsymbol{\mu}|\psi_0^2\rangle + \langle \psi_0^2|\boldsymbol{\mu}|\psi_0^0\rangle\,,\end{aligned} \tag{14.83}$$

where

$$\langle \boldsymbol{\mu}^{CP}\rangle_1 = 2\sum_{kn} \frac{H_{0k}^{P,T}}{E_0 - E_k}\boldsymbol{\mu}_{kn}\frac{H_{n0}^{ext}}{E_n - E_0}, \tag{14.84}$$

$$\langle \boldsymbol{\mu}^{CP}\rangle_2 = 2\sum_{kn} \boldsymbol{\mu}_{0k}\frac{H_{kn}^{P,T}H_{n0}^{ext}}{(E_0 - E_k)(E_0 - E_n)}, \tag{14.85}$$

$$\langle \boldsymbol{\mu}^{CP}\rangle_3 = 2\sum_{kn} \boldsymbol{\mu}_{0k}\frac{H_{kn}^{ext}H_{n0}^{P,T}}{(E_0 - E_k)(E_0 - E_n)}. \tag{14.86}$$

Special relativity enforces a double restriction on these matrix elements. First of all, the operator $H^{P,T}$ is intrinsically relativistic, as we know. However, even if this were not the case, in the non-relativistic (NR) limit the magnetic dipole operator cannot change principal quantum numbers, and thus cannot connect occupied and excited orbitals. Therefore, in this limit, $\langle \boldsymbol{\mu}^{CP}\rangle_2 = \langle \boldsymbol{\mu}^{CP}\rangle_3 = 0$. This leaves only $\langle \boldsymbol{\mu}^{CP}\rangle_1$ but it can be shown that the latter matrix element vanishes as well, unless one takes into account the spin-orbit interaction (a relativistic effect) in the matrix elements of H^{ext}. As a result the Z dependence of $\langle \boldsymbol{\mu}^{CP}\rangle$ is much more pronounced than the $Z^3\alpha^2$ dependence of paramagnetic atom enhancement factors; indeed β^{CP} is roughly proportional to $Z^5\alpha^4$. Using the Dirac–Hartree–Fock method, Ravaine, Kozlov, and Derevianko have employed the arguments sketched above to calculate β^{CP} for all of the rare gas atoms from helium through radon. The results are shown in Table 14.4. The most favorable case from a practical viewpoint is Xe (since Rn is radioactive). However, estimates reveal that the limit achievable by an experiment employing liquid xenon with the most sensitive magnetometry currently available would fall short of the current limit on d_e by almost two orders of magnitude.

On the other hand, the outlook is not so bleak for a related effect, the CP-violating magnetic moment of a diatomic molecule. As we know,

Table 14.4. Theoretical values of β^{CP} for rare-gas atoms.

Atom	Z	β^{CP}/d_e
He	2	$-3.8 \cdot 10^{-9}$
Ne	10	$-2.2 \cdot 10^{-6}$
Ar	18	$-7.4 \cdot 10^{-5}$
Kr	36	$-3.6 \cdot 10^{-3}$
Xe	54	-.045
Rn	86	-1.07

such a molecule is characterized by the projection $\boldsymbol{\Omega}$ of the total electronic angular momentum $\boldsymbol{J}_e$ on the internuclear axis $\hat{n}$. For a molecular state with definite Ω the molecular magnetic moment is directed along $\hat{n}$:

$$\boldsymbol{\mu} = \mu_B G_{\|}(\boldsymbol{J}_e \cdot \hat{n})\hat{n} + \mu^{CP}\hat{n}. \tag{14.87}$$

The first term on the right-hand side is the ordinary (P,T-even) contribution; here $G_{\|}$ is a number of order unity analogous to the Lande g-factor for atoms. For a diamagnetic molecule, this first term on the right-hand side of (14.87) vanishes. (Also, it can be shown that the nuclear magnetic moments and the small nuclear rotational magnetic moment do not have any significant effect on the conclusions to be drawn here). The second term on the right-hand side of (14.87) is P,T odd, and like the corresponding P,T-odd atomic magnetic moment appearing in (14.81), it is proportional to the local electric field. This molecular electric field is strongest in the neighborhood of a nucleus with large Z; therefore diatomic molecules with at least one large-Z atom are favored. In the absence of an external electric field, the internuclear axis is randomly oriented in space. However, as we have previously noted, only relatively modest external electric fields $\boldsymbol{\mathcal{E}}_{\text{ext}}$ are required to polarize $\hat{n}$ along $\boldsymbol{\mathcal{E}}_{\text{ext}}$. Derevianko and Kozlov [33] have shown that one can then have a sample of polarized molecules with a small but macroscopic P,T-odd magnetization that reverses with $\boldsymbol{\mathcal{E}}_{\text{ext}}$. They have estimated μ^{CP} for several diatomic molecules. A particularly favorable case is BiF which has a $^1\Sigma$ ground state and a nucleus (Bi) with $Z = 83$. One finds:

$$\mu^{CP}(\text{BiF}) \approx 1.63 \cdot 10^{-17} \left(\frac{d_e}{10^{-27}\ e\ \text{cm}}\right) \mu_B. \tag{14.88}$$

Although this is an extremely small magnetic moment, application of $\mathcal{E}_{\text{ext}} \approx$ few kV/cm to a sample of BiF with number density $\approx 10^{21}$ cm^{-3} in a volume ≈ 0.3 cm^3 would result in a magnetic field $B \approx 10^{-15}$ G. Using the best available magnetometry, it might be possible to arrange a BiF experiment that could reach a limit on d_e competitive with those aimed for in paramagnetic molecule experiments. Finally, Kozlov and Derevianko have proposed an electron EDM experiment in which a molecular radical (e.g. HgH in the $^2\Sigma_{1/2}$ state) would be frozen in a rare gas matrix. Here, as in the GGG experiment described in Sec. 14.3.10, one would measure the EDM-induced magnetic field, when the EDM and hence the electron spin magnetic moment is polarized in an applied electric field [31].

Acknowledgments

We thank E. Cornell, D. Heinzen, E. Hinds, L. Hunter, J.D. Jackson, M. Kozlov, S. Lamoreaux, A. Leanhardt, R. Littlejohn, C.-Y. Liu, M. Romalis, Y. Sakemi, and N. Shafer-Ray for very helpful discussions and/or for making available valuable information concerning their EDM researches.

References

[1] L. Landau, Sov. Phys. JETP 5, 336 (1957).
[2] K. Kleinknecht, "Uncovering CP violation: experimental clarification in the neutral K meson and B meson systems", *Springer Tracts in Modern Physics* 195 (Springer Verlag 2003).
[3] D. Kirkby and Y. Nir, CP Violation in Meson Decays, in Review of Particle Physics, W.-M. Yao *et al.*, J. Phys. G 33, 1 (2006).
[4] B. C. Regan, E. D. Commins, C. J. Schmidt, and D. DeMille, Phys. Rev. Lett. 88, 071805 (2002).
[5] F. J. Gilman, K. Kleinknecht, and B. Renk, The Cabibbo–Kobayashi–Maskawa Quark-Mixing Matrix, in Review of Particle Physics, S. Eidelman *et al.*, Phys. Lett. B 592, 1 (2004).
[6] C. Jarlskog, Phys. Rev. Lett. 55, 1039 (1985); Z. Phys. C 29, 491 (1985).
[7] M. Pospelov and I. B. Khriplovich Sov. J. Nucl. Phys. 53, 638 (1991).
[8] B. Kayzer, "Neutrino Mass, Mixing, and Flavor Change", in Review of Particle Physics, W. -M. Yao *et al.*, J. Phys. G 33, 1 (2006).
[9] J. P. Archambault, A. Czarnecki, and M. Pospelov, Phys. Rev. D 70, 073006 (2004).
[10] J. Bailey *et al.*, J. Phys. G. 4, 345 (1978); Nucl. Phys. B 150, 1 (1979).
[11] G. W. Bennett *et al.*, Phys. Rev. D73, 072003 (2006); G.W. Bennett, *et al.*, arXiv:0811.1207v2 [hep-ex], July 2009, to be published in Phys. Rev. D.
[12] F. J. M. Farley, K. Jungmann, J. P. Miller, W. M. Morse, Y. F. Orlov, B. L. Roberts, Y .K. Semertzidis, A. Silenko, and E. J. Stephenson, Phys. Rev. Lett. 93, 052001 (2004).
[13] R. Akers *et al.* (OPAL Collaboration), Z. Phys. C 66, 31 (1995).

[14] D. Buskelic *et al.* (ALEPH Collaboration), Phys. Lett. B 346, 371 (1995).
[15] G. Abbiendi *et al.*, Phys. Lett. B565, 61 (2003).
[16] H. P. Nilles, Phys. Reports 110, 1 (1984).
[17] M. V. Romalis, W. C. Griffith, and E. N. Fortson, Phys. Rev. Lett. 86, 2505 (2001).
[18] P. G. Harris *et al.*, Phys. Rev. Lett. 82, 904 (1999); see also: C. A. Baker *et al.*, Phys. Rev. Lett. 97,131801 (2006).
[19] S. M Barr, Int. J. Mod. Phys. A 8, 209 (1993).
[20] J. C. Pati and A. Salam, Phys. Rev. D. 10, 275 (1974).
[21] R. N. Mohapatra and J. C. Pati, Phys. Rev. D 11, 566 (1975); Phys. Rev. D 11, 2558 (1975).
[22] R. N. Mohapatra and G. Senjanovic, Phys. Rev. Lett. 44, 912 (1980); Phys. Rev. D 23, 165 (1981).
[23] D. F. Nelson, A. A. Schupp, R. W. Pidd, and H. R. Crane, Phys. Rev. Lett. 2, 492 (1959).
[24] L. I. Schiff, Phys. Rev. 132, 2194 (1963).
[25] P. G. H. Sandars, Phys. Lett. 14, 194 (1965).
[26] P. G. H. Sandars, Phys. Lett. 22, 290 (1966).
[27] Z. W. Liu and H. P. Kelly, Phys Rev. A 45, R4210 (1992).
[28] I. B. Khriplovich and S. K. Lamoreaux, *CP violation without strangeness: electric dipole moments of particles, atoms, and molecules*, Springer Verlag, Berlin, 1997.
[29] S. K. Lamoreaux, Phys. Rev. A 66, 022109 (2002).
[30] C.-Y. Liu and S. K. Lamoreaux, Mod. Phys. A 19, 1235 (2004).
[31] M. G. Kozlov and A. Derevianko, Phys. Rev. Lett. 97, 063001 (2006).
[32] B. J. Heidenreich *et al.*, Phys. Rev. Lett. 95, 253004 (2005).
[33] A. Derevianko and M. G. Kozlov, Phys. Rev. A 72, 040101 (R) (2005).
[34] O. P. Sushkov, V. V. Flambaum, and I. B. Khriplovich, Sov. Phys. JETP 60, 873 (1984).
[35] V. V. Flambaum, I. B. Khriplovich, and O. P. Sushkov, Phys. Lett. B 162, 213 (1985).
[36] V. V. Flambaum, I. B. Khriplovich, and O. P. Sushkov, Nucl. Phys. A 449, 750 (1986).
[37] V. M. Khatsymovsky, I. B. Khriplovich, and A. S. Yelkhovsky, Annals of Phys. 186, 1 (1988).
[38] W. C. Haxton and E. M. Henley, Phys.Rev. Lett. 51, 1937 (1983).
[39] X. He and B. McKellar , Phys. Lett. B 390, 318 (1997).
[40] I. B. Khriplovich, Sov. Phys. JETP 44, 25 (1976).
[41] I. B. Khriplovich, *Parity Nonconservation in Atomic Phenomena*, Gordon & Breach, Philadelphia (1991).
[42] V. F. Dmitriev, I. B. Khriplovich, and V. B. Telitzin, Phys. Rev. C 50, 2358 (1994).
[43] C. Bouchiat, Phys. Lett. B 57, 284, (1975).
[44] E. Hinds, C. Loving, and P. G. H. Sandars, Phys. Lett. B62, 97 (1976).
[45] V. A. Dzuba, V. V. Flambaum, and V. V. Sylvestrov, Phys. Lett. B 154, 93 (1985).

[46] A. M. Martensson-Pendrill, Phys. Rev. Lett. 54, 1153 (1985).
[47] S. M. Barr, Phys. Rev. Lett. 68, 1822 (1992); Phys. Rev. D. 45, 4148 (1992); Phys. Rev. D 47, 2025 (1993).
[48] I. B. Khriplovich, Nucl. Phys. B 352, 385 (1991).
[49] R. S. Conti and I. B. Khriplovich, Phys. Rev. Lett. 68, 3262 (1992).
[50] See for example: H. A. Bethe and E. E. Salpeter, *Quantum Mechanics of One- and Two- Electron Atoms*, Academic Press, New York (1957); pp 50–51.
[51] J. D. Bjorken and S. D. Drell, *Relativistic Quantum Mechanics*, McGraw-Hill, New York (1964).
[52] J. D. Jackson, *Classical Electrodynamics*, 3rd Ed., Wiley, New York (1998); p.556.
[53] E. E. Salpeter, Phys. Rev. 112, 1642 (1958).
[54] P. G. H. Sandars, J. Phys. B 1, 499 (1968).
[55] P. G. H. Sandars, J. Phys. B 1, 511 (1968).
[56] P. G. H. Sandars and R. M. Sternheimer , Phys. Rev. A 11, 473 (1975).
[57] G. E. Brown and D. G. Ravenhall, Proc. R. Soc. A 208, 552 (1951).
[58] J. Sucher, Int. J. Quantum Chem. 25, 3 (1984).
[59] E. Lindroth, B. W. Lynn, and P.G.H. Sandars, J. Phys. B 22, 559 (1989).
[60] W. R. Johnson, D. S.Guo, M. Idrees, and J. Sapirstein, Phys. Rev. A 34, 1043 (1986).
[61] A. Shukla, B. P. Das, and J. Andriessen, Phys. Rev. A 50, 1155 (1994).
[62] V. V. Flambaum, Sov. J. Nucl. Phys. 24, 199 (1976).
[63] V. A. Dzuba, O. P. Sushkov, W. R. Johnson, and U. I. Safronova, Phys. Rev. A 66, 032105 (2002).
[64] E. D. Commins, J. D. Jackson, and D. P. DeMille, Am. J. Phys. 75, 532 (2007).
[65] M. A.Bouchiat and C. Bouchiat, Jour. Phys. (Paris) 36, 493 (1975).
[66] E. N. Fortson, Bull. Am. Phys. Soc. 28, 1321 (1983).
[67] V. V. Flambaum and I. B. Khriplovich, Phys. Lett. A 110, 121 (1985).
[68] A. M. Martensson-Pendrill and P. Oster, Phys. Scripta 36, 444 (1987).
[69] M. G. Kozlov *et al.* Phys. Rev. A 56, R3326 (1997).
[70] M. G. Kozlov, Sov. Phys. JETP, 62, 1114 (1985).
[71] N. Mosyagin, M. Kozlov, and A. Titov, J. Phys. B 31, L763 (1998).
[72] A. V. Titov, N. Mosyagin, and V. Ezhov. Phys. Rev. Lett. 77, 5346 (1996).
[73] Y. Y. Dmitriev *et al.*, Phys. Lett. 167A, 280, 1992.
[74] A. N. Petrov *et al.*, Phys. Rev. A 72, 022505 (2005).
[75] E. R. Meyer and J.L. Bohn, Phys. Rev. A 78, 010502(R) (2008).
[76] T. A. Isaev *et al.*, Phys. Rev. Lett. 95, 163004 (2005).
[77] E. R. Meyer, J. L. Bohn, and M. P. Deskevitch, Phys. Rev A 73, 062108 (2006).
[78] A. N. Petrov, N. S. Mosyagin, T. A. Isaev, and A. V. Titov, Phys. .Rev. A. 76, 030501 (2007).
[79] B. E. Sauer, J. Wang, and E. Hinds, J. Chem. Phys. 105, 7412 (1996).
[80] M. G. Kozlov and L. N. Labzowsky, J. Phys. B 28, 1933 (1995).
[81] C. T. Munger, Jr., Phys. Rev. A 72, 012506 (2005).

[82] S. K. Lamoreaux, Phys. Rev. A60, 1717 (1999).
[83] J. Nenonen, J. Montonen, and T. Katila, Rev. Sci. Instrum. 67, 2396 (1996).
[84] M. A. Player and P.G.H. Sandars, J. Phys. B3, 1620 (1970).
[85] E. D. Commins, Am. J. Phys. 59, 1077 (1991).
[86] J. M. Pendlebury, *et al.*, Phys. Rev. A 70, 032102 (2004).
[87] S. K. Lamoreaux and R. Golub, Phys. Rev. A 71, 032104 (2005).
[88] M. Rupasinghe and N.E. Shafer-Ray, Phys. Rev. A 78, 057702 (2008).
[89] M. V. Romalis and E.N. Fortson, Phys. Rev. A 59, 4547 (1999).
[90] N. E. Shafer-Ray, Phys. Rev. A 73, 34102 (2006).
[91] N. E. Shafer-Ray, private communication.
[92] S. A. Murthy, D. Krause, Z.L. Li, and L. Hunter, Phys. Rev. Lett. 63, 965 (1989).
[93] M. V. Romalis, private communication.
[94] T. W. Kornack and M. V. Romalis, Phys. Rev. Lett. 89, 253002 (2002).
[95] T. W. Kornack, R. K. Ghosh, and M. V. Romalis, Phys. Rev. Lett. 95, 230801 (2005).
[96] J. M. Amini, C. T. Munger, and H. Gould, Phys. Rev. A 75, 063416 (2007).
[97] C.Chin, V. Leiber, V. Vuletic, A. J. Kerman, and S. Chu, Phys. Rev. A 63, 033401 (2001).
[98] D. Heinzen, private communication.
[99] D. S. Weiss, F. Fang, and J. Chen, Bull. Am. Phys. Soc. APR03, J1.008 (2003).
[100] Y. Sakemi, private communication.
[101] J. J. Hudson, B. E. Sauer, M. R. Tarbutt, and E. A. Hinds, Phys. Rev. Lett. 89, 023003 (2002).
[102] M. R. Tarbutt *et al.*, J. Phys. B 35, 5013 (2002).
[103] J. J. Hudson, H.T. Ashworth, D.M. Kara, M.R. Tarbutt, B.E. Sauer, and E. A. Hinds, Phys. Rev. A 76, 033410 (2007).
[104] M. R. Tarbutt, J. J. Hudson, B. E. Sauer, and E. A. Hinds, arXiv:0811.2950v1 [physics.atom-ph] (2008).
[105] D. Egorov *et al.*, Phys. Rev. A 63, 030501(R) (2001).
[106] D. DeMille *et al.*, Phys. Rev. A 61, 052507 (2000).
[107] L. R. Hunter *et al.*, Phys. Rev. A 65, 030501(R) (2002).
[108] D. Kawall, F. Bay, S. Bickman, Y. Jiang, and D. DeMille, Phys. Rev. Lett. 92, 133007 (2004).
[109] G. Herzberg, *Spectra of Diatomic Molecules*, 2nd Ed., Van Nostrand, New York.
[110] G. Edvinsson and A. Lagerqvist, J. Mol. Spectrosc. 113, 93 (1985); and references therein.
[111] V. Goncharov, J. Han, L. A. Kaledin, and M. C. Heaven, J. Chem. Phys. 122, 204311 (2005); and references therein.
[112] J. Paulovic *et al.*, J. Chem. Phys. 119, 798 (2003).
[113] S. E. Maxwell *et al.*, Phys. Rev. Lett. 95, 173201 (2005).
[114] D. Patterson and J.M. Doyle. J. Chem. Phys. 126, 154307 (2007).
[115] E. Cornell and co-workers, private communication.
[116] F. L. Shapiro, Sov. Phys. Usp. 11, 345 (1968).

[117] B. V. Vasil'ev and E. V. Kolycheva, Sov. Phys. JETP 47, 243 (1978).
[118] M. Mercier, Magnetism 6, 77 (1974).
[119] C.-Y. Liu, private communication.
[120] T. N. Mukhamedjanov, V.A. Dzuba, and O.P. Sushkov, Phys. Rev. A 68, 042103 (2003).
[121] B. Ravaine, M. G. Kozlov, and A. Derevianko, Phys. Rev. A 72, 012101 (2005).

Chapter 15

The Neutron Electric Dipole Moment: Yesterday, Today and Tomorrow

Steve K. Lamoreaux

Department of Physics
Yale University
P.O. Box 208120
New Haven, CT 06520 U.S.A.
steve.lamoreaux@yale.edu

Robert Golub

Department of Physics
North Carolina State University
Riddick Hall
2401 Stinson Drive
Raleigh, NC 27695 U.S.A.
rgolub@ncsu.edu

The possibility for the existence of an electric dipole moment (EDM) of the neutron has been of interest for nearly 60 years. In this review, we provide a brief discussion of the history of the neutron EDM both in regard to theory and experiment. We also discuss the motivation and rationale for new experiments that are under construction or planned, and the prospects for attaining a new level of sensitivity that will severely challenge theoretical extensions to the so-called Standard Model of electroweak interactions.

Contents

15.1. Introduction

The astute reader might be wondering why a discussion of an electromagnetic moment of the neutron is included in a volume on lepton moments (if you don't know, don't worry). The silly answer is that we were invited to write this; the serious answer is that the electron EDM together with the neutron EDM provide some of the most stringent constraints on extensions to the Standard Model of electroweak interactions, in particular those that fall under the general heading of supersymmetry (SUSY). It should further be noted that CP non-invariance effects, and hence T non-invariance, have only been observed in mesonic systems; originally (1964) in the decay of the K_0 and more recently at so-called B factories (BaBar, Belle), these effects being fully accounted for within the framework of the Standard Model.

K_0- and B-meson studies involve changes in the flavor quantum numbers "strangeness" and "beauty", while the electron and neutron EDMs are flavor conserving, and thus relatively suppressed in the Standard Model where they are induced at the multi-loop quantum level. The Standard-Model prediction for the neutron EDM is about 10^{-31} e cm, while the electron is about 10^{-40} e cm. Both predictions are impossibly small by today's experimental standards. But this is in fact an advantage and implies that we need not be encumbered by the usual CP violation in the Standard Model, and are thus granted free reign to explore new sources of CP

violation, unhindered by the need of arbitrary-accuracy QCD perturbative calculations of the effects of CP violation in the Standard Model, as in the case of B decay.

Of course, the long-standing observation of the matter-antimatter asymmetry in the universe provides additional impetus to discover new sources of CP violation, as those known within the context of the Standard Model are too small to account for this asymmetry.

Noting that serious interest in the possibility of a neutron or other EDM began only in 1964 with the observation of CP non-invariance in K_0 decay, the general question of the symmetry properties of fundamental forces was first put forward in 1950 by Ramsey and Purcell. The development of the ideas of symmetry, or lack thereof, in subatomic forces is described in the following quotes from Norman Ramsey's autobiography [1] (p. xxx).

> In 1950, when the assumption of parity (P) symmetry was universally accepted, I was lecturing on molecular beams with Ed Purcell sitting in. I was about to give the then-standard proof that in the absence of degeneracy a particle whose orientation was determined by its spin angular momentum could have no electric dipole moment, a proof that depends on the assumption of P symmetry for nuclear forces. I had already discovered that if I lectured on a topic I did not thoroughly understand, I could count on Purcell asking me an astute question that would reveal my ignorance. In anticipation of such a question I tried to find experimental evidence for the assumption of parity symmetry in nuclear forces but found none. An electric dipole moment interacting with an external electric field, for example, would show a P failure but most experiments were on charged particles which would be accelerated by an electric field into a zero field region or out of the apparatus. On the Military Principle that, if about to be attacked, counterattack, I asked Purcell for the evidence for P symmetry with nuclear forces, and after some effort he decided he could find none either. Thus, in 1950, we published Paper 3.1, [64] pointing out the absence of experimental evidence for the assumption of P symmetry, and we collaborated with a graduate student J. Smith, to test P by looking for a neutron electric dipole moment we found none, but did set a low limit to its value (Paper 3.2). [65]
>
> Most physicists then attributed this result to the assumed validity of P symmetry, but I continued to believe a failure of P symmetry in nuclear forces might be found with a sufficiently sensitive experiment At that time such forces were usually discussed together as nuclear forces rather than the two separate categories of weak and strong forces. Our electric dipole experiment

> was a very sensitive test for P failure in the strong, but not in the weak, force. As a result, in 1956, when I heard C. N. Yang in a colloquium suggest the possibility of a P failure in the weak force, I immediately wanted to test the weak interaction and proposed, in the colloquium discussion and subsequent correspondence, a method for doing so. I knew that L. Roberts of Oak Ridge had cryogenic facilities for polarizing nuclei and that radioactive ^{60}Co had been polarized, so I arranged to do an experiment with him to see if more decay electrons came out in one direction relative to the spin than in the opposite. Roberts agreed to provide the polarized ^{60}Co and the electron-counting equipment. Unfortunately, Roberts soon discovered that the angular distribution of the neutrons in fission was different than expected theoretically so the theory advisory group at Oak Ridge urged him to concentrate on exploiting this new discovery and postpone our highly speculative test of parity. By the time I was told of this decision, C. S. Wu and B. Ambler had already started preparing their own experiment which later showed a maximal failure of parity symmetry in the radioactive decay of ^{60}Co. The absence of an electric dipole moment in our neutron experiment and the forced postponement of our ^{60}Co experiments were the greatest disappointments in my research career. But by then, I had realized that research scientists have both good and bad luck and productive scientists do not allow the bad luck to discourage them from further research.

What do we learn from this? Never accept the advice of an advisory group? Continuing [1] (p. xxv),

> Soon after the experimental demonstration of the failure of parity symmetry in the weak interaction, most of the leading theorists, including L. Landau, T. D. Lee, C. N. Yang and E. Wigner, predicted there could still be no neutron electric dipole moment because of CP conservation, which would imply T symmetry if there were CPT symmetry as was generally believed. I then wrote the theoretical Paper 3.4 [66] pointing out that CP and T symmetries were assumptions which required experimental testing and that a search for a neutron electric dipole moment provided such a test. Subsequently various collaborators and I have carried out successive searches with increased sensitivity, but we have yet to find a non-zero electric dipole moment. The attitude of theorists toward these experiments changed markedly in 1964 with the discovery of CP non-invariance in the decay of the long-lived neutral kaon [K_0]. Most theories soon after that discovery predicted neutron electric dipole moments of sizes comparable to our limits.

> As our experimental limit lowered, most of these early theories were abandoned and replaced by new theories predicting smaller neutron electric dipole moments.

And with that said, we could close this book and all go home. But alas, as is often said, the devil is in the details; these details are to follow.

15.2. Theoretical Motivation

Much is presented elsewhere in this volume concerning the theoretical motivation for EDM searches. Here we present our unique overview of the problems that are being addressed. Given the difficulty with the experimental searches, we present the following to highlight the true excitement and relevance of these experiments, which are much more that simple exercises in precision spectroscopy. The progression of experimental limits compared with theoretically expected EDM ranges of values is shown in Fig. 15.1.

15.2.1. *The Higgs field, supersymmetry (SUSY) and all that*

The ability to calculate lepton moments within the framework of QED is well-known, with initial successes in the 1940s. These successes are due in large part to the smallness of the electromagnetic coupling parameter $\alpha \approx 1/137$ which can be thought of as the ratio between the Coulomb energy and the rest mass energies of two electrons separated by a Compton wavelength; the smallness of this parameter means that it makes sense to perform perturbative calculations for the effects of virtual field excitations on the ground state of the electron or muon. Of course, such calculations diverge, but these divergences can be handled through renomalization procedures, yielding finite and meaningful calculations of ground state properties. Also important is the fact that the photon field is massless.

In the case of massless vector fields associated with the Z_0 and $W_{\pm}$, introduced in the Weinberg–Salam–Glashow model of electroweak interactions that unified the weak and electromagnetic forces, higher order virtual calculations diverge and are non-renormalizable. Nonetheless, we consider this theory as completely successful, for it provides the basis of the Standard Model. These divergences are handled by introducing a scalar field, the Higgs field, which interacts with the massless vector field, making it appear as massive. Extra virtual processes associated with the Higgs field cancel those associated with the massless vector field, making the theory

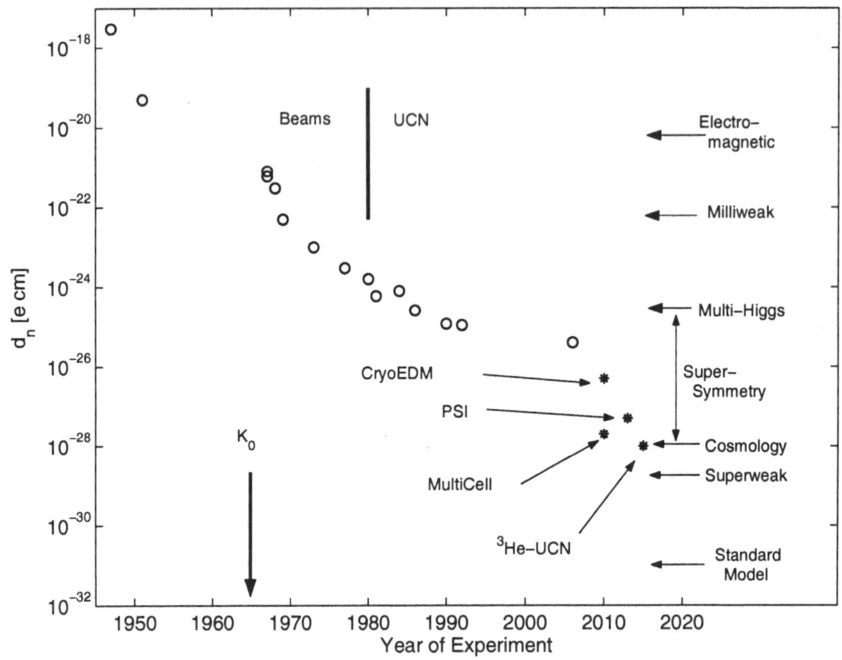

Fig. 15.1. Historical development of the neutron EDM experimental limit along with expectations from various theoretical models. The points marked with * are the anticipated limits from experiments presently under development or proposed, and will be discussed in this review.

renormalizable and therefore theoretically acceptable. This theory has had its most spectacular success in applications to measurements at the so-called Z-pole in e^+e^- collisions where the following successes, among many, have been achieved: Proof of three families of neutrinos; indirect measurement of the top quark mass; upper and lower limits on the mass of the Higgs boson.

Despite the successes of the Standard Model, in the context of the CP violation in the Higgs field, it is difficult to reconcile the large CP violation observed in K_0 decay with the small values of the neutron and electron EDMs. Furthermore, the Standard Model does not provide a mechanism with sufficient rate to generate the universe baryon asymmetry. Briefly, the problem can be outlined and a naive estimate of the neutron EDM can be obtained as follows:

$$d_n/e \approx m_p^{-1} \times G_F\, m_\pi^2 \times \eta \tag{15.1}$$

$$\approx (2 \times 10^{-14}\text{cm})(2 \times 10^{-7})(2 \times 10^{-3})$$

$$\approx 10^{-24} - 10^{-23}\,\text{cm}. \tag{15.2}$$

The first term here, the Compton wavelength of a proton, is the usual scale of a nucleon magnetic moment. The second one is the characteristic relative magnitude of weak interactions in hadronic processes (the typical momenta used to construct a dimensionless factor from the Fermi weak interaction constant G are assumed to be comparable to the pion mass m_π). Finally, one might expect that the CP-odd part of the weak interaction is somehow suppressed as compared to the CP-even one. As a natural value of this suppression factor, the third term η, is the ratio of the CP-odd and CP-even amplitudes in the decays of K_0 and B mesons.

This estimate is about two orders of magnitude larger than the experimental limit for the neutron EDM. However, the value of $\eta \sim 10^{-3}$ is larger than one might expect; its large value is due to the very small difference in mass between K_L and K_S mesons and their relatively small decay rates, which are at the level of 10^{-15} of the mass of either state, so the mass denominator in the perturbation expansion of the states is small. The K_0 and B mesons appear as unique systems to explore CP non-invariance. Although the K_0 and B decay properties can be described within the formalism of the Standard Model, the existence of these decays can be interpreted as the existence of new particle interactions, for example a milliweak interaction, which has a strength $10^{-3}G_F$ and changes strangeness by one so K decay is a second-order process, or Wolfenstein's superweak interaction which has coupling $10^{-9}G_F$ and changes strangeness by two, so K decay is a first order process. [54] The milliweak interaction leads to a neutron EDM of 10^{-22} e cm and is thus ruled out, whereas the superweak interaction leads to a neutron EDM of 10^{-29} e cm, and would thus appear as viable, at least in the context of the neutron EDM. On the other hand, the Standard Model neutron EDM prediction is of order 10^{-31} e cm.

The specific origin of CP non-invariance is not known; however, if CP non-invariance is incorporated into the Higgs field in a simple way (e.g., three doublets of complex Higgs fields), it is impossible to have a small neutron EDM and significant CP asymmetry in K_0 decay simultaneously. The Higgs bosons can be made as heavy as possible, and the CP odd phases tuned to be small to accommodate the neutron EDM limit, but this means that the CP odd effect observed in K_0 decay has a different origin.

This problem provides motivation for supersymmetric extensions of the Standard Model. In SUSY models, the Higgs field can become very complicated. In addition, each massive fermionic field is assigned a bosonic field, in much the same way that the Higgs field was introduced for the intermediate vector bosons. This has a several salubrious effects, including renormalizability in calculations, and the introduction of a multitude of new phases, some of which imply CP non-invariance. These additional particles and phases make it possible to accommodate the CP odd characteristic of K_0 and B decay with small neutron and electron EDMs, and provides a mechanism for the generation of the matter-antimatter asymmetry of the universe.

The conclusion is that there are likely multiple Higgs. The phases and masses are of course completely unknown, but EDMs much larger than predicted by the bare Standard Model are entirely possible, and might be expected. It is likely that EDM experiments will provide the only window into CP non-invariance in the Higgs sector, and will thus complement LHC and other studies where it is expected that the Higgs will be discovered. Given the difficulty of the LHC experiments, the goals of which are to detect the Higgs, little beyond the masses will be learned, and certainly exploration of the symmetry properties of the Higgs remains a remote possibility at present. This underscores the importance of the continuation of all EDM work, particularly given the modest budget of most experiments.

15.2.2. *The strong CP problem and the axion*

It is well-known that the usual Lagrangian of the electromagnetic field

$$-\frac{1}{4}F_{\mu\nu}F_{\mu\nu} = \frac{1}{2}(\vec{E}^2 - \vec{B}^2)$$

can in principle be supplemented by another Lorentz scalar [56]

$$F_{\mu\nu}\tilde{F}_{\mu\nu}, \quad \tilde{F}_{\mu\nu} = \frac{1}{2}\epsilon_{\mu\nu\kappa\lambda}F_{\kappa\lambda}.$$

This pseudoscalar violates both P and T invariance, which can most easily be seen from its three-dimensional form:

$$F_{\mu\nu}\tilde{F}_{\mu\nu} = -4\vec{E}\cdot\vec{B}.$$

However this scalar generates no observable effects in electrodynamics as its net contribution to the action can be shown to vanish because the fields fall off rapidly at infinity. The electromagnetic field is considered an Abelian gauge theory, which is known to high accuracy, e.g. the photon mass is zero.

In quantum chromodynamics (QCD) the situation is quite different. Due to the self-interaction of the gluon vector potential A^a_μ, the field configurations that do not fall off rapidly enough at infinity play a prominent role in the theory. Therefore, an analogous Lorentz scalar is no longer inconsequential. The corresponding possible P and T non-invariant term in the QCD Lagrangian is usually written as

$$L_\theta = -\theta\,(\alpha_s/8\pi)\,\tilde{G}^a_{\mu\nu}G^a_{\mu\nu} \tag{15.3}$$

and is called the θ term. Here α_s is the gluon coupling constant, the QCD analog of the fine structure constant α in electrodynamics.

The neutron EDM generated by a non-zero θ can be calculated within the framework of the Standard Model, and the present experimental limit implies

$$\theta < 10^{-10}. \tag{15.4}$$

The smallness of this parameter is referred to as the strong CP problem: Why should something that can be anything (to order unity) be so small?

There are several possible explanations. The calculation of the neutron EDM shows that is is proportional to θ times the product of the quark masses; if one of the quark masses is zero, there would be no neutron EDM generated by a non-zero θ. This possibility appears as ruled out by experimental measurements of the quark masses and their ratios.

A favored explanation is the so-called Peccei–Quinn (PQ) mechanism, where θ is considered as a field (particle). In this mechanism, a new global chiral symmetry is introduced to the Standard Model that becomes spontaneously broken. This leads to a new particle, a Goldstone boson of the new field, called the axion, with zero bare mass. However, due to instanton effects, the axion acquires a mass, and thereby couples linearly to gluon fields, thus generating a finite θ_{eff} which is then canceled by adjusting the location of the potential minimum in the pseudoscalar field. The original PQ mechanism has been ruled out, but has been recast with very light "invisible" axions, which remain of current experimental interest.

It should be noted that the PQ mechanism was introduced in the mid to late 1970s, just at the time that the Standard Model was being developed, to solve the θ puzzle. Introduction of the Higgs field solved another puzzle (perturbative covergence of the weak interaction), and it is unclear whether these two solutions are mutually exclusive or mutually compatible. It should be further noted that θ can be calculated directly within the context of the Standard Model. Its value is small, $\theta \approx 10^{-19} - 10^{-18}$ [38]. It is generated at the same $\alpha_s G^2$ (fourth order) as the neutron EDM. A more complete analysis shows that when θ is calculated within the context of the Standard Model, it is renormalized by other CP-odd interactions, and in general this renomalization can be infinite. In particular, the induced contributions to θ appears to diverge logarithmically starting at high order (fourteenth) in the electroweak coupling constant. Introduction of the axion solves this specific problem.

In the context of SUSY, $\theta = 0$ implies that CP is a good symmetry in the $SU(3)_c$ sector of these models. So in fact the strong CP problem can be inverted: θ is small precisely because the neutron EDM is small. However, there is no denying that the overall mystery persists.

15.2.3. *Matter-antimatter asymmetry of the universe*

Sakharov, in 1967, was first to propose a mechanism whereby an imbalance between matter and antimatter could be generated from an initially homogeneous mixture. This mechanism requires: 1. Non-equilibrium conditions; 2. Baryon number nonconservation; and 3. CP non-invariance which allows matter and antimatter to have different reaction rates. The first condition is met in the earliest stages of the Big Bang. At sufficiently high energy, the Standard Model does allow baryon number nonconservation, and it is now generally assumed that at some point the universe went through a phase transition from where the energy density was high enough to allow easy changes between baryon and antibaryon number, to a relatively stable state. This phase transition might have occurred in bubbles or more complicated regions, with expanding domain walls. CP non-invariance would allow baryons vs. antibaryons to have different reflection/penetration properties at the moving domain walls, so an excess matter vs. antimatter could be collected into the stable regions. On the other hand, in the very early high temperature dense universe, it is likely that there was an abundance of primal black holes, and as baryons or antibaryons were swallowed up by these, the knowledge of the baryon number was lost, providing another mechanism of baryon number nonconservation. As these primal black holes evaporated under non-equilibrium conditions, the net rate of baryon vs. antibaryon generation could be different, leading into a very slight imbalance in their numbers. As the universe expanded, this imbalance was maintained, with most of the antibaryons annihilating against the more abundant baryons.

These models appear as viable, except that the CP non-invariance observed in the decay of K_0 and B mesons, appears inadequate to give the observed matter excess, which can be estimated [7]

$$\frac{n_B}{n_\gamma} < \frac{d_{cp}}{T} \sim 10^{-19} \tag{15.5}$$

where $d_{cp} = 10^{-17} M_W^{12}$ (with $M_W = 80$ GeV) is the magnitude of the Standard Model Lagrangian describing CP non-invariance in the reactions at energies above the electroweak phase transition temperature, $T = 100$ GeV. Although this estimate is open to criticism, it appears as reasonable in that it is an estimate of a difference in the reaction rates between different CP states in a quasi-equilibrium situation, which is not to be confused with a calculation of the opposite CP state mixing in the K_0 system for example.

15.3. Comparison of Experimental Techniques

Over the last 60 years or so, a number of experimental techniques have been put forward to measure the neutron EDM; the only ones that have set significant limits have been based on magnetic resonance measurements. However, interest remains in the possibility to detect a neutron EDM in a scattering experiment, the idea that the neutron can interact with an atomic-scale electric field which is five orders of magnitude larger than any conceivable laboratory field.

All EDM searches are based on the application of an electric field and then searching for an appropriate response. In the case of magnetic resonance measurements, the value of the electric field is obvious, while in scattering experiments, determining the effective electric field is challenging. However, all experiments can ultimately be cast as measurement of the effective interaction energy of a neutron with an electric field: In the presence of a non-zero EDM, an electric Zeeman effect occurs in addition to the usual magnetic Zeeman effect, and the Hamiltonian of the system is

$$H = -(\mu\vec{B} + d_n\vec{E}) \cdot \frac{\vec{s}}{|s|}, \tag{15.6}$$

where $\vec{B}$ and $\vec{E}$ are applied static magnetic and electric fields, μ is the magnetic moment, $s = 1/2$ is the net angular momentum of the neutron, and d_n represents the EDM.

The art of all EDM measurements is in the separation of spurious electric field effects from a true EDM effect. The spurious effects can be made quite small; this illustrates an advantage of EDM experiments over T violation study involving β decay or neutron transmission [11] where the sought T violation signal cannot be turned on and off and appears alongside other allowed processes.

In writing Eq. (15.6), we have ignored, for example, changes in the internal structure of the neutron due to the application of the electric field (electric tensor polarizability). Also ignored is a possible static electric quadrupole moment; these two possible effects indicate some of the advantages of working with spin-1/2 systems where the only possible (PT even) electromagnetic moment is the magnetic dipole. For a spin-1/2 system, there is no energy shift between $m_{\rm F} = \pm 1/2$ due to application of an electric field, and therefore no directly observable effect. Also we have assumed that the net species charge is zero, supposedly this is exactly true for the neutron.

A typical experimental observable is the change in Larmor precession frequency associated with a reversal of $\vec{E}$ relative to $\vec{B}$; this is an energy shift correlated with the quantity $\vec{E}\cdot\vec{B}$, a P- and T-odd quantity. An EDM of 1×10^{-26} e cm would produce a relative change in precession frequency, on reversal of $\vec{E}$ relative to $\vec{B}$, of 1×10^{-7} Hz when $E = 10$ kV/cm. This frequency shift corresponds to a magnetic field of about 2×10^{-11} Gauss for a neutron or diamagnetic atom, or about 10^{-13} Gauss for a paramagnetic atom. Given that the Earth's magnetic field is of order 0.5 Gauss, we see immediately that magnetic field control is crucial for any EDM experiment.

Other EDM observables are changes in position or momentum of a neutron interacting with an electric field gradient; such effects have been sought in neutron scattering experiments. The magnitude of the force $\vec{f}$ is simply given by the gradient of (15.6), and therefore detection of an EDM force can ultimately be associated with an energy shift.

This leads to the definition of a figure of merit F for EDM measurements. From the uncertainty principle, the accuracy with which an energy change can be measured is inversely proportional to the time that the neutron interacts with a given electric field. The magnitude of the energy change is proportional to the effective electric field. Finally the shot noise of the measurement is proportional to the square root of the neutron current I, leading to

$$F = ET\sqrt{I} \tag{15.7}$$

where, in the case of a neutron storage experiment, $I = N/T = \rho V/T$ where N is the total number of neutrons (product of density ρ and volume

Table 15.1. Comparison of neutron EDM experimental sensitivities, where the systematic limit represents the control required to attain the full fundamental shot noise sensitivity.

Technique	E [kV/cm]	T [s]	I [n/s]	Sys. Lim.	$ET\sqrt{I}$
Bragg reflection	1×10^9	2×10^{-7}	10^4	$\theta_{\mathrm{EB}} < 10^{-4}$	2×10^4
Neutron beam magnetic resonance	2×10^5	1.5×10^{-2}	1×10^6	$\theta_{EB} < 10^{-5}$	3×10^6
Ultracold neutron	1×10^4	100	250	$\delta E/E < 0.1$	2×10^7
Pendellösung (α-quartz)	2×10^8	2×10^{-3}	2×10^3	$\theta_{\mathrm{EB}} < 10^{-7}$	2×10^7
UCN-^{3}He	5×10^4	500	5×10^3	$\delta E/E < 0.1$ $\partial B_0/\partial z < 0.01\ \mu$G/cm	2×10^9

V) counted at the end of the measurement of duration T, assumed to be dominated by the coherence time. This factor allows us to compare different experimental techniques, as shown in Table 15.1.

Because Bragg scattering experiments require the setting of an angle to a degree of precision that appears as experimentally impossible, we will not review this work here. Because the electric field in the crystal cannot be turned on and off, and detection and discrimination of an EDM effect requires absolute alignment of the crystal axes, the problem with these experiments are similar to those expected in neutron absorption or spin rotation experiments; see [11] for a discussion of the experimental issues.

15.4. Systematic Effects in Magnetic Resonance Experiments

15.4.1. $E \times v$ *effects in beam experiments*

Another spurious effect is the so-called motional magnetic field, first addressed in relation to a Cs atomic EDM experiment. Its effects are most severe for atomic beam experiments [12]. When one moves relative to the sources of a static electric field $\vec{E}$, according to special relativity, a magnetic field $\vec{B}_{\mathrm{m}}$ is generated in the co-moving frame which to first order in v/c is

$$\vec{B}_{\mathrm{m}} = \vec{E} \times \frac{\vec{v}}{c}. \tag{15.8}$$

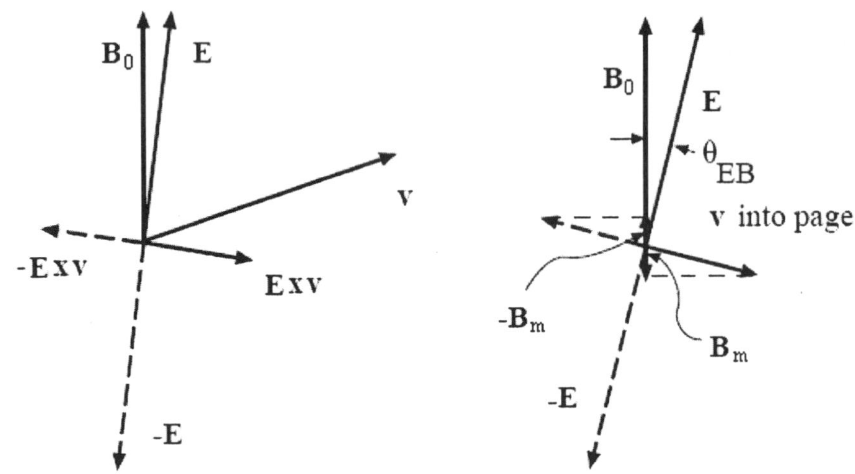

Fig. 15.2. Geometrical picture of the $\vec{E} \times \vec{v}$ effective magnetic field.

For a typical cold neutron velocity of $v = 1000$ m/s in an electric field of 100 kV/cm, $B_{\rm m} = 1$ mG. Now consider an experiment where there is a large applied magnetic field $\vec{B}_0$ and an EDM is sought by measuring the shift in Larmor precession frequency on reversal of a electric field $\vec{E}$, as implied by (15.6).

If $\vec{E}$ and $\vec{B}_0$ are nearly parallel as shown in Fig. 15.2, and $B_{\rm m} \ll B$, the effective magnetic field strength is given by the magnitude of $\vec{B} = \vec{B_0} + \vec{B_m}$

$$B = B_0 + \theta_{\rm EB} B_{\rm m} + \frac{1}{2}\frac{B_{\rm m}^2}{B_0}, \tag{15.9}$$

giving a change in Larmor frequency of

$$\Delta\omega = \frac{\gamma\theta_{\rm EB} v}{c} E + \frac{\gamma v^2}{2c^2}\frac{E^2}{B_0}, \tag{15.10}$$

where γ is the gyromagnetic ratio, and $\theta_{\rm EB}$ is the angle between $\vec{E}$ and $\vec{B}_0$ in the plane perpendicular to $\vec{v}$. If B is substituted into Eq. (15.6), it can be readily seen that if $\theta_{\rm EB} \neq 0$, a spurious shift correlated with $\vec{E} \cdot \vec{B}$ is generated, which has the same signature as an EDM. However, this is not a true T-violating effect, for under T, $\vec{v}$ reverses sign, and therefore so does $\vec{B}_{\rm m}$. The important point is that reversing $\vec{E}$ relative to $\vec{B}$ does not create the time-reversed Hamiltonian; $\vec{v}$ must also be reversed. Even in the case where $\theta_{\rm EB} = 0$, there is a relative shift quadratic in $B_{\rm m}$, which may require that the magnitude of E does not change significantly on reversal.

The limit on $\theta_{\rm EB}$ in the final beam experiment, performed in 1977 using the Oak Ridge apparatus that had been moved to the Institut Laue-Langevin, can be estimated as follows. The neutron velocity was about 100 m/s, implying a motional field of 0.1 mG. The reported uncertainty for this experiment is 1.5×10^{-24} e cm, implying a limit on the shift in resonance frequency of about 20 μHz in 100 kV/cm, further implying a magnetic field control of about 10 nG, or a part in 10^5 of the motional field. Thus, the requirement on $\theta_{\rm EB}$ for the last neutron beam EDM experiment was $\theta_{\rm EB} < 10^{-5}$ radians. Although much effort was expended in dealing with this effect, which included mounting the entire experiment on a Navy Surplus gun turret in an attempt to cancel the $E \times v$ systematic by reversing $\vec{v}$ through the apparatus, it remained the ultimate limiting factor for neutron beam experiments. These techniques were abandoned in favor of ultracold neutron storage experiments which have the advantage that $\langle v \rangle = 0$ so the motional field is expected to have very small effects. We address these effects later in this section.

15.4.2. *Electric-field correlated magnetic effects*

Whenever high voltages are applied to a system, small leakage currents invariably flow through insulators, and these currents generate magnetic fields which are correlated with the electric field direction and are indistinguishable from an EDM. The leakage current magnetic field is a function of the electric field and adds a term to the Hamiltonian (15.6)

$$H = -\left[\mu(\vec{B} + \hat{z}(\beta\hat{z} \cdot \vec{E}) + d\vec{E}\right] \cdot \vec{s}/|s| \,, \tag{15.11}$$

where β represents the average projection of the magnetic field generated by the current density

$$\vec{j} = \sigma\vec{E} \tag{15.12}$$

along the static magnetic field direction ($\hat{z}$), with σ representing the electrical conductivity. A non-zero β implies some helicity of $\vec{j}$ along $\hat{z}$.

The apparent T-odd character of this new term is the result of the irreversible "macroscopic" process(es) which lead to (15.12). Also, under parity reversal, the helicity of the leakage path, hence the effective magnetic field, changes sign; under parity reversal, $\vec{E}$ also changes sign, so we see that the leakage current effect is even under parity reversal, unlike a true EDM.

For beam experiments where the insulators and conductors leading to the high voltage plates can be relatively well spatially separated from the sensitive measurement area, leakage current fields are not so troublesome as in the case of storage experiments where the cell walls generally serve a second purpose as the high voltage electrode spacer.

A worst-case scenario is when all the leakage current flows in a closed loop around the cell; given that such a current flow is highly unlikely, to estimate a possible systematic effect, one-quarter of the field at the center of a loop is sometimes taken. Note that if such a helical current existed due to some imperfection in the cell walls, reversing the cell orientation does not distinguish this effect because the leakage current helicity and hence magnetic field direction is a fixed property of the cell.

It is also important that the leads which supply the high voltage are coaxial with the leakage current return leads, otherwise the leakage currents, charging displacement currents when the electric field magnitude/direction is changed, or impulse currents associated with sparks can cause a systematic magnetization of the magnetic shields.

This discussion might appear as academic, but in fact two neutron EDM experiments, one at the Institut Laue-Langevin (ILL) in Grenoble, France [20], and one at the Petersburg Nuclear Physics Institute (PNPI) in Gatchina, Russia [63], reported results in the mid-1980s that were affected by systematics apparently of this type.

Of interest was the announcement of results from these two groups that were non-zero, agreed in sign and magnitude, and were at the 90% statistical

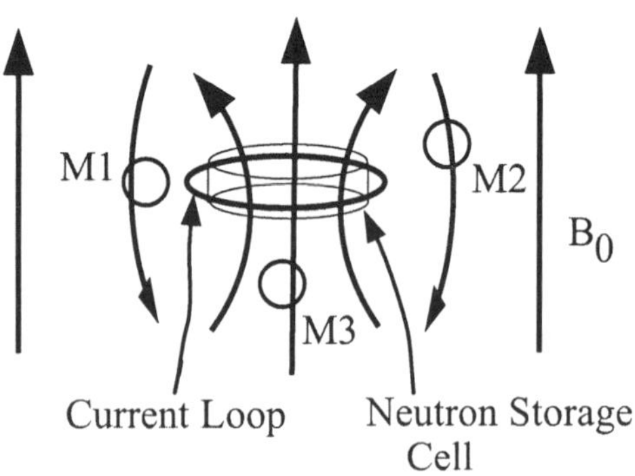

Fig. 15.3. The external magnetometer problem. Leakage currents associated with the application of a high voltage to the measurement cell can flow in a loop (or some fraction thereof) around the cell, creating a magnetic field that is correlated with the direction of the electric field. Depending on the location of a magnetometer, the field from the loop can add or subtract to the applied static field B_0.

confidence level. This announcement created a world-wide sensation. The problems with the data were found and reported in later publications [47, 58].

The quality of the data from the ILL experiment described in Ref. [47] is best illustrated in Fig. 15.4. In 1998, data from an improved neutron EDM experiment [57] (to be discussed later) was combined with the older data from 1990 [47] that was contaminated by magnetic field systematics associated with the application of voltage to the neutron storage cell. This figure was prepared for a subsequent analysis of the validity of combining

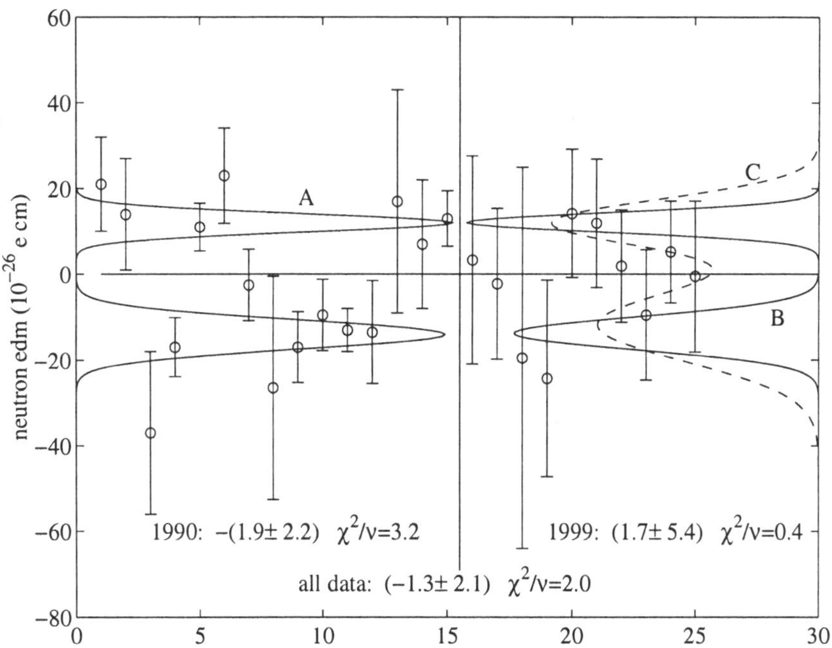

Fig. 15.4. Distribution of neutron EDM values over the course of running the 1990 ILL experiment, shown in the left-hand plot labeled A. Each point represents several weeks of running, usually a complete reactor fuel cycle. The plot on the right side includes subsequently acquired data using the Hg comagnetometer system. [4] Given that the previous data set was demonstrably contaminated by systematic effects, the combining of the data sets was not statistically valid, as discussed in Ref. [2]. Curve B is the distribution assuming the errors of the earlier data are systematic free, while curve C increases the error of the earlier data to reflect the systematic uncertainty, and there was no evident increase in sensitivity by combining the data sets. (This figure was reprinted with permission from Ref. [2]. Copyright 2000 by the American Physical Society.)

new, systematic free data, with old data that were limited in accuracy due to systematic effects. As can be seen in this plot, the data from the early experiment have a marked bi-modal character. In the data set, it was evident that major changes in the systematic EDM occurred whenever the measurement apparatus was disassembled and reassembled, usually for general maintenance. In the early analysis of the data of Ref. [47], it was assumed that if the average of the systematic magnetic field in the three rubidium magnetometers, shown schematically in Fig. 15.3, had zero average, the data set was systematic free. In fact, data selected by this criterion showed the largest neutron EDM. This is not surprising, as this selection of the data, where the average was near zero, made the experiment most sensitive to leakage currents in the neutron storage cell as the systematic field, due to helical leakage currents in the bottle, at the closest magnetometer was expected to be a factor of two larger than the outer two magnetometers, and of opposite sign.

The problem was uncovered when the correlations between the individual magnetometer systematic magnetic fields (associated with application of the high voltage), and the neutron frequency were shown to be statistically significant, invalidating any possibility of reliably detecting a non-zero neutron EDM at the level of the intrinsic sensitivity of the experiment. The correlation technique is described in detail in Ref. [26] (Sec. 7.3.4) and in [10] (Sec. 4.5.2).

The point in discussing this problem in such detail is that there are several planned or ongoing experiments that do not employ a so-called "comagnetometer." Results from these experiments will need to be considered with utmost caution, for as we know the past is usually prologue.

15.4.2.1. *The Ramsey comagnetometer*

With the increased sensitivity offered by the use of ultracold neutrons (UCN) in neutron EDM experiments, it became evident very early that magnetic field noise and systematic effects would ultimately limit the experimental sensitivity. Although the ideas had been discussed, the first published analysis of an *in situ* magnetometer was given by Ramsey [17]. The idea is that a spin polarized atomic gas can be stored along with the UCN in the same volume, and serve as a magnetometer. To very high accuracy, the atomic magnetometer can provide a measure of the magnetic field directly experienced by the UCN. The ambiguity associated with external magnetometers is thus eliminated.

Such an *in situ* magnetometer is referred to as a comagnetometer. This term appears to have been invented in the late 1980s by Prof. N. Fortson's group at the University of Washington. Identification of a comagnetometer for a specific EDM experiment is a sort of experimental Holy Grail. Finding such a magnetometer provides a measure of guarantee for the success of an experiment.

Ramsey's analysis addressed the use of polarized ^{3}He atoms. In his analysis, he shows that the direct effects due to field gradients and non-resonant radiofrequency pulses on the spin precession frequency are small for conditions normally found in UCN EDM experiments. Because these background effects are not correlated with application of the high voltage, they produce no intrinsic systematic effect. It was believed that the $\vec{E} \times \vec{v}$ field would be zero for a stored UCN experiment because the average value of v is effectively zero. However, recently it was realized that a quadratic effect persists which places requirements on the accuracy with which the applied voltage must be reversed. More important is the very recent realization that a quantum interference between a magnetic gradient and the $\vec{E} \times \vec{v}$ field can cause a systematic effect. These problems are discussed later in this section.

Ramsey discusses the requirement that the comagnetometer species does not have an EDM of its own. For light diamagnetic atoms, a nuclear EDM is suppressed by $\alpha^2 Z^2$ ($\alpha = 1/137$ is the fine structure constant, and Z is the atomic number) due to shielding by the electron cloud.

Despite several years of research with promising results at the University of Sussex, a practical ^{3}He magnetometer did not appear as feasible. Eventually, optically pumped ^{199}Hg was successfully employed as a comagnetometer in the ILL experiment, to be discussed later in this review.

Because UCN have velocity less than 7 m/s or so, their spatial density in a finite size storage cell is significantly modified by the earth's gravitational field. There is a considerable shift in the center of mass between a UCN gas and an atomic gas in the gravitational field, due to the difference in their effective temperatures [17]. Although the UCN gas does not strictly represent an equilibrium system, we can estimate the downward displacement by assuming an effective UCN temperature of 2.5 mK;

$$\Delta h = -\frac{h}{2} + \frac{1}{h}\int_0^h x e^{-m_{\rm n} g x/kT} {\rm d}x \approx -\frac{m_{\rm n} g h^2}{3kT}, \tag{15.13}$$

where h is the cell height. For a 20 cm cell, the displacement is on the order of 6 mm. For the higher temperatures of the magnetometer gas, the shift is comparatively insignificant.

Although this displacement represents an imperfection in the monitoring of the exact magnetic field as seen by the UCN, the discrepancy is small. A systematic magnetic shift that failed to be corrected would most certainly be evident in the direct UCN or atomic magnetometer signal, provided that the background magnetic field noise is sufficiently small.

It is generally assumed that for a comagentometer to be useful, the time to average the magnetic field over the entire storage volume is shorter than the spin coherence or total measurement time. If this averaging time is too long, there will be a relaxation (decoherence) effect unless the magnetic gradient is sufficiently small. Quantitatively, the experiment should be operated in the "motional narrowing" limit, where the inverse of the gradient-induced frequency shift is small compared to the time for a spin to diffuse across the storage cell. The dimensionless parameter d [6],

$$d = \left[\frac{2D}{L^2}\right]\left[\frac{\gamma G L}{2\pi}\right]^{-1} \gg 1 \tag{15.14}$$

in the motional narrowing limit, where D is the diffusion coefficient, L is a maximum characteristic length in the system, γ is the gyromagnetic ratio, and G is the magnetic field gradient. The first term in brackets is the rate that a spin moves diffusively through the entire cell, and the second term is the characteristic dephasing time associated with a spatial static magnetic field gradient. As will be discussed later, when the gradient is small enough for the "geometric phase" EDM to be small, d will tend to be large, for any imaginable D. However, a large d is a necessary but not sufficient requirement to reduce a possible EDM systematic gradient magnetic field (generated when an electric field is applied), and any particular system will require evaluation of its immunity to such effects. As will be discussed later, the fluctuating magnetic field due to the $E \times v$ motional field also can lead to relaxation. [8]

A finite averaging time can lead to a systematic effect if the application of an electric field creates a voltage-polarity-dependent magnetic field gradient and if there is a position dependent detection/measurement sensitivity. These types of problems have been discussed in relation to the ^{199}Hg EDM experiment. [3] The effect can be visualized as follows: Consider a cell of long spatial extent, with the spins detected only at one end (the "near" end) of the cell. Assume also that the systematic gradient only appears at the "far" end. In the limit where the comagnetometer diffusion time becomes extremely long, the comagnetometer, because of the spatial sensitivity to the detection, will not register magnetic fluctuations at the "far" end of the cell, while UCN, not being hindered by diffusion, will sample the entire cell relatively rapidly. Thus a second criterion for a comagnetometer to be effective is that the combination of diffusion time and of detection position sensitivity variations both be small so that that the cell is uniformly averaged by both the UCN and comagnetometer species to an adequate degree of precision.

Another comagnetometer imperfection results from the so-called pseudomagnetic field. This field results from the spin-dependent coherent scattering cross section, which leads to an energy shift for the UCN that is spin dependent and thus appears as a magnetic field. The pseudomagnetic field is not directly affected by the application of an electric field, but can be the source of precession frequency fluctuations and hence extra noise in the system. The magnitude of the pseudomagnetic field can be reduced by ensuring that the magnetometer spins have no component along the static magnetic field, which is possible by careful control of the spin flip pulses.

Such pseudomagnetic fields have appeared in other EDM experiments, for example, a ^{129}Xe experiment [49] where the field was of order 1 mHz due to the presence of spin polarized rubidium that was used to polarize and detect the ^{129}Xe spin precession. This frequency, as an EDM in a 5 kV/cm field that was used in the experiment, corresponds to 10^{-22} e cm, while the final experiment sensitivity is in the 10^{-26} e cm range. This level of discrimination results simply from the fact that the electric field does not directly affect the pseudomagnetic field, and the spin of the rubidium was approximately orthogonal to the applied static magnetic field.

A final concern is the possibility that the magnetometer atom could stick to the wall for a significant period of time compared to the time that a UCN interacts with a wall (i.e. the time for quantum reflection, which is of order 10^{-8} s). For a heavy atom like Hg, together with the known binding energy of 0.1 eV on typical surfaces, implies a sticking time of order 10^{-6} s, which can be calculated by considering the density of states on the two-dimensional surface compared to the density of states for the atom freely propagating in the storage cell. Estimates for the ILL Hg comagnetometer experiment suggest this effect, which would lead to a difference in the spatial averaging of the magnetic field by the UCN compared to the Hg, is very small. However, improvements in the experimental limit for the neutron EDM using this technique beyond 10^{-27} e cm will require careful study. Also of concern is a modification to the diamagnetic correction of the atom during its dwell on the wall, the idea being that the electron density at the nucleus is affected by the interaction with the storage container walls. Since the diamagnetism results from the inner electrons, this effect is expected to be quite small.

As a final note in this section, the use of a comagnetometer vs. external magnetometry offers a final and critical advantage. For external magnetometry, as shown in Fig. 15.3, the total magnetic noise registered by the magnetometer is the combined noise due to the magnetometer itself, and that due to external noise fields resulting from imperfections in the magnetic shielding. This total noise must be low enough so that a high voltage correlated shift in the magnetic field can be detected with the same accuracy as the neutron resonance signal. Therefore, it is difficult to use external magnetometers as "an extra layer of magnetic shielding" to compensate for the limited performance of a magnetic shield, as this requires a nearly impossible level of magnetic shielding to attain a level of accuracy for correlated magnetic field measurements below an EDM sensitivity of 10^{-26} e cm.

As an example, the ILL UCN experiment [47] employed three Rb magnetometers near the UCN storage cell. These magnetometers had net sensitivity just at the limit to be useful to detect and eliminate a systematic magnetic field change. The sensitivity was limited by the intrinsic magnetometer sensitivities, but mostly by magnetic field noise due to the finite shielding ability of the magnetic fields. When this experiment was rebuilt, incorporating a ^{199}Hg magnetometer, the innermost magnetic shield layer was removed. As a consequence, the magnetic field noise due to external sources was so large that the Rb magnetometers were useless in detecting possible small field changes, at the level of sensitivity of the neutron EDM frequency change, that one would hope to potentially detect with application of the high voltage. However, the ^{199}Hg magnetometer could be used to correct for field fluctuations, and even if the high voltage fluctuations could not be discriminated from the external noise, there is a reasonable degree of assurance that the systematic fields were corrected along with the other fluctuations. In fact the degree of correction can be tested by applying arbitrarily pathological magnetic gradients to the system, and then scaled to what could be reasonably expected from leakage currents, etc. Up to now, no specific studies have been performed, but the apparent performance of the ^{199}Hg magnetometer suggests at the present limit, the degree of perfection is adequate.

In the next sections, we will describe a newly discovered systematic effect due to the $\vec{E} \times \vec{v}$ generated magnetic field that affects mostly the comagnetometer atoms and represents the final known imperfection. This is an effect that can be controlled, but as we will discuss, requires a careful experimental design.

15.4.3. $E \times v$ *effects in storage experiments*

15.4.3.1. *Quadratic effect*

Although the motional field is most significant in the case of beam experiments, examples of which are the early neutron EDM experiments and the more recent thallium EDM experiment [13], there can be some subtle effects in other cases. EDM experiments using optically-pumped atoms or neutrons contained in a cell have on average $\vec{v} = 0$ simply because the atoms are free to rattle about the cell, so one might expect that there is no net motional effect. However, as we will show, the fluctuating field associated with the random velocity can in fact lead to sizable systematic effects; the term quadratic in v in $B = \sqrt{B_0^2 + B_{\mathrm{m}}^2}$ in the effective magnetic field persists even if the average velocity is zero, and one may wonder why it is possible to measure EDMs to the achieved levels of sensitivity. If we consider a case where $B_0 = 10$ mG, and $v = 120$ m/s, $E = 10$ kV/cm as in the case of the ^{199}Hg comagnetometer used in the current Institut Laue-Langevin experiment, the quadratic term amounts to about 50 nG, corresponding to a shift of 35 μHz for the optically pumped and detected ^{199}Hg. The experimental accuracy is at the level of 10^{-7} Hz, which implies a magnetic field of about 0.1 nG, and would seem to require an electric field magnitude reversal symmetry of 1 part in 10^3 for an apparent ^{199}Hg EDM to be below the experimental limit.

An important point has been neglected in this estimate. In fact the motional magnetic field is randomly fluctuating, and it simply is not correct to take the average square of this field. The motional field has a definite magnitude only for a time interval τ_c, the time between substantial velocity changes due to, for example, collisions with buffer gas molecules or cell walls. The parameter τ_c depends on the system geometry, nature of the collisions, and velocity of the particles.

For a spin-1/2 system, the net effect of the randomly fluctuating field can be readily quantitatively calculated in the context of the density matrix. [16] The Hamiltonian can be separated into static and time-dependent components

$$H = H_0 + H(t) = -2\pi\gamma\sigma_z B_0/2 - 2\pi\gamma f(t) B_{\mathrm{m}}\sigma_x/2, \qquad (15.15)$$

where γ is the gyromagnetic ratio (Hz/G), $\sigma_{x,z}$ Pauli matrices, and $f(t)$ represents the fluctuating character of $B_{\rm m}$. Here we only consider the possibility of an x component of $B_{\rm m}$, but this doesn't change the result significantly. Eventually, both time and ensemble averages of the effect of this Hamiltonian must be determined.

By transforming into a rotating frame, the static component of the Hamiltonian can be eliminated

$$H' = e^{i\omega t\sigma_z} H(t) e^{-i\omega t\sigma_z} = -2\pi\gamma f(t) B_{\rm m} D_z(\omega t)\sigma_x D_z(-\omega t), \qquad (15.16)$$

where $\omega = 2\pi\gamma B_0$ and D_z is the spin-1/2 axial rotation matrix.

The effect of H' on the system is most readily calculated in a density matrix formalism, as discussed in Refs. [14] and [15]

$$\frac{d\rho}{dt} = \Gamma\rho = -\left\langle \int_0^\infty [H'(t), [H'(t-\tau), \rho]] d\tau \right\rangle_{\rm av}, \qquad (15.17)$$

where ρ is the 2×2 spin-1/2 density matrix and the average is over a time much longer than $\tau_{\rm c}$; also assumed is an average over the statistical ensemble represented by the subscript "av". This result comes from the second-order perturbative approximation to the density matrix evolution (see Ref. [14], Chap. VIII, Eqs. (28)–(32)). Γ is referred to as the relaxation matrix.

The double commutator in the integrand is proportional to the autocorrelation function of $f(t)$, which can be taken as a simple form

$$f(t)f(t-\tau) = \begin{cases} 0, & \text{if } \tau > \tau_{\rm c} \\ 1 - \tau/\tau_{\rm c} & \text{otherwise} \end{cases} \qquad (15.18)$$

where $\tau_{\rm c}$ is the time between velocity changing collisions.

Ignoring exponential terms with arguments $\omega(\tau + 2t)$ gives

$$\Gamma_{11} = \Gamma_{22} = -\left\langle \frac{(2\pi\gamma B_{\rm m})^2}{2} \frac{1 - \cos\omega\tau_{\rm c}}{\omega^2\tau_{\rm c}} \right\rangle_{\rm av} \qquad (15.19)$$

for the diagonal elements of the relaxation matrix, and for the off-diagonal elements,

$$\Gamma_{12} = \Gamma_{21}^* = -\left\langle \frac{(2\pi\gamma B_{\rm m})^2}{2} \left(\frac{1 - \cos\omega\tau_{\rm c}}{\omega^2\tau_{\rm c}} + i\,\frac{\omega\tau_{\rm c} - \sin\omega\tau_{\rm c}}{\omega^2\tau_{\rm c}} \right) \right\rangle_{\rm av}. \qquad (15.20)$$

The real components of Γ represent the spin relaxation, while the imaginary components of the off-diagonal elements represent a frequency shift; it is

$$\Delta\omega = 2\pi f_{\rm m} = \frac{1}{2}\left\langle (2\pi\gamma B_{\rm m})^2 \, \frac{\omega\tau_{\rm c} - \sin\omega\tau_{\rm c}}{\omega^2\tau_{\rm c}} \right\rangle_{\rm av}. \qquad (15.21)$$

It is interesting to consider the limiting forms of (15.21). When $\omega\tau_c \gg 1$, the term $\sin\omega\tau_c$ has zero ensemble average (given a reasonably broad velocity distribution). Furthermore, taking into account the fact that $\vec{v}$ is not constrained to lie in a plane perpendicular to $\vec{E}$, $B_m \to B_m \sin\theta$ must be averaged over all possible directions on a sphere, giving a mean square effect $2B_m^2/3$. Thus, in the limit $\omega\tau_c \gg 1$,

$$f_m = \frac{1}{3}(\gamma B_m)^2/f_0 = \frac{1}{3}(\gamma v E/c)^2/f_0, \tag{15.22}$$

where $f_0 = \gamma B_0$. It should be noted that in this limit the shift does not depend on τ_c, and is the average quadratic expansion of the sum of the motional and applied magnetic fields. The ultracold neutron storage experiment operates in this regime.

In the case where $\omega\tau_c \ll 1$, the $\sin\omega\tau_c$ term can be expanded

$$f_m = \frac{(2\pi)^2}{9}(\gamma B_m)^2 f_0 \tau_c^2 = \frac{(2\pi)^2}{9}(\gamma v E/c)^2 f_0 \tau_c^2, \tag{15.23}$$

where B_m and τ_c represent appropriate ensemble averages. The behavior here is rather unexpected in that the shift increases with f_0, which is opposite to the previous case. Any EDM experiment which employs a buffer gas operates in this regime.

The behavior in the two limiting cases can be qualitatively understood. The time evolution operator for a spin-1/2 system is

$$U = e^{iHt} = \cos\gamma|B|t - i\,\frac{\vec{\sigma}\cdot\vec{B}}{|B|}\,\sin\gamma|B|t, \tag{15.24}$$

and when $\omega\tau_c \gg 1$, the system simply responds to the quadrature sum of all the fields in the problem, as we already knew. The case of $\gamma|B|t \ll 1$ is more subtle. It is useful to work in the rotating frame; the effect of the random field is determined by the magnitude of its static or slowly varying components in that frame. Since the power spectrum of the fluctuations is proportional to the cosine transform of the autocorrelation function, for the rectangular correlation function, the effective random field power is simply the second factor in the average as shown in (15.21). Pictorially, the slowly varying field components lead to a random walk of the spin vector. For small angles, the change in net spin direction is given by the vector sum of all the angular displacements. This gives qualitatively the same answer.

The conclusion for the most recent ILL experiment is that the quadratic motional effect for the ultracold neutrons (velocity 5 m/s) is small enough to be of no concern. The effect on the ^{199}Hg is suppressed by a factor $(f_0\tau_c)^2 = (10\text{ mG } 0.759\text{Hz/mG } 0.4\text{ m}/120\text{ m/s})^2 \approx 10^{-3}$ so is also negligible.

For the ^{3}He comagnetometer experiment that will be discussed later, the spin relaxation rate (real part of the relaxation matrix) is large enough to be of some concern for that experiment, and provides a limit on the static magnetic field and coefficient of diffusion, D [8].

15.4.3.2. *"Geometric phase" effect*

It is interesting to note that after more than 50 years of searching for an EDM of elementary particles an unknown effect can emerge. While this effect was unimportant for earlier searches it proved to be critical for the most recent experiment and will be crucial for the next generation of experiments.

First discovered and analyzed by Commins [39] in connection with a beam experiment, the effect was rediscovered in the most recent UCN storage experiment at the ILL [40]. Commins gives a very clear description of the effect valid for slowly varying fields, showing it can be understood as a manifestation of Berry's geometric phase [41].

Another approach to the basic idea can be seen as follows [40]: Any non-resonant time-varying magnetic field (say rotating around the dc field at frequency ω_r (note that most perturbations can be expressed as a superposition of such fields) will induce a frequency shift of

$$\delta\omega = -\frac{1}{2}\frac{\omega_1^2}{\omega_o - \omega_r} \tag{15.25}$$

with $\omega_{o,1} = \gamma B_{o,1}$, where B_1 is the magnitude of the off-resonant field. This equation can be derived by considering the effective field in a frame rotating with ω_r. So, if the perturbing field is the sum of two fields the square of the field magnitude will contain a term linear in each field, and if one of the magnetic fields is proportional to E, as is the case with the $\vec{B}_m = \vec{E} \times \vec{v}/c$ field discussed above, $\delta\omega$ will also contain a term linear in E, which will be difficult to distinguish from that due to an EDM unless one can manipulate the properties of the extraneous fields in such a way as to bring out the differences. In order for the cross term between the two fields to have a non-zero time average there needs to be some degree of coherence between the two fields. Unfortunately this is rather easy to achieve as the particle's velocity, and hence $\vec{B}_m$, is to some extent correlated with its position, and in the presence of field gradients there will be a term in the magnetic field, also correlated with position. Thus if the particles of velocity, v, are moving in a cylindrical vessel, radius R, as shown in Fig. 15.5, making specular reflections with the walls [40], the particle's velocity will make a step-wise

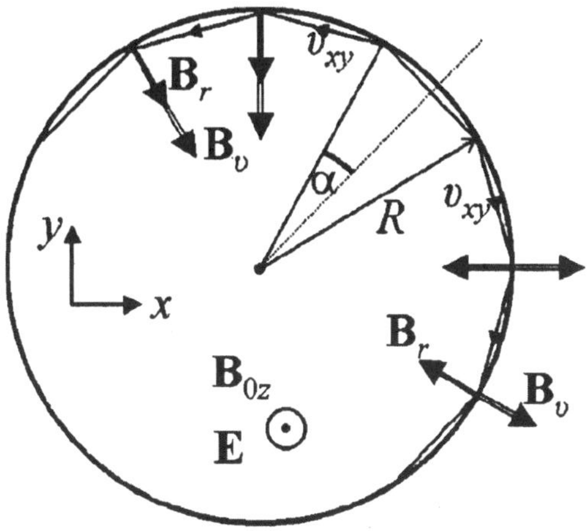

Fig. 15.5. Trajectory in a cylindrical cell with specular wall reflection. The frequency shift depends only on the component of the trajectory in the plane perpendicular to the axis. (This figure was reprinted with permission from Ref. [40]. Copyright 2004 by the American Physical Society.)

revolution with a fundamental frequency

$$\omega_r \approx \frac{v}{R} \tag{15.26}$$

and the field $\vec{B}_m$, directed perpendicular to $\vec{v}$ ($\vec{B}_m$ is perpendicular to the plane of the figure) will move with the same frequency. If, now, in addition the dc field, $\vec{B}_0$, has a radial component, B_r, that component, as seen by the particle will rotate with the same frequency so that the total perturbing field varying at ω_r will be

$$\omega_1 = \gamma \left(B_m + B_r \right) \tag{15.27}$$

which according to Eq. (15.25) will produce (among others) a frequency shift

$$\frac{\delta\omega}{\gamma^2} = -\frac{B_m B_r}{\omega_o - \omega_r} = -\frac{B_r v E}{c\left(\omega_o - v/R\right)}. \tag{15.28}$$

Assuming mechanical equilibrium there will be an equal number of particles with $\pm v$, so averaging the terms for the two directions we find

$$\delta\omega = -\frac{\gamma^2}{2} \frac{\left(\partial B_o/\partial z\right) E}{c} \frac{v^2}{\omega_o^2 - \omega_r^2} \tag{15.29}$$

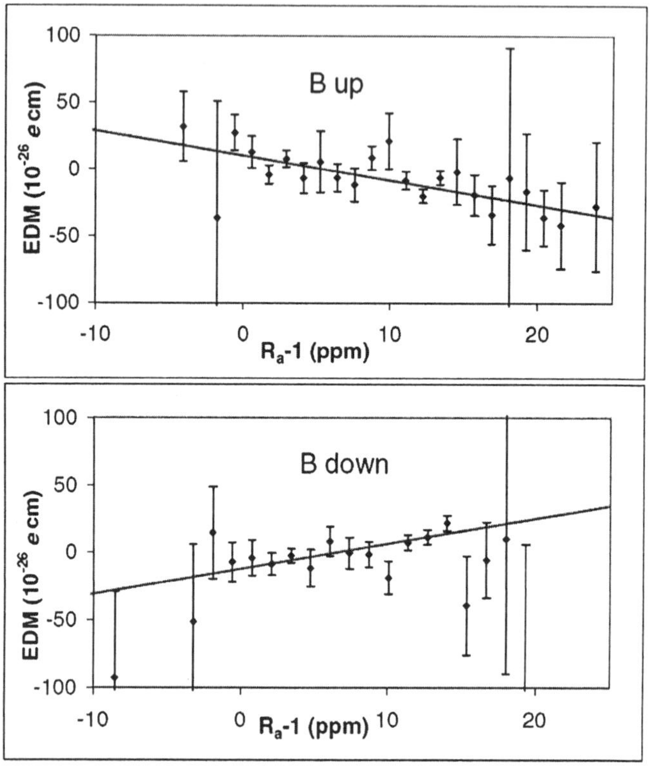

Fig. 15.6. Apparent neutron EDM vs. UCN-Hg frequency deviation from Ref. [48]. (This figure was reprinted with permission from [48]. Copyright 2006 by the American Physical Society.)

where the radial field component follows from the assumption of a constant dc field gradient, $\partial B_o/\partial z$, and $\vec{\nabla} \cdot \vec{B} = 0$. The effect was rediscovered in the context of EDM searches using stored UCN by Pendlebury *et al.* [40], who found a correlation between the values of an apparent EDM in their data and the ratio of the precession frequencies of the UCN and the Hg comagnetometer. The data are shown in Fig. 15.6.

To understand the way the systematic effect manifests itself in this data it is necessary to recognize that the center of mass of the UCN and Hg distributions are displaced by gravity by $\Delta h \lesssim 3$ mm (see Eq. (15.13)) along the z axis so that in the presence of a gradient $\partial B_o/\partial z$, the two species will see slightly different average magnetic fields and have slightly different Larmor frequencies. Defining, as in Ref. [40],

$$R_a = \frac{\omega_n/\gamma_n}{\omega_{Hg}/\gamma_{Hg}} \tag{15.30}$$

it is easy to see that

$$\frac{\partial B_o/\partial z\,|\Delta h|}{B_o} = \pm\,(R_a - 1) \tag{15.31}$$

where the plus sign is for B_o pointing down. Thus $(R_a - 1)$ is a measure of the z-gradient and according to (15.29) we should expect an EDM signal proportional to this quantity. The slope in Fig. 15.6 agrees very well with this equation. After discovering the effect in their data Pendlebury *et al.* undertook a detailed study of the effect in order to understand and deal with it, using the low and high frequency limits of Eq. (15.29). They then went on to solve the Bloch equations for the motion of the spin in the combined electric and gradient magnetic fields, for the case of a single specularly reflecting orbit as shown in Fig. 15.5 with no collisions, giving an expression for the systematic effect. They also studied the effects of collisions by means of extended numerical simulations of the Bloch equations.

Further study of the problem [42] led to the recognition that the frequency shift is given by the spectrum of the velocity-position correlation function and hence can be derived from a knowledge of the velocity correlation function averaged over the ensemble of particle trajectories:

$$\delta\omega = \frac{\gamma^2\,(\partial B_o/\partial z)}{4}\frac{E}{c}\int_0^t d\tau \cos\omega_o\tau R\,(\tau) \tag{15.32}$$

$$R\,(\tau) = \langle y\,(t)\,v_y\,(t-\tau) + x\,(t)\,v_x\,(t-\tau) - y\,(t-\tau)\,v_y\,(t) - x\,(t-\tau)\,v_x\,(t)\rangle \tag{15.33}$$

$$R\,(\tau) = 2\int_0^\tau dx\psi\,(x) \tag{15.34}$$

where $\psi\,(x)$ is the velocity-velocity correlation function.

This result is obtained in two ways [42], the first working out the relaxation matrix Eq. (15.17), as in many nmr applications, or, by solving the classical Bloch equations, as in [40], but applying the solution to arbitrary time dependence of the perturbing fields, B_m, B_r, rather than to the variation seen by a specific trajectory.

Equation (15.32) can be rewritten, using Eq. (15.34) as [42]

$$\delta\omega = -\frac{\gamma^2\,(\partial B_o/\partial z)}{2}\frac{E}{c}\int_{-\infty}^{\infty}\frac{\psi\,(\omega)}{(\omega_o^2-\omega^2)}d\omega, \tag{15.35}$$

i.e. the single frequency Bloch–Siegert result (15.29) summed over the freqency spectrum of the velocity auto-correlation function.

Figure 15.7 (dotted lines) shows the correlation function for a cylindrical measurement cell with specular wall reflections for different collision mean free paths obtained from numerical simulations. The solid curves show the same results obtained from an analytical form of $\psi(x)$ [43].

For a single trajectory (single α, see Fig. 15.7) without collisions the analytic solution agrees with the result obtained by direct solution of the Bloch equation (Eq. (78), Ref. [40]) for this case. For values of $\omega' \gg 1$,

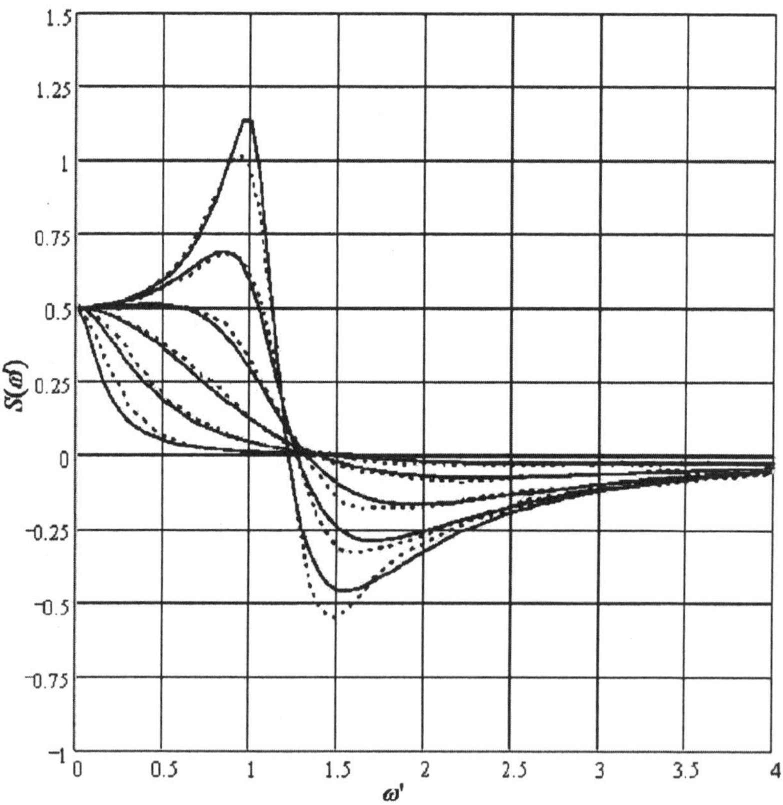

Fig. 15.7. Normalized frequency shift for a constant velocity as a function of normalized applied frequency $\omega' = \omega_o R/v$, for different values of the damping parameter $r_o = R/\lambda$. Solid curves: Results of the analytic function given in Eqs. (43) and (44) of Ref. [43]. Dotted lines: Numerical simulations from Ref. [43]. Starting at highest peak, r_0 = 0.2, 0.5, 2, 4, 10. (Cylindrical cell).

($\omega_o \gg \omega_r = R/v$) i.e., the short time region of the correlation function, the velocity auto-correlation function is given by $\psi(\tau) = e^{-\tau/\tau_c}$ with τ_c the collision time. This region, where the shift $\propto 1/\omega_o^2$, is appropriate for stored UCN. For the opposite limit (long time limit of the correlation function) appropriate for heavier comagnetometer atoms, e.g. Hg, the Diffusion theory applies. The effect is generally larger in this region as can be seen from Fig. 15.7. The analytic result obtained in Ref. [43] is valid in the intermediate region as well. It is interesting to explore the possibility of using the zero-crossing in this region to reduce the effect. In the case of a comagnetometer consisting of He^3 atoms moving in superfluid He^4, as in the experiment under development for operation at the Oak Ridge National Laboratory Spallation Neutron Source, the collision mean free path is strongly temperature dependent and this can be used to tune the effect around the zero crossing. Fig. 15.8 shows the frequency shift averaged over

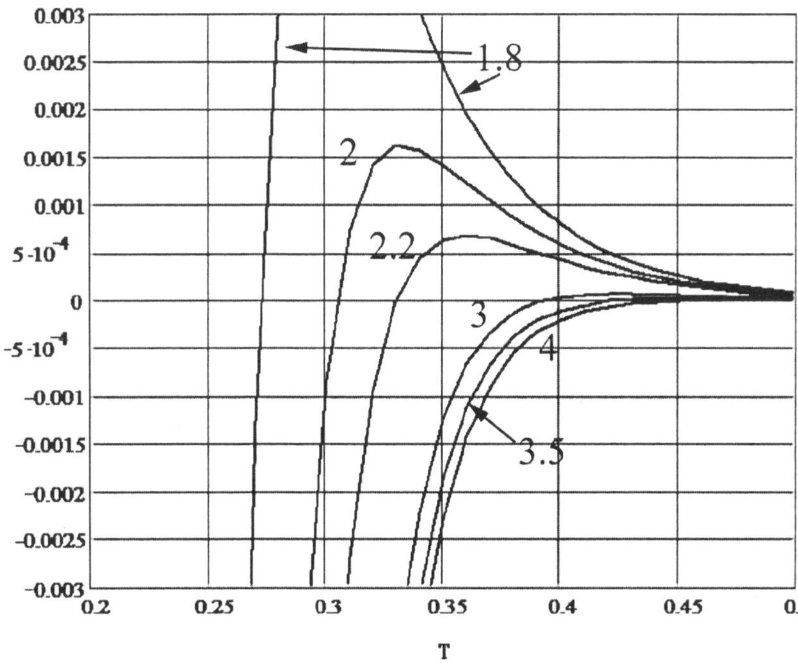

Fig. 15.8. Normalized, velocity averaged, frequency shift, $\Psi(\omega^*, T)$, vs temperature T for various reduced frequencies $\omega^* = \omega_o R/\beta(T)$ using the temperature dependent mean free path for ^3He in ^4He. (Cylindrical cell) [43].

the Maxwell distribution of the He^3 velocities as a function of temperature and Larmor frequency [43], ($\beta(T)$ is the most probable velocity at temperature T, calculated using the effective mass of the ^3He in ^4He). Reductions of more than 10^3 seem possible.

In the low frequency (long time) limit the diffusion theory can be used to calculate the shift for arbitrary geometry. At the moment we only have an analytic solution valid for all times for the cylindrical cell, for other geometries we must calculate the correlation function by numerical simulation of the trajectories as was done in Ref. [42]. Nevertheless this is much more efficient than simulating the spin dynamics directly.

It is amusing to note that since the magnetic moments of the neutron and Hg atoms have opposite signs the two species are precessing in opposite directions during the ILL measurement. It follows that the Earth's rotation will shift the two precession frequencies in opposite directions [44]. The laboratory is essentially a rotating frame, rotating with $\omega_\oplus = 2\pi/24\,(3600) = 2\pi\,(11.6\mu Hz)$, so that a term

$$-\left(\frac{\omega_\oplus \sin\theta_L}{B_o\gamma'}\right) \tag{15.36}$$

should be added to the right side of (15.31), which would correspond to an EDM shift of $d_{n\oplus} = -2.57 \times 10^{-26}$ e cm, a non-negligible shift given the limit fixed by the experiment $\left(|d_n| < 2.8 \times 10^{-26}\ e\right)$ cm. However it turns out that this effect was fortuitously canceled to better than 15% by the change of stray quadrupole magnetic fields on reversing the field, B_o. [45, 46]

15.5. Ultracold Neutron Magnetic Resonance Experiments: Current Experimental Limits

15.5.1. *Ultracold neutrons*

Brief mentions of ultracold neutrons (UCN) were made in previous sections of this Review. The UCN idea has its origin in so-called Neutron Optics: To describe the interaction of slow neutrons with bulk material, Fermi developed the concepts of the "pseudo-potential" and the neutron index of refraction. His idea is as follows. Although the range of nuclear forces is small, they are quite strong within that range so one cannot in general apply perturbation theory to a collision between a neutron and a nucleus. However, the amplitude for scattering of neutron of wavelength large compared to the nucleus is a constant independent of the velocity.

The constant amplitude can be obtained if we describe the interaction of the neutron with the nucleus by the point interaction

$$U(\vec{r}) = \frac{2\pi\hbar^2}{M} a\delta(\vec{r}) \tag{15.37}$$

where M is the reduced mass (or the neutron mass for rigidly bound nuclei) and a is the coherent scattering length. When this potential is substituted into the Born approximation,

$$f(\theta) = f = -\frac{M}{2\pi\hbar^2} \int U(\vec{r}) e^{-i\vec{q}\cdot\vec{r}} dV = -a, \tag{15.38}$$

the delta function makes the integral independent of the momentum transfer $\vec{q}$. Now consider many scatterers bound in a piece of bulk matter such that the distance between the scatters is much less than the neutron wavelength. As a slow neutron approaches the boundary, it will see an average potential

$$U = \frac{2\pi\hbar^2}{M} a\rho \tag{15.39}$$

where $a = -f$ is the coherent scattering length and ρ is the density of scattering points. This potential appears as a step as the neutron enters the bulk material. Thus, for nuclei with $a > 0$, the neutron loses kinetic energy and the wavelength increases on entering the bulk material; the index of refraction is less than 1.

Although the possibility of storing neutrons with low kinetic energy in material bottles is usually attributed to Fermi, Zeldovich was the first to take the idea seriously enough to put it into print. The idea is that neutrons with kinetic energy $E < U$ will be reflected from the material surface for all incidence angles, and thus a storage bottle can be constructed. The reflection from the material surface is analogous to the total internal reflection of light. The storage lifetime can be long because the time that a UCN interacts with the wall (10^{-8} s) compared to the time between wall collisions (0.05 s) is very small.

Neutrons with such low velocities ($v < 7$ m/s for most materials corresponding to U of order 100's of nano electron volts) are referred to as UCN because their average kinetic energy, as a temperature, is 5 mK or less. UCN production and storage are now well-developed technologies after some intense and difficult research over a 20-year period starting in the late 1960s. [26]

UCN can be transported as a gas through pipes of high potential materials (stainless steel, for example). In addition, UCN can be polarized by transmission through a thin magnetically saturated foil; the foil material, typically a Fe-Co alloy, is chosen so that the saturation flux Zeeman shift just cancels the UCN potential for one spin state; that spin state passes easily through the foil while the other is reflected. It is also possible to polarize UCN by applying a magnetic field in a region of the guide; one spin state will gain energy on entering the field region, while the other state will not have enough kinetic energy to pass the region. A field of several Tesla is sufficient to fully polarize a UCN current, as has been demonstrated in a number of experiments.

15.6. Present Experimental Limit: UCN Experiment with ^{199}Hg Comagnetometer

Since the both the ILL [47] and the PNPI [58] UCN EDM experiments were no longer limited by counting statistics but by magnetic systematics, it was decided to rebuild the ILL apparatus and include a comagnetometer, as we have discussed already, and thus provide a nearly exact spatial and temporal average of the magnetic field affecting the neutrons over the storage period. The use of polarized ^{3}He had already been considered [17], but the extreme difficulty in the detection of the ^{3}He polarization makes its use impractical.

The use of ^{199}Hg was suggested in 1986 [18], and is described in Ref. [19]. The advantage is that ^{199}Hg can be readily directly optically pumped and its polarization optically detected with 254 nm resonance radiation. Because ^{199}Hg is a $^1S_0$ atom, its ground state polarization is specified by the nuclear angular momentum, which is 1/2 for ^{199}Hg. In addition, the room temperature vapor pressure of Hg is more than adequate to provide the necessary density.

Of course, to be useful for a comagnetometer, it must be demonstrated that the chosen atomic species does not have an EDM of its own which could possibly mimic or mask a neutron EDM; in the case of ^{199}Hg, experimental limits were set at the level of sensitivity needed [21]. In these experiments, ground state spin-polarization lifetimes in excess of 100 s. were routinely achieved in cells of about 5 cm^3 volume, even in the presence of electric fields up to 15 kV/cm. However, these cells included 250 torr of nitrogen to improve the high voltage stability.

An unfortunate disadvantage of ^{199}Hg is that the walls of the container must be specially prepared to have long spin relaxation times. In all previous experiments, hydrocarbon waxes were used; these of course would be unusable with UCN. In addition, the wall coating has to be stable under the application of high voltage in vacuum since a high-pressure background gas cannot be used with the UCN. A fused silica insulating ring 20 cm high separating two diamond-like carbon coated aluminum plates were used as the storage cell, with total volume of 20 liters.

A schematic of the experimental apparatus is shown in Fig. 15.9. To increase the experimental sensitivity through storage time and UCN number increases, a 20 liter volume storage bottle was constructed, compared to 5 liters in the earlier version. The magnetic shields were the same as those used in the previous ILL experiment, only the innermost layer was removed. The loss in shielding factor was be made up for by the improved volume comagnetometry.

In addition, there was a safety consideration for the use of Hg, and it was necessary to isolate the experiment with a gas-tight window which can withstand atmospheric pressure. The thin foil polarizer was redesigned to

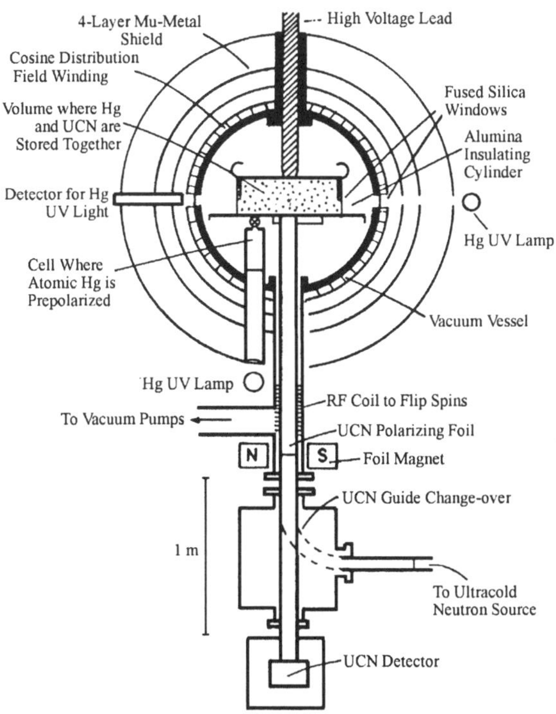

Fig. 15.9. Schematic of the ILL UCN EDM experiment incorporating a ^{199}Hg comagnetometer.

also serve as the window, with iron evaporated onto an aluminum foil. To account for the fairly high effective potential of the aluminum, after passing through the foil the UCN rise about one meter. Tests were performed to determine the optimum height to maximize the number of UCN left in a test bottle after a 100 s storage period.

Provisions were included for polarizing the atomic vapor; an optical pumping cell was connected to an isotopically enriched Hg reservoir, a few mg of HgO powder in a tube held at about 250°C. This provided a current of ^{199}Hg atoms and could be controlled with a valve. The Hg was optically pumped to the appropriate spin state, parallel to the static field, with circularly polarized light from a Hg discharge lamp. After the Hg was polarized, and after the storage vessel was filled with polarized UCN, the neutron valve was closed; then polarized Hg was admitted to the neutron bottle. $\pi/2$ pulses were applied first for the ^{199}Hg and then for the UCN (the ^{199}Hg magnetic moment is about one third of the neutron magnetic moment). The free precession of the Hg spin was observed with a beam of circularly polarized resonance light which propagates across the bottle diameter, through the fused silica insulating cylinder. The storage vessel spatial average magnetic field, averaged over the measurement time, could be determined from the free precession signal.

At the end of the storage period, the second neutron pulse was applied, the bottle door opened, and the neutrons were counted and the final polarization state, hence resonant frequency, was determined as before. The Hg was pumped away during the UCN counting period. While the storage was in progress, more Hg had been admitted to the optical pumping cell and polarized; the process was thus ready to be repeated.

An EDM would be evident from a change in the ratio of the magnetic moments between reversals of the electric field. Although the sensitivity of the Hg to a magnetic field is only 1/3 that of the neutron, the high signal to noise inherent in the free precession signal was a compensating factor, and the determination of the average field was a factor of 3 to 10 higher in sensitivity than the neutron accuracy and hence contributed very little noise to the measurement. The final uncertainty for this experiment, based on the shot-noise, is about 3×10^{-26} e cm (95% conf.).

The figure of merit (see Eq. (15.7)) $F = \alpha E\sqrt{\rho V T}$, where T is the coherence time, V is the storage volume, ρ the UCN density, E the applied electric field, and α the polarization factor, for this version of the experiment compared to the 1990 version is

$$\frac{F'}{F} = \frac{(.5 \times 4.5 \times \sqrt{0.7 \times 20000 \times 120})}{(.6 \times 10 \times \sqrt{3 \times 5000 \times 80})} = 0.45$$

with the reduction in F largely due to the reduction in electric field strength. The larger volume compensated for the loss of UCN number density due to the relatively low potential of the fused silica/diamond-like carbon storage cell of 110 neV (due to the fused silica), compared to 240 neV for Be/BeO used in the previous experiment, representing a loss in density by a factor of $(110/240)^{3/2} = 1/3$, compared to a factor of 1/4 in the experiment, with the additional loss due in part to less transmissive polarizer.

15.7. Present Experimental Development

15.7.1. *Hg comagnetometer experiment at PSI*

The success of the Hg comagnetometer suggests that this technology should not be abandoned. Thus, the present plan is to upgrade the experiment and move it to a more intense neutron source at the Paul Scherrer Institut (PSI) in Switzerland. [55]

It is anticipated that the coating technology can be improved, leading to a factor of three improvement in the figure of merit. It is also anticipated that the storage lifetime could be improved to perhaps 200 s. However, the principal advantage is to increase the UCN density by use of a solid deuterium spallation driven ultracold neutron source.

The idea that solid deuterium can be used as a UCN source is due to Golub and Böning who first discussed its use in the context of a thin film source. [53] Pokotilovskii suggested a configuration that would work at a pulsed neutron source, with the UCN being produced during the short duration of an intense neutron pulse and conducted to a UCN storage vessel, and then isolated from the UCN storage vessel by a fast valve so that the produced UCN would not be lost in the deuterium which has a relatively high loss cross section. [52] The advantages of enclosing a neutron spallation target and solid deuterium in a flux trap is discussed in Ref. [51], and this discussion led to the construction of a prototype source at Los Alamos National Laboratory (LANL) that produced 140 UCN/cc in a storage vessel above the source. [59] This is to be compared to the output of the ILL UCN source which, in a similar storage volume, produced about 40 UCN/cc.

For the PSI source, it is planned to use 30 liters of solid deuterium, compared to 1 liter or less for the LANL source. Comparisons between the configurations are difficult as the LANL source uses cold polyethylene as the flux trap and spallation neutron moderator, while the PSI source uses heavy water as the moderator, with the solid deuterium apparently serving as the UCN converter and the cold moderator. One might expect a larger density from the polyethylene moderator for the moderation length is about 2 cm in hydrogenous materials, compared to 25 cm in heavy water. However, the absorption rate is relatively low, with a neutron lifetime of about 0.2 s in heavy water compared to 0.16 millisec in polyethylene. So the moderated neutrons occupy a 1000 times larger volume in the heavy water system, with a lifetime 1000 times longer, and is limited by the diffusion time out of the heavy water central region. With all these factors canceling overall in the comparison, the principal means of increased density of the PSI source is the increase of the current in the spallation source, 2 mA compared to 100 μa in the LANL prototype source. Simply scaling by the currents suggests a density of about 3000 UCN/cc for the PSI source, comparable to their own estimates.

In evaluating the experimental sensitivity, it is assumed that the EDM experiment can be filled with UCN with 50% efficiency. This is ambitious in that for the ILL experiment, given a source density of 40/cc, produces a net UCN density of 5/cc for the 1990 experiment, which is due in part that only the neutrons surviving after the storage period contribute to the measurement. Applying these factors to the Hg comagnetometer experiment, along with an anticipated increase in storage time to 200 s, and an increase in electric field to 15 kV/cm, shows an increase in figure of merit compared to the present comagnetometer experiment (which produced a limit of 3×10^{-26} e cm) by a factor of 100; therefore a limit of 3×10^{-28} e cm appears as imminently feasible.

However, attaining this level of accuracy will require very careful control of the geometric phase effect. The magnetic shields presently in use do not have an adequately small gradient to eliminate this effect. The plan is to install a number of discrete alkali atom magnetometers that will allow control of the field gradients in real-time. Other comagnetometer issues that were discussed earlier also need to be addressed.

The apparatus was moved to PSI in early 2009, and it is anticipated that the refurbished apparatus will be taking data in late 2010.

15.7.2. *PNPI experiment at ILL*

A multicell experiment being built under the direction of A. P. Serebrov is nearing completion at the ILL [62]. This ambitious project employs 13 20-liter storage volumes, with an anticipated voltage of 15 kV/cm. The systematic magnetic fields will also be monitored with 16 discrete magnetometers. The experiment will be operated in exactly the same fashion as the earlier ILL and PNPI experiments. In some sense, the experiment is equivalent to running 13 copies of the earlier experiments together, although having the storage bottles in the same apparatus allows detection and cancellation of common mode magnetic field fluctuations. The storage cells will be oriented with their long dimension vertical, so the offset in the center of mass from the geometrical mean might be problematic. The general magnetometer problem remains.

The figure of merit, compared to the comagnetometer experiment, is 150 times larger, suggesting that from a statistical point a view, a sensitivity at the 10^{-28} e cm level is possible. The geometric phase effect will be important for the UCN at this level of sensitivity, and the correlation function for the UCN in the gravitational field requires study as the electric field is perpendicular to the gravitational field of the Earth, hence the UCN trajectories are significantly affected.

15.8. The Future: Superfluid ^{4}He

15.8.1. *The production of UCN in superfluid ^{4}He*

To increase the sensitivity of UCN EDM experimental searches, an increase in UCN density is required. Both planned and existing UCN sources, based on extraction of UCN from a cold moderator, are limited by the phase space density of low energy neutrons in the moderator. At most, one could expect a factor of perhaps ten over the density at the ILL reactor, but this requires the extension of reactor technology by about an order of magnitude in regard to radiation fluxes in the core. Spallation sources might eventually give a factor of 100 increase in density, but this remains to be proven.

There is another way to produce UCN; the idea is to inelastically scatter cold neutrons in a suitable material. As the neutron wavelength increases, the inelastic scattering efficiency of solids and liquids decreases. Thus, the rate of scattering from high to low energy can exceed the inverse, that is, scattering from low to high energy. In a suitable material, the density of low energy neutrons can be enhanced over what is expected from the source phase space density.

An ideal material for such a UCN source is superfluid ^{4}He [23]. ^{4}He has zero neutron absorption because it is the most tightly bound nucleus. Thus, if the superfluid bath is sufficiently free of ^{3}He (which has a rather large absorption cross section), UCN can be stored in the bath until β decay, wall absorption, or upscattering occur; it is expected that with modest effort, β decay can become the dominant loss mechanism.

The production of UCN by the downscattering of 8.9 angstrom neutrons in superfluid He has now been demonstrated and well studied [24, 25]. The process is nicely described in [26]. Fig. 15.10 shows the free neutron dispersion curve along with the dispersion curve for elementary excitations in superfluid ^{4}He [the Landau–Feynman (L–F) dispersion curve]. The dispersion relation of the free neutron, relating the energy to the momentum, is a parabola:

$$\omega = \hbar k^2/2m. \tag{15.40}$$

This curve crosses the L–F dispersion curve at $2\pi/k^* = 8.9$ angstrom, and $E^* = \hbar\omega = 11$ K. The crossing point is in the quasi-linear region of the L–F curve. (The curves also intersect at $k = 0$). Neutrons in this range of wavelength are readily produced by a liquid deuterium or liquid hydrogen moderator.

Because both energy and momentum are conserved in the scattering process, neutrons at or near rest can only absorb phonons of energy E^*,

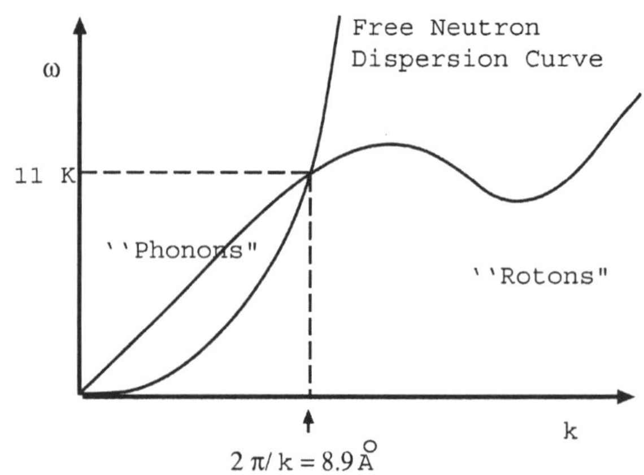

Fig. 15.10. Free neutron and superfluid ^{4}He elementary dispersion curves.

where the dispersion curves cross. This process is strongly suppressed by the Boltzmann factor, $e^{-E^*/T}$, when the superfluid temperature is less than 1 K. By the same argument, only neutrons with energy near E^* can scatter into the UCN energy region by emission of a single excitation.

A UCN source based on this process operates by the following principle. Neutrons of wavelength 8.9 angstrom can easily penetrate the walls of the storage container, and enter the superfluid bath. These neutrons then downscatter, producing UCN which are trapped in the container. (U_F for liquid ^{4}He is about 20 neV, much less than the potential for most solid materials, so we can assume it is zero in the following discussion.) UCN produced in this way will remain in the superfluid He bath until they are lost through one of the possible loss mechanisms, which include β decay, absorption by ^{3}He, and loss in the wall. The UCN will reach a saturation density

$$\rho_\mathrm{UCN} = P\tau \tag{15.41}$$

where τ is the total loss rate,

$$\tau^{-1} = \tau_\mathrm{wall}^{-1} + \tau_\beta^{-1} + \tau_{^3\mathrm{He}}^{-1} + \ldots \tag{15.42}$$

and P is the UCN production rate [UCN/(cm^3 s)] due to the above-mentioned downscattering process [27]:

$$P = 7.2 \frac{\mathrm{d}^2\Phi^*}{\mathrm{d}\lambda\,\mathrm{d}\Omega} \frac{1}{\lambda_\mathrm{u}^3} \delta\Omega \ \mathrm{UCN\ cm^{-3}\ s^{-1}}, \tag{15.43}$$

where the the neutron spectral density is specified at 8.9 angstrom, λ_u is the shortest UCN wavelength that can be stored in the container, and $\delta\Omega$ is the source solid angle subtended at the superfluid bath. A UCN source based on this principle is referred to as a "superthermal source". Multiphonon processes increase the production rate by about 30%.

The neutron-superfluid ^{4}He system is in some sense a two-level quantum system, and the production of UCN by the emission of a phonon can be compared to the spontaneous emission of radiation by an excited atom.

Cold neutrons of wavelength 8.9 angstrom have an attenuation length of order 100 m in superfluid ^{4}He at temperatures at around 1 K. Thus, for any conceivable experiment, the production rate will be constant (except for beam divergence) independent of position along the incident neutron beam.

The increase in neutron density near zero energy can be understood by the following argument. If we take a linear dispersion relation for the liquid He elementary excitations, $\omega = ck$ where c is the phonon (sound) velocity, we have the following condition, by conservation of energy and momentum, limiting the region around $k = k^* + \delta k$ which can scatter to an UCN with momentum k_{UCN}:

$$c|\vec{k} + \vec{k}_{\mathrm{UCN}}| = \frac{\hbar}{2m}(k^2 - k_{\mathrm{UCN}}^2). \tag{15.44}$$

The maximum and minimum of $|\vec{k} + \vec{k}_{\mathrm{UCN}}|$ are $k \pm k_{\mathrm{UCN}}$. We thus arrive at

$$\delta k = 2k_{\mathrm{UCN}}. \tag{15.45}$$

This is a remarkable result, and shows that Liouville's theorem, which was previously briefly mentioned, is apparently violated by this system. Incident neutrons occupy a (momentum) phase space volume of $4\pi k^{*2}\delta k$ whereas the UCN occupy a volume $\frac{4\pi}{3}\ k_{\mathrm{UCN}}^3$, which represents a factor of $\frac{1}{3}(k_{\mathrm{UCN}}/k^*)^2$ decrease in phase space volume, corresponding to an increase in phase space density. Given an arbitrarily long storage lifetime of the UCN, for any non-zero production rate P, the real space density will simply continue to increase as the incident neutrons downscatter, at least until the UCN density is so high that all the states of the Fermi gas are occupied, at which point no more downscattering can occur. This is possible because the produced phonons occupy a very large phase space, and these phonons are continually removed from the system by a refrigerator which keeps the superfluid bath cold. In this regard, the system is analogous to a heat powered refrigerator.

We have not addressed upscattering of UCN by phonons which leads to additional losses. The one phonon process is easy to calculate. By using microscopic reversibility [28] the production and upscattering processes can be related:

$$\frac{\sigma(E^* \to E_{\mathrm{UCN}})}{E_{\mathrm{UCN}}e^{-E_{\mathrm{UCN}}/T}} = \frac{\sigma(E_{\mathrm{UCN}} \to E^*)}{E^* e^{-E^*/T}}. \tag{15.46}$$

which implies that the reverse process is exponentially small, as was previously mentioned. However, this simple treatment does not include higher order processes, and in fact the dominant process below 1 K is two-phonon upscattering [29], which gives a loss time of about $\tau = 100T^{-7}$ s.

If the incident neutrons are polarized, the UCN that are produced will also be polarized because there are no magnetic process in the scattering interaction. This suggests that experiments where polarized UCN are required, a considerable improvement can be gained by using a polarized cold beam which can be polarized to a very high level with negligible loss.

15.8.2. *SNS superfluid helium experiment*

The possibility of a new neutron EDM experiment employing spin polarized ^{3}He stored together with UCN in a superfluid bath was first described in detail in Ref. [37]. Detailed reports describing the current status of efforts to implement this system are available. [60, 61] Here we will give an overview of the advantages of this experimental technique, and describe some of the special features of the system that tend to get buried in detailed reports.

Attempts to make a UCN source based on the superthermal process in ^{4}He all encountered technical difficulties, primarily in regard to extraction of the UCN from the bath. Invariably and inevitably, the thin material windows used to contain the liquid He but allow the UCN to pass, become covered by condensed, frozen gases (O_2 or N_2), increasing U_F and/or the UCN absorption. Typically, extracted densities have been a factor of 10 to 100 below that expected [25, 30]. Indirect measurement through the upscattering rate has confirmed that the expected density does indeed exist within the bath [31]. Recently, workers at Munich have constructed a helium source without a UCN exit window that shows the expected UCN density [67].

The extraction problems can be avoided by performing an EDM search directly in the liquid helium of the superthermal source. Such a system has a number of advantages; for example, because of the excellent dielectric properties of liquid helium, increasing the applied electric field by nearly an order of magnitude might be possible.

We can estimate the figure of merit of superfluid helium experiment operated at the Spallation Neutron Source (SNS) presently under construction at Oak Ridge National Laboratory. A polarized UCN production rate of 5/cc/sec, with a storage lifetime of 500 s, or a UCN density of 2500/cc in the experimental apparatus is anticipated. Applying an electric field of 50 kV/cm appears as possible, and comparing with the comagnetometer experiment, we see an increase in figure of merit by a factor of 1200, indicating that a level of sensitivity approaching 2×10^{-29} e cm appears as feasible.

The important point is that the UCN are produced in the experiment in the polarized state, so the usual losses associated with transport from a source to the experiment, and with polarization, are eliminated. The other advantage is that the electric field can be significantly increased, and at low temperatures enhanced storage times can be expected.

It is interesting to note that the figure of merit is linear in the storage time T (taken as equal to the coherence time) because the UCN density is proportional to T.

Also, it appears feasible to use a dilute solution of polarized ^{3}He as a UCN spin analyzer, detector, and magnetometer. ^{3}He only absorbs neutrons when the total spin is zero because the reaction occurs via the 0^+ excited state [32] as follows

$$^3\mathrm{He} + \mathrm{n} \rightarrow \mathrm{p} + \mathrm{T} + 764 \ \ \mathrm{keV}. \tag{15.47}$$

The polarization and cryogenic transport of polarized ^{3}He have also been studied [33]. Furthermore, energetic charged particles produce ultraviolet scintillations in liquid helium with about 4 photons per keV of deposited energy. The reaction between ^{3}He and neutrons in the liquid helium are thus easily detected, giving a detection of reactions with nearly 100% efficiency. See Ref. [34] for an application of these techniques to a measurement of the neutron β decay lifetime which may lead to a factor of 100 improvement in accuracy.

The ^{3}He serves as a UCN polarizer by absorbing neutrons in the singlet state. To be effective, this rate of absorption should be slightly higher than other UCN loss mechanisms in the system. This implies a ^{3}He concentration of 10^{-10}, or about 10^{12} atoms/cm^3. Such low densities of polarized ^{3}He can be produced with a hexapole state selector, with essentially perfect polarization, which has been demonstrated at Los Alamos. Other techniques for producing higher densities have polarization limited to 70%; since the ^{3}He serves three functions, it is expected that the experimental sensitivity varies as at least the square of the ^{3}He polarization. This agrees with detailed calculations.

An EDM experiment based on these ideas, as originally proposed, could be sensitive to a neutron EDM by looking at the scintillation rate at the end of a double-pulse sequence, as a function of electric field polarity. [35] It has been shown by solving the Schrödinger equation with a spin dependent absorption probability that this technique is slightly less sensitive than the conventional bottle technique, however, this loss of sensitivity is more than made up for by elimination of the extraction losses and increase in electric field.

In the following discussion, let the subscript 3 refer to the ^{3}He atoms, and subscript n refer to the UCN. In the case where both species are polarized, the spin-dependent loss rate can be written

$$\frac{1}{\tau_{\mathrm{abs}}} = \frac{1}{\tau_{^3\mathrm{He}}}(1 - \vec{p}_{\mathrm{n}} \cdot \vec{p}_3) = (1 - p_{\mathrm{n}} p_3 \cos\theta_{\mathrm{n3}})/\tau_{^3\mathrm{He}}, \tag{15.48}$$

where θ_{n3} is the angle between the spin polarization vectors and $|\vec{p}_{n,3}| \leq 1$. Each loss (nuclear reaction) produces a scintillation pulse; the scintillation rate thus becomes a measure of the angle between the polarization vectors.

One could search for a neutron EDM by using the above UCN production/ polarization technique. After the UCN are polarized (along a static field of magnitude B_0), the UCN and ^{3}He spins could be flipped by $\pi/2$; the spins then precess about the static field and there will be a modulation in the scintillation rate:

$$\phi(t) \propto (1 - \vec{p}_3 \cdot \vec{p}_n) = 1 - p_3 p_n \cos[(\gamma_3 - \gamma_n) B_0 t + \Phi], \tag{15.49}$$

where $\phi(t)$ is the time-dependent scintillation rate, Φ is an arbitrary phase, and the gyromagnetic ratios are

$$\gamma_n/2\pi \approx -3 \text{ Hz/mG}$$

and

$$\gamma_3/2\pi \approx -3.33 \text{ Hz/mG}.$$

The EDM of ^{3}He is expected to be quite small (due to shielding as described in section 15.4.2.1); thus, if an electric field is applied along B_0 there will be a change in the frequency of the scintillation rate modulation. Unfortunately, the problem of measuring the magnetic field remains (although the effects are only 1/10 as large since the gyromagnetic ratios are nearly equal) and it has been demonstrated that experiments are presently limited by magnetic systematic effects. It might be possible to use SQUID magnetometers to detect the precessing ^{3}He magnetization, so that the ^{3}He could then serve as a direct magnetometer. Recent advances in SQUID technology make this a possible alternative to the dressed spin technique described below.

15.8.2.1. *Dressed spin magnetometry*

In the above description, it is evident that a perfect experiment would be possible if the magnetic moments of the ^{3}He and neutron were equal; the fact that the magnetic moments are equal to within 10% reduces the sensitivity to background field by an order of magnitude, and if the moments were exactly equal, there would be no effect at all. Unfortunately, we have no direct control over the physics responsible for the observed magnetic moments; however, these moments can be artificially modified by using "dressed atom" techniques, [36] and it is possible make them equal. [37]

In the presence of a strong oscillating magnetic field, the magnetic moment will be modified, or "dressed", yielding an effective gyromagnetic ratio

$$\gamma' = \gamma J_0(\gamma B_{\rm RF}/\omega_{\rm RF}) = \gamma J_0(x), \tag{15.50}$$

where γ is the unperturbed gyromagnetic ratio, $B_{\rm RF}$ and $\omega_{\rm RF}$ are the amplitude and frequency of an applied oscillating magnetic RF field, and J_0 is the zeroth-order Bessel function. This effect can be qualitatively understood by taking the average of the spin in an oscillating magnetic field. Consider a spin pointing along $\hat{z}$ at $t = 0$. Now apply an oscillating field along $\hat{x}$; the spin precession frequency is time dependent,

$$\omega(t) = \dot{\theta}(t) = \gamma B_x(t) = \gamma B_{\rm RF} \sin \omega_{\rm RF} t,$$

so that the angle relative to $\hat{x}$ is

$$\theta = \gamma(B_{\rm RF}/\omega_{\rm RF}) \cos \omega_{\rm RF} t.$$

The average spin projection $\langle P_z \rangle$ along $\hat{z}$ is given by

$$\langle P_z \rangle = \frac{1}{T} \int_0^T \cos(\gamma B_{\rm RF} t \cos \omega_{\rm RF} t) {\rm d}t = J_0(\gamma B_{\rm RF}/\omega_{\rm RF}) = J_0(x).$$

A more sophisticated treatment shows that a spin will respond to a small (compared to the oscillating field amplitude) static field along $\hat{x}$, with an average magnetic moment $\gamma' = \gamma J_0(x)$; our simple estimate gives a picture of how the oscillating field dilutes the magnetic moment.

In practice, the oscillating field is at right angles to the static field B_0 around which the spins are precessing. In the absence of the oscillating field, one would see scintillation due to reactions occurring at a rate given by (15.49). Thus, there is an oscillation in the scintillation rate at the difference in the precession frequencies $[\delta\omega = (\gamma_{\rm n} - \gamma_3)B_0]$. If the RF dressing field is now applied, the effective magnetic moments become modified, and

$$\delta\omega = [\gamma_{\rm n} J_0(\gamma_{\rm n} x) - \gamma_3 J_0(\gamma_3 x)] B_0. \tag{15.51}$$

This has the property that $\delta\omega = 0$ when $\gamma_{\rm n} x \approx 1.19$; this condition is referred to as "critical dressing". It can be achieved in practice with a dressing field frequency of order 1 kHz and amplitude of 100 mG, and a spin precession frequency on the order of a few Hz.

If the neutron EDM is non-zero, the neutron precession frequency will be shifted by an amount $2d_{\rm n} E J_0(\gamma_{\rm n} x)$ (since the dressing dilutes the net spin projection). Thus, the value of $x = x_{\rm c}$ to give $\delta\omega = 0$ is changed. By measuring the value of $x_{\rm c}$ vs. electric field direction, a neutron EDM would be evident. The important point is that the effect of static magnetic fields is canceled.

Experimentally, the neutron and helium spin vectors could be kept nearly parallel; the scintillation would increase or decrease as x is varied away from the value x_c such that $\delta\omega = 0$. Over the course of a storage period, x could be sinusoidally modulated at a low frequency ω_m and the value $x_c(\pm E)$ inferred from variations in the scintillation rate which occur at harmonics of ω_m. If the average value of $x \neq x_c$, there will be a first harmonic to the scintillation rate growing linearly in time. If $x = x_c$, there will be only a second harmonic component. In practice, a feedback system might be used to force the first harmonic signal to zero; the second harmonic then serves as a system calibration. (Note that the modulation in x and the subsequent modulation in the scintillation rate are 90° out of phase because the spin vectors must precess before the effects due to a change in x are manifest.)

A detailed analysis of this system is given in Ref. [37], in which many technical issues are addressed. It is shown that a factor of over 1000 improvement in the neutron EDM experimental limit is feasible. This improvement is based on a factor of 5 increase in electric field strength, an increase in the net UCN storage lifetime, and an increase by a factor of nearly 10^4 in UCN density. Magnetic field noise and systematic effects are eliminated by the dressed spin technique.

An important difference between the previous UCN storage experiments and one performed directly in the superfluid bath is that essentially all of the UCN stored in the liquid helium contribute to the measurement. As was shown earlier, the ILL experiment UCN-use efficiency was only 3%, giving an effective experimental density of 4 cm^{-3}. This number should be compared to the 2×10^5 cm^{-3} given above for the superthermal source.

15.8.2.2. *Analysis of the dressed spin system and systematic effects*

The motion of a spin under the application of static and nonresonant oscillating magnetic fields is quite complicated. In some sense, saying that the magnetic moment is modified (or dressed) is the "zeroth-order" approximation. In Ref. [37], the system was studied both by numerically integrating the equations of motion, and through quantum perturbation theory. The ^{3}He–n spin system was solved numerically under various conditions. This system is difficult to solve numerically since it involves two timescales: the RF field of 1 kHz, and the relatively slow precession (1 Hz) around the DC field. The accuracy obtained is set by the step size at the 1 kHz level. The results were identical to those obtained with the analytical treatment.

Briefly, in the quantum treatment, effects of static fields both along the RF field (B_0^x, which is a spurious field), and perpendicular to the RF field (applied field $B_0^z \gg B_0^x$), were studied. The unperturbed states are specified by $|\pm 1/2\rangle|n\rangle$ where n is the RF field photon number. The states are degenerate between $\pm 1/2$ before the static fields are applied. Using the formalism developed in Ref. [36], the following first-order correction (due to B_0^z) to the $\pm 1/2$ eigenvalues were found:

$$E^{(1)} = \pm\frac{1}{2}\gamma\sqrt{(B_0^x)^2 + [B_0^z J_0(\omega_1/\omega)]^2}, \tag{15.52}$$

where $\omega_1 = \gamma B_{\mathrm{RF}}$. If we require that $E_3^{(1)} = E_{\mathrm{n}}^{(1)}$, the critical dressing condition is obtained. Thus, effects of $B_0^x \neq 0$ enter only in second order. Carrying the perturbation expansion to higher order mixes in states of different n. The second-order corrections are zero, while the third-order gives $E^{(3)} \propto (\gamma B_0^z)^3/\omega^2$, which shows that fluctuations in the precession field $\delta\omega_0$ only alter the critical dressing condition to order

$$\delta x_{\mathrm{c}} = (\omega_0/\omega)^2(\delta\omega_0/\omega_0). \tag{15.53}$$

With $\omega_0/\omega \approx 10^2$, the system shows excellent rejection of static field fluctuations.

An important result of this analysis is that the spin/field state cannot be affected by the static electric field. The total system angular momentum could be greater than or equal to one; however, there is no way for the static electric field to couple to the constituent system states (RF photons or particles of spin-1/2).

15.8.3. *CryoEDM at ILL*

An experiment presently under construction at the ILL is described in the proposal [50]. This experiment will be conducted in nearly the same way as the 1990 ILL experiment, except that the apparatus will be filled with superfluid helium. The UCN will be produced by the superthermal process in a region away from the storage cell, and then conducted into the storage cell, with EDM measurements performed as in the 1990 experiment. The Rb magnetometers will be replaced with SQUID magnetometers that will be placed in the region around the storage cell. The problems associated with external discrete magnetometers remain, but the leakage currents associated with the application of high voltage should be orders of magnitude less than the room temperature experiment, which should decrease the potential systematic. The figure of merit compared to the comagnetometer result is about a factor of 20 greater, therefore it is anticipated that a level of 10^{-27} e cm can be obtained.

A novel UCN detector has been developed for this experiment that allows detection directly in a superfluid filled guide. The detectors employ ^{6}Li deposited directly on a large area PN photodiode detector. The detector can be further coated with a magnetic thin film polarizer, which will make for a complete polarization analysis system for the UCN.

15.9. Conclusions

The fundamental nature of CP non-invariance in fundamental interactions remains largely unknown. The search for the neutron EDM remains among the most sensitive ways to test theoretical notions regarding the nature of the interactions that led to, for example, the matter-antimatter asymmetry of the universe.

The current experimental limit is based on a UCN storage experiment that employs a ^{199}Hg atomic spin precession magnetometer. In this work, a new "geometric phase" systematic was discovered that results from an interference between the motional $\vec{E} \times \vec{v}$ magnetic field and a magnetic gradient. This effect was accounted for, and appears as controllable in improved experiments.

Current plans for improving the neutron EDM experimental limit include moving the Hg comagnetometer experiment to a new intense UCN source based on solid deuterium now under construction at PSI. With upgrades to the experiment, including reduction of gradients that lead to the geometric phase systematic, and increase in electric field and UCN storage lifetime, it is anticipated that an EDM limit in the 10^{-28} e cm range will be possible. This experiment will likely produce the first improved neutron EDM limit among the new experiments discussed in this review.

The superfluid helium EDM experiment planned for the SNS is the only experiment presently under discussion that has potential to attain a sensitivity in the 10^{-29} e cm range. This experiment is very ambitious and will likely not be producing data until after 2012.

Other experiments that do not employ a comagnetometer are being developed, or are under construction. Given the past problems with discrete external magnetometry in accounting for systematic magnetic fields, a non-zero result from any of these experiments should be considered with heavy but healthy skepticism. In fact, a zero result should be considered with similar skepticism, for we are now in a range of sensitivity where both zero and non-zero results have far-reaching theoretical implications.

Acknowledgments

We gratefully acknowledge permission from the American Physical Society to use Figures 15.4, 15.5 and 15.6, which come from Refs. [2], [40] and [48] respectively.

References

[1] Norman F. Ramsey, *Spectroscopy with Coherent Radiation* (World Scientific, Singapore, 1998).
[2] S. K. Lamoreaux and R. Golub, Phys. Rev. D **61**, 051301 (2000).
[3] J. P. Jacobs *et al.*, Phys. Rev. A **52**, 3521 (1995).
[4] P. G. Harris *et al.*, Phys. Rev. Lett. **82**, 904 (1999).
[5] V. Cirigliano, S. Profumo, and M.J. Ramsey-Musolf, ArXiv:hep-ph/0603246v2.
[6] W. Brian Hyslop and Paul C. Lauterbur, J. Mag. Res. **94**, 501 (1991).
[7] Glennys R. Farrar and M.E. Shaposhnikov, Phys. Rev. D **50**, 774 (1994).
[8] B. Filippone, Private Communication, 2008.
[9] J. A. Casas, J. R. Espinosa, and H. E. Haber, Nucl. Phys. B **526**, 3 (1998).
[10] I. B. Khriplovich, S. K. Lamoreaux: *CP Violation Without Strangeness* (Springer-Verlag, Berlin, 1997).
[11] S .K. Lamoreaux, R. Golub: Phys. Rev. D **50**, 5632 (1994).
[12] P.G.H. Sandars, E. Lipworth: Phys. Rev. Lett. **13**, 718 (1964).
[13] E. D. Commins, S. B. Ross, D. DeMille, B. C. Regan: Phys. Rev. A **50**, 2960 (1994).
[14] A. Abragam, *Principles of Nuclear Magnetism* (OUP, London 1962).
[15] W. Happer, Phys. Rev. B **1**, 2203 (1970).
[16] S. K. Lamoreaux, Phys. Rev. A 53, R3705 (1996).
[17] N. F. Ramsey, Acta Physica Hungarica **55**, 117 (1984).
[18] S. K. Lamoreaux, Nucl. Instr. Meth. A **284**, 43 (1989).
[19] J. M. Pendlebury, Nucl. Phys. A **546**, 359c (1992).
[20] J. M. Pendelbury, Proc. Ninth Symposium on Grand Unification (France, Aix-Les Bains, 1988).
[21] J. P. Jacobs, W. M. Klipstein, S. K. Lamoreaux, B. R. Heckel, E. N. Fortson, Phys. Rev. A **52**, 3521 (1995); Phys. Rev. Lett. **71**, 3782 (1993).
[22] S. K. Lamoreaux, Ph.D. Thesis (University of Washington, 1986) (unpublished); S. K. Lamoreaux *et al.*, Phys. Rev. A **39**, 1082 (1989).
[23] R. Golub and J. M. Pendlebury, Phys. Lett. A **62**, 3376 (1977).
[24] P. Ageron *et al.*, Phys. Lett. A **66**, 469 (1978).
[25] R. Golub *et al.*, Z. Phys. B **51**, 187 (1983).

[26] R. Golub, D. J. Richardson, and S. K. Lamoreaux, *Ultracold Neutrons* (Adam Hilger, Bristol 1991).
[27] S. K. Lamoreaux and R. Golub, Pis'ma ZhETF **58**, 844 (1995), Sov. Phys. JETP Lett. **58**, 792 (1993).
[28] V. F. Turchin, *Cold Neutrons* (Program for Scientific Translations, Israel 1965).
[29] R. Golub, Phys. Lett. A **72**, 387 (1979).
[30] H. Yoshiki *et al.*, Phys. Rev. Lett. **68**, 1323 (1992); R. Golub, S.K. Lamoreaux: Phys. Rev. Lett. **70**, 517 (1993).
[31] A. I. Kilvington *et al.*, Phys. Lett. A **125**, 416 (1987).
[32] L. Passell and R. Schermer, Phys. Rev. **150**, 146 (1960).
[33] C. G. Aminoff *et al.*, Rev. Phys. Appl. **24**, 827 (1989).
[34] J. M. Doyle and S. K. Lamoreaux, Europhys. Lett. **26**, 253 (1994).
[35] R. Golub, J. Physique **44**, L321 (1983); *Proc. 18th Inter.Conf. on Low-Temperature Physics, Part 3; Invited Papers* (Kyoto 1987) p.2073.
[36] Polonsky, N. and Cohen-Tannoudji, C. (1965), Jour. de Phys. **26**, 409; Cohen-Tannoudji, C. and Haroche, S. (1969), Jour. de Phys. **30**, 153.
[37] R. Golub and S.K. Lamoreaux, Phys. Rep. **237**, 1 (1994)
[38] I. B. Khriplovich, Phys. Lett. B **173**, 193 (1986); Yad. Fiz. **44**, 1019 (1986) [Sov. J. Nucl. Phys. **44**, 659 (1986)].
[39] E. D. Commins, Am. J. Phys **59**,1077 (1991).
[40] J. M. Pendlebury *et al.*, Phys. Rev. **A70**, 032102 (2004).
[41] M. Berry, Proc. Roy. Soc. London, Ser. A **392**, 45 (1984).
[42] S. K. Lamoreaux and R. Golub, Phys. Rev. **A 71**, 032104 (2005).
[43] A. L. Barabanov, R. Golub, and S. K. Lamoreaux, Phys. Rev. **A74**, 02115 (2006).
[44] S. K. Lamoreaux and R. Golub, Phys. Rev. Lett. **98**,149101 (2007).
[45] C. A. Baker *et al.*, Phys. Rev. Lett. **98**, 149102 (2007).
[46] P. G. Harris and J. M. Pendlebury, Phys. Rev. A **73**, 014101 (2006).
[47] K. F. Smith, *et at.*, Phys. Lett. **234B**, 33 (1990).
[48] C. A. Baker *et al.*, Phys. Rev. Lett. **97**, 131801 (2006).
[49] T. G. Vold *et al.* Phys. Rev. Lett. bf 52, 2229 (1984).
[50] S. N. Balashov *et al.*, arXiv:0709.2428.
[51] S. K. Lamoreaux, ArXiv:nucl-ex/0103005.
[52] Y. N. Pokotilovskii, Nucl. Inst. Meth. A **356**, 412 (1995).
[53] R. Golub and K. Bonig, Z. Phys. B **51**, 95 (1983).
[54] L. Wolfenstein, Phys. Rev. Lett. **13**, 562 (1964).
[55] K. Kirch, Cape Code Lepton Moments Meeting, 2006, see: http://g2pc1.bu.edu/lept06/program.html.
[56] L. D. Landau and E.M. Lifshitz, *The Classical Theory of Fields* (Nauka, Moscow, 1988).
[57] P. G. Harris *et al.*, Phys. Rev. Lett. **82**, 904 (1999).
[58] I. S. Altarev *et al.*, Phys. At. Nucl. **59**, 1152 (1996).
[59] A. Saunders *et al.*, Phys. Lett. B **595**, 55 (2004).
[60] http://p25ext.lanl.gov/edm/edm.html
[61] http://nedm.bu.edu/

[62] A. P. Serebrov, Int. Conf. Prec. Meas. with Slow Neutrons, NIST, Gaithersburg, April 5–7 2004.
[63] I. S. Altarev et al, Sov. Phys. JETP Lett. **44**, 460 (1986).
[64] E. M. Purcell and N.F. Ramsey, Phys. Rev. **78**, 807 (1950).
[65] J. H. Smith, E.M. Purcell, and N.F. Ramsey, Phys. Rev. **108**, 120 (1957).
[66] N. F. Ramsey, Phys. Rev. **109**, 225 (1958).
[67] O. Zimmer *et al.*, Phys. Rev. Lett. **99**, 104801 (2007).

Chapter 16

Nuclear Electric Dipole Moments

W. Clark Griffith*, Matthew Swallows†, and Norval Fortson‡

Department of Physics, University of Washington
Seattle, WA, USA
‡*fortson@phys.washington.edu*

Permanent electric dipole moment (EDM) searches in diamagnetic atoms provide important bounds on nuclear EDMs. Such EDMs would most likely originate from CP-violating interactions between nucleons. On-going experiments in Hg, Xe, Ra, and Rn atoms are discussed, and a thorough description of the most sensitive experiment to date, in ^{199}Hg, is given. Improved bounds on new sources of CP violation based on the ^{199}Hg result are presented.

Contents

*Present address: NIST, 325 Broadway, Boulder, CO.
†Present address: JILA, University of Colorado, Boulder, CO.

16.1. Introduction

In this chapter we discuss experimental searches for a permanent electric dipole moment (EDM) of an atomic nucleus. Such an EDM might originate from an intrinsic EDM of the constituent protons and neutrons, but unlike measurements on the bare neutron (see Chapter 15), nuclear EDM searches are also sensitive to CP-violating interactions between nucleons. Since an underlying CP-violating theory can have very different contributions to a nucleon-nucleon interaction, an EDM of the electron, or a neutron EDM, it is essential that experimental searches be carried out in all three sectors. At present, results from each sector contribute comparable and complementary bounds on new sources of CP violation [1]. The tightest constraint on the nucleon-nucleon sector comes from a search for the EDM of the ^{199}Hg atom carried out at the University of Washington in Seattle [2]. The new limit on the atomic dipole, $|d(^{199}\text{Hg})| < 3.1 \times 10^{-29}\, e\,\text{cm}$, presently is the tightest experimental bound on the EDM of any system and improves on the CP violation limits quoted in Ref. [1].

After giving an overview of the general features of nuclear EDM searches, a brief summary of current experiments will be given, followed by a more detailed discussion of the ^{199}Hg experiment.

16.2. Measuring an EDM

The general strategy used in almost all EDM searches is to place the particle (or atom, or molecule) of interest in an electric ($\boldsymbol{E}$) and magnetic ($\boldsymbol{B}$) field. If the system under consideration possesses a non-zero EDM, the usual magnetic Zeeman effect is modified by an electric field-dependent term, giving the interaction:

$$H = -(\mu \boldsymbol{B} + d\boldsymbol{E}) \cdot \frac{\boldsymbol{F}}{|F|}, \tag{16.1}$$

where μ is the magnetic moment and d is the electric dipole moment of the particle. It is advantageous to study a system with $F = 1/2$, because then the only possible moments in the ground state are the magnetic and electric dipole moments shown in Eq. (16.1). For a higher spin system, the presence of higher order moments can add additional interactions that complicate the measurement and open up additional avenues for systematic effects to enter.

The EDM can then be measured by comparing the Larmor precession frequency of $\boldsymbol{F}$ about $\boldsymbol{B}$ with $\boldsymbol{E}$ parallel and anti-parallel to $\boldsymbol{B}$,

$\omega_L = (\mu B \pm dE)/(\hbar F)$. Most of the experimental effort is then associated with controlling magnetic field fluctuations, and ensuring that there are no magnetic-like effects associated with the electric field application. For example, any electrical currents flowing due to the electric field application generate magnetic fields that are likely to reverse direction when the electric field direction is changed, leading to a possible systematic effect.

The fundamental limitation on the sensitivity of such an experiment comes from the uncertainty in measuring ω_L, which goes as $\delta\omega_L = 1/\tau$ for a single particle, where τ is the length of time the spin precession can be measured, or the spin coherence time. The sensitivity is improved by performing the measurement on a large number of particles (N) simultaneously, and repeating the measurement a large number of times (N_m) during a total time $T = N_m\tau$, so that the uncertainty in d is

$$\delta d = \frac{\hbar F}{E\sqrt{N\tau T}}. \tag{16.2}$$

Experiments are designed to maximize the electric field strength, spin coherence time, and number of particles, in order to increase the statistical sensitivity of the measurement. The total time that the measurement takes can range from several months to many years.

16.3. Diamagnetic Atom EDM Searches

All nuclear EDM searches carried out to this point have used nuclei that are part of an electrically neutral atomic or molecular system. Although this somewhat complicates the theoretical effort involved in extracting the nuclear effect from the measurement, it enables large electric fields to be applied without accelerating the particle out of the apparatus. Charged particle EDM experiments in storage rings have recently been proposed, as is discussed in Chapter 17, including a search for the deuteron EDM which would be sensitive to CP-violating nucleon-nucleon interactions. In this chapter we will focus on nuclear EDM searches using neutral atoms.

16.3.1. *Shielding and the Schiff theorem*

At first glance it might not seem useful to apply the method described in Sec. 16.2 on a neutral atom when looking for the EDM of the nucleus. If placed in an external electric field, the constituents of the neutral system should arrange themselves such that the average electric field seen by the charged particles of the system is zero, otherwise the system would be accelerated by the electric field. According to the Schiff theorem [3] this shielding is perfect for nonrelativistic point charges with only electrostatic interactions.

In the case of paramagnetic atoms where there is a non-zero electron spin, though, the Schiff theorem is evaded due to relativistic effects that actually enhance the effect of an EDM of the electron, especially in heavier atoms. These systems are generally considered to be overwhelmingly sensitive to an electron EDM compared to any nuclear effects. Diamagnetic atoms used in EDM searches have zero electronic ground state angular momentum ($^1S_0$) and a non-zero nuclear spin. In this case the shielding of the external field is violated due to the finite size of the nucleus, giving sensitivity to an EDM of the nucleus. The violation is greater the larger the nuclear charge Z, rising as Z^2 due to the effect on the electronic wave function near the nucleus. Thus it is advantageous to use heavier atoms. There is also an additional enhancement for octupole-deformed nuclei such as Ra, compared to spherical nuclei in Xe and Hg [4, 5].

16.3.2. *Advantages and disadvantages of diamagnetic atoms*

Although compared to an electron EDM, a nuclear EDM generally induces a much smaller atomic EDM, there are experimental advantages for nuclear EDMs in diamagnetic atoms that partly redress the balance. The nuclear spin is much less sensitive to external magnetic field fluctuations than an electron spin, since the electron magnetic moment is much greater than a nuclear magnetic moment ($\mu_B/\mu_N \approx 2000$). This allows diamagnetic atoms to achieve better statistical sensitivity compared to paramagnetic atoms since they are less susceptible to magnetic noise, and they are also less susceptible to magnetic systematic effects such as leakage current and $\boldsymbol{v} \times \boldsymbol{E}$ fields.

In vapor cell experiments, diamagnetic atoms can also achieve better sensitivity than paramagnetic atoms since they tend to have much longer spin coherence times. In a cell, paramagnetic spin coherence times are typically tens of milliseconds, limited by spin-exchange and spin-destroying collisions with other atoms, while diamagnetic atoms are practically free of these interactions. Diamagnetic atoms with $I = 1/2$ do not couple to electric field gradients at the cell walls and so the spin polarization can survive many collisions with the cell walls, especially if an appropriate wall coating is used, making it possible to achieve spin coherence times over 1,000 s in some cases.

16.3.3. *Interpretation of diamagnetic atom edms*

The nuclear EDM associated with a diamagnetic atom is conventionally parameterized in terms of the Schiff moment (S), which can be considered the lowest order nuclear moment unaffected by the shielding discussed in Sec. 16.3.1. Atomic calculations are necessary to relate the Schiff moment to an atomic EDM [6]. Numerical factors for several diamagnetic species are given in Table 16.1.

Table 16.1. Comparison of diamagnetic atoms used in current nuclear EDM searches. The k value corresponds to atomic calculations relating the atomic EDM (d_a) to the Schiff moment (S), $d_a = k \times 10^{-17}(S/e\,\mathrm{fm}^3)\ e\,\mathrm{cm}$ [6]. References are given in the table for the estimation of S in terms of η.

Species	I	half-life	k	$S[10^{-8}\eta\, e\,\mathrm{fm}^3]$	Ref.	$d_a[10^{-25}\eta\, e\,\mathrm{cm}]$
^{129}Xe	1/2	stable	0.38	1.75	[9]	0.7
^{199}Hg	1/2	stable	2.8	1.4	[9]	4
^{225}Ra	1/2	15 days	8.5	300	[4]	2500
^{223}Rn	7/2	24 min.	3.3	1000	[4]	3300

The Schiff moment might arise from an EDM of the proton or neutron. For example, calculations for ^{199}Hg give [7]

$$S(^{199}\mathrm{Hg}) = (0.2d_p + 1.9d_n)\ \mathrm{fm}^2. \tag{16.3}$$

There is a stronger dependence on the neutron EDM since the valence nucleon in the ^{199}Hg nucleus is a neutron. Combined with the experimental limit [2], an upper bound on the neutron EDM of $|d_n| < 5.8\times10^{-26}\ e\,\mathrm{cm}$ can be obtained, within a factor of two above direct measurements on the neutron [8]. The ^{199}Hg EDM limit currently allows the best constraint on the proton EDM, $|d_p| < 7.9 \times 10^{-25}\ e\,\mathrm{cm}$, where a 30% theoretical uncertainty is included as is described in [7].

However, it is expected that a CP-violating nucleon-nucleon interaction is most likely to be the dominant contribution to the Schiff moment. The magnitude of the CP-violating interaction is commonly parameterized in terms of a dimensionless constant, η, and then the Schiff moment is calculated in terms of η. Table 16.1 shows theoretical estimates for the relative sensitivities to CP-violating nucleon-nucleon interactions of diamagnetic atoms used in current nuclear EDM searches. The strength of the CP-violating nucleon-nucleon interaction can be related to chromoelectric dipole moments of the quarks [10], which in turn can be estimated in supersymmetric models [1] or other extensions to the Standard Model.

The CP-violating phase in the QCD Lagrangian, $\bar{\theta}_{QCD}$ can also contribute to η.

Besides strictly nuclear effects, the EDM of a diamagnetic atom might originate from semileptonic interactions between the atomic electrons and the nucleus, typically parameterized in terms of the dimensionless constants C_S, C_P, and C_T. There is also the possibility of a contribution from the electron EDM through the hyperfine structure coupling between the nuclear and electron spins.

16.3.4. *Experiments with diamagnetic atoms*

As of this writing, the most sensitive measurement of the EDM of a diamagnetic atom is the experiment using ^{199}Hg atoms at the University of Washington in Seattle. Fig. 16.1 shows the upper bound on the ^{199}Hg EDM obtained from different iterations of the experiment. The measurement is carried out by comparing the nuclear spin precession frequencies in two vapor cells in parallel magnetic and antiparallel electric fields. While the first versions used Hg discharge lamps to provide 254 nm light for optical pumping and spin precession probing [11–13], the last two measurements used a frequency-quadrupled semiconductor laser [2, 14]. The most recent measurement added two additional magnetometer cells which have no electric field applied to them. These additional cells serve to cancel noise due to fluctuating magnetic field gradients, and allow monitoring for possible

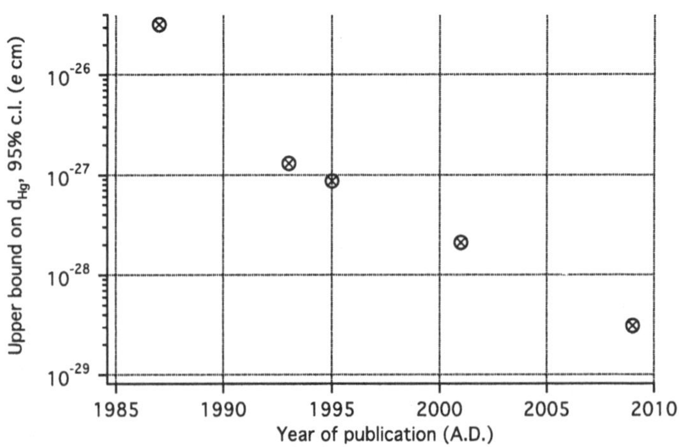

Fig. 16.1. Improvement in the ^{199}Hg EDM upper bound over time.

magnetic field systematic effects. Further information about the UW EDM measurement can be found in Sec. 16.4.

Several other experiments with diamagnetic atoms have been proposed or are currently underway (some relevant information is summarized in Table 16.1). The Romalis group at Princeton is planning to measure the EDM of ^{129}Xe [15, 16], using a cryogenic liquid sample. Xenon gas can be polarized by spin-exchange with optically-pumped rubidium. The polarized ^{129}Xe can then be liquefied in bulk quantities without significant relaxation, and the spin precession can be detected using SQUID magnetometers. Xenon is significantly lighter than Hg, and thus the Schiff shielding is more effective in this system. However, the increased shielding should be compensated by the large density of xenon atoms in the liquid phase ($\sim 10^{22}$ cm^{-3}), the long transverse relaxation times achievable with liquid xenon ($\sim$ 1300 s), and the large electric fields that can be applied ($\sim$ 400 kV/cm). From this information, the shot-noise limited sensitivity of such an EDM experiment can be estimated as $\sim 10^{-36}\, e\,\mathrm{cm}$ for one day of integration. Even if other noise sources restrict the Princeton experiment to a small fraction of the quantum-noise-limited sensitivity, the potential for progress seems evident.

Certain isotopes of radium, radon, and other heavy atoms are promising candidates for EDM experiments, because of the collective enhancement of their Schiff moments generated by the octupole deformation of these nuclei [4, 5]. The advantage of these exotic systems is that the experiments need only achieve a fraction of the sensitivity of the mercury or xenon experiments in order to produce comparable limits on new sources of CP-violation. Experiments with optically trapped ^{225}Ra are under development at Argonne National Laboratory in the United States, and at the Kernfysisch Versneller Instituut in the Netherlands. Such experiments can benefit from the large electric fields and long coherence times achievable with optical traps, but trapping light-induced noise and possible systematic shifts of the Larmor frequency must be carefully considered [17, 18]. Magneto-optical trapping of ^{225}Ra has been demonstrated by the Argonne group [19], and an EDM measurement is being actively pursued. With a brighter ^{225}Ra source, the Argonne group estimates that a statistical sensitivity of $\sim 1 \times 10^{-26}\, e\,\mathrm{cm}$ could be achieved in a first-generation experiment, which together with the enhancements of the radium system should provide a sensitivity to CP-violating effects that might exceed that of the 2009 mercury result.

An EDM experiment with ^{223}Rn has been proposed at TRIUMF in Canada [20]. While ^{225}Ra is produced as an α-daughter of relatively long-lived ^{229}Th, the short half-life of ^{223}Rn requires that the experiment be based at an accelerator facility. ^{223}Rn should also exhibit nuclear octupole deformation, and can be polarized through spin-exchange optical pumping. The spin-polarized radon would be collected in specialized vapor cells, and the spin precession would then be observed by detecting the asymmetry of the gamma or beta rays produced in the decay of these radioactive atoms. The collaboration expects to reach a sensitivity to the atomic EDM of $\sim 1 \times 10^{-26}\, e\,\text{cm}$, allowing a sensitivity to CP-violating effects similar to that of the radium system.

In addition to the above atomic systems, experiments have also been proposed to measure the EDM of the bare deuteron [21], and the nuclear Schiff moment of ^{207}Pb using a sample of the ferroelectric crystal $PbTiO_3$ [22]. These experiments will employ methods that are very different than those used in the atomic experiments, but would be sensitive to the same CP-violating nuclear interactions. A measurement of the deuteron EDM, in particular, would be of interest because of the simplified nuclear theory involved in the interpretation of the experimental result.

16.4. The ^{199}Hg EDM Measurement in Seattle

The authors are directly involved in the experimental search for the ^{199}Hg EDM at the University of Washington, and in this section we describe this experiment in greater detail.

16.4.1. *Experimental technique*

The ^{199}Hg EDM apparatus used in the most recent measurement [2] is shown in Fig. 16.2. The main improvement to the experiment was the construction of an apparatus that incorporates a stack of four vapor cells (see the cutaway view in Fig. 16.4 below). Previous versions of the experiment had all compared the spin precession frequency between two vapor cells, where the cells were in a common magnetic field and oppositely directed electric fields. In the current experiment the two additional cells are at zero electric field and are used as magnetometers above and below the EDM sensitive cells. They help to improve statistical sensitivity by allowing magnetic field gradient noise cancellation, and they are also used to look for possible magnetic systematic effects.

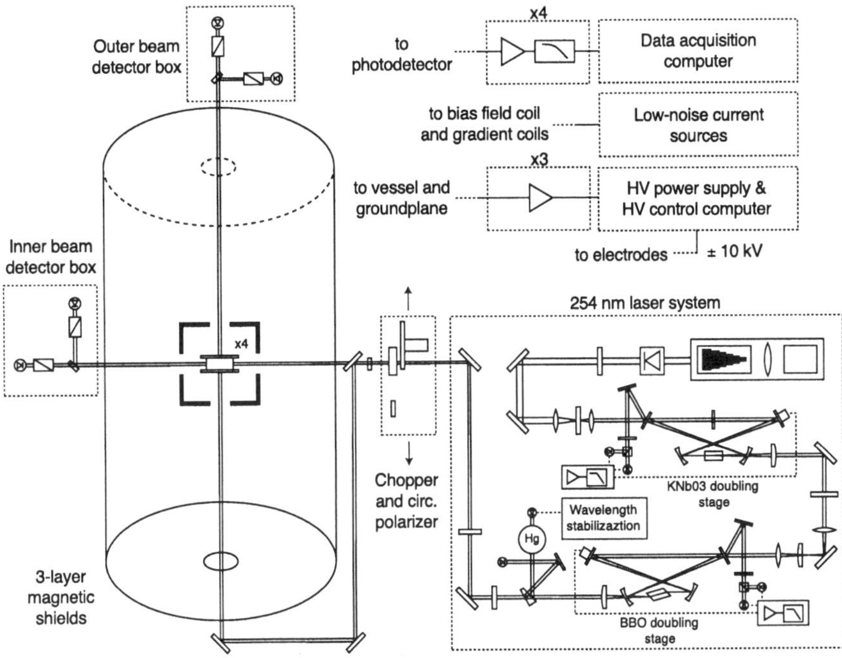

Fig. 16.2. Simplified diagram of the ^{199}Hg EDM apparatus.

As before, to search for an EDM, the Larmor spin precession frequency of ^{199}Hg is measured. A common magnetic field produces Larmor precession in a vapor of spin-polarized mercury in each cell, and a strong electric field applied in opposite directions in the middle two cells modifies the precession frequency by an amount proportional to the electric dipole moment. An EDM would cause a frequency shift of $2Ed/h$, with opposite sign in the two cells; so the magnitude of the EDM is given by $d = h\delta\nu/(4E)$, where $\delta\nu$ is the difference in precession frequency between the two middle cells.

The ^{199}Hg nuclei are spin polarized by optical pumping on the 253.7 nm absorption line in mercury. Since the light beam is transverse to the precession axis, the circularly-polarized pumping light is modulated at the Larmor frequency to synchronously pump the precessing spins [23, 24]. The optical rotation of a linearly-polarized off-resonant probe beam is used to detect the spin precession. Polarization rotation is converted to amplitude modulation using high-quality polarizers, and the resulting signals (see Fig. 16.3) are fit to extract the Larmor frequency. The ultraviolet light for

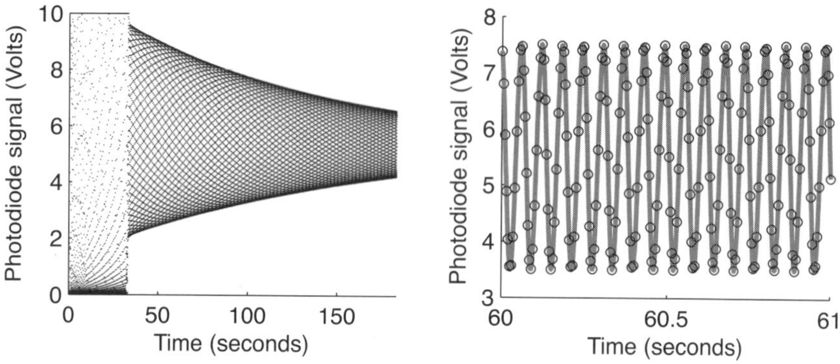

Fig. 16.3. Pump-probe cycle showing the Larmor precession frequency expanded in the right figure. The detectors saturate during the optical pumping phase of the experiment (first 30 seconds).

this transition is obtained by quadrupling the output of an infrared diode laser in a master oscillator, power amplifier (MOPA) configuration. The laser system produces several milliwatts of stable, tunable UV radiation with good spatial characteristics. This system has operated continuously and problem-free for several years, and requires only occasional maintenance. The laser frequency is locked to absorption lines in a separate vapor cell containing mercury at natural isotopic abundances.

The cells are held as shown in Fig. 16.4 inside a sealed vessel filled with about 1 bar of SF_6 or N_2 gas to reduce leakage currents. The vessel and electrodes are constructed of conductive polyethylene, which was chosen due to its low magnetic impurity content. The vapor cells were upgraded slightly after the 2001 measurement, containing a 100% CO buffer gas, instead of the 95% N_2 / 5% CO mixture used previously. Studies of spin relaxation in mercury vapor cells [25] indicated that the wax coating on the interior of the cells could be damaged by collisions with excited metastable mercury atoms. The CO buffer gas efficiently quenches these metastable states and thus helps prevent damage to the coating. The end result is that polarization lifetimes can be achieved that are a factor of 1.5 longer than was possible with the old vapor cells.

16.4.2. *4-cell data*

Larmor precession measurements are made simultaneously in the four vapor cells as described in the previous section, and the electric field direction in the middle two cells is reversed between each measurement. The

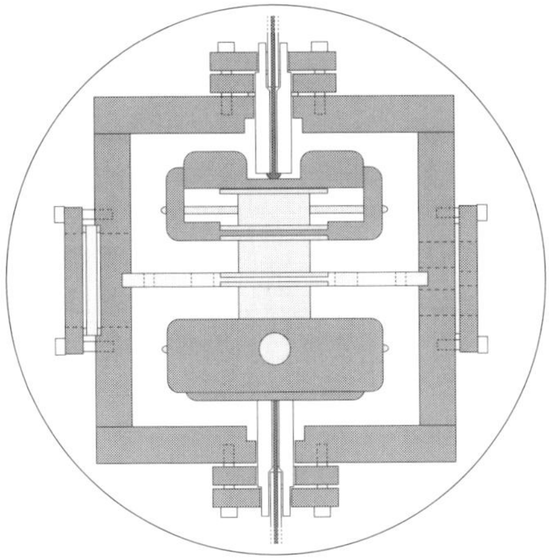

Fig. 16.4. Cutaway view of the EDM cell-holding vessel. High voltage (± 10 kV) is applied to the middle two cells with the ground plane in the center, so that the electric field is opposite in the two cells. The outer two cells are enclosed in the HV electrodes (with light access holes as shown here for the bottom-most cell), and are at zero electric field. A uniform magnetic field is applied in the vertical direction.

pump/probe cycle lasts between two and four minutes, and the measurement is repeated several hundred times in overnight data runs lasting between 12 and 24 hours. Fig. 16.5 shows the measured EDM signal from a typical data run. Between data runs various experimental parameters are changed, such as the magnetic field direction, the high voltage charging current, and the probe light polarization direction.

The overnight runs are grouped into "sequences," usually consisting of 12 consecutive data runs. In the ideal data taking schedule, eight of these runs would be optimized to detect an EDM, and four runs are dedicated to checking for systematic effects. Between sequences, individual Hg vapor cells may be reoriented, exchanged with each other, or swapped out for other cells, and likewise for individual electrodes. Also, two versions of the cell holding vessel have been used during data collection. The swapping of these nominally identical components is meant to randomize the effects of possible ingrained leakage current paths or embedded magnetic impurities.

16.4.2.1. *Frequency combinations*

With the four cell measurement it is possible to monitor a variety of linear combinations of the precession frequencies from the four positions, providing varying degrees of magnetic gradient noise cancellation and EDM

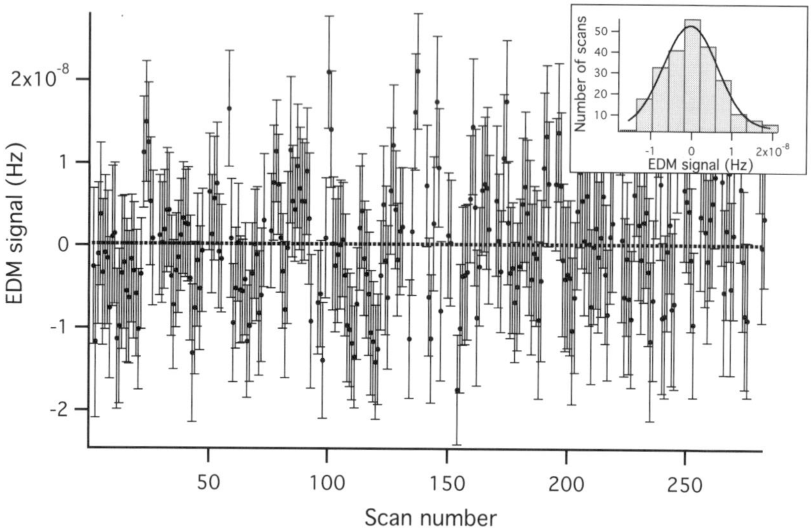

Fig. 16.5. EDM signal derived from the EDM sensitive 4-cell frequency combination from an overnight data run. Each data point corresponds to a 200 second spin precession measurement. The weighted mean of the EDM signal from this data set was $1.7 \pm 5.0 \times 10^{-10}$ Hz, and the reduced χ^2 was 0.78.

sensitivity. Table 16.2 shows several of the frequency combinations that are regularly monitored. The middle difference gives the same information available to previous two-cell versions of the experiment. The EDM-combination has the same EDM sensitivity as the middle difference, but has the advantage that it cancels up to second order magnetic field gradient noise. Combinations such as the outer difference and the magnetic systematic (MS) combination potentially give important information about systematic effects since they have zero sensitivity to an EDM, but might register signals due to magnetic fields from leakage currents or trace magnetic impurities. The MS-combination cancels up to first order magnetic field gradient noise.

16.4.2.2. *Statistical sensitivity*

The 4-cell dataset contributing to the most recent result [2] consisted of 166 overnight data runs taken over a span of about two years. The weighted average of these runs gives a statistical error on the ^{199}Hg EDM of 1.29×10^{-29} $e\,$cm, corresponding to a 0.1 nHz uncertainty on the difference

Table 16.2. Frequency combinations. Frequencies are labeled by cell position (OT = outer top, MB = middle bottom, etc.). The third column gives the relative sensitivity to an EDM. The EDM combination removes magnetic field gradient noise of up to second order, while retaining maximal sensitivity to an EDM. The MS combination is insensitive to an EDM, but is useful for revealing systematic effects due to electric field-correlated magnetic fields.

Name	Combination	EDM sens.
Middle difference	$\omega_{MT} - \omega_{MB}$	1
Outer difference	$\omega_{OT} - \omega_{OB}$	0
EDM-combination	$\omega_{MT} - \omega_{MB} - \frac{1}{3}(\omega_{OT} - \omega_{OB})$	1
MS-combination	$\omega_{OT} + \omega_{OB} - (\omega_{MT} + \omega_{MB})$	0

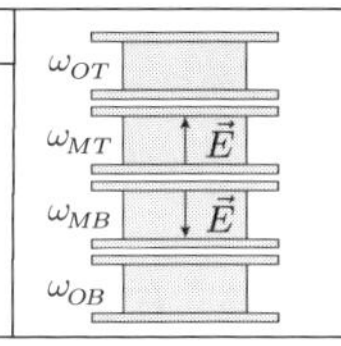

between the frequencies of the two middle cells. This is a factor of four improvement over the previous 2-cell version of the experiment [14]. The improvement is the result of roughly equal contributions from gradient noise cancellation, improved stability of the vapor cell spin lifetimes, and a longer total integration time.

It is hoped that the statistical sensitivity of the 4-cell measurement can be further improved, and efforts to fully understand the noise performance of the system are ongoing. The photon shot noise contribution is modeled using computer simulations, which show that individual Larmor frequency measurements are within a factor of three of the shot noise limit. However, the scatter among a series of such measurements is often larger than can be explained by the single-shot uncertainty. While the modeling also shows that further improvements to reduce the shot noise itself are possible, the current extraneous noise limiting the experiment must first be eliminated. This noise is worse in frequency difference channels that are sensitive to field gradient fluctuations, indicating that the effect is most likely magnetic in origin, although no such noise is apparent in individual Larmor frequency measurements when the measurement time is extended for several hundred seconds. As this apparently rules out long time scale background magnetic field fluctuations, that noise is most likely generated by the pump-probe cycling and might be related to the stability of the light beam steering.

16.4.3. *Systematic effects*

As discussed above, a statistical uncertainty at the $1 \times 10^{-29}\, e\,\text{cm}$ level has been reached thus far in the experiment and it is possible that new improvements may reduce this uncertainty further. To take advantage of such sensitivity, comparable bounds must be placed on any systematic effects in the measurement.

Among the most important possible systematics are:

(1) HV leakage currents and sparks,
(2) Ferromagnetic contaminants,
(3) The ^{199}Hg Stark interference effect,
(4) Motional $\vec{v} \times \vec{E}$ magnetic fields seen by the mercury atoms as they move inside the cells,
(5) HV pick-up on coils or other magnetic field sources.

The last two are examples of effects believed to be under control to well below $10^{-29}\,e\,\mathrm{cm}$. The first two are the systematic problems of greatest concern; they and number (3) will be discussed individually.

16.4.3.1. *Leakage currents and sparks*

When an electric field is applied to the vapor cells during the EDM experiment, small leakage currents flow across or through the cell body. If these leakage currents have a circumferential component, then they will produce a magnetic field that will add linearly to the bias field B_0. If these currents reverse when the high voltage polarity is reversed, they will generate an EDM-like signal that can be difficult to distinguish from a real EDM.

The measured cell leakage currents are typically less than 0.5 pA. If a current of this size followed a path that made a half turn about the circumference of a cell – an extreme case – it would generate a signal equivalent to an EDM of about $1 \times 10^{-29}\,e\,\mathrm{cm}$. This level is already quite small, and any leakage bias resulting from a particular cell/electrode combination is expected to be averaged to a yet smaller level by using different vapor cells and electrodes, and cycling each one through all possible positions in the vessel.

A part of the leakage currents can consist of sparks or microdischarges, which show up as spikes on the measured HV leakage currents. Sparks can produce not only the effects described above due to their average magnetic field but also other possible systematic effects. For example, the peak magnetic fields of sparks might be large enough to magnetize ferromagnetic contaminants (contaminants are discussed in the next subsection) and in that way generate HV-correlated Larmor frequency shifts that can mimic an EDM.

Although it is now understood how to reduce the occurrence of sparks to a negligible level by proper control of the HV vessel fill gas, a portion of the earlier data in the recent EDM measurement was contaminated with sparks, and correlations were observed between the EDM signal and the appearance of sparks. To address this problem the data with sparks were cut in two different ways: 1. Cutting out entire sequences containing any nights with significant sparks; or 2. Cutting only the individual pump/probe cycles that contained sparks, thus retaining more data. Fortunately, the two cutting schemes yielded EDM values in agreement with each other, and also with the spark-free data.

16.4.3.2. *Ferromagnetic contaminants*

Trace amounts of ferromagnetic material near or in the vapor cells could be disturbed in some way (moved, magnetized, etc.) by the HV, and therefore are a possible source of EDM-like signals. Ultra-precise tests have been made to ensure that any materials used near the vapor cells are as free as possible of any contaminants, and all parts are cleaned in an HCl acid solution, but it is difficult to exclude the possibility that an effect due to any remaining contaminants could bias the data. Thus it is important to detect such biases, or rule out their presence if possible. If a ferromagnetic source is large enough, then frequency combinations that are not sensitive to an EDM (e.g. the outer cell difference or the MS-combination) can be used to detect it independently of its contribution to the EDM-sensitive channels. An example of the utility of the MS-combination in actual data is shown in Fig. 16.6. Thus far, the presence of such detectable contamination has been traced to electrodes that were not flame polished, and seems to have been eliminated by flame polishing the electrodes.

The question remains whether there are other contaminants or similar sources of bias that are too small to be detected by the MS-combination but are large enough to be significant in the EDM data, since the MS-combination is somewhat less sensitive than the cell combinations that are used to extract an EDM. If small contaminants come and go frequently, then their effects will tend to cancel out and may not bias the data. If there is a more stable contaminant (as would be the case if it were a speck of ferrous metal embedded in the electrode body, for example), some bounds on it can be set with a method similar to the one used to set better bounds on leakage current systematics (see above). In this case the ferromagnetic source may be considered a property of a particular cell or electrode and

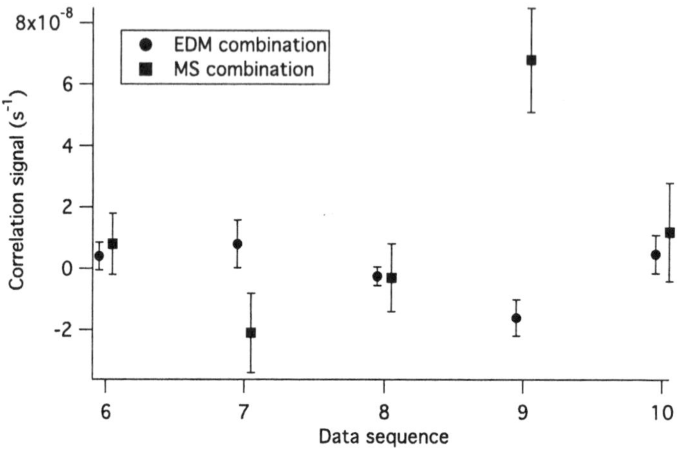

Fig. 16.6. Identifying occasional magnetic contaminants. In the 30 data sequences thus far, sequences 9 and 17 have a certain anomalous effect, shown here for sequence 9. Sequence 9 looks suspicious in the EDM cell combination, and clearly also has a serious offset in the magnetic systematic (MS) combination, which is insensitive to a true EDM. In both sequences 9 and 17, an electrode that had not been flame-polished was installed in the cell vessel; the observed deviations did not reappear when flame-polished electrodes were used subsequent sequences.

it could be detected by analyzing the data for signals correlated with the presence of that cell or electrode in the various possible positions within the vessel. The groundplane could also harbor stable contaminants, and this possibility is addressed by swapping in a different groundplane and comparing the data from before and after the change. Such comparisons have been made using the data not rejected because of the MS combination signal, the appearance of sparks, or other data filters. The absence of any spurious EDM-like Larmor frequency changes under the various changes of the cells, electrodes and ground plane is evidence that there are no serious contaminant (or leakage current) problems in the filtered data at the current sensitivity.

16.4.3.3. *The ^{199}Hg Stark interference effect*

A static electric field applied to an atom with an $E1$ (electric dipole) optical transition induces $M1$ (magnetic dipole) and $E2$ (electric quadrupole) transitions. The presence of these additional transitions leads to an interference effect of a particular vector character. For a $F = \frac{1}{2} \rightarrow F = \frac{1}{2}$ $E1$ transition, such as the one used in the ^{199}Hg EDM search, the fractional change in the absorptivity α is of the form,

$$\frac{\delta\alpha}{\alpha} = a(\hat{\boldsymbol{\epsilon}} \cdot \mathbf{E}_S)(\hat{\mathbf{k}} \times \hat{\boldsymbol{\epsilon}}) \cdot \hat{\boldsymbol{\sigma}}, \tag{16.4}$$

where a is a factor denoting the strength of the effect, $\hat{\boldsymbol{\epsilon}}$ is the direction of the electric field vector of the light driving the transition, $\hat{\mathbf{k}}$ is the propagation direction of the light, $\mathbf{E}_S$ is the static electric field, and $\hat{\boldsymbol{\sigma}}$ is the atomic spin polarization direction of the ground state. The factor a was originally estimated years ago [26] and recently was evaluated in a thorough many-body calculation that yielded the result [27]: $a = 0.80 \times 10^{-8}$ $(\text{kV/cm})^{-1}$ for the 254 nm $E1$ transition in ^{199}Hg. This Stark interference effect is of interest for the EDM search because it can lead to a *light shift* (also called an ac-Stark shift), an apparent Larmor frequency shift that is linear in the strength of the applied electric field; in other words, it can mimic an EDM. The shift is zero if the optical linear polarization direction $\hat{\boldsymbol{\epsilon}}$ is aligned parallel to the spin precession axis, and hence the average polarization is held close to this position during an EDM measurement. At the current level of EDM sensitivity, the linear polarization need not be controlled precisely to suppress the Stark interference sufficiently, provided the value of a is as small as calculated. The Stark interference can be measured with the present EDM apparatus with only minor modifications, and a preliminary result is in agreement with the calculated value of a above. A more precise measurement is currently underway.

If necessary there are additional ways to guard against the Stark interference appearing as a systematic effect. One way is to use the probe laser at two different wavelengths where the Stark interference light shift has opposite sign, and average the results to cancel out the Stark interference. A way to completely eliminate the Stark interference problem, and any other shifts due to the light beam, is to measure the Larmor frequency "in the dark" between two probe laser pulses (which establish the Larmor phase at the beginning and end of the dark period).

16.4.3.4. *Blind analysis*

Because of the need to cut some data (for example, when sparks or magnetic impurities do appear) while at the same time guarding against human bias in decisions about making data cuts, blind analysis has been used to hide the actual value of the EDM signal for all data taken after March 2006. The analysis program adds a fixed, blind HV correlated offset to the middle cell fitted frequencies, $+\delta/2$ to the middle top cell and $-\delta/2$ to the middle

Table 16.3. Limits on CP-violating parameters based on the experimental bound on $d(^{199}\mathrm{Hg})$ (95% C.L.) compared to limits from the Tl (90% C.L.) [28], neutron (90% C.L.) [8], or TlF (95% C.L.) [29] experiments. Values that improve upon (complement) previous limits appear above (below) the horizontal line. Relevant theory references for the Tl, neutron, and TlF limits are given in the last column. (This table was reprinted with permission from Ref. [2]. Copyright 2009 by the American Physical Society.)

Parameter	^{199}Hg bound	Hg theory	Best alternate limit		
$\tilde{d}_q(\mathrm{cm})$ [a]	6×10^{-27}	[6, 10, 30]	n:	3×10^{-26}	[1]
$d_p(e\,\mathrm{cm})$	7.9×10^{-25}	[6, 7]	TlF:	6×10^{-23}	[31]
C_S	5.2×10^{-8}	[32]	Tl:	2.4×10^{-7}	[33]
C_P	5.1×10^{-7}	[32]	TlF:	3×10^{-4}	[34]
C_T	1.5×10^{-9}	[32]	TlF:	4.5×10^{-7}	[34]
$\bar{\theta}_{QCD}$	3×10^{-10}	[6, 30, 35]	n:	1×10^{-10}	[1]
$d_n(e\,\mathrm{cm})$	5.8×10^{-26}	[6, 7]	n:	2.9×10^{-26}	[1]
$d_e(e\,\mathrm{cm})$	3×10^{-27}	[36, 37]	Tl:	1.6×10^{-27}	[38]

[a] For ^{199}Hg: $\tilde{d}_q = (\tilde{d}_u - \tilde{d}_d)$, while for n: $\tilde{d}_q = (0.5\tilde{d}_u + \tilde{d}_d)$.

bottom cell, which gives an artificial EDM-like signal of size δ, randomly generated between $\pm 2\times 10^{-28}\,e\,\mathrm{cm}$ (the 2001 upper bound). This range is large enough to insure the analysis is blind, but small enough to reveal any large spurious signals that might appear due to the changes made when the blind analysis began. Once selected, the blind offset can remain fixed throughout a number of data sequences, and therefore will not interfere with tests for systematic effects (e.g. correlations with leakage currents, sparks, etc) in which different sequences are compared. For example, a look back at Fig. 16.6 demonstrates that, while the EDM-combination data have a common offset for the 4 sequences shown, the change associated with sequence 9 is evident.

16.4.4. *Recent resuslt*

The most recent result for the 4-cell version of the ^{199}Hg EDM experiment is [2]:

$$d(^{199}\mathrm{Hg}) = (0.49 \pm 1.29_{stat} \pm 0.76_{syst}) \times 10^{-29} e\ \mathrm{cm}\,. \tag{16.5}$$

The main contributions to the systematic error are the leakage current error and the spark analysis error discussed in Section 16.4.3, and in addition a contribution from analysis of correlations between the EDM signal and a large number of experimental parameters. This result gives an upper bound,

$$|d(^{199}\text{Hg})| < 3.1 \times 10^{-29}\ e\,\text{cm}\ \ (95\%\,\text{confidence level})\,, \tag{16.6}$$

which improves over the 2001 limit by a factor of 7 and provides correspondingly tighter constraints on CP-violating parameters as shown in Table 16.3. The parameters are defined in Section 16.3.3. Of special note is the improved limit on the quark chromo-electric dipole moment, $\tilde{d}_q$, which has major implications for new sources of CP violation in supersymmetry and other theories [1]. Additional upgrades to the apparatus are expected to lead to another factor of 3 to 5 improvement in sensitivity.

Acknowledgments

The authors wish to thank colleagues Blayne Heckel, Tom Loftus, and Mike Romalis for information and discussions. This work was supported by NSF Grant PHY 0457320. Fig. 16.4 is taken from Ref. [39], copyright World Scientific, and is reproduced with permission from World Scientific.

References

[1] M. Pospelov and A. Ritz, *Annals of Physics* **318**, 119 (2005).
[2] W. C. Griffith, M. D. Swallows, T. H. Loftus, M. V. Romalis, B. R. Heckel and E. N. Fortson, *Physical Review Letters* **102**, 101601 (2009).
[3] L. I. Schiff, *Physical Review* **132**, 2194 (1963).
[4] V. V. Flambaum and V. G. Zelevinsky, *Physical Review C* **68**, 035502 (2003).
[5] J. Engel, M. Bender, J. Dobaczewski, J. H. de Jesus and P. Olbratowski, *Physical Review C* **68**, 025501 (2003).
[6] V. A. Dzuba, V. V. Flambaum, J. S. M. Ginges and M. G. Kozlov, *Physical Review A* **66**, 012111 (2002).
[7] V. F. Dmitriev and R. A. Sen'kov, *Physical Review Letters* **91**, 212303 (2003).
[8] C. A. Baker *et al.*, *Physical Review Letters* **97**, 131801 (2006).
[9] V. V. Flambaum, I. B. Khriplovich and O. P. Sushkov, *Physics Letters* **162B**, 213 (1985).
[10] M. Pospelov, *Physics Letters B* **530**, 123 (2002).
[11] S. K. Lamoreaux, J. P. Jacobs, B. R. Heckel, F. J. Raab and N. Fortson, *Physical Review Letters* **59**, 2275 (1987).
[12] J. P. Jacobs, W. M. Klipstein, S. K. Lamoreaux, B. R. Heckel and E. N. Fortson, *Physical Review Letters* **71**, 3782 (1993).
[13] J. P. Jacobs, W. M. Klipstein, S. K. Lamoreaux, B. R. Heckel and E. N. Fortson, *Physical Review A* **52**, 3521 (1995).

[14] M. V. Romalis, W. C. Griffith, J. P. Jacobs and E. N. Fortson, *Physical Review Letters* **86**, 2505 (2001).
[15] M. V. Romalis and M. P. Ledbetter, *Physical Review Letters* **87**, 067601 (2001).
[16] M. P. Ledbetter, *Progress Toward a Search for a Permanent Electric Dipole Moment in Liquid* ^{129}Xe, Ph.D. thesis, Princeton University (2005).
[17] M. V. Romalis and E. N. Fortson, *Physical Review A* **59**, 4547 (1999).
[18] C. Chin, V. Leiber, V. Vuletic, A. J. Kerman and S. Chu, *Physical Review A* **63**, 033401 (2001).
[19] J. R. Guest *et al.*, *Physical Review Letters* **98**, 093001 (2007).
[20] E. R. Tardiff *et al.*, *Nuclear Instruments and Methods in Physics Research A* **579**, 472 (2007).
[21] Y. F. Orlov, W. M. Morse and Y. K. Semertzidis, *Physical Review Letters* **96**, 214802 (2006).
[22] T. N. Mukhamedjanov and O. P. Sushkov, *Physical Review A* **72**, 034501 (2005).
[23] W. E. Bell and A. L. Bloom, *Physical Review Letters* **6**, 280 (1961).
[24] S. K. Lamoreaux, in *Particle Astrophysics, Atomic Physics and Gravitation: Proceedings of the XXIXth Recontre de Moriond*, ed. J. Tran Thanh Van, G. Fontaine, and E. Hinds, (Editions Frontieres, Gif-sur-Yvetter, France 1994) p. 271.
[25] M. V. Romalis and L. Lin, *Journal of Chemical Physics* **120**, 1511 (2004).
[26] S. K. Lamoreaux and E. N. Fortson, *Physical Review A* **46**, 7053 (1992).
[27] K. Beloy, V. A. Dzuba and A. Derevianko, *Physical Review A* **79**, 042503 (2009).
[28] B. C. Regan, E. D. Commins, C. J. Schmidt and D. DeMille, *Physical Review Letters* **88**, 071805 (2002).
[29] D. Cho, K. Sangster and E. A. Hinds, *Physical Review A* **44**, 2783 (1991).
[30] J. H. de Jesus and J. Engel, *Physical Review C* **72**, 045503 (2005).
[31] A. N. Petrov, N. S. Mosyagin, T. A. Isaev, A. V. Titov, V. F. Ezhov, E. Eliav and U. Kaldor, *Physical Review Letters* **88**, 073001 (2002).
[32] J. S. M. Ginges and V. V. Flambaum, *Physics Reports* **397**, 63 (2004).
[33] B. K. Sahoo, B. P. Das, R. K. Chaudhuri, D. Mukherjee and E. P. Venugopal, *Physical Review A* **78**, 010501 (2008).
[34] I. P. Khriplovich and S. K. Lamoreaux, *CP Violation Without Strangeness* (Springer, Berlin, 1997).
[35] R. J. Crewther, P. Di Vecchia and G. Veneziano, *Physics Letters* **88B**, 123 (1979), 91B:487(E), 1980.
[36] V. V. Flambaum and I. B. Khriplovich, *Soviet Physics – JETP* **62**, 872 (1985).
[37] A.-M. Martensson-Pendrill and P. Öster, *Physica Scripta* **36**, 444 (1987).
[38] Z. W. Liu and H. P. Kelly, *Physical Review A* **45**, R4210 (1992).
[39] M. D. Swallows *et al.*, in *Proceedings from the Institute for Nuclear Theory - Vol. 16: Rare isotopes and fundamental symmetries*, ed. B.A. Brown, Jonathan Engel, W. Haxton, M. Ramsey-Musolf, M. Romalis, and G. Savard, World Scientific, 2009.

Chapter 17

Search for a Permanent EDM of Charged Particles Using Storage Rings

B. Lee Roberts and James P. Miller

Department of Physics, Boston University
Boston, MA 02215, USA
roberts@bu.edu; miller@bu.edu

Yannis K. Semertzidis

Department of Physics, Brookhaven National Laboratory
Upton, NY 11973-5000, USA
yannis@bnl.gov

In a storage ring it is possible to search for a permanent electric dipole moment of a charged particle. While direct EDM searches have been carried out on the neutron, the limits on the EDM of the electron, and the ^{199}Hg atom have been carried out on atomic systems, as discussed in other articles in this volume. In this chapter we describe the storage ring technique to search for the EDM of charged particles. This technique was first used to place a limit on the EDM of the muon, which made use of the motional electric field $E_m \propto \vec{\beta} \times \vec{B}$ felt by a muon circulating in a muon $(g-2)$ storage ring. This technique can be expanded by the "frozen spin" method which would significantly reduce the systematic errors of the past muon experiments, and provide a powerful new technique to search for an EDM of charged particles such as the proton and deuteron.

Contents

17.1. Introduction

In his famous paper on the relativistic theory of the electron, Dirac [1] first mentioned the possibility of an electric dipole moment, which like the magnetic dipole moment would be directed along the electron spin direction. The magnetic dipole (MDM) and electric dipole (EDM) moments are given by

$$\vec{\mu} = g_s \left(\frac{q}{2m}\right) \vec{s}; \qquad \vec{d} = \eta \left(\frac{q}{2mc}\right) \vec{s}. \tag{17.1}$$

The quantity η in the EDM expression is analogous to the g value for the magnetic dipole moment. As discussed in Chapter 1, the presence of an EDM violates both P and T symmetries.

EDM searches started with the suggestion by Purcell and Ramsey [2] that a permanent EDM of the neutron would show parity violation in nuclear interactions. Because of the difficulty of placing a charged particle in an electric field region for an adequate time to be sensitive to its EDM, they proceeded with the neutron. Their subsequent experiment reached a sensitivity of 10^{-20} e-cm [3].

EDM searches have been carried out almost continuously since the 1950s, with significant progress during each decade. In the 1950s the first neutron EDM experiment was carried out at Oak Ridge National Laboratory. In the 1960s EDM searches in atomic systems were proposed and begun. In the 1970s the storage-ring EDM method was applied for the first time, and a limit on the muon EDM was obtained. In the 1980s theoretical studies on molecules with large enhancement factors were carried out. In the 1990s the first significant experimental attempts to search for an EDM with molecules began, and independently the dedicated storage-ring EDM method was developed. In the present decade, next-generation neutron EDM experiments are going forward in three separate laboratories, and the proposal to search for the deuteron EDM in a dedicated storage ring received scientific approval at the Brookhaven National Laboratory (BNL).

Thus every decade has seen major advances in developing more sensitive EDM methods for both hadronic and leptonic systems. EDM limits currently set the limits on many beyond the Standard Model (SM) models, e.g. they set the most strict limitations on SUSY parameters.

The important elements in an EDM experiment are:

(1) *Polarization:* Preparation of the system of interest with a well defined spin direction with as high intensity and polarization as possible.
(2) *Interaction with an electric field:* The effective electric field needs to be the highest possible for the longest possible time, thus requiring long spin coherence times (SCT).
(3) *Analysis of the spin direction:* A high-efficiency analyzer with large analyzing power is needed to observe the spin evolution with time.
(4) *Interpretation of the result.* This can be difficult, or require additional theory in the case of atomic or molecular systems.

As an example, measuring the EDM of the neutron involves the presence of both an electric (E) and magnetic (B) field, parallel to each other. The interaction energy is given by

$$H = -\vec{\mu} \cdot \vec{B} - \vec{d} \cdot \vec{E}. \tag{17.2}$$

The spin precession frequency is compared with the $\vec{E}$-field parallel to $\vec{B}$, (ω_1), and anti-parallel, (ω_2):

$$\hbar\omega_1 = 2\mu B + 2dE\,, \quad \hbar\omega_2 = 2\mu B - 2dE\,. \tag{17.3}$$

The EDM is determined from the difference between these two frequencies,

$$d = \frac{\hbar(\omega_1 - \omega_2)}{4E}\,, \tag{17.4}$$

which for an EDM value of $d_n = 10^{-28}$ e-cm and electric field strength of $E = 100\text{kV/cm}$ would result in a frequency change of $\delta\omega = 6 \times 10^{-8}\text{rad/s}$.

Three types of EDM experiments are presently underway: searches for a neutron EDM; searches for a permanent EDM in atoms or molecules; and the search for a permanent EDM of a *charged particle* using polarized particles trapped in a storage ring. The first two techniques are covered in other articles in this volume.

In this chapter we present the storage ring technique of measuring an EDM. The strong *motional* electric field present in the rest frame of relativistic particles circulating in a magnetic storage ring provides a new and unique tool to search for an EDM of a charged particle. The combination of electric and magnetic fields in a dedicated storage ring will permit measurements of the EDMs of the deuteron, proton, and the muon.

17.2. Motivation for EDM Searches

The physics at the frontier of science is accomplished by pursuing two different approaches: The energy frontier, and the precision frontier. The energy frontier is moving soon from the Fermilab Tevatron collider to the Large Hadron Collider (LHC) at CERN. The LHC, with a mass scale reach of about 1 TeV has the potential to discover the Higgs particle, and possibly new physics like supersymmetry (SUSY) and/or extra dimensions. The precision frontier provides a complementary approach in the search for physics beyond the Standard Model, which in many cases has a sensitivity orders of magnitude beyond that accessible using the direct approach. Certainly if New Physics is discovered at LHC, the precision experiments will play a significant role in constraining the interpretation of these new results.

The deuteron EDM experiment discussed below has a physics reach of 300 TeV or, if there is New Physics at the LHC scale, probing *CP*-violating phases in the models of New Physics at the level of 10μrad, an unprecedented sensitivity level. Thus far, *CP* violation has only been seen in neutral kaon and B meson decays, and it is well described phenomenologically by the existence of one *CP*-violating phase in the CKM mixing matrix. Since EDMs are not very sensitive to this source of *CP* violation, any observation of an EDM of a fundamental particle would either mean the existence of physics beyond the SM, or for baryons the existence of *CP* violation through the θ term in the standard-model QCD Lagrangian.

A non-zero permanent EDM violates both the time (T) and parity (P) symmetries. This is evident when one considers the interaction energy (H given in Eq.(17.2). The magnetic moment interaction is *even* under C, P and T symmetries. On the other hand, the electric field term is only even under C, and has the transformation properties:

$$\mathcal{P}(-\vec{d}\cdot\vec{E}) \to +\vec{d}\cdot\vec{E} \tag{17.5}$$

and

$$\mathcal{T}(-\vec{d}\cdot\vec{E}) \to +\vec{d}\cdot\vec{E}\,. \tag{17.6}$$

Thus the EDM interaction is odd under both of these symmetries, and if the EDM exists, T, P are not good symmetries of the interaction Hamiltonian Eq. (17.2).

Induced EDMs *are* permitted because they are proportional to the applied electric field, $\vec{d}_{ind} = d_{ind}\vec{E}$:

$$\mathcal{T}(-d_{ind}\vec{E}\cdot\vec{E}) \to -d_{ind}\vec{E}\cdot\vec{E}\,, \tag{17.7}$$

and

$$\mathcal{P}(-d_{ind}\vec{E}\cdot\vec{E}) \to -d_{ind}(-\vec{E})\cdot(-\vec{E})\,, \tag{17.8}$$

showing that induced EDMs do not violate P or T.

The T-violation when combined with the the assumption of conservation of the combined *CPT* symmetries implies *CP* violation. *CP* violation is very important because it is one of the three conditions required to enable the universe containing equal amounts of matter and anti-matter to evolve into the matter-dominated universe we observe today [4].

CP violation was first discovered [5] in the neutral kaon system at the Brookhaven Laboratory in 1964, and more recently in the B-system at SLAC and KEK. However, the *CP* violation incorporated into the CKM matrix that describes the observed standard-model *CP* violation in the weak interactions is insufficient by nine orders of magnitude to account for the observed baryon asymmetry of our universe. The observed baryon number density over the observed photon number density is of order of 10^{-9}. Theoretical models based on the Standard Model *CP* violation produce an asymmetry of only 10^{-18}. Hence a new, stronger source of *CP* violation is required which might also produce EDMs at measurable levels. While Standard-Model CKM physics will produce a neutron EDM of $\simeq 10^{-32}$ e-cm, the present experimental limit of $d_n < 1.6\times10^{-26}$ e-cm leaves a rather large window through which to search for a new-physics contribution to the baryon EDMs.

Standard-Model extensions such as SUSY, Multi-Higgs, Left-Right Symmetric, etc., easily accommodate new sources of *CP* violation, and predict EDM values within the sensitivity of current or planned experiments. This possibility combined with the fact that the weak interaction *CP* violation predicts negligible EDMs, makes EDM searches an ideal place to search for non-CKM *CP* violation.

17.2.1. *Hadronic EDMs*

A significant EDM in hadrons can rise from various sources: Quark electromagnetic (EM) or Color (chromo) EDMs; and/or from the *CP*-violating parameter θ-QCD ($\bar{\theta}$). The first two contributions would have to be beyond the SM sources, e.g. SUSY, while the third one is part of the strong interactions within the SM.

The QCD Lagrangian includes a *CP*-violating parameter, θ-QCD:

$$\mathcal{L}_{CPV} = \bar{\theta}\frac{\alpha_s}{8\pi}G\tilde{G} \tag{17.9}$$

from which we can estimate the neutron EDM within an order of magnitude

$$d_n(\bar{\theta}) \approx \bar{\theta}\frac{e}{m_n}\frac{m_*}{\Lambda_{QCD}} \approx \bar{\theta} \cdot (5 \times 10^{-17})\, \mathrm{e} \cdot \mathrm{cm} \tag{17.10}$$

with

$$m_* = \frac{m_u m_d}{m_u + m_d} \tag{17.11}$$

the reduced mass of the up and down quarks. Λ_{QCD} is the QCD scale and m_n the neutron mass. When the estimation is done more precisely (Refs. [6, 7] and Chapter 13) it becomes

$$d_n(\bar{\theta}) \approx \bar{\theta} \cdot (3.6 \times 10^{-16})\, \mathrm{e} \cdot \mathrm{cm}. \tag{17.12}$$

The present neutron EDM limit [12] of 2.9×10^{-26} e-cm results in a limit on theta-QCD: $\bar{\theta} \leq 10^{-10}$. It is estimated [6, 7, 10] that the deuteron EDM has one third the neutron sensitivity (for the same nominal EDM limit) to θ-QCD and at 10^{-29} e-cm the deuteron would be sensitive down to $\bar{\theta} \leq 10^{-13}$.

On the other hand the quark EM and Color (chromo) EDM Lagrangian is

$$\mathcal{L}_{CPV} = -\frac{i}{2}\sum_q \bar{q}\left(d_q \sigma_{\mu\nu} F^{\mu\nu} + d_q^c \sigma_{\mu\nu} G^{\mu\nu}\right)\gamma_5 q \tag{17.13}$$

and the neutron and deuteron EDM values are [6, 7, 10]

$$d_n \approx 1.4(d_d - 0.25 d_u) + 0.83e(d_d^c + d_u^c) + 0.27e(d_d^c - d_d^c) \tag{17.14}$$

and

$$d_D \approx (d_d + d_u) - 0.2e(d_d^c + d_u^c) + 6e(d_d^c - d_d^c)\,; \tag{17.15}$$

i.e., the deuteron and neutron EDM are different combinations of quark- and chromo-EDMs, and thus complementary. Regarding the isovector part of the quark-chromo EDM, the deuteron has 20 times the neutron sensitivity. This has to do with the special structure of the deuteron where a neutron and proton are held together by T-odd nuclear forces, as shown in Fig. 17.1.

Suppose the neutron EDM experiments discover a non-zero EDM value, let's say at 10^{-28} e-cm, if the source is θ-QCD the expected deuteron EDM value would be $d_D \approx 3 \times 10^{-29}$ e-cm. However, if SUSY is the EDM source and in particular the isovector part of the interaction, then the expected value would be $d_D \approx 2 \times 10^{-27}$ e-cm. In Chapter 2, Czarnecki and Marciano make the point that the neutron, proton and deuteron experiments,

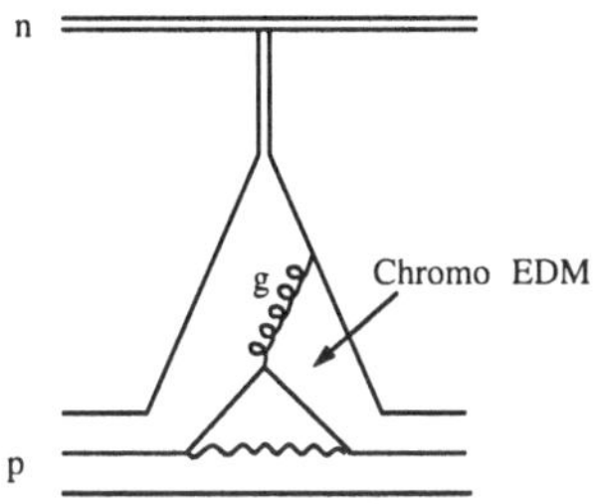

Fig. 17.1. The T-odd nuclear forces (shown here as exchange of a pion) between the proton and neutron constituents of the deuteron nucleus is shown here. The loop shown in the bottom may include SUSY particles carrying *CP*-violating phases.

together with an EDM sensitivity of 10^{-28} *e*-cm each, can pinpoint the *CP*-violating source should a non-zero EDM be observed.

There are three main physics reasons to carry out a deuteron EDM experiment (dEDM) at the 10^{-29} *e*-cm level. The present sensitivity on $\bar{\theta}$ is $\bar{\theta} \leq 10^{-10}$, which would become $\bar{\theta} \leq 10^{-13}$ with the proposed dEDM measurement. Such an experiment would have a sensitivity to new contact interactions at the 3000 TeV level. Furthermore there is a sensitivity to SUSY-type New Physics,

$$dEDM \approx 10^{-24}\,\mathrm{e\cdot cm}\ \times \sin\delta \times \left(\frac{1\,\mathrm{TeV}}{M_{\mathrm{SUSY}}}\right)^2 , \qquad (17.16)$$

where δ is a *CP*-violating phase. A deuteron EDM measurement at 10^{-29} *e*-cm sensitivity has a reach of about 300 TeV for SUSY-type New Physics or, if New Physics exists at the LHC scale, it has significant sensitivity to a *CP*-violating phase in models such as supersymmetry of 10^{-5} rad.

Other hadronic systems under study are the ^{199}Hg and the ^{129}Xe atoms. However, due to the Schiff shielding of the nucleus by the atomic electrons, their sensitivities to nuclear EDMs are significantly reduced (see Chapter 16). Table 17.1 shows the current limit, future goal and the neutron equivalent of the future goal [11]. The physics reach of the various hadronic systems depends on the underlying *CP*-violating source: such as θ-QCD, quark electro-magnetic or quark-color EDM. Different systems have different sensitivities to various combinations of *CP*-violating sources. In Table 17.1 we give the range of physics reach as neutron equivalent, i.e. what the neutron EDM experimental sensitivity should be to match the same physics reach.

Thus searches for the deuteron and proton EDMs are complementary to the neutron EDM ones, and under certain circumstances (isovector part of

Table 17.1. The physics strength comparison for a few hadronic EDM systems showing the current limit, future goal and the neutron equivalent of the future goal all in (e-cm) units.

System	Current limit	Future goal	Neutron equivalent
Neutron [12]	$< 2.8 \times 10^{-26}$	10^{-28}	10^{-28}
^{199}Hg atom [13]	$< 3.1 \times 10^{-29}$	$\leq 10^{-29}$	$10^{-26} - 10^{-27}$
^{129}Xe atom [14]	$< 6.6 \times 10^{-27}$	$10^{-30} - 10^{-33}$	$10^{-26} - 10^{-29}$
Deuteron nucleus		10^{-29}	$3 \times 10^{-29} - 5 \times 10^{-31}$
Proton nucleus [13, 15]	$< 7.9 \times 10^{-25}$	1×10^{-29}	$4 \times 10^{-29} - 2.5 \times 10^{-30}$

the T-odd nuclear forces) the deuteron has better sensitivity to *CP* violation by an order of magnitude for the same nominal EDM value. Together the deuteron, proton and neutron can pinpoint the *CP*-violating source. The physics reach of the proposed deuteron and proton EDM measurements typically extends well beyond the LHC scale, and along with all of the present and proposed EDM experiments is complementary to it.

17.3. Storage Ring EDM Method

The detection of an EDM requires measuring the interaction of the EDM with an electric field which is as large as possible. It is a challenge to place a charged particle in an electric field for an extended period of time, since it will be accelerated away. The storage ring method circumvents this difficulty; the Lorentz force that holds the particle in circular motion is accompanied by a large motional electric field, $\vec{\beta} \times \vec{B}$, in the particle rest frame. This motional field can be equal to or greater than electric fields obtainable in the laboratory [9]. With a storage ring, it is possible to search directly for an EDM of the proton and deuteron, although the storage rings needed might be quite different. The muon presents a unique opportunity to search for an EDM in a second-generation particle using (yet a different) dedicated storage ring. We consider each of these options below.

17.3.1. *The search for an EDM of the muon*

To understand how to measure an EDM in a storage ring, it is necessary to understand how the muon $(g-2)$ experiment works. First we review the muon $(g-2)$ experiment described in Chapter 11 and emphasize the role of the focusing electric field. We discuss how the spin equation is modified by the electric field, and then expand this equation to include the presence of a muon electric dipole moment as well as the magnetic dipole moment.

The first element of the measurement is the production of a polarized beam of muons by the pion decay, $\pi^- \rightarrow \mu^- + \bar{\nu}_\mu$ in flight. Since the antineutrino is right-handed, the muon is left-handed to conserve angular momentum. The pions decay isotropically in the pion rest frame, but in a moving beam of pions, the highest-energy (forward) muons, or the lowest-energy (backward) muons are highly polarized. By selecting the highest- or lowest-energy muons, a polarized beam of muons can be produced.

The spin precession in a magnetic field for a muon with a magnetic moment (see Eq.(17.2)) (in SI units) is given by

$$\vec{\omega}_s = -g\frac{q\vec{B}}{2m} - \frac{q\vec{B}}{\gamma m}(1-\gamma), \tag{17.17}$$

where $\gamma = (1-\beta^2)^{-\frac{1}{2}}$, $\beta = v/c$, the muon charge is $q = \pm e$, with e a positive number. For a non-relativistic particle, $\gamma \rightarrow 1$ and the second term vanishes. If the particle is stored in a magnetic storage ring its cyclotron angular frequency is given by

$$\omega_c = -\frac{qB}{\gamma m}, \tag{17.18}$$

which is the frequency that the momentum vector goes around in a circle.

The difference between the spin precession rate and the cyclotron precession rate, namely the rate that the spin turns *relative to the momentum vector*, is

$$\omega_a = \omega_s - \omega_c = -\left(\frac{g-2}{2}\right)\frac{qB}{m} = -a\frac{qB}{m} \tag{17.19}$$

where a is the anomalous magnetic moment (anomaly) of the particle. Note that the expression for ω_a is the same for the non-relativistic case, $\gamma \rightarrow 1$, and for any other value of γ. Because $g > 2$, the spin will advance faster than the momentum vector, which is shown in cartoon form in Fig. 17.2. The frequency, ω_a is *independent* of the momentum, for a specific particle, only depending on the magnitude of the magnetic field. As we will see below, when an externally applied electric field is involved, the spin motion can depend strongly on momentum.

The muon lifetime at rest is about $\tau = 2.2\,\mu$s, and is boosted to $\gamma\tau$, with γ the Lorentz relativistic factor. The muon decays through the parity violating weak force to an electron and two neutrinos (see Chapter 11, Section 11.2.1). In the muon rest frame, the energy of the electron is highest when the neutrino and antineutrino are emitted parallel to each other, with opposite helicities, and antiparallel to the direction of the electron.

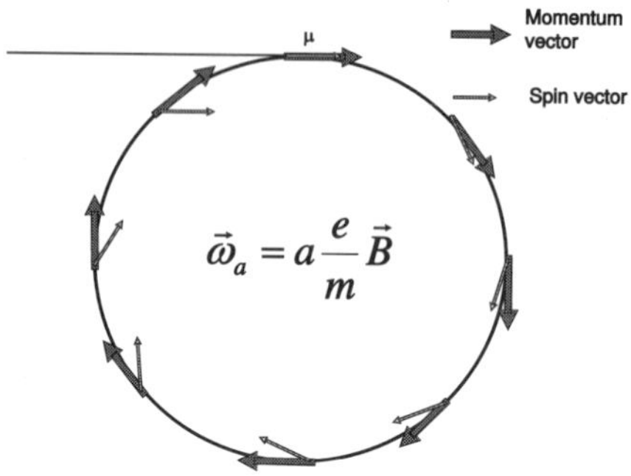

Fig. 17.2. A longitudinally polarized muon beam is injected into the ring. When the beam enters the storage ring the spin and momentum are aligned. As the particle travels around the ring, the spin vector advances ahead of the momentum vector as a function of time according to Eq.(17.19)

By conservation of angular momentum, the electron will have the same spin direction as the muon. If the electron were a zero mass particle, its helicity would be left-handed. Since the electron is a very light particle, in the $V - A$ decay of the electron, production of a left-handed electron is more probable than a right-handed one. (Recall that neutrinos, which have almost zero mass are almost purely left-handed.) Therefore, in the limit when a maximum energy decay electron is produced in the muon rest frame, it is more likely directed anti-parallel to the muon spin. The opposite is true for the μ^+.

In the lab frame, the energy of the electron is largest when its CM energy is maximum and when its direction is parallel to the muon lab momentum. There are more of these electrons when the spin is antiparallel to the muon momentum than when it is parallel – thus the number of high-energy electrons from muon decay oscillates at the spin precession frequency, Eq. (17.19), where the number of *high-energy electrons* above an energy threshold of E_{th} as a function of time is given by

$$N(t, E_{th}) = N_0(E_{th})e^{-t/\gamma\tau}[1 + A(E_{th})\cos(\omega_a t + \phi(E_{th}))]. \qquad (17.20)$$

Through this weak decay process, the spin direction at the decay time can be determined.

The simplicity of Eq.(17.19) has a lot to do with the success of the muon $(g-2)$ experiment (see Ref. [16] and Chapter 11).

The accuracy with which the anomalous magnetic moment can be determined depended only on the accuracy of the determination of precession frequency ω_a and the magnetic field, each quantity being averaged over the muon ensemble. The B-field is determined by NMR, therefore there is also a dependence on the ratio of the magnetic moments of the muon and proton. The average magnetic field determination is easier if the B-field where the muons circulate is as uniform as possible.

A storage ring without a field gradient does not have a good capture efficiency, and of course, the particles will travel in helical trajectories and quickly be lost. As explained in Chapter 11, an electric quadrupole field can be used to provide vertical focusing to the muon beam. When the E-field is included, Eq.(17.19) becomes

$$\vec{\omega}_a = -\frac{q}{m}\left[a\vec{B} + \left[a - \left(\frac{m}{p}\right)^2\right]\frac{\vec{\beta}\times\vec{E}}{c}\right] \tag{17.21}$$

when $\vec{\beta}\cdot\vec{B} = 0$, which is equivalent to Eq.(11.6) of Chapter 11. At the so called "magic" momentum $p = m/\sqrt{a} = 3.1$ GeV/c the effect of the E-field on the muon spin and momentum are equal and cancel.

This technique produced a 7.3 ppm (part per million) measurement of the muon anomaly at CERN [18] and a 0.54 ppm measurement at Brookhaven. The small uncertainty obtained at BNL was possiblly due to the ability to store a substantial number of muons in a storage ring with a magnetic field uniformity of ± 1 ppm, when averaged over azimuth. Essential to the measurement was the fact that at the "magic" muon momentum the effect of the motional magnetic field, $\vec{\beta}\times\vec{E}$ on the spin motion relative to the momentum cancels. This cancellation is easy to understand with a simple example. Suppose that the $\vec{B}$-field is up, the charge of the muon is +, and the $\vec{E}$-field is directed radially outward. The E-field will decrease the cyclotron frequency. In general, transformation of the E-field from the lab frame to the particle rest frame will lead to both magnetic and electric fields. The B-field in the muon rest frame due the lab frame E-field is proportional to $-\vec{\beta}\times\vec{E}$; this is the so-called "motional B-field". At low β, the electric field does not contribute an appreciable magnetic field in the particle rest frame and produces little change in ω_s. Therefore, $\omega_a = \omega_s - \omega_c$ increases when such an E-field is applied. As β increases, $\vec{\beta}\times\vec{E}$ increases, in our example leading to a decrease in ω_s which can be larger than the decrease in ω_c, leading to a decrease in ω_a. There is a particular momentum, the magic momentum, where the effect on ω_a is zero.

17.3.1.1. *Effect of a radial electric field on spin precession*

Consider the general case where the momentum is fairly far removed from the magic momentum. In contrast to a magnetic field orthogonal to the momentum vector, where the $(g-2)$ precession rate is independent of the muon relativistic γ-factor, the radial electric field effect on the $(g-2)$ precession rate is strongly dependent on it. This is a purely relativistic effect, the radial E-field is partially transformed into a magnetic field in the muon's rest frame depending on the muon's velocity.

While the last two muon $(g-2)$ experiments [16, 18] were operated at the muon "magic" momentum of 3.1 GeV/c, the finite muon momentum spread introduces a small correction of order 0.5 ppm, with a negligible uncertainty [16, 20] that must be applied to the experimental value obtained from the observed muon $(g-2)$ frequency. If the experiments were performed at a momentum not equal to the "magic" one, the electric field correction would be large, with an uncertainty that would be significant compared with the other errors in the experiment.

Just as an electric field can affect the spin motion of a relativistic muon through the motional magnetic field $\vec{B}_m \propto \vec{\beta} \times \vec{E}$, if the muon possesses an electric dipole moment the motional electric field $\vec{E}_m \propto \vec{\beta} \times \vec{B}$ will create a torque on the electric dipole $\vec{N} = \vec{d} \times \vec{E}_m$ that introduces a spin precession about the motional electric field $\vec{E}_m$.

The spin precession frequency given in Eq.(17.21) must be modified to account for this additional torque. If a static electric field is present, one obtains an additional term $\vec{d} \times \vec{E}$. The net result of B- and E-fields on the spin precession if the muon has both a magnetic and electric dipole moment is (to first order) in the lab frame

$$\vec{\omega}_{a\eta} = -\frac{q}{m}\left[a\vec{B} + \left(a - \left(\frac{m}{p}\right)^2\right)\frac{\vec{\beta} \times \vec{E}}{c}\right] - \eta\frac{q}{2m}\left[\frac{\vec{E}}{c} + \vec{\beta} \times \vec{B}\right] \quad (17.22)$$

where η plays the same role for the EDM as the g-factor plays in the magnetic dipole moment, and it is equal to

$$\eta = \frac{m}{e}\frac{4dc}{\hbar} \text{ for spin } \frac{1}{2}\,; \text{ and } \eta = \frac{m}{e}\frac{2dc}{\hbar} \text{ for spin } 1\,. \quad (17.23)$$

The first term in square brackets results from the torque on the magnetic dipole moment from the static and motional *magnetic* fields, while the second term comes from the torque on the electric dipole moment from the static and motional *electric* fields.

At the magic momentum, Eq.(17.22) becomes

$$\vec{\omega}_{a\eta} = -\frac{q}{m}\left[a\vec{B} + \frac{\eta}{2}\left(\frac{\vec{E}}{c} + \vec{\beta} \times \vec{B}\right)\right], \quad (17.24)$$

which if $|\vec{E}| \ll c|\vec{\beta} \times \vec{B}|$, becomes

$$\vec{\omega}_{a\eta} \simeq -\frac{q}{m}\left[a\vec{B} + \frac{\eta}{2}\left(\vec{\beta} \times \vec{B}\right)\right]. \tag{17.25}$$

The observed frequency $\vec{\omega}$ is the vector sum of two orthogonal angular frequencies, $\vec{\omega}_{a\eta} = \vec{\omega}_a + \vec{\omega}_\eta$. The first term comes from the anomalous magnetic moment, a, and the second from the electric dipole moment. These two frequencies are shown in Fig. 17.3, where the EDM related frequency ω_η is greatly exaggerated.

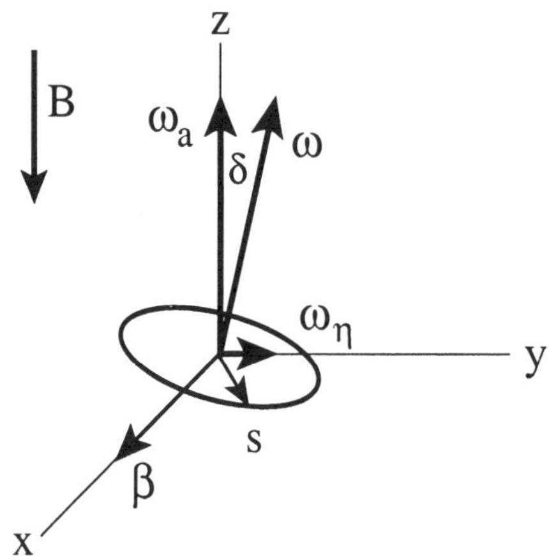

Fig. 17.3. The two frequencies present if the muon has both a magnetic and electric dipole moment (not to scale). Note that the EDM ω_η is much smaller than ω_a. The muon spin precession plane is tilted by an angle proportional to the particle's EDM value. The tilt is highest for small $(g-2)$ frequencies.

Thus there are *two* effects due to an electric dipole moment. The observed frequency is the vector sum of ω_a and ω_η so the magnitude of the observed frequency is increased from ω_a to

$$\omega_{a\eta} = \omega_a \sqrt{1 + \left(\frac{\eta\beta}{2a}\right)^2} \tag{17.26}$$

and, the spin precession plane is tilted by a (very small) angle

$$\delta = \tan^{-1}\frac{\omega_\eta}{\omega_a} = \tan^{-1}\left(\frac{\eta\beta}{2a}\right) \tag{17.27}$$

as shown in Fig. 17.3.

Thus the spin precession plane is tilted everywhere around the ring, very much like there is a net radial magnetic field which when integrated around the ring is not zero. In a ring with a purely magnetic field, the average radial B-field for a stored particle is zero, since the particle adjusts its vertical position in the focusing system to ensure this. However, in the presence of other forces, like vertical E-fields, gravity, etc., this is not strictly true and must be taken into account for systematic error estimation. A major tool against these types of systematic errors would be the ability to inject into the storage ring both in a clockwise (CW) and counter-clockwise (CCW) sense, where the non-magnetic forces are kept the same while the EDM signal changes sign.

The tipping of the plane of precession results in an up-down oscillation of the muon spin which is out of phase by $\pi/2$ with the $(g-2)$ precession. It was this effect which was searched for in the third $(g-2)$ experiment at CERN, and in E821 at Brookhaven. At CERN one detector station was outfitted with two scintillators, one just above the mid-plane, one just below.

Assuming the gain and acceptance of the upper and lower detectors are equal and the storage ring and vertical detector mid-plane are identical, the number of electrons above (+) or below (-) the mid-plane is given by [17]

$$N^{\pm}(t) \propto [1 \mp A_{\eta} \sin(\omega t + \phi) + A_{\mu} \cos(\omega t + \phi)] \tag{17.28}$$

where A_{η} is proportional to d_{μ}. A major source of systematic error arises if there is an offset between the average vertical position of the beam and the position of the boundary between the upper and lower detectors.

In E821, three separate methods were used to search for the up-down oscillations [17]. Five-element hodoscopes were placed in front of about half of the 24 electron calorimeters, and the vertical centroid of the decay electron distribution was fit as a function of time. Five calorimeter stations had finer-grained hodoscopes which also provided the vertical electron distribution of decay electrons as a function of time. One of the stations was equipped with a straw tube array that gave both x and y information, so that the electron tracks could be fit. These "traceback" chambers were primarily designed to provide information on the muon distribution in the storage ring [17], but turned out to be a powerful tool to search for the EDM signal. No evidence for an up-down oscillation was seen, and the result is [17]

$$d_{\mu} = (0.1 \pm 0.9) \times 10^{-19} e\text{–cm};\ |d_{\mu}| < 1.9 \times 10^{-19} e\text{–cm}\ (95\%\ \text{C.L.})\,, \tag{17.29}$$

a factor of five smaller than the previous limit.

Since the traceback chambers provide the average vertical angle, they avoid much of the systematic error due to vertical misalignment between the beam and the detectors, and therefore they potentially provide a much more sensitive way to improve the measurement at the next level. If a new $(g-2)$ experiment is done, either at Fermilab or at J-PARC, many detector stations should be equipped with a traceback system to improve on the EDM limit, perhaps 12 to 16 out of 24 detector stations. A simple estimate shows one to two orders of magnitude improvement might be possible, and is being studied in the preparation of a proposal to Fermilab.

17.3.2. *Frozen spin method and d_μ*

It is clear from Fig. 17.3 that the EDM signal is very difficult to see. The problem is that the very small EDM signal is masked by the large $(g-2)$ precession, introducing very large systematic errors into the measurement [17]. If ω_a were close to zero, then the motional electric field would cause the spin to rise steadily out of the plane instead of oscillating up and down, with the net effect being a large amplification of the signal.

The idea of the frozen spin method is to employ the proper combination of radial electric field and vertical magnetic field to cancel the spin precession due to the magnetic moment (see Eq.(17.21)). As emphasized above, the momentum is chosen for the E821 measurement so that the electric field has no effect on the spin, so clearly a different momentum choice is required. If the beam momentum were to be lowered, a radial electric field could be arranged such that the spin precession from the magnetic moment interaction could be canceled [8], and the spin *"frozen"*. If the radial field is used to cancel the $(g-2)$ precession, then the applied electric field $\vec{E}$ and the motional electric field $\vec{E}_m \propto \vec{\beta} \times \vec{B}$ are parallel, and ω_η is radial, transverse to both $\vec{B}$ and $\vec{\beta}$. The E-field required to freeze the muon spin is

$$E = \frac{aBc\beta\gamma^2}{1 - a\beta^2\gamma^2} \simeq aBc\beta\gamma^2. \qquad (17.30)$$

The torque on the EDM will cause the spin to move out of the plane of the storage ring, so one measures an up-down asymmetry in the decay electrons proportional to the EDM which will steadily build up with time. It would be measured using detectors placed above and below the stored beam.

Use of the frozen spin method to measure the muon EDM was discussed in Ref. [9] using $p_\mu = 500$ MeV/c. More recently, a suggestion by Adelmann et al., [21] proposed using a small ring with a 0.42 m radius, operating at $p_\mu = 125$ MeV/c ($\gamma = 1.57$). The uncertainty in η is given by

$$\sigma_\eta = \frac{2}{\gamma\tau(e/m)\beta BAP\sqrt{N}} = \frac{2ac\gamma}{\tau(e/m)EAP\sqrt{N}} \qquad (17.31)$$

where the right-hand expression uses the E-field expression in Eq. (17.30). The right-hand expression implies that low γ is preferred. However, the small γ experiment only works if there is no injection related flash in the EDM detectors, since the short lifetime would drastically reduce the size of the data sample if it were necessary to gate the detectors off at injection and wait some number of μs after injection to begin data collection. Adelmann *et al.* [21], get around this problem by injecting one muon at a time. If a bunched muon beam were to be available for this small ring, the muon intensity per fill will be limited by this constraint. The two suggestions are compared in Table 17.2.

Table 17.2. Muon storage ring parameters suggested in Refs. [9] and [21] for a muon EDM measurement. R_0 is the central radius of the storage ring, and r_0 is the beam aperture.

$\lvert\vec{E}\rvert$ MV/m	r_0 mm	$\lvert\vec{B}\rvert$ T	p_μ MeV/c	γ	$\gamma\tau$ μs	R_0 m	σ_{d_μ}	Ref.
2	100	0.25	500	5	11	7	$\simeq 2\times 10^{-16}N^{-\frac{1}{2}}$	[9]
0.64	20	1	125	1.57	3.5	0.42	$\simeq 1\times 10^{-16}N^{-\frac{1}{2}}$	[21]

As in all EDM experiments, systematic errors are of paramount importance. A careful discussion of types of systematic errors is given in Ref. [9] along with methods of canceling them. One powerful tool is to inject into the ring in clockwise, and then counterclockwise directions. We return to this topic below in the discussion of the deuteron EDM. It is important to understand the level to which the $(g-2)$ precession is canceled, so any experiment would need detectors in the mid-plane, as were used in the $(g-2)$ experiment, to determine when the spin is frozen, or to measure how well it is frozen.

Dedicated muon storage ring experiments have been discussed for the Paul Scherrer Instituter (PSI) [21], and the Japan Proton Accelerator Complex (J-PARC) [22]. The PSI experiment could reach a sensitivity of $\sigma_{d_\mu} \simeq 5\times 10^{-23}$ e-cm in one year of operation. The J-PARC sensitivity could potentially reach the $< 10^{-24}$ e-cm level. At a very intense muon source, such as at the front-end of a neutrino factory, one could probe yet another order of magnitude or so in sensitivity. We give a schematic of sensitivities as a function of year in Fig. 17.4.

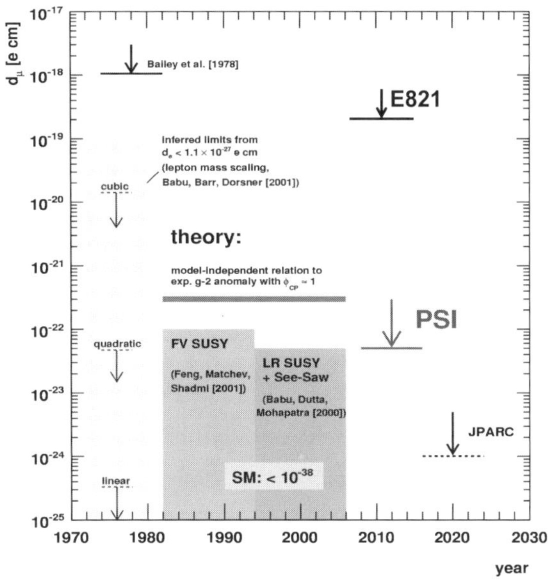

Fig. 17.4. The history and projections for measurements of the muon EDM. (Figure courtesy of T. Schietinger.)

17.3.3. *Frozen spin method with radioactive beams*

We end the discussion of measuring EDMs of unstable particles by mentioning the suggestion of Khriplovich [23] that one could use polarized beams of isotopes that β-decay to search for an EDM of a bare nucleus. A number of nuclei with appropriate half-lives and small anomalies have been tabulated in Ref. [23]. The β-decay asymmetry provides the analyzing power just as the muon weak decay does in the d_μ discussion above. In principle these experiments are appealing, if difficult, since they avoid the issue of Schiff screening in neutral atoms such as ^{199}Hg which complicates the interpretation of a result in terms of a nuclear EDM.

17.3.4. *The deuteron EDM using the frozen spin method*

The lightest complex nucleus, the deuteron, is a prime candidate for a storage ring EDM measurement. It has a small anomaly, $a_d = -0.143$, and thus the spin can easily be frozen with a radial electric field. High intensity polarized deuteron beams ($\sim 10^{11}$/measurement cycle) exist at a number of facilities in the world. The spin precession frequency can be well measured using proton or deuteron scattering on ^{12}C, where the analyzing power for 1 GeV/c (250 MeV kinetic energy for the deuteron) is very high, close to 50% for a detection efficiency of 1%. Furthermore, long spin coherence times in accelerators are possible using well understood techniques.

The radial electric field needed to freeze the spin is $E \simeq aBc\beta\gamma^2$, and the effective E-field acting on the EDM is $\vec{E} + c\vec{\beta} \times \vec{B}$. If the particle's g-factor were exactly equal to 2, i.e. $a = 0$, then a radial E-field alone in a storage ring could be used to probe the EDM of the particle. The electric field will always be radial at every ring place and this will be true for any particle momentum. A combination of E- and B-fields can be used to probe the particle EDM when $a \neq 0$.

For the deuteron, both the applied radial E-field and the motional field $E_m \propto \vec{\beta} \times \vec{B}$ enter, so the rest frame E-field divided by γ is equal to

$$E_1^* = E + \beta Bc \, , \tag{17.32}$$

due to the negative sign of the anomalous magnetic moment of the deuteron, i.e. the radial electric field *reduces* the effective E-field. Note that the rest frame E-field is multiplied by the relativistic factor γ. However due to time dilation this factor is lost and we drop this extra factor of γ in the estimation of the rest frame E-field.

Taking into account (the full expression) for the applied electric field needed to freeze the spin (Eq. (17.30)) the effective E-field becomes

$$E_1^* = E\left[\frac{1}{|a|\gamma^2}(1+a)\right] = 4.7E \tag{17.33}$$

where the right-hand result is for 1 GeV/c deuterons with $a_d = -0.143$. There is an added advantage that the effective rest frame E-field is enhanced by the factor in square brackets over the applied field in the laboratory, thereby reducing the measurement error by this factor (see Eq.)17.50)).

A longitudinally polarized deuteron beam will be stored in the EDM ring with combined dipole magnetic and radial electric fields (BE-sections). The fields will be tuned so that the spin will remain frozen in the horizontal plane during the storage time of about 10^3 s. Small horizontal spin precession will be allowed for systematic error studies. If there is an EDM, the motional electric field, i.e. the rest frame electric field will act on it and will precess the spin out of plane.

Since the deuteron does not decay, the spin motion will have to be monitored by a polarimeter based on elastic nuclear scattering off ^{12}C nuclei. This polarimeter, shown schematically in Fig. 17.5, will continuously monitor the spin precession in both the vertical and horizontal planes. The

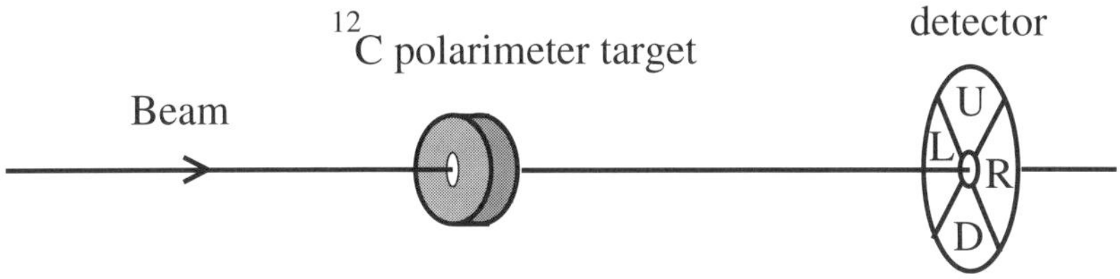

Fig. 17.5. The polarimeter consists of a solid target made out of ^{12}C, where the deuterons elastically scatter before they are captured by the detector at the end with the labels U (up), D (down), and L (left), R (right). The deuteron emittance is slowly increased by electric field kicks from a stripline system located in a straight section of the ring.

scattering target will be about 5 cm long placed at one specific azimuthal location in the storage ring, and will be the limiting aperture in the storage ring. A polarimeter based on elastic nuclear scattering off ^{12}C nuclei has an average efficiency better than 1%, and an asymmetry of $\simeq 40\%$ [30].

Two asymmetries, horizontal and vertical, can be formed from the polarimeter,

$$\varepsilon_H = \frac{L-R}{L+R} \text{ and } \varepsilon_V = \frac{D-U}{D+U}\,. \tag{17.34}$$

The horizontal asymmetry carries the EDM signal and would slowly build up with time. The vertical asymmetry carries the in-plane precession signal. A controlled mechanism for increasing the emittance of the beam as a function of time will be used to slowly drive the beam onto the scattering target. One way to analyze and extract the beam is by adding white noise on the beam emittance using stripline electrodes mounted in a straight section of the ring.

An experiment with scientific approval at Brookhaven proposes to use an electric field of 120 kV/cm across a 2 cm aperture with a magnetic field of 0.5 T for a beam energy of 1 GeV/c deuterons (see Eq. (17.30)). Several straight sections will be interleaved between the BE-sections for focusing and de-focusing magnetic quadrupoles, as well as magnetic sextupoles to prolong the spin coherence time of the beam. Two long straight sections, about 9 m in length, will be located on either side of the ring for the injection kickers, polarimeters and a beam transfer focusing de-focusing (FODO) quadrupole magnet system. A working lattice is shown in Fig. 17.6.

A normal-conducting RF-cavity will be used to cancel the first-order momentum dispersion which will increase the spin-coherence time (SCT) to about 1 s. Second-order effects originating from finite transverse motion

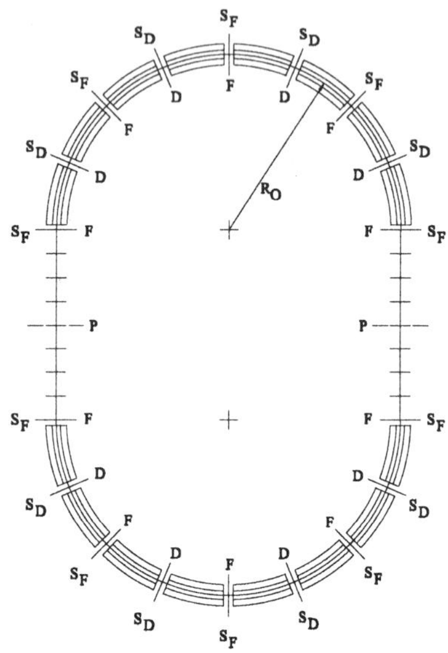

Fig. 17.6. The working lattice for the deuteron EDM consists of the combined *BE*-sections where a dipole magnetic field of 0.5 T and a radial *E*-field of 120kV/cm are used to freeze the spin precession in the horizontal plane. The focusing (F) and de-focusing (D) quadrupoles magnets form a typical strong-focusing FODO lattice. The polarimeters (P) and injection kickers are located in the long straight sections. The sextupole magnets (denoted as S_D and S_F) are used to prolong the spin coherence time. The two bunches will have opposite polarizations for polarimeter systematic error minimization.

and second-order momentum related effects will be corrected for by using sextupole magnets located in specific places around the ring, which should be able to increase the SCT to $\sim 10^3$ s, based on similar experimental work at Novosibirsk [27]. The vertical spin polarization as a function of time depends on ω_η, the EDM component of the precession frequency,

$$\Delta P_V = P\frac{\omega_\eta}{\Omega}\sin(\Omega t + \theta_0) \tag{17.35}$$

where

$$\Omega = \sqrt{\omega_\eta^2 + \omega_a^2} \tag{17.36}$$

and θ_0 is the initial angle between the spin direction and momentum vector. Clearly, the vertical polarization development is maximum when the $(g-2)$ frequency is minimized and θ_0 is either 0 or π.

The main ingredients of the deuteron EDM experiment proposed at BNL are:

(1) A polarized deuteron source that is capable of producing high intensity (few 10^{11} particles/cycle), with a highly polarized beam ($> 80\%$). The beam will be accumulated and bunched in the booster synchrotron and accelerated to 1 GeV/c. In the AGS it will undergo modest cooling resulting in a vertical emittance (95%) of 10 mm-mrad, a horizontal emittance of 3 mm-mrad and a maximum momentum spread $\Delta P/P = 10^{-3}$. The bunch is then injected into the EDM ring where the beam polarization will be kept horizontal for maximum sensitivity.
(2) Two separate bunches with opposite polarization will be stored per ring. The EDM signals from the two bunches will be opposite and they will be used to minimize the polarimeter systematic errors.
(3) The spin coherence time of an un-bunched beam would be of order of 10 ms due to the momentum spread. As mentioned above, the use of an RF-cavity and sextupole magnets should increase the SCT to $\simeq 10^3$ s.

The average vertical E-field is a major systematic error. The force due to that field would be compensated by a radial magnetic field from the focusing system, which will also precess the spin out of plane resulting in an EDM-like signal. This effect will be canceled by clockwise (CW) and counter-clockwise (CCW) consecutive injections into the storage ring. CW and CCW will only work if the beam sees the same E-fields and this requirement sets the specifications on the vertical E-field uniformity and stability. The required E-field plate parallelism is of the order (on average) of 10^{-7} rad. We are planning to use a trolley that travels inside of the storage ring to measure the relative distance between the two plates with nm level resolution. It is currently possible to measure relative distances with sub-nm resolution, using capacitive measurements [28, 29].

Storing particles CW and CCW will require flipping the B-field direction while the E-field direction remains the same. The E-field plates will be monitored using very high resolution Fabry–Perot resonators to make sure that the plate distance is not influenced by the magnetic field direction [10].

The effect of geometrical phases that could arise from the non-exact local cancellation (in a single BE section) of the $(g-2)$ spin precession must be minimized. For this error to become small there is a requirement of very good E and B-field alignment and good local matching to reduce the $(g-2)$ precession in every BE-section. The local B and E-field cancellation requirement is of order of 10^{-4}, which can be accomplished by shimming the fields to match them along the azimuth. Storing particles CW and CCW also cancels this effect as long as they remain the same after reversal of the magnetic field.

17.3.5. *A proton EDM experiment*

The proton anomaly is large, $a_p = 1.79\cdots$, $g_p = 5.58\cdots$ so in the presence of a vertical magnetic field the electric field needed to freeze the spin is very large. However, the dipole bending magnets can be replaced by a radial electric field, with magnetic quadrupoles providing horizontal and vertical focusing. Eliminating the B-field from Eq. (17.22), it becomes

$$\vec{\omega}_{a\eta} = -\frac{q}{m}\left[\left[a - \left(\frac{m}{p}\right)^2\right]\frac{\vec{\beta}\times\vec{E}}{c} + \frac{\eta}{2}\frac{\vec{E}}{c}\right]. \tag{17.37}$$

The $(g-2)$ (i.e. in-plane) spin precession can be made zero at a momentum

$$p = \frac{m}{\sqrt{a}}. \tag{17.38}$$

The magic momentum for the proton in a radial electric field is 0.7 GeV/c. Recent advances in achieving large electric field gradients [25] using high pressure water rinsing (HPR), combined with the fact that proton beam emittance can be very effectively cooled using electron cooling, makes this method very promising.

HPR has been used in the past to enhance the effective E-field in RF cavities. The method has now been applied to enhance the E-field gradient in DC applications by a factor of two to three over previous limits. The electric field sustainable between two plates depends on the distance ℓ between them, and follows a $1/\sqrt{\ell}$ rule [26]. Assuming 15 MV/m for a 2 cm plate separation, the ring circumference (including the straight sections needed for instrumentation) would be of order of 200 m.

From Eq. (17.37) it is clear that at the magic momentum of 0.7 GeV/c, the proton spin will be frozen independent of the E-field value as long as the average momentum is kept constant to the correct value. In order to eliminate the vertical E-field background we will still have to inject CW and CCW. The focusing of the system is still based on magnetic quadrupoles since the elimination of small stray magnetic fields would be very strict otherwise and very expensive to achieve.

There are differences in running protons and deuterons. Clockwise and counterclockwise injection is necessary for both. A few contrasting requirements are:

- The proton storage ring needs a radial electric field whereas the deuteron ring needs combined E and B-field sections with their magnitudes well matched.
- For the deuteron measurement, the dipole magnetic field must be flipped between CW and CCW injections while the (bending) radial electric field in the proton ring does not change.
- A sensitive (state of the art) Fabry–Perot resonator is needed for the deuteron ring to ensure flipping the B-field does not influence the E-field direction in a systematic way. This is not needed for the proton ring.
- The local $(g-2)$ phase cancellation is much easier in the proton ring since one only needs to deal with the E-field plates.
- The proton polarimeter is simpler since the proton has only vector polarization, compared with the vector and tensor polarization of the deuteron.
- The estimated ring circumference for proton storage is about 200 m, much longer than the estimated 85 m for the deuteron ring.

While the experimental sensitivities of the proposed proton and deuteron EDM searches are comparable, their potential physics reach is model dependent. For some cases, such as the θ parameter, the proton may be about a factor of 3 better, while for the case of SUSY-induced color quark EDMS, the deuteron can be considerably more sensitive than the proton or neutron. The storage-ring EDM collaboration at Brookhaven is exploring both the deuteron and proton options, and in discussions with the Laboratory regarding resources and funding availability will decide which experiment to pursue first.

17.3.5.1. *Experimental sensitivity for d_p and d_d*

The statistical sensitivity of the experiment depends on the time dependence of the collected data and the time constants of the machine cycles compared to the spin coherence time. The signal $S(t)$ will be proportional to

$$S(t) = \frac{R(t) - L(t)}{R(t) + L(t)} = PA\theta(t) \tag{17.39}$$

where $R(t)$, $L(t)$ are the signals from the left and right detectors. P is the polarization, A the analyzing power and $\theta(t)$ the vertical spin angle as a function of time. For the (spin 1/2) proton $\theta(t)$ is

$$\theta(t) = \frac{d_p E_1^*}{\hbar/2} t\,, \tag{17.40}$$

and for the (spin 1) deuteron it is

$$\theta(t) = \frac{d_d E_1^*}{\hbar} t\,. \tag{17.41}$$

$d_{p,d}$ is the EDM of the particle, E_1^* is the rest frame electric field divided by the relativistic Lorentz factor γ, and t is the time in the lab frame.

The error in the ratio $S(t)$ is given by

$$\sigma_s = \sqrt{\frac{1-S^2}{L+R}} \simeq \frac{1}{\sqrt{L+R}} = \frac{1}{\sqrt{N_0 e^{-t/\tau}}} \tag{17.42}$$

under the assumption that $S(t)$ is small, and τ is beam lifetime.

The χ^2 is given by

$$\chi^2 = \sum_{i=1}^{n} \left[\frac{P\rho d t_i - N_i}{(\sigma_s)_i} \right]^2 \tag{17.43}$$

which is minimized with respect to d

$$\frac{\partial \chi^2}{\partial d} = 2 \sum_{i=1}^{n} \left[\frac{P\rho d t_i - N_i}{(\sigma_s)_i} \right] \frac{P\rho t_i}{(\sigma_s)_i} \tag{17.44}$$

with $\rho_p = (AE_1^*)(\hbar/2)$ for the protons and $\rho_d = (AE_1^*)/\hbar$ for the deuterons. Assuming there is a DC offset in the signal $S(t)$ at $t = 0$, which will be included in the fit, the error (per measurement cycle) is

$$\sigma_{d_p} = \frac{1}{2} \frac{\hbar}{\tau P A E_1^* \sqrt{N_{\rm tot,c}}} \tag{17.45}$$

for the proton and

$$\sigma_{d_d} = \frac{\hbar}{\tau P A E_1^* \sqrt{N_{\rm tot,c}}} \tag{17.46}$$

for deuterons. Here, $N_{\rm tot,c}$ is the total number of particles per measurement cycle.

The limiting measurement factor is the spin coherence time or polarization lifetime. The goal is to achieve a SCT of 10^3 s, much larger than the accelerator cycle time of ~ 1 s. The optimized beam extraction lifetime, with a rather broad minimum, is about half of the polarization lifetime, with the measurement length per cycle to be equal to the polarization lifetime.

The final errors, assuming that the extraction rate is proportional to the instantaneous stored beam intensity, are

$$\sigma_{d_p} = \frac{4\hbar}{PAE_1^* \sqrt{N_{\text{tot,c}} T_{\text{tot}} \tau_p}} \tag{17.47}$$

for the protons and

$$\sigma_{d_d} = \frac{8\hbar}{PAE_1^* \sqrt{N_{\text{tot,c}} T_{\text{tot}} \tau_p}} \tag{17.48}$$

for the deuterons. The total live-time of the experiment is T_{tot}, τ_p is the polarization lifetime, and $N_{\text{tot,c}}$ is the total number of particles accumulated per machine cycle.

For the proton, as explained above, the effective electric field is equal to the lab electric field the laboratory electric field. and Eq. (17.47) becomes:

$$\sigma_{d_p} = \frac{4\hbar}{PAE \sqrt{N_{\text{tot,c}} T_{\text{tot}} \tau_p}}. \tag{17.49}$$

For the deuteron, the enhancement factor of Eq.(17.33) enters, and Eq. (17.48) becomes

$$\sigma_{d_d} = \frac{8\hbar}{PAE \left[\frac{1}{|a|\gamma^2}(1+a)\right] \sqrt{N_{\text{tot,c}} T_{\text{tot}} \tau_p}}. \tag{17.50}$$

We assume the following parameters for the proton and deuteron EDM experiments:

- Polarization lifetime is 10^3 s.
- The asymmetry observed by the polarimeter $A = 0.5$ for 0.7 GeV/c protons and $A = 0.4$ for 1 GeV/c deuterons.
- The beam polarization at injection into the EDM ring $P = 0.8$.
- The number of particles per cycle $N_{\text{tot,c}} = 4 \times 10^{11} \times f$, with f the detector efficiency.
- The total measurement time $T_{\text{tot}} = 10^7$ s per year.
- The efficiency of the polarimeter $f = 0.01$, which will multiply the number of particles injected into the ring to obtain the number of detected particles.
- The lab frame electric field 15 MV/m for the proton and 12 MV/m for the deuteron[a].

[a]In the deuteron ring we assume the presence of the dipole magnetic field will restrict the maximum E-field possible between the plates, but this may not prove to be true in practice.

Table 17.3. A summary of the beam and ring parameters for the different storage ring EDM measurement.

	Goal (e-cm)	β	γ	Circumference	particles /fill	Running time (year)
μ	5×10^{-23}	0.78	1.6	2.6 m	1	1
μ	$< 10^{-24}$	0.98	4.8	44 m	10^9	?
d	10^{-29}	0.47	1.13	85 m	4×10^{11}	1
p	10^{-29}	0.60	1.25	200 m	4×10^{11}	1

The total statistical error then becomes $\sigma_{d_d} \simeq 5.5 \times 10^{-30}$ *e*-cm per year for the deuteron, assuming that the entire ring is filled with the combined E and B sections. This is true for 60% of the ring so the error becomes $\sigma_{d_d} \simeq 0.9 \times 10^{-29}$ *e*-cm per year. Similarly for the proton ring we would have $\sigma_{d_p} \simeq 7 \times 10^{-30}$ *e*-cm per year, which when corrected by this same efficiency gives $\sigma_{d_p} \simeq 1.2 \times 10^{-29}$ *e*-cm per year.

17.4. Conclusions and Outlook

The use of a storage ring permits direct searches for electric dipole moments of charged particles. Unlike the atomic experiments, which require significant additional information to extract an EDM of the electron or of the atomic nucleus, the storage ring technique will provide direct measurements that are significantly easier to interpret should evidence for an EDM appear. A summary of the storage-ring parameters discussed above is given in Table 17.3 along with the projected sensitivities of the storage ring EDM method for different particles.

The storage ring technique has already been used to determine a limit on the muon EDM, and the possibilities to extend this search in a dedicated frozen-spin experiment provides a unique opportunity to search for an EDM in the second generation. The next muon $(g-2)$ experiment could lower this limit by perhaps as much as two orders of magnitude by employing the decay-electron traceback technique. A dedicated frozen-spin experiment could improve by three to five orders of magnitude further.

The deuteron and proton experiments provide unique opportunities, which are complementary to the ongoing neutron and atom EDM searches. The full set of EDM experiments should pin down the *CP*-violating source, should a non-zero EDM value be found in any system. Even if the neutron EDM experiment does not discover a non-zero EDM, the storage ring experiments should be done since they are more sensitive in general by a couple of orders of magnitude, especially for interactions such as a T-odd component in the nuclear exchange force. Such efforts await funding, but have significant discovery potential, which makes a compelling case for their approval.

Acknowledgments

We wish to thank our colleagues on the muon $(g-2)$ experiment E821, as well as our colleagues on the deuteron and proton EDM proposal at Brookhaven for numerous useful discussions. Preparation of this manuscript was supported in part by NSF Grant PHY-0758603.

References

[1] P.A.M. Dirac, Proc. R. Soc. (London) **A117**, 610 (1928).
[2] E.M. Purcell and N.F. Ramsey, Phys. Rev. **78**, 807 (1950).
[3] J.H. Smith, E.M. Purcell, and N.F. Ramsey, Phys. Rev. **108**, 120 (1957) and references therein.
[4] A.D. Sakharov, JETP Lett., **5**, 24 (1967).
[5] J.H. Christenson *et al.*, Phys. Rev. Lett. **13**, 138 (1964).
[6] I.B. Khriplovich, R.A. Korkin, Nucl. Phys. A **665**, 365 (2000); O. Lebedev *et al.*, Phys. Rev. D **70**, 016003 (2004); M. Pospelov, A. Ritz, Ann. Phys. **318**, 119 (2005).
[7] C.P. Liu and R.G.E. Timmermans, Phys. Rev. C**70**, 055501 (2004).
[8] The frozen spin technique was first proposed by Y.K. Semertzidis, *et al.* at the AGS2000 workshop at BNL in 1996, and later at the Workshop on Frontier Tests of Quantum Electrodynamics and Physics of the Vacuum, Sandansky, Bulgaria, 9–15 Jun 1998, published in the proceedings, Sandansky 1998, Frontier tests of QED and physics of the vacuum, 369–376, ed. by E. Zavattini, D. Bakalov, C. Rizzo; and at the AGS2000 workshop at BNL, May 2000.
[9] F.J.M. Farley, K. Jungmann, J.P. Miller, W.M. Morse, Y.F. Orlov, B.L. Roberts, Y.K. Semertzidis, A. Silenko, E.J. Stephenson, Phys. Rev. Lett. **93**, 052001 (2004); Y.K. Semertzidis *et al.*, AIP Conf. Proc. **698**, 200 (2004); Y.K. Semertzidis *et al.*, High Intensity Muon Sources (HIMUS99), Tsukuba, Japan Dec. 1999, hep-ph/0012087.
[10] Deuteron Storage Ring EDM Proposal to the BNL PAC, March 2008, available at http://www.bnl.gov/edm/
[11] W. Marciano, HEP seminar at BNL on the theoretical aspects of deuteron, proton and neutron EDM.
[12] C.A. Baker *et al.*, Phys. Rev. Lett. **97**, 131801 (2006).
[13] W. C. Griffith, M. D. Swallows, T. H. Loftus, M. V. Romalis, B. R. Heckel and E. N. Fortson, Phys. Rev. Lett. **102**, 101601 (2009).
[14] M.A. Rosenberry and T.E. Chupp, Phys. Rev. Lett. **86**, 22 (2001).
[15] V.F. Dmitriev, R.A. Sen'kov, Phys. Rev. Lett. **91**, 212303 (2003).

[16] G.W. Bennett *et al.*, Phys. Rev. D**73**, 072003 (2006).
[17] G.W. Bennett, *et al.*, arXiv:0811.1207v2 [hep-ex], July 2009, to be published in Phys. Rev. **D**.
[18] J. Bailey *et al.*, J Nucl. Phys. B **150**, 1 (1979).
[19] W. Flegel and F. Krienen, Nucl. Instr. and Meth. **113**, 549 (1973).
[20] Y.K. Semertzidis *et al.*, Nucl. Instr. Meth. in Phys. Res. A **503**, 458 (2003).
[21] A. Adelmann, K. Kirch, C.J.G. Onderwater, and T. Schietinger, arXiv:hep-ex/0606034v2, Dec. 2008.
[22] J-PARC Letter of Intent L22, Search for a permanent Muon Electric Dipole Moment, J.P. Miller, Y.K. Semertzidis, Y. Kuno *et al.*, February 2003.
[23] I.B. Khriplovich, Phys. Lett. bf B 444, 98 (1998).
[24] Yuri Orlov was the first to suggest some 10 years ago to study the electron EDM at its magic momentum of P=15 MeV/c. Francis Farley suggested to use the muon at its magic momentum and others suggested other systems. They were all rejected as not practical, the electron for lack of an efficient polarimeter, the muon and the others due to the very large ring needed since at the time we only considered the then achievable very modest electric fields.
[25] B.M. Dunham *et al.*, Proceedings of PAC07, 1224 (2007).
[26] L. Cranberg, Journ. of Appl. Phys. **23**, 518 (1952).
[27] I.B. Vasserman *et al.*, Phys. Lett. B **198**, 302 (1987). Y. Orlov did the analytical work for the BNL proposal.
[28] Physik Instrumente, http://www.pi-usa.us/
[29] http://www.lionprecision.com/tech-library/appnotes/cap-0030-thickness-measurement.html
[30] Y. Satou *et al.*, Phys. Lett. B **549**, 307 (2002).

Chapter 18

Models of Lepton Flavor Violation

Yasuhiro Okada

Theory Group, Institute of Particle and Nuclear Studies, KEK,
and
Department of Particle and Nuclear Physics,
The Graduate University for Advanced Studies (Sokendai),
Tsukuba, Ibaraki 305-0801, Japan
yasuhiro.okada@kek.jp

Muon lepton flavor violation exists in many physics models beyond the Standard Model. Predictions for lepton flavor-violating processes such as $\mu \to e\gamma$, $\mu \to 3e$ and $\mu - e$ conversion in muonic atoms are discussed in various New Physics models.

Contents

18.1. Introduction

To date, charged lepton processes that violate the separate conservation of flavor for each generation have never been observed. When muons were discovered and thought to be a heavy state of the electron, it was natural to expect that a muon would decay to an electron by emitting a photon. Experimental efforts to find *lepton flavor violation* (LFV) were started in the very early days of muon experiments. The non-observation of LFV indicated that the muon is not a simple excited state of the electron, and led to the idea of the generation structure of elementary particles. The concept of generation was one of the foundations in formulating the Standard Model (SM) of elementary particle physics in 1970s. Absence of LFV is attributed to the zero neutrino mass in the SM because we can define conservation of electron, muon, and tau lepton numbers within renormalized interactions of the SM.

LFV in the charged lepton sector has received renewed attention since the discovery of neutrino oscillations. Since the SM assumes massless neutrinos, it has to be extended to accommodate massive neutrinos, and the separate conservation of the lepton number for each generation is likely to be violated. The simplest mechanism for neutrino mass generation, the seesaw neutrino model or the Dirac neutrino model, however, turns out to predict extremely small branching ratios for LFV processes due to the smallness of the neutrino masses. In other scenarios, new particles and/or new interactions associated with the neutrino mass generation can induce sizable LFV in the charged lepton sector. Patterns of LFV signals in various processes including tau decays would provide clues to help choose a correct model of neutrino mass generation.

In this chapter, various theoretical models are reviewed in connection with charged lepton LFV processes [1]. Some are directly related to the neutrino mass generation mechanism, and others are not. In any model that predicts branching ratios of LFV processes large enough to be accessible to near-term experimental searches, we can expect new particles and/or new phenomena in the TeV energy scale. Some of these new particles/phenomena may be seen at the LHC experiments. In such a case, LFV searches and the LHC experiments can play complementary roles to clarify the nature of the physics beyond the SM.

18.2. Supersymmetry

18.2.1. *Flavor problem in the SUSY models*

Supersymmetry (SUSY) was introduced as an extension of the Poincaré algebra that represents a symmetry of four-dimensional space-time. Unlike other symmetries, the SUSY transformation connects a boson and a fermion, and the existence of super partners is a requirement of this symmetry. A unique feature of SUSY is that it is related to space-time symmetry. In fact SUSY is most elegantly expressed as translation in superspace, which is, in a sense, an extension of the space-time concept [2].

Since SUSY is a general framework, this symmetry itself cannot constrain the structure of particle models very well. We can introduce gauge interactions and Yukawa interactions as we like. What SUSY can do, is to relate various interactions associated with particles and their super partners. As far as dimensionless couplings are concerned, the number of free parameters are essentially the same if a particle model is extended by SUSY.

In a realistic particle model, SUSY has to be realized as a broken symmetry because a super partner should have the same mass as the corresponding particle in the unbroken phase of SUSY. The number of new parameters in possible SUSY breaking terms is large, about one hundred, even for the minimal SUSY extension. Most of the new parameters are related to flavor mixings of the super partners of quarks and leptons (squarks and sleptons). Since the early days of phenomenological studies on SUSY models, it has been realized that the flavor mixings are strongly constrained by existing data on flavor changing neutral current (FCNC) processes, and LFV in the charged lepton sector. Roughly speaking, degeneracy in masses at a percent level is required for squarks and sleptons of different generations with the same gauge quantum numbers, if their masses are a few hundred GeV. Even if we take the new particle scale to be in a TeV range, which is a natural New Physics scale from the viewpoint of electroweak symmetry breaking, we need to require that these masses should be degenerate within a 10% level. This implies that there should be some physical reason that explains the pattern of squark and slepton mass terms. Flavor mixings together with the SUSY mass spectrum will provide us with a clue to the origin of the SUSY breaking mechanism.

There is an important difference between quark FCNC processes and LFV. In the quark case, FCNC processes are induced at one loop level even within the SM, so that we need to measure deviations from SM predictions to extract New Physics effects. Observation of LFV processes in the charged lepton sector is a direct evidence of existence of new flavor-mixing sources, which will be discussed in the following sections.

18.2.2. *SUSY seesaw neutrino model*

The SUSY seesaw neutrino model is an interesting example of physics models that predict large LFV branching ratios. There is a class of models of SUSY breaking in which the slepton mass matrix is generated in a flavor-blind way, i.e. all sleptons with the same gauge quantum numbers are degenerate at the scale where SUSY breaking terms are generated. Minimal supergravity is one example where all scalar quarks and leptons have universal masses at the Planck scale. Flavor off-diagonal terms in the slepton mass matrices are induced through renormalization effects due to the neutrino Yukawa coupling constants, even if one starts from a flavor-blind initial condition at the Planck scale. As a consequence, branching ratios of charged lepton LFV processes can be enhanced significantly [3]. In this sense, LFV processes are probes to interactions at a very high energy scale.

Typical predictions of branching ratios of LFV processes [4], $\mu \to e\gamma$, $\tau \to e\gamma$, and $\tau \to \mu\gamma$ in the SUSY seesaw model are shown in Fig. 18.1. Here we show three branching ratios as a function of the lightest slepton mass. In this model, there are two sources of flavor mixings in the lepton sector. One is the heavy Majorana mass matrix and the other is the neutrino Yukawa coupling matrix (or the neutrino Dirac mass matrix). The light neutrino masses and the mixing matrix, i.e. *Pontecorvo–Maki–Nakagawa–Sakata*

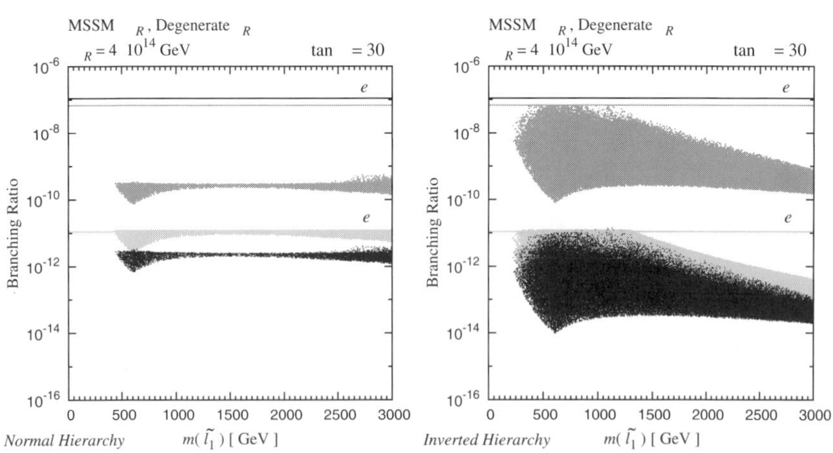

Fig. 18.1. Branching ratios of lepton flavor violation processes $\mu \to e\gamma$ (light-gray), $\tau \to \mu\gamma$ (gray), and $\tau \to e\gamma$ (black) as functions of the lightest charged slepton mass $m(\tilde{l}_1)$ for the SUSY seesaw model. Horizontal lines denote experimental upper limits. Left (Right) figure corresponds to the normal (inverse) hierarchy case for the light neutrino masses.

(PMNS) matrix, are determined through the seesaw mass relation, namely the light neutrino mass matrix is given by $m_\nu = m_D^T M_N^{-1} m_D$ where m_D is the neutrino Dirac mass matrix and M_N is the heavy Majorana mass matrix. Since the flavor mixing in the slepton mass matrix is related to the neutrino Yukawa coupling matrix, branching ratios of muon and tau LFV processes depend on both the light and heavy neutrino mass and mixing parameters. In Fig. 18.1, two cases (the normal and the inverse hierarchy cases of the light neutrino masses with three degenerate heavy Majorana masses) are shown. The Majorana mass is taken to be 4×10^{14} GeV corresponding to $O(1)$ neutrino Yukawa coupling constants. The ratio of two Higgs vacuum expectation values ($\tan\beta$) is taken to be 30 and other SUSY parameters are scanned taking account of various phenomenological constraints with the assumption of the minimal supergravity model. We can see that the branching ratio of $\mu \to e\gamma$ can be close to 10^{-11} even if the slepton mass is larger than 1 TeV. On the other hand, the branching ratio of tau LFV processes depends on the mass pattern of light neutrinos. Under the constraints from $B(\mu \to e\gamma)$, $B(\tau \to \mu\gamma)$ can be $O(10^{-8})$ for the inverse hierarchy case. This range of tau LFV branching ratios is within the reach of future B factory experiments [5].

There is an interesting special case which can be realized for large values of $\tan\beta$ [6–8]. In this case, SUSY loop correction to the Higgs-lepton vertex can generate a large LFV coupling. As a result, heavy Higgs boson exchange diagrams can be dominant, and the $\mu - e$ conversion process is enhanced relative to the $\mu \to e\gamma$ process [9]. For large values of $\tan\beta$, the Higgs exchange contribution to the $\mu - e$ conversion rate is proportional to $(\tan\beta)^6$, whereas photonic diagrams have $(\tan\beta)^2$ dependence. An example is shown in Fig. 18.2 [9]. For a smaller heavy Higgs boson mass, the two branching ratios can be closer. The dominance of scalar operators can be also confirmed through the atomic number dependence of the $\mu - e$ conversion rate [10]. The ratio of the $\mu - e$ conversion rate for heavy and light nuclei depends on which LFV operators are present in the interaction, for example the contribution of scalar operators reduces the ratio of the $\mu - e$ conversion rates for *Pb* and *Al* nuclei as is shown in Fig. 18.2. The ratio of branching ratios of $\mu \to 3e$ and $\mu \to e\gamma$ is also shown. In this case, the Higgs exchange effect is small because of the small electron Yukawa coupling constant.

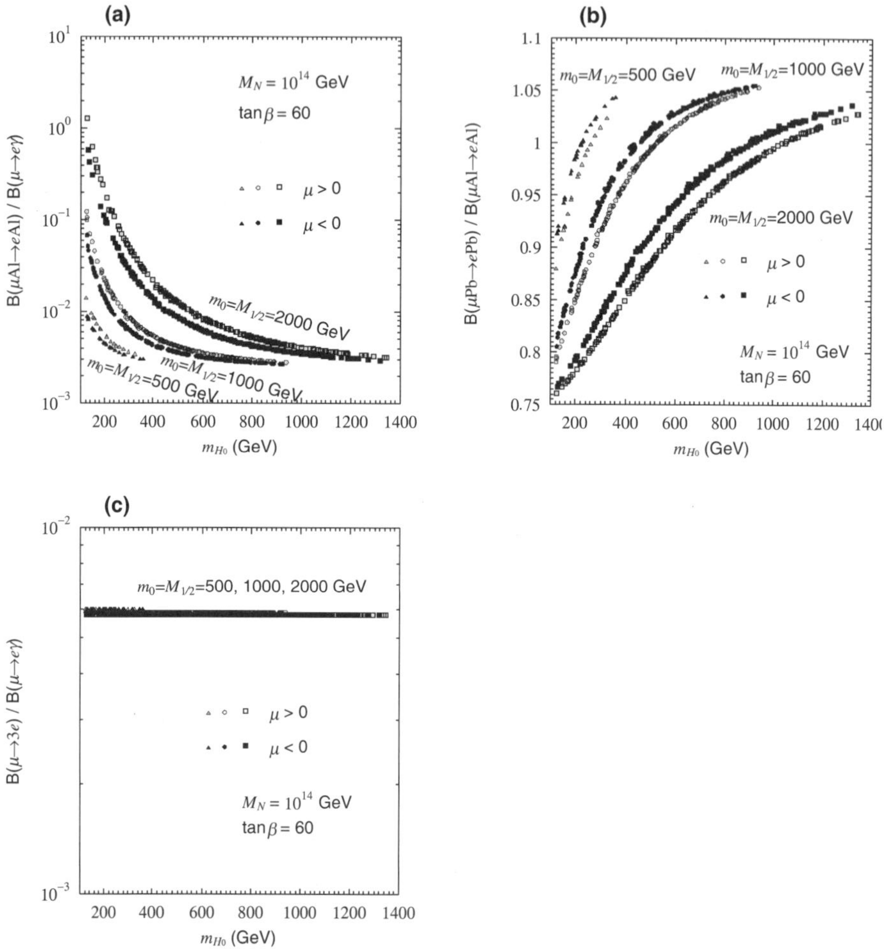

Fig. 18.2. Various ratios of branching ratios as a function of the heavy CP even Higgs boson mass in the SUSY seesaw model. (a) $B(\mu\text{Al} \to e\text{Al})/B(\mu \to e\gamma)$, (b) $B(\mu\text{Pb} \to e\text{Pb})/B(\mu\text{Al} \to e\text{Al})$, (c) $B(\mu \to 3e)/B(\mu \to e\gamma)$. We take the right-handed neutrino masses to be 10^{14} GeV, and $\tan\beta = 60$.

18.2.3. *SUSY GUT*

In the framework of a *grand unified theory* (GUT), flavor mixing in the the slepton mass matrices can be generated by renormalization effects due to Yukawa couplings related to quark masses. This is because quarks and leptons are assigned to be in the same representation with respect to the GUT gauge group. In particular, a large top Yukawa coupling constant can

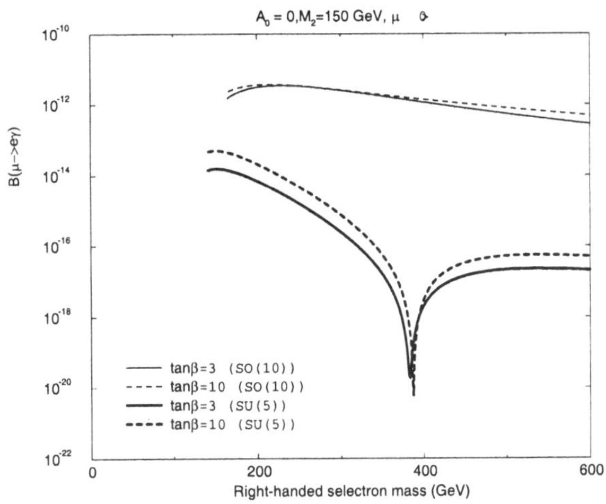

Fig. 18.3. The branching ratio of $\mu \to e\gamma$ processes for $SU(5)$ and $SO(10)$ SUSY GUT as a function of right-handed slepton mass. Solid (dashed) curves corresponds to $tan\beta$=3 (10), and the top quark mass is taken to be 175 GeV.

be a source of large branching ratios for LFV processes [11]. In the minimal $SU(5)$ SUSY GUT, the flavor mixing appears in the the right-handed slepton sector, because right-handed sleptons and up-type quarks are members of the same $SU(5)$ gauge representation, namely the **10** representation. The flavor mixing is essentially controlled by the *Cabibbo–Kobayashi–Maskawa (CKM) matrix*. The branching ratio of $\mu \to e\gamma$ process is shown in Fig. 18.3 for SUSY GUT models [1]. In the minimal SU(5) case, there could be cancellation between different diagrams, so that the branching ratio can be smaller than 10^{-14} [12]. The branching ratio of the $SO(10)$ SUSY GUT is also shown in this figure. No cancellation is expected in this case because both left-handed and right-handed slepton mass matrices have flavor mixings and both chargino-slepton and neutralino-slepton one-loop diagrams are involved. The branching ratio is typically $O(10^{-12})$. It should be noticed that the minimal $SU(5)$ is somewhat special because the branching ratio is enhanced in general if we try to accommodate realistic fermion mass spectrum within the context of $SU(5)$ SUSY GUT [13]. Furthermore, effects of neutrino Yukawa coupling constants can be a source of an additional large contribution to the branching ratio, just like the SUSY seesaw model, as long as the gauge-singlet Majorana mass scale is larger than 10^{13} GeV.

Polarized muons are useful to discriminate different theoretical models of LFV. In the case of the $\mu \to e\gamma$ search, there are two effective LFV operators corresponding to $\mu^+ \to e^+\gamma_L$ and $\mu^+ \to e^+\gamma_R$. We can distinguish contributions of the two operators by the angular distribution of the final particles with respect to the initial spin of the muon, if we use polarized muon decays. In the SUSY model, the two operators reflect whether the flavor mixing arises in the right-handed slepton sector or in the left-handed slepton sector. For example, the SUSY seesaw model predicts $\mu^+ \to e^+\gamma_R$ and the top Yukawa contribution in the minimal SU(5) SUSY GUT corresponds to $\mu^+ \to e^+\gamma_L$. If $\mu \to e\gamma$ is observed, a polarized muon experiment will be very important to determine nature of the LFV interaction.

Polarized muon decays provide further information on LFV interactions in the $\mu^+ \to e^+e^+e^-$ case [14]. We can define two P-odd and one T-odd asymmetries using the initial muon polarization direction and three final particle momentum directions. These asymmetries as well as the $\mu \to e\gamma$ asymmetry is useful to constrain possible contributions of various types of photonic dipole and four-fermion LFV operators. In the $SO(10)$ SUSY GUT case, for example, contributions from two photonic dipole operators are dominant and we can derive specific relations among two P-odd asymmetries and the $\mu \to e\gamma$ asymmetry. The T-odd asymmetry is particularly interesting because this is sensitive to CP-violating phases in the lepton sector. In order to obtain a sizable T-odd asymmetry, we need both photonic and four-fermion interactions which interfere with each other, and coupling constants of two operators should have a relative complex phase. Such examples exist in both $SU(5)$ SUSY GUT [14] and the SUSY seesaw model [15], in the case where the photonic dipole contribution is somewhat suppressed due to the cancellation of different Feynman diagrams.

18.2.4. *LFV and dipole moments*

There is an interesting relationship between $\mu \to e\gamma$ and the anomalous magnetic moment of the muon in SUSY models. These two processes are generated by similar loop diagrams with internal sleptons and charginos/neutralinos. The decay amplitude of the $\mu \to e\gamma$ process involves flavor off-diagonal elements of the slepton mass matrix, whereas the muon anomalous magnetic moment is related to diagonal elements. If these two processes are dominated by either left-handed or right-handed slepton loop diagrams we can derive a relation between two observables in terms of a relevant flavor mixing angle [16, 17]. For instance, in the SUSY seesaw model we obtain

$$B(\mu \to e\gamma) \sim 3 \times 10^{-5} \left(\frac{\delta a_\mu^{SUSY}}{10^{-9}} \right)^2 \left(\frac{(m_{\tilde{L}}^2)_{12}}{m_{SUSY}^2} \right)^2 \tag{18.1}$$

where $\delta a_\mu^{SUSY} \equiv (g-2)_\mu^{SUSY}/2$ is the SUSY contribution to the muon anomalous magnetic moment, $(m_{\tilde{L}}^2)_{12}$ is the 1-2 element of the left-handed slepton mass-squared matrix, and m_{SUSY} is a SUSY particle mass which is assumed to be the same for all SUSY particles. We can see that if the deviation of muon anomalous magnetic moment is $O(10^{-9})$, the present upper bound of $B(\mu \to e\gamma)$ puts a very stringent constraint on the flavor mixing angle of the slepton matrix at the level of 10^{-3}. On the other hand, if positive signals are obtained for both processes, we will be able to determine the flavor mixing angle of the slepton mass matrix.

The muon electric dipole moment is also generated by one-loop diagrams which involve sleptons and charginos/neutralinos. In this case, however, relationship to LFV processes is not straightforward, because there are many new sources of CP-violating complex phases in SUSY models.

18.2.5. *R-parity violation and LFV*

In the minimal supersymmetric Standard Model (MSSM), if we required only gauge invariance to write all possible superpotentials, the following interactions would also be allowed:

$$W = \lambda_{ijk} L_i L_j E_k^c + \lambda'_{ijk} L_i Q_j D_k^c + \lambda''_{ijk} U_i D_j^c D_k^c - \mu_i L_i H_2. \tag{18.2}$$

These interactions violate the baryon- or lepton-number conservation. To forbid proton decays that are too fast, a parity called the R-parity is often imposed. All superparticles are assigned to be odd under the R-parity whereas ordinary particles are assigned to be even. We can eliminate the above terms by R-parity. The R-parity has an important consequence for our universe, namely the lightest superparticle becomes stable, and is a good candidate for the dark matter particle.

Imposing R-parity is not the only way to prevent fast proton decays. Since proton decay requires both baryon and lepton number violations, we can consider the case where baryon number violating terms (λ'') are absent, but other lepton number violating terms are present. In such a case, the combination of two coupling constants can be constrained by LFV processes. Typical tree and loop diagrams for LFV processes are shown in Figs. 18.4 and 18.5 [1].

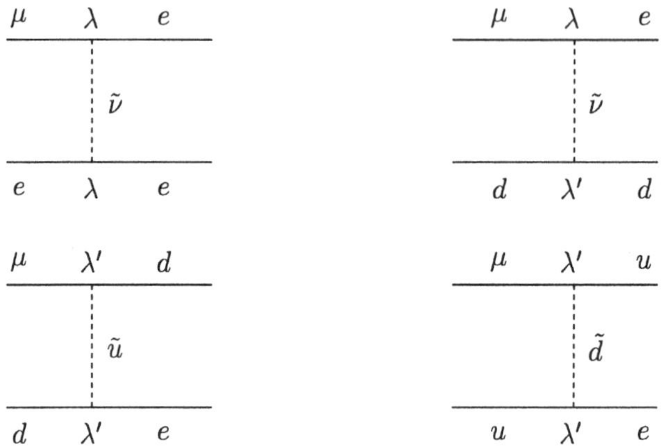

Fig. 18.4. Tree diagrams for LFV processes in SUSY models with R-parity violation.

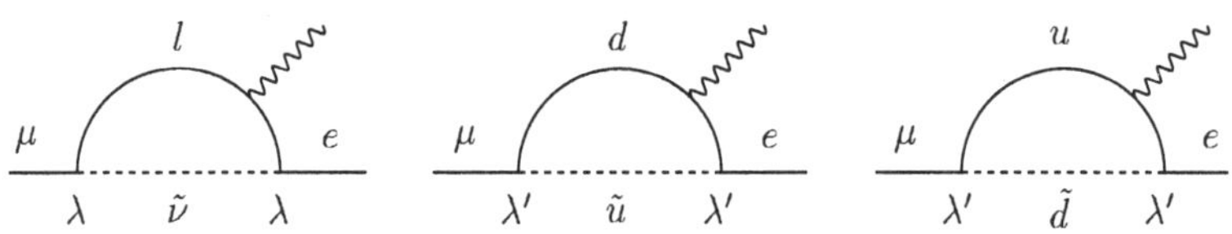

Fig. 18.5. Loop diagrams for LFV processes in SUSY models with R-parity violation.

We can distinguish three categories [18]. The first is where the $\mu \to 3e$ process occurs at the tree level. The branching ratio of $\mu \to 3e$ is three (two) orders of magnitude larger than that of $\mu \to e\gamma$ ($\mu - e$ conversion). Therefore, $\mu \to e\gamma$ is already strongly constrained ($< O(10^{-15})$), but future $\mu - e$ conversion experiments may reach the signal region. The second category is where tree diagrams induce the $\mu - e$ conversion. In such cases, the $\mu - e$ conversion is the most promising and the other two processes are very suppressed. All three processes are induced by one-loop diagrams for the last category. In the loop diagrams in Fig. 18.5 it is known that the $\mu \to 3e$ decay and $\mu - e$ conversion processes receive a logarithmic enhancement of a type $\ln m_\mu/m_{SUSY}$. As a result, these branching ratios can be comparable or even larger than the $\mu \to e\gamma$ branching ratio. Considering future prospects of $\mu - e$ conversion experiments, which aim to go below 10^{-16}, the $\mu - e$ conversion is the most promising process of LFV in SUSY models with R-parity violation. Furthermore, measurements of other processes and P- and T-odd asymmetries are useful to distinguish different cases of models with R-parity violation.

18.3. Little Higgs Models with T-parity

Although little is known regarding the Higgs sector experimentally, theoretical consideration may provide some hints that could lead to a correct mechanism of the electroweak symmetry breaking. Consider that the Higgs theory is a good description of physics of the electroweak symmetry breaking below some cutoff scale. Present knowledge of electroweak measurements indicate that the cutoff scale is already above a multi-TeV range, say 5 TeV. Such a high cutoff scale requires a fine tuning in the renormalization of the Higgs mass term to a significant degree because scalar mass terms have a quadratic dependence of the cutoff scale. This is called the little hierarchy problem [19].

Little Higgs models were proposed as a solution of the little hierarchy problem [20, 21]. The Higgs field is realized as a pseudo Nambu–Goldstone boson below the cutoff scale, presumably around 10 TeV. Quadratic divergence of the Higgs mass renormalization is absent at the one-loop level by a properly chosen structure of global and local symmetries. The most famous example is the littlest Higgs model which is based on an $SU(5)/SO(5)$ nonlinear sigma model [21]. In this model, heavy gauge bosons and a heavy top quark are introduced in order to cancel the quadratic divergence of the Higgs boson mass term at the one-loop level. Subsequent study on the littlest Higgs model, however, showed that the mass scale of the new particles should be unnaturally high in order to satisfy constraints from electroweak precision measurements [22].

The little Higgs model with T-parity is a modified version of the littlest Higgs model by requiring a discrete symmetry called T-parity [23]. In this model, contributions to electroweak precision observables from tree-level heavy gauge boson exchange diagrams are absent due to the assignment of T-parity, and the new particle scale can be below 1 TeV.

Although the little Higgs model was introduced to solve a problem associated to the electroweak symmetry breaking physics, the model with T-parity has an important implication to flavor physics. In order to assign the T-parity, we need to introduce heavy partners of the SM fermions (T-odd quarks and leptons). The vertex between the heavy gauge boson, heavy quark (lepton) and the SM quark (lepton) has a flavor mixing independent of the corresponding flavor mixing matrix, namely the CKM

(PMNS) matrix [24]. In fact, we can introduce one new 3×3 unitary matrix for the quark and the lepton sectors respectively. Quark FCNC processes and LFV processes receive contributions from heavy fermion and heavy gauge boson loop diagrams that depend on new unitary matrices.

LFV and FCNC processes have been studied in the little Higgs model with T-parity [25–28]. Present experimental data already put strong constraints on heavy fermion mass spectrum and flavor mixings if we assume the heavy mass scale is below 1 TeV. Roughly speaking, the present upper bounds on $\mu \to e\gamma$ and $\mu \to 3e$ processes can translate to the degeneracy of three heavy lepton masses up to the 10% level or alignment of the heavy lepton and the SM lepton flavor mixings at 10% level [28]. This implies that further experimental improvements on the limits for muon LFV processes will put more stringent constraints on model parameter space, or lead to an observation.

There is a clear difference in the prediction of muon LFV branching ratios between the little Higgs model and SUSY models [26]. In well-motivated SUSY models like the SUSY seesaw and SUSY GUT models, the photonic dipole operator is dominant for most of the parameter space, and Z-penguin and box diagrams are sub-dominant. As a result, the branching ratios of $\mu \to 3e$ and $\mu - e$ conversion are smaller than the $\mu \to e\gamma$ branching ratio by two orders of magnitude. (The ratio $B(\mu \to 3e)/B(\mu \to e\gamma)$ is about 6×10^{-3}.) On the other hand, the contribution of the photonic dipole operator is sub-dominant for $\mu \to 3e$ in the little Higgs model with T-parity, and $B(\mu \to 3e)/B(\mu \to e\gamma)$ is $O(1)$. Similarly, the ratio of the $\mu - e$ conversion and $\mu \to e\gamma$ branching ratio can be enhanced compared to the SUSY case. This implies that searches for the $\mu - e$ conversion and the $\mu \to 3e$ process become important to distinguish theoretical models, should $\mu \to e\gamma$ be discovered in the current MEG experiment at the Paul Scherrer Institute.

18.4. Neutrino Mass from TeV Physics and LFV

Various mechanisms of the neutrino mass generation have been proposed besides the simplest seesaw model and the Dirac neutrino model. In many of these cases, the interaction responsible for the neutrino mixings also induces LFV. The supersymmetric seesaw model discussed in the previous section is one of the examples. Other examples include the Zee model, [29] Dirac-type bulk neutrinos in the warped extra dimension, [30] the triplet

Higgs model, [31, 32] and the non-supersymmetric left-right symmetric model [33–36]. Since each model introduces the lepton flavor violation in a different way, phenomenological features can be quite different. These are important clues to identify the correct model of neutrino mass generation.

The triplet Higgs model provides a simple framework to generate neutrino masses from a small triplet vacuum expectation value. In this model, the triplet Higgs and lepton coupling for the neutrino mass also induces doubly charged Higgs boson and lepton coupling, and the neutrino mixing matrix is directly related to the LFV doubly charged Higgs boson coupling. LFV in the triplet Higgs model has been studied in detail [31, 32]. Since the doubly charged Higgs boson makes a tree-level contribution to the $\mu \to 3e$ process, this process has a larger branching ratio than $\mu \to e\gamma$ and the $\mu - e$ conversion in general. This is especially the case for the inverse hierarchy and degenerate cases of the neutrino mass spectrum, in which the former is larger by about two orders of magnitudes than the latter two. In the normal hierarchy case, there is a possibility that three processes have similar branching ratios due to partial cancellation in the $\mu \to 3e$ process.

The left-right symmetric model also has the triplet Higgs field. In this case, however, neutrino masses can be generated by the low-scale seesaw mechanism. The right-handed neutrino mass term arises in association with $SU(2)_L \times SU(2)_R \times U(1)_{B-L}$ symmetry breaking to the SM gauge groups. If this scale is close to the TeV scale, observable LFV effects are generated through the doubly charged Higgs boson and lepton couplings [33]. Unlike the triplet Higgs model, the relationship between the neutrino mixing and LFV is not straightforward. A generic feature is however that the $\mu \to 3e$ branching ratio is larger by two orders of magnitude compared to $\mu \to e\gamma$ and the $\mu - e$ conversion branching ratios [34]. There are various possibilities of tau LFV processes depending on where the origin of large neutrino mixings is attributed [36].

18.5. Model with Extra Dimensions

The idea of *extra dimensions* is quite old, and went back to the Kaluza–Klein theory, where gauge theory and the gravity interaction were unified in the five-dimensional space-time. The modern theory of extra dimensions is motivated by the superstring theory which can be formulated in special space-time dimensions, and compactification to the four-dimensional space-time is mandatory. Various models of particle physics have been proposed so far based on extra-dimensions: Some are intimately related to the superstring, and others are not.

In recent years, particle models in the context of extra-dimensional space-time are introduced to explain the weakness of the gravity interaction compared to the gauge interaction. In the model of the flat extra dimensions, [37] the size of extra-dimensional space is taken to be large and gravity is allowed to propagate to extra-space, so that there is no physical mass scale identified as the Planck scale. The physical fundamental scale of gravity lies in the range of the TeV scale, and the gravity interaction becomes as strong as other gauge interactions at the fundamental scale. In the warped extra-dimension, [38] the hierarchy between the Planck scale and the weak scale is attributed to the warped geometry. This model provides a solution of the hierarchy problem in the SM. In both cases, a variety of possibilities have been considered regarding which particles are allowed to propagate in extra-dimensions, and phenomenological consequences change from one model to another.

Flavor physics becomes relevant in models of extra-dimensions if one tries to understand the known pattern of the fermion mass hierarchy in terms of geometrical setup in extra-dimensions. For instance, if one allows fermions to propagate in (some of) the extra dimensions, their masses are determined by overlaps of relevant wave functions in the extra direction, rather than the size of the Yukawa coupling constants. LFV in this type of models was considered in the connection to neutrino mass generation in the model of the warped extra-dimension [30, 39]. Here, Dirac type neutrinos are allowed to propagate in the extra-dimension. A small Dirac neutrino Yukawa coupling is realized for the lowest mode of Kaluza–Klein neutrinos, but couplings of higher modes to the SM particles are not in general suppressed. In this model, LFV processes are not necessarily suppressed, but provide severe constraints to model parameters. Generalization to other phenomenological models based on a warped extra-dimension has been considered [40–42]. In recent works, an appropriate flavor symmetry is introduced to explain neutrino mixing patterns in this geometrical setup [43, 44]. These kinds of models open an interesting possibility of connecting flavor and energy frontier physics. The masses and mixings of the SM fermion sector is dynamically determined by the TeV scale physics, and therefore can be experimentally studied by the combination of new particle searches at colliders, along with FCNC and LFV processes at low energies.

18.6. Violation of Lorentz Invariance

A possible violation of the Lorentz invariance has been considered in connection with very high energy cosmic rays beyond the Greisen, Zatsepin, and Kuz'min (GZK) cutoff [45]. In this scheme the Lorentz transformation is not invariant, but only the translational and rotational symmetries are assumed to be exact in a preferred system. Thus the maximum attainable velocity could be different for each species of particles, and this would cause many unique phenomena in particle physics and cosmic-ray physics.

Muon LFV processes provide a good test of the violation of Lorentz invariance. If a small Lorentz-non-invariant interaction exists in the SM Lagrangian, flavor mixing couplings can be in general allowed in the photon-fermion interaction even at the renormalizable interactions of quantum electro dynamics. The current limit on the $\mu \to e\gamma$ branching ratio puts a strong constraint on the relevant coupling constants. Another interesting effect is a change of the muon lifetime at a high energy. Since the $\mu \to e\gamma$ decay width due to the Lorentz non-invariant interaction would increase with higher energy, it would eventually dominate over the ordinary muon decay. As a result, the muon lifetime might start decreasing at a sufficiently-high energy. The current limit on the energy dependence of the muon lifetime has been obtained by the experiment of the muon anomalous magnetic moment. The constraint on possible Lorentz violation parameters is similar to that obtained from the $\mu \to e\gamma$ branching ratio.

18.7. Summary of LFV in Various Models

As we have discussed, there are many New Physics models which predict sizable branching fractions of muon LFV processes. All of these models involve new particles and new interactions at the TeV scale. For each model, the relative importance of the three muon LFV processes, $\mu \to e\gamma$, $\mu \to 3e$, and $\mu - e$ conversion, is different.

- The branching ratio of $\mu \to e\gamma$ is larger by two orders of magnitude than those of $\mu \to 3e$ and the $\mu - e$ conversion for most of parameter space in R-parity conserving SUSY models. An exception is the Higgs-mediated LFV that is possible for large $\tan\beta$ cases, where the $\mu - e$ conversion branching ratio can be close to that of $\mu \to e\gamma$.
- There are a variety of possibilities in R parity-violating SUSY models. Depending on combinations of allowed coupling constants, $\mu - e$

conversion or $\mu \to 3e$ processes may be induced at the tree level and can be more important than the $\mu \to e\gamma$ process.

- In the littlest Higgs model with T-parity, the pattern of three branching ratios are different from SUSY models, even though LFV processes are induced by loop diagrams of new particles in both cases. The $\mu - e$ conversion and $\mu \to 3e$ can have a similar branching ratios to $\mu \to e\gamma$.
- Models with neutrino mass generation at the TeV scale have direct impact on LFV processes because interactions which determine neutrino mass matrices can also generate LFV processes. Many of these models involve doubly charged Higgs bosons with flavor off-diagonal lepton couplings. The $\mu \to 3e$ process is the most important constraining processes for such cases.
- In models of extra-dimensions, the phenomenology of LFV processes depends on details of model structure, in particular on which particles are allowed to propagate in extra-dimensional space.
- In a scheme of Lorentz violation the $\mu \to e\gamma$ process has a unique feature because this process is induced at the tree level within renormalizable interactions.

Relationships among the three muon LFV processes is a useful way to distinguish between different theoretical models. We have also discussed other observable quantities sensitive to the nature of LFV interactions. Angular distributions of polarized muon decays in $\mu \to e\gamma$ and $\mu \to 3e$ processes, atomic number dependence of the $\mu - e$ conversion branching ratios are examples. The relationship between muon and tau LFV processes is also important to explore the flavor structure of LFV interactions.

Acknowledgments

The work is supported in part by the Grant-in-Aid for Science Research, Ministry of Education, Culture, Sports, Science and Technology, Japan, No. 16081211 and by the Grant-in-Aid for Science Research, Japan Society for the Promotion of Science, No. 20244037.

References

[1] See for example Y. Kuno and Y. Okada, Rev. Mod. Phys. **73**, 151 (2001).
[2] For reviews on supersymmetry, see for example H. P. Nilles, Phys. Rept. **110**, 1 (1984).

[3] F. Borzumati and A. Masiero, Phys. Rev. Lett. **57**, 961 (1986). For recent references, see A. Masiero, S. K. Vempati and O. Vives, New J. Phys. **6**, 202 (2004).
[4] T. Goto, Y. Okada, T. Shindou and M. Tanaka, Phys. Rev. D **77**, 095010 (2008).
[5] T. Browder, M. Ciuchini, T. Gershon, M. Hazumi, T. Hurth, Y. Okada and A. Stocchi, JHEP **0802**, 110 (2008).
[6] K. S. Babu and C. Kolda, Phys. Rev. Lett. **89**, 241802 (2002).
[7] M. Sher, Phys. Rev. D **66**, 057301 (2002).
[8] A. Dedes, J. R. Ellis and M. Raidal, Phys. Lett. B **549**, 159 (2002).
[9] R. Kitano, M. Koike, S. Komine and Y. Okada, Phys. Lett. B **575**, 300 (2003).
[10] R. Kitano, M. Koike and Y. Okada, Phys. Rev. D **66**, 096002 (2002) [Erratum-ibid. D **76**, 059902 (2007)].
[11] L. J. Hall, V. A. Kostelecky and S. Raby, Nucl. Phys. B **267**, 415 (1986); R. Barbieri and L. J. Hall, Phys. Lett. B **338**, 212 (1994); R. Barbieri, L. J. Hall and A. Strumia, Nucl. Phys. B **445**, 219 (1995).
[12] J. Hisano, T. Moroi, K. Tobe and M. Yamaguchi, Phys. Lett. B **391**, 341 (1997) [Erratum-ibid. B **397**, 357 (1997)].
[13] J. Hisano, D. Nomura, Y. Okada, Y. Shimizu and M. Tanaka, Phys. Rev. D **58**, 116010 (1998).
[14] Y. Okada, K. i. Okumura and Y. Shimizu, Phys. Rev. D **58**, 051901 (1998); Phys. Rev. D **61**, 094001 (2000).
[15] J. R. Ellis, J. Hisano, S. Lola and M. Raidal, Nucl. Phys. B **621**, 208 (2002).
[16] D. F. Carvalho, J. R. Ellis, M. E. Gomez and S. Lola, Phys. Lett. B **515**, 323 (2001).
[17] J. Hisano and K. Tobe, Phys. Lett. B **510**, 197 (2001).
[18] A. de Gouvea, S. Lola and K. Tobe, Phys. Rev. D **63**, 035004 (2001).
[19] R. Barbieri and A. Strumia, Phys. Lett. B **462**, 144 (1999); arXiv:hep-ph/0007265.
[20] N. Arkani-Hamed, A. G. Cohen, E. Katz, A. E. Nelson, T. Gregoire and J. G. Wacker, JHEP **0208**, 021 (2002).
[21] N. Arkani-Hamed, A. G. Cohen, E. Katz and A. E. Nelson, JHEP **0207**, 034 (2002).
[22] C. Csaki, J. Hubisz, G. D. Kribs, P. Meade and J. Terning, Phys. Rev. D **67**, 115002 (2003); J. L. Hewett, F. J. Petriello and T. G. Rizzo, JHEP **0310**, 062 (2003); C. Csaki, J. Hubisz, G. D. Kribs, P. Meade and J. Terning, Phys. Rev. D **68**, 035009 (2003).
[23] H. C. Cheng and I. Low, JHEP **0309**, 051 (2003); H. C. Cheng and I. Low, JHEP **0408**, 061 (2004).
[24] J. Hubisz, S. J. Lee and G. Paz, JHEP **0606**, 041 (2006); M. Blanke, A. J. Buras, A. Poschenrieder, S. Recksiegel, C. Tarantino, S. Uhlig and A. Weiler, Phys. Lett. B **646**, 253 (2007).
[25] M. Blanke, A. J. Buras, A. Poschenrieder, C. Tarantino, S. Uhlig and A. Weiler, JHEP **0612**, 003 (2006); M. Blanke, A. J. Buras, A. Poschenrieder, S. Recksiegel, C. Tarantino, S. Uhlig and A. Weiler, JHEP **0701**,

066 (2007); M. Blanke, A. J. Buras, S. Recksiegel, C. Tarantino and S. Uhlig, Phys. Lett. B **657**, 081 (2007); M. Blanke, A. J. Buras, S. Recksiegel, C. Tarantino and S. Uhlig, JHEP **0706**, 082 (2007); M. Blanke, A. J. Buras, S. Recksiegel and C. Tarantino, arXiv:0805.4393 [hep-ph].
[26] M. Blanke, A. J. Buras, B. Duling, A. Poschenrieder and C. Tarantino, JHEP **0705**, 013 (2007).
[27] T. Goto, Y. Okada and Y. Yamamoto, Phys. Lett. B **670**, 378 (2009).
[28] F. del Aguila, J. I. Illana and M. D. Jenkins, arXiv:0811.2891 [hep-ph].
[29] K. Hasegawa, C. S. Lim and K. Ogure, Phys. Rev. D **68**, 053006 (2003).
[30] R. Kitano, Phys. Lett. B **481**, 39 (2000).
[31] E. J. Chun, K. Y. Lee and S. C. Park, Phys. Lett. B **566**, 142 (2003).
[32] M. Kakizaki, Y. Ogura and F. Shima, Phys. Lett. B **566**, 210 (2003).
[33] C. S. Lim and T. Inami, Prog. Theor. Phys. **67**, 1569 (1982).
[34] V. Cirigliano, A. Kurylov, M. J. Ramsey-Musolf and P. Vogel, Phys. Rev. D **70**, 075007 (2004).
[35] O. M. Boyarkin, G. G. Boyarkina and T. I. Bakanova, Phys. Rev. D **70**, 113010 (2004).
[36] A. G. Akeroyd, M. Aoki and Y. Okada, Phys. Rev. D **76**, 013004 (2007).
[37] N. Arkani-Hamed, S. Dimopoulos and G. R. Dvali, Phys. Lett. B **429**, 263 (1998); I. Antoniadis, N. Arkani-Hamed, S. Dimopoulos and G. R. Dvali, Phys. Lett. B **436**, 257 (1998); N. Arkani-Hamed, S. Dimopoulos and G. R. Dvali, Phys. Rev. D **59**, 086004 (1999).
[38] L. Randall and R. Sundrum, Phys. Rev. Lett. **83**, 3370 (1999).
[39] Y. Grossman and M. Neubert, Phys. Lett. B **474**, 361 (2000).
[40] S. J. Huber, Nucl. Phys. B **666**, 269 (2003).
[41] G. Moreau and J. I. Silva-Marcos, JHEP **0603**, 090 (2006).
[42] K. Agashe, A. E. Blechman and F. Petriello, Phys. Rev. D **74**, 053011 (2006).
[43] G. Perez and L. Randall, arXiv:0805.4652 [hep-ph].
[44] C. Csaki, C. Delaunay, C. Grojean and Y. Grossman, JHEP **0810**, 055 (2008).
[45] S. R. Coleman and S. L. Glashow, Phys. Rev. D **59**, 116008 (1999).

Chapter 19

Search for the Charged Lepton-Flavor-Violating Transition Moments $l \to l'$

Yoshitaka Kuno

Department of Physics, Osaka University
1-10-19 Machikane-yama, Toyonaka, Japan
kuno@phys.sci.osaka-u.ac.jp

This article describes the experimental status of the searches for processes in which charged lepton flavor is not conserved. Phenomenology, current experimental status and future prospects for selected processes of lepton flavor violation for muon and tau leptons are presented.

Contents

19.1. Introduction

Our understanding of elementary particle physics is based on the Standard Model (SM), which is a gauge theory of the strong and electroweak interactions. The SM has been scrutinized by numerous experimental tests, and to date it is still consistent with most precision measurements. In the minimal version of the SM where massless neutrinos are assumed, lepton flavor conservation is a natural consequence of the gauge invariance. Therefore, it provided a naive explanation for why lepton flavor violation (LFV) in charged leptons[a] is highly suppressed.

However, recently there has been firm evidence for the existence of non-zero neutrino masses and mixing based on the results of the neutrino oscillation experiments. Since neutrino oscillations indicate that lepton flavor is not conserved, LFV processes involving muon and tau leptons are also expected to occur. In the framework of the SM with massive neutrinos, however, the neutrino mixing introduces only small contributions to cLFV processes. For example, the branching ratio of $\mu^+ \to e^+\gamma$ is of the order of $O(10^{-54})$.

However, in extensions to the SM, cLFV could occur from various sources of New Physics beyond the SM. In fact, in many New Physics scenarios, one would expect cLFV at a sizeable level. One of the well-motivated theoretical models predicting cLFV is supersymmetry (SUSY). The resulting cLFV rates can be as large as the present experimental upper bounds. And therefore they could be accessible and will be tested at future experiments.

There has been much experimental progress in searching for cLFV with muons and taus. First of all, several new results have been obtained using the highly intense sources of muons and taus now available, and on-going and proposed experiments are aiming for further improvements. Furthermore, in the long-term future attempts to create new sources of muons and taus with even higher intensities have been initiated. With this increased muon flux, significant improvements in experimental searches can be anticipated.

[a]Lepton flavor violation for charged leptons will be referred to as "cLFV" from now on where appropriate.

In this article, we review the current experimental status of the field of searches for lepton flavor violation, and its potential for probing physics beyond the SM. We particularly emphasize the importance of low-energy cLFV searches with muons and taus. There have been many excellent review articles on muon decays and lepton flavor violation [1–9]. But in order to renew current interest in cLFV, this article has been written to bring this topic up to date. The phenomenology and experimental status of some of the selected important processes for muons and taus are described in detail.

This article is organized as follows. In Section 19.2, we give a short history on cLFV since the discovery of the muon in 1937. In Section 19.3, the physics motivation of cLFV is discussed in a model-independent approach and then the supersymmetric extension of the Standard Model is described. In Sections 19.4, 19.5 and 19.6, we describe the phenomenology and experimental status of the most recent experiments which have searched for $\mu^+ \to e^+\gamma$, $\mu^+ \to e^+e^+e^-$, $\mu^- - e^-$ conversion in a muonic atom, respectively. In Section 19.7, the searches for LFV in τ decays in low-energy e^+e^- colliders are briefly discussed, together with in-flight LFV processes to produce taus. In Section 19.8, the future outlook is presented.

19.2. History

In 1937, the muon was discovered by Neddermeyer and Anderson [10] in cosmic rays, with a mass which was found to be about 200 times the mass of the electron. The discovery of the muon was made just after Yukawa [11] postulated the existence of the π meson as a force carrier of the nuclear force in 1935. However, it was demonstrated by Conversi *et al.* [12] in 1947 that the muon did not interact through the strong interaction, and thus it could not be the π meson. Rabi made the famous comment "Who ordered that?", which indicates how puzzling the existence of a new lepton was.

At that time, it was believed that if the muon were simply a heavy electron it would decay into an electron and a γ-ray. In 1947, the first search for $\mu^+ \to e^+\gamma$ was made by Hincks and Pontecorvo by using cosmic-ray muons [14]. Its negative result set an upper limit on the branching ratio of less than 10%. This was the beginning of the search for cLFV. In 1948, the continuous spectrum of electrons from muon decay was established by Steinberger [13]. It suggested that a three-body decay of the muon gives rise to a final state of an electron accompanied by two neutral particles. Soon afterwards, in 1952, the search for the neutrino-less $\mu^- - e^-$ conversion process ($\mu^- N \to e^- N$, where N is a nucleus capturing the muon) was also carried out by Lagarrigue and Peyrou [15], who obtained a negative result. Such searches were significantly improved when muons became artificially produced at accelerators. In 1955, the upper limits of the branching ratios of $B(\mu \to e\gamma) < 2\times10^{-5}$ by Lokonathan and Steinberger [16] and $B(\mu^- Cu \to e^- Cu) < 5\times10^{-4}$ by Steinberger and Wolfe [17] were set using the Columbia University Nevis cyclotron.

After the discovery of parity violation in 1956, it was suggested by Feynman and Gell-Mann [18] that the weak interaction took place through the exchange of charged intermediate vector bosons. In 1958, Feinberg [19] pointed out that the intermediate vector boson, if it exists, would lead to $\mu^+ \to e^+\gamma$ at a branching ratio of 10^{-4}. The absence of any experimental observation of the $\mu^+ \to e^+\gamma$ process with $B(\mu \to e\gamma) > 2 \times 10^{-5}$ led directly to the two-neutrino hypothesis by Nishijima [20] and Schwinger [21], in which the neutrino that coupled to the muon differs from that coupled to the electron, and the $\mu^+ \to e^+\gamma$ process would be forbidden. The two-neutrino hypothesis was verified experimentally at Brookhaven National Laboratory (BNL) by observing muon production but not electron production from the scattering of neutrinos produced from pion decays by G. Danby *et al.* [22]. This introduced the concept of a separate conservation law for individual lepton flavors, electron number (L_e) and muon number (L_μ).

In the 1970s, three meson factories, SIN (Swiss Institute for Nuclear Research, Switzerland)[b], LAMPF (The Clinton P. Anderson Meson Physics Facility, New Mexico, USA) and TRIUMF (The TRI-University Meson Factory, Vancouver, Canada), which produced many muons and pions using highly intense, low-energy proton beams, were built. The searches for cLFV processes with muons were rapidly improved.

The historical progress in various cLFV searches in muon and kaon decays is shown in Fig. 19.1, from which it is seen that the experimental upper limits have been continuously improved at a rate of about two orders of magnitude per decade during the 50 years since the first cLFV experiment by Hincks and Pontecorvo in 1947.

The present upper limits of various cLFV decays are listed in Table 19.1. As seen in Table 19.1, the current searches for cLFV with muons are now sensitive to branching ratios of the order of $10^{-12} - 10^{-13}$. In general,

---

[b]It is now called the Paul Scherrer Institute (PSI).

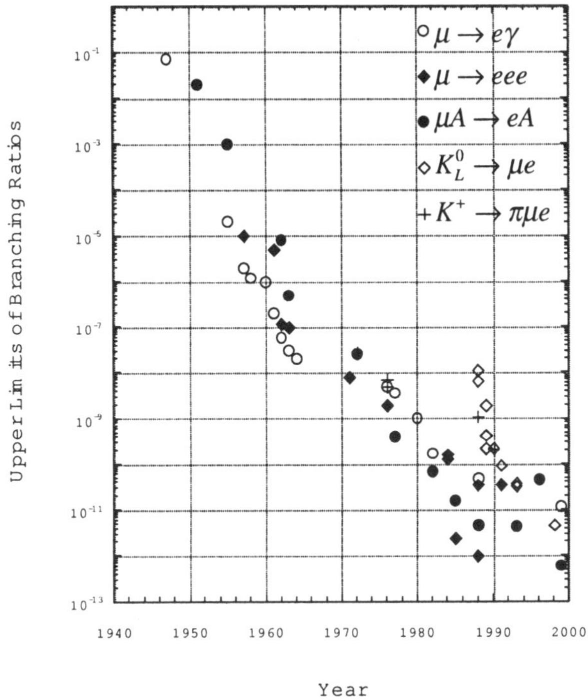

Fig. 19.1. Historical progress of lepton flavor violation (LFV) of charged leptons for various processes of muons and kaons. (Reprinted with permission from Ref. [8]. Copyright (2001) by the American Physical Society.)

searches for rare processes could probe new interactions mediated by very heavy particles. For example, in the four fermion interaction, the LFV branching ratios could be scaled by $(m_W/m_X)^4$, where m_X is the mass of a hypothetical heavy particle responsible for the LFV interaction and m_W is the mass of the W gauge boson. In such a scenario, the present sensitivities for cLFV searches in muon decays could probe m_X up to several 100 TeV, which is not directly accessible at present or planned accelerators. From Table 19.1, it can be seen that the cLFV sensitivity for the muon system is very high. This is mostly because of the large number of muons available for experimental searches nowadays (about $10^{14} - 10^{15}$ muons/year). Moreover, an even greater number of muons (about $10^{19} - 10^{20}$ muons/year) will be available in the future, if new highly intense muon sources are realized. Important LFV processes involving muons are $\mu^+ \to e^+\gamma$, $\mu^- - e^-$ conversion in a muonic atom ($\mu^- N \to e^- N$), $\mu^+ \to e^+e^+e^-$, and muonium (the

Table 19.1. Experimental limits for the lepton-flavor violating decays of muon, tau, pion, kaon and Z boson.

Reaction	Present limit	Reference
$\mu^+ \to e^+\gamma$	$< 1.2 \times 10^{-11}$	Brooks *et al.* [49]
$\mu^+ \to e^+e^+e^-$	$< 1.0 \times 10^{-12}$	Bellgardt *et al.* [55]
$\mu^- Ti \to e^- Ti$	$< 4.3 \times 10^{-12}$	C. Dohmen *et al.* [70]
$\mu^- Ti \to e^- Ti$	$< 6.1 \times 10^{-13}$	Wintz [72] *
$\mu^- Au \to e^- Au$	$< 7 \times 10^{-13}$	Bert *et al.* [73]
$\mu^- Pb \to e^- Pb$	$< 4.6 \times 10^{-11}$	Honecker *et al.* [71]
$\mu^+e^- \to \mu^-e^+$	$< 8.3 \times 10^{-11}$	Willmann *et al.* [23]
$\tau \to e\gamma$	$< 1.1 \times 10^{-7}$	Aubert *et al.* [24]
$\tau \to \mu\gamma$	$< 4.5 \times 10^{-8}$	Hayasaka *et al.* [25]
$\tau \to \mu\mu\mu$	$< 3.2 \times 10^{-8}$	Miyazaki *et al.* [26]
$\tau \to eee$	$< 3.6 \times 10^{-8}$	Miyazaki *et al.* [26]
$\pi^0 \to \mu e$	$< 8.6 \times 10^{-9}$	Edwards *et al.* [27]
$K_L^0 \to \mu e$	$< 4.7 \times 10^{-12}$	Ambrose *et al.* [28]
$K^+ \to \pi^+\mu^+e^-$	$< 2.1 \times 10^{-10}$	Lee *et al.* [29]
$K_L^0 \to \pi^0\mu^+e^-$	$< 3.1 \times 10^{-9}$	Arisaka *et al.* [30]
$Z^0 \to \mu e$	$< 1.7 \times 10^{-6}$	Akers *et al.* [31]
$Z^0 \to \tau e$	$< 9.8 \times 10^{-6}$	Akers *et al.* [31]
$Z^0 \to \tau\mu$	$< 1.2 \times 10^{-5}$	Abreu *et al.* [32]

*Not published.

μ^+e^- atom) (Mu) to anti-muonium ($\overline{\text{Mu}}$) conversion (Mu$-\overline{\text{Mu}}$ conversion).

19.3. Physics Motivation

19.3.1. *The Standard Model*

In the minimal Standard Model, lepton flavor conservation is built in by assuming vanishingly small neutrino masses. However, recently, neutrino mixing has been experimentally confirmed by the discovery of neutrino oscillations, and lepton flavor conservation is known to be violated. However, LFV of charged leptons has yet to be observed experimentally. The predicted branching ratio to the $\mu^+ \to e^+\gamma$ decay in the Standard Model with massive neutrinos and their mixing are given by

$$B(\mu \to e\gamma) = \frac{3\alpha}{32\pi}\left|\sum_l (V_{MNS})^*_{\mu l}(V_{MNS})_{el}\frac{m^2_{\nu_l}}{M^2_W}\right|^2 \qquad (19.1)$$

where $(V_{MNS})_{\alpha l})$ is the lepton flavor mixing matrix (the Maki–Nakagawa–Sakata matrix). m_{ν_l} and m_W are the masses of neutrino ν_l and of the W boson respectively. It is known that this contribution is extremely small, since it is proportional to $(m_{\nu_l}/m_W)^4$ (i.e. the GIM mechanism), yielding the order of $O(10^{-54})$ or less in its branching ratio, which depends on the neutrino mixing parameters and neutrino mass hierarchy. Therefore, discovery of cLFV would imply New Physics beyond "neutrino oscillations". As a matter of fact, any New Physics or interactions beyond the Standard Model would predict cLFV at some level. The physics motivation for studying the physics of cLFV throughout the next decade will be very robust.

19.3.2. *Model-independent approach*

In an extension of New Physics beyond the Standard Model, the effective Lagrangian for $\mu^+ \to e^+\gamma$ (of a dipole-interaction type) can be given by

$$\mathcal{L}_D = y_D \frac{em_\mu}{\Lambda_D^2} \bar{\mu}_R \sigma^{\mu\nu} e_L F_{\mu\nu} + h.c. + ..., \tag{19.2}$$

where Λ_D is an energy scale of New Physics and y_D is an effective coupling constant. By using this effective Lagrangian, the branching ratio of $\mu^+ \to e^+\gamma$ decay can be calculated as

$$B(\mu \to e\gamma) = (y_D)^2 \frac{3(4\pi)^3 \alpha}{G_F^2 \Lambda_D^4}. \tag{19.3}$$

When the new interaction operates at a tree level, then $y \sim 1$ and

$$B(\mu \to e\gamma) = (1 \times 10^{-11}) \times \left(\frac{400 \text{ TeV}}{\Lambda_D}\right)^4 \left(\frac{y_D}{1}\right)^2, \tag{19.4}$$

and a search for $\mu^+ \to e^+\gamma$ is sensitive to very high energy scale like several 100 TeV.

On the other hand, when the new interaction occurs at a loop level, then by defining $y_D \sim \theta_{\mu e} g^2/16\pi^2$, the branching ratio of $\mu^- - e^-$ is given by

$$B(\mu \to e\gamma) = (1 \times 10^{-11}) \times \left(\frac{2 \text{ TeV}}{\Lambda_D}\right)^4 \left(\frac{\theta_{\mu e}}{10^{-2}}\right)^2, \tag{19.5}$$

where g is the coupling of weak interaction and $\theta_{\mu e}$ is an effective coupling parameter of New Physics. It is sensitive to physics at the 1 TeV scale with a small effective coupling parameter $\theta_{\mu e}$ of 10^{-2} level. It would be the case for low-energy supersymmetry. From this investigation, it is known that a search for cLFV at a 10^{-11} level in branching ratios is sensitive to New Physics at TeV energy, which will be complementarily studied at the LHC, and other precision flavor mixing physics of 10^{-2}.

Based on this model-independent approach, a ratio of the branching ratios of $\mu^+ \to e^+e^+e^-$ to $\mu^+ \to e^+\gamma$ decays is given by

$$\frac{B(\mu \to eee)}{B(\mu \to e\gamma)} = \frac{\alpha}{3\pi}\left(\ln\frac{m_\mu^2}{m_e^2} - \frac{11}{4}\right) = 6 \times 10^{-3}. \tag{19.6}$$

Similarly, $\mu^- - e^-$ conversion in a muonic atom is also suppressed with respect to $\mu^+ \to e^+\gamma$ decay. It will be discussed in Section 19.6.1.1.

Next, let us examine the relation between cLFV and the muon $g-2$ value. The effective Lagrangian of New Physics for the muon $g-2$ can be given by

$$\mathcal{L}_{\mathrm{g}-2} = y\frac{em_\mu}{2\Lambda_D^2}\bar{\mu}\sigma^{\mu\nu}\mu F_{\mu\nu}, \tag{19.7}$$

where y is the flavor-conserving effective coupling for New Physics. The anomalous magnetic moment δa_μ is given by

$$\delta a_\mu = y\frac{2m_\mu^2}{\Lambda_D^2}, \tag{19.8}$$

and the relation between the muon $g-2$ and cLFV can be created if $y_D = y\theta_{\mu e}$ by

$$B(\mu \to e\gamma) = \frac{3(4\pi)^3\alpha}{4G_F^2 m_\mu^4}(\delta a_\mu)^2\theta_{\mu e} \tag{19.9}$$

$$\sim 0.6 \times 10^{-11}\left(\frac{\delta a_\mu}{10^{-9}}\right)^2\left(\frac{\theta_{\mu e}}{10^{-4}}\right)^2. \tag{19.10}$$

It would indicate that New Physics contribution to the muon $g-2$ at the level of $\delta a_\mu \sim 10^{-9}$ can also contribute to cLFV at the 10^{-11} level.

We can also consider the Lagrangian for an effective four fermion interaction with lepton flavor violation. It can be given, for example, by

$$\mathcal{L}_F = y_F\frac{1}{\Lambda_F^2}\bar{\mu}_L\gamma^\mu e_L\bar{f}_L\gamma_\mu f_L + h.c., \tag{19.11}$$

where y_F and Λ_F are an effective coupling and an energy scale of New Physics respectively. And f is any fermion, which could be an electron for $\mu^+ \to e^+e^+e^-$ decay or light quarks for $\mu^- - e^-$ conversion. If $\mu^+ \to e^+e^+e^-$ decay occurs at a tree level, then a ratio of the branching ratios of $\mu^+ \to e^+e^+e^-$ and $\mu^+ \to e^+\gamma$ decays is given by

$$\frac{B(\mu \to eee)}{B(\mu \to e\gamma)} = \frac{1}{12(4\pi)^2}\left(\frac{y_F}{y_D}\right)^2\left(\frac{\Lambda_D}{\Lambda_F}\right)^4. \tag{19.12}$$

In this case, the branching ratios of $\mu^- - e^-$ and $\mu^+ \to e^+e^+e^-$ could become comparable depending on the parameters.

We can combine the Lagrangian for a dipole interaction and that for an effective four fermion interaction as follows [76],

$$\begin{aligned}\mathcal{L} &= y_D \frac{em_\mu}{\Lambda_D^2} \bar{\mu}_R \sigma^{\mu\nu} e_L F_{\mu\nu} + y_F \frac{1}{\Lambda_F^2} \bar{\mu}_L \gamma^\mu e_L \bar{f}_L \gamma_\mu f_L + h.c. \\ &= \frac{m_\mu}{(\kappa+1)\Lambda^2} \bar{\mu}_R \sigma^{\mu\nu} e_L F_{\mu\nu} + \frac{\kappa}{(\kappa+1)\Lambda^2} \bar{\mu}_L \gamma^\mu e_L \bar{f}_L \gamma_\mu f_L + h.c.,\end{aligned} \tag{19.13}$$

where the parameter κ determines the relative magnitudes for the dipole interaction and the effective four fermion interaction. κ and Λ are given by

$$\kappa = \frac{y_F}{e y_D} \left(\frac{\Lambda_D^2}{\Lambda_F^2} \right) \tag{19.14}$$

$$\Lambda^2 = \frac{\Lambda_D^2 \Lambda_F^2}{y_F \Lambda_F^2 + y_D e \Lambda_D^2}. \tag{19.15}$$

The branching ratio of $\mu^- - e^-$ conversion in Ti can be calculated as a function of κ and is given in Fig. 19.2. The parameter κ interpolates between an effective dipole LFV interaction ($\kappa << 1$) and an effective four fermion LFV interaction ($\kappa >> 1$).

19.3.3. *Supersymmetry models*

It is known that cLFV has received sizable contributions from low-energy supersymmetry (SUSY). In particular it becomes significant if SUSY particles exist in the LHC energy range. In SUSY models, cLFV would occur through mixing of their SUSY partners, namely mixing of sleptons $\tilde{l}$. Fig. 19.3 shows one of the diagrams of SUSY contributing to a muon to electron transition, where the mixing of a smuon ($\tilde{\mu}$) and a selectron ($\tilde{e}$) is given by the off-diagonal slepton mass matrix element $\Delta m^2_{\tilde{\mu}\tilde{e}}$, where the slepton mass matrix ($m^2_{\tilde{l}}$) is given by Eq. (19.16).

$$m_{\tilde{l}} = \begin{pmatrix} m^2_{\tilde{e}\tilde{e}} & \Delta m^2_{\tilde{e}\tilde{\mu}} & \Delta m^2_{\tilde{e}\tilde{\tau}} \\ \Delta m^2_{\tilde{\mu}\tilde{e}} & m^2_{\tilde{\mu}\tilde{\mu}} & \Delta m^2_{\tilde{\mu}\tilde{\tau}} \\ \Delta m^2_{\tilde{\tau}\tilde{e}} & \Delta m^2_{\tilde{\tau}\tilde{\mu}} & m^2_{\tilde{\tau}\tilde{\tau}} \end{pmatrix}. \tag{19.16}$$

In one type of SUSY model called the supergravity model, the slepton mass matrix is assumed to be diagonal at the Planck mass scale ($\sim 10^{19}$ GeV), and no off-diagonal matrix elements exist (for instance, $\Delta m^2_{\tilde{\mu}\tilde{e}} = 0$). It is called the universal scalar mass hypothesis. However, non-zero off-diagonal matrix elements can be induced by radiative corrections from the Planck

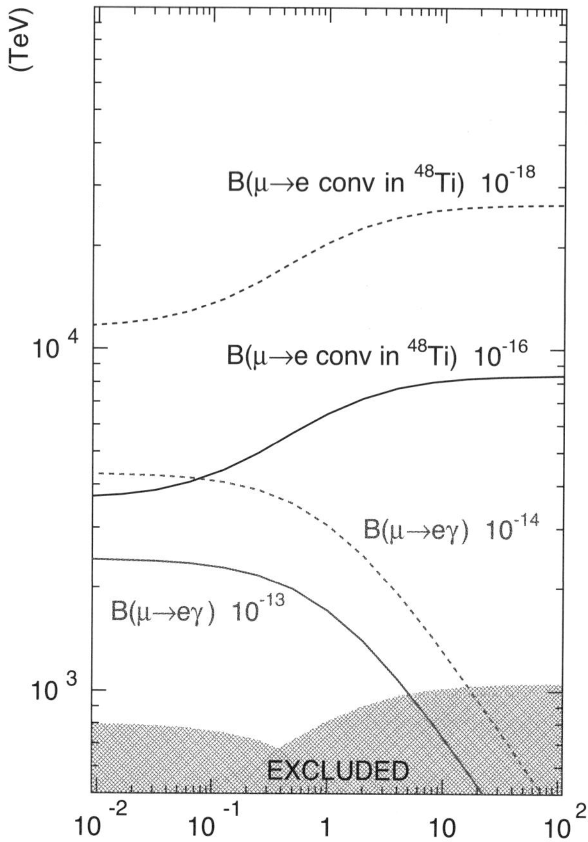

Fig. 19.2. Sensitivity of a $\mu^- - e^-$ conversion in Ti and $\mu^- - e^-$ decay to New Physics scale Λ as a function of κ. The parameter κ interpolates between an effective dipole LFV interaction ($\kappa << 1$) and an effective four-fermion LFv interaction ($\kappa >> 1$). The excluded region by the present experimental limits is also shown. From the Mu2e Proposal (2008).

scale to the weak scale ($\sim 10^2$ GeV), when New Physics mechanisms exist between the Planck scale and weak scale [33, 34].

One scenario of New Physics could be the Grand Unification Theory (GUT) models, where the Yukawa interactions at a GUT energy scale create non-zero off-diagonal elements of the slepton mass matrix [35]. This scenario with supersymmetry is called a SUSY-GUT model. Another scenario could be constituted by the neutrino seesaw mechanism, where the

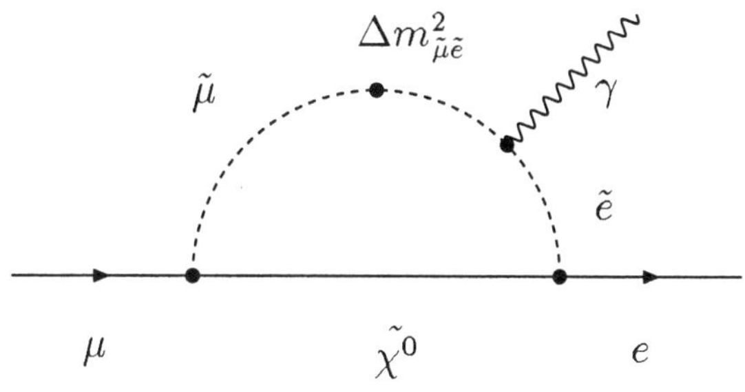

Fig. 19.3. One of the diagrams of SUSY contributions to a μ to e transition. $\Delta m^2_{\tilde{\mu}\tilde{e}}$ ($m^2_{\tilde{\mu}\tilde{e}}$ in the text) indicates the magnitude of the slepton mixing. (Reprinted with permission from Ref. [8]. Copyright (2001) by the American Physical Society.)

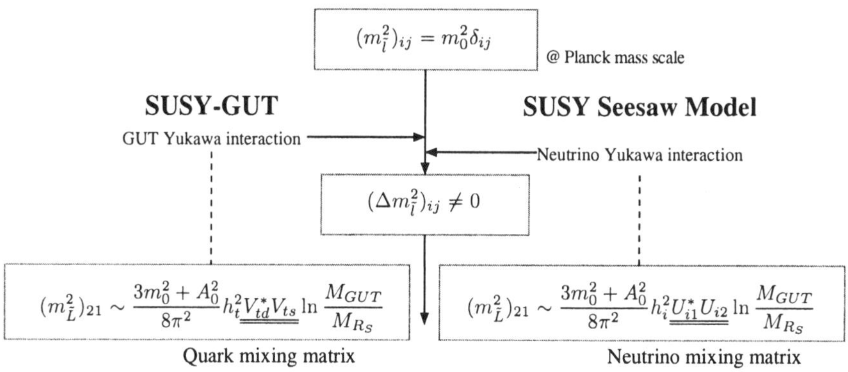

Fig. 19.4. Two physics scenarios in low-energy SUSY (SUSY-GUT and SUSY-Seesaw), introducing non-zero slepton mixing into the minimum supersymmetric Standard Model (MSSM).

neutrino Yukawa interaction has similar effects. This scenario is called the SUSY-seesaw models [36–38]. These two scenarios are illustrated in Fig. 19.4. Both of the models predict large branching ratios for cLFV, which are just a few orders of magnitude below the current experimental upper limits. Figs. 19.5 and 19.6 show the predictions of the cLFV branching ratios in the SU(5) SUSY-GUT and SUSY-seesaw models, respectively. If we could improve experimental sensitivity by a few orders of magnitude, this would provide a great potential for new discoveries.

Intensive calculations on cLFV predictions based on SO(10) SUSY GUT models have been done [39]. The SO(10) SUSY-GUT models naturally

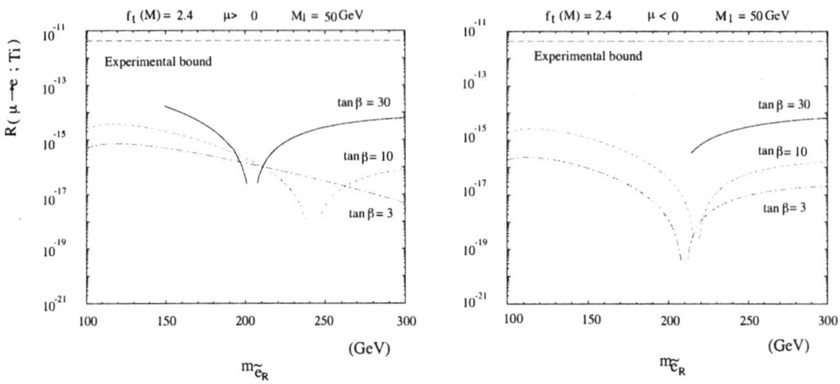

Fig. 19.5. Predicted branching ratios for $\mu^- - e^-$ conversion in the minimal SUSY SU(5) GUT model. (Reprinted with permission from Ref. [36]. Copyright (2001) by the American Physical Society.)

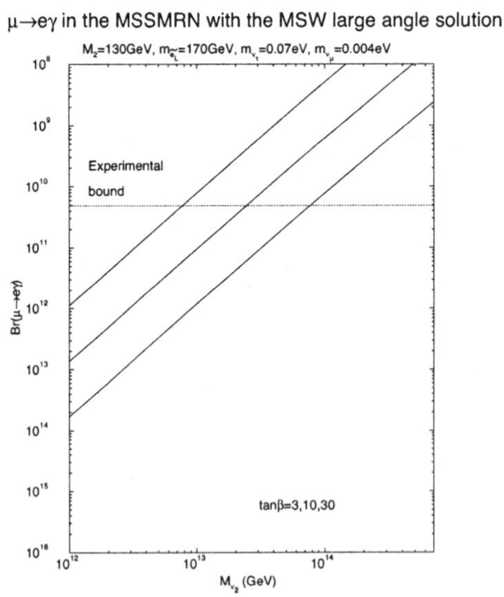

Fig. 19.6. Predictions of $\mu^+ \to e^+\gamma$ branching ratio in SUSY-seesaw models. The three lines correspond to the cases of $\tan\beta = 30, 10, 3$ from top to bottom, respectively. (Reprinted with permission from Ref. [8]. Copyright (2001) by the American Physical Society.)

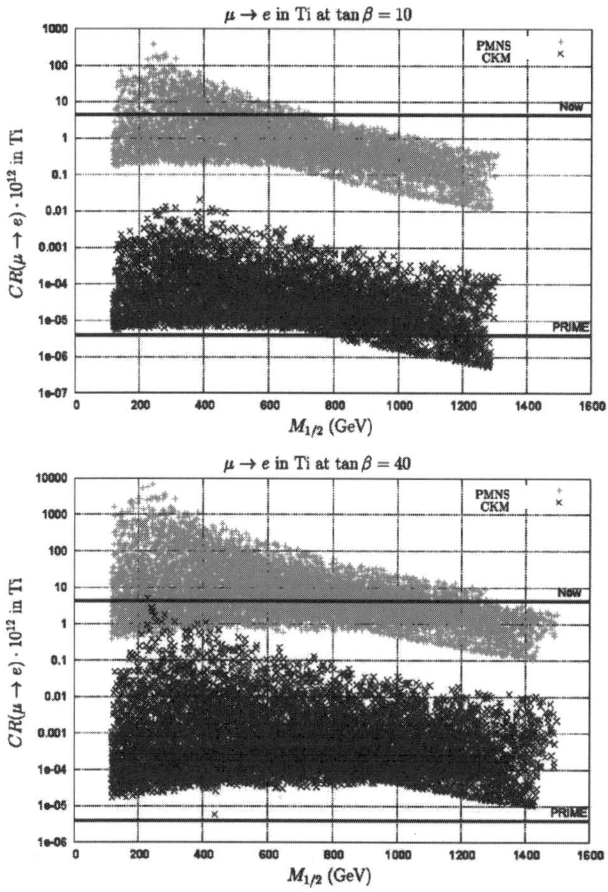

Fig. 19.7. Predictions of branching ratios of $\mu^- - e^-$ conversion in Ti in SO(10) SUSY-GUT models. Plotted points are obtained by scanning the LHC accessible parameter space. Gray and black points represent the maximal case where the neutrino Yukawa matrix is a MNS-type and minimal cases where the neutrino Yukawa matrix is a CKM-type, respectively. The top and bottom figures correspond to $\tan\beta = 10$ and $\tan\beta = 40$ respectively. The horizontal lines are the present limit from SINDRUM II and the sensitivity expected by the PRISM/PRIME. The latter would be able to cover most of the parameter spaces even for low $\tan\beta$ case. (Reprinted with permission from Ref. [39]. Copyright (2006) by the American Physical Society.)

include the seesaw mechanism. The off-diagonal slepton mass squared is given by

$$\Delta m^2_{\tilde{\mu}\tilde{e}} \simeq \sum \frac{M^2_{SUSY}}{16\pi^2} y^*_{\mu k} y_{ek} \log\Big(\frac{M_{GUT}}{M_k}\Big), \tag{19.17}$$

where $y_{\alpha k}(\alpha = e, \mu, \tau)$ are the neutrino Yukawa couplings. $M_k(k = 1, 2, 3)$ are the masses of right-handed Majorana neutrinos, and M_{SUSY} and M_{GUT} are a typical supersymmetric mass scale and a GUT scale, respectively. In Ref. [39], they considered the LFV contribution from only the seesaw mechanism, and studied two cases, where one is the "minimal case" that the neutrino Yukawa couplings are similar to those given by the Kobayashi–Maskawa (KM) quark mixing matrix, and the other is "maximal case" that the neutrino Yukawa couplings are similar to those given by the observed neutrino mixing matrix (the Maki–Nakagawa–Sakata (MNS) matrix). Figure 19.7 shows predictions of branching ratios of $\mu^- - e^-$ conversion in Ti for both the maximal and minimal cases. A planned future experiment like PRISM/PRIME aiming at a sensitivity of $B(\mu^- + Ti \to e^- + Ti) < 10^{-18}$, which will be described in Section 19.6, would cover the most SUSY parameter space that be explored at the LHC, even for the minimal case with low $\tan\beta$ values. In some case, cLFV with very high sensitivity would be able to test the SUSY framework for SUSY masses that are even beyond the LHC sensitivity reach.

19.3.3.1. *cLFV and the LHC*

If the LHC finds SUSY, cLFV might be likely to be observed in the current and planned experiments when either SUSY-GUT or SUSY-seesaw models are correct. Then, cLFV searches would provide the information of slepton mixing, which might not be measured at the LHC at high precision. If the LHC does not find any evidence for SUSY, two potential cases can be considered. One is that SUSY does not exist at all. The other is that SUSY particles exist for the mass region heavier than the LHC reach, such as in a multiple TeV scale. For the latter case, measurements with high precision for cLFV become very important, since such measurements are sensitive to a heavier mass scale than that which can be reached by high-energy accelerators. For heavier SUSY, if cLFV search has sufficient experimental sensitivity (such as 10^{-18} for $\mu^- - e^-$ conversion), it could be sensitive to the SUSY mass scale up to several TeV [39]. Therefore, the search for cLFV would be worth carrying out even if the LHC does not find any evidence for SUSY below the TeV energy scale.

19.3.3.2. *Other theoretical models*

It should be noted that besides SUSY, there are many other models that predict sizable effects of cLFV. These include extra-dimension models, little Higgs models, heavy neutrino models, leptoquark models, composite models, two Higgs doublet models, Z' models, and anomalous Z coupling. Additional discussion can be found in Chapter 18.

19.4. $\mu^+ \to e^+\gamma$ Decay

19.4.1. *Phenomenology of $\mu^+ \to e^+\gamma$ decay*

One of the most popular cLFV processes is the decay $\mu^+ \to e^+\gamma$. The Lagrangian for the $\mu^+ \to e^+\gamma$ amplitude is given by

$$\mathcal{L}_{\mu\to e\gamma} = -\frac{4G_F}{\sqrt{2}}\left[m_\mu A_R \overline{\mu_R}\sigma^{\mu\nu} e_L F_{\mu\nu} + m_\mu A_L \overline{\mu_L}\sigma^{\mu\nu} e_R F_{\mu\nu} + h.c.\right], \quad (19.18)$$

where A_R and A_L are coupling constants that correspond to the processes of $\mu^+ \to e_R^+\gamma$ and $\mu^+ \to e_L^+\gamma$, respectively, and $e^+_{R(L)}$ is a right-handed (left-handed) positron. This Lagrangian presents a dipole-type interaction with photons, but changing lepton flavor. The differential angular distribution of $\mu^+ \to e^+\gamma$ decay is given by

$$\frac{dB(\mu^+ \to e^+\gamma)}{d(\cos\theta_e)} = 192\pi^2\left[\left|A_R\right|^2(1 - P_\mu\cos\theta_e) + \left|A_L\right|^2(1 + P_\mu\cos\theta_e)\right], \quad (19.19)$$

where θ_e is the angle between the muon polarization and the e^+ momentum vectors. P_μ is the magnitude of the muon spin polarization. The branching ratio is given by

$$B(\mu^+ \to e^+\gamma) = \frac{\Gamma(\mu^+ \to e^+\gamma)}{\Gamma(\mu^+ \to e^+\nu\overline{\nu})} = 384\pi^2(|A_R|^2 + |A_L|^2). \quad (19.20)$$

From Eq. (19.19), one can consider that when spin-polarized muons are used, an angular distribution of $\mu^+ \to e^+\gamma$ decay with respect to the muon polarization vector would be useful to determine A_R and A_L [42].

19.4.2. *Event signature and backgrounds*

The event signature of $\mu^+ \to e^+\gamma$ decay at rest is a positron and a photon moving back-to-back in coincidence, with their energies equal to half that of the muon mass ($m_\mu/2 = 52.5$ MeV). The searches in the past were made using positive muons at rest to fully utilize its kinematics. Negative muons have not been used because they are captured by a nucleus when they are stopped in matter.

There are two major backgrounds to the search for $\mu^+ \to e^+\gamma$ decay. One of them is a physics (prompt) background from radiative muon decay, $\mu^+ \to e^+\nu\overline{\nu}\gamma$, when e^+ and photon are emitted back-to-back with the two neutrinos carrying off a small amount of energy. The other background is an accidental coincidence of an e^+ in a normal muon decay, $\mu^+ \to e^+\nu\overline{\nu}$, accompanied by a high energy photon. Possible sources of the latter would be either $\mu^+ \to e^+\nu\overline{\nu}\gamma$ decay, annihilation-in-flight or external bremsstrahlung of e^+s from a normal muon decay. They will be explained in detail in the following.

19.4.2.1. *Physics background*

One of the major physics backgrounds is radiative muon decay, $\mu^+ \to e^+\nu\overline{\nu}\gamma$ (branching ratio = 1.4% for $E_\gamma > 10$ MeV), when the e^+ and photon are emitted back-to-back with the two neutrinos carrying off a small amount of energy.

The differential decay rate of this radiative muon decay was calculated as a function of the e^+ energy (E_e) and the photon energy (E_γ) normalized to their maximum energies, namely $x = 2E_e/m_\mu$ and $y = 2E_\gamma/m_\mu$ [40, 41]. The kinematic case when $x \approx 1$ and $y \approx 1$ is important as a background to $\mu^+ \to e^+\gamma$. Given the detector resolutions of δx and δy, the sensitivity limitation from this physics background can be estimated by integrating the differential decay rate over the signal box [42].

Figure 19.8 shows the fraction of the $\mu^+ \to e^+\nu\overline{\nu}\gamma$ decay for the given δx and δy values with unpolarized muons. From Fig. 19.8, it can be seen that both δx and δy of the order of 0.01 are needed to achieve a sensitivity limit at the level of 10^{-15}. Radiative corrections to radiative muon decay for the case of the physics background to $\mu^+ \to e^+\gamma$ decay have been calculated to be of the order of several percent, depending on the detector resolution [43].

19.4.2.2. *Accidental background*

The accidental background becomes more important than the physics background for a very high rate of incident muons. It is present in current experiments, and is expected to become more serious at future ones. The effective branching ratio ($B_{\rm acc}$) which is an event rate of the accidental

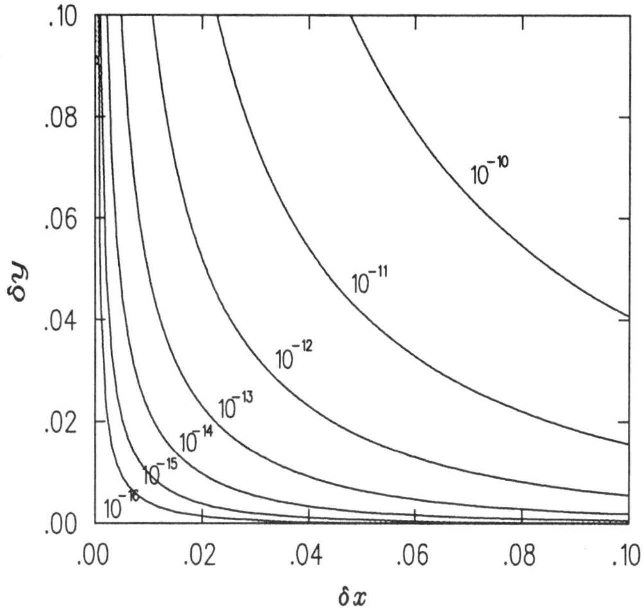

Fig. 19.8. Effective branching ratio of the physics background from the $\mu^+ \to e^+\nu\bar{\nu}\gamma$ decay as a function of the e^+ energy resolution δx and photon energy resolution δy. (Reprinted with permission from Ref. [8]. Copyright (2001) by the American Physical Society.)

background normalized to the total decay rate, is given by

$$B_{\rm acc} = R_\mu \cdot f_e^0 \cdot f_\gamma^0 \cdot (\Delta t_{e\gamma}) \cdot (\frac{\Delta\omega_{e\gamma}}{4\pi}), \tag{19.21}$$

where R_μ is the instantaneous muon intensity. f_e^0 and f_γ^0 are respectively the integrated fractions within the signal region of the spectrum of e^+ in the normal muon decays ($\mu^+ \to e^+\nu\bar{\nu}$) and photons in radiative muon decays ($\mu^+ \to e^+\nu\bar{\nu}\gamma$) or e^+e^- annihilation. They include their corresponding branching ratios. $\Delta t_{e\gamma}$ and $\Delta\omega_{e\gamma}$ are respectively the full widths of the signal regions for timing coincidence and angular constraint of the back-to-back kinematics.

Given the sizes of the signal region, B_{acc} can be evaluated. When we take δx, δy, $\delta\theta_{e\gamma}$, and $\delta t_{e\gamma}$ to be respectively the half width of the signal region for e^+, photon energies, angle $\theta_{e\gamma}$, and relative timing between e^+ and photon, the effective branching ratio of the accidental background is given by

$$B_{acc} = R_\mu \cdot (2\delta x) \cdot \left[\frac{\alpha}{2\pi}(\delta y)^2(\ln(\delta y) + 7.33)\right] \times \left(\frac{\delta\theta_{e\gamma}^2}{4}\right) \cdot (2\delta t_{e\gamma}). \tag{19.22}$$

Table 19.2. Historical progress of searches for $\mu^+ \to e^+\gamma$ since the era of meson factories with 90% C.L. upper limits. The resolutions quoted are given as a full width at half maximum (FWHM).

Place	Year	ΔE_e	ΔE_γ	$\Delta t_{e\gamma}$	$\Delta\theta_{e\gamma}$	Upper limit	Ref.
TRIUMF	1977	10%	8.7%	6.7 nsec	–	$< 3.6 \times 10^{-9}$	[45]
SIN	1980	8.7%	9.3%	1.4 nsec	–	$< 1.0 \times 10^{-9}$	[46]
LANL	1982	8.8%	8%	1.9 nsec	37 mrad	$< 1.7 \times 10^{-10}$	[47]
LANL	1988	8%	8%	1.8 nsec	87 mrad	$< 4.9 \times 10^{-11}$	[48]
LANL	1999	1.2%*	4.5%*	1.6 nsec	15 mrad	$< 1.2 \times 10^{-11}$	[49]
PSI	2008	0.9%	5%	0.1 nsec	23 mrad		[50]

*Shows an average of the numbers given in Brook *et al.* (1999) [49].

To evaluate the accidental background, detector resolutions are needed. The detector resolutions in the past and current $\mu^+ \to e^+\gamma$ experiments are summarized in Table 19.2. For instance, let us take some realistic values such as $\delta x = 0.5\%$ for the e^+ energy resolution, a photon energy resolution of $\delta y = 3\%$, $\delta\omega_{e\gamma} = 1.5 \times 10^{-4}$ steradians, $\delta t_{e\gamma} = 0.5$ nsec, and $R_\mu = 3 \times 10^8$ μ^+/s, B_{acc} would be 3×10^{-13}. This indicates that the accidental background could be very difficult to beat. Therefore, it is critical to make significant improvements in the detector resolution in order to reduce the accidental background.

19.4.2.3. *Muon polarization*

The use of polarized muons has been found to be useful in suppressing backgrounds for $\mu^+ \to e^+\gamma$ searches [42, 44]. For the physics background, the angular distribution of radiative muon decay ($\mu^+ \to e^+\nu\overline{\nu}\gamma$) with respect to the muon spin direction is given by[c]

$$dB(\mu^+ \to e^+\nu\overline{\nu}\gamma) = \frac{\alpha}{16\pi}\Big[J_1 \cdot (1 - P_\mu \cos\theta_e) + J_2 \cdot (1 + P_\mu \cos\theta_e)\Big] d(\cos\theta_e), \quad (19.23)$$

where the coefficients J_1 and J_2 are given by

$$J_1 = (\delta x)^4(\delta y)^2 \quad \text{and} \quad J_2 = \frac{8}{3}(\delta x)^3(\delta y)^3, \quad (19.24)$$

and δx, δy are half widths of the $\mu^+ \to e^+\gamma$ signal region for x and y, respectively. Experimentally, the resolution of the e^+ energy is better than that of the photon energy, i.e. $\delta x < \delta y$. Thereby, J_2 is much larger than J_1 in most cases. Therefore, the angular distribution of the physics background

[c]Here, only the case when the angular correlation between e^+ and γ is poorly measured is considered.

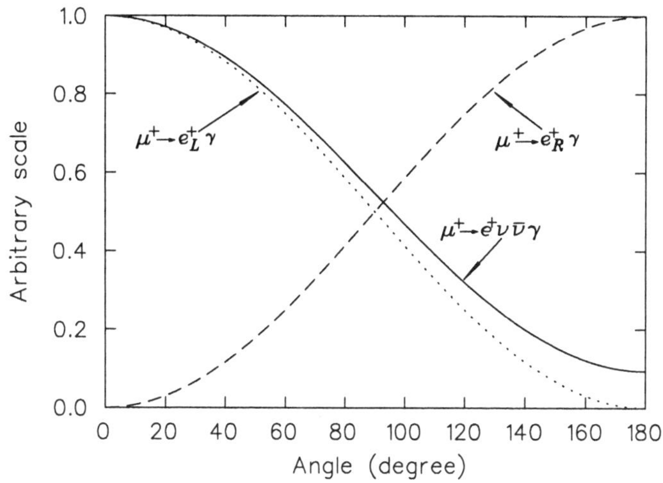

Fig. 19.9. Angular distribution of e^+ from the physics background of the $\mu^+ \to e^+\nu\bar{\nu}\gamma$ decay from polarized muons with respect to the muon polarization direction (in solid line). The dotted and dashed lines are for $\mu^+ \to e_L^+\gamma$ and $\mu^+ \to e_R^+\gamma$ decays, respectively. (Reprinted with permission from Ref. [8]. Copyright (2001) by the American Physical Society.)

follows approximately $(1+P_\mu \cos\theta)$ as long as $\delta y > \delta x$. Figure 19.9 shows the angular distribution of $\mu^+ \to e^+\nu\bar{\nu}\gamma$ with, for instance, $\delta y/\delta x = 4$. If we selectively measure the e^+s in $\mu^+ \to e^+\gamma$ which move opposite to the muon-polarization direction, the background from $\mu^+ \to e^+\nu\bar{\nu}\gamma$ would be significantly reduced in the search for $\mu^+ \to e_R^+\gamma$. Furthermore, by varying δx and δy, the angular distribution of the $\mu^+ \to e^+\nu\bar{\nu}\gamma$ background can change according to Eq. (19.23), thus providing another means to discriminate the signal from the backgrounds.

The use of polarized muons would also provide suppression of the accidental background [44]. This is due to the sources of accidental backgrounds having a specific angular distribution when a muon is polarized. For instance, the e^+s in normal Michel μ^+ decay are emitted preferentially along the muon spin direction, following a $(1 + P_\mu \cos\theta_e)$, whereas the inclusive angular distribution of a high-energy photon (e.g. ≥ 50 MeV) from $\mu^+ \to e^+\nu\bar{\nu}\gamma$ decay follows a $(1 + P_\mu \cos\theta_\gamma)$ distribution, where θ_γ is the angle of the photon direction with respect to the muon spin direction. This inclusive angular distribution of a high-energy photon in $\mu^+ \to e^+\nu\bar{\nu}\gamma$ implies that the accidental background could be suppressed for $\mu^+ \to e_L^+\gamma$, where high-energy photons must be detected at the opposite direction to the muon polarization. A similar suppression mechanism of accidental background can be seen for $\mu^+ \to e^+_R\gamma$ when high-energy positrons are detected in the opposite direction to the muon polarization. As a result, the selective measurements of either e^+s or photons antiparallel to the muon spin direction would give the same accidental background suppression for $\mu^+ \to e^+_R\gamma$ and $\mu^+ \to e^+_L\gamma$ decays, respectively. The suppression factor, η, is calculated for polarized muons by

$$\begin{aligned}\eta &\equiv \int_{\cos\theta_D}^{1} d(\cos\theta)(1+P_\mu\cos\theta)(1-P_\mu\cos\theta)/\int_{\cos\theta_D}^{1} d(cos\theta)\\ &= (1-P_\mu^2)+\frac{1}{3}P_\mu^2(1-\cos\theta_D)(2+\cos\theta_D),\end{aligned} \tag{19.25}$$

where θ_D is a half opening angle of detection with respect to the muon polarization direction. η is shown in Fig. 19.10 as a function of θ_D. For instance, for $\theta_D = 300$ mrad, an accidental background can be suppressed to the level of 1/20 (1/10) when P_μ is 100 (97)%.

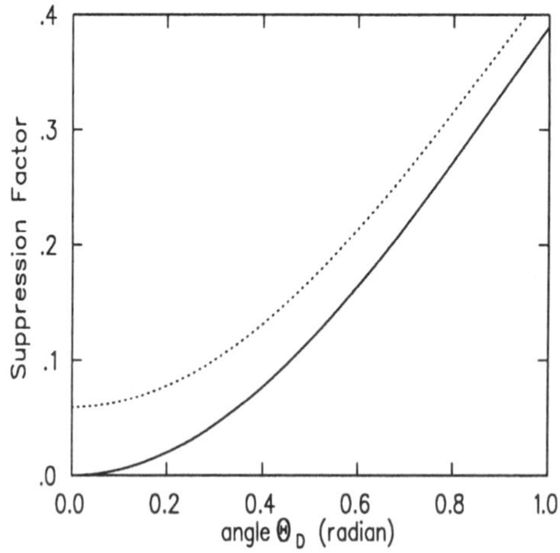

Fig. 19.10. Suppression factor of the accidental background in a $\mu^+ \to e^+\gamma$ search as a function of half of the detector opening angle. The solid line is 100% muon polarization and the dotted line is 97% muon polarization. (Reprinted with permission from Ref. [44]. Copyright (1997) by the American Physical Society.)

19.4.3. *Experimental status of $\mu^+ \to e^+\gamma$ decay*

Experimental searches for $\mu^+ \to e^+\gamma$ decay has a long history. The history of the searches for $\mu^+ \to e^+\gamma$ after the era of meson factories is summarized in Table 19.2. It is a history of how the detection resolutions are improved to eliminate background events.

The present experimental upper limit for $\mu^+ \to e^+\gamma$ is 1.2×10^{-11}, which was obtained by the MEGA experiment [49] at Los Alamos National Laboratory (LANL) in the US. A schematic layout of the MEGA detector is shown in Fig. 19.11. The MEGA detector consisted of a magnetic spectrometer for positron detection and three concentric pair spectrometers for photon detection. These detectors were placed inside a superconducting solenoid magnet of 1.5 Tesla. The positron spectrometer is comprised of eight cylindrical wire chambers and scintillators for timing. The average positron-energy resolution was about 1.2% (FWHM) and the photon resolution was 3.3% and 5.7% for the inner and outer converters respectively. A pulsed surface μ^+ beam of 29.8 MeV/c was introduced and stopped in the muon-stopping target made of a thin tilted Mylar foil. The intensity of the muon beam was 2.5×10^8/sec with a macroscopic duty factory of 6%. The total number of muons stopped was 1.2×10^{14}.

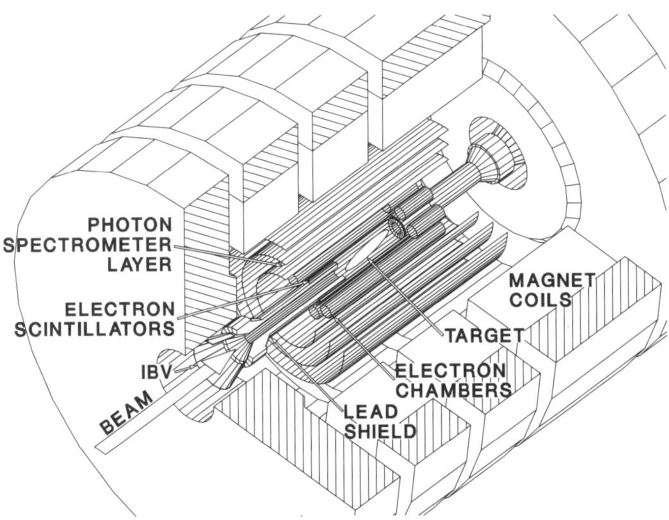

Fig. 19.11. Schematic layout of the MEGA detector at LANL. (Reprinted with permission from Ref. [8]. Copyright (2001) by the American Physical Society.)

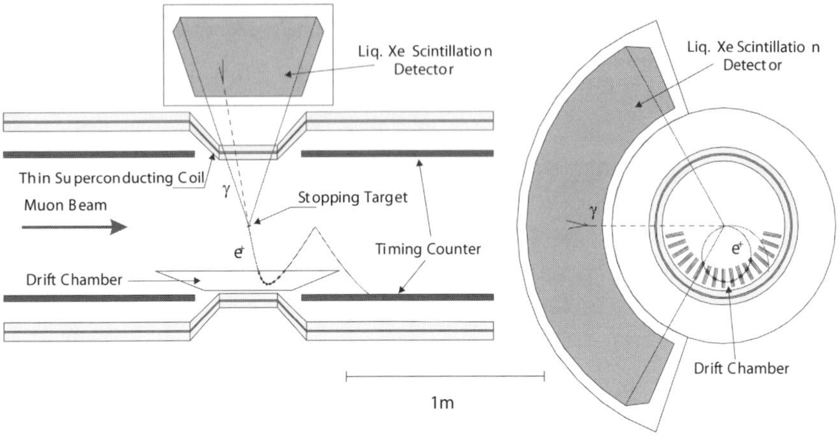

Fig. 19.12. Side and end views of the MEG detector. The magnetic field is shaped so that positrons of 52.8 MeV could have the same radius independently of their emission angles. This shaped magnetic field also sweeps positrons out of the tracking region, thus minimizing the detector rates. The liquid Xe photon detector of 0.8 m^3 volume is viewed by 846 PMTs immersed inside. (Figure courtesy of T. Mori and reproduced by permission of the MEG Collaboration.)

A new experiment at PSI called MEG [50], which aims to achieve a single event sensitivity of 10^{-13} in the $\mu \to e\gamma$ branching ratio, was built and started data-taking in 2008. A significant improvement in the $\mu^+ \to e^+\gamma$ sensitivity is expected from the use of a continuous muon beam (100% duty factor) at PSI. Using the same instantaneous beam intensity as MEGA, the total number of muons available can be increased by a factor of about 16.

A schematic view of the MEG detector is shown in Fig. 19.12. The MEG spectrometer uses a COBRA (COnstant Bending RAdius) scheme, in which a magnetic field is graded so that the radius of the 52.8 MeV positrons is constant, independently of their emission angles (θ) within $|\cos\theta| < 0.35$, and at the same time all positrons are swept away faster than in a straight solenoid field. Another improvement is the use of a novel liquid xenon scintillation detector of the "Mini-Kamiokande" type, which is a 0.8-m^3 volume of liquid Xenon, viewed by an array of 846 photomultipliers immersed inside liquid Xenon from all the sides. This system allows not only detection of photon energy but also reconstruction of the photon conversion point and its direction. Physics data taking has already started in 2008.

19.5. $\mu^+ \to e^+e^+e^-$ Decay

19.5.1. *Phenomenology of* $\mu^+ \to e^+e^+e^-$ *decay*

In a similar way to the process of $\mu^- - e^-$ conversion described in Section 19.6, the $\mu^+ \to e^+e^+e^-$ decay could have not only photonic (dipole) contributions but also non-photonic contributions. If only the photon-penguin diagrams contribute to $\mu^+ \to e^+e^+e^-$ decay, a model-independent relation between the two branching ratios can be derived, as follows:

$$\frac{B(\mu^+ \to e^+e^+e^-)}{B(\mu^+ \to e^+\gamma)} \simeq \frac{\alpha}{3\pi}(\ln(\frac{m_\mu^2}{m_e^2}) - \frac{11}{4}) = 0.006. \tag{19.26}$$

When muons are polarized, the T-odd asymmetry in $\mu^+ \to e^+e^+e^-$ decay can be made as follows

$$A_T \propto \vec{s}_\mu \cdot (\vec{p}_{e1} \times \vec{p}_{e2}) \tag{19.27}$$

where $\vec{s}_\mu$ is a muon spin vector, and $\vec{p}_{e1}$ and $\vec{p}_{e2}$ are momentum vectors of the decay positron of a higher energy and that of a lower energy respectively. The T-odd asymmetry can arise from interference between the photonic (dipole) diagrams and the four-fermion interaction diagrams. This T-odd asymmetry could become sizable in supersymmetric models [58, 59].

19.5.2. *Event signature and backgrounds*

The event signature of the decay $\mu^+ \to e^+e^+e^-$ is kinematically well constrained, since all particles in the final state are detectable. Muon decay at rest has been used in all past experiments. In this case, the conservation of momentum sum ($|\sum_i \vec{p}_i| = 0$) and energy sum ($\sum_i E_i = m_\mu$) could be effectively used together with the timing coincidence between two e^+s and one e^-, where $\vec{p}_i$ and E_i ($i = 1 - 3$) are respectively the momentum and energy of each of the e's.

One of the physics background processes is the allowed muon decay $\mu^+ \to e^+\nu_e\overline{\nu}_\mu e^+e^-$ which becomes a serious background when ν_e and $\overline{\nu}_\mu$ have very small energies. Its branching ratio is $(3.4\pm0.4)\times10^{-5}$. The other background is an accidental coincidence of an e^+ from normal muon decay with an uncorrelated e^+e^- pair, where a e^+e^- pair could be produced either from Bhabha scattering of e^+, or from the external conversion of the photon in $\mu^+ \to e^+\nu_e\overline{\nu}_\mu\gamma$ decay. Since the e^+e^- pair from photon conversion has a small invariant mass, it could be removed by eliminating events with a small opening angle between e^+ and e^-. This, however, causes a loss in the signal sensitivity, in particular for theoretical models in which $\mu^+ \to e^+e^+e^-$ decay occurs mostly through photonic diagrams.

The other background, which comes mainly at the trigger level, comprises of fake events with an e^+ curling back to the target, which mimics an e^+e^- pair. For this background, an e^+e^- pair forms a relative angle of 180°, and can therefore be rejected.

As has been discussed for the $\mu^+ \to e^+\gamma$ searches in Section 19.4, the search for $\mu^+ \to e^+e^+e^-$ decay is also limited by accidental backgrounds in a high rate of incident muons. To reduce accidental background events, an instantaneous rate of incident muons should be kept low, and thus a continuous muon beam should be utilized.

19.5.3. *Experimental status of $\mu^+ \to e^+e^+e^-$ decay*

The historical progress of the search for $\mu^+ \to e^+e^+e^-$ decay is summarized in Table 19.3. In 1976, the pioneering measurement using a cylindrical spectrometer gave an upper limit of $B(\mu^+ \to e^+e^+e^-) < 1.9 \times 10^{-9}$ [51]. Since then, various experiments to search for $\mu^+ \to e^+e^+e^-$ decay have been carried out. In particular, a series of experimental measurements with the SINDRUM I magnetic spectrometer at SIN [53–55] were carried out. A surface μ^+ beam with $5 \times 10^6 \mu^+/\mathrm{s}$ was used, and the muons were stopped in a hollow double-cone target. The e^+s and e^-s were tracked by the SINDRUM spectrometer, which consisted of five concentric multi-wire proportional chambers (MWPC) and a cylindrical array of 64 plastic scintillation counters under a solenoid magnetic field of 0.33 T. The momentum resolution was $\Delta p/p = (12.0 \pm 0.3)\%$ (FWHM) at $p = 50$ MeV/c. This experiment gave a 90% C.L. upper limit of $B(\mu^+ \to e^+e^+e^-) < 1.0 \times 10^{-12}$, assuming a constant matrix element for the $\mu^+ \to e^+e^+e^-$ decay [55]. They also observed 9070 ± 10 events of $\mu^+ \to e^+\nu_e\overline{\nu}_\mu e^+e^-$ decay. A detailed analysis

Table 19.3. Historical progress and summary of searches for $\mu^+ \to e^+e^+e^-$ decay.

Place	Year	90%C.L. upper limit	Reference
JINR	1976	$< 1.9 \times 10^{-9}$	[51]
LANL	1984	$< 1.3 \times 10^{-10}$	[52]
SIN	1984	$< 1.6 \times 10^{-10}$	[53]
SIN	1985	$< 2.4 \times 10^{-12}$	[54]
LANL	1988	$< 3.5 \times 10^{-11}$	[48]
SIN	1988	$< 1.0 \times 10^{-12}$	[55]
JINR	1991	$< 3.6 \times 10^{-11}$	[56]

of the differential decay rate of $\mu^+ \to e^+\nu_e\overline{\nu}_\mu e^+e^-$ decay was studied, and was found to be consistent with the $V - A$ interaction [57].

Another recent experiment to search for $\mu^+ \to e^+e^+e^-$ was performed at the Joint Institute for Nuclear Research (JINR), Dubna, Russia [56]. A magnetic 4π spectrometer with cylindrical proportional chambers was used. They obtained an upper limit of 90% CL of $B(\mu^+ \to e^+e^+e^-) < 3.6\times10^{-11}$, where the matrix element of $\mu^+ \to e^+e^+e^-$ was assumed to be constant.

19.6. $\mu^- - e^-$ Conversion in a Muonic Atom

19.6.1. *Phenomenology of $\mu^- - e^-$ conversion*

Another prominent muon cLFV process is the coherent neutrino-less conversion of a negative muon to an electron ($\mu^- - e^-$ conversion) in a muonic atom. When a negative muon is stopped in some material, it is trapped by an atom, and a muonic atom is formed. After it cascades down energy levels in the muonic atom, the muon is bound in its $1s$ ground state. The fate of the muon is then either decay in orbit ($\mu^- \to e^-\nu_\mu\overline{\nu}_e$) or nuclear muon capture by a nucleus $N(A, Z)$ of mass number A and atomic number Z, namely, $\mu^- + N(A,Z) \to \nu_\mu + N(A, Z-1)$. However, in the context of lepton flavor violation in physics beyond the Standard Model, the exotic process of neutrino-less muon capture, such as

$$\mu^- + N(A,Z) \to e^- + N(A,Z), \tag{19.28}$$

is also expected. This process is called $\mu^- - e^-$ conversion in a muonic atom. This process violates the conservation of lepton flavor numbers, L_e and L_μ, by one unit, but the total lepton number, L, is conserved.

The final state of the nucleus (A, Z) could be either the ground state or one of the excited states. In general, the transition to the ground state, which is called coherent capture, is dominant. The rate of the coherent capture over non-coherent capture is enhanced by a factor approximately equal to the number of nucleons in the nucleus, since all of the nucleons participate in the process.

The branching ratio of $\mu^- - e^-$ conversion is defined as

$$B(\mu^-N \to e^-N) \equiv \frac{\Gamma(\mu^-N \to e^-N)}{\Gamma(\mu^-N \to \text{all})} \tag{19.29}$$

where Γ is the decay width. The time distribution of $\mu^- - e^-$ conversion follows a lifetime of a muonic atom. The lifetime of a muonic atom depends on a nucleus. A list of mean lifetimes for typical muonic atoms is given in Table 19.4.

Table 19.4. Lifetimes of various muonic atoms.

Nucleus	H	C	Al	Fe	Cu	W	Pb
Z	1	6	13	26	29	74	82
Lifetime (nsec)	2195	2027	880	200	164	78	74

19.6.1.1. *Photonic and non-photonic contributions*

The $\mu^- - e^-$ conversion process can have two possible contributions, which are the photonic (dipole) contribution and the non-photonic contribution. In principle, this process could have a non-photonic contribution that does not contribute to $\mu^- - e^-$ decay. The photonic contribution in the $\mu^- - e^-$ conversion process has some definite relation to that in $\mu^+ \to e^+\gamma$ decay as a function of the mass number (A) and the atomic number (Z). It can be parametrized as

$$\frac{B(\mu^+ \to e^+\gamma)}{B(\mu^-N \to e^-N)} = \frac{96\pi^3\alpha}{G_F^2 m_\mu^4} \cdot \frac{1}{3 \times 10^{12} B(A,Z)} \sim \frac{428}{B(A,Z)} \tag{19.30}$$

where $B(A,Z)$ represents the rate dependence on the mass number (A) and the atomic number (Z) of the nucleus. The values of $B(A,Z)$ are calculated based on various approximations. Some of them are tabulated in Table 19.5. For instance, by using $B_{CMK}(A,Z)$, the ratios B($\mu^+ \to e^+\gamma$)/B($\mu N \to eN$) of 389 for ^{27}Al, 238 for ^{48}Ti, and 342 for ^{208}Pb are obtained.

Table 19.5. Z dependence of the photonic contribution in the $\mu^- - e^-$ conversion estimated by various theoretical models (after Czarnecki *et al.*, (1997)).

Models	Al	Ti	Pb	Reference
$B_{WF}(A,Z)$	1.2	2.0	1.6	Weinberg and Feinberg (1959) [60]
$B_S(A,Z)$	1.3	2.2	2.2	Shanker (1979) [61]
$B_{CMK}(A,Z)$	1.1	1.8	1.25	Czarnecki *et al.* (1997) [62]

If the non-photonic contribution dominates, $\mu^- - e^-$ conversion could be sufficiently large to be observed, even if $\mu^+ \to e^+\gamma$ decay is small. It might be worth noting that if a $\mu^+ \to e^+\gamma$ signal is found, a $\mu^- - e^-$ conversion signal should also be found. If no $\mu \to e\gamma$ signal is found, there will still be an opportunity to find $\mu^- - e^-$ conversion signals because of the potential existence of non-photonic contributions.

19.6.1.2. *Dependence on muon-stopping target material*

Recently, the rates of coherent $\mu^- - e^-$ conversion processes for general effective LFV interactions (such as dipole, scalar and vector interactions) were calculated for various nuclei [63]. The calculations also took the relativistic wave functions and the proton and neutron distributions with their ambiguities into account. Their results, which are shown in Fig. 19.13, indicate that the branching ratios for $\mu^- - e^-$ conversion increase for light nuclei up to the atomic number of $Z \sim 30$. and high for the region of $Z = 30 - 60$, and decrease for heavy nuclei of $Z > 60$. It is also pointed out that the atomic number dependence of the $\mu^- - e^-$ conversion rate would be useful to distinguish different effective LFV interactions.

19.6.2. *Signal and background events*

The event signature of coherent $\mu^- - e^-$ conversion in a muonic atom is a mono-energetic single electron emitted from the conversion with an energy

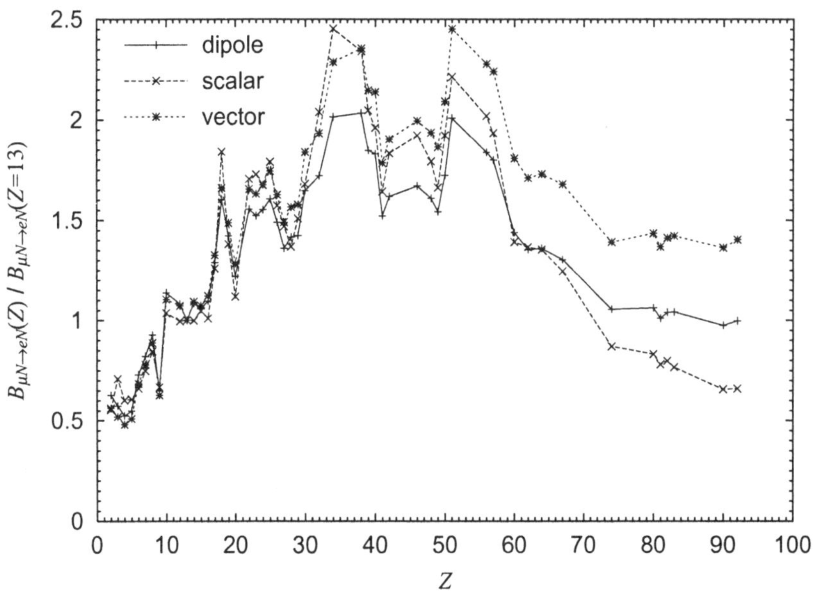

Fig. 19.13. The $\mu^- - e^-$ conversion ratios for various general LFV interactions are plotted as a function of the atomic number Z. The $\mu^- - e^-$ conversion rates are normalized by those for aluminum nuclei ($Z = 13$). The solid, long-dashed and dashed lines represent the cases of photonic (dipole), scalar and vector interactions, respectively. (Reprinted with permission from Ref. [63]. Copyright (2002) by the American Physical Society.)

($E_{\mu e}$) of

$$E_{\mu e} = m_\mu - B_\mu - E_{recoil} \sim m_\mu - B_\mu, \qquad (19.31)$$

where m_μ is the muon mass, and B_μ is the binding energy of the $1s$ muonic atom. E_{recoil} is the nuclear recoil energy which is small and can be ignored. Since B_μ varies for various nuclei, $E_{\mu e}$ could be different. For instance, $E_{\mu e}$ = 104.3 MeV for titanium (Ti) and $E_{\mu e}$ = 94.9 MeV for lead (Pb).

From an experimental point of view, $\mu^- - e^-$ conversion is a very attractive process for the following reasons:

- The energy of the signal electron of about 105 MeV is far above the endpoint energy of the normal muon decay spectrum ($\sim$ 52.8 MeV).
- Since the event signature is a mono-energetic electron, no coincidence measurement is required. The search for this process has the potential to improve sensitivity by using a high muon rate without suffering from accidental background events, which would be serious for other processes, such as $\mu \to e\gamma$ and $\mu \to eee$ decays.

There are several potential sources of electron background events in the energy region around 100 MeV, which can be grouped into three categories as follows. The first group is intrinsic physics backgrounds which come from muons stopped in the muon-stopping target. The second is beam-related backgrounds which are caused by beam particles of muons and other contaminated particles in a muon beam. The third is other backgrounds which are, for instance, cosmic-ray backgrounds, and fake tracking events, and so on.

19.6.2.1. *Intrinsic physics backgrounds*

The intrinsic physics background events are caused by muons stopped in a muon-stopping target. One of the major backgrounds in this category is muon decays in orbit (DIO) in a muonic atom, in which the e^- endpoint energy is close to the energy of the signal electron, owing to a nuclear recoil effect. Energy distributions for DIO electrons have been calculated for a number of muonic atoms [64, 65]. Since the energy distribution of DIO falls steeply as the fifth power of ($E_{\mu e} - E_e$) toward its endpoint, where $E_{\mu e}$ and E_e are the energy of the signal electron and that of DIO electrons respectively. Experimentally, the momentum resolution of e^-

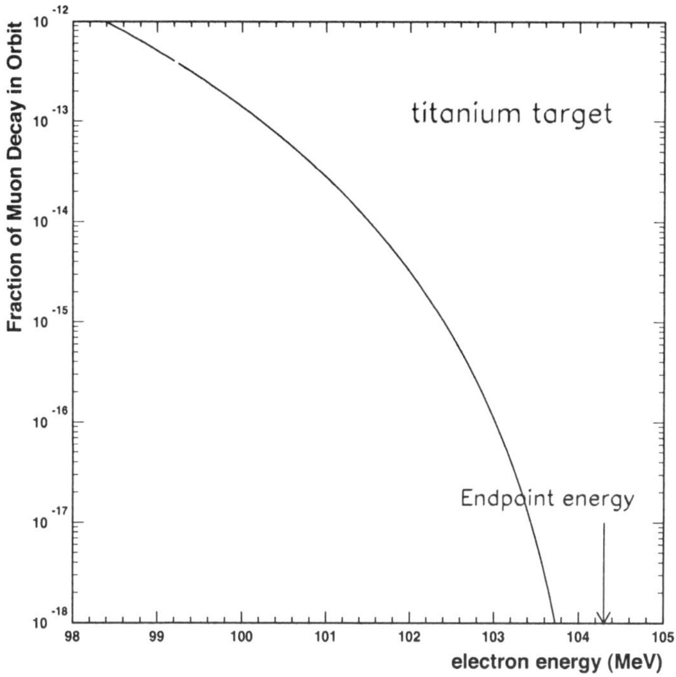

Fig. 19.14. Energy distribution of electrons from muon decay in orbit (DIO), normalized to the total nuclear muon capture rate for a titanium target. This represents an effective branching ratio of muon decay in orbit as a background to the $\mu^- - e^-$ conversion. (Reprinted with permission from Ref. [8]. Copyright (2001) by the American Physical Society.)

detection must be improved to eliminate any DIO background events. For a resolution better than 0.5%, the contribution from DIO occurs at a level of below 10^{-16}.

The other intrinsic physics backgrounds are radiative muon capture (RMC), given by

$$\mu^- + N(A, Z) \to \nu_\mu + N(A, Z-1) + \gamma, \tag{19.32}$$

followed by internal and/or external asymmetric e^+e^- conversion of the photon ($\gamma \to e^+e^-$). The kinematical endpoint (E^{end}_{RMC}) of radiative muon capture is given by

$$E^{end}_{\mathrm{RMC}} \sim m_\mu - B_\mu - \Delta_{Z-1} \tag{19.33}$$

where Δ_{Z-1} is the difference in a nuclear binding energy of the final $N(A, Z-1)$ from the initial $N(A, Z)$ nuclei in radiative muon capture.

Therefore, a muon-stopping target with a large Δ_{Z-1} should be selected to keep a wide background-free region. The other intrinsic physics background events are particle emission (such as protons and neutrons) of after nuclear muon capture.

19.6.2.2. *Beam-related backgrounds*

Beam-related background events may originate from muons, pions or electrons in a beam. Muon decays in flight with the muon momentum greater than 75 MeV/c may create electrons in the energy range of 100 MeV. Pions in a beam may produce background events by radiative pion capture (RPC) given

$$\pi^- + N(A,Z) \to N(A,Z-1) + \gamma \tag{19.34}$$

followed by internal and external asymmetric e^+e^- conversion of the photon ($\gamma \to e^+e^-$). The others are electrons in the beam scattering off the target. To eliminate the backgrounds from pions and electrons, the purity of the beam is crucial.

19.6.2.3. *Other backgrounds*

The other sources of background events are (i) cosmic rays, and (ii) tracking errors. To eliminate cosmic-ray backgrounds, passive and active shielding with high efficiency is needed.

19.6.3. *Present experimental status*

The experimental status of searches for $\mu^- - e^-$ conversion processes is presented. Table 19.6 summarizes the history of searches for $\mu^- - e^-$ conversion.

The latest search for $\mu^- - e^-$ conversion was performed by the SINDRUM II collaboration at PSI. A schematic view of the SINDRUM II spectrometer is shown in Fig. 19.15. It consisted of a set of concentric cylindrical drift chambers inside a superconducting solenoid magnet of 1.2 Tesla. Negative muons with momenta of about 90 MeV/c were stopped in a muon-stopping target located at the center of the magnet after passing through an energy degrader. Charged particles with transverse momenta above 80 MeV/c originating from the target were detected in the spectrometer. A momentum resolution of about 2.8% (FWHM) was achieved for 100 MeV/c. Figure 19.16 shows their result on $\mu^- + Au \to e^- + Au$. The main

Table 19.6. Past experiments on $\mu^- - e^-$ conversion. (*Reported only in conference proceedings.)

Year	Location	Process	Upper Limit	Reference
1972	SREL	$\mu^- + Cu \to e^- + Cu$	$< 1.6 \times 10^{-8}$	[66]
1982	SIN	$\mu^- +^{32} S \to e^- +^{32} S$	$< 7 \times 10^{-11}$	[67]
1985	TRIUMF	$\mu^- + Ti \to e^- + Ti$	$< 1.6 \times 10^{-11}$	[68]
1988	TRIUMF	$\mu^- + Ti \to e^- + Ti$	$< 4.6 \times 10^{-12}$	[69]
1988	TRIUMF	$\mu^- + Pb \to e^- + Pb$	$< 4.9 \times 10^{-10}$	[69]
1993	PSI	$\mu^- + Ti \to e^- + Ti$	$< 4.3 \times 10^{-12}$	[70]
1996	PSI	$\mu^- + Pb \to e^- + Pb$	$< 4.6 \times 10^{-11}$	[71]
1998*	PSI	$\mu^- + Ti \to e^- + Ti$	$< 6.1 \times 10^{-13}$	[72]
2006	PSI	$\mu^- + Au \to e^- + Au$	$< 7 \times 10^{-13}$	[73]

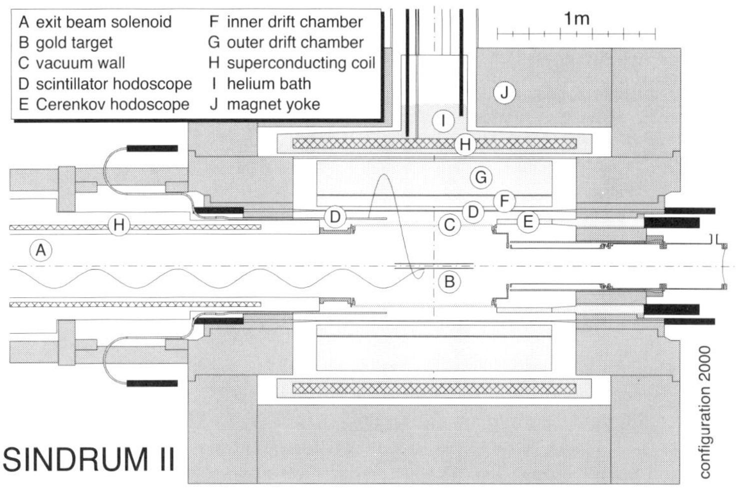

Fig. 19.15. Schematic layout of the SINDRUM II detector. (Reprinted with permission from Ref. [8]. Copyright (2001) by the American Physical Society.)

spectrum shows the steeply falling distribution expected from muon DIO. Two events were found at higher momenta, but just outside the region of interest. The agreement between measured and simulated positron distributions from μ^+ decay means that confidence can be held in the accuracy of the momentum calibration. At present there are no hints concerning the nature of the two high-momentum events: They might have been induced by cosmic rays or RPC by pions in a beam, for example.

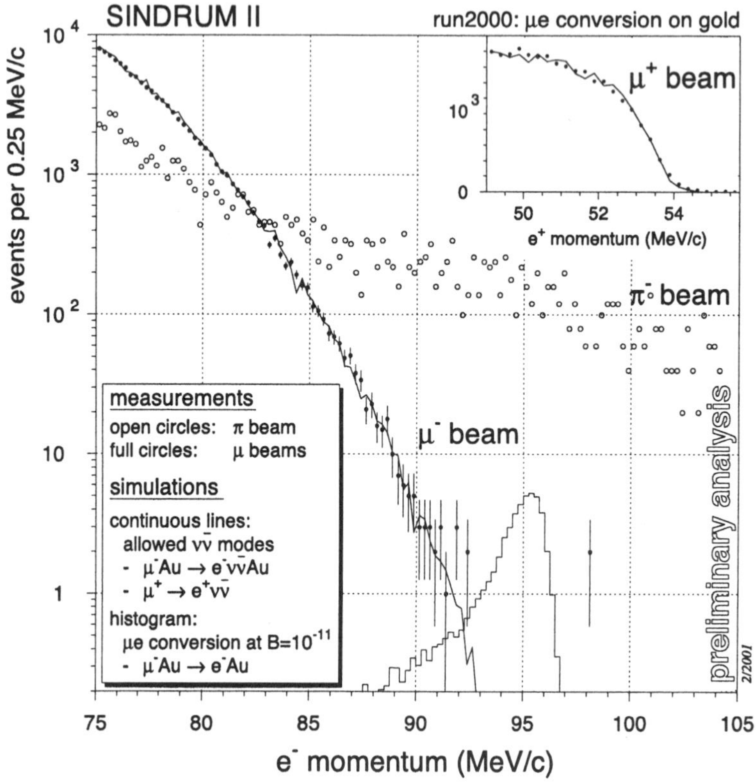

Fig. 19.16. Recent results of $\mu^- + Au \to e^- + Au$ by SINDRUM II. Momentum distributions for three different beam momenta and polarities: (i) 53 MeV/c negative, optimized for μ^- stops, (ii) 63 MeV/c negative, optimized for π^- stops, and (iii) 48 MeV/c positive, optimized for μ^+ stops. The 63 MeV/c data were scaled to the different measuring times. The μ^+ data were taken using a reduced spectrometer field.

19.6.4. *Future experimental prospects*

Considering its marked importance to physics, it is highly desirable to consider a next-generation experiment to search for cLFV with muons. There are three muon cLFV processes to be considered; namely, $\mu^+ \to e^+\gamma$, $\mu^+ \to e^+e^+e^-$ decays and $\mu^- - e^-$ conversion.

The three muon LFV processes have different experimental issues that need to be solved to realize improved experimental sensitivities. They are summarized in Table 19.7. The processes of $\mu^+ \to e^+\gamma$ and $\mu^+ \to e^+e^+e^-$ decays are limited by accidental backgrounds. If the incident muon beam rate is increased by a factor N, background suppression has to be

Table 19.7. A list of major backgrounds, beam requirement and issues for various cLFV processes with muons.

Process	Backgrounds	Beam Requirement	Issue
$\mu^+ \to e^+\gamma$	accidentals	continuous beam	detector resolutions
$\mu^+ \to e^+e^+e^-$	accidentals	continuous beam	detector resolutions
$\mu^- - e^-$ conversion	beam-associated	pulsed beam	beam qualities

improved by a factor of N^2. To achieve this, the detector resolutions have to be significantly improved, which is in general very challenging. In particular, improving the photon energy resolution for $\mu^+ \to e^+\gamma$ is difficult. On the other hand, for $\mu^- - e^-$ conversion, there are no accidental background events, and thus an experiment with higher rates can be performed. If a new muon source with a higher beam intensity and a better beam quality for suppressing beam-associated background events can be constructed, measurements of higher sensitivity can be performed.

Furthermore, it is known that there are more physics processes contributing to $\mu^- - e^-$ conversion and a $\mu^+ \to e^+e^+e^-$ decay than a $\mu^+ \to e^+\gamma$ decay. Namely, the dipole interaction of photon-mediation can contribute to all the three processes, but the box diagrams and four-fermion contact interaction can contribute to only $\mu^- - e^-$ conversion and $\mu^+ \to e^+e^+e^-$ decay. In summary, in consideration of the experimental and theoretical aspects, a search for $\mu^- - e^-$ conversion would be a natural next choice to accomplish significant improvements in the future.

Future experimental projects to search for $\mu^- - e^-$ conversion with a higher sensitivity are being pursued in the USA and Japan. To suppress background events, in particular beam-related backgrounds, the following key elements have been proposed. They are based on the ideas developed in the MELC proposal at the Moscow Meson Factory [74].

- *Beam pulsing*:
  Since muonic atoms have lifetimes of the order of 1 μsec, a pulsed beam with its width that is short compared with these lifetimes would allow one to remove prompt background events by performing measurements in a delayed time window. To eliminate prompt beam-related backgrounds, proton beam extinction is required during the measurement interval.
- *High Field Solenoids for Pion Capture*:
  Superconducting solenoid magnets of a high magnetic field surround a proton target to capture pions in a large solid angle. It

leads to a dramatic increase of muon yields by several orders of magnitude.

- *Curved Solenoids for Muon Transport*:
  The solenoid system for muon transport has high transmission efficiency, resulting a significant increase of muon flux. The curved solenoids select charges and momenta of muons as well as removing neutral particles in a beam. The principle is as follows. In a curved solenoidal magnetic field, a center of the helical trajectory of a charged particle is shifted perpendicular to the curved plane. The shift, whose amount is given as a function of momentum and its charge, makes a dispersive beam. By placing appropriate collimators, charges and momenta of muons can be selected.

One proposal in the USA was the MECO experiment at BNL [75]. It was mostly based on the MELC design and aimed to search for $\mu^- - e^-$ conversion at a sensitivity of less than 10^{-16}. A schematic layout of the MECO beamline and detector is shown in Fig. 19.17. It consists of the production

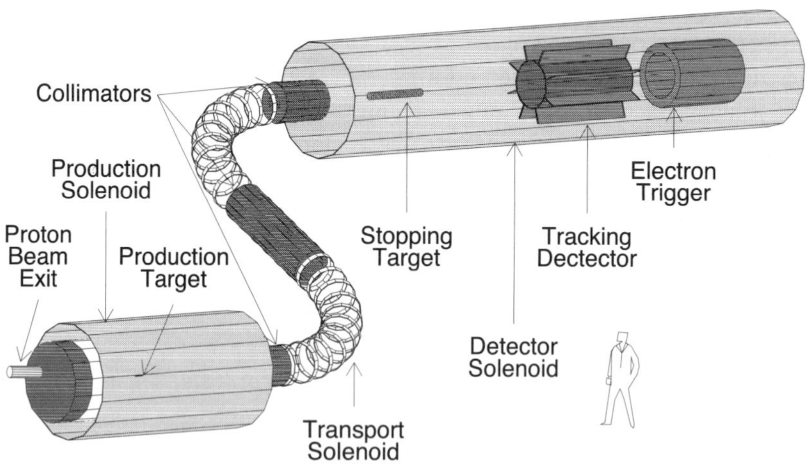

Fig. 19.17. Schematic layout of the MECO experiment at BNL. Protons hit a (pion) production target to produce pions, which decay to muons. They are transported through the transport solenoid system, and brought to a (muon) stopping target. The signal electrons are detected by a tracking detector and an electron calorimeter in the detector solenoid system. (Reprinted with permission from Ref. [8]. Copyright (2001) by the American Physical Society.)

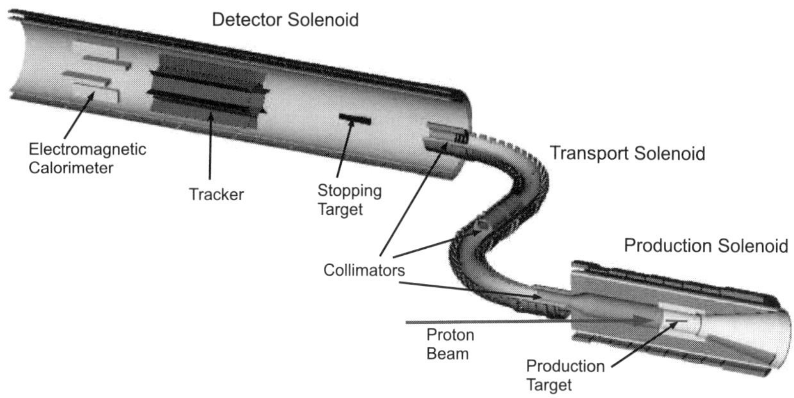

Fig. 19.18. Schematic layout of the Mu2e experiment at FNAL.

solenoid system, the transport solenoid system and the detector solenoid system. Unfortunately, the MECO proposal was canceled in 2005, due to funding problems.

However, in 2008 a new initiative at Fermi National Accelerator Laboratory (FNAL), which is called the Mu2e experiment, has been made to perform a MECO-type experiment [76]. The Mu2e experiment is planned to combat beam-related background events with the help of a 8 GeV/c proton beam from the Booster machine at FNAL. Figure 19.18 shows the proposed layout of the Mu2e experiment. Pions are produced by 8 GeV/c protons, and they are captured by surrounding superconducting solenoid magnets in the production solenoid system. Muons from the decays of the pions are collected efficiently with the help of a graded magnetic field. Negatively charged particles with 20–70 MeV/c momenta are transported by a curved solenoid to the experimental target. In the spectrometer magnet, a graded field is also applied. A major challenge has to be made to meet the requirement for proton extinction in between the proton bursts. In order to maintain the pion coming rate in the pulsed beam interval, a beam extinction factor better than 10^{-9} is required.

The other experimental proposal to search for $\mu^- - e^-$ conversion, which is called COMET (COherent Muon to Electron Transition), is being prepared for the Japan Proton Accelerator Research Complex (J-PARC), Tokai, Japan [77]. The aimed sensitivity at COMET is less than 10^{-16}, which is almost the same as that of Mu2e at FNAL. A schematic layout of the COMET experiment is presented in Fig. 19.19. The differences of the

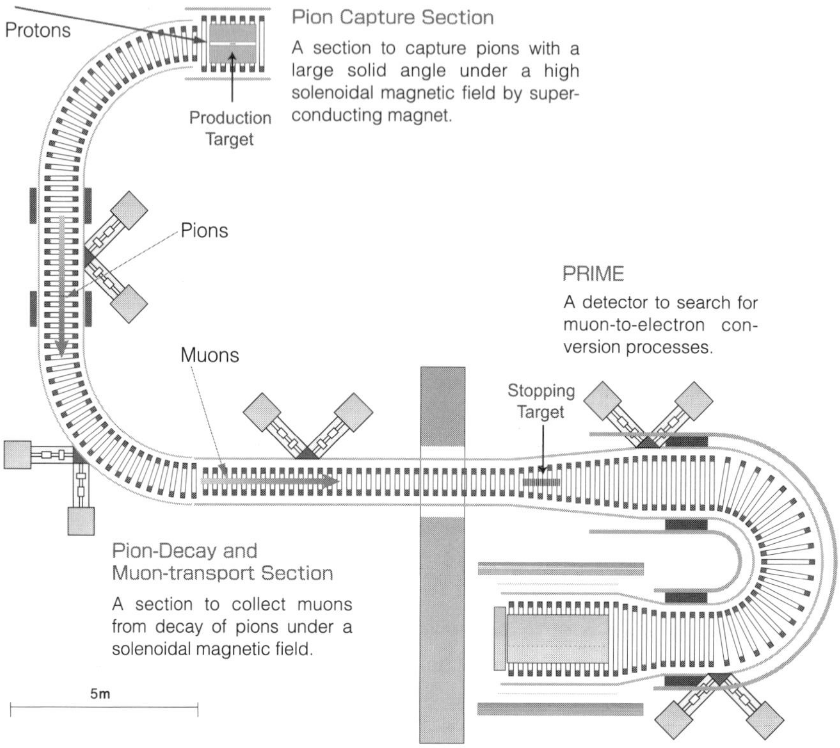

Fig. 19.19. Schematic layout of the COMET beamline and detector.

designs between Mu2e and COMET exist in the adoption of C-shape curved solenoid magnets for a muon beamline and a e^+ spectrometer in COMET. First of all, in Mu2e, after the first 90-degree bending, the muons of their momenta of interest are necessarily shifted back to the median plane in the second 90-degree bending with opposite bending direction (therefore a S-shape), whereas in COMET, by applying a vertical correction magnetic field, the muons of interest can be kept on the median curved plane. From this fact, any opposite bending direction is not needed and a 180-degree bending in COMET would provide larger dispersion to give a better momentum selection. Secondly, a curved solenoid spectrometer in COMET is useful to eliminate low-energy DIO events before going into the detector, resulting in lower single counting rates in the detectors. To eliminate beam-related backgrounds at this sensitivity, both experiments, Mu2e and COMET, place a stringent requirement on the beam extinction of 10^{-9}

during the measurement interval. To meet the requirement, additional kicker magnets in the accelerator ring as well as in the extracted proton beam line is being considered.

In the long-term future, significant improvements to aim at an experiment with a 10^{-18} sensitivity could be considered. Potential key requirements for the improvement are the following.

- *Beam purity:*
  A low-momentum (< 70 MeV/c) μ^- beam with no pion contamination ($< 10^{-20}$) would keep prompt background events at a negligible level. This could be achieved by adopting a muon storage ring, where pions decay out during their flight of many turns in the ring. An additional advantage of the method is that heavy muon-stopping targets such as gold, whose muonic-atom lifetimes is around 100 nsec, can be studied.
- *Narrow energy spread:*
  The e^- energy resolution is determined by multiple scattering and energy straggling in the muon-stopping target. To improve the resolution, a thinner muon-stopping target is required. To keep a good muon-stopping efficiency, a narrow energy spread of a muon beam is needed.
- *Extinction of a muon beam:*
  As discussed, requirements on beam extinction are very stringent. In addition to the proton beam extinction, the extinction of a muon beam in low energy, which might be easier than high-energy protons, would be needed. To achieve this, fast kicker magnets are needed.

In consideration of these requirements, the PRISM (Phase Rotated Intense Slow Muon source) project is being developed in Japan [78]. In the PRISM project, a muon storage ring, which comprises a fixed field alternating gradient (FFAG) ring, is considered. The FFAG ring has large aperture to accept a muon beam of a large size and allows fast acceleration due to a fixed magnetic field. To achieve narrow energy spread, phase rotation, where fast muons are decelerated and slow muons are accelerated by RF fields in the muon storage ring, is adopted. Furthermore, the kicker magnets for injection and extraction to the muon storage ring would serve the muon-beam extinction. A schematic layout of PRISM and its PRIME detector is shown in Fig. 19.20.

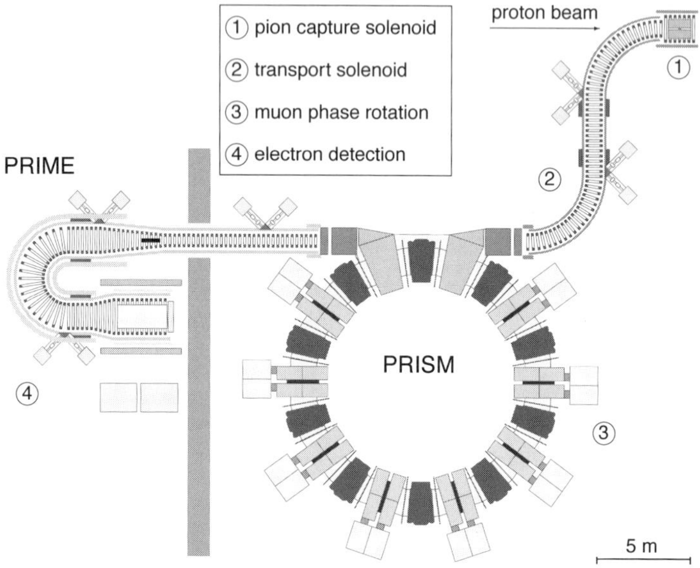

Fig. 19.20. Schematic layout of the PRISM/PRIME detector.

19.7. Lepton Flavor Violation in τ Leptons

Recently lepton flavor violation in τ decays has been extensively studied. The present B factories, which are operating at the $\Upsilon(4S)$ resonance, can produce many τs, since the production cross sections for $\sigma_{\tau^+\tau^-} = 0.9$ nb whereas $\sigma_{b\bar{b}} = 1.05$ nb at the center of mass energy of 10.58 GeV. Almost as many as τ pairs as b pairs are produced and thus the B factories serve as τ factories. Moreover, the jet-like topology of $\tau^+\tau^-$ pairs can be easily distinguished from the spherical event shape of $B\bar{B}$ events. As a result, the B factories represent an optimal framework for the search for LFV in τ decays due to high statistics and the clean environment. In particular, the KEKB have achieved the highest luminosity of $1.7 \times 10^{34}/\text{cm}^2/s$.

19.7.1. *Signature and background events*

The analysis is mostly carried out as follows. Firstly, one τ lepton in SM decays, which are either 1-prong (of its branching fraction of about B ~85%) or 3-prongs (of B ~14%), is reconstructed, and LFV decays of the other τ lepton is studied. The former is called "tag" side, while the latter is called "signal" side.

The signal events of LFV decays of the τ leptons can be extracted by the following requirements. They are (1) the measured energy of τ decay products (E_{rec}) that should be close to a half of the CM beam energy, and (2) the total invariant mass (M_{rec}) of the τ decay products that should be the mass of the τ lepton. Namely,

$$E_{rec} = E_{beam} \tag{19.35}$$

$$M_{rec} = m_{\tau}. \tag{19.36}$$

The distributions of E_{rec} and M_{rec} might have non-Gaussian tails due to initial and final state radiations.

Potential sources for background events come from radiative QED events (such as dimuon events and Bhabha processes) and continuum ($q\bar{q}$) events. There is hard initial-state radiation which contributes a background photon in the search for $\tau \to l\gamma$ ($l = e, \mu$). A blind analysis is usually adopted, in which the signal region is defined in advance in the energy-mass plane of the τ decay products and various selection criteria are considered to optimize a signal sensitivity and background rejection by using control samples, sideband data and Monte Carlo simulation data.

19.7.2. *Present experimental status*

BELLE and Barbar analyzed the data of integral luminosities of $\mathcal{L} \sim 535$ and 376 fb^{-1}, respectively. It is as many as about 10^9 τ decays. No signal events have been observed yet and thus upper limits on the branching ratios at 90% C.L. have been set. They are shown in Table 19.8. The $\tau \to lll$ modes have no background events and $B(\tau \to lll) < (2.0 - 4.1) \times 10^{-8}$ at 90% C.L. [26].

19.7.3. *Future experimental prospects*

An upgraded project such as a super B-factory is planned either in Japan and/or Italy. At a Super B-factory, an increase of the luminosity ($\mathcal{L}$) of about 10–100 times is expected and thus about 10^{10} τ pairs can be produced. Thereby further significant improvement in sensitivities are anticipated. However, future projection on the sensitivity improvement depends on the nature of background events. There are two extreme cases: (1) If there is no expected background event, the sensitivities would be scaled by $1/\mathcal{L}$, yielding sensitivities of 10^{-8} level. It is the case for 3-prong events such as $\tau \to \mu\mu\mu$ where the expected background level is still very low at the present. (2) If there are background events, the sensitivities would

Table 19.8. Upper limits on the branching ratios at 90% C.L. and the corresponding luminosities for the searches for LFV in τ decays.

	BELLE			Barbar		
Decay modes	B (10^{-8})	$\mathcal{L}$ (fb^{-1})	Ref.	B (10^{-8})	$\mathcal{L}$ (fb^{-1})	Ref.
$\tau \to \mu\gamma$	5	535	[25]	6.8	232	[24]
$\tau \to e\gamma$	12	535	[25]	11	232	[24]
$\tau \to \mu\eta$	7	401	[79]	15	339	[81]
$\tau \to e\eta$	9	401	[79]	16	339	[81]
$\tau \to \mu\eta'$	13	401	[79]	13	339	[81]
$\tau \to e\eta'$	16	401	[79]	24	339	[81]
$\tau \to \mu\pi^0$	12	401	[79]	15	339	[81]
$\tau \to e\pi^0$	8	401	[79]	13	339	[81]
$\tau \to lll$	2.0–4.1	535	[26]	3.7–8.0	376	[82]
$\tau \to lhh$	20–160	158	[80]	7–48	221	[83]

be scaled by $1/\sqrt{\mathcal{L}}$, yielding sensitivities of 10^{-8} level. It is the case, for instance for the $\tau \to \mu\gamma$ decay, where $\tau \to \mu\nu\bar{\nu}$ decay accompanied by initial-state radiation contributes irreducible background events. At a future Super B-factory, one can consider further optimization of background rejection and improvement of the mass resolution, and future extrapolation could be better than $1/\sqrt{\mathcal{L}}$.

19.7.3.1. *In-flight LFV Processes for tau lepton production*

Another method to study cLFV with tau leptons (either $\mu - \tau$ or $e - \tau$ transitions) has been proposed using in-flight scattering processes. In this method, tau leptons are directly produced by either incident electrons or muons scattering off a nuclear target [84], namely

$$e(\mu) + N \to \tau + X \tag{19.37}$$

where N is an initial-state nucleus and X includes a final-state nucleus with all other particles produced in this reaction. Since the cross section of the reactions in Eq. (19.37) increases as the energy of incident particles becomes higher, these reactions should be considered at high incident energy regions such as in the deep inelastic scattering (DIS) regime. A potential advantage of these reactions is as follows. A future increase of the number of tau leptons expected in future low-energy e^+e^- colliders (such as super B-factories) is limited to be about one order of magnitude, but a future international e^+e^- linear collider (ILC) or a neutrino factory/muon collider would produce huge numbers of electrons and muons respectively. Therefore, even if a scattering process is in general less efficient than a

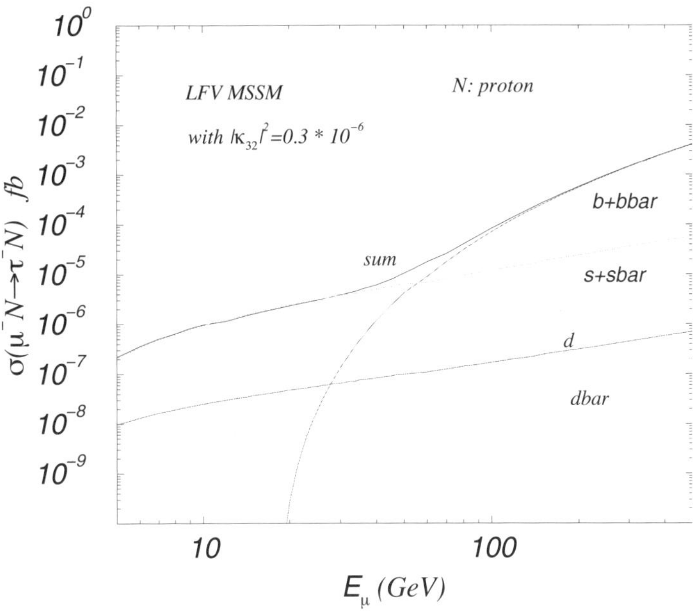

Fig. 19.21. Cross section of the $\mu^- N \to \tau^- X$ DIS process as a function of the muon energy for the Higgs mediated interaction. It is assumed that the initial muons are purely left-handed. CTEQ6L is used for the parton distribution function. (Reprinted with permission from Ref. [84]. Copyright (2005) by Elsevier.)

decay process, the reaction processes in Eq. (19.37) might have a better sensitivity if such future accelerator facilities are available. The prediction of the cross section is estimated by supersymmetric models with the Higgs mediation, where the theoretical parameters in the model were determined to be maximally-allowed values from the experimental upper limits in rare τ decays. For instance, the evaluated cross section of $\mu^- + N \to \tau^- + X$ DIS reaction is shown in Fig. 19.21. Here N is a proton, and CTEQ6L [85] is used for the parton distribution function. In Fig. 19.21, the cross section sharply increases above $E_\mu \sim 50$ GeV. This enhancement comes from a consequence of the b-quark contribution in addition to the d and s quark contribution in nuclei. The coupling for the b quark to the Higgs particle is enhanced by a factor of m_b/m_d over the d quark contribution.

For the $\mu N \to \tau X$ reaction, with the intensity of 10^{20} muons per year and the target mass of 100 g/cm^2, about 10^4 (10^2) events could be expected for the cross section $\sigma(\mu X \to \tau X) = 10^{-3}$ (10^{-5}) fb, which corresponds to E_μ= 300 (50) GeV respectively. Such a muon intensity could be available at a future muon collider or a neutrino factory. This would provide good potential to improve the sensitivity by four (two) orders of magnitude from the present limit from $\tau \to \mu\eta$ decay. For the $eN \to \tau X$ reaction, at ILC with the center of mass energy of 500 GeV and the luminosity of 10^{34} $\text{cm}^{-2}sec^{-1}$, about 10^{22} electrons (or positrons) with the energy of 250 GeV would be available. Assuming a fixed target of the mass of 10 g/cm^2, it is expected to have about 10^{4-5} of the $eN \to \tau X$ events for $\sigma(eN \to \tau X) \sim 10^{-3}$ fb. Experimental issues such as the signal identification and the background rejection have not been discussed in detail yet. However, significant improvements by several orders of magnitude can be expected with further future developments.[d]

19.8. Conclusions and Outlook

The search for cLFV has large potential to obtain hints for New Physics beyond the Standard Model. The physics of cLFV has received glowing attention from both theorists and experimentalists. In this article, the phenomenology, the experimental status and future experimental prospects for various cLFV processes with muon and tau leptons are described. For the muon sector, the MEG experiment at PSI is now running to aim at a sensitivity of 10^{-13} for $\mu^+ \to e^+\gamma$ decay. Next-generation experiments to search for $\mu^- - e^-$ conversion at a sensitivity of better than 10^{-16}, both Mu2e and COMET, are being proposed. An ultimate search for $\mu^- - e^-$ conversion at a sensitivity of 10^{-18} at PRISM is also being studied. For the tau sector, super B-factories are planned to accomplish an order of magnitude increase in their luminosities and improve sensitivities of cLFV with τ decays. Thus, various experimental attempts for cLFV are being made. These attempts would offer extraordinary opportunities for exploring new phenomena which would otherwise be directly inaccessible at future high energy colliders.

[d]Flavor-changing charged current (FCCC) LFV processes, which produce directly tau leptons in the final state by incident neutrinos such as $\nu_\mu(\nu_e) + N \to \tau + X$, can also be considered. This kind of experiment should be carried out at a near-detector location to suppress the contribution from neutrino oscillation. These processes might be cleaner from the experimental point of view, but no enhancement of the b quarks existing in nuclei can be expected even from SUSY models with the Higgs-particle mediation.

Acknowledgments

The author would like to thank Prof. Y. Okada for many discussions on cLFV. He would like to thank Prof. K. Tobe for the discussions in the model-independent approach for LFV physics in Section 3.2. This work is supported in part by a Grant-in-Aid, Creative Scientific Research 15GS0211 "A Study of A Super Muon Beam for New Initiative on Muon Physics" by the Japan Society for the Promotion of Science (JSPS).

References

[1] S. Frankel, in *Muon Physics II: Weak Interaction*, ed. V.W. Hughes and C.S. Wu (Academic, New York), p. 83 (1975).
[2] F. Scheck, *Phys. Rep.* **44**, 187 (1978).
[3] J.D. Vergados, *Phys. Rep.* **133** 1 (1986).
[4] R. Engfer and H.K. Walter, *Ann. Rev. Nucl. Part. Sci.* **36**, 327 (1986).
[5] P. Depommier, in *Neutrinos* edited by H.V. Klapdor, (Springer, Berlin), 265 (1987).
[6] A. Van der Schaaf, *Progress in Particle and Nuclear Physics* **31**, 1 (1993).
[7] P. Depommier, and C. Leroy, *Reports on Progress in Phys.* **58**, 61 (1995).
[8] Y. Kuno and Y. Okada, *Rev. Mod. Phys.* **73**, 151 (2001)
[9] W.J. Marciano, T. Mori, and J.M. Roney, *Ann. Rev. Nucl. Part. Sci.* **58** 315 (2008).
[10] S.H. Neddermeyer and C.D.Anderson, *Phys. Rev.* **51**, 884 (1937).
[11] H. Yukawa, *Prog. Phys. Math. Soc. Japan* **17**, 48 (1935).
[12] M. Conversi, E. Pancini, and O. Piccioni, *Phys. Rev.* **71**, 209 (1947).
[13] J. Steinberger, *Phys. Rev.* **74**, 500 (1948).
[14] E.P. Hincks and B. Pontecorvo, *Phys. Rev. Lett.* **73**, 246 (1947).
[15] A. Lagarrigue and C. Peyrou, *Acad. Sci. Paris*, **234**, 873 (1952).
[16] S. Lokonathan and J. Steinberger, *Phys. Rev.* **98**, 240 (1955).
[17] J. Steinberger and H.B. Wolfe, *Phys. Rev.* **100**, 1490 (1955).
[18] R.P. Feynman and M. Gell-Mann, *Phys. Rev.* **109**, 193 (1958).
[19] G. Feinberg, *Phys. Rev.* **116**, 1482 (1958).
[20] K. Nishijima, *Phys. Rev.* **108**, 907 (1957).
[21] J. Schwinger, *Ann. Phys.* **2**, 407 (1957).
[22] G. Danby, J.M. Gaillard, K. Goulianos, L.M. Lederman, N. Mistry, M. Schwartz, and J. Steinberger, *Phys. Rev. Lett.* **9**, 36 (1962).
[23] L. Willmann *et al.*, *Phys. Rev. Lett.* **82**, 49 (1999).
[24] B. Aubert *et al.* (Barbar Collaboration), *Phys. Rev. Lett* **96** 041802 (2006).
[25] K. Hayasaka *et al.* (BELLE Collaboration), *Phys. Lett.* **B666**, 16 (2008)
[26] Y. Miyazaki *et al.* (BELLE Collaboration), *Phys. Lett.* **B660**, 154 (2008)
[27] P. Krolak *et al.*, *Phys. Lett.* **B320**, 407 (1994).
[28] D. Ambrose *et al.* (BNL E871 Collaboration), *Phys. Rev. Lett.* **81** 5734 (1998).

[29] A.M. Lee *et al.*, *Phys. Rev. Lett.* **64**, 165 (1990).
[30] K. Arisaka *et al.*, *Phys. Lett.* **B432**, 230 (1998).
[31] R. Akers *et al.* (OPAL Collaboration), *Z. Phys.* **C67**, 555 (1995).
[32] P. Abreu *et al.* (DELPHI Collaboration), *Z. Phys.* **C73**, 243 (1997).
[33] F. Borzumati and A. Masiero, *Phys. Rev. Lett.* **57**, 961 (1986).
[34] L.J. Hall, V.A. Kostelecky and S. Raby, *Nucl. Phys.* B **267**, 415 (1986).
[35] R. Barbieri, L.J. Hall, and A. Strumia, 1995a, Nucl. Phys. B **445**, 219 (1995) : R. Barbieri, L.J. Hall, and A. Strumia, 1995b, Nucl. Phys. B **449**, 437 (1995).
[36] J. Hisano, T. Moroi, K. Tobe, and M. Yamaguchi, *Phys. Rev.* **D53**, 2442 81996).
[37] J. Hisano, D. Nomura, and T. Yanagida, *Phys. Lett.* **B437**, 351 (1998).
[38] J. Hisano and D. Nomura, *Phys. Rev.* **D59**, 116005 (1999).
[39] L. Calibbi, A. Faccia, A. Masierro, and S.K. Vempati, Phys. Rev. **D74** 116002 (2006).
[40] C. Fronsdal and H. Überall, *Phys. Rev.* **118**, 654 (1959).
[41] S.G. Eckstein and R.H. Pratt, *Ann. of Phys.* **8**, 297 (1959).
[42] Y. Kuno and Y. Okada, *Phys. Rev. Lett.* **77**, 434 (1996).
[43] A.B. Arbuzov, O. Krehl, E.A. Kuraev, E.N. Magar, B.G. Shaikhatdenov, *Phys. Lett.* **B432**, 421 (1998).
[44] Y. Kuno, A. Maki, and Y. Okada, *Phys. Rev.* **D55**, 2517 (1997).
[45] P. Depommier and C. Leroy, *Reports on Progress in Phys.* **58**, 61 (1995).
[46] A. Van der Schaaf *et al.*, *Nucl. Phys.* **A340**, 249 (1980).
[47] W.W. Kinnison *et al.*, *Phys. Rev.* **D25**, 2846 (1982).
[48] R.D. Bolton *et al.*, *Phys. Rev.* **D38**, 2077 (1988).
[49] M.L. Brooks, *et al.* (MEGA Collaboration), *Phys. Rev. Lett.* **83**, 1521 (1999): M. Ahmed, *et al.* (MEGA Collaboration), *Phys. Rev.* **D65**, 112002 (2002).
[50] T. Mori *et al.*, "Search for $\mu^+ \to e^+\gamma$ down to the 10^{-14} branching ratio", PSI proposal R-99-5, May 1999, available as UT-ICEPP 00-02; A. Baldini *et al.*, Research Proposal to INFN, Sep.2002. These documents and the updated information of the MEG experiment are available at http://meg.web.psi.ch/docs/.
[51] S.M. Korenchenko, *et al.*, *Sov. Phys.* JETP **43**, 1 (1976).
[52] R.D. Bolton, *et al.*, *Phys. Rev. Lett.* **53**, 1415 (1984).
[53] W. Bertl, *et al.*, *Phys. Lett.* **B140**, 299 (1984).
[54] W. Bertl, *et al.*, *Nucl. Phys.* **B260**, 1 (1985).
[55] U. Bellgardt, *et al.*, *Nucl. Phys.* **B229**, 1 (1988).
[56] V.A. Baranov, *et al.*, *Sov. J. Nucl. Physics* **53**, 802 (1991).
[57] A. Kersch, N. Kraus, and R. Engfer, *Nucl. Phys.* **A485**, 606 (1988).
[58] Y. Okada, K. Okumura and Y. Shimizu, *Phys. Rev.* **D58**, 051901 (1988).
[59] Y. Okada, K. Okumura, and Y. Shimizu, *Phys. Rev.* **D61**, 094001 (2000).
[60] S. Weinberg and G. Feinberg, *Phys. Rev. Lett.* **3**, 111, and Erratum: 1959, *ibid.*, 244 (1959).
[61] O. Shanker, *Phys. Rev.* **D20**, 1608 (1979).
[62] A. Czarnecki, W.J. Marciano, and K. Melnikov, 1997, in *Proceedings of*

Workshop on Physics at the First Muon Collider and at the Front End of the Muon Collider, edited by S.H. Geer and R. Raja, Fermilab, U.S.A., AIP Conference Proceedings 435, p. 409.
[63] R. Kitano, M. Koike and Y. Okada, *Phys. Rev.* **D66**, 096002 (2002) [Erratum-ibid. **D76**, 059902 (2007)]
[64] O. Shanker, *Phys. Rev.* **D25**, 1847 (1982).
[65] R. Watanabe, M. Fukui, H. Ohtsubo, and M. Morita, *Prog. Theor. Phys.* **78**, 114 (1987).
[66] D.A. Bryman, M. Blecher, K. Gotow and R.J. Powers, *Phys. Rev. Lett.* **28** 1469 (1972).
[67] A. Badertscher *et al. Nucl. Phys.* **A 377** 406 (1982).
[68] D.A. Bryman *et al. Phys. Rev. Lett.* **55** 465 (1985).
[69] S. Ahmad *et al. Phys. Rev.* **D38** 2102 (1988).
[70] C. Dohmen *et al.* (SINDRUM II Collaboration) *Phys. Lett.* **B317** 631 (1993).
[71] W. Honecker*et al.* (SINDRUM II Collaboration) *Phys. Rev. Lett.* **76** 200 (1996).
[72] P. Wintz 1998 in *Proceedings of the First International Symposium on Lepton and Baryon Number Violation*, ed. H.V. Klapdor-Kleingrothaus and I.V. Krivosheina (Institute of Physics Publishing, Bristol and Philadelphia), p.534. Unpublished.
[73] W. Bertl *et al.*, *Eur. Phys. J.* **C47**, 337 (2006).
[74] R.M. Dzhilkibaev and V.M. Lobashev, *Sov. J. Nucl. Phys.* **49** 384 (1989) [*Yad. Fiz.* **49** 622 (1989)].
[75] M. Bachman *et al.*, MECO, BNL Proposal E940 (1997).
[76] R.M. Carry *et al.* (Mu2e collaboration), "Proposal to Search for $\mu^- - e^-$ with a Single Event Sensitivity Below 10^{-16}, FNAL proposal (2008).
[77] Y. Kuno *et al.* (COMET collaboration), J-PARC 50 GeV Proton Synchrotron Proposal P21 "A Experimental Search for Lepton Flavor Violating $\mu^- - e^-$ Conversion at Sensitivity of 10^{-16} with A Slow-Extracted Bunched Proton Beam", unpublished, 2007. see http://j-parc.jp/NuclPart/pac_0801/pdf/Kuno.pdf
[78] Y. Kuno *et al.* (PRISM collaboration), J-PARC 50 GeV Proton Synchrotron LOI P20 " An Experimental Search for a $\mu^- - e^-$ Conversion at Sensitivity of the Order of 10^{-18} with a Highly Intense Muon Source: PRISM", unpublished, 2006. see http://j-parc.jp/NuclPart/pac_0606/pdf/p20-Kuno.pdf
[79] Y. Miyazaki *et al.* (BELLE Collaboration), *Phys. Lett.* **B648**, 341 (2007).
[80] Y. Yusa *et al.* (BELLE Collaboration), *Phys. Lett.* **B640**, 138 (2006).
[81] B. Aubert *et al.* (Barbar Collaboration), *Phys. Rev. Lett.* **98** 061803 (2007).
[82] B. Aubert *et al.* (Barbar Collaboration), *Phys. Rev. Lett.* **99** 251803 (2007).
[83] B. Aubert *et al.* (Barbar Collaboration), *Phys. Rev. Lett.* **95** 191801 (2005).
[84] S. Kanemura, Y. Kuno, M. Kuze, and T. Ota, *Phys. Lett.* **B607** 165 (2005).
[85] J. Pumplin it et al., *JHEP* **0207** 012 (2002).

Epilogue

Physicists often have a love affair with numbers. Those feelings may stem from early abilities in arithmetic calculations, skills that may have led them to scientific careers. Some numbers, such as alpha (1/137), hold a special fascination, one that beckons to ever more precise determinations. Toward that end, the incredible measurements of the electron anomalous magnetic moment and impressive high-order perturbative calculations of the QED prediction (in powers of α), which have been reviewed in this book, provide its currently most precise value

$$\alpha = 1/137.035999084(51).$$

Comparison of that value with other (somewhat less precise) determinations from atomic and condensed-matter physics confronts the validity of QED, arguably making it the best tested theory in physics. Further improvements in these measurements and quantum calculations are expected. The quest for a better α will continue.

In the case of the muon anomalous magnetic moment, experimental and theoretical advances have also occurred and been reviewed in this book. Although the muon anomaly is not measured as precisely as the electron's, the current experimental value

$$a_\mu = 116\,592\,080(63) \times 10^{-11}$$

is truly impressive. Furthermore, the muon anomaly, because of its larger mass, is about 43,000 times more sensitive to hadronic and electroweak quantum loops, further challenging our computational skills. It is similarly more sensitive to effects from New Physics. The current, roughly $3\,\sigma$ difference between theory and experiment hints at the presence of New Physics, with supersymmetry the leading candidate explanation. Alternatively, it could indicate problems with the hadronic loop calculations, but a significant change there would have implications for other Standard-Model predictions such as the Higgs mass. Further improvements in theory and experiment are highly warranted.

If the muon is unveiling New Physics in its magnetic moment, it could be a harbinger of other exciting discoveries on the horizon. Supersymmetry with its plethora of new particles might populate the LHC with rich spectroscopy. Alternatively, new strong dynamics or manifestations of extra dimensions may appear.

On the low-energy frontier, a New Physics effect in the muon magnetic moment could also imply dramatic consequences. It suggests that the electric dipole moment searches described in this book may come to fruition, finally observing non-zero values. That would culminate an extraordinary effort that started with the pioneering insights of Purcell and Ramsey. Such an observation would imply a new, relatively strong source of *CP* violation, one that might help explain the matter-antimatter asymmetry of our Universe and the chain of cosmological events that led to our existence.

New Physics in the muon anomaly, combined with a new source of lepton flavor violation which is likely to accompany it, suggests the existence of transition dipole moments and the occurrence of rare decays such as $\mu \to e\gamma$. Searches for charged lepton flavor violating decays, as described in this book, have been carried out ever since the muon's discovery, but as yet are unobserved. Their sensitivity continues to improve, and further dramatic experimental advances are envisioned.

The above experimental goals are all strongly motivated and could lead to major discoveries. They represent examples of exciting fundamental research. Where they ultimately lead is to be seen. However, we can assert with some confidence that their quest will continue to develop and refine our technological and computational skills, further expanding our understanding of nature.

William J. Marciano and B. Lee Roberts

Subject Index